Synoptic Climatology

Methods and Applications

R.G. Barry
and A.H. Perry

Methuen & Co Ltd
11 New Fetter Lane London EC4

First published 1973
by Methuen & Co Ltd
11 New Fetter Lane, London EC4

Printed in Great Britain by
William Clowes & Sons, Limited
London, Beccles and Colchester

SBN 416 08500 8

Distributed in the U.S.A. by
HARPER & ROW PUBLISHERS, INC.
BARNES & NOBLE IMPORT DIVISION

Synoptic Climatology

Methods and Applications

to Valerie and Vivien

Contents

Acknowledgments

The final version of this book has benefited greatly from the suggestions and critical comments of several individuals. The authors are particularly indebted to Professor F. Kenneth Hare, University of Toronto, (presently Director-General, Research Coordination Directorate, Department of Environment, Ottawa) who found time amidst his numerous other commitments to read the entire manuscript. Chapters 2A and 4 have also been improved as a result of suggestions by Dr Julius London, University of Colorado, and both Dr Yuk Lee, University of Colorado (Denver Centre) and Dr Erat Joseph, Keystone College, Pennsylvania, respectively. Any remaining errors or shortcomings are our own.

The greater part of the typing has been cheerfully borne by Mrs Jennifer Skwiot, Institute of Arctic and Alpine Research.

The authors and publishers wish to thank the following individuals, societies and organizations for permission to reproduce material:

Academic Press Inc., New York, for fig. 3.29 from *Atmospheric Circulation Systems* by E. Palmen and C. Newton; fig. 2.33 by J. L. Holloway from *Advances in Geophysics*; fig. 5.28 by H. H. Lamb and A. Woodroffe from *Quaternary Research*.

Air Force Cambridge Research Laboratories, Bedford, Mass., for fig. 4.21 from *Atmospheric Flow Patterns and their Representation by Spherical-Surface Harmonics* by B. Haurwitz and R. A. Craig.

American Geophysical Union, Washington, D.C., for fig. 5.48 by G. W. Brier from *Reviews of Geophysics*; fig. 4.12 by I. Rodriguez-Iturbé and C. F. Nordin from *Water Resources Research*.

American Meteorological Society, Boston, Mass., and the authors for fig. 3.15 by R. D. Elliott from *Compendium of Meteorology* edited by T. F. Malone; fig. 2.17 from *An Introduction to the Study of Air Mass and Isentropic Analysis* by J. Namias; fig. 5.1 by A. Ehrlich, fig. 5.3 by R. A. Bryson and W. P. Lowry, and fig. 3.30 by D. L. Bradbury, from *Ameri-*

can Meteorological Society Bulletin; figs. 2.34 and 2.35 by W. R. Gommel, fig. 4.22 by J. A. Leese and E. S. Epstein, figs. 2.27 and 2.28 by L. C. Clarke and R. J. Renard, fig. 4.25 by C. K. Stidd, fig. 3.49 by J. E. Kutzbach, figs. 2.5 and 2.8 by J. J. Taljaard, fig. 4.14 by F. A. Huff and W. L. Shipp, and fig. 4.18 by R. R. Dickson, from *Journal of Applied Meteorology*; fig. 2.32 by J. Namias and P. F. Clapp, and figs. 2.18 and 2.19 by E. F. Danielsen, from *Journal of Meteorology*; fig. 3.16 by W. C. Jacobs, fig. 5.15 by T. C. Yeh, and fig. 3.42 by J. Namias, from *Meteorological Monographs.*

Edward Arnold (Publishers) Ltd, London, for fig. 4.33 from *Network Analysis in Geography* by R. Chorley and P. Haggett.

The Association of American Geographers, Washington, D.C., for fig. 3.32 by P. E. Lydolph, and fig. 4.6 by L. H. Horn and R. A. Bryson, from the *Annals of the Association of American Geographers.*

The Canadian Association of Geographers, Montreal, for fig. 4.23 by J. N. Rayner from *Canadian Geographer.*

Deutscher Wetterdienst, Offenback am Main, Germany, for figs. 3.13 and 5.33 by P. Hess and H. Brezowsky, fig. 4.4 by O. Essenwanger, fig. 5.17 by H. Bürger, figs. 4.3 and 4.13 by H. Wachter, and fig. 4.11 by R. Dobertiz, from *Berichte.*

Editions J. Duculot à Gembloux, Belgium, and the author for fig. 4.19 by S. Gregory from *Mélanges de géographie offerts à M. Omer Tulippe*, I edited by J. A. Spork.

The English Universities Press Ltd, London and The American Elsevier Publishing Co. Inc., New York, for fig. 4.7 from *Statistics in the Computer Age* by J. Craddock.

Entomological Society of Canada, Ottawa, and the author, for figs. 5.62 and 5.63 by W. G. Wellington from *Canadian Entomologist.*

Environment Canada, Atmospheric Environment Service, Downsview, Ontario, for fig. 2.25 from *The Analysis and Interpretation of Precipitation Patterns* by E. C. Jarvis, and figs. 3.4 and 3.5 from *Fourier Analysis Applied to Hemispheric Waves of the Atmosphere* by B. W. Boville and M. Kwizak.

Erdkunde, Bonn, for fig. 2.14 by R. A. Bryson and P. M. Kuhn.

Fachliche Mitteilungen, Porz-Wahn, Germany, for fig. 5.36 by I. Weiss and H. H. Lamb.

W. de Gruyter & Co., Berlin, for figs. 3.12 and 3.24 from *Allgemeine Klimageographie* by J. Blüthgen.

The Controller, Her Majesty's Stationery Office, London, for figs. 2.1, 3.45 and 3.46 by J. Sawyer from *Royal Meteorological Society Quarterly Journal*; fig. 3.8 by J. Sawyer and figs. 3.9 and 3.10 by R. F. M. May from Met. Office Professional Notes Nos. 118 and 98; figs. 5.44 and 3.47 by J. Craddock and R. Ward from Met. Office Sci. Paper No. 12; fig. 5.61 by G. W. Hurst and R. P. Rumney from Met. Office Agric. Mem. No.

291; fig. 5.13 from *Aviation Meteorology of the Azores*; figs. 3.50 and 3.51 by R. Murray, fig. 2.6 by J. Findlater, fig. 2.21 by J. G. Lockwood, and fig. 5.56 by A. H. Perry, from *Meteorological Magazine.*

Institute of British Geographers for fig. 5.26 by B. W. Atkinson, and fig. 4.26 by A. H. Perry, from *Transactions.*

Institute of Geophysics, Mexico, for figs. 5.20 and 6.3 by P. A. Mosiño from *Geofisica Internacional.*

International Glaciological Society, Cambridge, and the author for fig. 5.32 by H. C. Hoinkes from *Journal of Glaciology.*

Japan Meteorological Agency, Tokyo, for fig. 5.2 by T. Nagao from *Geophysical Magazine.*

The Longman Group, Harlow, Essex, and the editors, for figs. 5.23 and 5.24 by R. G. Barry from *Liverpool Essays in Geography* edited by R. W. Steel and R. Lawton.

McGraw-Hill Book Co., New York, for fig. 3.3 from *Weather Analysis and Forecasting* by S. Petterssen.

Macmillan & Co. Ltd, London, for fig. 2.23 from *British Weather in Maps* by J. Taylor and R. Yates.

Meteorological Research Institute, Tokyo, for fig. 4.16 by T. Izawa from *Papers in Meteorology and Geophysics.*

Naturwissenschaftlicher Verein für Karnten, Klagenfurt, Austria, for fig. 3.23 by F. Fliri from *Carinthia II.*

Norwegian Universities Press, Oslo, for fig. 3.41 by T. Hesselberg, and fig. 5.55 by S. Petterssen, D. L. Bradbury and K. Pedersen, from *Geofysiske Publikasjoner.*

Physical Medicine Library, New Haven, Conn., for fig. 5.65 by H. Brezowsky from *Medical Climatology* edited by S. Licht.

Prentice-Hall Inc., Englewood Cliffs, N.J., for fig. 1.1 from *Introduction to Astronomy* by C. Payne-Gaposchkin and K. Haramundanis.

Les Presses de l'Université de Montréal for fig. 5.22 by R. G. Barry from *Revue de Géographie de Montréal.*

The Rand Corporation, Santa Monica, California, for fig. 3.26 by A. A. Girs and fig. 4.9 by J. M. Mitchell from Rand Report RM–5233–NSF edited by J. O. Fletcher.

Rivista di Meteorologia Aeronautica, Rome, for fig. 3.20 by A. Gazzola.

Royal Meteorological Society, Bracknell, Berkshire, for figs. 2.4 and 2.30 by S. Petterssen from *Centenary Proceedings*; fig. 5.29 by H. H. Lamb, R. P. W. Lewis and A. Woodroffe from *World Climate from 8000 to 0 B.C.* edited by J. S. Sawyer; fig. 4.10 by C. B. Williams, fig. 2.37 by B. Saltzmann and J. Peixoto, fig. 3.19 by R. G. Barry, fig. 5.47 by G. W. Brier and J. Simpson, fig. 3.40 by K. A. Browning and T. W. Harrold, fig. 2.16 by T. N. Krishnamurti, and figs. 5.51 and 5.52 by R. A. S. Ratcliffe and R. Murray, from *Quarterly Journal*; fig. 5.30 by H. H. Lamb, fig. 2.43 by G. Sutton, fig. 5.5 by R. H. Frederick, fig. 4.29 by

A. H. Perry, and fig. 5.45 by R. Murray and B. J. Moffit, from *Weather*.

Schweizerische Meterologische Zentralanstalt, Zurich, for fig. 5.21 by M. Schüepp and F. Fliri from *Veröffentlichungen der Schweizerischen Meteorologischen Zentralanstalt*.

Società Italiana di Geofisica e Meteorologia, Genoa, for fig. 2.9 by M. Schüepp, and fig. 5.18 by H. Kern, from *Geofisica e Meteorologia*.

Springer-Verlag, Berlin, for fig. 5.19 by H. Flohn and J. Huttary, fig. 3.14 by M. Yoshino, and figs. 5.11 and 3.37 by R. W. James, from *Meteorologische Rundschau*.

Superintendent of Documents, Government Printing Office, Washington, D.C., for fig. 5.4 by R. T. Duquet, fig. 4.17 by W. H. Klein, figs. 2.38 and 2.39 by J. F. O'Connor, fig. 2.36 by L. P. Stark, fig. 3.39 by W. I. Christensen and R. A. Bryson, fig. 3.38 by W. H. Klein, D. L. Jorgensen and A. F. Korte, fig. 5.57 by S. Manabe and K. Bryan, fig. 4.15 by P. R. Julian, and fig. 5.35 by J. E. Kutzbach, from *Monthly Weather Review*.

Tellus, Stockholm, for fig. 2.41 by H. Riehl, fig. 5.49 by H. Wexler, fig. 5.50 by J. Namias, and fig. 5.46 by H. van Loon and R. L. Jenne.

Thomas Nelson & Sons Ltd, London, for fig. 3.17 by S. Gregory from *The British Isles* edited by J. Wreford-Watson and J. B. Sissons.

UNESCO, Paris, for fig. 3.27 by B. L. Dzerdzeevski and fig. 5.58 by J. Bjerknes from *Arid Zone Research, 20: Changes of Climate, Proceedings of the Rome Symposium 1963*.

The University of Chicago Press for fig. 2.12 from *Principles of Meteorological Analysis* by W. Saucier.

Van Nostrand Reinhold Co., New York, for figs. 5.41 and 5.42(a) by W. H. Klein from *Humidity and Moisture, 2* edited by A. Wexler. Copyright 1965 by Litton Educational Publishing.

Wetter und Leben, Vienna, for fig. 4.27 by F. Fliri.

World Meteorological Organization, Geneva, for fig. 5.64 by C. I. H. Aspliden and R. C. Rainey from *W.M.O. Bulletin*; figs. 5.40 and 5.42(b) by W. H. Klein from Tech. Note No. 66; figs. 2.10 and 6.2 by D. H. Johnson from Tech. Note No. 69; fig. 4.1 by C. L. Godske from Tech. Note No. 71; fig. 3.28 by N. G. Davidova from Tech. Note No. 87; figs. 5.59 and 5.60 by W. H. Hogg, C. E. Hounam, A. K. Mallik and J. C. Zadoks from Tech. Note No. 99.

Preface

The origins of this book lie in the long-standing interest of the senior author in the climatological application of meteorological concepts. Synoptic climatology provides a meeting ground for meteorological and climatological viewpoints, concerned as it is with the relationships between the atmospheric circulation and local (or regional) climate. The term originated in the 1940s but studies that can be designated as synoptic climatology have a history of almost a century. It is surprising, therefore, that no book to the authors' knowledge has been written which deals specifically with synoptical climatological methods. The renewal of interest in synoptic climatology in recent years and the opportune development of the research interests of the junior author along similar lines at the University of Southampton prompted collaboration in undertaking this book. Access to diverse library facilities in the United States (particularly at the National Center for Atmospheric Research, Boulder) and in the British Isles (especially the Meteorological Office Library, Bracknell) has undoubtedly helped to compensate for the authors' spatial separation.

Beyond the introductory level in climatology, courses usually concentrate on either physical climatology or regional aspects of climate. In advanced regional treatments such as G. Trewartha's *The Earth's Problem Climates* it is assumed that synoptic climatological concepts and procedures are understood. However, most of the pertinent literature is widely scattered and much of it is not in English. The present book is intended to fill this gap for advanced climatology courses at the senior undergraduate or beginning graduate level. Although the book is particularly intended for climatology students in geography, anyone concerned with data on the atmospheric environment should find the study of synoptic climatological methods worthwhile.

Synoptic climatology is by no means a neat coherent field. Indeed, such is the variety of interpretation given to the term that it is in danger of becoming unusable. Hopefully this book may, as a byproduct, help to restore it as a recognized technical term. The basic consideration in synoptic climatology is the nature of the relationships between atmospheric circulation systems and weather conditions, especially in particular geographical locations. This problem has led to the formulation of numerous classifications of weather systems. The various approaches to such classification are examined systematically in Chapter 3. Many workers regard this topic as the principal aspect of synoptic climatology but it should certainly not be considered as the end product. The variety of possible applications of such classifications and related analyses, which is elaborated in Chapter 5, underlines the present and potential value of the field. Some background work in introductory meteorology and in statistical methods has been assumed in the presentation of the book although the basic meteorological parameters and related analytical techniques are reviewed in Chapter 2. The statistical concepts discussed in Chapter 4 are treated from the standpoint of the user. The organization of the material, as presented will probably require modification for use in a teaching sequence. For example, the analytical and statistical techniques may be dealt with in parallel courses or worked into the synoptic climatological topics.

Throughout the book themes and ideas are stressed which seem to have been somewhat neglected. Conventional material that is adequately covered elsewhere is given less space. Also, while early contributions to particular problems are not overlooked, the presentation is in a systematic rather than a historical framework. The reader will see that much remains to be accomplished in synoptic climatology. This is true with respect to making fuller use of available statistical techniques and incorporating the analysis of a wider range of meteorological parameters. More fundamental problems remain, however, in our endeavour to link understanding of the global circulation of the atmosphere with our knowledge of local and regional scale phenomena.

Roger G. Barry
Institute of Arctic and Alpine Research
and Department of Geography,
University of Colorado,
Boulder, Colorado.

Allen H. Perry
Geography Department,
University College,
Swansea, Glamorgan
(formerly at University
College, Dublin).

1 An introduction to synoptic climatology

In recent years there has been much argument about the proper status, scope and methods of climatology. This argument shows no sign of being settled. . . . (*F. K. Hare, 1957*)

A Historical antecedents

Agricultural necessity has made man long aware of the nature and regularity of the seasons and of the significance of the different winds. Chinese farmers of the third century B.C. divided the year into twenty-four intervals related to differing sequences of weather phenomena or phenological events (Chu, 1962), while an Aztec 'calendar' – the Great Stone of the Sun (fig. 1.1) – based on Maya antecedents, illustrates early knowledge of seasonal regime in the New World. Biblical references show recognition of the importance of wind characteristics. We read, for example, the following:

Out of the south cometh the whirlwind
And cold out of the north (Job 37:9)

Fair weather cometh out of the north (Job 37:22)

When the east wind toucheth it, it shall wither (Ezekiel 17:10)

The Greeks of the second century A.D. constructed an octagonal Tower of the Winds, which still stands in Athens (Karapiperis, 1951; see also D'Arcy Thompson, 1918). This depicts the wind from each cardinal point by a bas-relief figure personifying the characteristic type of weather associated with such a wind.

In most countries popular weather lore reflects awareness of, or at least interest in, typical weather patterns and seasonal tendencies or anomalies. Early meteorological knowledge of this type is reviewed by Hellman (1908), Shaw (1926), Schneider-Carius (1955), Heninger (1960), and in R. L. Inward's book *Weather Lore* (1950, first published 1869). Among the more remarkable writings are those of the Greek Theophrastus, who in the fourth century B.C. published a compendium of 'weather signs' relating to the appearance of the sun, moon and sky, and to the behaviour of birds and animals. A comparable collection of medieval weather lore *Die Bauern-Praktik* appeared in Germany in 1508 and ran to no fewer than sixty editions. Rather later in England we find *The Shepherd of Banbury's Rules* (Claridge, 1748) based largely on the writings of Pointer (1723).

Fig. 1.1 The Great Stone of the Sun – an Aztec representation based on earlier Maya calendars (from C. Payne-Gaposchkin and K. Haramundanis, *Introduction to Astronomy*, 2nd ed., © 1970, Prentice-Hall Inc., Englewood Cliffs, N.J.). The inner circle round the sun symbol gives the glyphs for the 20 calendar days. The year had 18 months – relating to agriculture and the weather – each of 20 days and a separate 5-day period.

Calendar periods which tend to have cold or warm, wet or dry, conditions were identified in many European countries and often related to saints' days. Among the best known (but by no means most reliable!) are the forty days of rain (or fine weather) that supposedly follow a wet (or fine) St Swithun's day (15 July) in England and the 'Ice Saints' spell of frosts in Europe in mid-May. An old French saying, which is related to the latter occurrences, cautions that 'in the middle of May comes the tail of the winter'. In the early seventeenth century Francis Bacon wrote a natural history of the winds in which he pointed to their main characteristics in England – cold from the north, dry from the east warm and often wet from the south, and generally rainy from the west. Such ideas were greatly elaborated in the nineteenth century by meteorologists such as Dove in Germany and Abercromby and Buchan in Great Britain.

Fig. 1.2 The Tower of the Winds, Athens.

Following these early studies there have been numerous attempts to investigate the seasonal structure of climate and the relationships between the character of the atmospheric circulation and weather and climate on a local and a regional scale. Such work has intensified greatly during this century and in the last twenty to thirty years it has become evident that, in seeking explanations for many of the observed characteristics, we are led to consider the atmospheric circulation on scales up to the global one.

B Synoptic Climatology

1 LEVELS OF CLIMATOLOGICAL SYNTHESIS

Synoptic climatology has not, hitherto, generally been regarded as a coherent and discrete field of study. The term is applied rather loosely, usually in the narrow sense pertaining to investigations of regional weather and circulation types. Occasionally it is used to refer to any climatological analysis which makes some reference to synoptic weather phenomena.

Table 1.1 Levels of climatic synthesis (modified from Godske, 1966)

Order	*Variables*	*Description*	*Representation*
0	$f(e_0s_0t_0)$	Instantaneous value of one element at a point.	Synoptic observation of one element
1	$f(e_0t_0s)$	Spatial distribution of one element at one instant	'Synoptic map' of one element; satellite cloud photo
1	$f(e_0s_0t)$	Time variation of one element at a point	Time series (autographic record)
1	$f(s_0t_0e)$	Instantaneous, point value of several elements	Station weather report
2	$f(e_0st)$	Changes in the distribution of one element with time	Series of pressure maps, etc.
2	$f(t_0es)$	Spatial covariance of weather elements	Synoptic weather map
2	$f(s_0et)$	Time covariance of weather elements	Climogram
3	$f(est)$	Covariance of weather elements in space and time	Series of synoptic weather maps

In the words of Durst (1951) 'climate is but the synthesis of weather'. As an introduction to the definition of synoptic climatology let us, therefore, examine the basis of climatological synthesis. Our basic data, or zero order level, consists of the instantaneous value of one weather element (such as temperature) at a given location. Different levels of synthesis may then be built up by considering the variation of the number of weather elements (e), the time variation (t) and the space variation (s), as shown in table 1.1. Notice that space is regarded as having one dimension. In Godske's original table (1966), the zero order synthesis in climatology was characterized as the frequency distribution of a single element at a particular point for a series of observations at a given hour of the day. We consider this to be incorrect, because such a frequency distribution can only be derived from a time series. An analogous frequency distribution can be envisaged relating to the spatial variation of one element, at a given instant as determined from point observations on a microclimatic or topoclimatic scale. The apparent distinction between point values and continuous series in time, on the one hand, and in space, on the other, is being removed by new observational techniques such as satellite photography and imagery, which permit the determination of continuous fields. It needs to be recognized, however, that the time variation may refer to any number of time scales; for example, the diurnal or annual variation, year-to-year differences of daily or monthly averages, etc.

Climatic analysis and synthesis are not made completely explicit in table 1.1 For this purpose some statistical considerations are necessary. The description and summary of a climatic regime (sometimes called climatography) involves a knowledge of the frequency of occurrence of the complete range of values of each weather element. Statistics of variance (s^2), or the standard deviation (s), and the higher moments of a frequency distribution (see p. 215) are still calculated too infrequently although it is generally recognized that the traditional climatic means are often of little value. In determining time (or space) averages the basic variability is filtered out of the data and it has only been realized relatively recently that this process may have unexpected effects on our final results. This subject will be examined later, but we can note here that mean values may be more difficult to comprehend and interpret than information at a lower level of synthesis.

2 COMPARISON OF SYNOPTIC AND DYNAMIC CLIMATOLOGY

The field of *synoptic climatology* is concerned with obtaining insight into local or regional climates by examining the relationship of weather elements, individually or collectively, to atmospheric circulation processes. Some definitions of this field in the literature are as follows:

> The description and analysis of the totality of weather at a single place or over a small area, in terms of the properties and motion of the atmosphere over and around the place or area. (*Court, 1957*)

> The method of arriving at inferences about local and regional climates and about the distribution of climatic elements from synoptic weather maps and from the weather patterns depicted on these maps. (*Landsberg, 1957*)

> The study and analysis of climate in terms of synoptic weather information primarily in the form of synoptic charts. (*Huschke, 1959*)

> The analysis of the structure of climate with interest directed to geographical distributions and to practical prediction for different parts of the world. (*Sutcliffe, 1964*)

According to Jacobs (1947) the term synoptic climatology was first proposed in 1942 in connection with wartime studies of historical synoptic weather data at the Headquarters of the American Air Force, and Hare (1955) notes that it was used in the same context about this time by Durst in Britain. Jacobs states that 'synoptic' is employed in the meteorological connotation of simultaneous or synchronous events although it also implies that the large-scale state of the atmosphere is being considered. This rather conflicts with some interpretations since, according to Hare, synoptic climatology 'deals specifically with regions small enough for the recognized circulation types to be interpreted in the ordinary weather elements'. He contrasts this with *dynamic climatology* which 'is broader in scale, being in effect a regional or global synthesis of daily circulation types'. Subsequently he defines dynamic climatology as: 'the explanatory description of world climates in terms of the circulation or disturbances of the atmosphere' (Hare, 1957), but admits that distinctions between the two fields in meteorological writing are often not clear. Certainly, if we read the objectives of dynamic climatology as originally set out by Bergeron, it is clear that much of the field is now regarded as being encompassed within synoptic climatology:

> . . . a dynamic climatology should describe the frequencies and intensities of well-defined systems that are more or less closed in a thermodynamic sense. This method can easily be applied to the stable weather types of the tropical zone or to weather types directly caused by mountain ranges. In order to obtain a dynamic climatology for the latitudes of non-periodic weather changes, semi-stationary weather types or their constituents must be isolated analytically and can then be studied as well-defined, unambiguous total systems or processes, in the same way as those of monsoons and trade winds in the zones of periodic or stable weather. As in the tropics, these systems can probably be best understood by their characteristic components. (*Bergeron, 1930*)

However, Hesselberg (1932) came nearer to present-day views: 'Dynamic climatology must be concerned with the quantitative application of the laws of hydrodynamics and thermodynamics . . . to investigate the general circulation and state of the atmosphere, as well as the average state and motion for shorter time intervals.' The emphasis here is on the physical mechanisms. Nevertheless, as we shall see in Chapter 3.C, even some studies of hemispheric circulation patterns can be regarded as part of synoptic climatology in terms of their methodology. In respect of scale, therefore, the topics included in this book go beyond Hare's definition, but it would in any case be pedantic to try and circumscribe too precisely the relevant subject matter. As the subsequent chapters show it is difficult, if not impossible, to assign research studies into neat methodological pigeonholes.

The basic aim of synoptic climatology, then, is to relate local or regional climates to a meaningful frame of reference – the atmospheric circulation – instead of using an arbitrary time base for assessing the average or modal values. There are essentially two stages to a synoptic climatological study:

1 The determination of categories of atmospheric circulation type.
2 The assessment of weather elements in relation to these categories.

Under the first topic there is a wide range of problems relating to the spatial scale with which one is dealing, and to the way in which one views the circulation – as an instantaneous feature, or as a system which may be averaged in its time and space coordinates. The question of the relationship of weather elements to circulation patterns has involved the development of many familiar synoptic models – the Bjerknes polar-front cyclone, the easterly wave, and so on – as well as studies of weather and airflow, or air mass type, for particular areas. However, it is with the latter type of approach that we are primarily concerned here.

Beyond this broad category of synoptic climatological studies are other, more specific, topics which nevertheless considerably enlarge the scope of the field we shall consider. The original purpose of most synoptic climatological analyses was forecasting and, although the early attempts were quite inadequate, present day long-range forecasting methods employ many synoptic climatological concepts. Indeed, Sutcliffe (1964) refers to practical prediction as a major function, as already noted. Related to this problem on the time scale of years and decades (super long-range, or climatic, forecasting) is the analysis of climate and its variations in the historical period. These two themes are closely interrelated but there are other areas in which synoptic climatological methods can be of value. For example, it is now recognized that elementary climatic data are inadequate in much biometeorological work. In entomology and epidemiology, in particular, the importance of a synoptic climatological framework is apparent from recent research although this term is seldom used explicity.

All synoptic climatological analysis involves the use of various types of synoptic meteorological data. We begin, therefore, by examining the major synoptic parameters and then review the characteristics of the global circulation of the atmosphere since this is the setting within which the smaller systems ultimately have to be understood.

2 The basic data and their analysis

Climatology . . . was not so long ago . . . accepted as defining statistical properties of some of the first moments of the atmospheric state variables at the earth's surface. The aerological era exposed the three-dimensional facet of climatology [and] soon . . . the higher statistical moments . . . were added to the repertoire of a complete climatic specification of the atmosphere. (*J. Smagorinsky, 1970*)

Meteorological observations and derived physical parameters form the basis for any climatological analysis. In general, only a small fraction of the available meteorological information has been put to use in climatology up to the present time, but with the advent of new computational techniques it seems probable that synoptic climatological methods of analysis will be extended to parameters that are, as yet, rather unfamiliar to climatologists. For this reason it is worthwhile examining the essential characteristics of the major items of data and the ways in which they are commonly presented.

A Synoptic data

We consider first synoptic data and the related maps, discussing in turn the pressure and wind fields and parameters derived from them, then isentropic charts, maps of weather elements and frontal analysis. The purpose here is only to emphasize the salient points, not to provide a thorough or rigorous coverage of the physical principles involved, although an outline of some of the basic equations is given in the Appendix, pp. 531–35. For further details of techniques of meteorological analysis and related topics the reader is referred to Petterssen (1956), Saucier (1955), Sawyer (1964) and World Meteorological Organization (1961), in addition to the specific references cited below.

1 THE PRESSURE FIELD

Surface pressure is undoubtedly the most frequently used synoptic parameter. Analysis of its horizontal distribution is simple since this is a scalar field; that is, for a given time each point in space has a specific value of pressure. Pressure data are, for synoptic purposes, reduced to the theoretical value at a given time at mean sea level (MSL) and since the necessary correction involves the observed temperature at the station, 'fictitious' MSL pressures may be reported over montane regions. The apparent intensity of the winter Siberian anticyclone is certainly due, in part, to this correction (Berry, Bollay and Beers, 1945, p. 87; Walker, 1967). In synoptic analyses MSL isobars are usually drawn at 4 or 5 mb intervals and local features are deliberately smoothed out of the pattern by the analyst. Fig. 2.1 illustrates the effects of this procedure. In mesoscale studies, on the other hand, it is precisely such

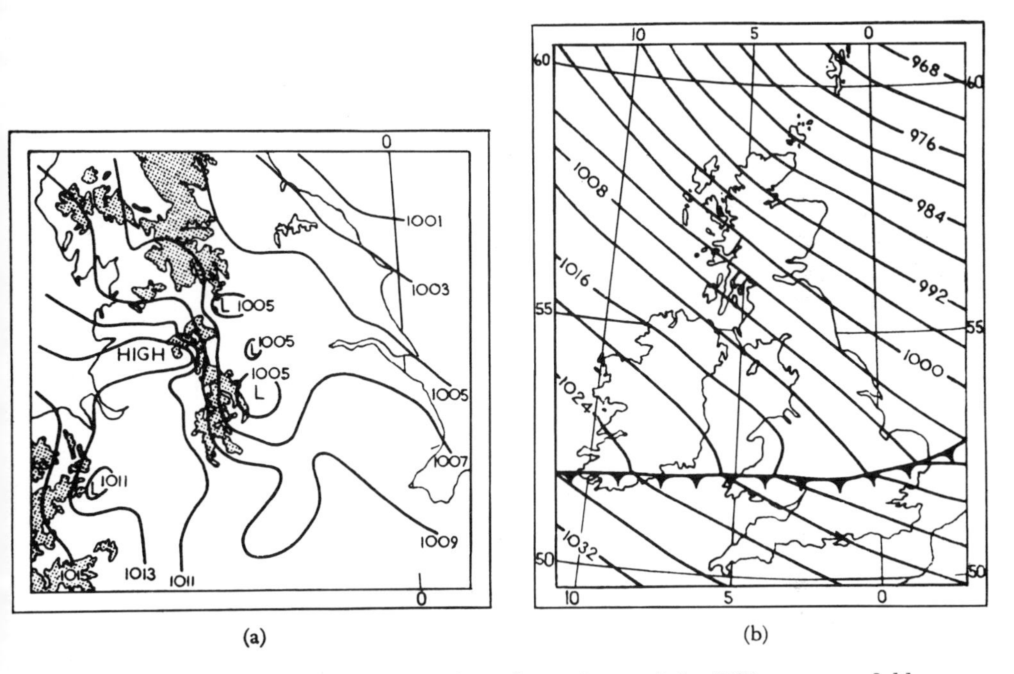

Fig. 2.1 (*a*) Local and (*b*) synoptic scale analyses of the MSL pressure field, 0900 GMT, 16 February 1962 (from Sawyer, 1964).

features that are of interest. Maps of MSL pressure at a given time have been prepared on at least a daily basis for Europe and most of North America since the 1870s, but in parts of the tropics and the polar regions the availability of historical weather-map series is limited at best only to some ten to twenty years. A useful guide to the basic world-wide sources of data and maps has been prepared by the World Meteorological Organization (1965).

Prior to 1945, pressure maps were prepared at constant levels in the free atmosphere, such as 10,000 feet or 3 km. Subsequently, *contour charts* for constant pressure surfaces have become the internationally accepted means of depicting the horizontal pressure fields in the free

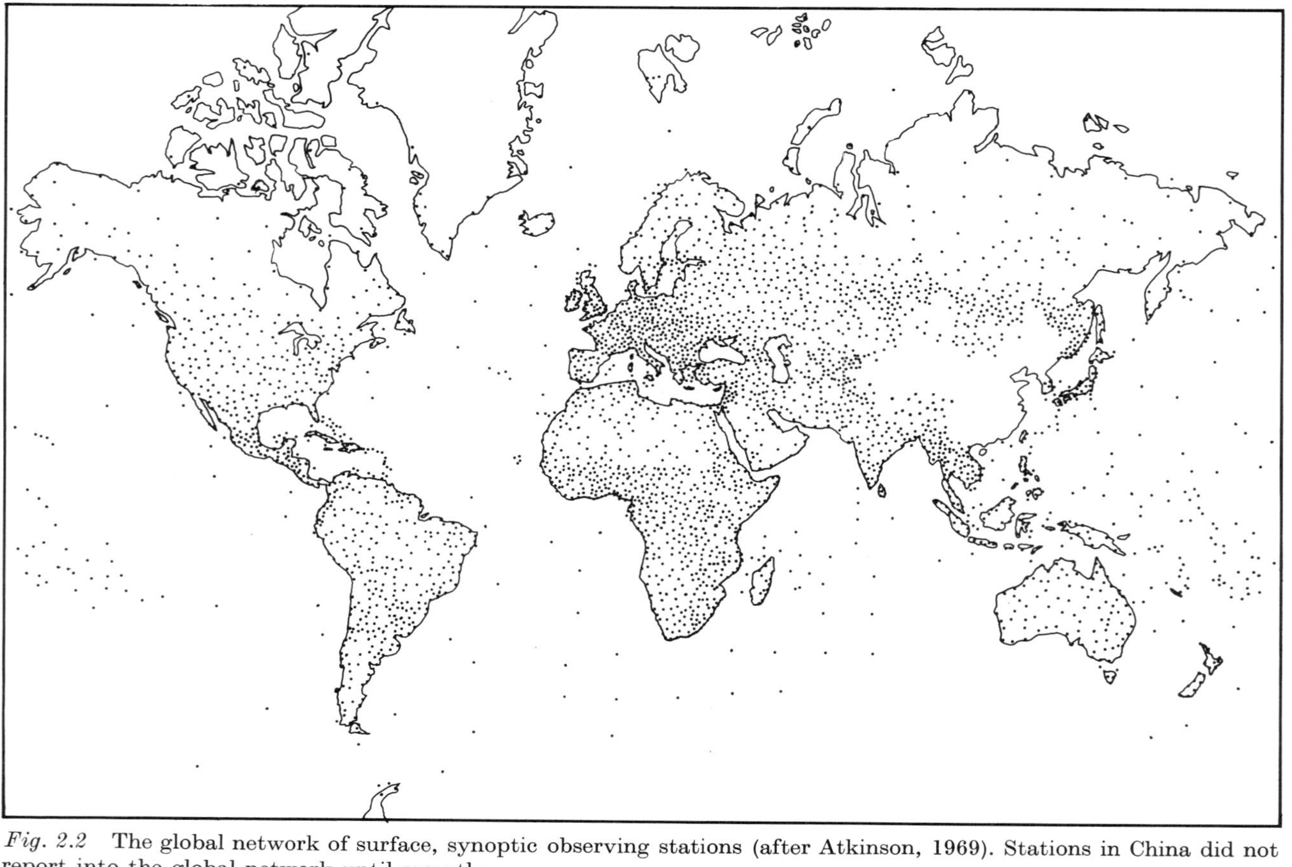

Fig. 2.2 The global network of surface, synoptic observing stations (after Atkinson, 1969). Stations in China did not report into the global network until recently.

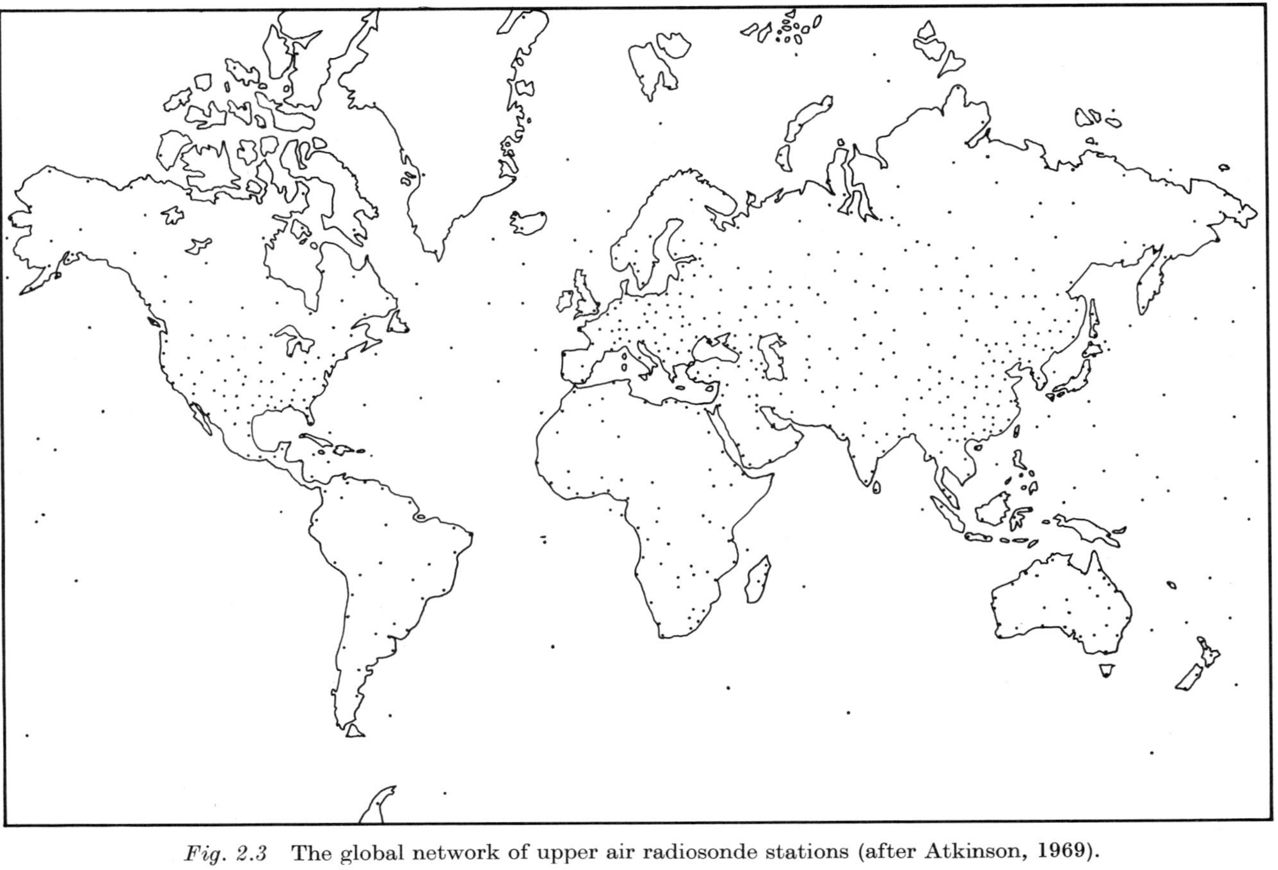

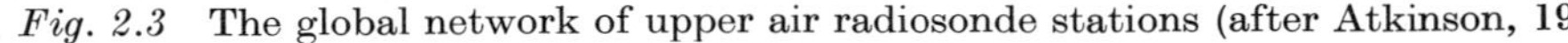

Fig. 2.3 The global network of upper air radiosonde stations (after Atkinson, 1969).

atmosphere.[1] The most commonly available upper air chart is that for the 500 mb surface (about 5·5 km elevation in middle latitudes), which approximates to the mean level of non-divergence in middle and higher latitudes. Johnson and Mörth (1960) regard it as an unsatisfactory level for the analysis of tropical weather systems, however. Other standard surfaces (mandatory reporting levels for upper air data) are the 1000, 850, 700, 300, 200 and 100 mb levels (see Jacobs, 1958). Height is expressed in *geopotential* rather than geometric units:

$$1 \text{ g.p.m. (geopotential metre)} = \frac{g}{9{\cdot}81} \text{ (geometric metre)}$$

where g = acceleration due to gravity (m s^{-2}) (see Appendix).

At this point it is convenient to discuss *thickness* (relative topography) maps since the representation is by simple isolines (*isohypses*). The thickness of an air layer $(z_2 - z_1)$ between two pressure surfaces is proportional to its mean temperature (see Appendix):

$$z_2 - z_1 = \frac{R\overline{T}_v}{g} \ln \frac{p_1}{p_2}$$

where $\overline{T}_v$ = the mean virtual temperature of the layer in °K (virtual temperature is the temperature of dry air at the same pressure and density as the given (moist) air sample).

R = the specific gas constant for dry air ($2{\cdot}867 \times 10^2$ m^2 s^{-2} K^{-1})

g = the acceleration due to gravity ($9{\cdot}81$ m s^{-2}).

Thickness charts for the 1000–500 mb or 1000–700 mb layers have been prepared routinely in European weather offices since the 1940s although this parameter was in fact discussed in 1910 by V. Bjerknes. The application of thickness charts in forecasting has been discussed and illustrated by Sutcliffe (1948) and Sutcliffe and Forsdyke (1950). Climatological applications have almost entirely referred to mean thickness charts (Van Loon and Taljaard, 1958).

It is vital to recognize those areas of the world where analyses are likely to be unreliable owing to inadequate data. Figs. 2.2 and 2.3 illustrate the present coverage provided by surface and upper air station networks. Numerous additional stations are due to be added for the international World Weather Watch programme, together with tethered buoys in the oceans, constant-volume horizontal sounding balloons (like those of the GHOST programme: Lally *et al.*, 1966) and increased satellite capability.

[1] Constant-level analysis began in Britain and North America in 1933 and constant-pressure analysis started simultaneously in Germany. This method was later adopted in the Soviet Union (1940) and Britain (1941) and found international acceptance in 1945 (see Fulks, 1945).

Derived data

The frequency of high and low pressure centres. Maps showing the occurrence of both highs and lows over the northern hemisphere in January and July have been presented by Petterssen (1950) and more extensively by Klein (1957, 1958), based on the 'Historical Weather Maps' of the United States Weather Bureau for 1899–1939. Similar studies for more limited areas have been undertaken by Evjen (1954) for Europe; Weickmann (1960) for the Middle East; Keegan (1958a), Reed and Kunkel (1960), and Mackay, Findlay and Thompson (1970) for the Arctic; Lamb (1959) for the southern hemisphere westerlies; and Taljaard (1967) for the whole of the hemisphere south of 15°S. Figs. 2.4 and 2.5 show the results of these studies.

Such data are an indispensable complement to mean pressure maps (see p. 64), but certain difficulties in their construction must be noted. The delimitation of the centre is often ambiguous in the case of high pressure areas since the pressure gradient is much slacker than in lows. Also, there may be several weak maxima within the last closed isobar, as illustrated in fig. 2.6, and these may shift irregularly over short time periods. Hence statistics of high centres are not particularly accurate. Recent work by Dodds (1971) shows that depressions over the northeast Atlantic are deeper and have lower central pressures than those over Europe, so that any frequency study based on a critical pressure value may be subject to regional bias unless various categories are used. The question of the reliability of the original synoptic analysis is also crucial. Petterssen's analysis, for example, considerably underestimated the frequency of cyclone centres in high latitudes due to the paucity of data north of the Arctic Circle. In the words of Keegan (1958b, p. 25), 'the effect of the presence or absence of even one station in the central Arctic cannot be overestimated'.

Apart from daily synoptic map series, certain data catalogues and atlases of synoptic features have been prepared. Bulinskaia (1963) presented an atlas of the frequency of the main pressure features for 1943–57 over Europe and the eastern North Atlantic. The intensity of each system and the number of isobars comprising it were taken into account. Over the period 1960–4 Vitels (1968) examined the pressure field over the 'European natural synoptic region' in terms of nine features:

0 strong anticyclone, central pressure $\geqslant$ 1035 mb
1 moderate anticyclone, central pressure 1025–1035 mb
2 weak anticyclone, central pressure 1019–1024 mb
3 ridge, or margin of anticyclone
4 diffuse anticyclonic field
5 diffuse cyclonic field
6 trough, or depression margin

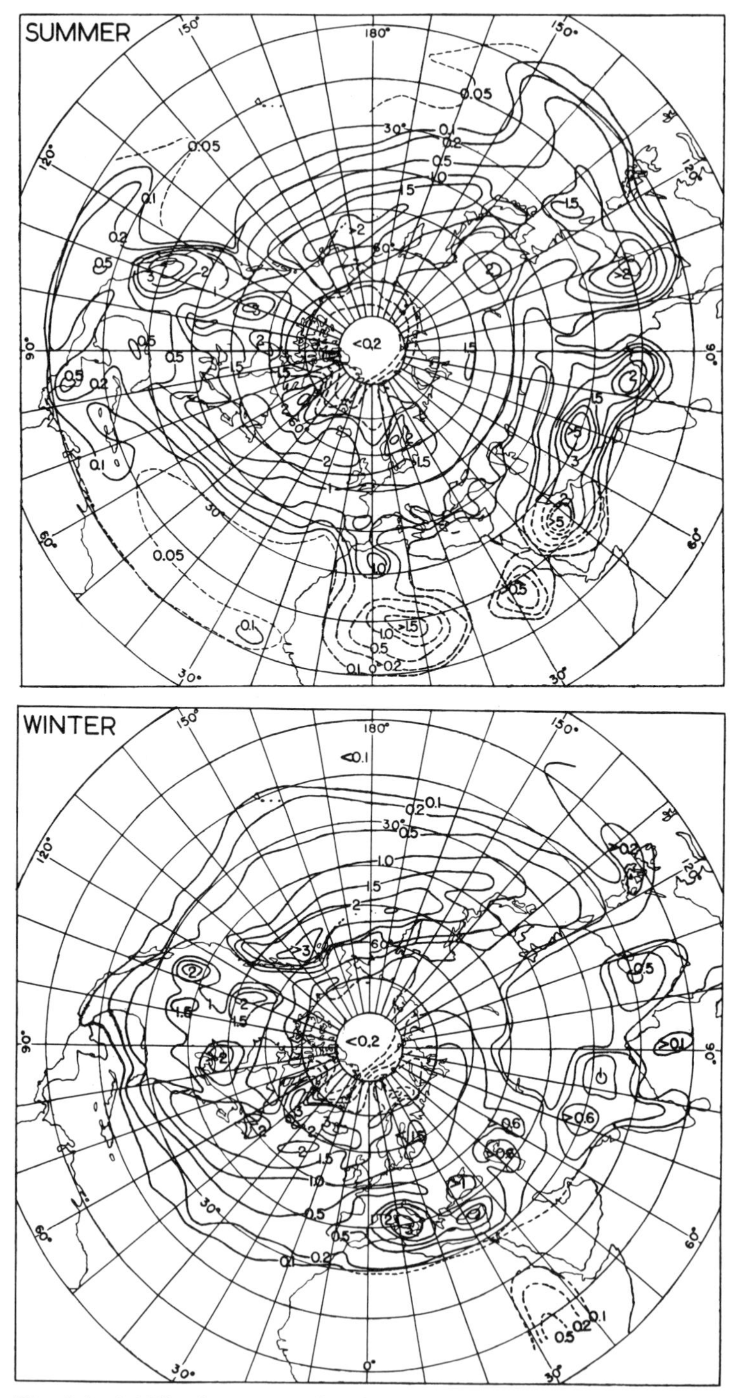

Fig. 2.4 (*a*) The frequency of cyclone centres per 100,000 km² in the northern hemisphere 1899–1939 (from Petterssen, 1950).

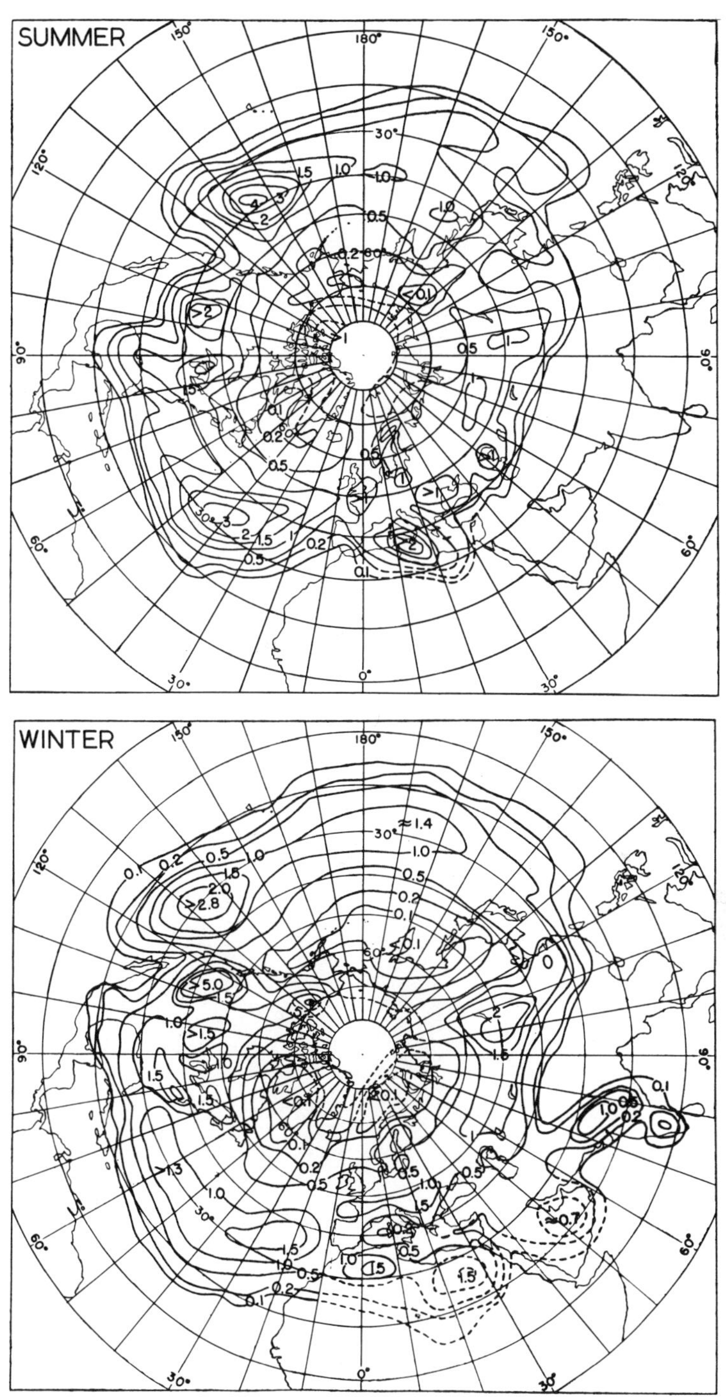

Fig. 2.4 (*b*) The frequency of anticyclone centres per 100,000 km² in the northern hemisphere 1899–1939 (from Petterssen, 1950).

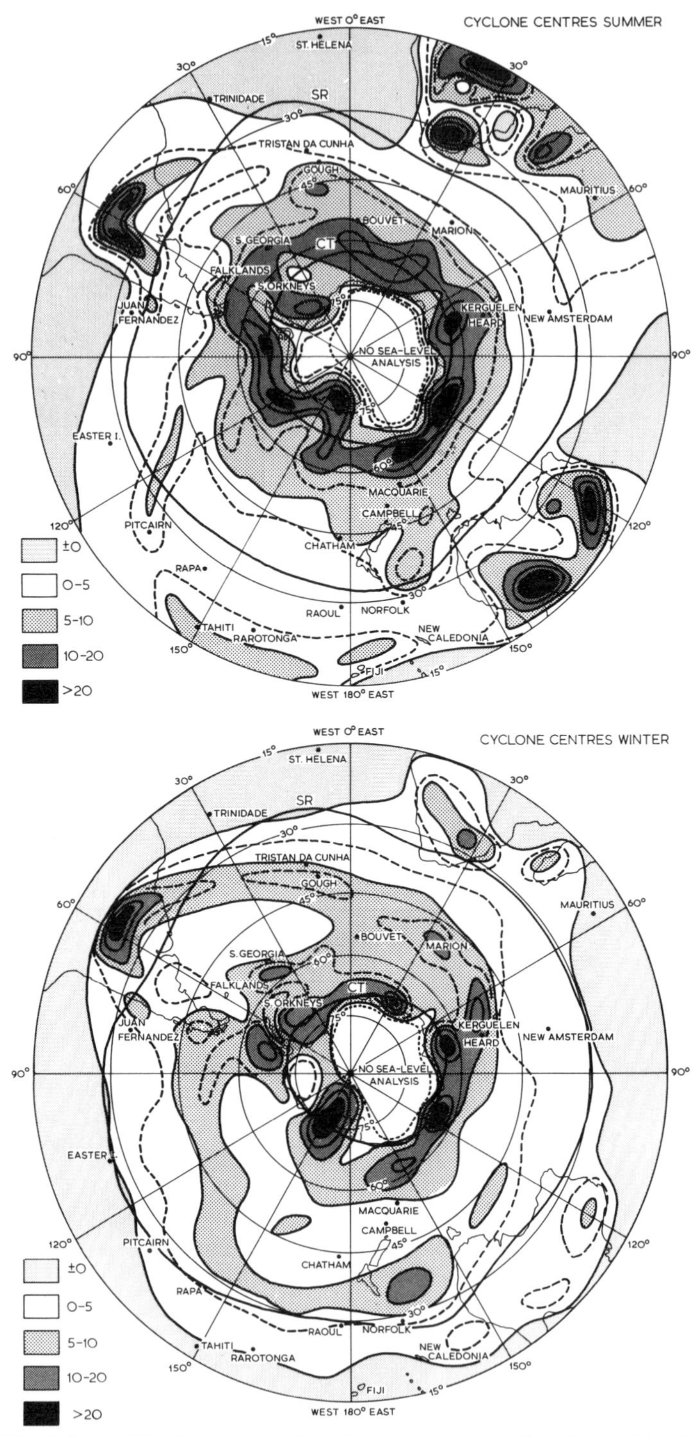

Fig. 2.5 (*a*) The frequency of cyclone centres per 438,000 km^2 in the southern hemisphere in the IGY 'year' (from Taljaard, 1967).

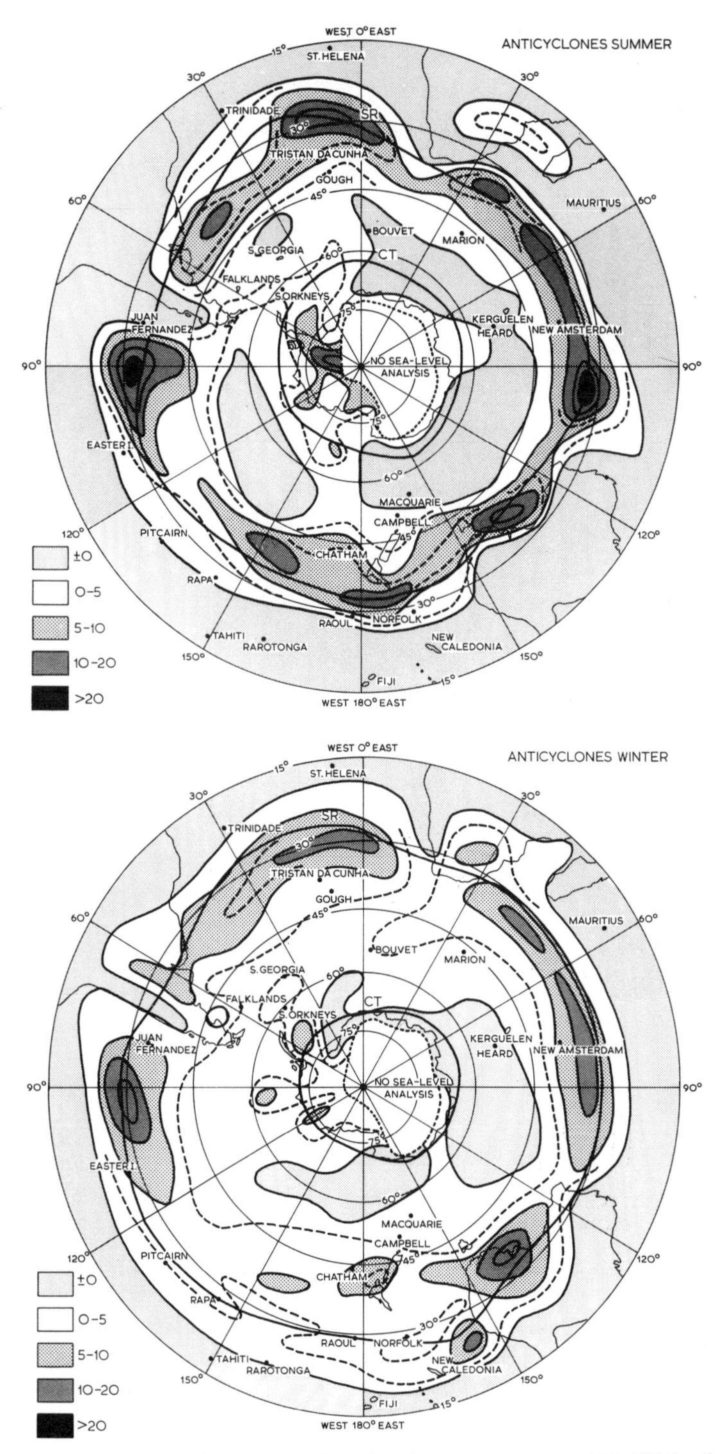

Fig. 2.5 (*b*) The frequency of anticyclone centres per 438,000 km^2 in the southern hemisphere in the IGY 'year' (from Taljaard, 1967).

Winter = June–September Summer = December–March
CT = circumpolar trough SR = subtropical ridge

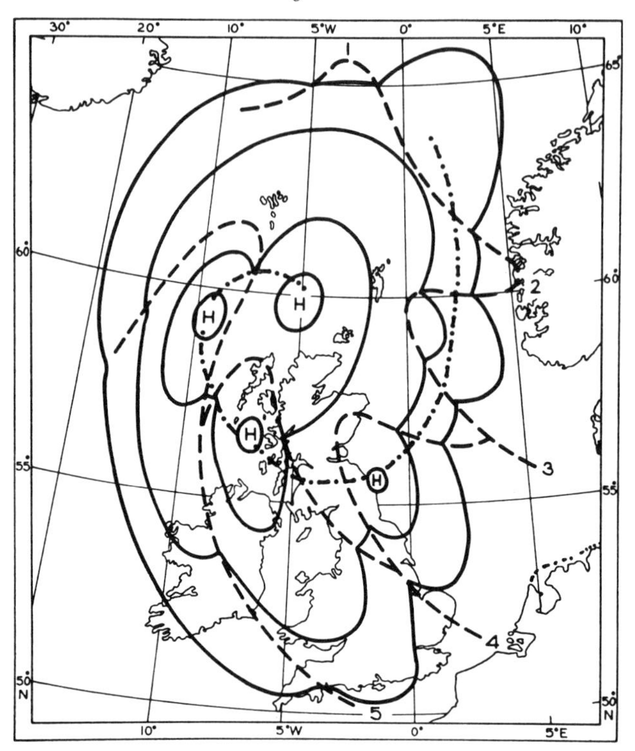

Fig. 2.6 A model of the detailed structure of an anticyclone composed of a cluster of cells (from Findlater, 1967). Solid line = isobars (schematic), broken line = trough lines, dot-dashed line = trajectory of an anticyclone cell.

7 weak depression, central pressure 1001–1005 mb
8 moderate depression, central pressure 990–1000 mb
9 deep depression, central pressure $\leqslant 990$ mb

and provided monthly, seasonal and annual frequency tables.

Relatively little work of a similar nature has been performed for daily upper air charts although Duquet and Spar (1957) have mapped the frequency of 500 mb cyclone centres in each season over North America. Damman (1960) has also analysed the occurrence of 500 mb features on hot summer days (centres > 5800 d.m.)[1] and on cold days (centres < 5600 d.m.) in relation to surface centres for the sector 35°–65°N, 15°W–30°E. Considerable attention has been devoted to five-day and thirty-day mean maps and these are discussed on p. 66. In an attempt to introduce greater objectivity into descriptions of synoptic patterns, Lenova (1970) has determined the 'centres of gravity' of cyclones and anticyclones with respect to a threshold pressure of 1017 mb. The over-

[1] Dynamic metre (d.m.) $\simeq 0{\cdot}98$ m (an absolute unit).

lying mass of the atmosphere was calculated for an area of $1 \cdot 28 \times 10^6$ km^2 on a daily basis for 1951–60. Such an analysis provides detailed information on the intensity of the systems.[1]

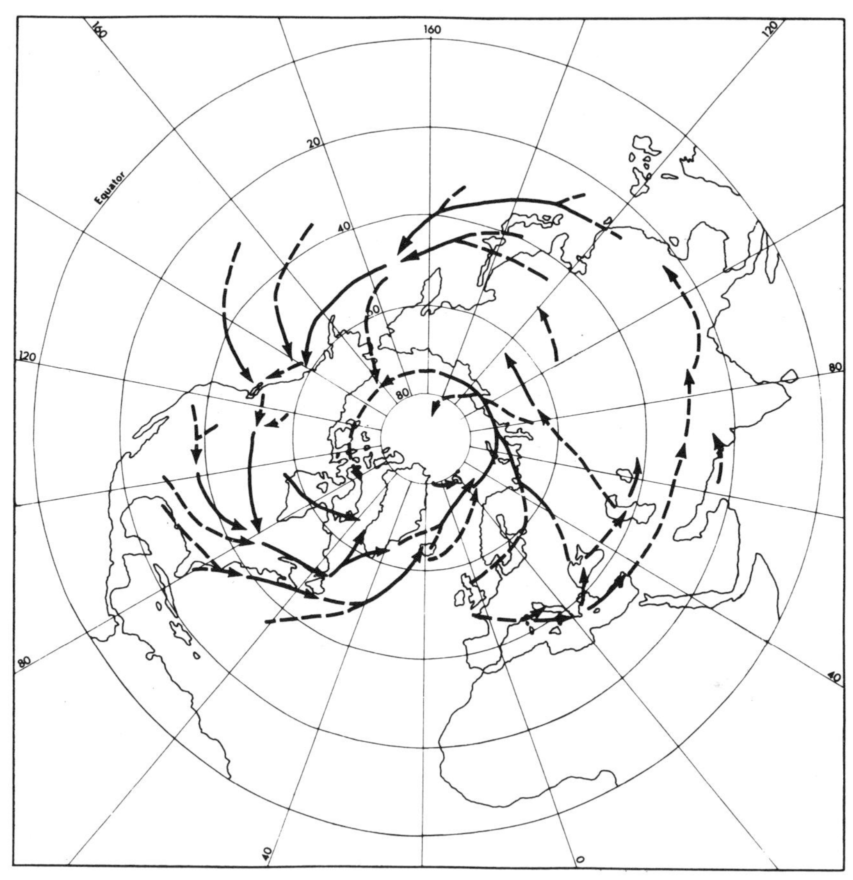

Fig. 2.7 Cyclone tracks in the northern hemisphere in January 1899–1939 (after Klein, 1957; from Barry and Chorley, 1971). Broken line = secondary tracks.

Tracks of highs and lows. Track charts of the movement of high and low pressure centres are prepared by noting the position of the identified centre on successive daily synoptic maps. The earliest work of this type was carried out for Europe and the North Atlantic by Köppen (1880) and Van Bebber (1891) and their depression track 'Vb', from the Mediterranean to the Black Sea, is still referred to in European meteorological literature (Flohn and Huttary, 1950; Schwarzl, 1965). Klein

[1] Godske *et al.* (1957, p. 784) note that a 'centre of gravity' method was used at Massachusetts Institute of Technology in 1943 for determining map analogues. The 'centre of gravity' was defined with respect to indices of zonal and meridional pressure gradient divided by the total pressure over the area.

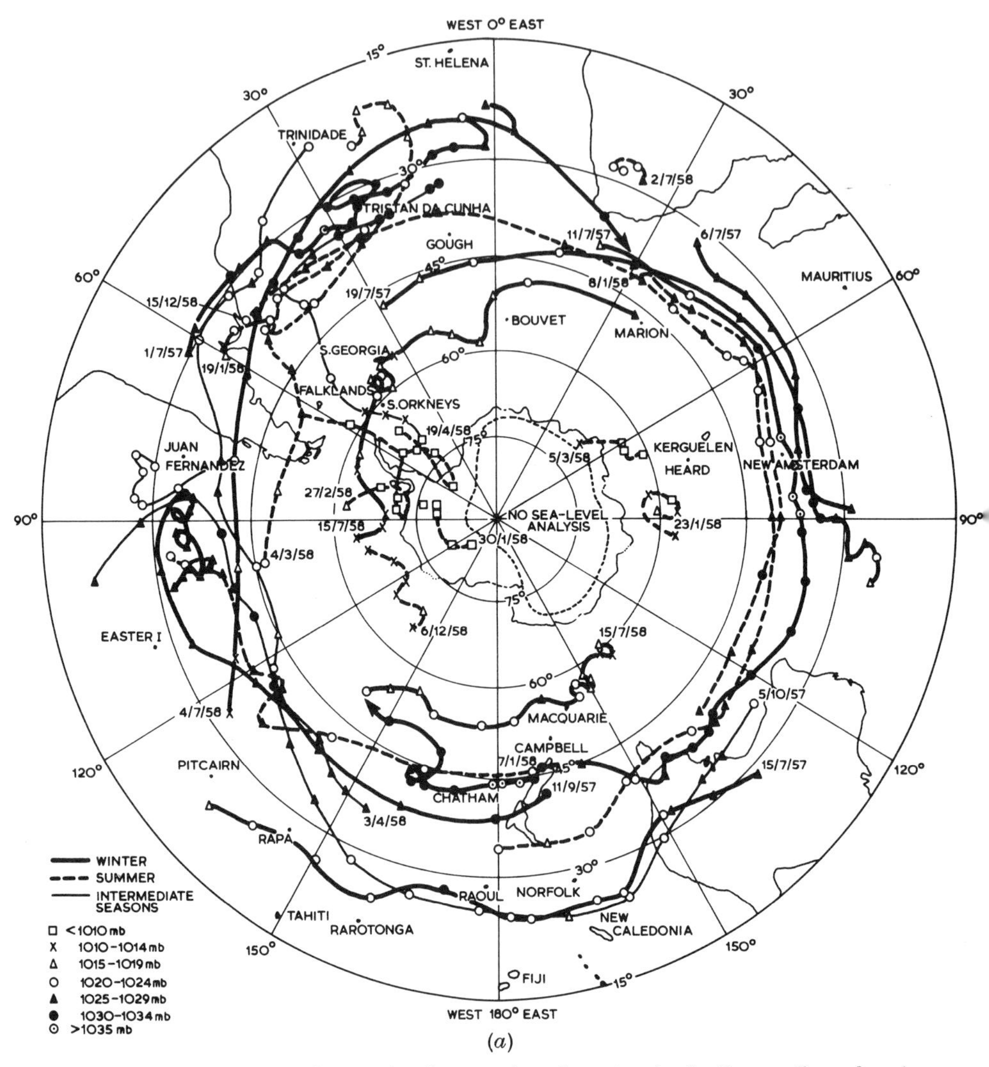

Fig. 2.8 Representative anticyclone and cyclone tracks in the southern hemisphere (from Taljaard, 1967). Tracks are numbered by date of first appearance.

(1957) made a very extensive survey of the movement of pressure systems in the northern hemisphere (fig. 2.7), distinguishing major and minor tracks and thereby identifying preferred areas for the genesis (or intensification) and decay of highs and lows. A similar study has recently been carried out in the Soviet Union by Kryzhanovskaia (1968, 1969). Other analyses of note are those of Keegan (1958a) for the arctic region in winter and of Reinel (1960) for anticyclones over Europe. Information for the southern hemisphere is much more limited, but Taljaard (1967)

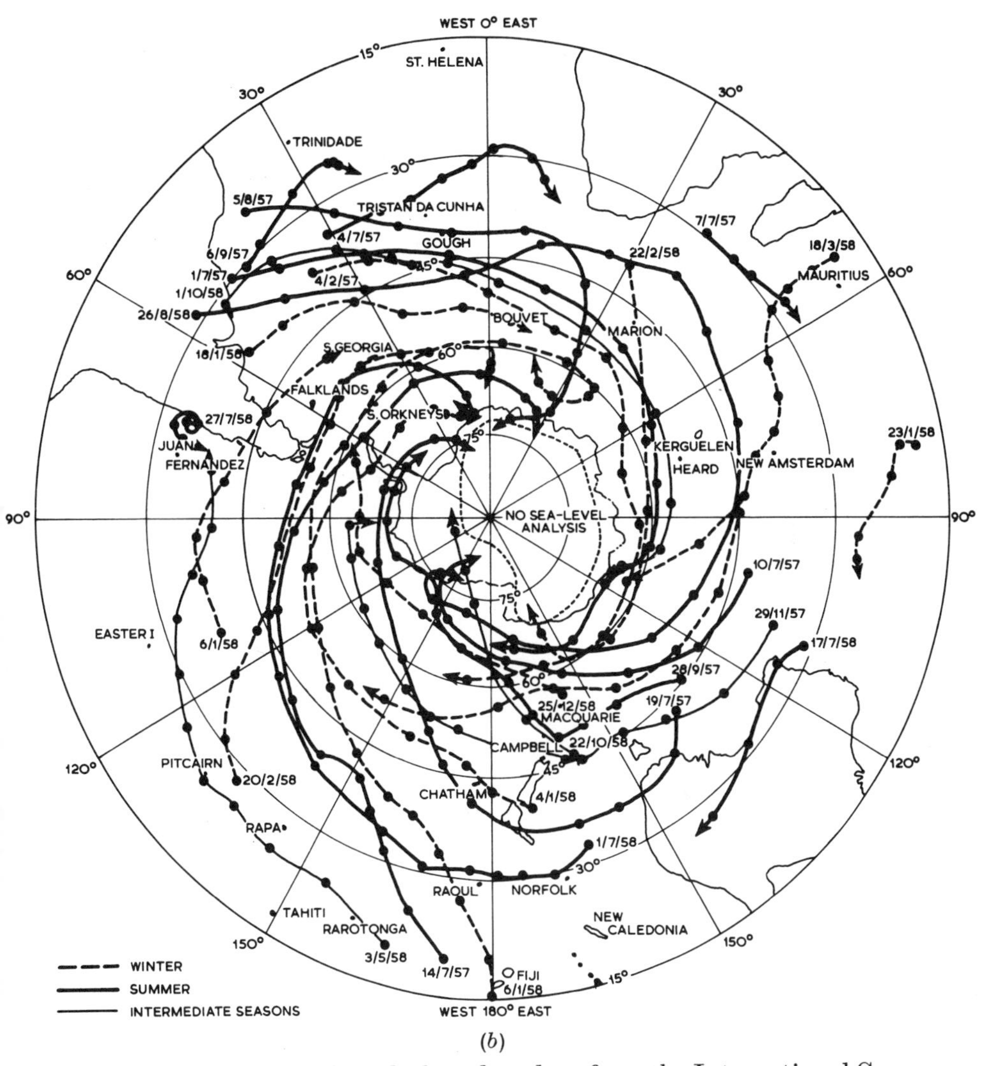

(*b*)

shows 'representative' tracks based on data from the International Geophysical Year 1957–8. Fig. 2.8 illustrates the results of these studies.[1]

As an extension of this type of analysis, Petterssen (1950) also determined the frequency of formation of new centres of high and low pressure, anticyclogenesis and cyclogenesis, respectively. The criterion adopted was the appearance of the first closed 5 mb isobar, provided that the cyclone or anticyclone was identifiable on the next day's chart.

[1] A problem of such analyses is that no distinction can usually be made as to intensity of the system, and also the speed of its movement is ignored. Both of these characteristics may nevertheless be of considerable significance in terms of synoptic climatology (see Chapter 3.8, pp. 102 and 113).

Isallobaric charts. These show the pressure change at given points over a specified time interval by means of lines of constant pressure change (isallobar). They were first used by Ekholm (1904). In low latitudes, where the semi-diurnal pressure oscillations may mask synoptic changes, it is necessary to take twenty-four hour pressure differences. Fields of pressure change, or height change on an isobaric surface, are primarily used in forecasting (Kagawa and Lee, 1967). They are used to indicate changes occurring in the pressure (or height) gradient and therefore the geostrophic winds, as well as to determine the movement of lows and their tendency to deepen or fill. Such charts have, so far, found little climatological application although twenty-four-hour isallobars were used by Dunn (1940) to identify wave disturbances in the tropical easterlies.

2 THE WIND FIELD – KINEMATIC PROPERTIES

Wind velocity, like any vector quantity, may be represented by a directional arrow (showing the direction from which the wind is blowing) proportionate in length to the magnitude. For many purposes this type of presentation is cumbersome although it is commonly used to show the flux of properties such as water vapour, for example.

It is sometimes useful to consider the westerly (u) and southerly (v) components of the horizontal wind velocity ($\mathbf{V}_H$). In cartesian coordinates the u component (positive for west wind) is given by the projection of $\mathbf{V}_H$ on the west–east (x) axis and the v component (positive for south wind) by the projection of $\mathbf{V}_H$ on the south–north (y) axis. Where the wind direction θ is the azimuth determined from north (360°) corresponding to the $+y$ axis of a cartesian graph

$$u = -\mathbf{V}_H \sin\theta$$
$$v = -\mathbf{V}_H \cos\theta.$$

In evaluating the sign it is helpful to remember that in the sector

0°–90°	sin +	cos +
90°–180°	sin +	cos −
180°–270°	sin −	cos −
270°–360°	sin −	cos +.

Component fields are rarely plotted for synoptic analyses, but we shall return to them again in section C of this chapter (see pp. 72–75) in connection with the global circulation.

In general the most convenient means of analysis is to examine the scalar quantities of direction and speed individually. These can be depicted in the form of *isogon* maps of constant direction and *isotach* maps of constant speed (see fig. 2.9). It is more usual, however, to determine

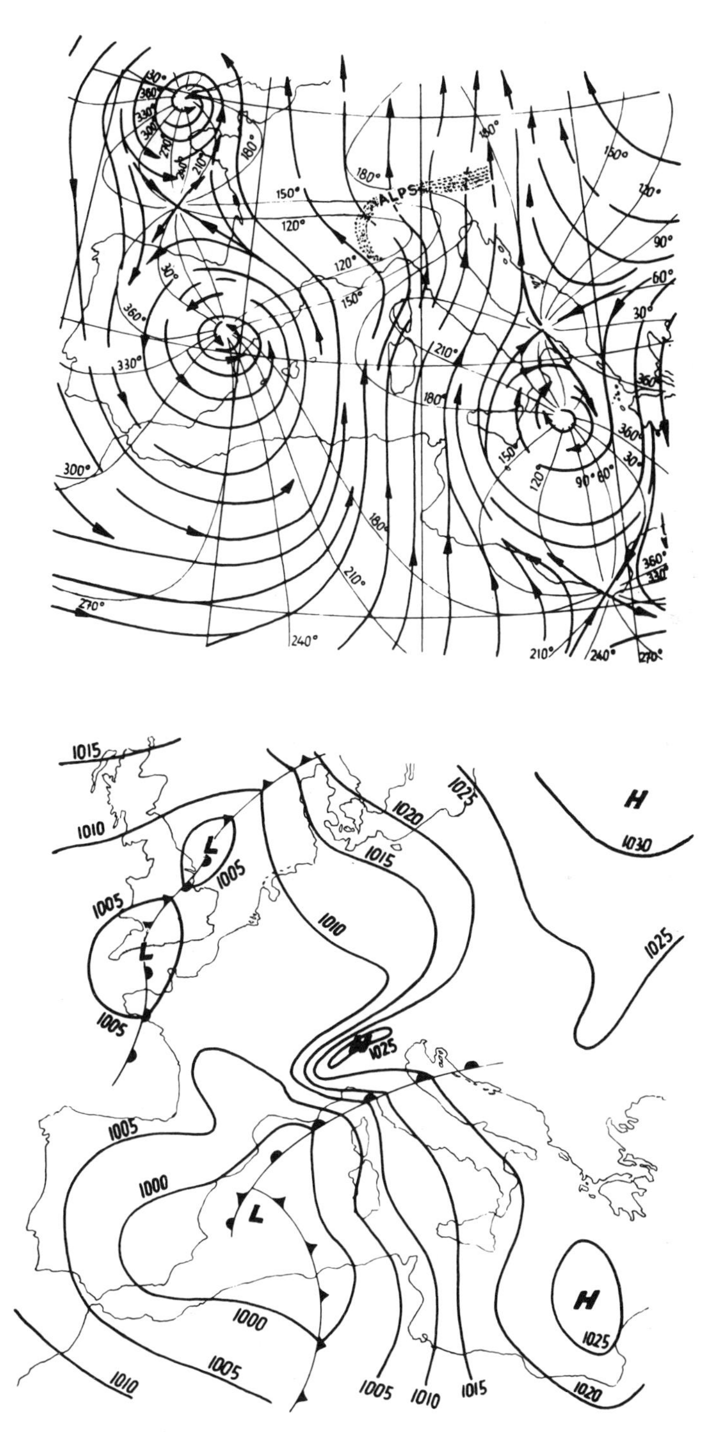

Fig. 2.9 An example of isogon and streamline analysis at 850 mb, 17 April 1962, and the corresponding MSL pressure map (from Schüepp, 1963).

streamlines, or lines tangent to the instantaneous motion at each point. These can be constructed by sketching directly from the velocity vectors. A more precise method using an isogon map is to draw line segments parallel to the wind direction across each isogon and then connect the segments by tangent curves as shown in fig. 2.10. This approach is particularly valuable in that wave motions in the flow are readily detected by isogons. The wind speed may be shown by superimposed isotachs with the spacing of the streamlines independent of the wind speed (Palmer, 1952) or, alternatively, the spacing of the streamlines is made inversely proportional to the speed (Watts, 1955). Streamlines may respectively converge into, or diverge from, centres of inflow and outflow known as 'singular points'. Inflow may also occur along a singular line or asymptote of convergence. The rate at which flow is converging or diverging with respect to an axis perpendicular to the flow is referred to as confluence and diffluence, respectively.

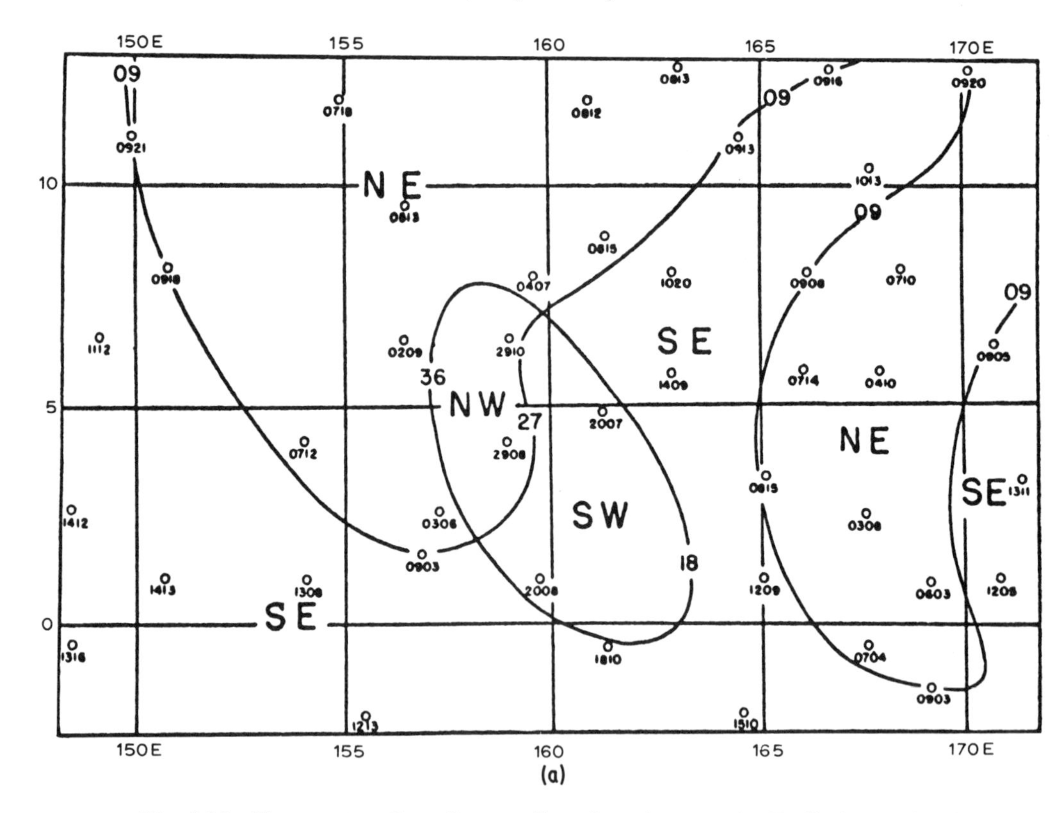

Fig. 2.10 The construction of streamlines from isogons. (*a*) Preliminary sketch of cardinal isogons – the numbers are wind direction in tens of degrees and speed in knots. (*b*) Complete isogon analysis (tens of degrees) with singular points. (*c*) Streamlines constructed from isogon line elements (from Johnson, 1965).

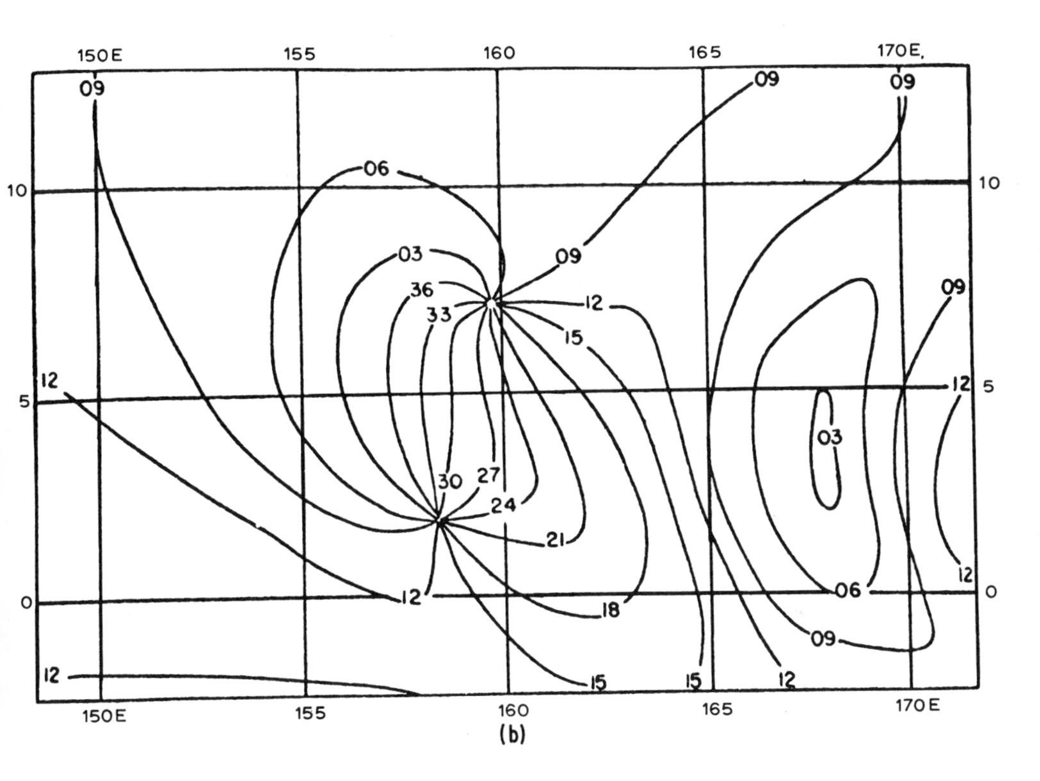
150E
155
160
165
170E
10
5
0
09
06
03
36
33
30
27
24
21
18
15
12
(b)

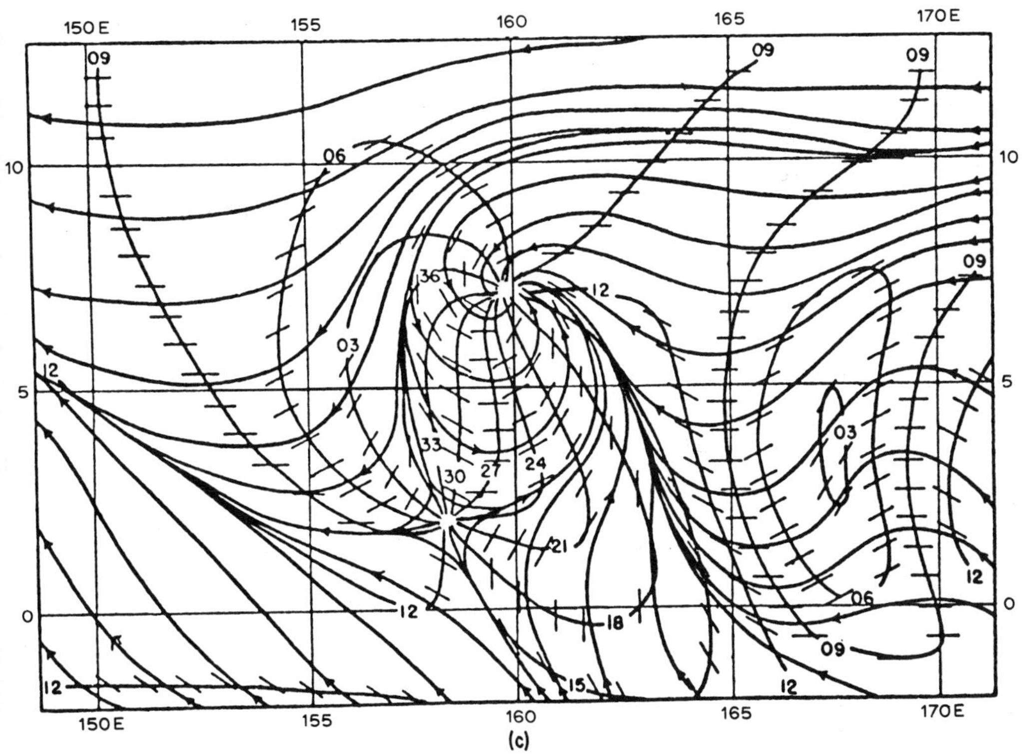
150E
155
160
165
170E
10
5
0
09
06
03
36
33
30
27
24
21
18
15
12
(c)

The basic patterns of streamline curvature and diffluence or confluence are illustrated for the anticyclonic cases in fig. 2.11. The corresponding cases of cyclonic curvature give rise to patterns which are the exact reverse of the anticyclonic ones.

Streamline analysis is used most frequently in low latitude analysis since pressure gradients are generally small and consequently the geostrophic wind field is not readily determined. Illustrations of such streamline maps may be found in Palmer (1952), Watts (1955) and Sadler (1965). Because small variations in the wind vector may be significant in the tropics, and the reliability and representativeness of observations at a single level is sometimes in doubt, it has become common to use *mean layer* winds for 3000–10,000 feet (*c.* 930–3080 m). These averages are routinely reported in tropical RAWIN-sonde ascents (Zipser and Colon, 1962).

Recently, interest in streamline analysis has been revived in the Canadian Meteorological Service. Jarvis (1967), for example, studied the occurrence of boundary layer cloud over eastern Canada in relation to

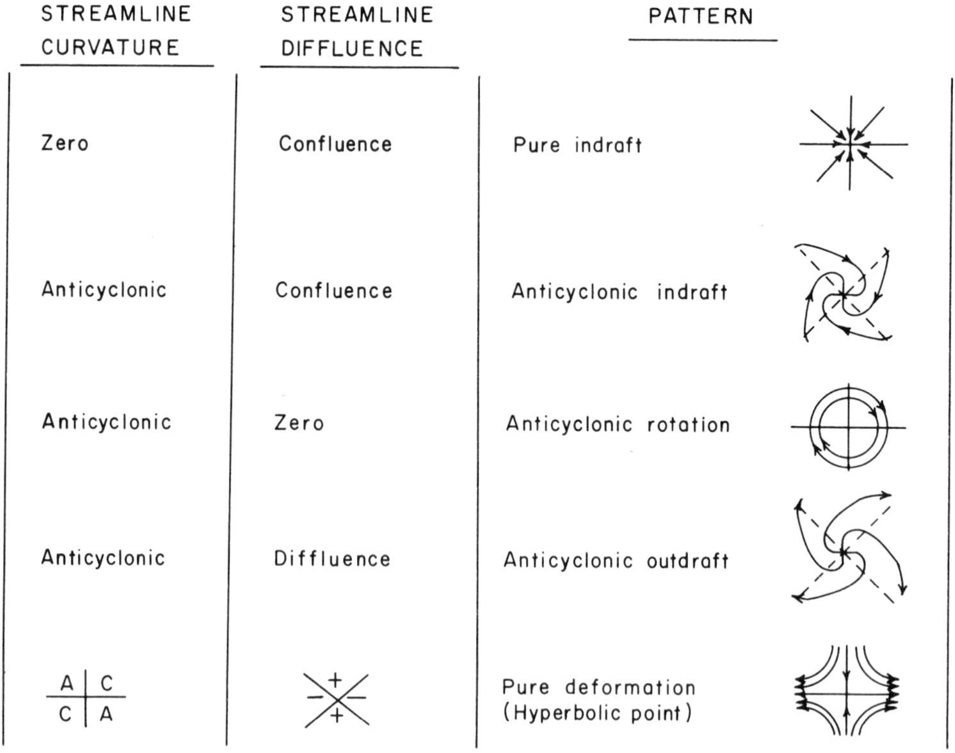

STREAMLINE CURVATURE	STREAMLINE DIFFLUENCE	PATTERN
Zero	Confluence	Pure indraft
Anticyclonic	Confluence	Anticyclonic indraft
Anticyclonic	Zero	Anticyclonic rotation
Anticyclonic	Diffluence	Anticyclonic outdraft
A C C A	+ − − +	Pure deformation (Hyperbolic point)

Fig. 2.11 Basic patterns of (anticyclonic) streamline curvature and diffluence (after Jarvis, 1967).

streamline curvature and confluence zones and Thompson (1969) examined the variation in snowfall patterns around the Great Lakes due to confluence zones.

Derived data

Horizontal derivatives. Four primary characteristics, which involve combinations of the horizontal derivatives of the velocity components, are derived from the wind field (fig. 2.12). They are:

$$\frac{\partial u}{\partial x}+\frac{\partial v}{\partial y} = \text{horizontal divergence, } \nabla_H.\mathbf{V} \text{ in vector notation}$$

$$\frac{\partial u}{\partial x}-\frac{\partial v}{\partial y} = \text{`stretching' deformation}$$

$$\frac{\partial v}{\partial x}+\frac{\partial u}{\partial y} = \text{`shearing' deformation}$$

$$\frac{\partial v}{\partial x}-\frac{\partial u}{\partial y} = \text{relative vorticity about a vertical axis, } \nabla_H \times \mathbf{V}.$$

A word must be said about vector notation (see Appendix). The gradient operator ∇ (del) is defined as:

$$\nabla \equiv \mathbf{i}\frac{\partial}{\partial x}+\mathbf{j}\frac{\partial}{\partial y}+\mathbf{k}\frac{\partial}{\partial z}$$

where **i**, **j** and **k** are unit vectors in the x, y and z directions respectively. In the horizontal case (∇_H) the vertical term $\partial/\partial z$ is of course omitted. By definition, $\nabla.(\ )$ is the divergence of, $\nabla \times (\ )$ is the curl (or vorticity) of, the term following the operator.

The horizontal divergence may be estimated by using a finite difference grid. That is to say, $\partial u/\partial x$ and $\partial u/\partial y$ are approximated by finite values $\Delta u/\Delta x$ and $\Delta v/\Delta y$ of the velocity components and lengths. Thus, with reference to the coordinates of Appendix fig. 1,

$$\nabla_H.\mathbf{V} = \frac{u_x - u_{-x}}{L}+\frac{v_y - v_{-y}}{L}$$

where L=a unit length in a rectangular grid (Miller, 1948; Panofsky, 1951),[1] $L/2=x=-x=y=-y$ (see Appendix fig. 1). By definition divergence is positive. Divergence (convergence) which measures the overall

[1] An alternative means of estimating divergence and vorticity from streamlines and isogons is given by Graham (1953). For mesoscale studies techniques have been developed, using two Doppler radar sets, to map streamline divergence directly (Browning and Wexler, 1968).

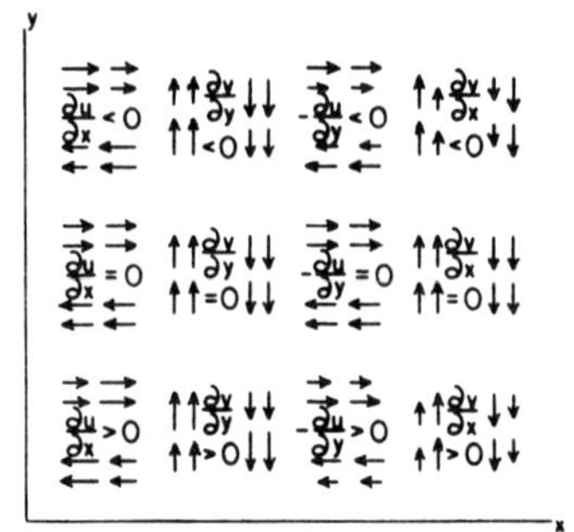

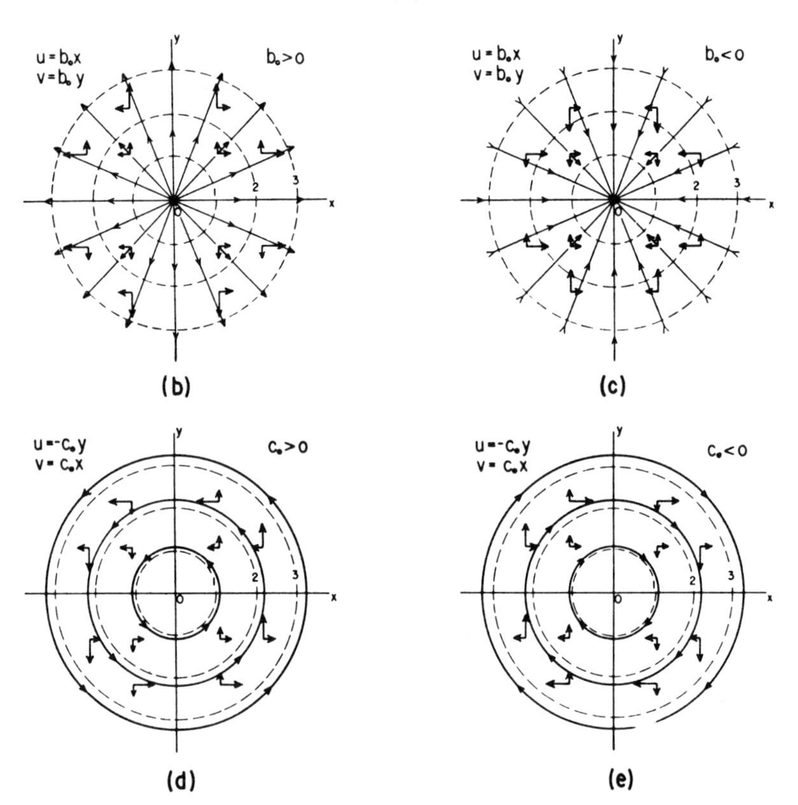

Fig. 2.12 Schematic models of kinematic properties of the wind field (from Saucier, 1955).

(*a*) Sign conventions. Length of arrow indicates relative speed. The sign of each partial derivative refers to each pair of cases shown (above and below, or left and right of the expression).

(*b*) Divergence (positive).

(*c*) Positive vorticity (cyclonic in the northern hemisphere).

(*d*) Stretching deformation.

(*e*) Shearing deformation.

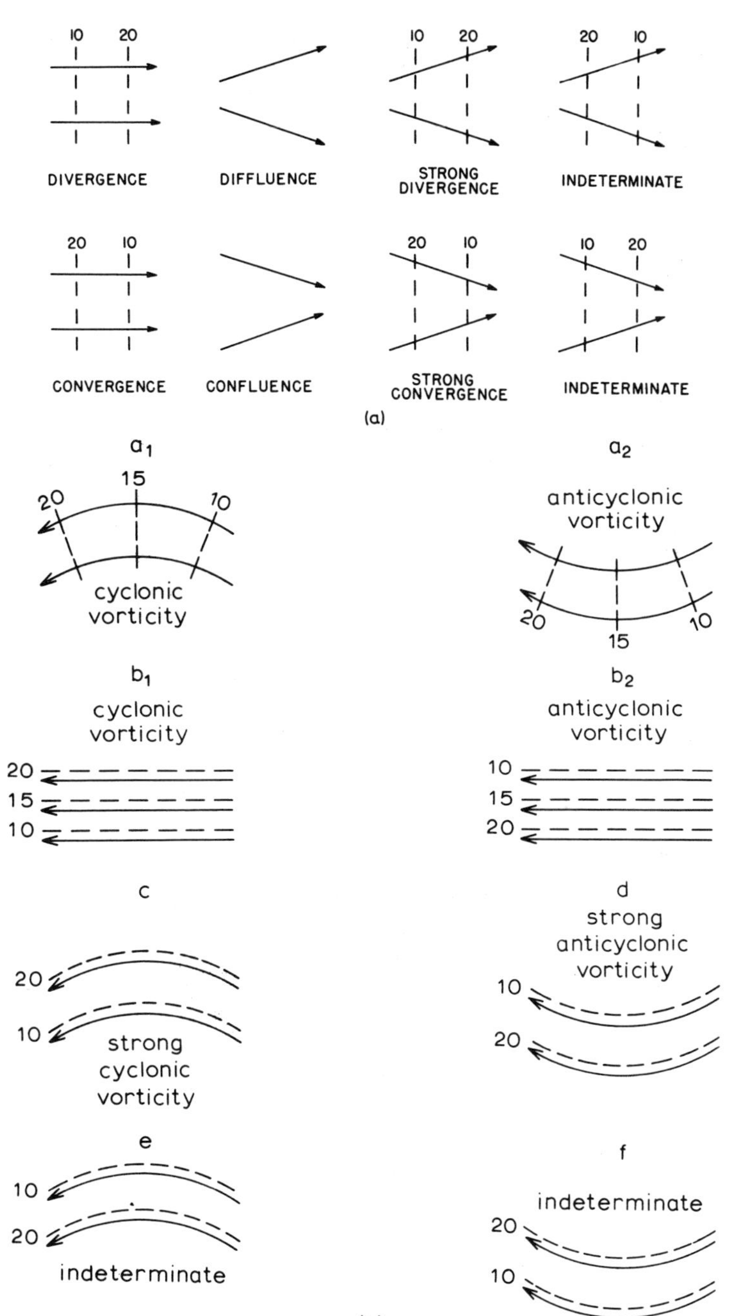

Fig. 2.13 Schematic models illustrating (*a*) divergence and convergence, (*b*) vorticity due to streamline curvature and lateral shear. The dashed lines are isotachs ($m\ s^{-1}$) (from Barry and Chorley, 1971).

expansion (contraction) of the wind velocity field must not be confused with confluence (diffluence). Diffluent streamlines may, for example, be associated with decreased wind speed so that the two effects tend to cancel out (fig. 2.13(*a*)).

In practice, wind data are usually inadequate for obtaining very reliable estimates of the divergence because of inaccuracy in the basic measurements and because the wind is nearly geostrophic above the friction layers. The horizontal divergence is zero for geostrophic flow[1] (when the horizontal pressure force is exactly balanced by the Coriolis acceleration).

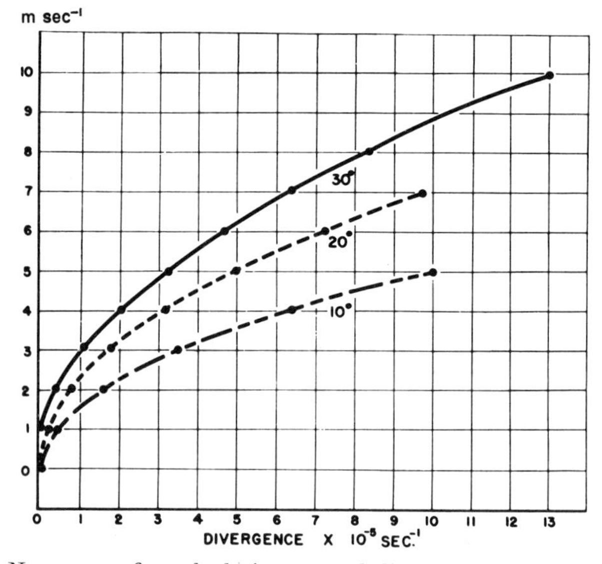

Fig. 2.14 Nomogram for calculating coastal divergence. Enter with the wind components (m s^{-1}) parallel to and normal to the coast against latitude. The sum of these components gives the divergence (10^{-5} s^{-1}) on the abscissa (from Bryson and Kuhn, 1961).

The special case of divergence induced by differential stress at a coastline has been examined by Bryson and Kuhn (1961). Onshore winds produce convergence through the slowing down of the flow due to increased friction over land. Winds parallel to the coast, with high pressure over the land (in the northern hemisphere), also produce convergence as a result of the shear set up by the reduced speed over land. The air is thus forced up the pressure gradient. Bryson and Kuhn determine the divergence by examining the frictional drag across a coastal strip. Fig. 2.14 shows their nomogram for direct computation of the divergence in low latitudes.

[1] This is strictly true only when the geostrophic wind has no north or south component (and therefore the Coriolis parameter is constant).

The vertical component of relative vorticity, ζ or $\nabla_H \times \mathbf{V}$, which is that due to the local rotation about an axis vertical to the earth's surface, is determined by finite differences (see Appendix fig. 1) from:

$$\nabla_H \times \mathbf{V} = \frac{v_x - v_{-x}}{L} - \frac{u_y - u_{-y}}{L}.$$

The vertical relative vorticity in plane polar coordinates is made up of two elements – lateral shear and streamline curvature (Scorer, 1957, 1958). We can write

$$\zeta = \frac{\partial V_s}{\partial r} + \frac{V_s}{r}$$

where V_s = horizontal velocity along a streamline
r = radius of curvature of the streamline.

These elements may reinforce one another or tend to cancel out, as illustrated in fig. 2.13(*b*). By definition, vorticity in the same sense as the earth's rotation, cyclonic in the northern hemisphere, is positive (fig. 2.12). Relative vorticity is a most important synoptic parameter, but its climatological usage has been minimal up to the present.

For some purposes it is necessary to consider *absolute vorticity*. The vertical component of absolute vorticity is made up of the sum of the local value of the Coriolis parameter, f, and the relative vorticity ζ determined by the circulation pattern. The Coriolis parameter which is due to the earth's rotation has a value $2\omega \sin \phi$, where ϕ = latitude angle and ω = the earth's angular velocity. It increases from zero at the equator to a maximum of $1{\cdot}458 \times 10^{-4}\ \mathrm{s}^{-1}$ at the poles.

The rate of change of absolute vorticity (following the motion) and divergence are related through the 'vorticity equation'

$$\frac{d(f+\zeta)}{dt} \simeq -(f+\zeta)\nabla_H \cdot \mathbf{V},$$

if we neglect the effects of baroclinicity, tilting of the vortex axis, and friction. This relationship, which shows that horizontal convergence is associated with increased absolute vorticity, is important in analysis since finite difference estimates of ζ are rather more reliable than those of divergence since they are less commonly close to zero. However, the neglected terms and the effects of time changes in intensity and in the relative motion of a circulation system all lead to inaccuracy in the estimation of $d(f+\zeta)/dt$.

The deformation terms are indicators of zones where frontogenesis is likely to occur. The two elements of deformation are often considered together, although in modern synoptic practice new parameters relating to frontal zones are being used (see p. 59). Finally, we may note that if there is no net divergence, vorticity and deformation of the motion, then the streamlines are straight and the velocity is unchanged. This distribution is referred to as pure translation.

Stream function. A horizontal wind vector, $\mathbf{V}_H$ can be separated into a non-divergent part and an irrotational part, thus:

$$\mathbf{V}_H = (\mathbf{k} \times \nabla\psi) + \nabla\chi$$

where ψ = a horizontal stream function
χ = a horizontal velocity potential
$\mathbf{k}$ = a unit vertical vector
∇ = a gradient operator.

For horizontal, non-divergent flow and $\mathbf{V}_H$ parallel to lines of constant ψ as shown in fig. 2.15

$$\frac{\partial\psi}{\partial x}dx + \frac{\partial\psi}{\partial y}dy = 0$$

so that ψ is defined by

$$u = -\frac{\partial\psi}{\partial y}, \qquad v = \frac{\partial\psi}{\partial x}.$$

If variations in the Coriolis parameter (f) are ignored, ψ in an isobaric surface is proportional to gz, i.e. $\psi = gz/f$.

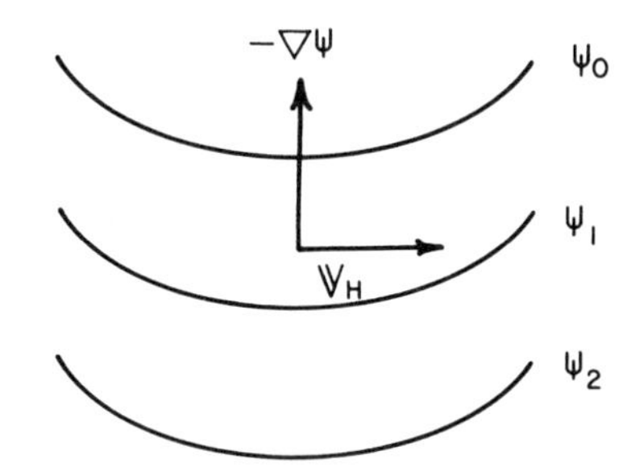

Fig. 2.15 Stream function and wind velocity for two-dimensional non-divergent flow.

Fig. 2.16 illustrates a streamline field for the 200 mb level and the equivalent streamfunctions of the rotational non-divergent wind. In the tropics at 200 mb this comprises the major part of the total wind (Krishnamurti, 1971).

The velocity potential, χ, is defined by:

$$u = -\frac{\partial\chi}{\partial x}, \qquad v = -\frac{\partial\chi}{\partial y}.$$

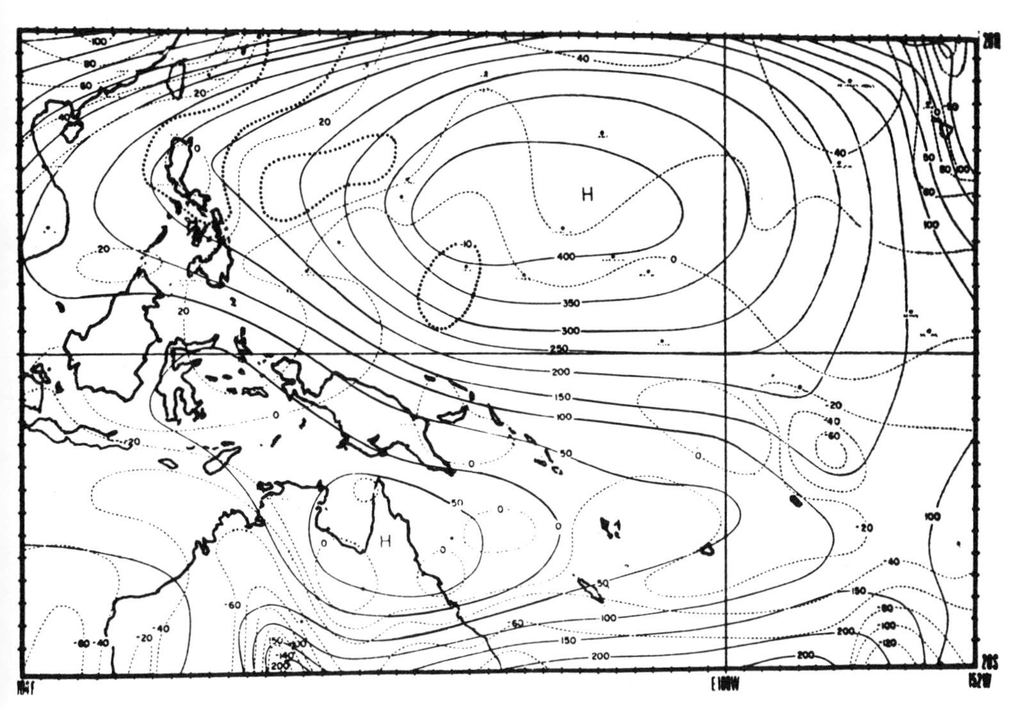

Dotted lines (above) show absolute vorticity.

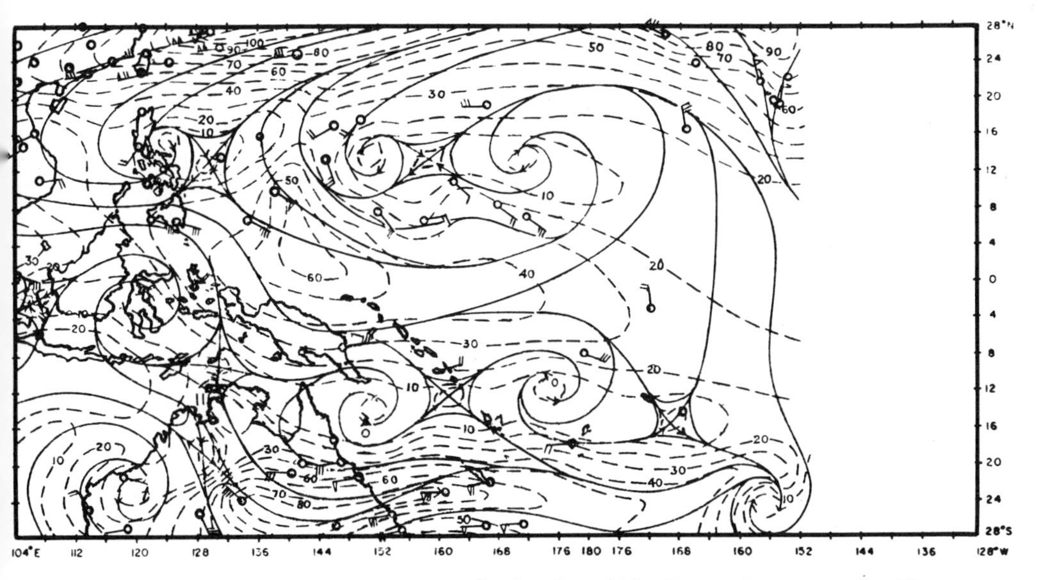

Fig. 2.16 A stream function field (10^5 m^2 s^{-1}) at 200 mb over the western Pacific Ocean at 1200 GMT, March 1965 (above), and the corresponding streamline (solid) and isotach (dashed in knots) analysis (below) (from Krishnamurti, 1969). The streamlines show two anticyclonic centres in each hemisphere, the stream function field only one.

It follows from the above that:

$$\text{Horizontal divergence} = \frac{\partial^2 \chi}{\partial x^2} - \frac{\partial^2 \chi}{\partial y^2} \equiv \nabla^2 \chi$$

$$\text{Relative vertical vorticity} = \frac{\partial^2 \psi}{\partial x^2} - \frac{\partial^2 \psi}{\partial y^2} \equiv \nabla^2 \psi$$

where ∇^2 = the two-dimensional Laplacian operator (see Appendix, p. 535).

The point of such procedures is often the elimination of some undesirable characteristic of the kinematic properties of the wind field. For instance, the deformation, divergence or vorticity can be altered, or removed, for purposes such as smoothing, or in order to obtain conformity with a particular synoptic model in the estimation of wind flow from satellite photographs. Unfortunately, the solution of the appropriate equations in ψ and χ raises serious difficulties, and for this reason Endlich (1967) proposes an alternative procedure. By iterative methods, he produces for example a non-divergent wind field which still retains the original vorticity. Hence the irrotational wind is the difference between the original and the non-divergent patterns.

At the present time the application of these methods is particularly connected with numerical modelling but, with the advent of computer programs to perform the necessary computations and to map the fields, their possible implications for climatological analysis should not be overlooked. Petterssen and Calabrese (1959), for example, isolated the rotational component of the pressure field over the Great Lakes for two synoptic intervals in winter in order to examine the effects of the local heat source. A detailed discussion of stream functions in relation to general types of horizontal motion is given by Godske *et al.* (1957, p. 387–98).

Trajectories. A trajectory is the actual path followed by an air parcel (in general, any fluid particle) over a given period of time. This path will only coincide with a pattern of streamlines if these are constant with time and this is rarely the case. As remarked by Dove (1862, p. 79), 'Most winds are liars as they do not come from the regions from which they appear to blow.' Direct 'observations' of trajectories are now made in the upper troposphere by horizontal sounding balloons, or TRANSOSONDES, such as those of the GHOST (Global Horizontal Sounding Technique) programme. These balloons are adjusted to travel along a constant pressure level with the upper tropospheric winds and are tracked by radio transmissions (Angell, 1961). Such flights have provided valuable information on the wave structure in the upper westerlies of the southern hemisphere.

It is usually necessary to compute approximate trajectories for synoptic studies. Some of the earliest work of this type on trajectories in depressions by Shaw and Lempfert (1906) was a major stimulus to synoptic meteorology. Such analysis of motion where the coordinate system moves with the individual particle considered is referred to as a Lagragian method of description, whereas the study of motion fields with respect to a fixed reference frame is a Eulerian method (see Appendix). In practice, however, a time sequence of fields is used to calculate selected trajectories. Suppose we wish to trace the horizontal movement of an air parcel now at point P (at some level above the friction layer) during the preceding twenty-four hours using six-hourly charts. The simplest, but least accurate, method is as follows:

1 Determine the geostrophic velocity, V_0 (km hr^{-1}), from the contour spacing at a selected point, P_0, at the terminal time, T.
2 Six hours earlier ($T-6$) the air parcel was $6 \times V_0$ km in the upwind direction, based on the streamline at P_0. Call this point P_1.
3 Determine the geostrophic velocity, V_1, at $T-6$ at point P_1. Based on the streamline at the point the air parcel was $6 \times V_1$ miles upwind at $T-12$. Call this point P_2, and so on.

An assessment of the errors due to approximating the trajectory over a time period by a streamline based on isobars or contours in this manner has been made by Hogben (1946). The displacement of synoptic systems with time is a major source of error in estimating trajectory curvature and distortion of the isobars creates further uncertainty. The accuracy may be improved if we use the mean speed at the beginning and end of each time interval, i.e.:

$(V_0 + V_1)/2$ for the period T to $T-6$
$(V_1 + V_2)/2$ for the period $T-6$ to $T-12$, etc.

Petterssen (1956, p. 27–28) gives an account of a more detailed procedure using successive approximations based on the *vector* mean for each time interval.

Near the surface, allowance has to be made for the effects of friction. On average, the angle between the wind direction and an isobar is about 10°–15° over oceans and 20°–25° over land, although in some situations the deviations may be much greater. More serious is the effect of vertical motion on the trajectory. This problem is considered below.

The determination of trajectories is time consuming, and while there are many limited case studies (Burbidge, 1951), there have been few climatological analyses. One major investigation of trajectories in relation to air masses over North America during July 1945–51 and 1954–6 has recently been reported by Bryson (1966). Case studies of air masses over the British Isles are given by Taylor and Yates (1967).

3 ISENTROPIC CHARTS

An isentropic chart shows meteorological elements on a surface of constant potential temperature (an isentropic surface). Potential temperature (θ) is the temperature an air parcel attains if brought dry adiabatically to a reference pressure, usually 1000 mb:

$$\theta = T\left(\frac{1000}{p}\right)^{\kappa}$$

where κ (Poisson constant) $= (c_p - c_v)/c_p = 0{\cdot}288$

c_p = the specific heat at constant pressure

and c_v = the specific heat at constant volume (for dry air).

It may be noted that the thermodynamic diagram known as the pseudo-adiabatic chart has $p^{0\cdot288}$ as ordinate so that dry adiabats (isentropic surfaces) are straight lines. The tephigram chart is similar in this respect.

In the atmosphere isentropic surfaces slope upward towards the poles, i.e. towards cold air, as long as the air is stable. The slope is about 1/500 in air masses and 1/100 in frontal zones.[1]

The technique of isentropic analysis was first suggested by Sir Napier Shaw (1930, p. 259) and was used extensively in the United States during the 1930s and 1940s by C. G. Rossby and his associates (1937), although it was abandoned for operational purposes at the end of the Second World War. The isentropic surfaces suggested for analysis over the United States are 290–295°K in winter and 310–320°K in summer (Namias, 1940). Three sets of lines are necessary for airflow analysis on a selected isentropic surface:

1 Isobars, usually drawn at 50 mb intervals (these also give the saturation mixing ratio for the given temperature).
2 Condensation pressure (the pressure at the lifting condensation level, also called the 'characteristic point') – which gives the potential wet-bulb temperature (θ_w) and the humidity mixing ratio (r).
3 The isentropic stream function, $\psi_\theta = c_p T_\theta + g z_\theta$ (Montgomery, 1937).[2] $c_p T$ is referred to as the specific enthalpy, gz is the geopotential, subscript θ denotes reference to the isentropic surface.

The magnitude of ψ_θ is given by:

$$\psi_\theta = (1{\cdot}0046 T_\theta + 9{\cdot}806\, z_\theta) \times 10^3 \text{ m}^2\text{ s}^{-2}$$

[1] These values may be compared with general slopes in the range $1/10^4$–$1/2 \times 10^5$ for isobaric (constant pressure) surfaces (Wilson, Beaudoin and Donais, 1968).

[2] This function is also known as Montgomery's acceleration potential.

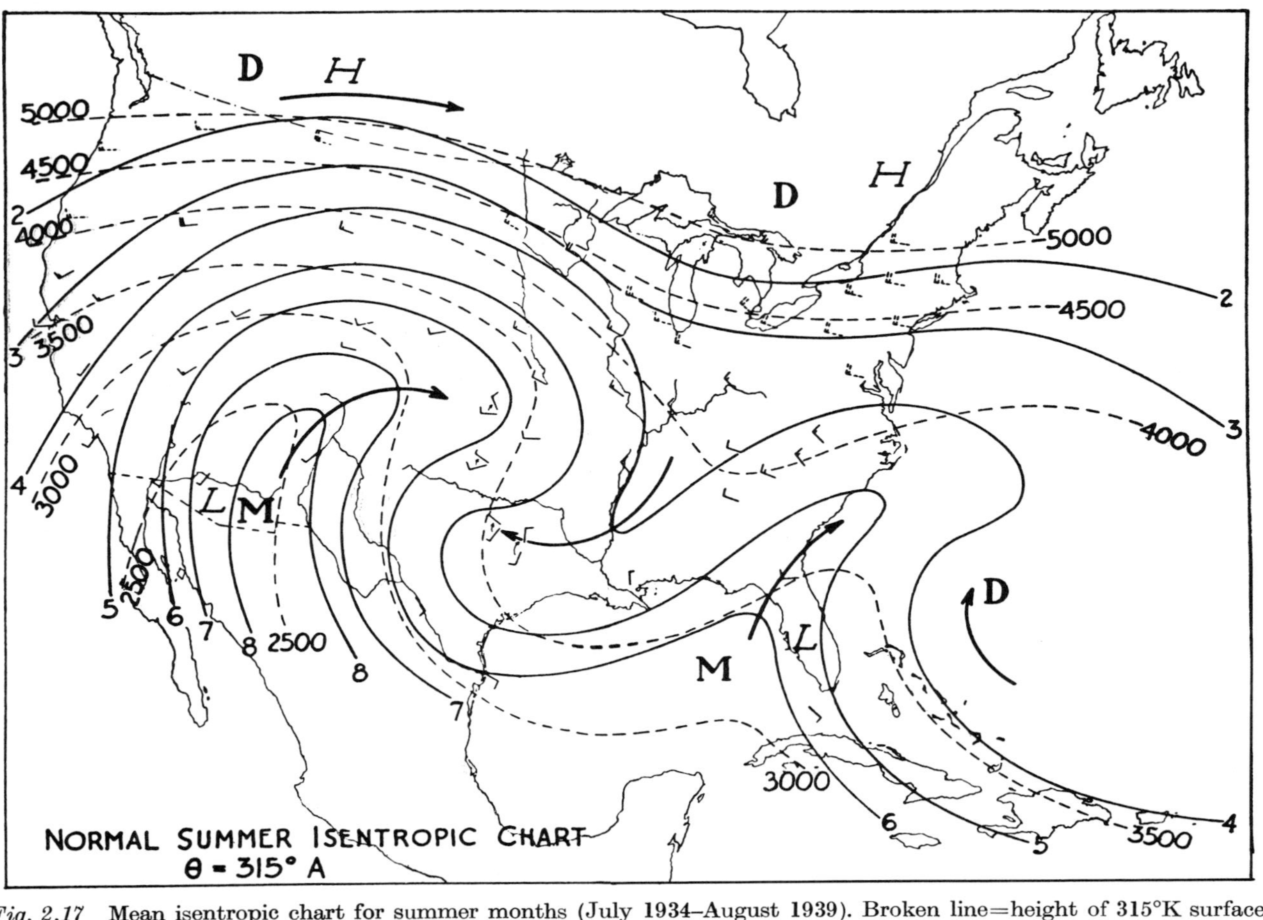

Fig. 2.17 Mean isentropic chart for summer months (July 1934–August 1939). Broken line=height of 315°K surface (*m*), solid line=specific humidity (g kg^{-1}); winds are resultants for the isentropic level; D=dry, M=moist (from Namias, 1940).

where $T_\theta = °K$

z_θ = geopotential km.

For x, y, θ, t coordinates streamlines of ψ_θ are equivalent to isobars in a cartesian system, so that $\nabla\psi_\theta$ on an isentropic surface defines the geostrophic wind. Thus,

$$fu_g = -\left(\frac{\partial\psi}{\partial y}\right)_\theta, \qquad fv_g = \left(\frac{\partial\psi}{\partial x}\right)_\theta,$$

An analysis of the errors in computing the stream function has been made by Danielsen (1959) who showed that T_θ and z_θ must be determined as dependent functions of p_θ to keep the error small (Reiter, 1972).

A major application of isentropic analysis has been made with respect to moisture patterns. Wexler and Namias (1938) demonstrated, for example, that isentropic surfaces show a semi-permanent pattern of adiabatic saturation pressure values over North America in summer (see fig. 2.17). There are isentropic ridges, with dry tongues of air, off the

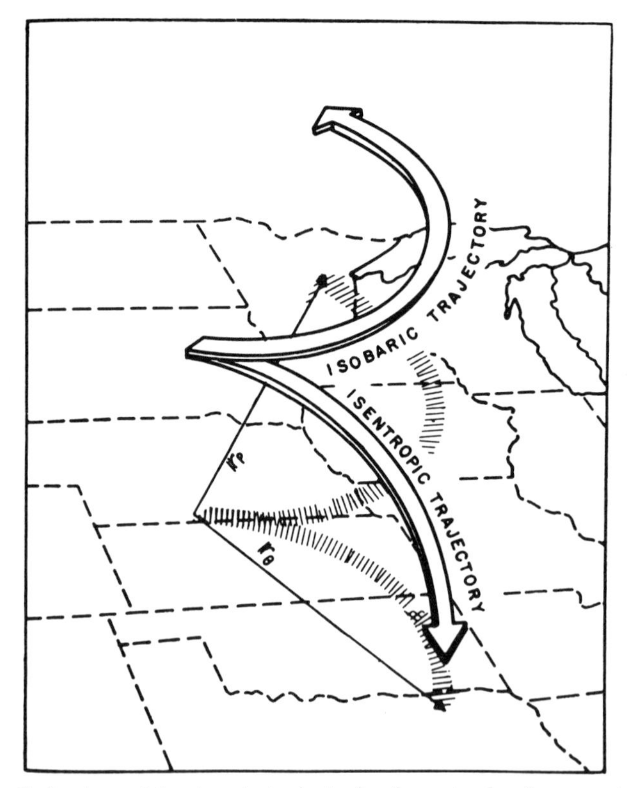

Fig. 2.18 Isobaric and isentropic trajectories for a twelve-hour period originating at the 700 mb level, 0300 GCT, 28 March 1956. The terminal pressure on the isentropic trajectory is 880 mb (from Danielsen, 1961).

east coast and from the central United States southward to the Gulf of Mexico. These tend to curve westward with decreasing latitude, while the intervening moist tongues curve eastward with increasing latitude. Several hypotheses have been advanced to explain this anticyclonic curvature, but the problem does not appear to have been answered unequivocally. The moist tongue from the Gulf of Mexico accounts for the summer shower activity over eastern Arizona and New Mexico.

In so far as vertical motion is adiabatic, isentropic charts indicate vertical displacements of air and, since an isentropic surface is in this respect conservative, they can be used to estimate three-dimensional trajectories. The vertical movement, in pressure units, is obtained from the difference between the initial and final points on the trajectory, and the geostrophic movement by the trajectory on the selected isentropic surface.

A comparison of isobaric and isentropic trajectories has been made by Danielsen (1961) and fig. 2.18 illustrates the completely different picture that these methods may give over a twelve-hour period. The horizontal deviation is about 1300 km, the curvatures are of opposite sign and the isentropic trajectory descends from 700 mb to 880 mb! Other cases may be less dramatic but the value of isentropic analysis is evident.

A computation method for determining isentropic trajectories has been detailed by Danielsen (1961). It is based on the total-energy equation, assuming hydrostatic equilibrium and frictionless flow.

$$\frac{\partial\psi}{\partial t} \approx \frac{d}{dt}\left(\frac{V^2}{2}+\psi\right) \quad \text{or} \quad \int \frac{\partial\psi}{\partial t_\theta}\,dt \approx \left(\frac{V_2{}^2-V_1{}^2}{2}\right)_\theta + (\psi_2-\psi_1)_\theta.$$

The first term on the right represents the change in the kinetic energy of the horizontal wind between the first and last positions (neglecting diabatic effects) and the second is the total change in the stream function which must be approximated. The integral term denotes the change in enthalpy and potential energy which is not converted into kinetic energy. If the trajectory starts at a data point (so that ψ_1 and V_1 are known) it is then necessary to find a terminal point (ψ_2 and V_2) which satisfies the equations:

$$\psi_2-\psi_1+\left(\frac{V_2{}^2-V_1{}^2}{2}\right) \approx \frac{\Delta\psi_1+\Delta\psi_2+2\psi_M}{4}$$

and
$$D \approx \left(\frac{V_2+V_1}{2}\right)\Delta t$$

$$\left[\text{more accurately, } D = \left(\frac{V_2+V_1+2V_M}{4}\right)\Delta t\right]$$

where D = the distance travelled in the time interval Δt
$\Delta\psi$ = the difference in ψ at a fixed point during Δt
M = the mid-point value along the trajectory.

Backward extrapolation can also be performed by determining a locus of points which satisfy the distance criterion (D) for Δt and then substituting values of ψ_1 and V_1 at these points into the first equation until

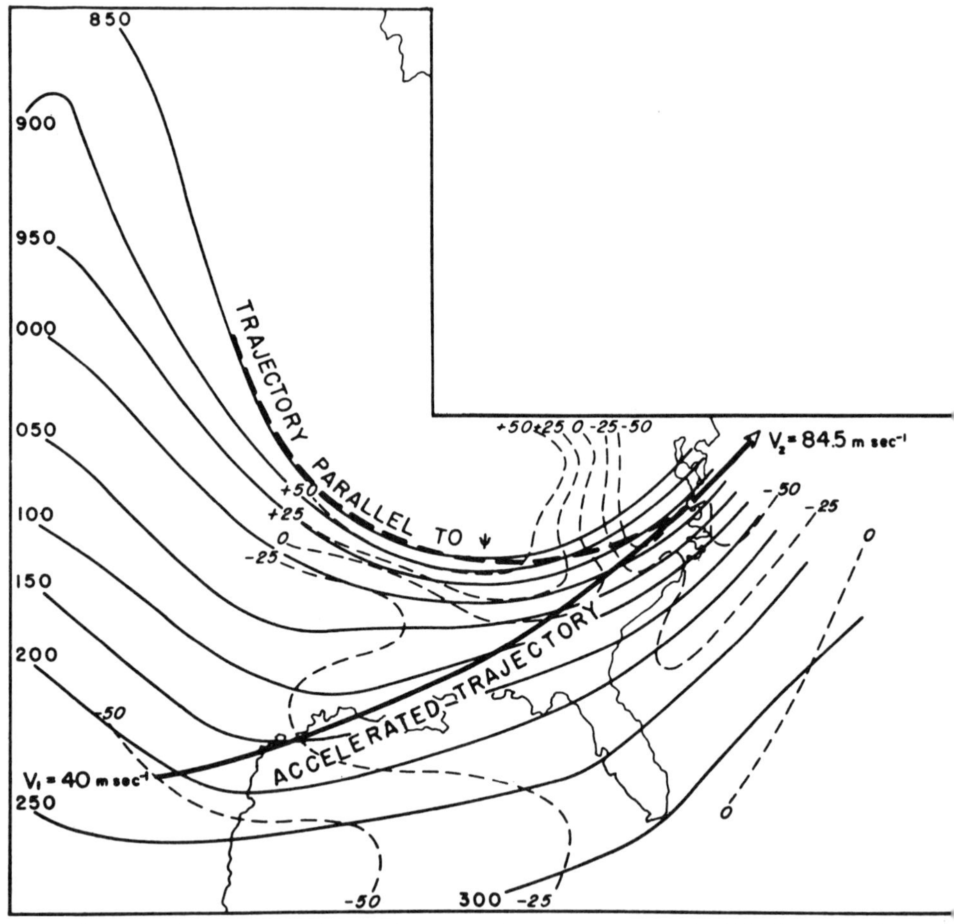

Fig. 2.19 Example of the construction of a trajectory for the 323°K isentropic surface, 0000 GCT, 2 January 1958. Continuous lines are isopleths of stream function, ψ (last three significant figures in units of 10^5 cm^2 s^{-2}), dashed lines are twelve-hour changes of ψ (last two significant figures), V_2 denotes the wind speed over Norfolk, Virginia, at 1200 GCT at 323°K level. Extrapolation upstream for 12 hours along constant ψ (heavy dashed line) satisfied the distance equation but not the energy requirements. Both equations are satisfied by the solid line where $V_1 = 40$ m s^{-1} and ψ_1 ($32{,}217 \times 10^5$ cm^2 s^{-2}) is much larger than ψ_2 ($31{,}910 \times 10^5$ cm^2 s^{-2}) (from Danielsen, 1961).

this identity is also satisfied. Fig. 2.19 illustrates the construction of a trajectory by this method. Christie and Ritchie (1969) have applied the method in a study of pollen transport in central Canada. Subsequently Danielsen and Bleck (1966) have investigated flows and trajectories on surfaces of equivalent potential virtual temperature (θ_{ve}) in a wave cyclone situation.

$$\theta_{ve} = \theta_v \exp\left(\frac{qL}{c_p T_c}\right)$$

where $\theta_v = T_v\left(\frac{1000}{p}\right)^{\kappa}$

T_v = virtual temperature (the temperature of dry air with the same pressure and density as a given moist air parcel)

T_c = condensation temperature.

Equivalent temperature (T_e) denotes the sum of the actual air temperature and the increment due to latent heat of the water vapour it holds.

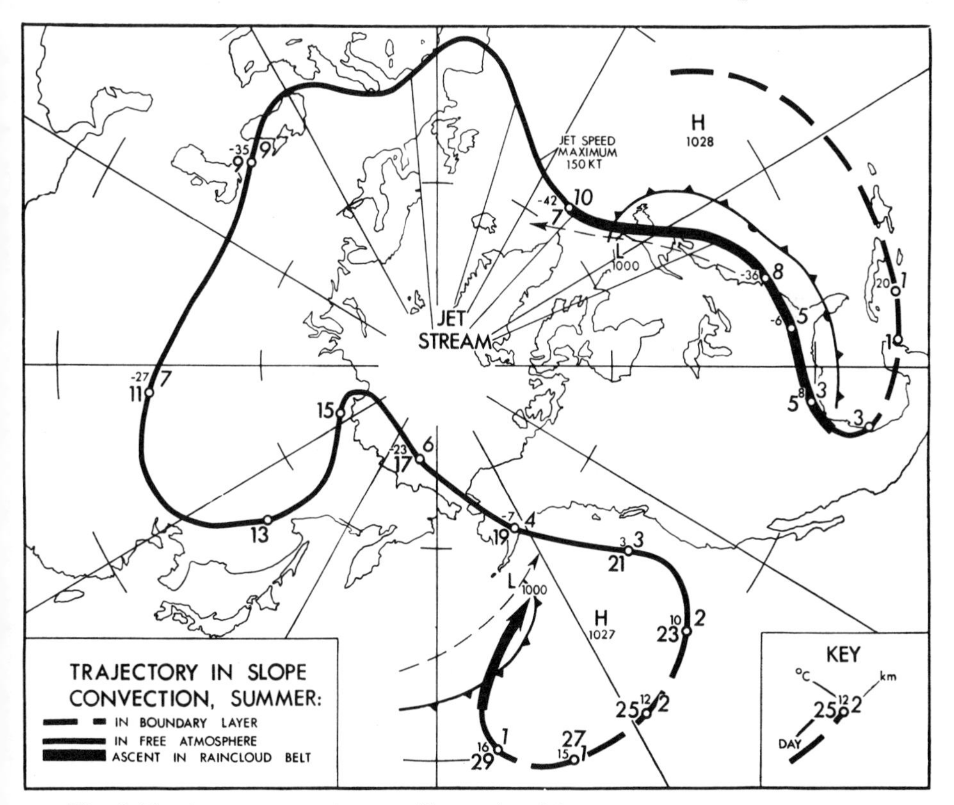

Fig. 2.20 A summer trajectory illustrating 'slope convection' (from Ludlam, 1966).

The complexity of the necessary calculations for these trajectory determinations make their computerization essential.

Isentropic methods have also been applied to trajectory analysis in synoptic and mesoscale systems by Green, Ludlam and McIlveen (1966). Isentropic maps are prepared for a time when, for about twenty-four hours before and after, the large-scale motion systems are not changing rapidly. Streamlines are computed relative to the system, so that these are approximate trajectories wherever the adiabatic and steady-state assumptions are appropriate. The results of such analyses demonstrate that trajectories from the trade wind zone rise into the upper troposphere through vertical motion in cloud systems along a cold front and there enter the associated jet stream. Fig. 2.20 shows such a trajectory, emphasizing the complexity of actual air movements.

A somewhat similar analysis has been carried out by Lockwood (1968) who prepared maps for surfaces of constant wet-bulb potential temperature (θ_w) – that is to say, constant total energy content. θ_w is the

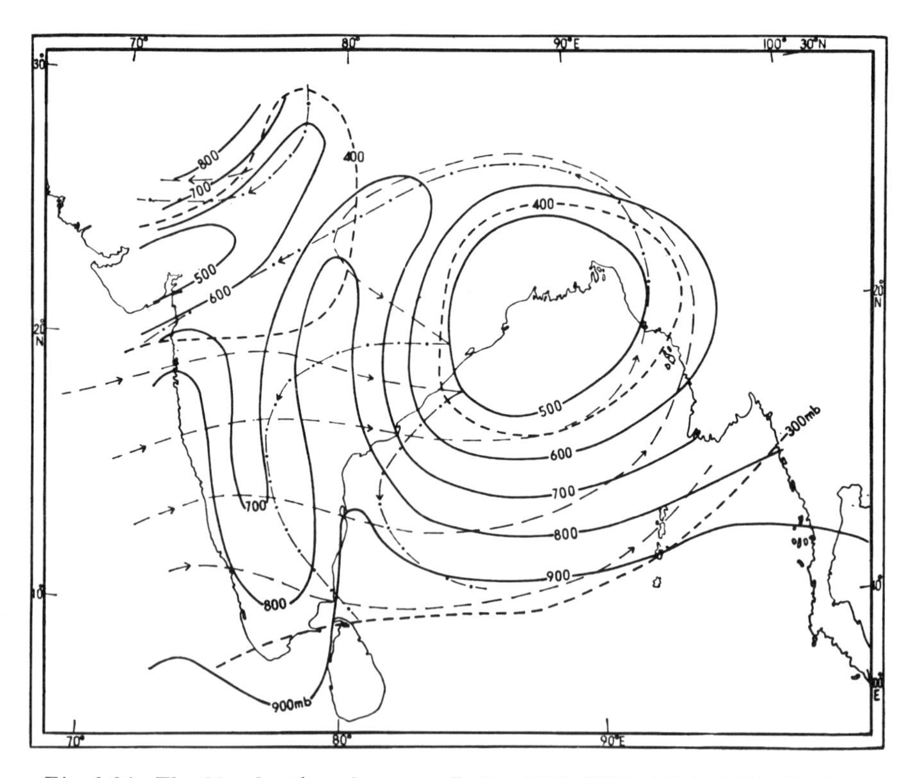

Fig. 2.21 The 81 cal g^{-1} surface over India, 0000 GMT, 4 July 1964. Contours (continuous), streamlines (fine dashed line). The structure gives a higher surface (heavy dashed line) with a different flow (streamlines shown dot-dashed) (from Lockwood, 1968).

wet-bulb temperature of air brought to 1000 mb pressure by a saturated-adiabatic process (see Hewson, 1936–8). Surfaces of θ_w are not affected by the condensation process so that they may be drawn through areas where condensation is occurring. The total energy content per unit mass of moist air, Q, is:

$$Q = \frac{V^2}{2} + gz + c_p T + Lq$$

where V = wind velocity
g = acceleration due to gravity
z = height above MSL
c_p = specific heat at constant pressure
T = temperature (°K)
q = specific humidity
L = latent heat of condensation.

The four terms on the right-hand side of the equation are, respectively, kinetic energy, geopotential, enthalpy and latent energy. The total energy content of air parcels tends to be conservative over several days in the free atmosphere. In fact, Q is constant for steady, adiabatic flow without friction, so that it can be used as a tracer of air parcels. Contours (or isobars) of constant Q, in conjunction with streamlines, show approximately the instantaneous three-dimensional motion. Fig. 2.21 illustrates the airflow relative to the 81 cal g^{-1} surface over India on 4 July 1964. At 0000 GMT there was a tropical storm over the northern Bay of Bengal and a minor disturbance, giving little or no rainfall, over West Pakistan. The map shows marked ascent over northeastern India from 850 mb to 500 mb, but little or none over West Pakistan. There is also ascent in the air crossing the Western Ghats from the Arabian Sea.

Analyses of θ_w have also been given considerable attention recently in investigations of cyclonic precipitation and air motion over the British Isles (Browning and Harrold, 1969; Harrold and Browning, 1969). These researchers combine isentropic analysis with detailed Doppler radar measurements.

Yet a further parameter in use is the *static energy* (σ), per unit mass

$$\sigma = gz + c_p T + Lq$$

where the small kinetic energy contribution is omitted. This parameter is conservative for dry and moist adiabatic displacements, like θ_w, but is a more direct indicator of energy changes. Kreitzberg and Brown (1970) make use of it in a study of the thermal structure of a mature extra-tropical depression.

4 VERTICAL VELOCITY

In terms of its direct significance in the production of weather phenomena, vertical air motion is undoubtedly the most important single parameter. Unfortunately, it cannot in general be directly measured to provide routine information. Instead it has to be estimated by one of a variety of techniques, depending on the scale of motion which is of interest and the applicability of their various assumptions in specific instances. Large-scale vertical motion is normally almost imperceptible (see table 2.1) whereas the vertical velocities associated with local storm systems may be dramatically evidenced by the build-up of cumulonimbus heads. The variety of approaches which may be used in estimating vertical motion makes it best considered as a special topic.

Table 2.1 Relative magnitudes of vertical motion in different scales of system

System	*Velocity* (cm s^{-1})	*Time scale*
Thunderstorm	10^3	1 hour
Tropical storm; subsynoptic (frontal zone)	$10^2 - 5 \times 10^2$	6 hours
Intense depression	10	6–12 hours
Average depression	5	1–2 days
Planetary wave	1	1 week

The simplest means of determining synoptic or larger-scale vertical velocity is based on the continuity equation relating the local rate of density change to the mass divergence per unit volume.

$$\frac{\partial \rho}{\partial t} = -\nabla . (\rho \mathbf{V})$$

$$= -\left[\frac{\partial(\rho u)}{\partial x} + \frac{\partial(\rho v)}{\partial y} + \frac{\partial(\rho w)}{\partial z}\right]$$

where ρ = density
$\mathbf{V}$ = wind velocity
w = vertical velocity.

From this equation it can be shown that

$$w_z = -\frac{\bar{\rho}}{\rho_z}(z\nabla_H . \mathbf{V})$$

where z = an arbitrary height and the bar denotes a vertical average between the surface and z.

For a steady state

$$w_z = z\nabla_H.\mathbf{V}.$$

In practice, pressure is used as the vertical coordinate and the equivalent of vertical velocity (ω) at an arbitrary pressure level, p, is determined by integration. Thus,

$$\omega_p = \omega_{p-1}+\int_p^{p-1} \nabla_H.\mathbf{V}\,dp$$

where $\omega = dp/dt$:

It is generally assumed that $\omega = 0$ at the surface pressure value. This is satisfactory as long as the topography has little slope. The integral is approximated by summation over a number of isobaric layers (Rex, 1958; Vaisanen, 1961).

The horizontal divergence is calculated by the cartesian grid or streamline methods outlined on p. 27 or by the objective triangle method of Bellamy (1949). Examples of vertical motion computations for triangles formed by the upper air stations in the British Isles are cited by Sheppard (1949). The triangle method is somewhat unsatisfactory in that the value does not refer to any particular point.

These kinematic methods are subject to considerable inaccuracies even when the available winds are numerous (Landers, 1955) and smoothing of any computations is essential (Palmén and Holopainen, 1962, for example). An advanced method of smoothing based on spatial autocorrelation has been described by Eddy (1964).

A second method of computing vertical velocity relates to temperature changes in the free atmosphere. The temperature field is assumed to be altered by horizontal advection and vertical advection only. Observations of the former and of the local temperature change allow the vertical motion necessary for balance to be computed.

$$w\frac{\partial T}{\partial z} = \frac{dT}{dt}-\frac{\partial T}{\partial t}-\mathbf{V}.\nabla_H T$$

where $\partial T/\partial t$ denotes the local rate of temperature change and dT/dt that following the motion of a particle. $\mathbf{V}.\nabla_H T$ is the advection of the horizontal temperature field. For adiabatic changes

$$\frac{dT}{dt} = -w\Gamma$$

where Γ = the dry adiabatic lapse rate. Thus,

$$w = -\left(\frac{(\partial T/\partial t)+\mathbf{V}.\nabla_H T}{\Gamma-\gamma}\right)$$

$$= -\left(\frac{\delta T/\delta t}{\Gamma-\gamma}\right)$$

where γ = the environmental lapse rate
$\delta T/\delta t$ = temperature change along a horizontal trajectory.

Alternatively, using potential temperature on isobaric surfaces,

$$\omega = -\left[\frac{(\partial\theta/\partial t)+\mathbf{V}(\partial\theta/\partial s)}{\partial\theta/\partial p}\right]$$

where V = horizontal wind speed
$\partial\theta/\partial t$ = local change of potential temperature on an isobaric surface
and $\partial\theta/\partial s$ = variation of potential temperature along a streamline.

This approach is particularly suited to analysis based on records from constant-level balloons, provided diabatic effects (particularly radiative ones) can be ignored. Alternatively, trajectories can be estimated from geostrophic winds. It should be noted also that in slant ascent in a changing pressure field the adiabatic lapse rate itself may be overestimated by up to 1°C km^{-1} (Staley, 1966). Such errors could seriously affect vertical velocities estimated by this method.

Computations of vertical velocity in conjunction with numerical forecasting studies have been based on the vorticity equation (Collins and Kuhn, 1954), the vorticity and thickness tendency equation (Sawyer, 1949; Bushby, 1952; Knighting, 1960), the 'omega equation' (Petterssen, Bradbury and Pedersen, 1962; Danard, 1964) and the 'primitive equations' of motion (see, for example, Kasahara and Washington, 1967). It is beyond the scope of this book to do more than outline the basis of the first two of these methods. They all involve heavy computational demands and the details are, in any case, subject to continual improvement in terms of the degree of resolution possible and of the mathematical procedures.

The approximate vorticity equation (see p. 31) is

$$\frac{\partial\zeta_a}{\partial t} \simeq -\mathbf{V}.\nabla\zeta_a - \frac{\omega\,\partial\zeta_a}{\partial p} + \zeta_a\frac{\partial\omega}{\partial p}$$

where ζ_a = the vertical component of absolute vorticity if the effects of friction and the turning of vortex lines are disregarded. The latter is important in frontal zones, however. Approximate integration of the above equation leads to the following expression:

$$\left(\frac{\omega}{\zeta_a}\right)_p = \left(\frac{\omega}{\zeta_a}\right)_{p_{1000}} - \frac{1}{2}\left[\left(\frac{(\partial\zeta_a/\partial t)+\mathbf{V}.\nabla\zeta_a}{\zeta_a^{\,2}}\right)_{p_{1000}} + \left(\frac{(\partial\zeta_a/\partial t)+\mathbf{V}.\nabla\zeta_a}{\zeta_a^{\,2}}\right)_p\right]$$

where p_{1000} = 1000 mb level
p = an arbitrary pressure level.

Collins and Kuhn computed ζ_a from $\zeta+f$. The relative vorticity, ζ, can be determined graphically using a method based on space-averaged contour fields (see section B of this chapter, p. 69) introduced by Fjørtoft (1952; see also Haltiner and Martin, 1957, p. 395), while f is tabulated. Charts at six- or twelve-hourly intervals are used to give $\partial\zeta_a/\partial t$. The vorticity advection $\mathbf{V}.\nabla\zeta_a$ can also be calculated by graphical methods. Knowing these terms and assuming $\omega=0$ at the 1000 mb level we can determine (ω/ζ_a) and therefore ω at any level p from the hydrostatic assumption, $w=-\omega/\rho g$. This method was used by Collins and Kuhn in conjunction with a precipitation 'forecast' over North America. It provides good estimates of vertical velocity on a synoptic scale so long as the absolute vorticity is not too small.

Penner (1963) has developed this type of approach in terms of thickness advection and vorticity advection. The original method related to charts of space-mean vorticity at 500 mb and was modified for use with the 500 mb absolute vorticity analyses subsequently adopted by the Canadian Weather Service (Harley, 1964, 1965b; Harley *et al.* 1964).

At the level of non-divergence (600 mb), vertical velocity $(-\omega_6)$ can be determined from an expression:

$$-\omega_6 = k_1(A_{\zeta a})_5 + k_2(A_{\Delta z})$$

where $(A_{\zeta a})_5$ = horizontal advection of absolute vorticity at 500 mb
$A_{\Delta z}$ = horizontal advection of 1000–500 mb thickness.

k_1 and k_2 incorporate scale factors including the Coriolis parameter. ω_6 is expressed in units of 10^{-3} mb s^{-1} ($\simeq$ cm s^{-1}). The two terms of the equation can be determined graphically using a special geostrophic advection scale (Ferguson, 1961, 1963) and their individual contributions are evaluated in vertical velocity units before adding.

A further operational method has been devised by Harley (1965b) utilizing computer-produced maps of 500 mb absolute vorticity and 1000–500 mb thickness advection. Vertical velocity can be read directly from graphs which solve the above equation for locations within 42–56°N. Extension to other latitudes is made by a weighting of the appropriate values of the Coriolis parameter. Graphs for the separate components are given by McPherson *et al.* (1969) covering latitudes 30–90°.

The basis of the 'omega equation' approach used by Petterssen *et al.* is that the advection of temperature and vorticity will disturb geostrophic balance unless compensated for by horizontal divergence and therefore the vertical motion field. The relationship is formulated in terms of divergence and the advection of vorticity by the thermal wind.[1]

[1] Thermal wind is a theoretical mean vector wind in geostrophic balance with the thickness lines (and therefore with the gradient of mean temperature in the layer between the chosen isobaric surfaces). This wind shear vector is parallel to the thickness lines with cold air on the left, viewed downwind, in the northern hemisphere.

Petterssen *et al.* show that vertical velocity can be separated into a component due to dry adiabatic motion and one due to diabatic heating by the input of sensible and latent heat. Patterns of thickness tendency associated with these components were analysed for typical stages of development of mid-latitude cyclones. This question has been examined further by Danard (1964) who showed that the heating term, primarily released latent heat, is very important with respect to the computed vertical velocity in precipitation areas. Indeed, vertical velocity is only 25% of kinematic estimates if this effect is not incorporated in the computations.

The most recent generation of numerical prediction models uses the 'primitive equations' of motion (Lorenz, 1967). These are essentially one step removed from the exact hydrodynamic equations incorpora-

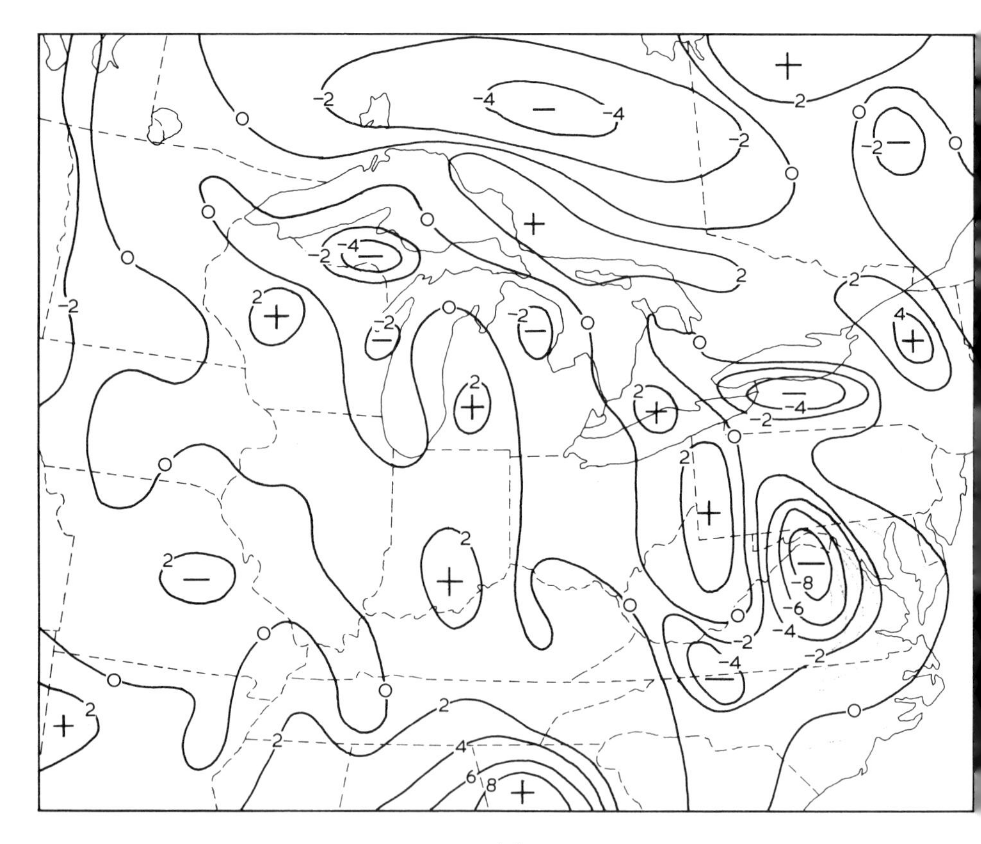

(*a*)

Fig. 2.22 Terrain-induced vertical velocity over eastern North America, for winds of (*a*) 225°, 10 knots, (*b*) 315°, 10 knots (after Jarvis and Leonard, 1969). Positive values denote rising air.

ting, for example, hydrostatic equilibrium. The general circulation model developed at the National Center For Atmospheric Research (Kasahara and Washington, 1967; Oliger *et al.*, 1970) computes vertical motion on this basis from an equation first formulated by Richardson (1922) involving horizontal divergence and pressure change.

The determination of vertical motion on various scales remains a major problem in synoptic meteorology.

Vertical velocity at the lower boundary (in practice the surface) is usually assumed to be zero in order to simplify the computations. However, the influence of terrain (and also of friction) on airflow has been stressed by Graystone (1962), Haltiner *et al.* (1963) and Jarvis and Agnew (1970). Quantitative assessment of terrain-induced vertical velocity, on a synoptic scale over a time period of about one day, can be performed as follows:

$$w_H = \mathbf{V}_g . \nabla H$$

where $\mathbf{V}_g$ = the horizontal geostrophic wind

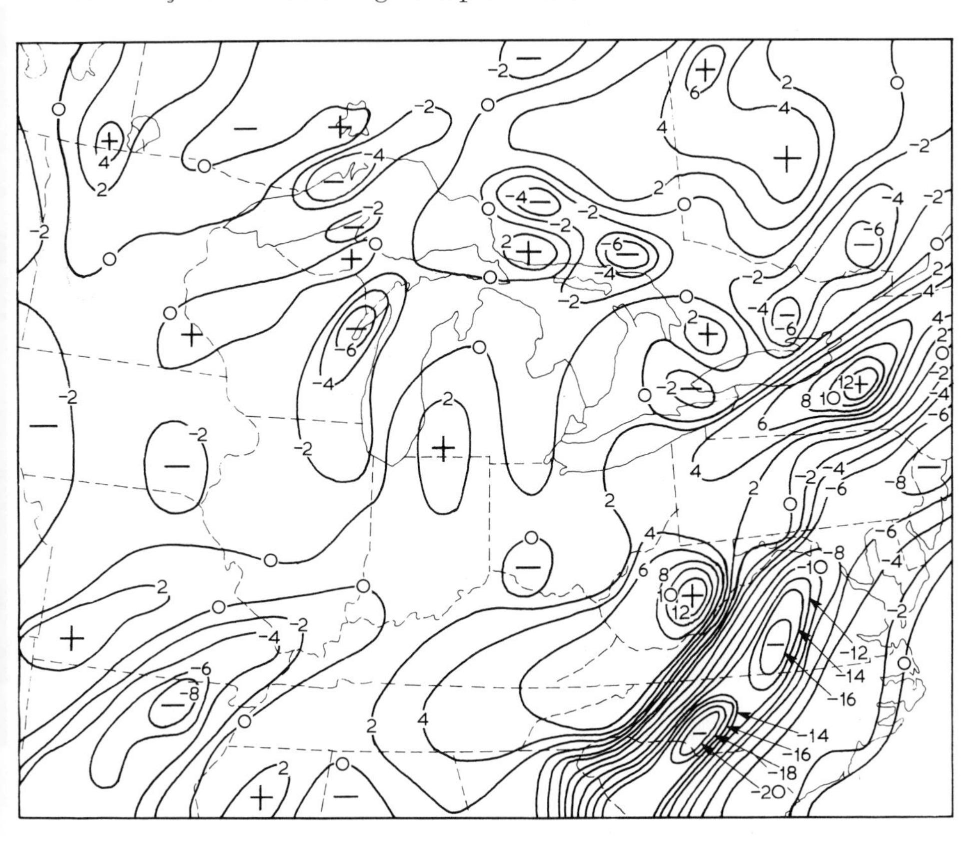

(*b*)

and H = vertical relief, smoothed over a grid length of the order of 200–300 km.

Using this approach Jarvis and Leonard (1969) have prepared maps of w_H over central and eastern North America for 10 knot winds from each cardinal point. These are illustrated for winds of 225° and 315° in fig. 2.22. The generation of similar base maps for other suitable areas of the world would provide an invaluable research tool. It should of course be noted that in the case of major topographic barriers the flow may be partially diverted or more or less wholly blocked. Friction effects are now incorporated in numerical models but, where necessary, a simple approach could be made along the lines of Bryson and Kuhn's (1961) estimates of frictional divergence.

5 WEATHER ELEMENTS

Synoptic maps of individual weather elements are uncommon, in spite of Godske's first-order level of synthesis, $f(e_0t_0s)$ category (see table 1.1). (Until recently routine weather office practice has been limited mainly to the depiction on synoptic charts of fog areas by yellow shading and precipitation belts by green shading!) Taylor and Yates (1967) illustrate a more complete range of examples, for different air masses and frontal situations, in their textbook on British weather. They use such parameters as cloud cover, precipitation occurrence, visibility, wind velocity, temperature and dew-point depression.

These parameters comprise the larger part of what is commonly regarded as weather and, in the longer term, climate. They are, therefore, important with respect to the many applications of weather data (some of which are considered in Chapter 5). Furthermore, they reflect various fundamental properties of the atmosphere, directly or indirectly. Thus, for example, cloud and precipitation are indicators of vertical motion. Knowledge of the space and time distribution of these parameters is therefore of value in furthering our understanding of synoptic systems.

Let us consider some specific weather elements. The distribution of humidity is of key importance in the production of 'weather' and it is therefore rather surprising that it is not featured more prominently in synoptic analyses. The dew point at the surface, or some isobaric level, is readily shown by isopleth maps, although it must be noted that humidity data in the free air are unreliable at low temperature and low humidity. Nevertheless, upper air values often clearly demonstrate patterns associated with subsidence and ascent of air (Sawyer, 1964, p. 233). Isopleths of dew-point depression are probably more useful than of the dew point itself. Fig. 2.23 illustrates a typical pattern associated with a cold front over the British Isles. An alternative parameter is precipitable water content w of the air column, between P_s and P_0:

$$w = \frac{1}{g} \int_{P_0}^{P_s} q \, dp$$

where P_0 is a convenient upper level, q = specific humidity. In practice, humidity data are rarely available above about 400 mb. Sutcliffe (1956) advocated the routine calculation of moisture transport at RAWIN-sonde stations and precipitable moisture contents, at least, are now available for the United States via the weather map facsimile network of the Weather Bureau. The Canadian Meteorological Service also publishes humidity mixing ratio data in tabulations of RAWIN-sonde soundings in the *Monthly Upper Air Bulletin*.

Until the advent of satellite photography, such information as was available on cloud cover was little used. There was nevertheless recognition of the significance of cloud data. Ludlam and Miller (1959) proposed that areas of various cloud features should be mapped, including: cloud-free sky, large amounts of upper cloud and of stratocumulus, stratus or fog, cumuliform cloud, precipitation systems.

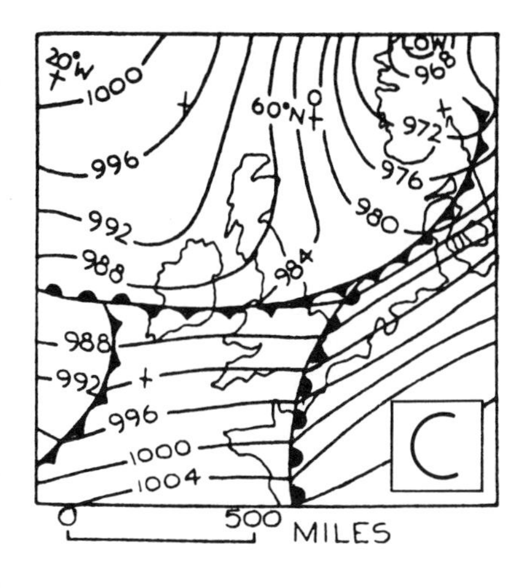

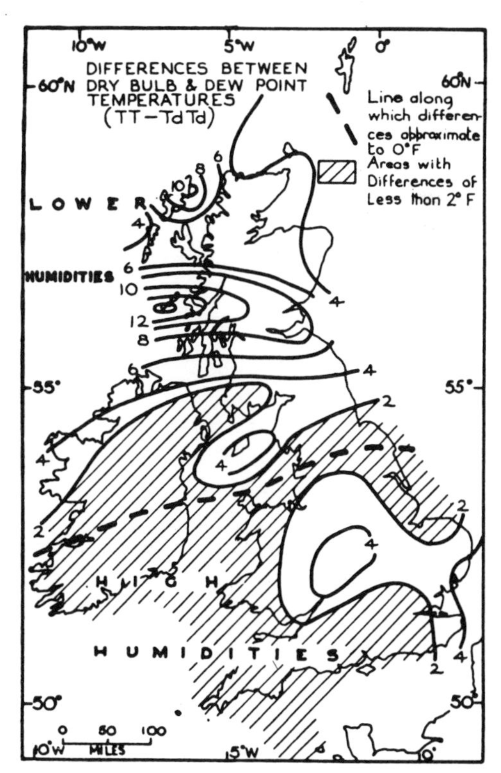

Fig. 2.23 Distribution of dew-point depression associated with the passage of a cold front, 1800 GMT, 10 January 1955 (after Taylor and Yates, 1967).

Table 2.2 A system of nephanalysis from surface observations (after Viaut, 1961)

General classes	*Types of sky*	*Major characteristics*
Travelling extratropical depression systems	Forward	Organized cirrus and cirrostratus progressively invading the sky.
	Lateral (*a*) cold	Partial or complete cirrostratus veil.
	(*b*) warm	Isolated altocumulus bands.
	Central	Continuous altostratus/nimbostratus usually with precipitation; fractostratus also common.
	Rear	Alternation of gaps and confused, threatening sky.
	Connecting zone	Extensive stratocumulus or stratus, often with fog; little precipitation.
Thundery systems	Pre-thundery	Dense cirrus and cirrostratus with altocumulus (often castellanus) and commonly cumulus.
	Thundery	Overcast, chaotic sky usually giving showers and thunderstorms.
Stationary cloud systems	Thermal origin	Cumulus types.
	Orographic origin	Wave clouds and related skies.
Interval		Mainly cloud-free areas between various systems.

A highly developed scheme of *nephanalysis* was evolved in France (Viaut, 1961). The classification of the major types of sky is shown in table 2.2; in addition, areas of stratiform and instability (cumulus) cloud are distinguished separately. In spite of their detail, schemes of this type, including the more specific depression models (Bergeron, 1951), have been demonstrated to have limited applicability in many individual situations. However, this reflects inadequacies of the early models and not an absence of pattern in cloud systems. The major problem has always been patchy information. With the NIMBUS and ESSA satellites, global coverage is now available on a daily basis and nephanalyses are prepared from the photography.

The cloud characteristics that are used in indentification are, according to Conover (1962):

1 Cloud brightness, relating particularly to the depth and composition.
2 Texture – smooth, fibrous, opaque, mottled, etc.
3 Form of cloud elements – regular, irregular.
4 Pattern of elements – relating to topographic effects, airflow, vertical and wind shear.
5 Size of elements and patterns.
6 Vertical structure – sometimes evident from cloud shadows, etc.

The basis of classification, however, has primarily been concerned with pattern. Hopkins (1967) details a classification with the following groups:

1 Vortical features:
  (*a*) Circular and/or spiral bands
  (*b*) Crescent or comma-shaped masses (on a synoptic scale)
  (*c*) Quasi-circular non-banded masses
  (*d*) Curved or linear bands in a vortical pattern.

2 Major cloud bands:
  A linear organization where the length is much greater than the breadth.

3 General features not included above:
  (*a*) Minor lines and bands (e.g. cloud streets, jet stream bands, lee waves)
  (*b*) Cumuliform features – cellular patterns (polygonal, actiniform) or unorganized
  (*c*) Stratiform features – sheets, fog areas.

In the first two groups distinctions can also be made according to tropical or extratropical features and the latter can then usually be differentiated into frontal or non-frontal systems. A genetically based classification along these lines has been proposed by Barrett (1970b) and is shown in fig. 2.24. It is from such methodology that new models of synoptic and large-scale systems have been developed, including tropical perturbations (Merritt, 1964; Fett, 1964), the intertropical convergence zone (Sadler, 1965; Hubert, *et al.*, 1969; Barrett, 1970a) and extratropical depressions (Boucher and Newcomb, 1962). In addition to the basic photography, a variety of mosaics and composites are prepared by the National Satellite Center, Washington, either routinely or for special projects. Taylor and Winston (1969), for example, have analysed monthly and seasonal average brightness for February 1967 to February 1968.

The analysis of precipitation areas is also a recent innovation. The motivation has been primarily a concern for improved short-range weather forecasts. Jarvis (1966) shows, for example, that on the synoptic

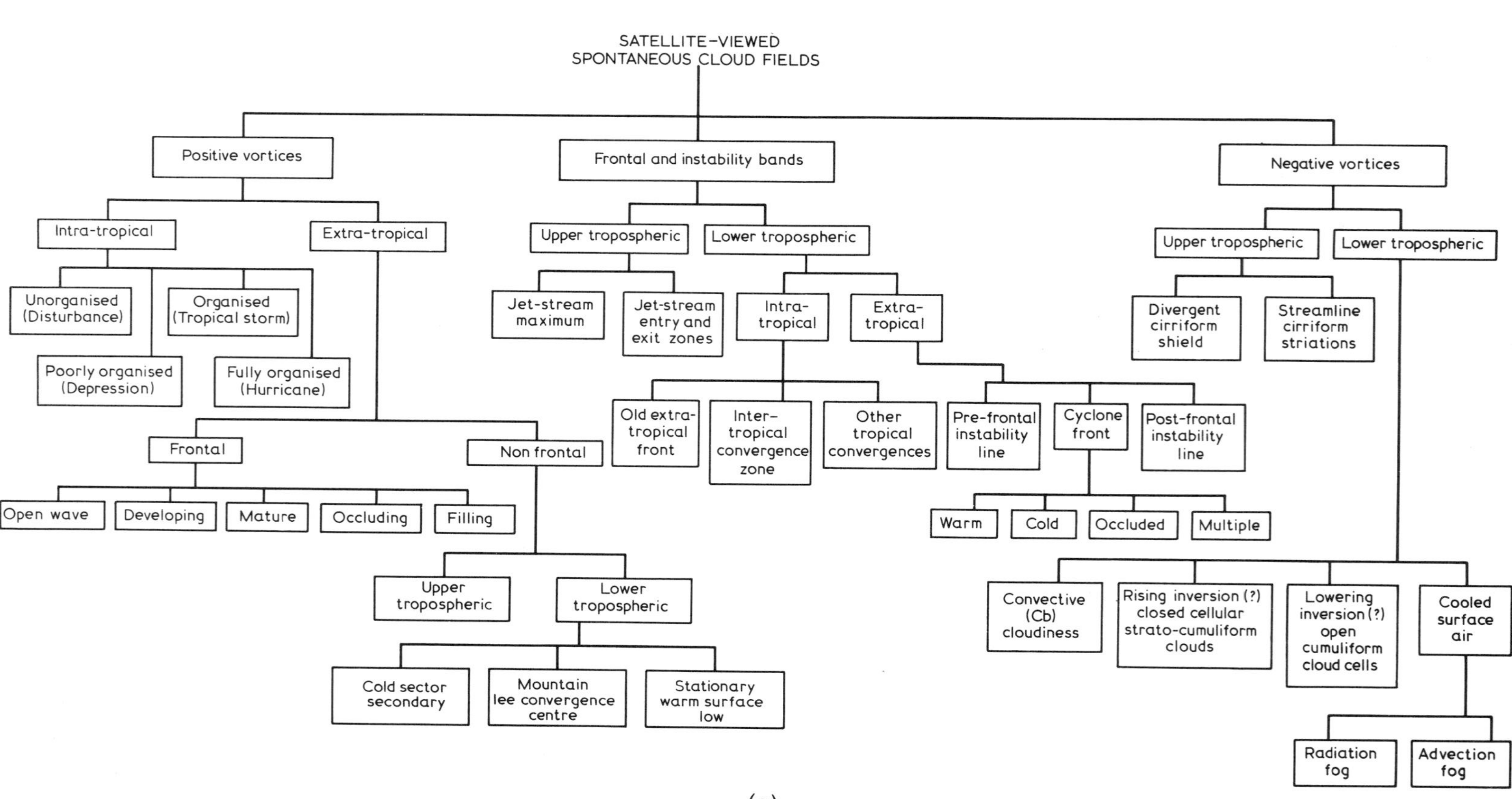
SATELLITE-VIEWED
SPONTANEOUS CLOUD FIELDS
Positive vortices
Frontal and instability bands
Negative vortices
Intra-tropical
Extra-tropical
Unorganised (Disturbance)
Organised (Tropical storm)
Poorly organised (Depression)
Fully organised (Hurricane)
Frontal
Non frontal
Open wave
Developing
Mature
Occluding
Filling
Upper tropospheric
Lower tropospheric
Cold sector secondary
Mountain lee convergence centre
Stationary warm surface low
Upper tropospheric
Lower tropospheric
Jet-stream maximum
Jet-stream entry and exit zones
Intra-tropical
Extra-tropical
Old extra-tropical front
Inter-tropical convergence zone
Other tropical convergences
Pre-frontal instability line
Cyclone front
Post-frontal instability line
Warm
Cold
Occluded
Multiple
Upper tropospheric
Lower tropospheric
Divergent cirriform shield
Streamline cirriform striations
Convective (Cb) cloudiness
Rising inversion (?) closed cellular strato-cumuliform clouds
Lowering inversion (?) open cumuliform cloud cells
Cooled surface air
Radiation fog
Advection fog
(a)

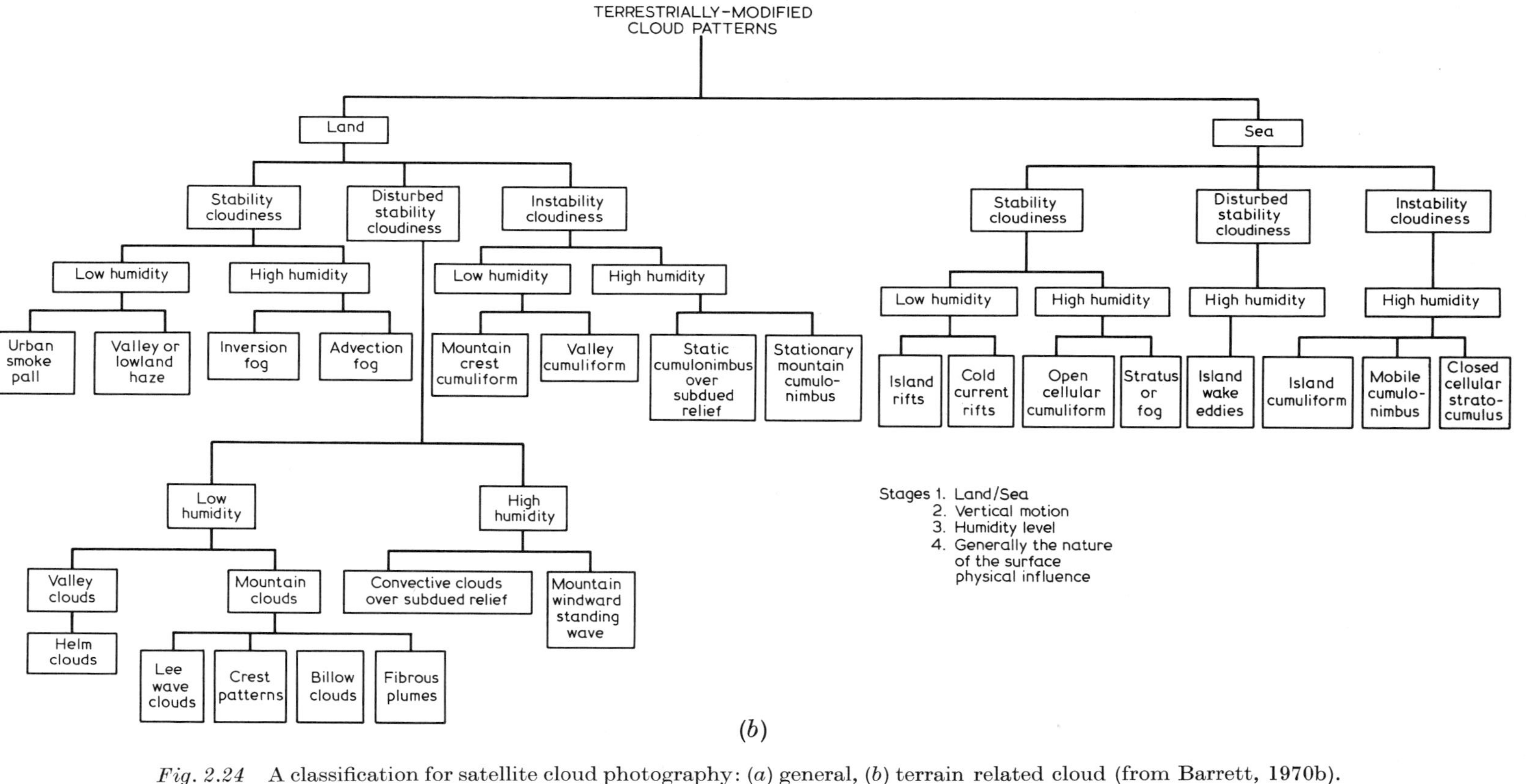

(*b*)

Fig. 2.24 A classification for satellite cloud photography: (*a*) general, (*b*) terrain related cloud (from Barrett, 1970b).

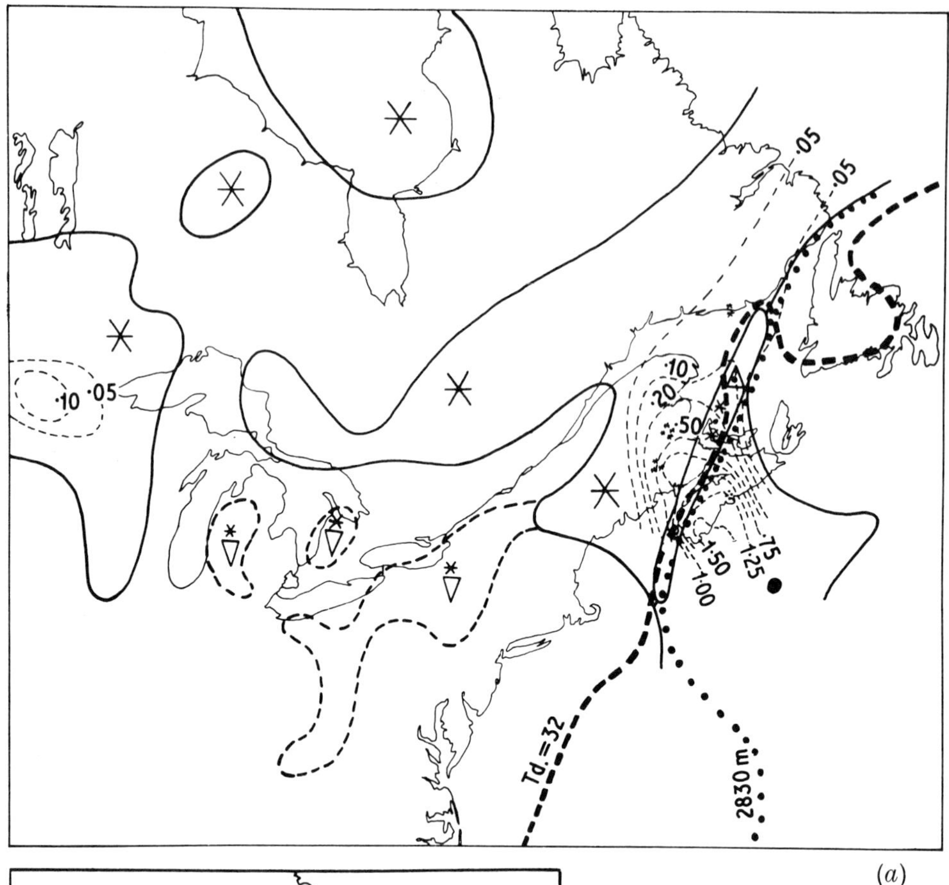

(a)

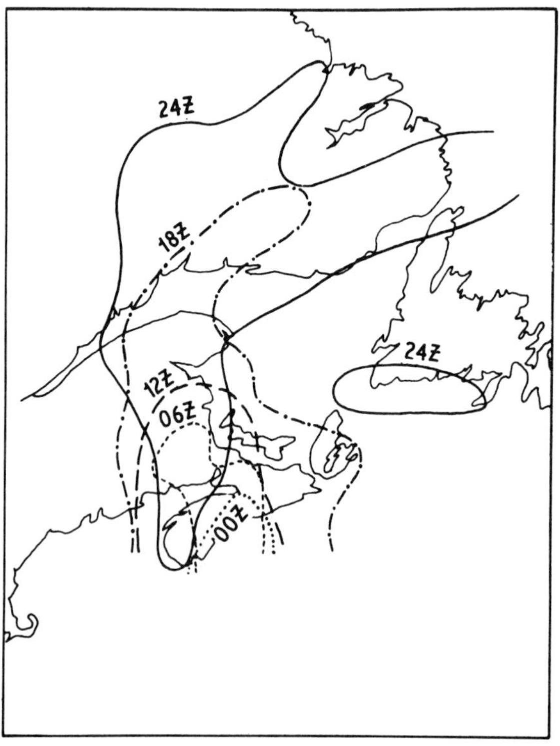

(b)

Fig. 2.25
(*a*) Precipitation chart for eastern North America, 1200 GMT, 1 December 1964, showing shower (▽) areas (dashed lines) and areas of snow (✳), rain (•), ice pellets (△) (full lines). 0°C surface dew-point temperature (heavy dashed line) and 2830 m 1000–700 mb thickness line (heavy dotted line). Thin broken lines indicate six-hour rate of precipitation as water equivalent (inches) (from Jarvis, 1966).
(*b*) Isochrones at six hour intervals for a precipitation rate of 0·20 in/6 hr, 1 December 1964 (from Jarvis, 1966).

and subsynoptic scales it is helpful to construct isochrones, at three- or six-hourly intervals, enclosing areas of particular categories of precipitation. Synoptic parameters provide a guide to certain boundaries (of freezing precipitation, or snow, for example) in regions of sparse data. Isopleths of the rate and intensity of precipitation assist in differentiating the physical processes responsible for the various precipitation areas. Fig. 2.25(*a*) illustrates a precipitation chart for eastern North America and fig. 2.25(*b*) shows the isochrones for a specified precipitation rate. The latter delineates the region influenced by the cyclone system. The light snow falling in a belt from the Great Lakes to southeast Labrador (fig. 2.25(*a*)) began as a separate boundary layer feature, for instance.

Another mapping technique, which uses conventional data of daily precipitation totals, deserves note. Johnson (1962) in a study of east African rainfall classifies rain areas 0–10 according to the number of stations reporting a measurable fall. The areas are delimited in descending order of 'concentration', neglecting small-scale (10 km^2) inhomogeneities. Divisions into separate areas are made whenever higher or lower indices extend over more than a one-degree square. The rain areas are then related to synoptic models of upper flow.

Finally, mention should be made of the mesoscale analysis and mapping of precipitation systems which can now be carried out using radar. Such work has been invaluable in the study of severe storms and of frontal systems (Newton, 1966; Browning and Harrold, 1969; Mason, 1969).

We may note in conclusion that while 'storage parameters' such as temperature and specific humidity are routinely analysed it is not yet possible to examine flux properties such as evapotranspiration, convective heat flux, soil heat flux, etc., in terms of their space–time characteristics. Generally, the best that can be done is to obtain climatological estimates, or in some cases averages, for a common synoptic situation. This topic is taken up again in Chapter 5.E, pp. 419–422.

6 FRONTAL ANALYSIS

In traditional terms, a front is a sharp transition zone between air masses with differing temperature (and often humidity) conditions. The warm air boundary of the maximum temperature gradient is identified as the actual frontal surface. The principles of subjective methods of frontal analysis are fully described in texts such as Petterssen (1956) and Godske, *et al.* (1957), and will be omitted here. A thorough summary of frontal structure is given in an excellent recent book by Palmén and Newton (1969, Chapter 7).

Subsequent to the 'classical' ideas of the Norwegian school (Bjerknes and Solberg, 1922) on the polar front and and related air mass concepts developed by Bergeron (1928, 1930) – see Chapter 3.D, pp. 178–80 – the first major attempt to introduce greater objectivity into the analysis was made by Canadian meteorologists. Frontal contour charts (Crocker, Godson and Penner, 1947), first devised by Bjerknes and Palmén

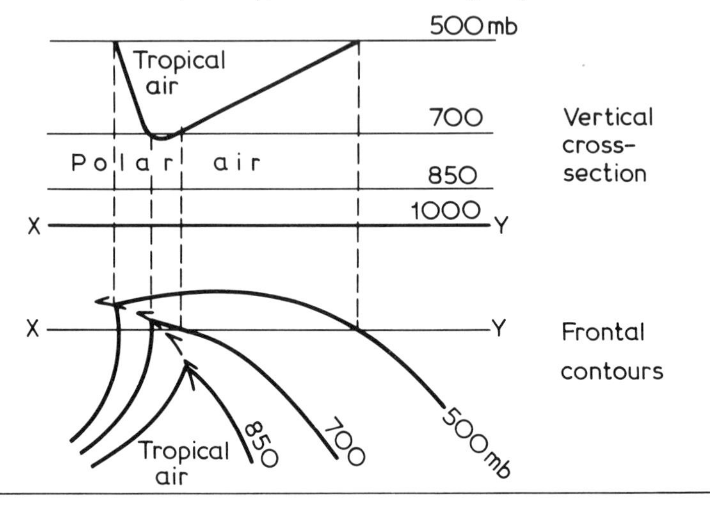

⟵ ⟵ ='Trowal' (trough of warm air aloft)

Fig. 2.26 The depiction of upper fronts by means of frontal contour charts (after Crocker, Godson and Penner, 1947).

(1937), were used to obtain a three-dimensional view of frontal zones. Fig. 2.26 illustrates their mode of construction. A frontal scheme proposed by Godson (1950), which reflected the classical Norwegian viewpoint, has four frontal zones:

Air mass	*Front*
Continental Arctic	
	Inter-Arctic
Maritime Arctic	
	Arctic
Cold Maritime Polar	
	Inter-Polar
Warm Maritime Polar	
	Polar
Tropical	

Later (Penner, 1955; Anderson, Boville and McClellan, 1955; Galloway, 1958; McIntyre, 1958) this was modified for the winter season in North America to a four air mass, three front model:

Continental Arctic	
	Continental Arctic Front
Maritime Arctic	
	Maritime Arctic Front
Maritime Polar	
	Polar Front
Maritime Tropical	

Penner noted that there is no source of continental arctic air in summer, but in fact, all three fronts still appear on most Canadian frontal contour analyses over North America (Barry, 1967).

The following criteria were adopted by Anderson *et al.* for defining a front:

1 It is a three-dimensional zone of pronounced baroclinicity[1] with a first-order discontinuity (i.e. the magnitude of the gradient is discontinuous) in the temperature and wind fields.
2 It has space and time continuity and is displaced with the airflow.

The most useful parameters were found to be wet-bulb potential temperature (the application of which was evaluated statistically by Harley, 1962) and the saturated wet-bulb potential temperature.

Recently, techniques have been developed for identifying fronts by quantitative field analysis. The parameters used here relate to temperature (Kirk, 1966b), potential temperature (Renard and Clarke, 1965), and wet-bulb potential temperature (Creswick, 1967). Kirk (1966a, 1966b) shows that a frontal zone may be identified from the vertical rate of change of geostrophic vorticity, which is proportional to the Laplacian of the temperature field ($\nabla^2 T$). For quasi-geostrophic flow:

$$f\frac{\partial \zeta_g}{\partial p} = -\frac{R}{p}\nabla^2 T$$

where ζ_g = geostrophic vorticity
f = Coriolis parameter
R = gas constant for air

$$\nabla^2 T = \frac{\partial^2 T}{\partial x^2} + \frac{\partial^2 T}{\partial y^2}$$
$$\approx (T_x + T_{-x} + T_y - 4T_0)/L^2,$$

with reference to the coordinates illustrated in Appendix fig. 1. This term, the divergence of the temperature gradient, is determined by the spacing and curvature of the isobars (see Kirk, 1970). Evidently, discontinuities develop in the temperature field as a result of differential concentrations of vorticity in the vertical plane. Kirk also showed that discontinuities in the geostrophic wind and in the gradient of pressure – both of which are criteria used in the subjective identification of fronts – are related closely to certain characteristics of the vorticity field.

The approach developed by Renard and Clarke (1965; Clarke and Renard, 1966) is somewhat different, but is related mathematically to that of Kirk. The divergence of the potential temperature gradient is

$$-\nabla^2\theta = -\nabla|\nabla\theta|\cdot\mathbf{n}_\theta - |\nabla\theta|\nabla\cdot\mathbf{n}_\theta$$

[1] Godson (1951) uses the term 'hyperbaroclinic'. It implies that surfaces of constant pressure intersect sharply those of constant density (and temperature).

where $\mathbf{n}_\theta$ = a unit vector in the direction of $\nabla\theta$. The first term on the right-hand side is the directional derivative of the magnitude of the gradient of potential temperature along its gradient. It is proportional to the horizontal shear of the thermal wind. The second represents the tan-

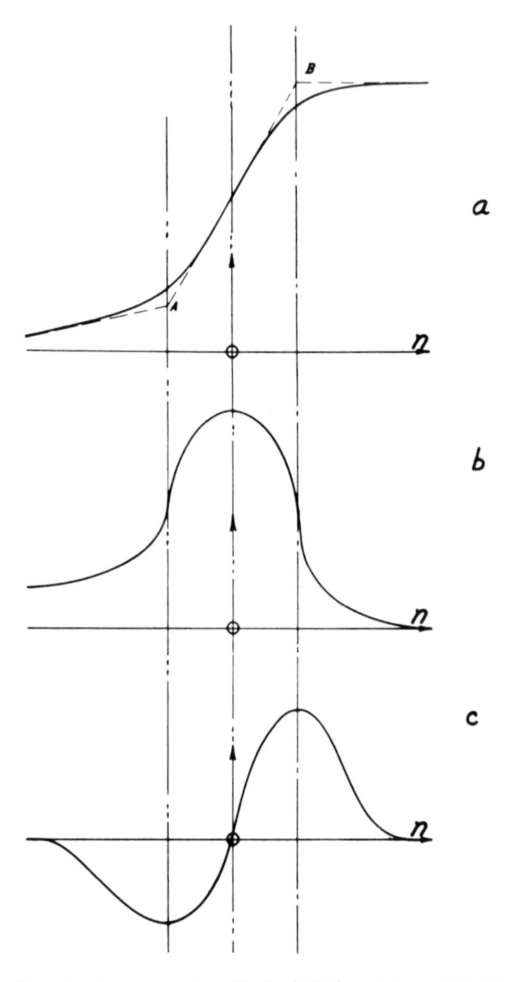

Fig. 2.27 The relationship between (*a*) X, (*b*) $|\nabla X|$ and (*c*) $-\nabla|\nabla X|\mathbf{n}_x$ in the one-dimensional case where parameters vary with $\vec{\mathbf{n}}_x$ (from Clarke and Renard, 1966).

gential curvature of the potential isotherms. Renard and Clarke use the axis of maximum $|\nabla\theta|$, which is coincident with a zero value for the first term, as the locus of the baroclinic zone. The meaning of these terms can be clarified diagramatically. Fig. 2.27 illustrates for a one-dimensional case the variation of X, $|\nabla X|$ and $-\nabla|\nabla X|.\mathbf{n}_x$, respectively, with $\mathbf{n}_x$.

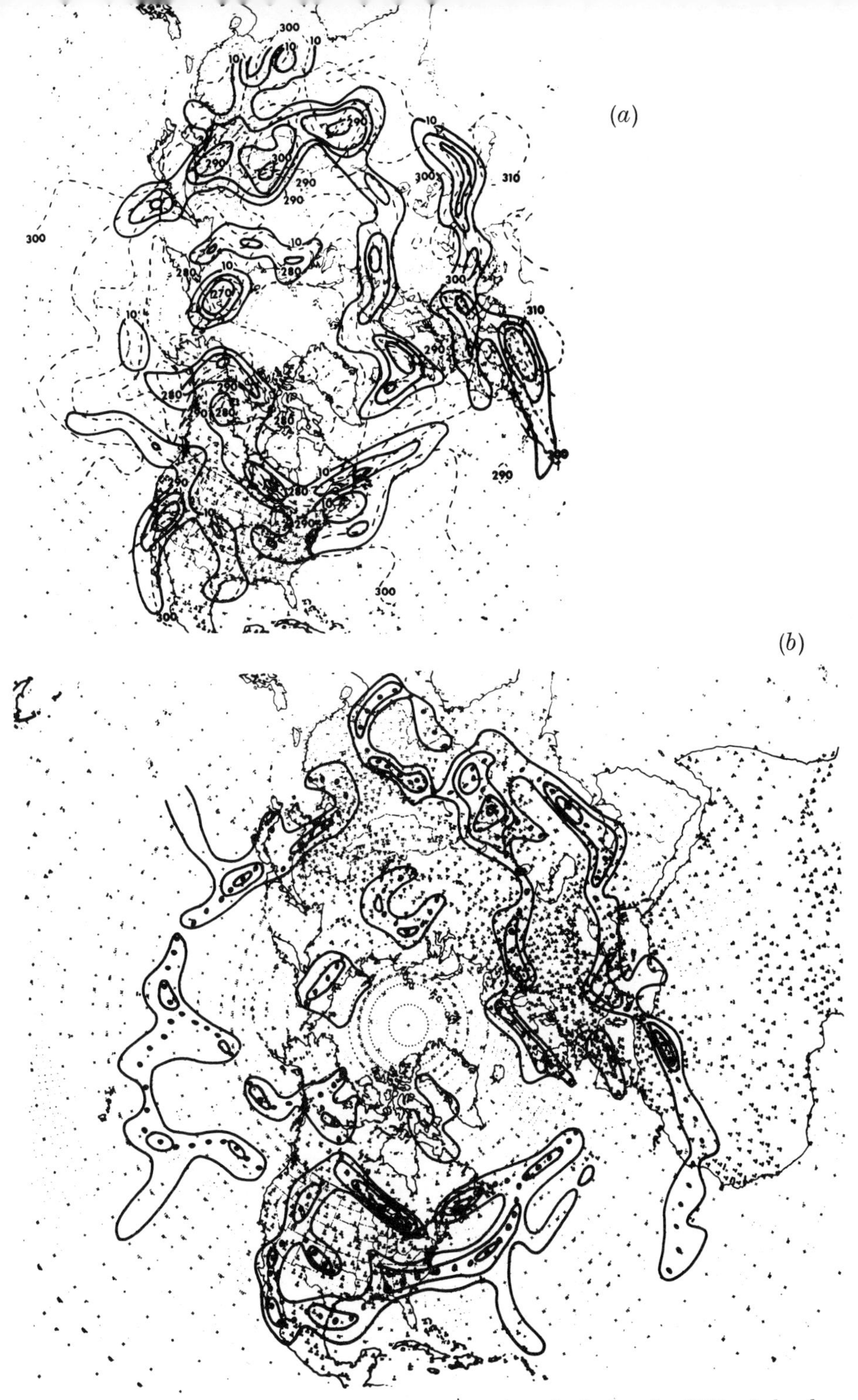

Fig. 2.28 An example of an objective frontal analysis for the 1000 mb level, 1200 GMT, 24 August 1965 (from Clarke and Renard, 1966).

(*a*) θ (dashed lines) °K at 5°K intervals·

$|\nabla\theta|$ (solid lines) 10^{-1} C (100 km)$^{-1}$ at 10, 15, 20, 30.

(*b*) $-\nabla|\nabla\theta|.\mathbf{n}_\theta$ 10^{-2} C (100 km)$^{-2}$ at 5, 15, 25; major positive axis dotted.

Note that $\nabla|\nabla X|.\mathbf{n}_x = 0$ corresponds to both maximum $|\nabla X|$ and minimum $|\nabla X|$. On isentropic maps the latter is the centrum of a barotropic zone (where isothermal and isobaric surfaces are parallel). The distance from the maximum to minimum value of $-\nabla|\nabla\theta|.\mathbf{n}_\theta$ indicates the width of the frontal zone. The occurrence of the maximum and minimum can be determined from a further operator which is the directional derivative of a parameter along the gradient of $-\nabla|\nabla X|.\mathbf{n}_x$ (see fig. 2.27). An important advantage of these procedures is that this information together with an index of frontal intensity (maximum $|\nabla\theta|$) is provided. $|\nabla\theta|$ may be of the order of 1°C/100 km and $-\nabla|\nabla\theta|.\mathbf{n}_\theta = \pm 0{\cdot}2°\mathrm{C}/(100\ \mathrm{km})^2$. Fig. 2.28 provides a synoptic picture of these parameters.

Comparisons of $-\nabla^2\theta$ and $-\nabla|\nabla\theta|.\mathbf{n}_\theta$ (Clarke and Renard, 1966) show that the former provides no significant improvement in the definition of the frontal pattern. Occluded fronts in particular present difficulties because isotherms commonly intersect them at close to a right angle. It appears that different combinations of thermal shear and curvature may be needed, according to the magnitude of these terms and also perhaps geographical area, in order to achieve a consistently satisfactory result. Creswick (1967) has extended the work of Renard and Clarke using wet-bulb potential temperature (θ_w). The objective results showed some marked divergences from the routine subjective analysis of frontal locations, and Creswick emphasized that fronts should not be regarded as surfaces of constant θ_w. Moreover, he demonstrated that air masses do not have constant θ_w in the vertical and concluded that air mass analysis 'should not be considered as synonymous with frontal analysis'.

It is evident from the above that the determination of frontal frequencies is rather unsatisfactory if the fronts are only identified subjectively. Nevertheless, many climatological studies of this type have been made in order to delimit mean zones of frequent frontal activity (Reed, 1960; Taljaard, Schmitt and Van Loon, 1961; Van Loon, 1965; Yoshimura, 1969) and to distinguish between frontal and other lows (Hare, 1955). It is useful, however, to depict not only the average frontal position but also its variability. This has been shown for the 850 mb Arctic Front over North America (Barry, 1967) and for the Arctic Front over Eurasia (Krebs and Barry, 1970) using the median, quartiles and deciles. Fig. 2.29 illustrates the latter map for July 1952–6. Yoshino (1969) plots percentages frequencies of surface and 850 mb fronts based on 2° latitude–longitude 'squares'. One of the few 'objective' approaches to identifying frontal zones in the troposphere is to be found in Kurashima's (1968) analysis of zones with a larger meridional gradient of 1000–500 mb thickness, although even here the critical gradient must be chosen arbitrarily.

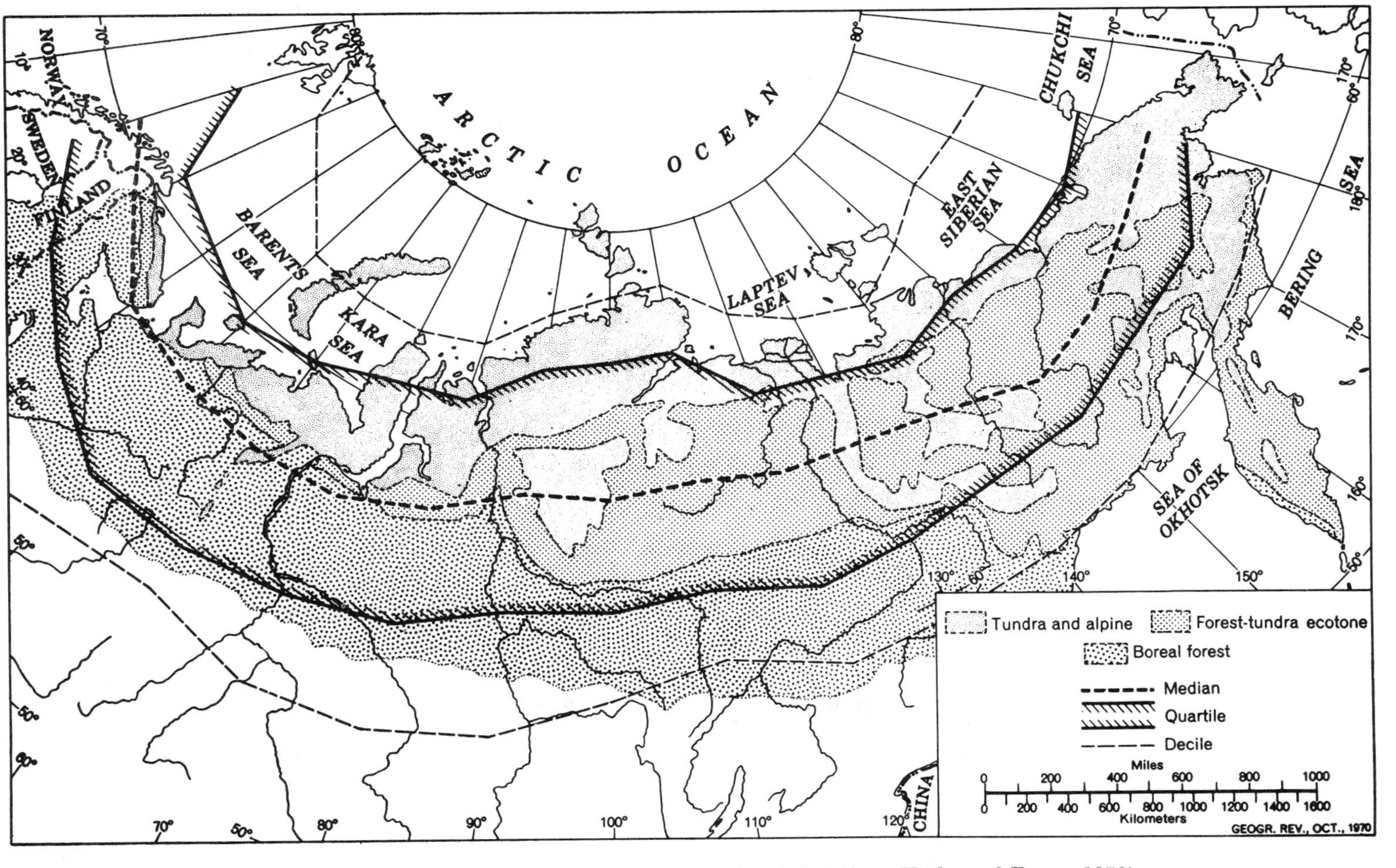

Fig. 2.29 The MSL Arctic front over Eurasia, July 1952–6 (from Krebs and Barry, 1970).

B Climatological maps

In this section we are concerned with maps of time- and/or space-averaged meteorological parameters and of deviations from such averages. Distribution maps of conventional climatic elements are not discussed.

1 MEAN PRESSURE AND WINDS

Maps of mean pressure (or mean isobaric contours) are a basic climatological tool, but they are nevertheless open to misinterpretation. It is essential first of all to recognize the very different significance of mean highs and mean lows. The former indicate areas where anticyclones tend to remain *in situ*, whereas the latter delineate areas into which deep depressions frequently tend to move (Petterssen, 1950).[1] The contrast between the regions of the Siberian winter high pressure and the Icelandic low is but one obvious example. Mean pressure maps therefore need to be examined with reference to charts of the frequency and tracks of cyclones and anticyclones for proper understanding of the mean features that they show. An additional tool which is useful for this purpose is a map of the rate of alternation of high and low pressure centres. Petterssen uses an index of alternation: Fc/Fa (when $Fc < Fa$) or Fa/Fc (when $Fa < Fc$), where Fa and Fc denote the percentage frequencies of anticyclones and cyclones, respectively, in a given square. Areas of low or high mean pressure have low indices of alternation while the 'perturbation ducts' along which such cells travel most frequently are depicted by high indices (fig. 2.30).

Another basic problem with mean pressure maps is that frequency distribution (in time) of atmospheric pressure patterns is not Gaussian. Consequently, the physical interpretation of average pressure maps is open to question. As Hare (1957) points out, a mean map for an individual month may incorporate segments of several distinct 'populations' of atmospheric pressure patterns (see also section C of this chapter, p. 86). In studying mean maps for individual months it is often useful to examine the pattern of departure from the long-term term. Hesselberg (1962) provides an extensive study of this type for Norway, using seasonal means. The lines of pressure deviations indicate the 'mean additional gradient wind' and different directions of this wind component can be related systematically to deviations of temperature and precipitation. This topic is discussed further in Chapter 3.D, pp. 194–6.

[1] At the same time, it must be pointed out that the mean Aleutian and Icelandic lows and also the Siberian high have been shown to be maintained by the thermal structure of the atmosphere (Spar, 1950; Smagorinsky, 1953). This topic is discussed further on p. 86.

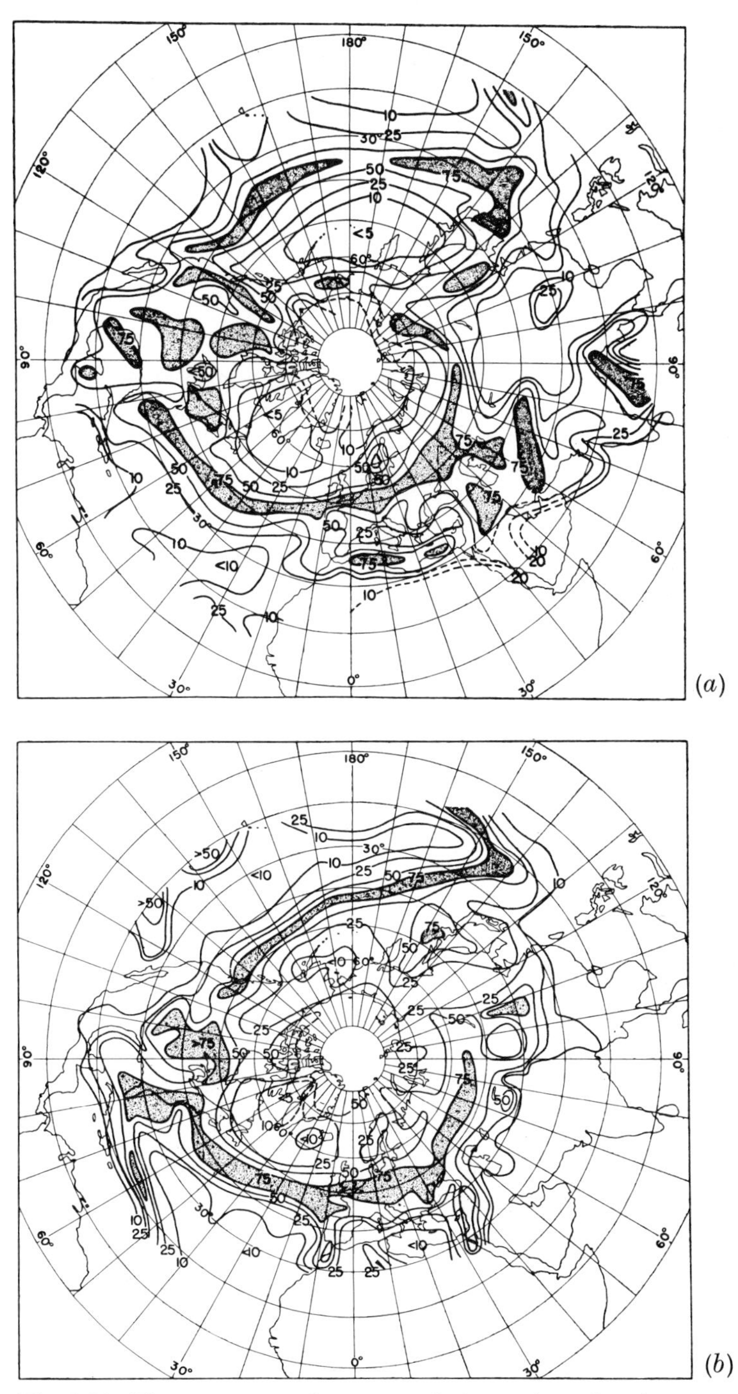

Fig. 2.30 The percentage frequency of alternation of cyclones and anticyclones (see text) (from Petterssen, 1950). (*a*) Winter, (*b*) summer.

In the case of upper air charts, where the patterns are mainly open troughs and ridges, mean maps greatly blur even the most significant features. Stark (1965) illustrates this point by maps of the positions of monthly mean troughs and ridges at 700 mb over the northern hemisphere during 1949–63 (see fig. 2.36). The type of detail that is required for synoptic climatological studies is exemplified by the atlas of five-day mean heights, standard deviations and changes between consecutive pentads at 700 mb over the northern hemisphere prepared by Wahl and Lahey (1969).

The wind velocity determined from a mean pressure (or isobaric contour) map is the *geostrophic resultant* (or mean vector). This, however, does not allow discrimination between strong, but variable, pressure gradients and predominantly weak pressure gradients. Over the Arctic Basin, for example, the January mean MSL pressure map conceals the fact that the circulation is at times dominated by a polar anticyclone and at times by slow-moving cyclones. This problem with the resultant can be avoided to some extent if maps are also prepared of the standard vector deviation. Tucker (1960) provides such information for winds up to the 100 mb level:

$$\textbf{Standard vector deviation, } \sigma = \left(\frac{\sum V^2}{N} - \bar{V}^2\right)^{1/2}$$

where $\sum V^2/N$ = the mean square velocity,
$\bar{V}$ = mean resultant velocity.

An alternative procedure is to compute wind constancy, q:

$$q = \frac{100}{\bar{V}_s}\bar{V}$$

where $\bar{V}_s$ = mean speed (regardless of direction).

Global maps for comparison with mean MSL streamlines and resultant winds have been prepared by Mintz and Dean (1952; see also Trewartha, 1968, p. 89).

Mean streamlines are sometimes drawn, based either on the mean pressure field, or on the vector mean wind field. Among the most extensive analyses of this type are the publications of the British Meteorological Office on global winds at several levels (Heastie and Stephenson, 1960; Frost and Stephenson, 1965).

One application of mean streamlines is in the determination of mean zones of confluence. Bryson (1966), for example, has prepared maps of resultant surface streamlines for each month over North America east of the Rocky Mountains (based on rather heterogeneous data) and showed that in summer the zone of confluence between Arctic and Pacific airstreams (fig. 2.31) corresponds closely with the modal position

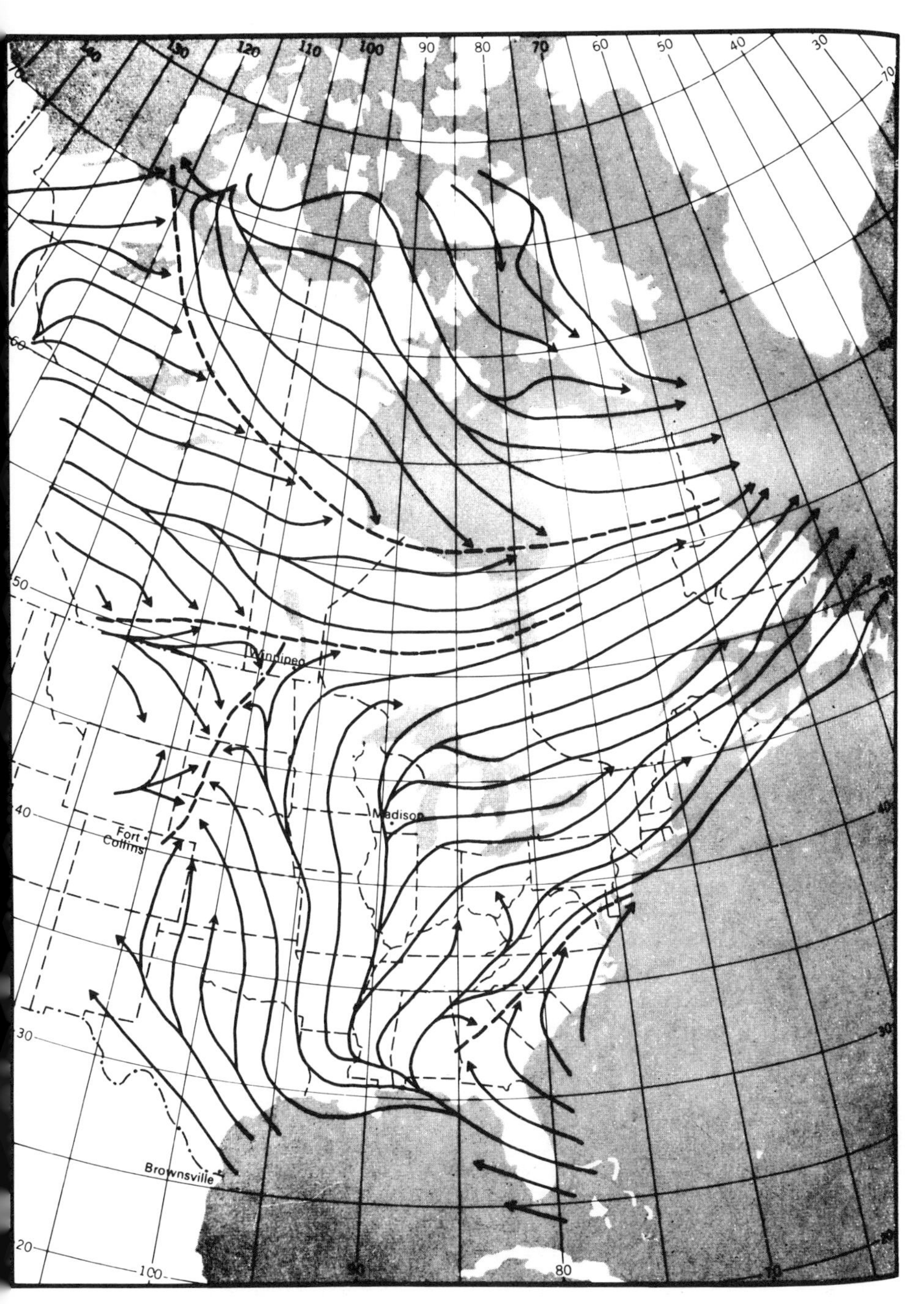

Fig. 2.31 Confluence of mean resultant streamlines at the surface for August demarcating various frontal zones (from Bryson, 1966).

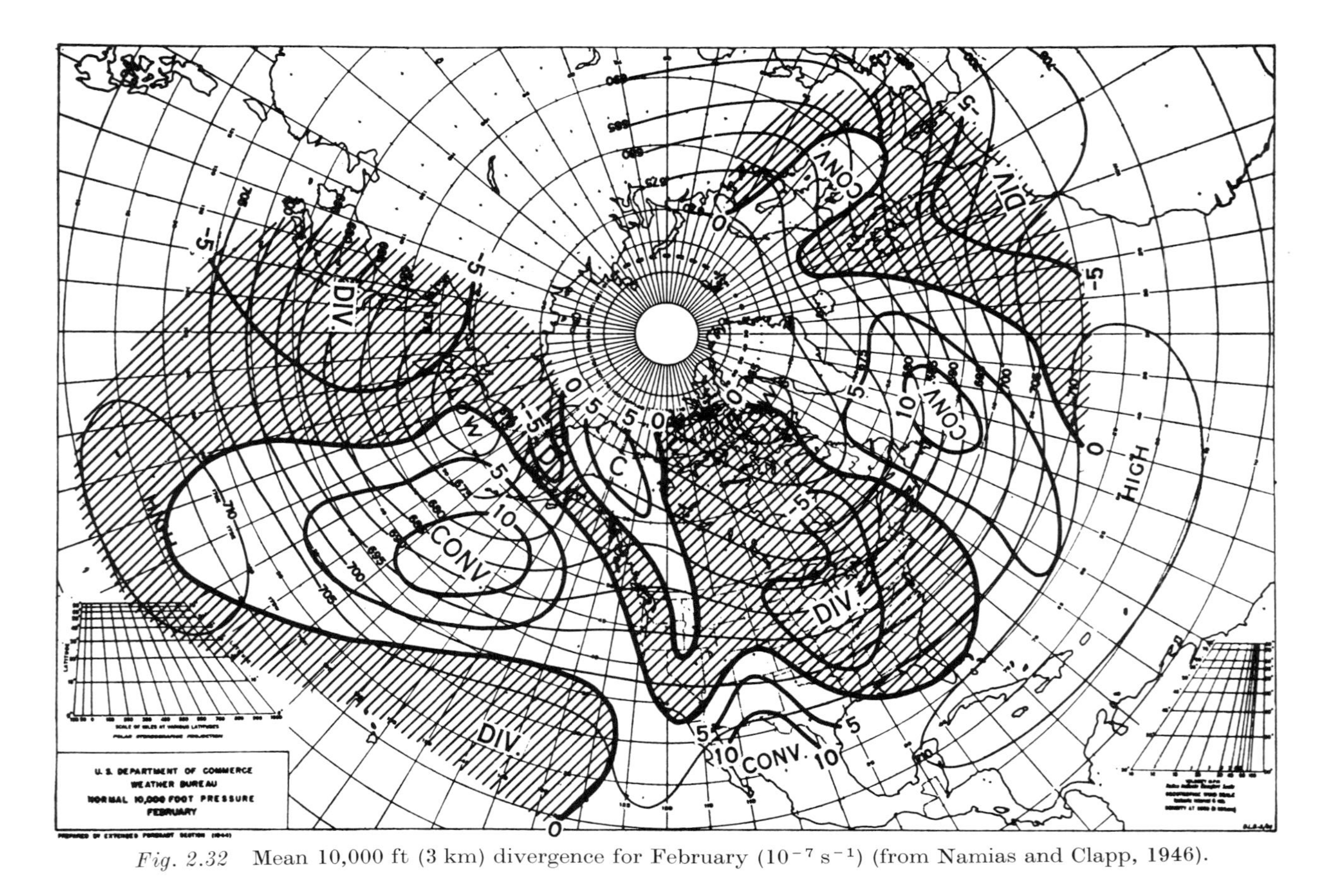

Fig. 2.32 Mean 10,000 ft (3 km) divergence for February ($10^{-7}\,s^{-1}$) (from Namias and Clapp, 1946).

of the Arctic frontal zone as defined by air masses. Thompson (1951) and Watts (1955) make considerable use of mean streamlines in southeast Asia, although the confluence zones they recognize show considerable disparities.

2 MEAN DIVERGENCE, VERTICAL VELOCITY AND RELATIVE VORTICITY

Few studies have been made of quantities such as mean divergence or vorticity fields, probably because of the problem of the unknown character of their frequency distribution as well as their rather limited value for general circulation studies. Namias and Clapp (1946) estimated the mean monthly fields of divergence over the northern hemisphere at the 10,000 feet (*c.* 3 km) level. Fig. 2.32 illustrates the computed mean divergence field at 10,000 feet for February. Convergence occurs just east of the major troughs and divergence to the west of them. This pattern, which reverses in the upper troposphere, reflects the occurrence of precipitation and cyclonic development ahead of the troughs with subsidence in their rear sector. Divergence also occurs in associations with the subtropical anticyclone cells over the eastern parts of the oceans. Analysis of divergence related to the mean surface wind fields in January and July has been performed by Mintz and Dean (1952; see Trewartha, 1968). More recently, Bleeker (1960) determined mean divergence for mid-season months over the Mediterranean and Romanov (1965) prepared mean maps of resultant wind divergence and vorticity for the Indian Ocean.

Sela and Clapp (1971) have attempted to compute mean vertical velocities for each mid-season month using a thermodynamic model. The patterns resemble the divergence maps of Mintz and Dean, and of Namias and Clapp except in the western oceans where topographic effects are important. However, they conclude that climatological fields of cloudiness and precipitation provide a more accurate picture although the relationship is imprecise due to the complexity of the scales of vertical motion which produce precipitation.

The most extensive analysis of climatological fields of relative vorticity is that of O'Connor (1964) for the 700 mb level on a seasonal basis for 1947–58. These maps are discussed in section C of this chapter (pp. 82–86).

3 SPACE AVERAGING

Climatologists are familiar enough with time averaging. A comparable analysis, with quite different aims, is that of space averaging. Space-mean charts are sometimes used in large-scale studies to filter synoptic

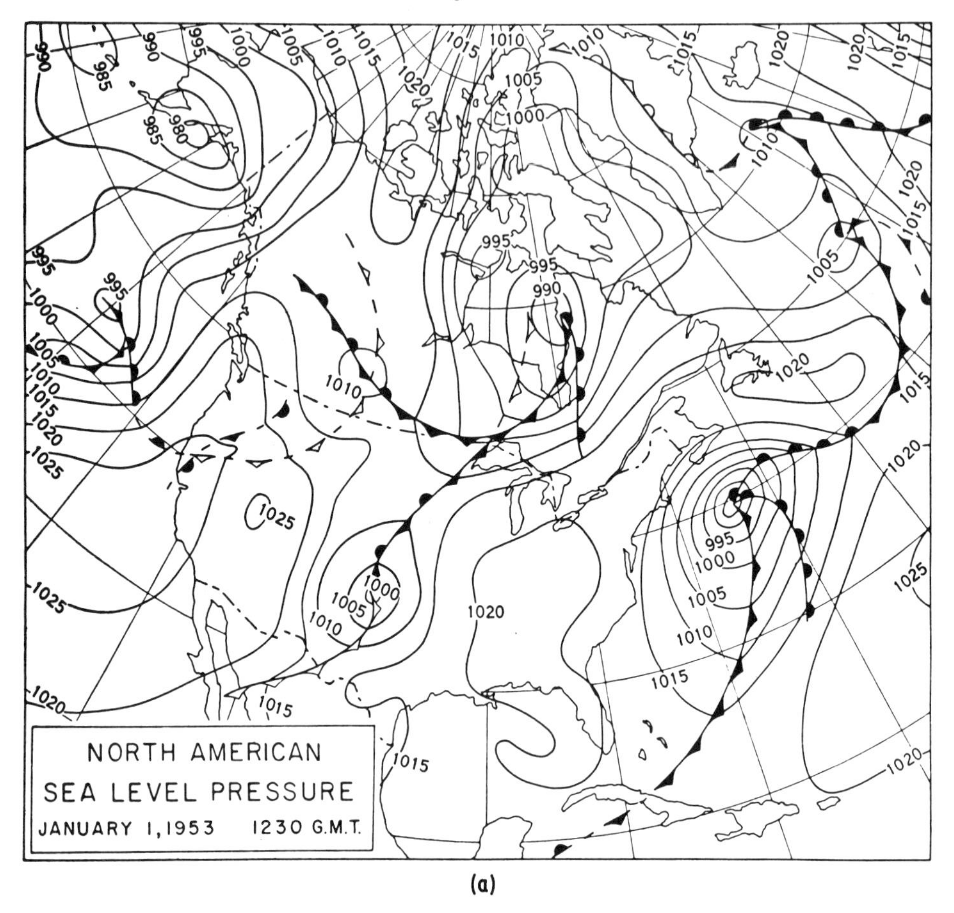

(a)

variability out of upper air contour patterns and clarify the major long-wave structure (Berry, Haggard and Wolff, 1954). A typical smoothing function is for a diamond grid with Z_0 at its centre:

$$Z_0 = (Z_1 + Z_2 + Z_3 + Z_4)/4$$

A 20° latitude spacing will eliminate cyclones, for example. Composite smoothing functions are developed by iteration. Holloway (1958) illustrates the procedure, as shown in table 2.3 from left to right.

Table 2.3 Space smoothing functions (Holloway, 1958)

1/3 1/3 1/3	1/9 1/9 1/9 3/9 1/9 1/9 1/9	1/81 2/81 1/81 2/81 8/81 8/81 2/81 1/81 8/81 15/81 8/81 1/81 2/81 8/81 8/81 2/81 1/81 2/81 1/81

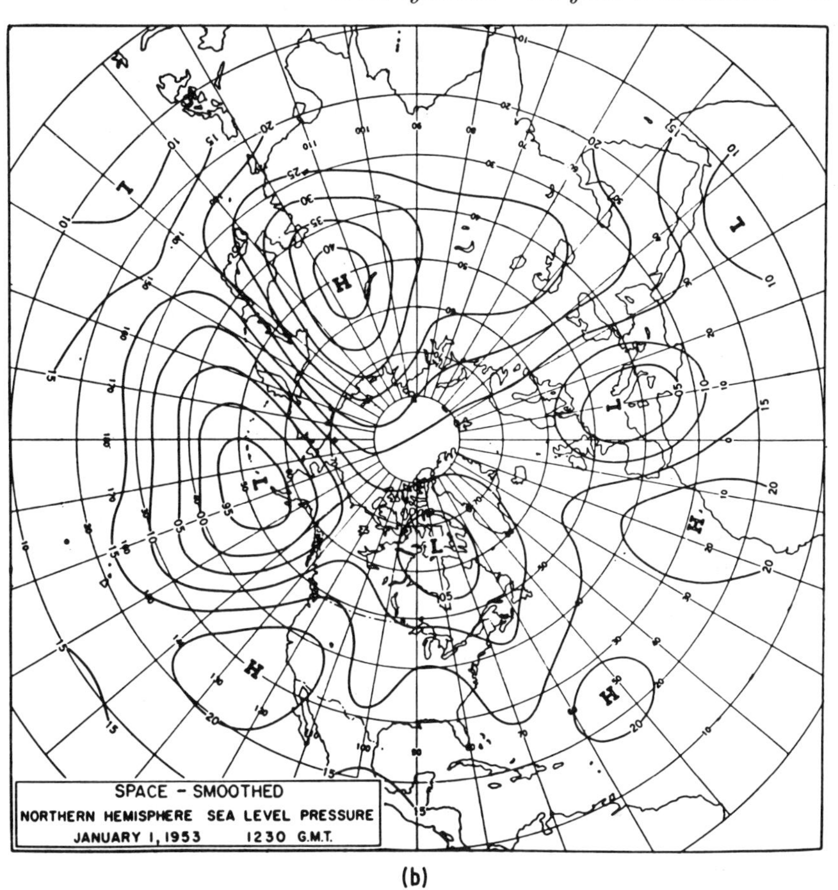

Fig. 2.33 Space smoothing of MSL pressure for 1 January 1953 (from Holloway, 1958). (*a*) Original map over North America, (*b*) Space-smoothed hemispheric map.

Fig. 2.33 shows a surface map and the smoothing obtained by a function similar to that at the right of table 2.3. Features of 1500 km scale are almost completely attenuated whereas those of 4000 km scale are retained with a reduction in amplitude of approximately 25%. If the smoothed map pattern is subtracted from the original the smaller-scale feature can be isolated. Procedures for determining smoothing functions are detailed by Asselin (1966).

C Data synthesis – the general circulation

To provide a framework for the synoptic climatological material, we need to review the principal characteristics of the general circulation.

In the words of Corby (1970), 'we . . . recognize that a comprehensive knowledge and understanding of the atmospheric general circulation is an essential prerequisite for an understanding of the climate of the earth including its variations on all time and space scales'. For the most part the studies we shall discuss are synthetical in character but it is convenient to present them at this point since it is statistical aspects of the circulation that are principally involved. Only those aspects that are pertinent to the main themes of the book are considered. Detailed treatments of dynamic meteorology and climatology may be found in recent publications by Lorenz (1967), Palmén and Newton (1969), Rex (1969) and Riehl (1970).

1 TRANSPORT MECHANISMS

The features of the global circulation, which have been intensively studied in recent years under the heading of what we may call dynamic climatology, can be grouped into four categories with respect to the degree of averaging of the variables that is involved. They are:

1 Time- and longitude-averaged features (e.g. the zonal westerlies and trade winds, mean meridional cells).
2 Longitude-averaged features (such as variations in the strength of the westerlies indicated by the zonal index).
3 Time-averaged features (e.g. the Asian monsoon regimes).
4 'Instantaneous' features such as the migratory synoptic-scale perturbations occurring in the westerlies and the tropical easterlies.

A major focus of attention has been the energy transformations and the transport mechanisms whereby the circulation achieves the necessary climatic balances of energy, momentum and water in the various latitude zones. The analysis method uses a concept developed by O. Reynolds in the late nineteenth century with reference to fluid flow in pipes. The total flow past a point, u, can be regarded as comprising the time–mean flow, $\overline{u}$, and a deviation, u', from the time average

$$u = \overline{u} + u'.$$

We can also consider departures with respect to space averages

$$u = [u] + u^*$$

where [ ] denotes an average along a latitude circle and * a spatial deviation from this average. Thus, considering joint time and space averages

$$u = [\overline{u}] + \overline{u}^* + [u]' + u^{*\prime}$$

where $[\bar{u}]$ = zonal winds averaged along a latitude circle and over a time interval
$\bar{u}^*$ = local departure from the longitudinal average, averaged in time
$[u]'$ = time departures from the longitudinal average
$u^{*\prime}$ = instantaneous local departures from the time and longitudinal averages.

It follows that:

$$\bar{u} = [\bar{u}] + \bar{u}^* \quad \text{and} \quad \bar{v} = [\bar{v}] + \bar{v}^*$$

and, by expansion, the mean poleward transport of relative angular momentum per unit mass at one level across a latitude circle is:

$$[\overline{uv}] = [\bar{u}][\bar{v}] + \overline{[u]'[v]'} + [\bar{u}^*\bar{v}^*] + \overline{[u^{*\prime}v^{*\prime}]} \tag{1}$$

where the terms on the right refer, respectively, to transports by standing (meridional) cells, transient cells, standing (zonal) eddies and transient eddies.

When the analysis is extended to the vertical dimension the terms corresponding to those in equation (1) represent, respectively, mean and instantaneous cell circulation along a meridian, and mean and transient circulations in the horizontal plane across a latitude circle.

In the literature, two different schemes may be encountered. These are compared by Starr and White (1952):

$$\overline{[uv]} = \overline{[u]}\,\overline{[v]} + \overline{[u]'[v]'} + \overline{[u^*v^*]} \tag{2}$$

and

$$[\overline{uv}] = [\bar{u}][\bar{v}] + [\bar{u}^*\bar{v}^*] + [\bar{u}'\bar{v}']. \tag{3}$$

Note the difference in the order of averaging:

$$\overline{[uv]} = [\overline{uv}]$$

since space and time averaging are commutative. Similarly:

$$\overline{[u]}\,\overline{[v]} = [\bar{u}][\bar{v}].$$

The other equivalent terms in the two equations are not equal, however. $\overline{[u]'[v]'}$ is due to time variations in the meridional flow, while $[\bar{u}^*\bar{v}^*]$ is due to the asymmetry of the mean streamlines; both terms appear in equation (1) above.

$\overline{[u^*v^*]}$ is associated with the asymmetry of instantaneous streamlines and $[\bar{u}'\bar{v}']$ corresponds to time variations of wind at a given point. These two terms are combined in the transient eddy term in equation (1).

In general, the poleward transport of any atmospheric property, S (such as heat, water vapour), is obtained by replacing u in the above equations by S. Following the early studies of Priestley (1949) and Starr and White (1951, 1952, 1954), in particular, numerous investigations have been carried out along these lines. Comprehensive reviews of the various findings have recently been prepared by Lorenz (1967) and Reiter (1969). Most of this work is concerned with the problem of the maintenance of the general circulation itself – the central problem of dynamic climatology.

For present purposes it is sufficient to summarize the salient findings of these studies very briefly. The picture that emerges is a complex one. A poleward transport of energy is necessary to maintain balance in latitudes poleward of about 35° where the net loss of long-wave radiation exceeds the incoming short-wave radiation. About 80% of this energy is transferred in the atmosphere in the form of latent heat and sensible heat (enthalpy), the remainder is transported by warm ocean currents. In low latitudes the heat transport is accomplished primarily by a mean meridional cell known as the Hadley cell, with equatorward flow at low levels in the trades, poleward flow aloft. The upward arm of this cell, which was originally regarded as being associated with a simple equatorial heat source, in fact originates mainly in sub-synoptic scale cloud bands especially within tropical storm systems where the 'hot towers' of giant cumulonimbus clouds extend up to the tropopause. Moreover, the poleward branch is concentrated in preferred longitudes at the western margins of the subtropical anticyclones aloft. In middle latitudes the poleward heat transport is almost entirely a result of horizontal eddies, particularly the travelling disturbances in the westerlies.

In addition, westerly angular momentum must be transported poleward in order to maintain that which the westerlies transfer to the earth through the mechanisms of friction and the mountain torque. The latter arises from horizontal pressure differences across mountain barriers. The poleward transport is at a maximum in the upper troposphere of the subtropics and is related to eddy circulations across the axis of the mean subtropical jet stream. The mean cell component is negligible. The upward transport of angular momentum in low latitudes appears to derive mainly from the mean cell. Since the absolute angular momentum (per unit mass) increases towards the equator, the rising arm of the Hadley cell transfers more upward than is transferred downward at the tropics, thus creating a net upward transport.

While horizontal moisture transport (and surface runoff) is necessary for the global water balance it is not a direct constraint on the atmospheric circulation. Nevertheless, there is a proportional relationship between this transport and that of latent heat. In low latitudes the moisture flux is directed equatorward due to the low-level branch of the Hadley cell whereas in mid-latitudes the poleward transport is determined by the low-level eddies.

It should be emphasized that in several instances the eddy transfer is counter to the mean gradient. Thus, they are generative rather than dissipative as in classical turbulence theory. The flux of relative angular momentum towards zones of higher angular velocity in lower middle latitudes – referred to as 'negative viscosity' (Starr, 1968) – is one illustration of this. The implication is that kinetic energy is fed from the eddies to the mean zonal circulation. Time- and longitude-averaged features, such as the meridional cells, are not of particular significance from the synoptic climatological standpoint and neither are time variations in meridional flow. Time variations of longitudinally averaged features such as the zonal wind $[u]'$ are of more interest but we shall reserve discussion of these for a later section (Chapter 3.C, pp. 165–75). Most conventional climatological study has dealt with the spatial variation of time-averaged features, like $[\bar{u}^*\bar{v}^*]$ in equation (3), while the migratory systems which primarily contribute to the transient eddy component of transfer have been the concern of synoptic meteorology. In view of our interest in the statistical properties of the perturbations in a spatial framework and their relationship to time–mean features of the atmosphere, these two categories must be discussed further. It is convenient to look first at the horizontal and vertical structure of the atmosphere and then to examine the eddy motions in the context of scale considerations.

2 THE HORIZONTAL DISTRIBUTION OF PRESSURE AND GEOPOTENTIAL

The distribution of average mean sea level (MSL) pressure in January and July is shown in fig. 2.34. The major high and low centres, often referred to as the *centres of action* (Teisserenc de Bort, 1883), are the Icelandic and Aleutian lows, the Azores and North Pacific subtropical highs, and the Siberian winter high, in the northern hemisphere. In the southern hemisphere, where the land masses are less extensive, the subtropical high pressure and the subpolar low pressure areas form more or less continuous belts around the hemisphere. In contrast to the northern hemisphere, however, the subtropical highs are best developed in winter. In the summer season they are disrupted by heat lows over the continents (Taljaard *et al.*, 1969). This also occurs of course in the northern hemisphere, but there the anticyclone cells over the oceans are particularly extensive in summer. The equatorial zone is characterized by a low pressure trough throughout the year although the location of this shifts seasonally, especially over the continents. In the Arctic the mean pressure pattern is weak and indeterminate in summer while in winter there is a trough in the Norwegian Sea–Barents Sea sector and a ridge across the Beaufort Sea to northeast Siberia. Only in spring, and then over the Canadian Arctic Archipelago, is there a true 'Polar' anticyclone.

Fig. 2.34
Mean MSL pressure over the northern hemisphere, 1950–9 (from Gommel, 1963). (*a*) January, (*b*) July.

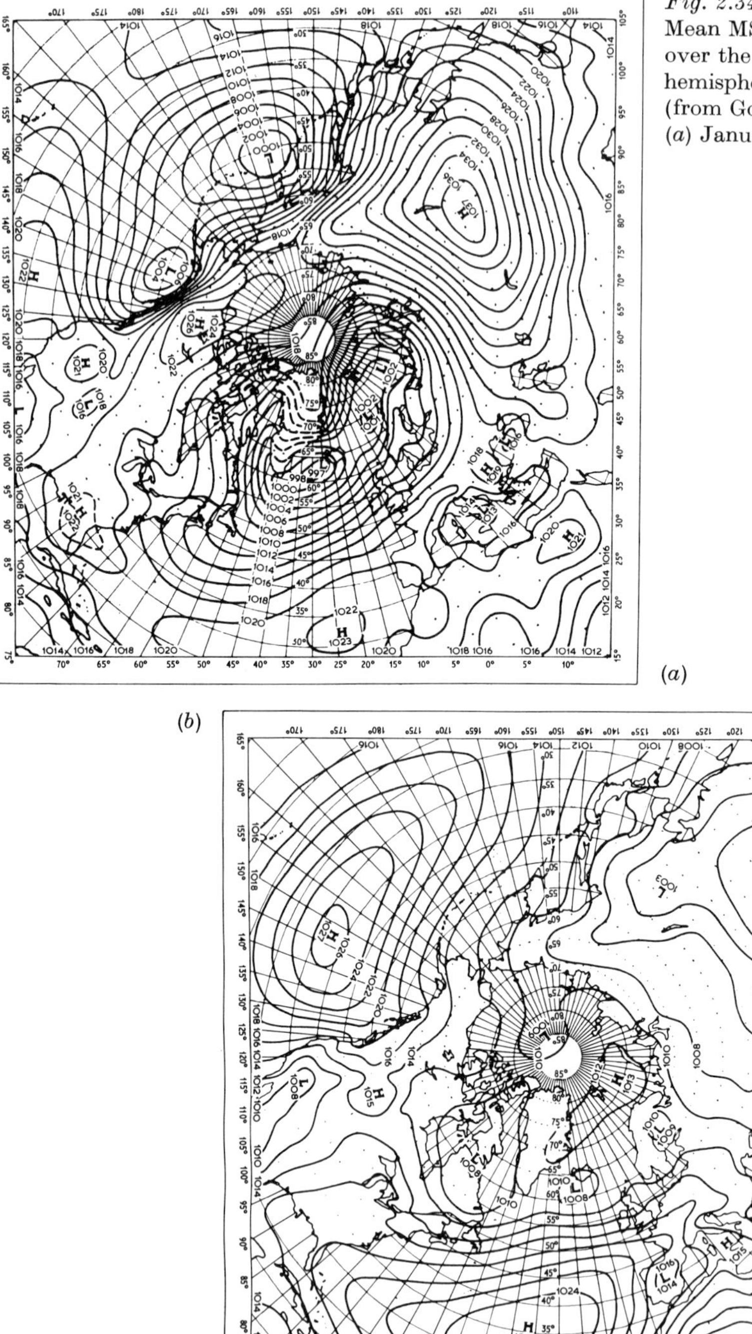

(*a*)

(*b*)

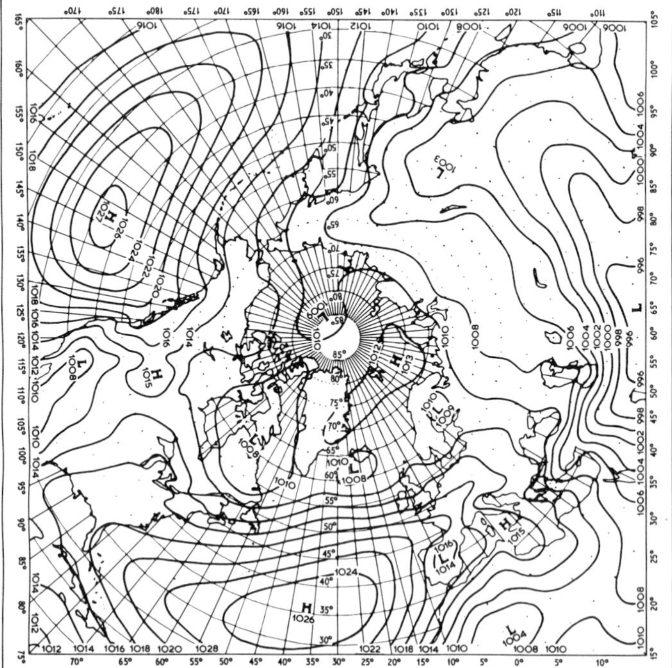

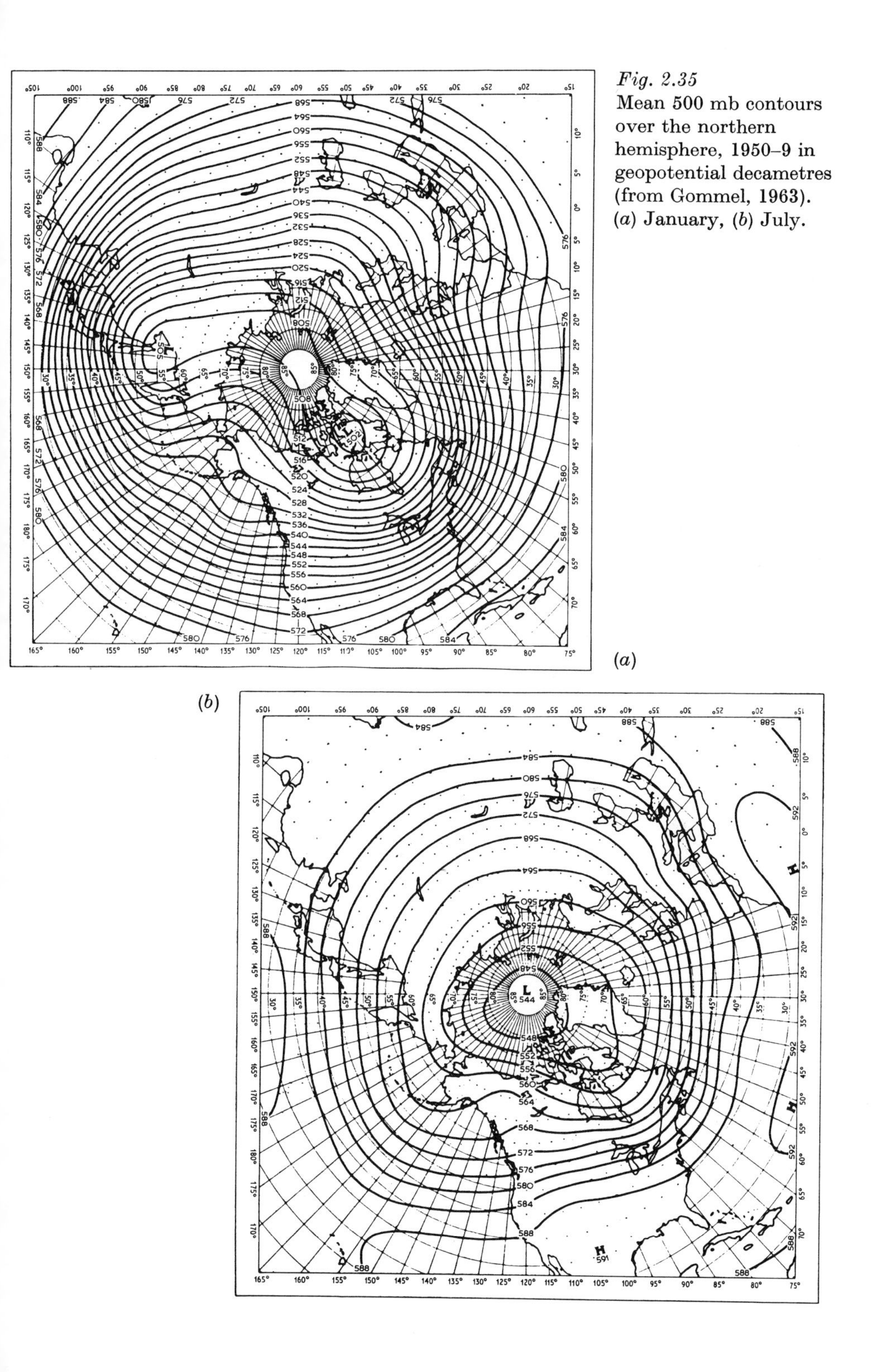

Fig. 2.35
Mean 500 mb contours over the northern hemisphere, 1950–9 in geopotential decametres (from Gommel, 1963). (*a*) January, (*b*) July.

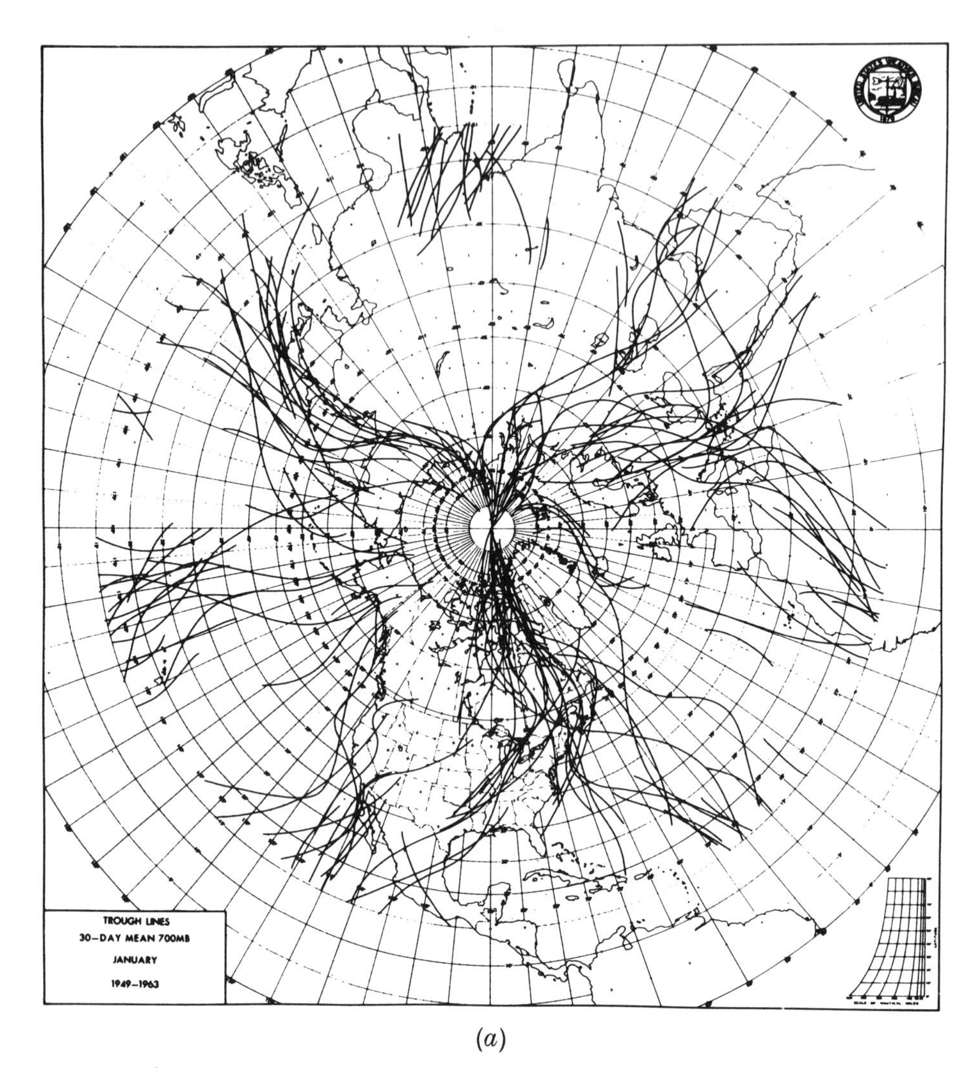

(*a*)

Fig. 2.36 Trough lines for thirty-day mean 700 mb charts, 1949–63 (from Stark, 1965). (*a*) January, (*b*) July.

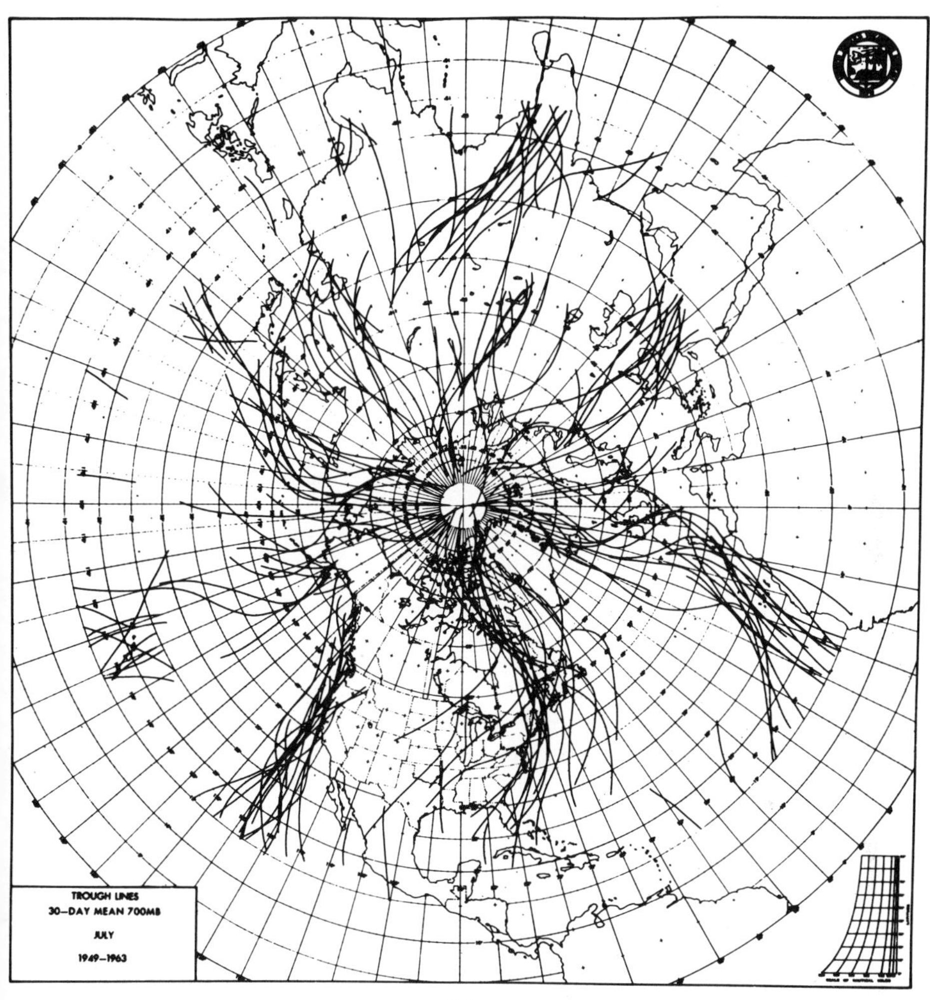

(*b*)

The extent of the major oceanic wind systems at the surface can be gauged from table 2.4. This basis of comparison eliminates the monsoonal effect of the continents. It should be noted that the northern hemisphere 'Polar Easterlies' are essentially restricted to the polar side of the Aleutian and Icelandic lows (fig. 2.34) rather than forming a coherent system. Hence, a hemispheric average for these latitudes will show easterly zonal flow only at times when the subpolar lows are well-developed and persistent features.

In the middle troposphere, adequately represented by the 500 mb surface (fig. 2.35), there is much greater simplicity of pattern than at the surface with an extensive 'Ferrel' vortex centred close to each pole and

Table 2.4 The oceanic wind belts (Blüthgen, 1966)

		% of ocean surface area	*10^6 km²*
70–90°N	Polar easterlies	4·9	17·4
35–70°N	Westerlies (and subpolar low)	7·9	28·4
25–35°N	Subtropical anticyclone	7·2	25·75
10–25°N	Tropical easterlies	13·2	47·4
5°S–10°N	Intertropical convergence-zone	10·5	37·5
5–20°S	Tropical easterlies	22·4	79·4
20–35°S	Subtropical anticyclone	12·3	44·65
37–70°S	Westerlies (and subpolar low)	15·0	56·9
70–90°S	Polar easterlies	5·6	20·0

nearly continuous belts of high pressure in the tropics. The centres of the high cells are at about 25°N and 10–12°S in July and about 10°N and 20°S in January. The vortex in the northern hemisphere is markedly asymmetric with deep troughs in winter over eastern North America and eastern Asia, and a subsidiary one over eastern Europe, whereas the waves in the southern hemisphere are much more subdued features. This does not mean, however, that synoptic disturbances in the southern hemisphere are weakly developed. They are mobile and show much less meridional movement than their northern hemisphere counterparts and less tendency to intensify in particular longitudes. Thus, there is a zonal belt of subpolar low pressure rather than well-developed mean centres.

As we have already indicated (see p. 64) the interpretation of mean pressure or geopotential fields poses certain problems. Although no comprehensive survey of upper air features, comparable to that of Petterssen's (1950) for MSL pressure features, is available we can nevertheless draw on several studies. For the northern hemisphere the geographical distribution of the frequency of troughs and ridges has been

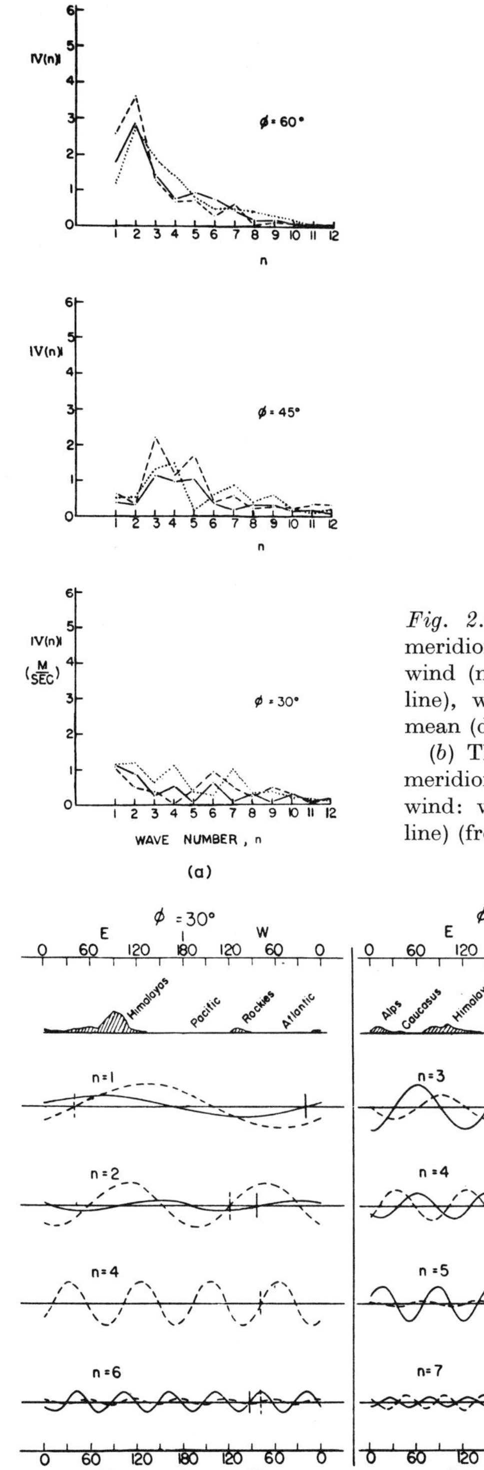

Fig. 2.37 (*a*) Amplitude spectra for the meridional component of the 500 mb mean wind (m s^{-1}) for 1950: annual mean (solid line), winter mean (dashed line), summer mean (dotted line).

(*b*) The dominant waves comprising the meridional component of the 500 mb mean wind: winter (solid line), summer (dashed line) (from Saltzmann and Peixoto, 1957).

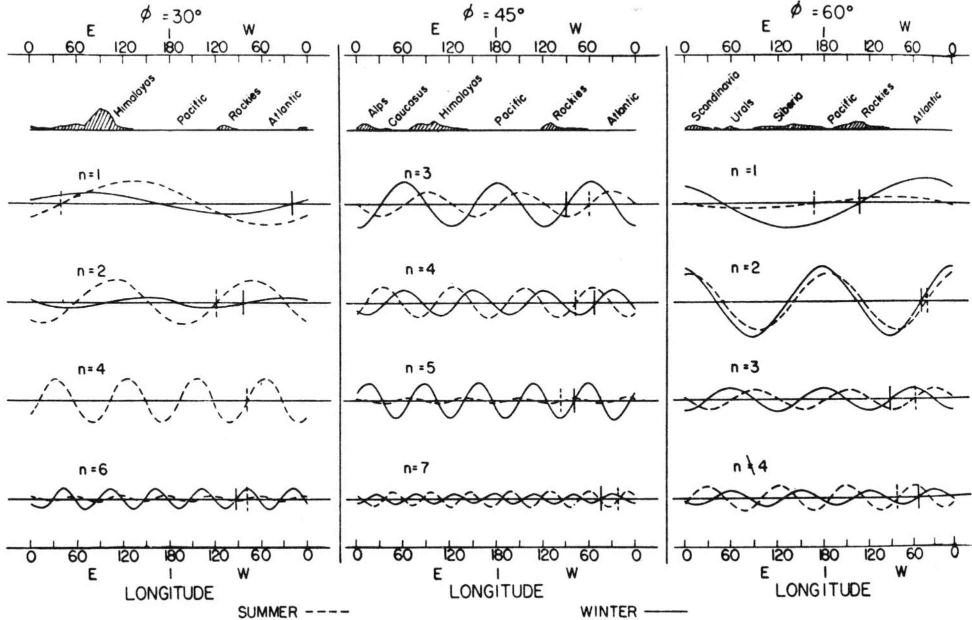

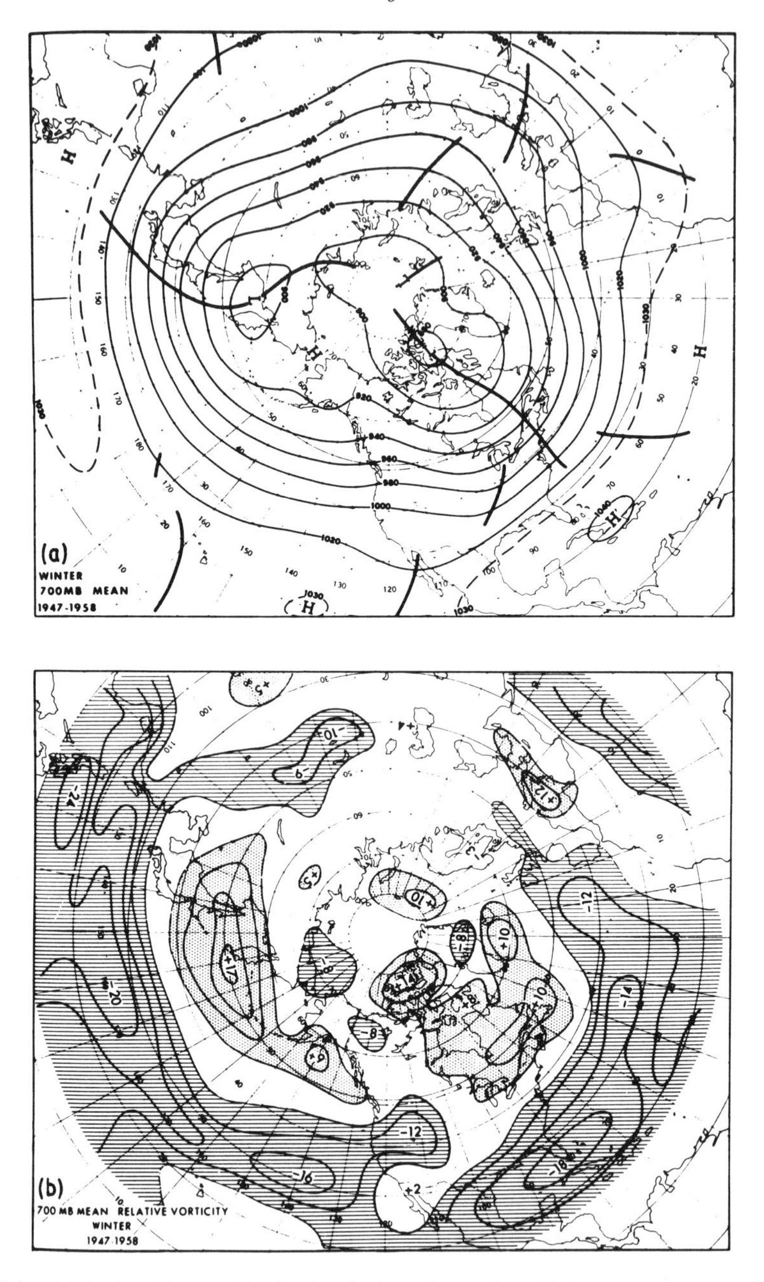

Fig. 2.38 (*a*) Mean 700 mb circulation, December–February, 1947–58. (*b*) Mean relative vorticity at 700 mb, December–February 1947–58. (*c*) Locations of five day mean low centres at 700 mb, January 1947–58. (*d*) Locations of five day mean high centres at 700 mb, January 1947–58. (From O'Connor, 1964.)

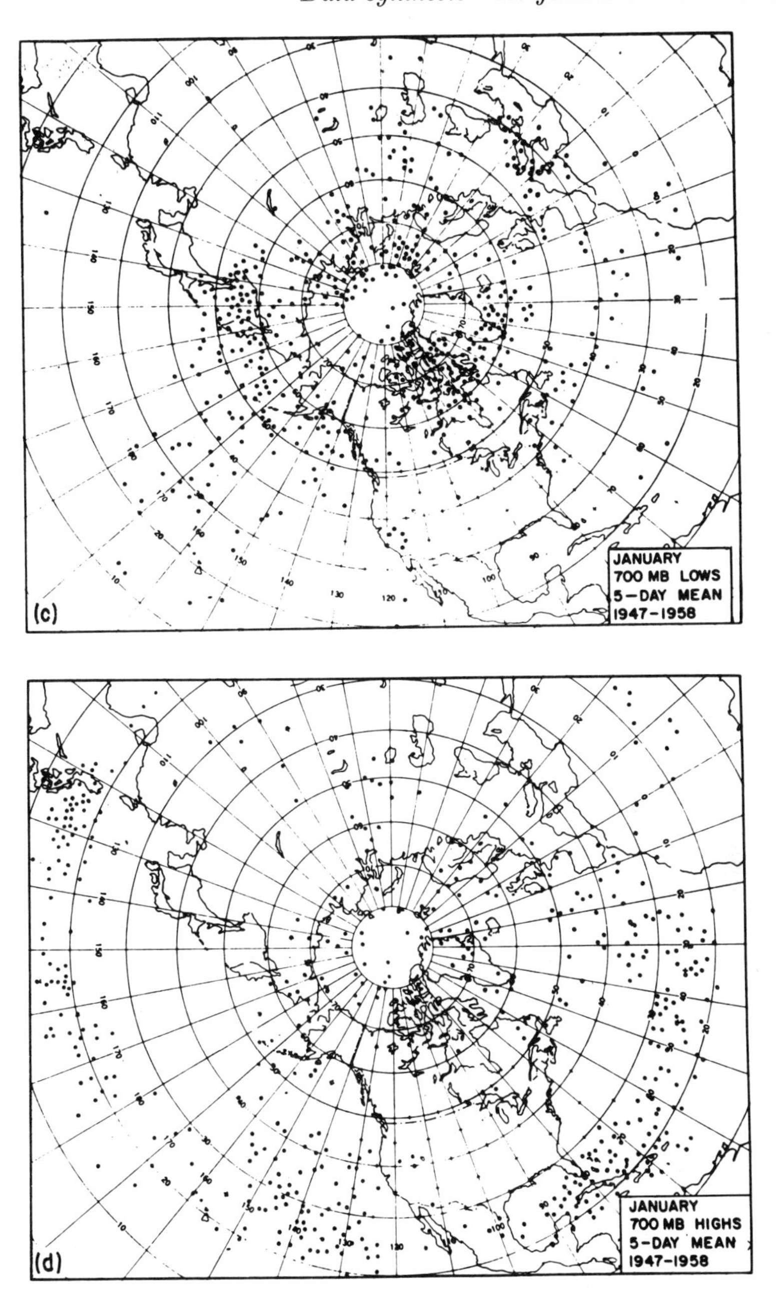
(c)
JANUARY
700 MB LOWS
5-DAY MEAN
1947-1958
(d)
JANUARY
700 MB HIGHS
5-DAY MEAN
1947-1958

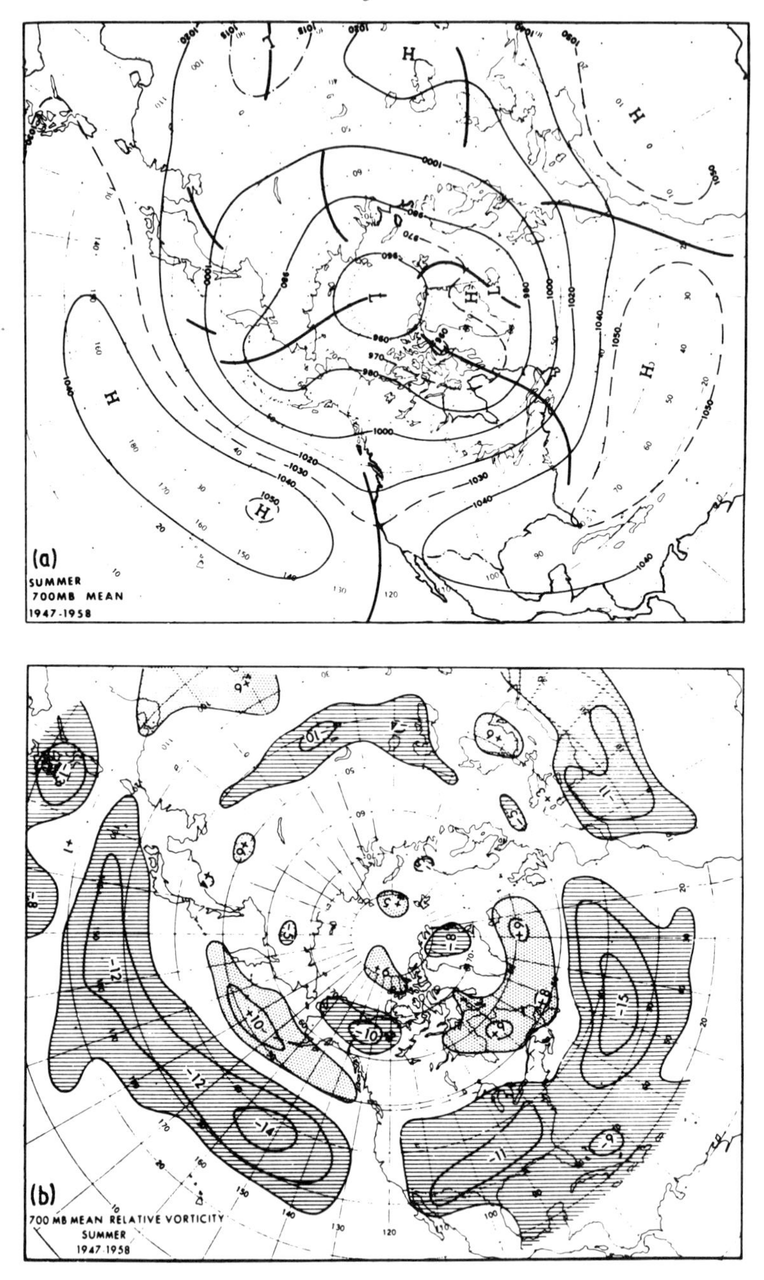

Fig. 2.39 (*a*) Mean 700 mb circulation, June–August 1947–58. (*b*) Mean relative vorticity at 700 mb, June–August 1947–58. (*c*) Locations of five day mean low centres at 700 mb, July 1947–58. (*d*) Locations of five day mean high centres at 700 mb, July 1947–58. (From O'Connor, 1964.)

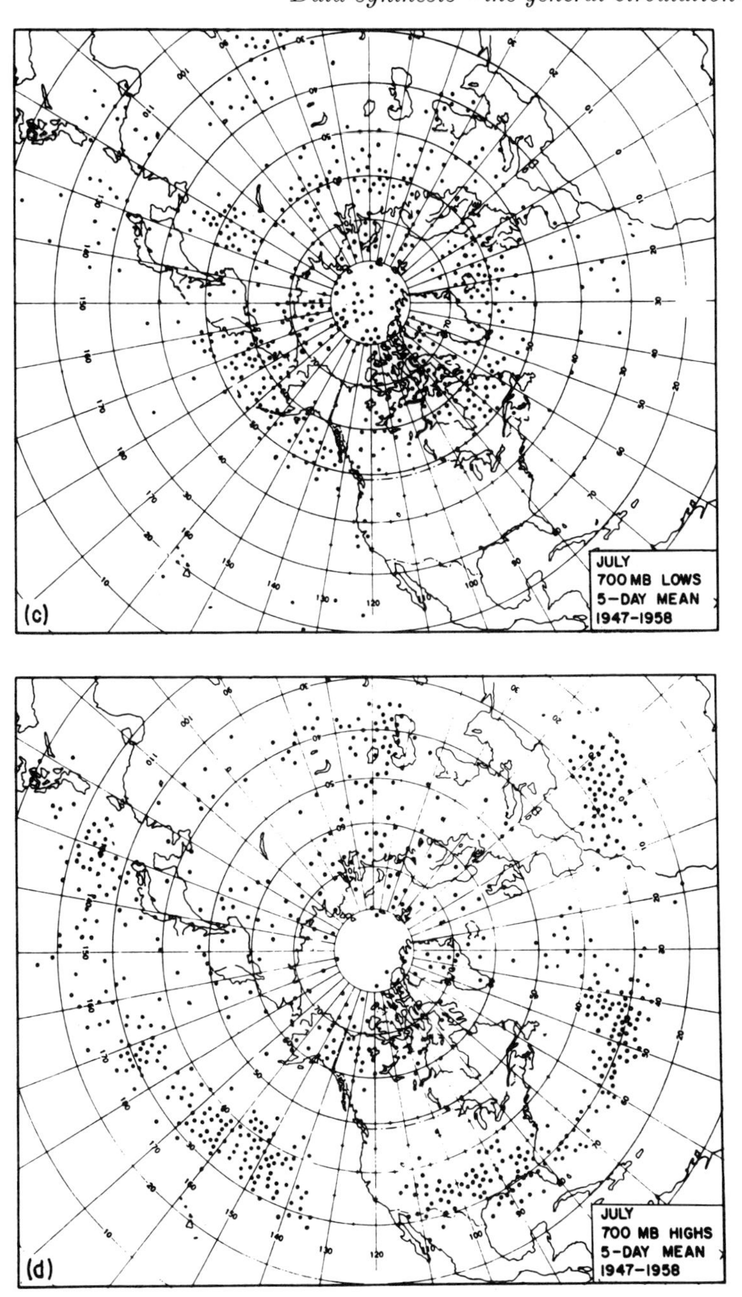
(c)
JULY
700 MB LOWS
5-DAY MEAN
1947-1958
(d)
JULY
700 MB HIGHS
5-DAY MEAN
1947-1958

analysed for the thirty-day mean 700 mb charts (Klein and Winston, 1958; Stark, 1965) and for five-day mean 700 mb charts (O'Connor, 1964). Fig. 2.36 illustrates the predominant three-wave pattern in middle latitudes and six- to seven-wave pattern in the subtropics in January. In July the troughs are much more variable in location in middle and high latitudes, except over eastern North America. Harmonic analysis of the 500 mb mean meridional wind components for 1950 (Saltzmann and Peixoto, 1957) provides objective evidence of the dominant wave patterns (fig. 2.37) and the joint role of orography and land–sea temperature differences in determining them. O'Connor's study, which includes analyses of the 700 mb seasonal mean relative vorticity fields in the northern hemisphere, illustrates the complexity of interrelationship between the mean circulation, mean relative vorticity and the frequency of high and low centres (figs. 2.38 and 2.39). Features such as the subtropical anticyclone cells and mean subpolar lows are readily identifiable. In winter, the vorticity minima and frequency maxima of high centres are concentrated over the western oceans. In summer continental centres are added and the oceanic centres are farther east, especially in the North Pacific. The patterns of mean circulation, vorticity maxima and occurrences of lows match well in winter although the maximum frequency of centres per unit area is near Kodiak Island in December, followed closely by the Baffin Island area, whereas the relative vorticity maximum is over Kamchatka. Other patterns demonstrate more subtle relationships. The 700 mb highs in winter over western Siberia (fig. 2.38), for example, are well to the west of the ridge in the mean circulation and its associated vorticity minimum.

These empirical results still leave unanswered the most crucial question in the present context concerning mean features. Namely, do they possess a physical reality or are they statistical averages only? Baur's (1951) concept of *Grosswetterlage* (steering patterns) is predicated on the concept of centres of action having a certain physical reality of their own and, as Namias (1953) points out, this is borne out to some extent by the degree of success achieved in monthly forecasts of the mean circulation pattern for the United States. On a day-to-day basis the steering of vortices is effected by the mean tropospheric flow in which they are embedded[1] and the 500 mb level flow is a good indicator of this (Veigas and Ostby, 1963). At the same time, the developing cyclone may itself modify the steering pattern through thickness and vorticity advection, particularly when it reaches its mature state. The

[1] A review of the history of the steering concept is given by Putnins (1962). He shows that over the Canadian Arctic Archipelago and the Greenland ice cap steering appears to be initiated in the lower troposphere rather than in the upper troposphere as is usual.

details of these aspects of cyclone development need not concern us here, however (see Palmén and Newton, 1969, ch. 11). On the time scale of weeks, steering reflects to a greater degree the thermodynamic effects of heat sources (sinks) which serve to generate cyclonic (anticyclonic) vorticity in geographically preferred locations. The long-wave pattern in the troposphere may be reinforced by this heating, or it may interact with a somewhat different pattern induced by heating, but whatever the net result, it is this that determines the broad steering pattern for a period of perhaps two to four weeks. Apart from interactions on these time and space scales, we also need to take account of hemispheric or global teleconnections. These are discussed more fully in Chapter 5.D. There are, for example, well-known links between the long-wave trough over eastern North America and blocking over Alaska. Moreover, seasonal shifts of the circulation regime over southern Asia are linked to the replacement of a trough which is near 90°E from October to May by one near 65–70°E from June to September. Such mean circulation centres, referred to by Flohn (1965) as 'centres of interaction', undoubtedly have important climatic roles and appear to be much more than statistical features. In addition, in the monsoon regions we need to take into account transequatorial fluxes (Vuorela and Tuominen, 1964). For example, the mean meridional Hadley cell is displaced across the equator so that rising air in the equatorial trough of low pressure forms its upward arm, being replaced at low levels by the oceanic monsoon flow.

3 VERTICAL STRUCTURE

The generalized vertical structure of the troposphere and lower stratosphere in the northern hemisphere in winter is shown schematically in fig. 2.40. The major features are the three nearly barotropic air masses and the intervening baroclinic zones (see Chapter 3.D, p. 185). Interchange between mid-latitude and tropical air is clearly no problem since the baroclinicity is confined to the upper troposphere of the subtropics, although in fact the poleward transport of relative angular momentum occurs primarily at the jet stream level. Air mass exchange must also take place across the polar arctic fronts to effect the poleward transfer of momentum and energy necessary for balance. We must recognize that frontal zones and, in particular, mean fronts are not substantial surfaces. Further discussion of the relationship between air masses and atmospheric structure is presented in Chapter 3.D (pp. 178–187).

Turning to longitudinal variations in the vertical structure, a primary characteristic is the slope with height of the axes of pressure centres. It follows from the hydrostatic equation (p. 529) that lows slope towards cold air and highs towards warm air so that in both instances their axes slope westward with height. Thus the Icelandic low at the surface is represented by the mean 500 mb trough over eastern North America. The subtropical anticyclone cells have their surface centre over the eastern part of the North Atlantic and North Pacific and their upper tropospheric centre (on average) over the western sector of these oceans. The Siberian winter anticyclone is something of a special case. The pressure on the MSL chart are largely hypothetical because the topography is montane and the valley stations experience very low

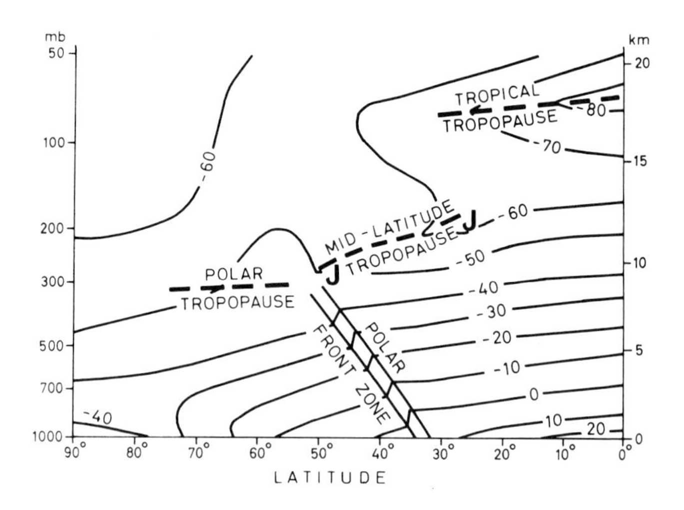

Fig. 2.40 Schematic cross-section of the troposphere and lower stratosphere in the northern hemisphere in winter (after Defant and Taba, 1957; from Barry and Chorley, 1971). Temperatures in °C; J = jet stream core.

temperatures rendering the reduction of barometric pressure to sea level much more problematic than usual for mountain stations (Walker, 1967). The 1000–500 mb thickness field is subject to similar interpretation problems. Nevertheless, the anticyclone, although usually exaggerated in intensity on synoptic maps, is not fictitious. High pressure centres are observed to move out of this region, across the arctic coastline especially, and the upper flow pattern is one indicative of subsidence over the area of Mongolia.

A very significant aspect of the structure of the atmosphere in different regions of the world is the vertical wind shear, $\partial V/\partial Z$. It is in fact a key to the occurrence of the various structures of weather disturbances. Fig. 2.41 illustrates the major types of wind profile in middle and low latitudes. The primary contrast is between the strong vertical shear in the Ferrel westerlies and the generally weak or even reversed shear in low latitudes. However, there is considerable variety within the tropics according to season and location. In winter, the easterly trades are

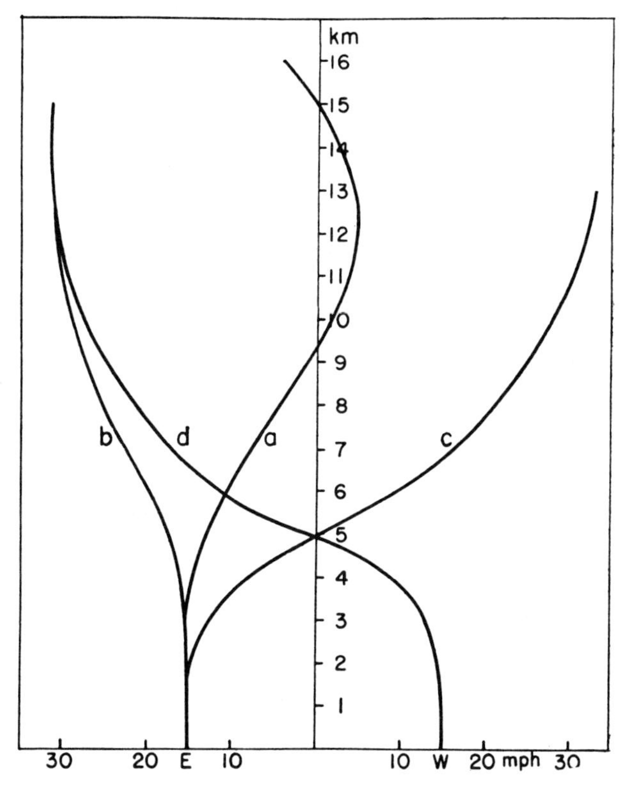

Fig. 2.41 Schematic models of the vertical profile of the zonal wind in low latitudes (from Riehl, 1950). a, b = the trades in summer. c = the trades in winter with westerlies aloft. d = summer monsoon areas.

overlain by the Ferrel westerlies in the upper troposphere, whereas in summer there may be westerlies or easterlies at high levels depending on the direction of the meridional temperature gradient. In addition, in the areas affected by the low-level equatorial westerlies (the monsoon regions, particularly) the upper tropospheric flow is easterly. These contrasts must certainly be borne in mind when attempting to develop synoptic climatological classification schemes in different parts of the world.

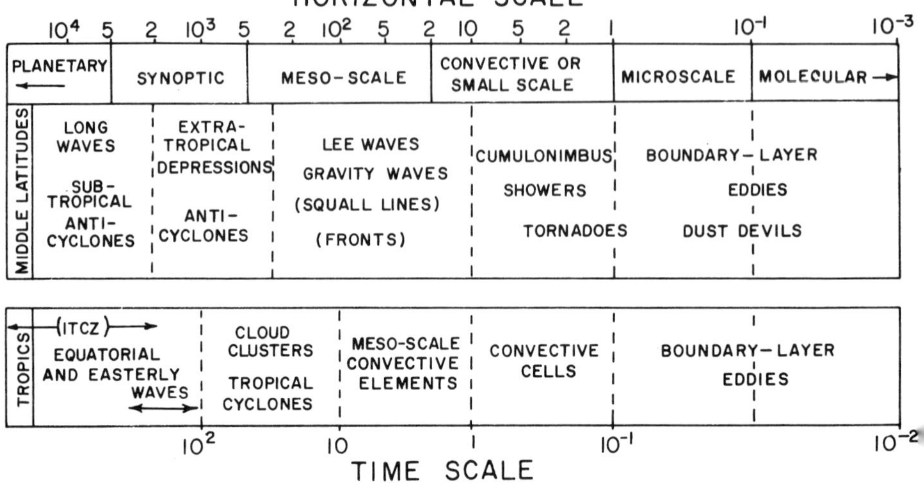

Fig. 2.42 The space and time dimensions of atmospheric systems (after Mason, 1970).

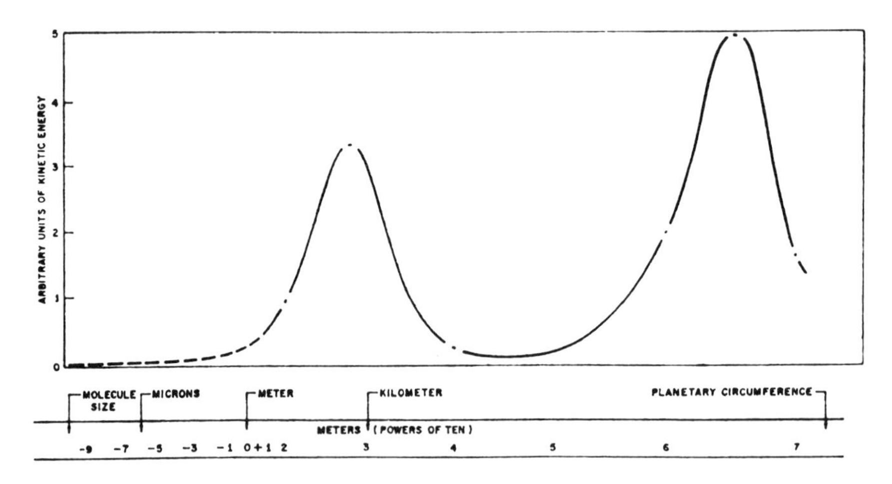

Fig. 2.43 The spectrum of kinetic energy in different scales of motion system (National Science Foundation; from Sutton, 1965).

4 SCALE CONSIDERATIONS

Finally we need to consider the various scales of motion in the atmosphere since it is the characteristics of the eddies that are modified by climatological averaging procedures. Fig. 2.42 summarizes the space and time dimensions of the systems which are usually identified. These are somewhat arbitrary as a result of the observational networks and associated techniques of analysis used to determine them. Although the range of scales is continuous, the energy spectrum is not uniform. There are known to be preferred scales of motion. Fig. 2.43 illustrates the concentration of energy in the synoptic and microscale systems. However, Fiedler and Panofsky (1970) indicate that the microscale energy is dependent on frequent, strong turbulence (mechanical and/or convective). Over the oceans and smooth terrain a mesoscale gap may be less apparent since microscale turbulence is weak. Mesoscale wind systems related to diurnal variations and topography may also help to fill the gap over land (Tyson, 1968). In the free atmosphere mesoscale motions are common, including gravity waves and convective up- and down-draughts, especially in the tropics (Zipser, 1970). There is, nevertheless, a quite frequent spectral gap between the micro- and mesoscale in the free atmosphere. The principal practical significance of these gaps is in the parametrization of small-scale processes with respect to large-scale properties of the atmosphere for purposes of numerical modelling. In this area synoptic climatological methods can provide one means of examining the small-scale processes in relation to synoptic and planetary flow characteristics and other factors.

As we have seen, the geographical distribution and seasonal occurrence of planetary and synoptic scale motion systems are fairly well known, but this is still far from the case at the mesoscale. Moreover, interactions between the larger scale of motion are to some extent determined by geographical controls. Thus, the planetary wave pattern is related to heat sources and sinks as well as to large-scale orography, and the formation and movement of frontal depressions is in part governed by these planetary waves. Mesoscale systems, however, seem to be related to geographical features and the larger motion systems only in special circumstances (Barry, 1970). There are diurnal wind regimes, for example, which usually require a slack regional pressure gradient for optimum development and there are also phenomena such as the topographically influenced 'tornado alleys' of Arkansas, but many mesoscale motion systems seem to be independent of such interactions.

A regional macroclimate may be accounted for in general terms by reference to the planetary wave pattern and the particular regime of synoptic and mesoscale disturbances on seasonal and shorter time scales. But the year-to-year variability in circulation regimes means that precise specification of a macroclimate in terms of meteorological properties is rarely possible. By providing a vital link between hourly and daily climatological data, on the one hand, and characteristics of motion systems, on the other, and also by its attention to long-term synoptic variability, synoptic climatology can play a significant role in furthering our understanding of world climates.

3 Synoptic climatological analysis

Synoptic systems have many characteristics making them very complex and so are not easily classified in terms of physical variables. (*H. C. S. Thom, 1970*)

A Introduction

In Chapter 2 we examined the basic types of meteorological and climatological data together with some of the more common derived characteristics. These are obtained by a variety of analytical procedures, but in nearly all cases a single parameter is being handled. We turn now to the major theme of this book: the treatment of assemblages of meteorological data. The approach of the dynamic meteorologist is to develop models based on the hydrodynamic and thermodynamic equations and to seek their verification or revision by detailed case studies. The climatologist is concerned with statistical generalization and for many purposes it is found more useful to group the data into classes rather than treating the whole frequency distribution or selected statistics of it such as the mean. Among the central problems of synoptic climatology are the choice of appropriate meteorological parameters and the formulation of an effective classification. The description of the state of the atmosphere, which for our purposes is mainly represented by a synoptic weather map, presents special difficulties. No two weather maps are identical; apart from the almost infinite variety of synoptic patterns, synoptic systems differ markedly in size and intensity throughout their individual life cycle and from one sequence to the next. These factors should be borne in mind in the following discussion. Their implications are examined in section B of this chapter (pp. 101–2) after a review of subjective classifications of circulation types.

A brief chronological perspective of the development of synoptic climatology is worthwhile, as the remainder of the chapter is arranged as systematically as the subject matter permits. The points that are noted below are discussed at greater length, either in the body of the chapter or in Chapter 5 which deals with practical applications of synoptic climatology.

1 THE HISTORICAL DEVELOPMENT OF SYNOPTIC CLIMATOLOGY

The first serious investigations of synoptic patterns were made in the nineteenth century. The study of winds and associated station weather received much attention from the German meteorologist Dove. However, as Leighly (1949) notes, once the relationship of wind to pressure systems was appreciated a new foundation was provided for synoptic climatology. Köppen (1874) studied the effects of the surface airflow on weather at St Petersburg (Leningrad) using six classes: straight, anticyclonically or cyclonically curved isobars, an anticyclone or cyclone centre, and indeterminate situations. This classic approach has been followed, albeit independently, by many subsequent workers. The most complete early attempt to develop a synoptic climatology is to be found in the writings of Abercromby (1883, 1887). He analysed the weather conditions over the British Isles in terms of airflow from the four major directions (N, S, E and W). The next significant step was the classification of pressure patterns over England by Gold (1920), again relating weather conditions to them. This type of approach was extended by Baur (1931, 1936a, 1936b, 1947, 1948) in Germany, who examined large-scale circulation patterns (*Grosswetter*) over Europe and the eastern North Atlantic. The emphasis of the classification was on tropospheric patterns that remain more or less steady for several days, and the concept of large-scale *steering* of individual depressions was introduced. About the time Baur was developing his concepts a very different approach was suggested by Bergeron (1928, 1930; see Willett, 1931). In laying a framework for 'dynamic climatology' he formulated the concept of air masses and proposed that studies be made of their frequency and intensity. Numerous investigations of the general characteristics of air masses and their local effects on weather and climate – which would now be properly regarded as part of synoptic climatology – were in fact carried out in many parts of the world by workers such as Dinies (1932), Willett (1933), Arakawa (1937), Showalter (1939) and Tu (1939).

Synoptic climatology developed considerably during and after the Second World War. The immediate practical purpose was the assessment of likely weather conditions for military operations. The proposed master plan for this type of study, outlined by Jacobs (1947) is shown in table 3.1. The examples described by Jacobs relate to the secondary level in the scheme and, in particular, to the spatial distribution of a particular meteorological element for the various airflow patterns over a selected area. The large-scale approach through analogue patterns for a period of days, which was also developed about this time, is discussed by Elliott (1949, 1951).

In the post-war period a notable achievement was the work of Lamb

(1950) in preparing an airflow pattern catalogue for the British Isles back to 1873, since extended to 1861. Synoptic climatological studies were carried out for many parts of the world in the 1950s, and in the Soviet Union classifications for the entire hemispheric circulation were developed. In the late 1950s and 1960s objective means of classifying weather maps were devised but, for the most part, their application has

Table 3.1 Scheme for synoptic climatology studies (after Jacobs, 1947)

Order	*Classification feature*	*Characterized with reference to:*
Primary	Large-scale synoptic types (analogue or index)	– Area, month (season), hour.
	Airflow pattern	– Area, month (season), hour; time sequence and duration; also spatial coupling of patterns.
Secondary	Gradient wind speed	– Geographical subregions
	Vorticity	– Isobar curvature
	Air mass	– Vertical structure

been on an experimental basis. Computer data handling has increased the scope of synoptic climatological projects by many orders of magnitude, but to date an immense range of possibilities remains to be explored. Indeed, it is not improbable that with the planned Global Atmospheric Research Programme (G.A.R.P.) the 'banks' of data awaiting thorough climatological analysis will continue to grow.

2 CLASSIFICATION APPROACHES

We turn now to the various ways in which pressure maps or the weather elements themselves may be classified. Considering first an individual map of the pressure field (or a contour chart), the *static* pattern may be classified into categories from two main standpoints: (i) the identification of individual 'features of circulation' such as high and low pressure cells, ridge and trough, in specified locations; (ii) the description of the complete pressure pattern, or its most significant features, by subjective or objective methods. This second approach in fact covers an immense range of methods.

The pressure (or streamline) map may also be regarded in a *kinematic* framework, and two broad lines of work of comparable importance have developed here. They are: (iii) classification of cyclone and anticyclone paths, and (iv) the classification of circulation or airflow patterns. Inevitably there is some overlap between these four approaches in practice since the type of flow pattern is a function of the pressure distribution and its changes.

Finally, meteorological and climatological parameters may be treated directly. Three approaches to classification can be recognized: (v) the combination of selected parameters in 'weather types'; (vi) the identification of air masses on the basis of selected parameters; (vii) the abstraction of combined weather 'factors' by means of mathematical techniques. These different approaches will be examined in turn following a brief rationale for such classifications.

3 CONCEPTUAL FRAMEWORK

Some meteorologists hold the view that, while the classification of weather patterns may be a necessary practical convenience, the physical validity of such a procedure is dubious. The work of Baur (1944, 1963) is particularly relevant here. By investigating a large number of series of air pressure, temperature, cloudiness, precipitation and spatial pressure differences for locations in Europe (for fifty days each in midwinter and midsummer over a fifty-year period) he has shown that the probabilities of occurrence of particular weather characteristics are different from year to year. Baur argues that these differences must originate from large-scale weather patterns whose physical existence is thereby essentially proven.

The statistical basis of this result involves assessing whether the probability of an event, W (i.e. a weather characteristic), is equal from trial to trial (day to day), but different from series to series (year to year). Baur shows that to compare these alternatives an appropriate 'coefficient of deviation' for meteorological data series which are highly serially correlated is

$$D = s_h/s_m$$

where s_h = the standard deviation of the frequencies of event W for a particular series of observations

s_m = the Markov deviation of the series; that is, the standard deviation of the frequencies of the event W for a first-order Markov series (where each term depends on the preceding term) (see also pp. 239 and 241).

$$s_m \approx s_B \sqrt{\frac{1+\Delta}{1-\Delta}}$$

s_B, standard deviation for constant probability (Bernoulli probability), is

$$s_B = \sqrt{np(1-p)}$$

where p = probability of event W. (See also 4.A.4.) $\varDelta$ is a measure of persistence ($0 < \varDelta < 1$ for positive persistence). It is the determinant of the matrix of transition probability for event W.

$$1 - \varDelta = W / E(W)$$

where $E(W)$, the expected value of W, in N cases $= 2N\bar{p}(1-\bar{p})$, the bar denoting a mean value.

$$\varDelta = 1 - \frac{W}{2N\bar{p}} - \frac{W}{2N(1-\bar{p})}$$

$$= 1 - \frac{W}{2N\bar{p}(1-\bar{p})}.$$

Further details may be found in Baur (1944) and Müller-Annen (1955).

If the probability of an event is constant from trial to trial (day to day), but varies from series to series (year to year), then D exceeds unity. Such statistics are referred to as Lexian (Anderson, 1967). The results which Baur obtained for Europe are all greater than unity; with one exception D exceeds 1·18. Statistically, at least, we may accept this as demonstrating the occurrence of large-scale weather patterns and proceed to a consideration of the various means of classifying weather and circulation patterns. Further aspects of this general question are discussed below and particularly in Chapter 5. We should at the same time note that the results of various studies of the centres of action and steering (see p. 122) provide more substantial evidence of the reality of large-scale patterns of circulation.

B Static view of the weather map

In this section we consider classifications which primarily emphasize the location of selected features of the pressure field. While the airflow is obviously closely related to the pressure field, it is not used as a principal element in defining the various types categories.

1 'FEATURES OF CIRCULATION'

This approach is rather surprisingly a very recent one (Sands, 1966). Attention is focused on individual components of the pressure field. Sands defines a feature of circulation as 'a recurring, dynamically or thermodynamically caused, pattern of flow over a limited portion (or specific adjacent portion) of a study area. It may be a surface feature or an upper air feature but not both. It may vary slightly in size and strength and location.' For the western United States, Sands identified the following four classes of feature on the MSL pressure map and seven classes on the 500 mb contour charts:

MSL	*500 mb*
Closed lows	Closed lows
Troughs	Meridional polar troughs Tilted polar troughs
Closed highs Ridges	Ridges and highs
	Easterly flows Zonal (westerly) flows Indeterminate weak flows

Sixty upper air and forty-five surface features were recognized and on the basis of a daily catalogue for 1958–63 it was found that three features per day were tabulated for the 500 mb charts and five per day for the MSL charts.

A major practical difficulty with this approach arises when features are displaced from their 'usual' defined location to a position midway between that and some other type location. This prevents identification of weather conditions at any selected point with reference to the features designated on individual days.

2 PRESSURE PATTERN CLASSIFICATION: SUBJECTIVE STUDIES

The earliest classification of the complete pressure pattern was that developed for the North Atlantic and western Europe by Van Bebber and Köppen (1895). They distinguished five main types (twenty subtypes) in relation to the position and movement of highs and lows for 1883–7, following from the earlier studies of Van Bebber on depression tracks.

The main point that emerges from examination of their study is the fact that several of the principal depression tracks occur with three or four of the major pressure pattern types and up to ten of the subtypes. Clearly, the grouping does not provide an adequate discrimination of patterns of movement in relation to the static pressure field.

More significant for British meteorologists was the work of Gold (1920). On the basis of MSL pressure maps for 1905–18 he defined fifteen types (twenty-eight including subtypes) of pressure distribution over the British Isles and adjacent areas. The orientation of the isobars and the positions of the dominant highs and lows were his primary criteria. The most frequent types were:

		Annual frequency
Type I.	High to the southwest, low over Scandinavia	10·2%
IV.	High to the south and ridge over the British Isles	6·7% (winter mainly)
II.	High to the south and low to the north	6·3% (mid-summer and also mid-winter)

Gold also includes notes on the general weather conditions associated with each type over northeast France. Pick (1929, 1930) later investigated the persistence of these types for summer and winter. Of the types noted above, type I shows spells of about three to six days whereas type IV is transient. The classification was later simplified by Bilham (1938) into seven types – anticyclonic, cyclonic, and five based on the orientation of the isobars – so that this scheme comes to resemble the airflow approach of Abercromby and subsequent workers discussed on pp. 136–42.

Newnham (1925) adopted a similar approach to that of Gold in another classification of pressure patterns over the North Atlantic and Europe. She divided the area between 30–80°N, 30°E and 70°W into ten sectors and for the period 1896–1910 differentiated ten types of pressure distribution, related to the occurrence of anticyclones with a pressure >1020 mb covering at least half of a sector. No maps of the patterns were included in the publication.

Some of the most recent work of this type has been carried out for Alaska by Putnins (1966) and for the eastern Canadian Arctic by Barry (1972). Putnins classifies the daily MSL pressure patterns for 1945–63 into twenty-two types (of which ten occurred on $<2\%$ occasions). The 500 mb flow was also taken into account with respect to anticyclonic, cyclonic and neutral curvature of the contours ('mixed' situations where none of these predominated over Alaska were also recognized). The most frequent pattern (annual occurrence 23%) is a low pressure system with several centres over Alaska and/or near the coast, and cyclonic circulation at 500 mb (fig. 3.1). This type has a maximum frequency of 39% in October and a minimum of 11% in January. The second most frequent type (8% annual frequency) is a south–north ridge over Alaska, with anticyclonic circulation aloft (fig. 3.2). There is a pronounced maximum occurrence in summer (19·5% in July). Two other types which deserve mention occur mainly in winter. The situation where there is a dominant high to the north or northwest (over the Bering Sea) and lows affecting southern Alaska occurs on 10–14% of

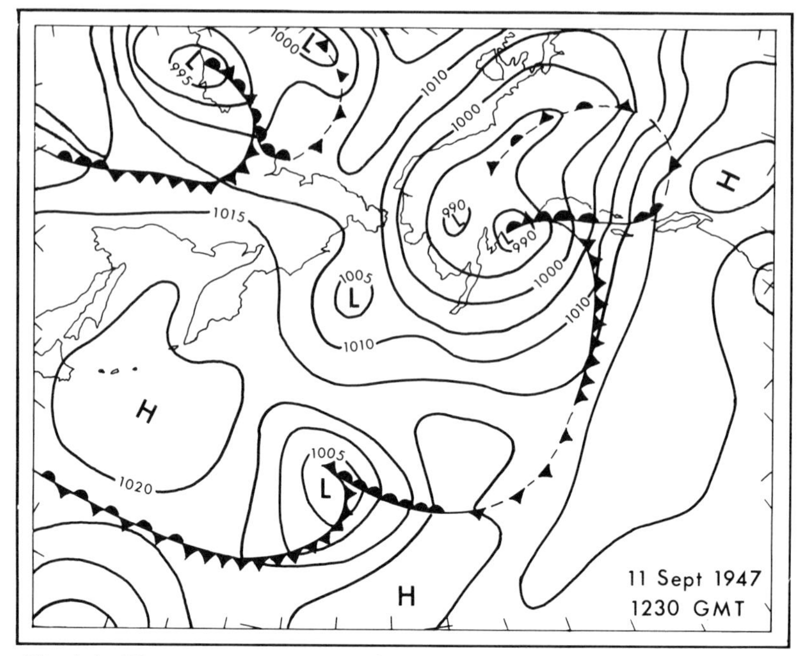

Fig. 3.1 The most frequent, A¹ (23%), MSL pressure pattern over Alaska (after Putnins, 1966).

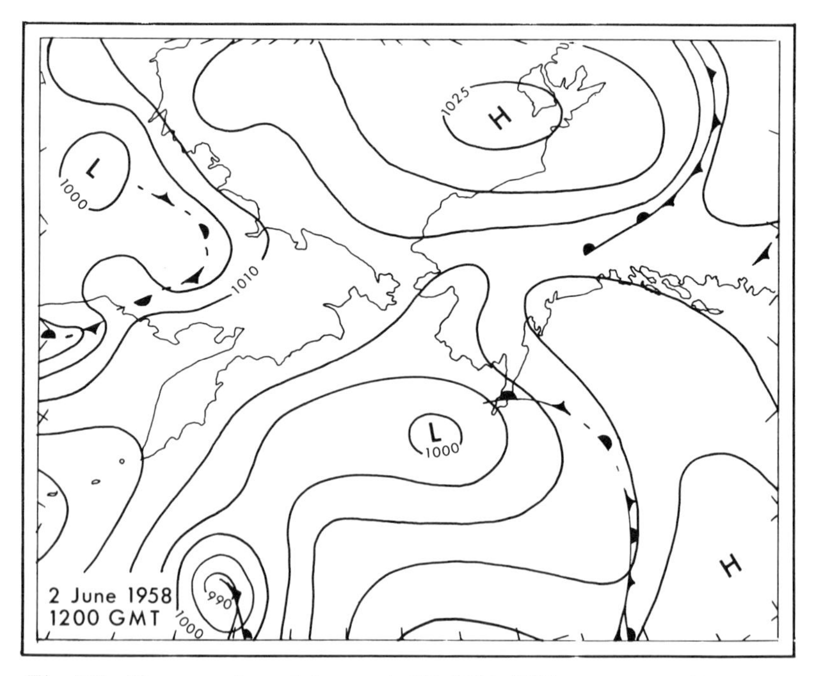

Fig. 3.2 The second most frequent, E¹ (8%), MSL pressure pattern over Alaska (after Putnins, 1966).

days in December to March but is almost absent in July–August. The other pattern – a high, or ridge, over or to the east of Alaska, with anticyclonic flow aloft – occurs on 18% of January days but is infrequent from May to December. It is interesting to compare these features with the mean monthly maps. The mean July pressure field has a strong subtropical anticyclone over the eastern North Pacific and a ridge northward from it towards Alaska. This is reflected in the one in five days resembling fig. 3.2. The mean January field has a deep low over the Aleutians and a ridge across the Bering Sea from northeast Siberia to northwest Canada.

For an area of similar dimensions, centred over Baffin Island, Barry (1972) identified forty-two types of MSL pressure patterns for January–February, April, and September–October for the period 1961–5 and July–August for 1961–70. The 500 mb pattern was also classified as to the dominance of cyclonic or anticyclonic curvature and, when meaningful, the flow direction over Baffin Island.

Examination of the static pattern is perhaps most appropriate in areas where a proportion of systems form and/or decay in situ. This is often the case off west Greenland, for instance, with Disko and Thule lows. In regions such as northwest Europe classifications which focus on airflow patterns and the movement of systems are generally preferable.

Problems of subjective identification of circulation types

The subjective identification of discrete categories of pressure patterns, or circulation types, presents several problems. First, and most important, is the fact that atmospheric modes are continuous so that the delimitation of any boundary between classes must be arbitrary and therefore somewhat unsatisfactory. With a large number of types this problem is multiplied and it is usually more convenient, and also more realistic, to define a limited number of broad groups. This procedure restricts the number of borderline cases and is also advantageous in providing larger samples of each type. Second, and related to the first point, is the problem of delimiting the termination of a sequence of a particular type. Atmospheric circulation patterns sometimes switch abruptly, but on other occasions a gradual evolution takes place. With respect to both of these questions the decisions of any two meteorologists may well differ, so that a unique classification cannot be obtained. Experience in developing a synoptic classification for the eastern Canadian Arctic (Barry, 1972) suggests that about 70% of cases present no difficulty, 20% might be assigned to one of two categories and the remaining 10% are virtually unclassifiable, mainly as a result of weak patterns. Lamb (1950) made use of an 'unclassifiable' group, but many other researchers prefer to accommodate all cases within one or other group. Putnins (1966, p. 2) states that 'a cetrain amount of "scientific imagination" was applied [in these cases] to preserve the continuity'.

Two further problems are often overlooked. One is the variable intensity of pressure systems and the other is the seasonal variation in type characteristics. In part the latter factor is related to the generally greater intensity of weather systems in winter, but it is also a result of the seasonal change in the heat budget of land and sea. The first point is rather more critical since it affects the classification categories themselves. If we are identifying 'features of circulation', or the general pressure pattern, there are various subjective decisions to be made as to when to include or neglect a small weak low, or ridge. In the Lamb classification, for example, Cyclonic type (code Z) applies whenever a low is present over or near the British Isles on at least one of the four six-hourly MSL synoptic maps. Some classifiers define limiting pressure values, however. For example, Barry (1960) defined anticyclonic situations over Labrador–Ungava as having anticyclonic curvature and a pressure $\geqslant$ 1016 mb.

In summary, the different sources of variation affecting airflow types are the type, intensity and size of the weather system, and the diurnal, season and local effects. It is interesting to record that Abercromby pointed these out in 1887 although they have not always been explicitly recognized by subsequent workers.

Apart from these practical considerations there are other methodological ones. Subjective classifications used in synoptic climatology are usually formulated *a priori* and then modified on the basis of experience in applying the categories. This creates problems in evaluating them since they are assessed only in terms of the classification problem for which they were set up. Hence a circular argument develops in any attempt to determine the adequacy and efficiency of the scheme for stratifying the basic data.

3 PRESSURE PATTERN CLASSIFICATION: OBJECTIVE STUDIES

In view of the many problems of defining characteristic pressure or flow patterns subjectively, much thought has been given in recent years to objective classification techniques. Three major approaches have been used – correlation methods, specification techniques, and empirical orthogonal functions (or eigenfunctions). These will be considered briefly in turn; fuller discussion of the methods is reserved until Chapter 4.

Correlation methods

Lund (1963) used correlation techniques to group MSL pressure patterns over eastern North America into types. For convenience, station pressures rather than grid point values were used.[1] The time period investigated by Lund covered 445 days so that a (445 × 445) correlation matrix was required. The procedure is as follows. The pattern of the day which has the largest number of maps correlated with it, using an arbitrary threshold of $r > +0{\cdot}7$, is selected as type A. After abstraction of these days the date with the next highest number of related maps is designated as type B, and so on. Lund found that 89% of the period was accounted for by ten such types. The same approach has been used by Houghton (1969) in a study for the Great Basin area using 700 mb heights with a correlation threshold of 0·85. An additional criterion, that the standard deviation of each daily set of heights did not differ by > 20 m from that of any other day in the same group, was included. Forty-two groups were obtained in this way for 1962–3, but it was found that these could be combined subjectively into seventeen types on the basis of the distribution of pressure and wind field characteristics and further reduced to six groups on the basis of the synoptic distributions of temperature, relative humidity and dew point. Clearly these procedures are not entirely free from subjectivity. As Lund pointed out, if the correlation threshold is raised, say to $r > 0{\cdot}9$, then more types result. Hartranft *et al.* (1970) state that a threshold of > 0·7 is most satisfactory for MSL maps and > 0·9 for 700 mb maps. Andersen (1967) uses Lund's method for forty-three stations in Europe and the eastern North Atlantic area. With a correlation threshold $r > 0{\cdot}9$, eleven types account for only 328 of the 541 days in the period December–February 1957–62. Although the resulting types are more homogeneous the omission of so many days is clearly unsatisfactory. In an unpublished trial for the British Isles, Long (1967) found that twelve types (including two hybrid groups) accounted for 75% of his stratified random sample of 180 days (three days per month) for 1961–5, but some days were initially placed in two or, in a few instances, three groups. It is apparent that misclassification may occur in such cases. Lund placed these situations in the group for which the correlation with the 'pattern day' was a maximum, but it may be preferable to define 'hybrid' groups, as Long suggests, if such duplicated classifications are numerous. The subjective study of Barry (1972) referred to above also highlights a less obvious problem. There it was found that superficially similar patterns with a low over the northern Labrador Sea might on a few occasions be associated with an unusual frontal pattern where the warm front extended northward into Baffin Bay instead of the normal more or less zonal orientation. The weather pattern in such a situation is drastically different from that

[1] Grid point pressure and 700 mb geopotential height data, which are available on magnetic tape for the northern hemisphere since 1947, provide a more suitable input for large-scale computer processing.

with the usual cyclonic pattern, but it is unlikely that a correlation classification would differentiate between the pressure fields.

Most important of all, it must be recognized that the correlation method relates each day of a given group to the 'pattern day', and does not consider the 'distances' to other members of the group. This clustering method may not produce the optimum grouping of the data, as we shall see in Chapter 4D. It may be instructive to determine the mean pressure field for each group although, of course, the features on this are weaker than those on the corresponding pattern day.

An objection to the Lund method sometimes raised, is that the correlation matrix has to be recomputed if the data series is extended. If useful groups have already been established, however, then additional days can simply be correlated with the pattern days already adopted. A large matrix makes the computations relatively slow and therefore expensive of computer time. This can be circumvented if months or seasons in a data series are treated individually but the pattern days thus determined are not comparable so that seasonal variations in type frequency cannot be evaluated.

Another factor affecting the computational problem is the area size and number of data points. Lund used 22 stations covering approximately the sector 35–45°N between the east coast of the United States and $82\frac{1}{2}$°W whereas Hartranft *et al.*, use only 12 grid points over sectors of 10° latitude by 15° longitude. They emphasize that a regional analysis will delineate synoptic scale flow patterns adequately and eliminate the problem of low correlations between maps which differ only in one small part of a larger grid. In practice, the trade-off between size of area, grid spacing and length of data series may be determined by practicalities of computer storage capacity and costs.

Specification by orthogonal functions and harmonic analysis

Wadsworth, Gordon and Bryan (1948) introduced the term *specification*, defining it as 'a mathematical representation of the distribution of some variable over a stated region of space'. The basic steps in this procedure are as follows:

1 The pressure field is expressed as an arbitrary spatial function. The coordinates may be cartesian or curvilinear (a cylindrical system was developed for polar analysis by Godson (see Hare *et al.*, 1957) for example).
2 Mathematical functions of the space coordinates are chosen so as to provide efficient specification and to allow physical interpretation of the functions.
3 Using grid point data, the pressure field is fitted to the functions by a 'least squares' method.

One way of examining the pressure field as a preliminary to classification is to regard it as comprising a number of individual surfaces. In the simplest case of straight, uniformly spaced isobars and low pressure to the northeast, this is equivalent to a linear surface inclined equally in the x (eastward) and y (northward) directions (fig. 3.3). A variety of basic plane and curved surfaces may be envisaged, involving either the x

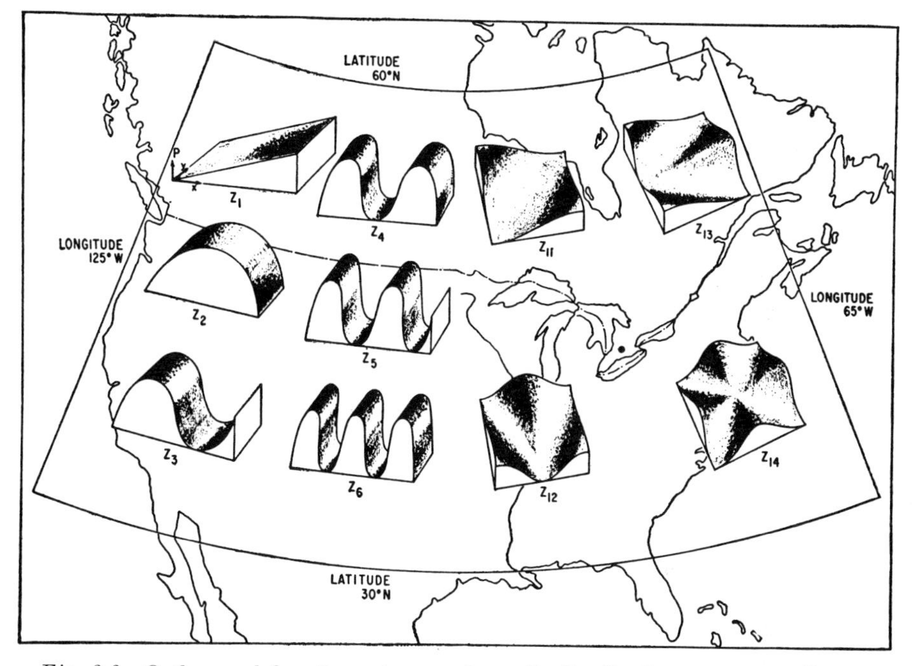

Fig. 3.3 Orthogonal functions shown schematically. Surfaces corresponding to Z_7–Z_{10} are equivalent to surfaces Z_1–Z_4 rotated 90° clockwise (after T. F. Malone; from Petterssen, 1958).

(or y) coordinate alone, or both together, as illustrated in fig. 3.3. Each surface is described by a polynomial function of the general type

$$y = a + bx + cx^2 + dx^3 + \cdots$$

and, by developing these as *orthogonal* equations, each surface is made independent of any other. That is to say, no correlation exists between the surfaces. Further discussion of the mathematical aspects of this approach is reserved for Chapter 4.C, pp. 262–3.

The earliest meteorological applications of this technique were made between 1942 and 1946 by Wadsworth, Gordon and Bryan (1948). Later it was applied to arctic pressure patterns by Hare and his associates (Hare *et al.*, 1957; Hare, 1958; Wilson, 1958) and to half-hemispheric 500 mb contours by Horn, Essenwanger and Bryson (1958). By reconstruction of a series of charts from orthogonal polynomials it was shown by Wadsworth, Gordon and Bryan that the major features of the pattern are adequately represented when 80% of the variance is accounted for by the polynomials. However, in order to achieve this reduction of variance some twenty to thirty surfaces may be necessary. This greatly complicates the subsequent development of classification categories. The work of Wilson (1958), for surface pressure fields over the Arctic, shows that the specification tends to be least satisfactory when the pressure field is flat or exhibits cellular patterns.

The application of these procedures to the classification of large-scale pressure patterns is based on the correlation of the pressure fields on successive pairs of days. The autocorrelation coefficient, which is determined from the correlation coefficients between the pressure pattern and each polynomial surface, provides an index of the persistence of the pattern. An arbitrary threshold ($r > 0{\cdot}7$ or $0{\cdot}8$) is used to delimit changes of type. Wilson notes that, as a result of the less satisfactory specification achieved due to cellular or flat patterns in the summer months over the Arctic, interdiurnal correlations of $+0{\cdot}6$ may be less significant than values of $+0{\cdot}4$ in the winter half-year when patterns are more pronounced.

The most fundamental objection to this approach is that the pressure field is fitted by surfaces of predetermined configuration which may tend to bias the interpretation of the results. The method of harmonic analysis avoids this particular problem although other difficulties occur.

The basis of harmonic analysis is discussed in Chapter 4.B (pp. 228–31) and its application to spatial problems in Chapter 4.C (pp. 263–5). Here it is sufficient to point out that, in the one-dimensional case, a continuous contour of an isobaric surface (say 500 mb) around the hemisphere can be described by a function consisting of several cosine terms (Graham, 1955). Alternatively, the wave pattern along a particular latitude circle can be specified in terms of phase angle – indicating the longitudinal interval between each wave – and the amplitude of the waves with respect to height departures of the selected isobaric surface from the latitudinal mean (Boville and Kwizak, 1959). Fig. 3.4 demonstrates the forms of the cosine terms of wave numbers 1 to 4 with an amplitude of 15° latitude and an origin at 120°W. We see that wave number 1 has one ridge and trough and corresponds to an eccentric polar vortex, wave number 2 has two ridges (180° apart) and 2 troughs, wave number 3 has three ridges (120° apart), and so on. It must be stressed that, since the wave number is an integer, an actual wave spacing of 150° in one sector of the hemisphere will be represented in the analysis by contributions to the variance reduction by wave numbers 2 *and* 3. The pattern obtained by combining the four waves is shown in fig. 3.5. Some actual computation for the 500 mb mean meridional wind

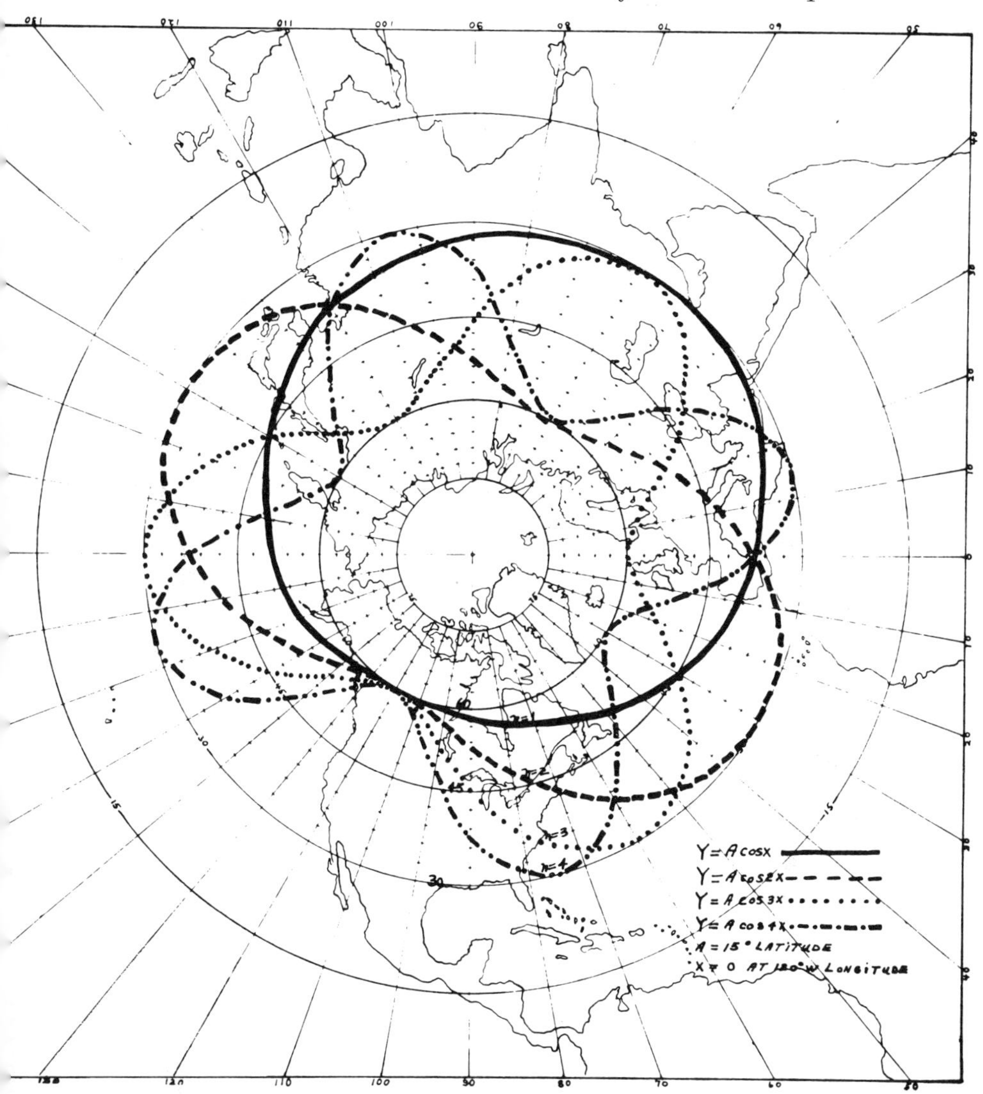

Fig. 3.4 The hemispheric form of the cosine terms of wave numbers 1–4 with an origin at 120°W and on amplitude of 15° latitude (from Boville and Kwizak, 1959).

component at latitudes 30°, 45° and 65°N during 1950 are given in fig. 2.37(*a*). (Saltzmann and Peixoto, 1957). The importance of very long waves is apparent in high latitudes although it must be recognized that the actual wavelength for a given wave number decreases with increasing latitude such that $n = 7$ at 30° is approximately equal to $n = 3$ at 60°. Fig. 2.37(*b*) provides a schematic picture of the longitudinal dis-

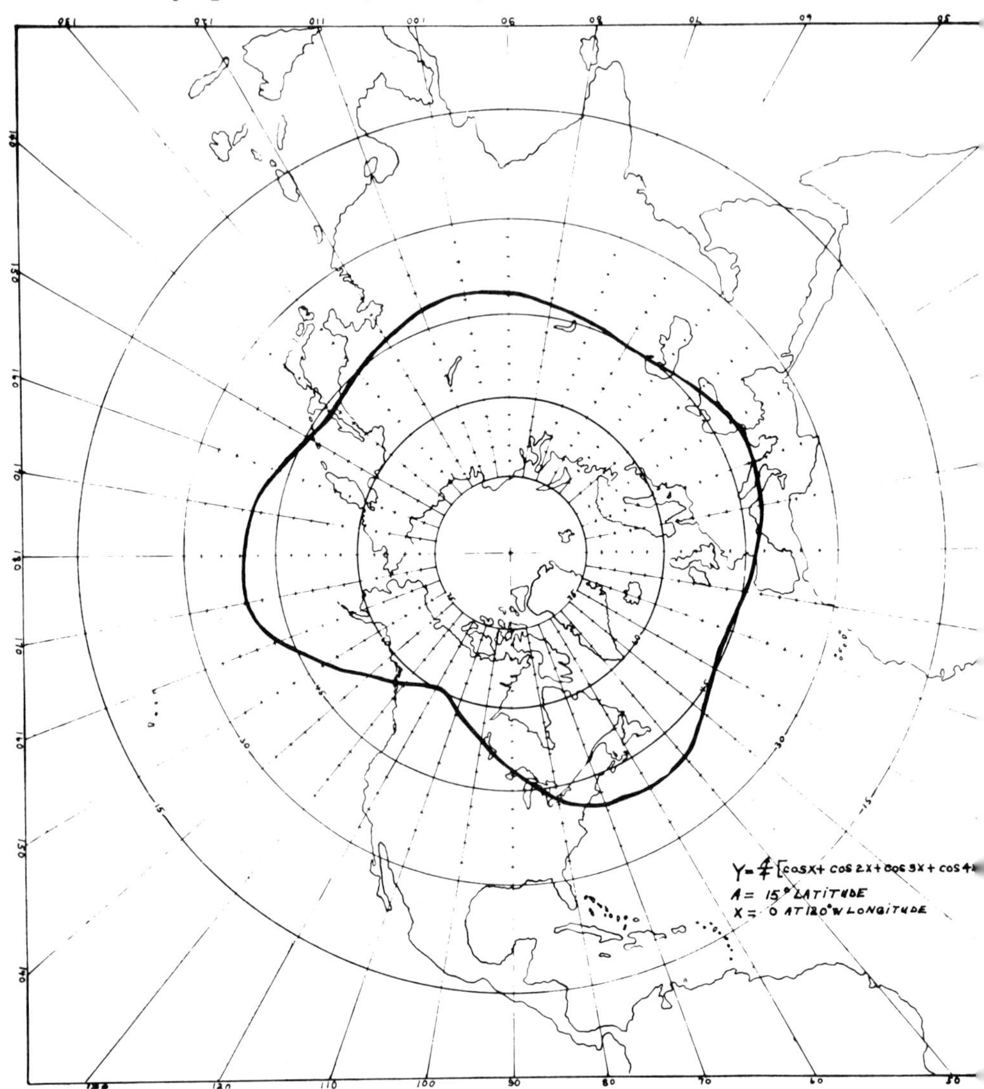

Fig. 3.5 The pattern formed by combining the four waves in fig. 3.4 (from Boville and Kwizak, 1959).

position of the waves in summer and winter illustrating the changes in phase and amplitude.

Harmonic analysis has also been applied to determinations of the spectra of energy and momentum transfers in the general circulation (Saltzmann, 1958; Saltzmann and Fleisher, 1962), but these applications do not concern us here.

The analysis of variations of periodic phenomena around a latitude circle is somewhat restrictive. The method of harmonic analysis has been greatly elaborated, however, to allow specification of the pressure field over the whole earth (or a selected latitudinal belt). Spherical-surface harmonics are computed (see Chapter 4.C, p. 263) but the results of this procedure, like that of orthogonal polynomial surfaces, do not provide an easy basis for developing a synoptic classification. Indeed there does not appear to be a single example of such an attempt in the meteorological literature. Actual approaches to hemispheric classification which are mainly subjective are reviewed in section C of this chapter (pp. 158–65).

Empirical orthogonal functions
We may also avoid the problem of predetermined functions referred to above by using the technique of principal components (eigenvector) analysis. Essentially, this reduces sets of intercorrelated data into some smaller number of mutually uncorrelated functions which provide the optimum specification, expressed as reduction of the original variance, with the least number of functions. These functions are entirely empirical being determined by the nature of the measurements, including the relative scaling of the different original variables. The principal use of eigenvectors up to now in meteorology has been in the analysis of mean fields and anomaly patterns but the wide availability of computer programs to perform the computations heralds their increasing application. Discussion of the procedures is deferred until Chapter 4.C (pp. 268–76) but it will be convenient to summarize the basic ideas involved here.

The mathematical technique involves the transformation of a set of intercorrelated variables into a new set, termed principal components (eigenvectors), which are mutually orthogonal. The representation of the original data by each component is expressed by a weighting factor on the original variables. Let us clarify these concepts with a specific example. Kuipers (1970) derived component patterns of 500 mb height departures from the mean for a 125-day sample over Europe and the North Atlantic. The three most important, in terms of their contribution to the total variance, are shown in fig. 3.6(*a*). It must be emphasized that these patterns do not necessarily represent actual circulation types. They are abstractions from reality by virtue of their orthogonality. The contribution of each component pattern to each grid point height anomaly varies from day to day and is dependent on the daily value of the weightings on each component. Thus the map on any particular day comprises some linear combination of the orthogonal patterns, as illustrated in fig. 3.6(*b*). It follows that this technique still leaves us with a classification problem. The solution adopted by Kuipers is discussed on p. 285 after further details of the procedures involved have been given.

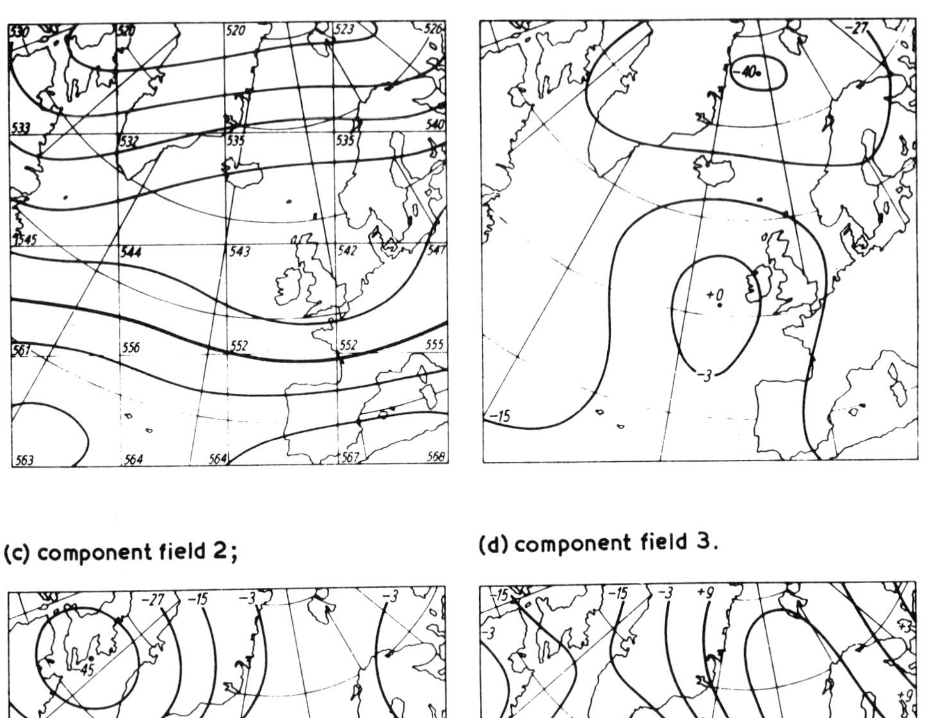

Fig. 3.6 (*a*) Mean field of 500 mb heights and three principal components fields of 500 mb height departures for 125 cases during January–June 1969 (from Kuipers, 1970).

C Kinematic view of the weather map

In the classifications discussed in the previous section the airflow characteristics were, for the most part, regarded as implicit functions of the (static) pressure distribution. We now examine classifications in which the synoptic weather map is viewed in terms of airflow and the

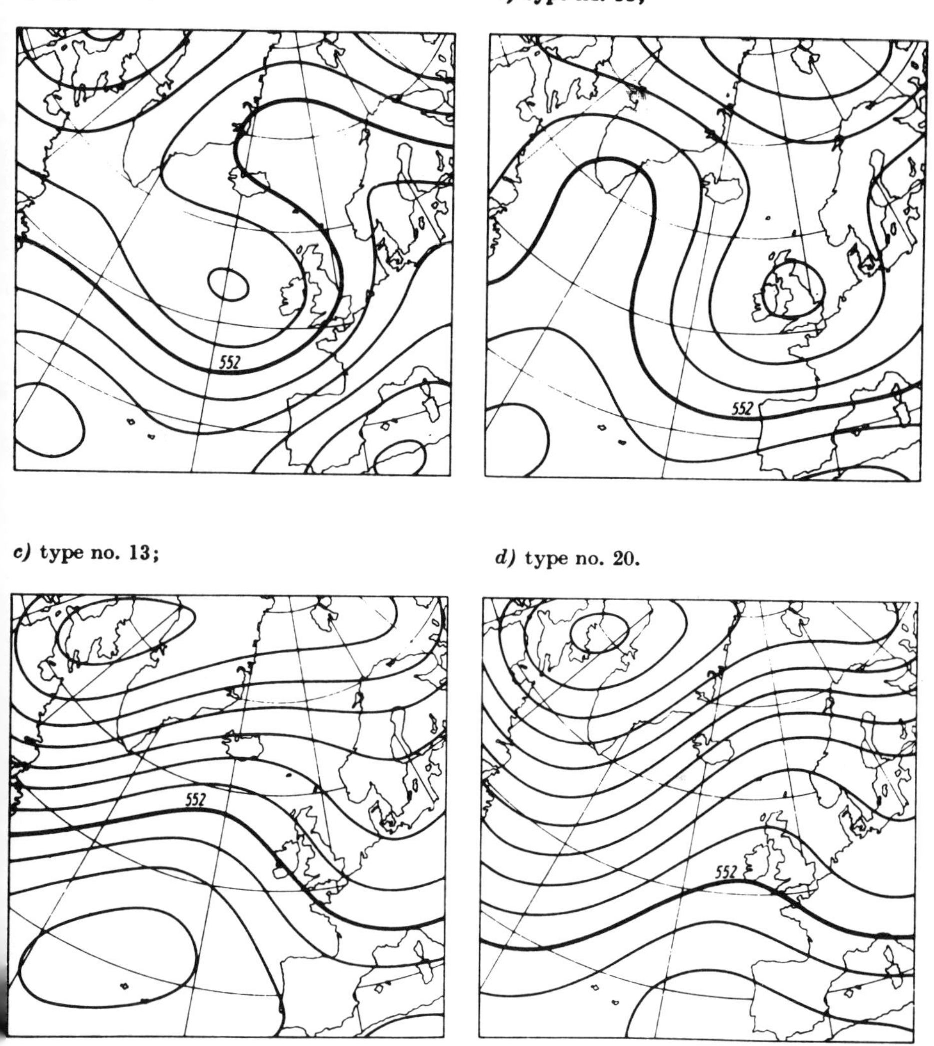

(*b*) Map based on different weightings of the three principal components, as follows:

Type 4: −54·5, 0, 19	Type 13: 16·5, −29·5, −19
Type 11: −16·5, 29·5, −19	Type 20: 54·5, −29·5, 19

movement of pressure systems, although it should be emphasized that any attempt to separate the various approaches into rigidly distinct groups would be unrealistic. The general aims of all these studies are similar, and differences between them are essentially a matter of emphasis.

1 PRESSURE FIELD CLIMATOLOGY

This term was introduced by Court (1957) to refer to studies which relate local weather conditions to 'the direction and distance to the nearest centres of high and low pressure, and their intensities and motions'. Court cited a few late nineteenth- and early twentieth-century references dealing with the first topic, but none of these has proved to be of any consequence in the development of synoptic climatology. Nevertheless, other studies which consider the motions of cyclones and anticyclones and their effect on weather can be placed appropriately in this category.

Most of the studies of depression and anticyclone tracks referred to in Chapter 2.A (p. 19) are concerned only with basic analysis. Two examples of very thorough investigations of depression movement involving classifications are the studies by Forsdyke (1955) for Labrador–Ungava and eastern North America to 40°N, and by Walden (1959) for Greenland. Forsdyke was primarily concerned with patterns of development and movement as a possible guide to subsequent behaviour of the systems. He concluded that although 28% of the systems move northward to Baffin Island or west to Greenland in association with slow-moving, large amplitude flow patterns aloft, the latter (as well as the thickness field) evolve with the surface pattern. In other words, no simple basis was found for differentiating such depressions from those which move out over the Atlantic. Walden was more interested in processes in the vicinity of Greenland itself. He distinguished the following major patterns:

Group A Low moves from between SW and WNW towards central Greenland (twelve subtypes).
B Low approaches south Greenland from SW to W (thirteen subtypes).
C Low moves into Davis Strait from the south (seven subtypes).
F Low moves to south Greenland from S to SE (seven subtypes).
G Low moves to north Greenland (six subtypes).
H Low approaches Greenland from an easterly direction (three subtypes).
J Stationary lows and cyclogenesis (four subtypes).

These patterns were illustrated both schematically and with reference to specific occurrences.

As early as 1882 Van Bebber analysed the movement of depressions over Europe for 1876–80 and found five major tracks (fig. 3.7). He also determined the seasonal weather conditions for each track at stations throughout Europe in terms of temperature, cloudiness, precipitation frequency and amount. Subsequently, many such studies have been made of weather at a point, or over an area, in relation to the most frequent tracks. Much attention has been given to precipitation in this respect (Hay, 1949; Sawyer 1956; Thomas, 1960b; Jackson, 1969), but there are various problems affecting the analysis. It is difficult to devise a satisfactory classification for primary and secondary depressions and troughs yet a large proportion of the prccipitation may derive from secondaries and troughs. Also the precipitation yield will be greatly

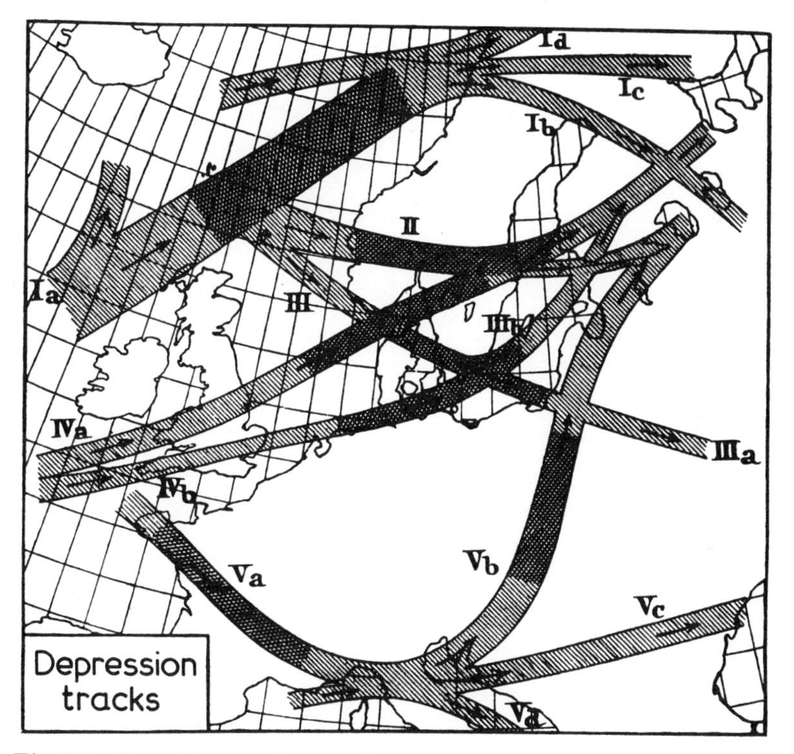

Fig. 3.7 Depression tracks over Europe according to Van Bebber (1882).

augmented if the system is slow-moving, or becomes quasi-stationary, but the classifications usually ignore the rate of movement. Jackson, however, separates out types where depressions over the British Isles and adjacent areas remain stationary for at least twenty-four hours.

Sawyer found a broad relationship between the distribution of rainfall and four depressions over the British Isles (fig. 3.8), but the variability in the totals associated with each track indicates that this approach is of limited value for prediction purposes. The earlier work of Hay identified twelve tracks and examined for each of them the

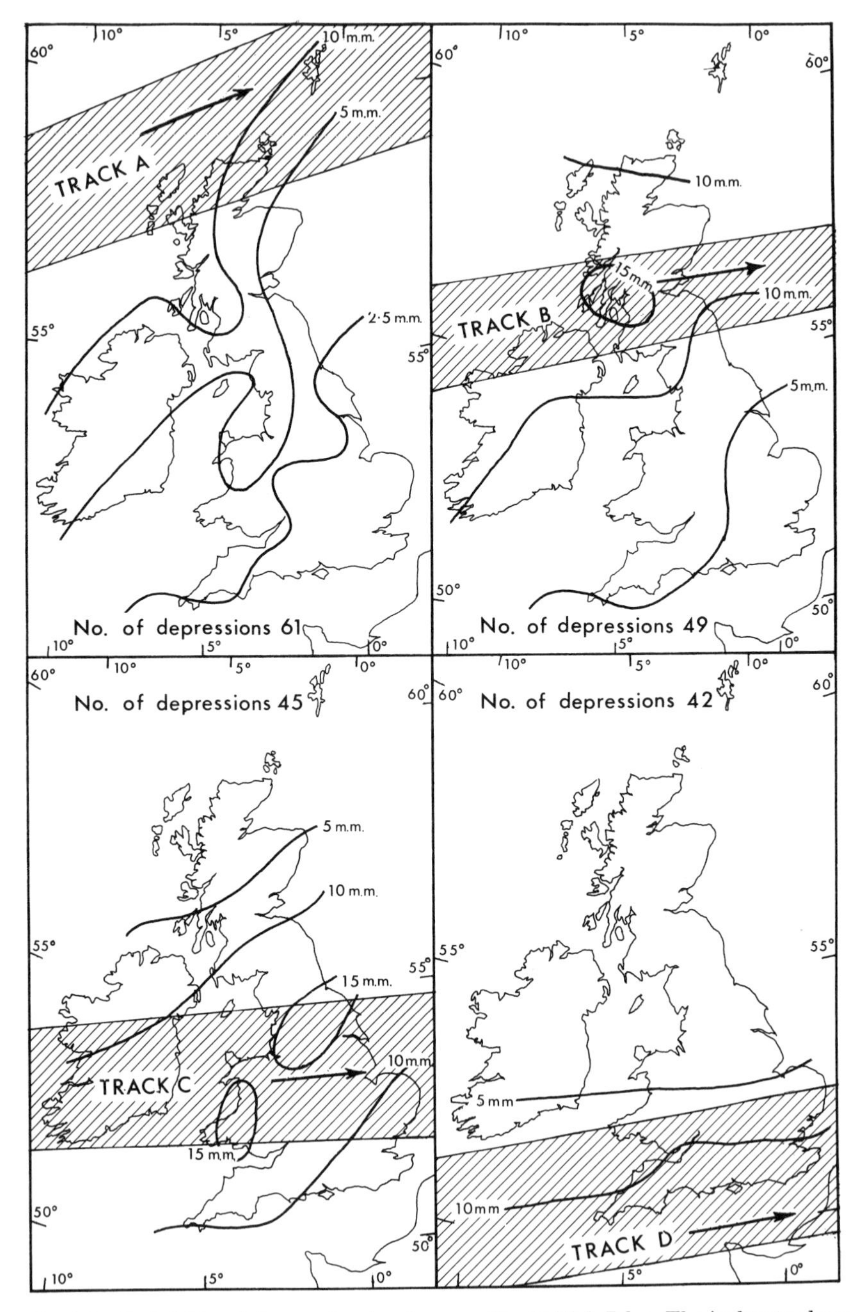

Fig. 3.8 Rainfall and depression tracks over the British Isles. The isohyets show average amounts for the number of depressions indicated (from Sawyer, 1956; Crown copyright reserved).

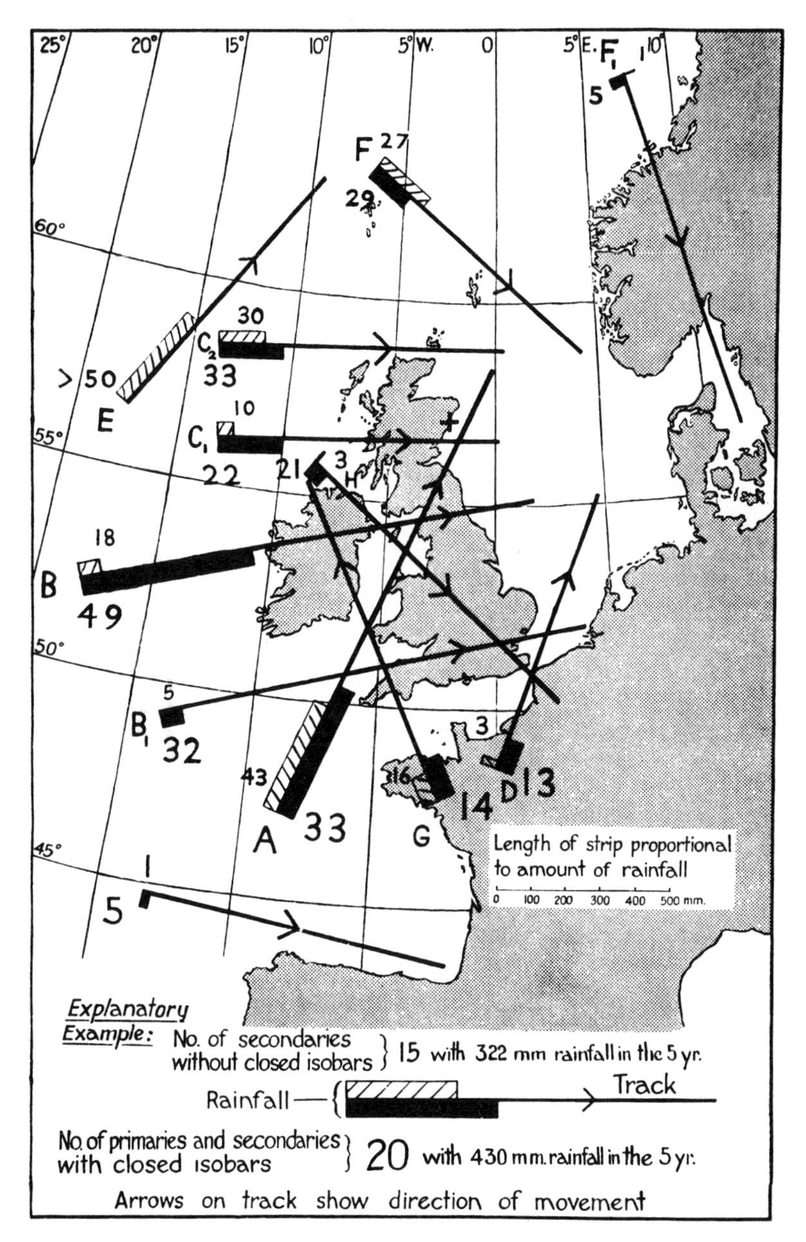

Fig. 3.9 Classification of depression tracks showing frequency for 1926–30 and precipitation at Aberdeen (from Hay, 1949; Crown Copyright reserved).

precipitation falling at Aberdeen during 1926–30 (fig. 3.9). Secondary waves were distinguished from closed primaries and secondaries. In addition he calculated the average seasonal precipitation yield of each system at Aberdeen (fig. 3.10).

Thomas (1960a, 1960b) performed a similar analysis for the eastern North Atlantic and British Isles, distinguishing seven main tracks and a number of subtypes (a total of fifteen). The major ones are illustrated in fig. 3.11. For a five-year period (unspecified) the approximate frequency of depressions on each track are as shown in table 3.2. The precipitation

Table 3.2 Approximate annual frequency of depressions on tracks over the eastern North Atlantic and British Isles (after Thomas, 1960a)

Track	A	B1	B2	B3	B4	B5	C	D	E	F	G	Annual total
Percent	6	17	14	7	2	2	6	5	15	13	14	160–180

occurring with each subtype A1 to D2 was considered in terms of frontal, orographic and instability types and with reference to its general duration. Table 3.3 gives the major findings. It is interesting to note that, while the results are necessarily generalized, Thomas considered that 70–80% of depressions of type A1, B1, B3 and D2 conformed reasonably closely with the type sequence of precipitation as to intensity, duration and distribution. For A2, C1 and C2 tracks the estimate was 50–60% of depressions in these categories.

A more specific study of stormy weather situations in central and east Europe led Sussebach (1968) to classify 291 patterns of cyclone movement over the North Atlantic and Europe during 1950–61. The most frequent tracks were those crossing northern Scotland towards Sweden (20%), those moving on a more northerly track from Iceland towards Scandinavia (17%) and the Vb type (see fig. 3.7) northeastward from the Adriatic Sea (11%). A particularly detailed analysis was performed with maps for each pattern of the mean 500 mb contours, MSL pressure, 1000–500 mb thickness and twenty-four hour change of 500 mb height and MSL pressure for the days of occurrence and each of the two preceding days.

Depressions on the Vb track, which was first identified by Van Bebber, are of particular interest since they may cause heavy rains in central Europe (see also pp. 334–5). Their northeastward track from the Adriatic region is generally associated with the eastern limb of an upper trough over the Alps. Reiter (1963, p. 371) points out that this pattern is often preceded by northwesterly flow over the western Alps which leads to cyclogenesis in the Gulf of Genoa as cold air enters the Mediterranean basin. Lee effects are augmented by the cyclonic vorticity created over the warm waters. This situation may now produce the Vb

Table 3.3 Precipitation effects of the more frequent depression types in the British Isles (after Thomas, 1960b)

DEPRESSION TYPE	FRONTAL PRECIPITATION		OROGRAPHIC PRECIPITATION		INSTABILITY PRECIPITATION	
	> 5 hours rainfall	*About 2 hours rainfall*	*Light rain/drizzle for 1–3 days*	*Intermitt. light rain 6–24 hours*	*Showers over 1–3 days*	*Showers over 12–24 hours*
A1	W, NW Scotland	Remainder of area except Cen., E, SE England			N England, N Wales, N Ireland and Scotland except SE	Remainder of area
A2	SW Scotland and rest of area except E, SE England	SE Scotland, E, SE England		SW Ireland, SW England, S Wales		NW Ireland, Ulster, NW, W, SW Scotland, NW, SW England, S Wales
B1	W, NW Scotland	NW Ireland, Ulster, SW Scotland, NW England	Wales, SW England, NW, SW Ireland	NW, W, SW Scotland, Ulster, NW England		NW Scotland, NW Ireland, NW England
B2	NW, W, SW Scotland, NE England, N Wales, NW Ireland and Ulster	Remainder of Scotland, and Ireland, and England except E, SE		Ireland, except Cen., E; SW Scotland, NW, SW England, Wales		NW, W, SW Scotland, Ireland except Cen., E; NE England, Wales
B3	All areas except SW, S, Cen., E, SE England	Remainder of England		SW Ireland, SW England	NW, W, SW Scotland, NW England, NW Ireland and Wales	E, SE Scotland, Wales, SW Ireland, S-Cen., SW England
C1	Ireland except NW; NW, W, SW, Cen. England, S Wales	Scotland, NW Ireland, E, NE England, N Wales				SW, SE Scotland, Ireland, England, except SW
C2	SW Scotland, Ireland, Wales, NW, SW, W, S-Cen. England	Remainder of area				SW, SE Scotland, Ireland, England, S Wales

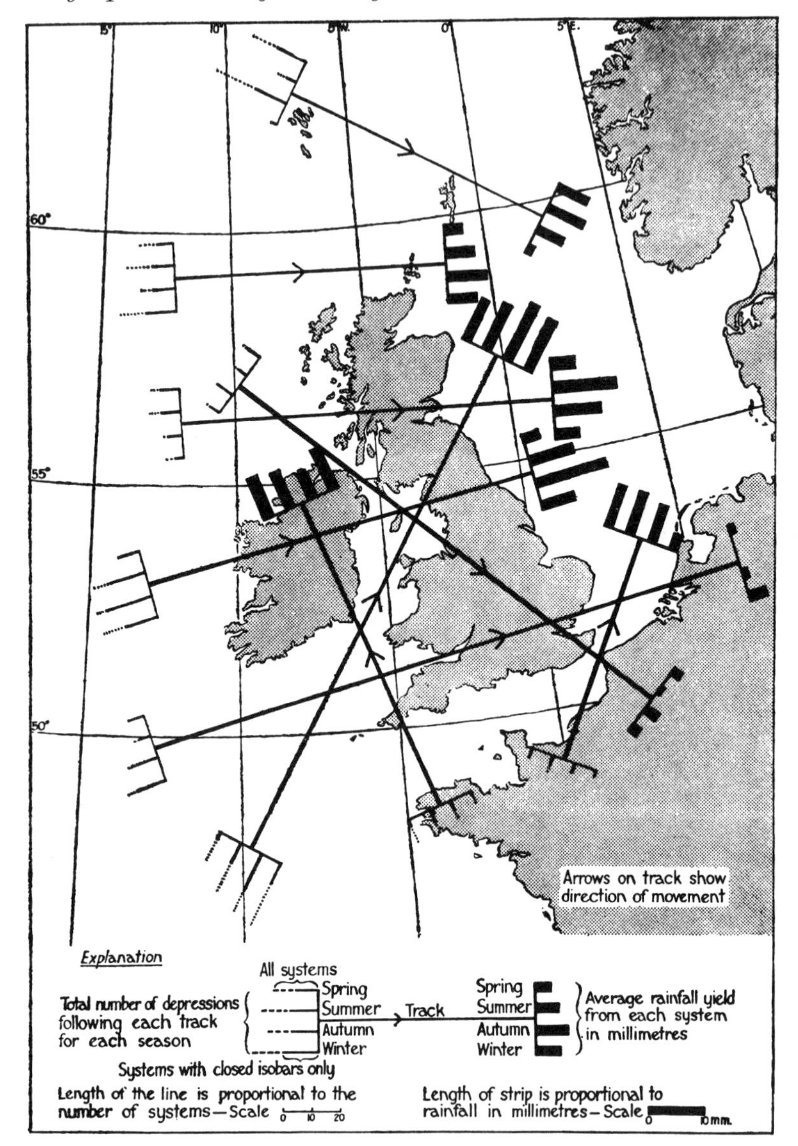

Fig. 3.10 Seasonal frequency of depression tracks, 1926–30, and precipitation at Aberdeen (from Hay, 1949; Crown Copyright reserved).

pattern, especially if a northwesterly jet stream maintains the cold air supply on the western limb of the trough.

Rather less analysis has been carried out for anticyclone tracks and their weather characteristics. In part this reflects the problem of defining anticyclone centres. Reinel (1960) identifies ten anticyclone tracks over Europe – four from west to southwest (Azores type), two from the west (Azores–subpolar), three from northwest to north (subpolar), and one from the east. The movement of individual centres during 1947–57 is mapped, as illustrated in fig. 3.12, for the most important Azores type 1 which moves into central Europe and becomes stationary. Reinel then provides case studies of the distribution of weather elements (temperature, sunshine, cloudiness and precipitation) in relation to each track.

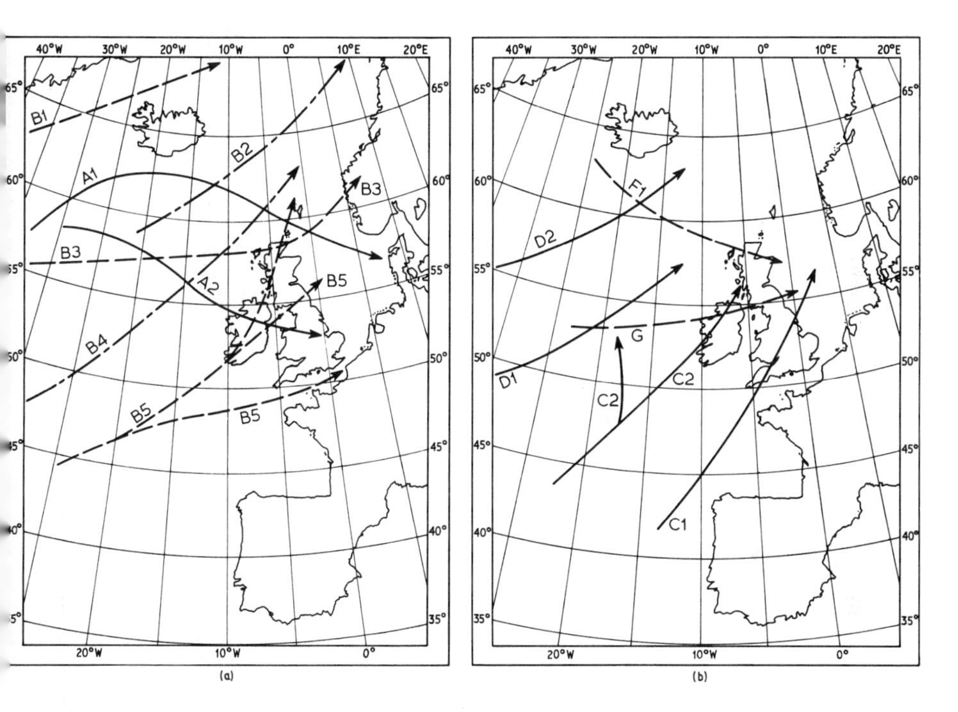

Fig. 3.11 Classification of depression tracks over the eastern North Atlantic and British Isles (after Thomas, 1960a).

One of the most ambitious endeavours of this type was the work of Multanovski (Aristov and Blumina, 1969).[1] In his first work, published in 1915, Multankovski stressed the significance of the polar 'centres of action' – a concept outlined by Teisserenc de Bort twenty years previously. These centres (in Denmark Strait, the American sector, and the Taimyr peninsula) and the Azores anticyclone were recognized as controls of the motion of cyclone and anticyclone tracks over European Russia, and a classification of the main tracks was developed for the

[1] Convenient summaries and commentaries are given by Schell (1940) and Neis (1949).

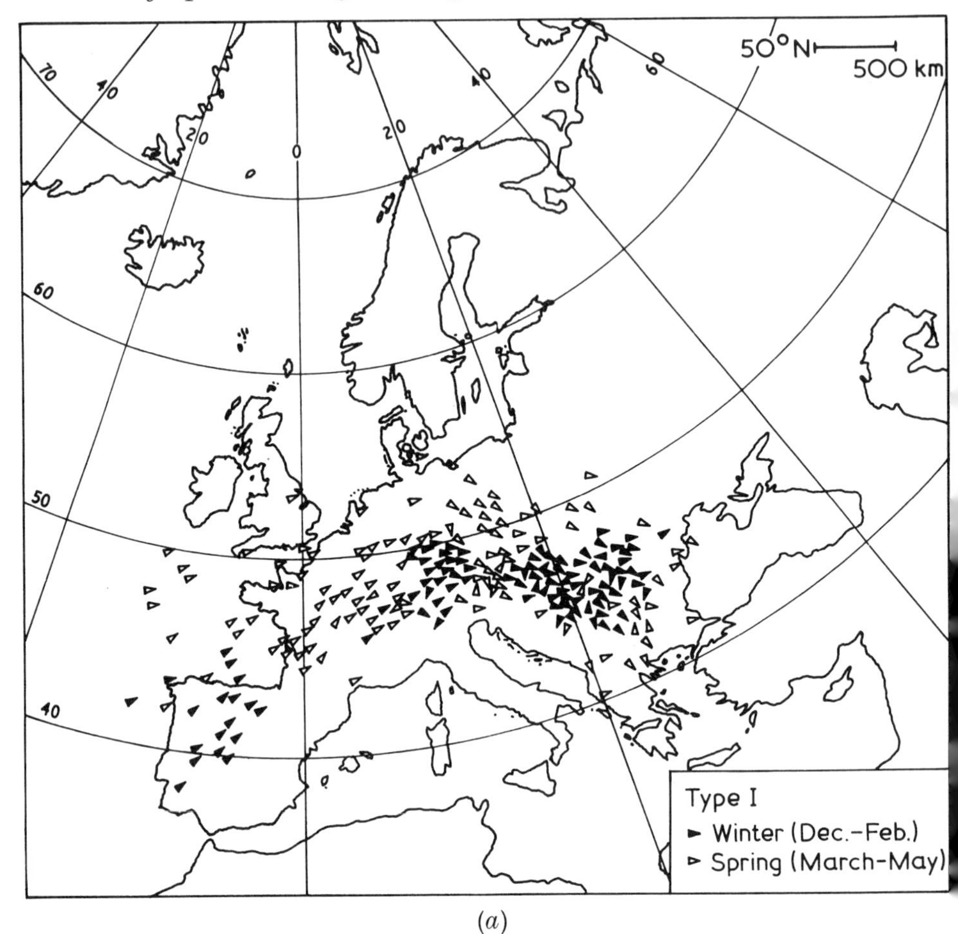

(*a*)

Fig. 3.12 Azores type anticyclones of Reinel's classification, 1947–57. (*a*) Winter and spring. (*b*) Summer and autumn (from Reinel, 1960; in Blüthgen, 1966).

purpose of long-range forecasting. It was claimed that movements along three major axes,

(*a*) from the northwest into Europe (Normal Polar),
(*b*) from the northeast in European U.S.S.R. (Ultra-Polar),
(*c*) along a narrow path from due west (Azores Normal),

tended to be quasi-persistent with a 'natural synoptic period' of ten to twelve days. Such axes, for the movement of both highs and lows, were grouped and plotted with the average pressure distribution on 'composite maps' in order to determine weather types, but these are not all clearly differentiated by the maps and the relationship of a given weather type to a natural period is far from consistent.

Although Multanovski's work had many shortcomings it undoubtedly

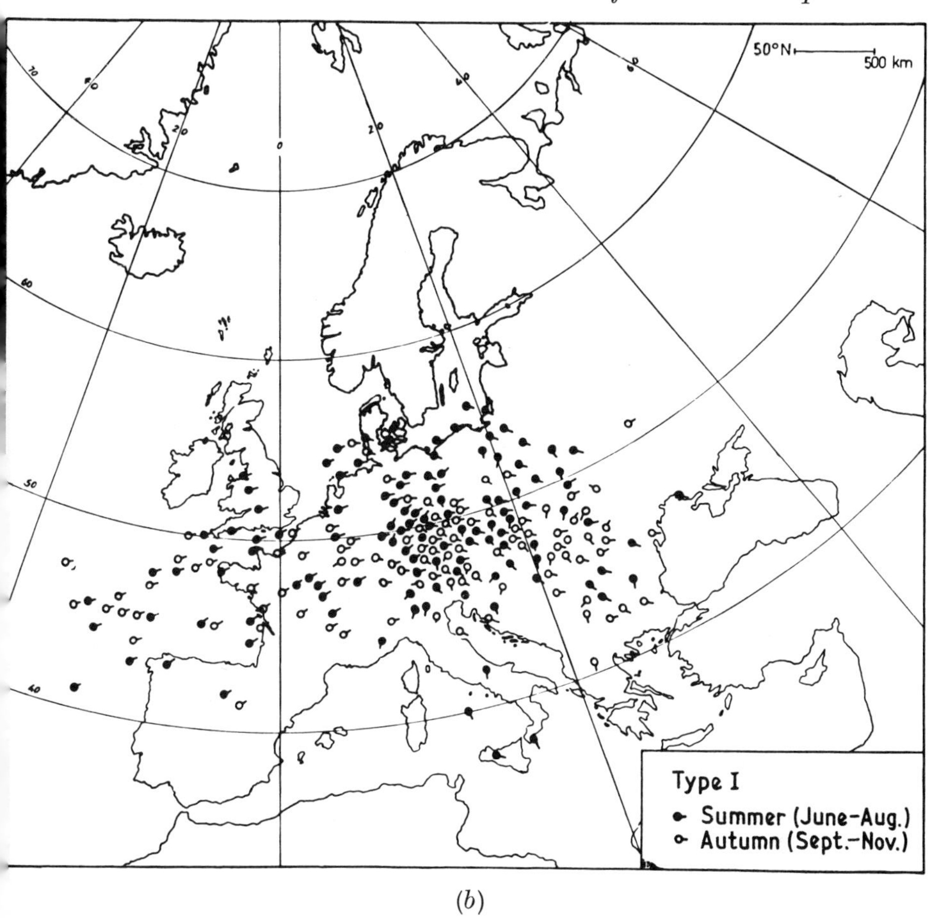

(*b*)

influenced many subsequent researchers. Studies along similar lines have been pursued by Vangengeim (1935), as discussed below, and more recently Irostnikov (1957) analysed the evidence of rhythmic activity of the ultrapolar category. Borsos (1952) developed a classification of cyclone and anticyclone tracks over Hungary with reference to donor and recipient areas and Kurashima (1957) used axes of anticyclone movement as a basis for distinguishing large-scale weather types and macrosynoptic processes over the Far East in winter. The investigations of Dzerdzeevski (1962, 1963) on hemispheric circulation types which are related to major lines of movement of perturbations (discussed on pp. 158–61), and particularly the work of Ragozin and Chukanin (1966), who examined the trajectories of highs and lows in relation to a hemispheric classification scheme, also reflect something of Multanovski's ideas and influence.

2 GROSSWETTER AND RELATED CLASSIFICATORY STUDIES

To obtain a clear appreciation of the *Grosswetter* concept, developed in Germany over a period of thirty years or more by Franz Baur (1936a, 1947, 1948, 1963), it is essential that we review the definitions which he proposed:

> *Weather* (*Wetter*) is the directly observable state of the atmosphere at a given instant or, at most, over twenty-four hours.
> *Weather character* (*Witterung*) refers to the basic pattern of weather conditions over a time interval of several days. This is essentially what is meant by 'weather type' although *Witterung* has no precise equivalent in English.
> *Large-scale weather* (*Grosswetter*) is a term which is also applied to a period of several days. 'Large-scale weather is not specifically uniform but consists of different weather characters which, because of simultaneity, are physically interrelated' (Baur, 1963, p. 27).

The concept of *Grosswetter* was applied to areas in Europe and adjacent parts of the eastern North Atlantic in terms of *large-scale weather patterns* (*Grosswetterlagen*). A *Grosswetterlage* identifies the major trends in atmospheric events over the region during several days of essentially similar weather character in the various parts of the region.[1] It is characterized by the pressure distribution at the surface and in the middle troposphere. In an earlier definition (Baur, 1947) the concept of 'steering' was incorporated: 'A *Grosswetterlage* is the mean pressure distribution (at sea level) during a time interval during which the position of the stationary (steering) cyclones and anticyclones and the steering within a special circulation region remain essentially unchanged.'

The steering refers to the direction of propagation of twenty-four-hour isallobars (Baur, 1951, p. 825) and hence the upper tropospheric pressure distribution is taken into account at least indirectly. The concept of steering was evolved primarily by the 'Frankfurt school' of meteorologists (Baur, 1936b; Ficker, 1938; Mügge, 1938; Putnins, 1962). They showed that the general airflow in the troposphere is closely linked to the pressure gradient in the *lower* stratosphere – Dines's compensation principle. Isallobaric fields are displaced in the direction of the airflow and hence perpendicular to the high level pressure gradient. The 500 mb level chart was used to represent the steering of tropospheric disturbances from the lower stratosphere. Modern concepts and terminology relate steering to events at the jet-stream level in the upper

[1] Blüthgen (1966, p. 30) correctly points out that *Grosswitterungslage* would be a more appropriate term.

Table 3.4 *The* Grosswetter *classification of Hess and Brezowsky* (*1952*)

CIRCULATION TYPE	ZONAL	MIXED	MERIDIONAL		
			N flow	*S–SE flow*	*E–NE flow*
Grosswetterlagen	W, BM	HM, SW, NW	HN, HB, N, TrM, TM	TrW, TB, S, SE	HF, HNF, NE, Ww
Witterung	Each *Grosswetterlage* is subdivided into cyclonic (Z) and anticyclonic (A) situations, with reference to surface isobars over central Europe.				

Zonal flow refers to an Azores subtropical anticyclone in the normal position, mixed flow indicates that it is displaced northward or northeastward towards 50°N, and meridional flow occurs with a blocking high located between 50–70°N.

The *Grosswetterlagen* are defined as follows:

- HM Closed anticyclone over central Europe.
- HN Closed anticyclone over the North Sea.
- HF Closed anticyclone over Scandinavia.
- HNF Closed anticyclone over the North Sea and Scandinavia.
- HB Closed anticyclone over the British Isles.
- BM Ridge of high pressure over central Europe.
- TM Closed depression over central Europe.
- TB Closed depression over the British Isles.
- TrM Trough of low pressure over central Europe.
- TrW Trough of low pressure over western Europe.
- Ww Blocked westerly type, with a blocking high over Russia and frontal zones orientated meridionally over Europe and the North Sea.

In addition to the Z, A, subdivisions of the *Grosswetterlagen*, W type also has a W_s subtype when the tropical high and the eastward-moving depressions are south of their normal positions.

troposphere. Depression tracks are closely related to the mean jet position, with maximum frequency of cyclones to the left of the mean jet viewed downwind. The surface frontal zone is continually being developed on the leading edge (the delta region) of a jet maximum by a transverse circulation of rising cold air and sinking warm air. The diffluence in the upper troposphere strengthens cyclogenesis by inducing pressure falls at the surface. To the rear of the jet maximum, in the entrance zone, a thermally direct transverse circulation dissipates the surface frontal zone although at the jet level frontogenetic tendencies are strengthened by confluence (see Reiter, 1963, pp. 167–174). No attempt has so far been made to incorporate these concepts directly into

Table 3.5 The percentage frequency of Grosswetterlagen *and* Grosswettertypen *(GT), 1881–1966 (Hess and Brezowsky, 1969)*

	Grosswetterlage	J	F	M	A	M	J	J	A	S	O	N	D	Year
	WA	6·1	4·7	4·2	4·6	4·2	6·1	7·3	8·7	8·1	6·5	5·5	3·6	5·8
	WZ	13·5	12·3	11·4	11·7	12·2	15·7	19·9	23·5	15·1	14·7	13·7	17·9	15·2
	WS	4·3	5·8	6·1	2·6	0·7	1·8	1·8	2·1	1·0	3·4	3·4	6·8	3·3
	WW	3·4	1·4	3·3	2·1	1·1	2·1	1·2	2·9	2·4	2·1	4·5	5·1	2·6
Zonal circulation	W (GT)	27·3	24·2	25·0	21·0	18·2	25·7	30·2	37·2	26·6	26·7	27·1	33·4	26·9
	SWA	2·8	3·0	2·1	1·6	1·3	1·0	1·2	1·4	1·4	2·9	2·7	2·9	2·0
	SWZ	3·9	2·5	1·1	1·4	1·1	0·5	0·4	0·8	0·8	2·3	3·0	2·1	1·6
	SW (GT)	6·7	5·5	3·2	3·0	2·4	1·5	1·6	2·2	2·2	5·2	5·7	5·0	3·6
	NWA	3·4	3·9	4·1	3·6	3·8	7·7	10·0	7·1	3·5	2·6	3·9	2·7	4·7
	NWZ	4·7	4·7	3·3	4·3	2·7	4·2	8·2	5·5	4·2	3·2	4·2	4·7	4·5
	NW (GT)	8·1	8·6	7·4	7·9	6·5	11·9	18·2	12·6	7·7	5·8	8·1	7·4	9·2
	HM	14·5	13·7	11·3	6·6	9·4	9·7	10·8	10·0	16·0	12·5	7·8	10·1	11·0
	BM	4·1	4·8	3·5	5·5	3·6	4·8	5·1	6·9	7·5	7·6	9·8	9·3	6·0
	HM (GT)	18·6	18·5	14·8	12·1	13·0	14·5	15·9	16·8	23·5	20·1	17·6	19·4	17·0
	TM	3·2	2·6	3·5	3·9	3·8	2·1	2·7	1·5	1·7	2·3	3·2	1·5	2·7

Mixed circulation		36·6	35·2	28·9	26·9	25·7	30·0	38·4	33·1	35·1	33·4	34·6	33·2	32·5
	NA	0·3	0·4	1·2	1·0	2·7	2·6	1·7	1·3	1·0	0·3	0·6	0·7	1·1
	NZ	2·3	2·1	3·7	4·1	5·7	5·5	2·4	1·8	2·9	2·0	1·6	1·2	2·9
	HNA	2·3	2·1	2·8	5·7	6·5	6·9	3·3	3·1	3·8	3·4	2·1	2·3	3·7
	HNZ	1·2	1·4	1·8	1·8	2·3	1·9	0·9	0·8	0·4	2·1	0·9	0·9	1·4
	HB	2·8	4·5	3·8	3·3	3·9	4·3	2·0	2·5	4·6	3·3	2·6	2·1	3·3
	TRM	3·5	4·4	4·1	5·4	3·6	2·9	3·8	2·6	4·4	3·7	5·5	3·3	3·9
	N (GT)	12·4	14·9	17·4	21·3	24·7	24·1	14·1	12·1	17·1	14·8	13·3	10·5	16·3
	NEA	1·7	2·5	2·9	2·6	4·8	5·1	4·5	3·0	2·4	1·1	0·8	1·1	2·7
	NEZ	2·0	1·7	2·3	4·4	3·5	3·1	2·4	2·1	2·6	1·6	0·6	1·8	2·3
	HFA	6·1	5·1	3·9	4·4	3·5	2·2	1·9	2·4	3·8	4·6	2·8	4·2	3·7
	HFZ	1·0	1·1	1·2	1·0	0·5	0·5	0·2	0·3	0·4	0·8	1·2	1·2	0·8
	HNFA	1·1	1·8	0·7	1·8	4·2	1·9	0·3	0·2	0·9	0·9	0·7	0·5	1·2
	HNFZ	1·1	1·8	3·9	2·8	3·3	0·8	0·7	0·8	0·6	0·6	1·5	0·6	1·5
	SEA	1·4	1·4	3·8	2·6	2·8	0·9	0	0·1	2·2	4·2	3·6	2·9	2·2
	SEZ	3·6	3·1	2·9	2·5	0·7	0	0	0·1	0·9	1·4	1·7	1·7	1·5
	E (GT)	18·0	18·5	21·6	22·1	23·3	14·5	10·0	9·0	13·8	15·2	12·9	14·0	15·9
	SA	3·3	1·5	2·7	1·8	1·0	0·1	0·2	0·3	2·8	2·9	5·0	2·5	2·0
	SZ	0·6	2·3	1·0	0·7	0	0	0	0	0·3	1·2	2·0	2·4	0·9
	TB	0·8	1·7	1·0	2·6	3·5	1·7	2·8	4·3	1·1	2·1	1·9	2·1	2·1
	TRW	1·1	1·5	1·8	2·8	2·8	2·4	3·8	3·7	2·7	3·2	2·9	1·8	2·5
	S (GT)	5·8	7·0	6·5	7·9	7·3	4·2	6·8	8·3	6·9	9·4	11·8	8·8	7·5
Meridional circulation		36·2	40·4	45·5	51·3	55·3	42·8	30·9	29·4	37·8	39·4	38·0	33·3	39·7
Unclassifiable		0	0·2	0·6	0·8	0·7	1·5	0·5	0·3	0·5	0·5	0·3	0	0·9

classifications of depression tracks or of the *Grosswetter* type, although it is long overdue.

The original Baur classification of twenty-one types of *Grosswetterlage* was modified by Hess and Brezowsky (1962) in the light of the increased amount of upper-air information, and a daily reclassification for 1881–1950 was prepared. This has subsequently been updated to 1966 (Hess and Brezowsky, 1969). The daily *Grosswetterlage* categories have been tabulated since 1952 in the monthly editions of *Die Grosswetterlagen Mitteleuropas* published by the Deutscher Wetterdienst. The revised classification is summarized in table 3.4. Figs. 3.13 (*a–e*) illustrate some of the common *Grosswetterlagen* patterns at the surface and 500 mb. Baur (1948) and Butzer (1960) claim that the *Grosswetterlagen* catalogue is valid for the area between 30°W–45°E, 24–70°N, but the type maps demonstrate that the overall emphasis is on central Europe and this tendency is rather more evident in the revised Hess–Brezowsky version.

Table 3.5 gives monthly frequency data for the major types. This is discussed further in Chapter 5 (see fig. 5.33). Hess and Brezowsky (1969) also analysed pentad frequencies for 1881–1966 and 1924–66.

In a study complementing the 'classical' *Grosswetterlagen* work, Maede (1965) has recently analysed upper air (500 mb) *Wetterlagen* over the eastern North Atlantic and Europe for 1949–62. He uses eight directional types combined with three classes of velocity or relative vorticity, as appropriate, and three non-directional categories, giving a total of twenty-seven types. The vorticity estimate is based on isohypse values at the four vertices of a diamond grid and their local mean (H_0):

$$\text{relative vorticity} = \tfrac{1}{4}(H_1+H_2+H_3+H_4)-H_0.$$

The seasonal trend of these categories is discussed.

A catalogue comparable with the *Grosswetter* calendar has been developed for eastern Asia by Yoshino (1968) although the daily catalogue for 1946–65 is at the present time unpublished. There are six main patterns and a total of fifteen subtypes. The main groups for the approximate area 110–150°E, 25–50°N, are:

1 High pressure to the west, low to the east (winter monsoon pattern).
2 Trough pattern. Four subtypes according to the position and direction of movement of the lows.
3 Migratory anticyclone pattern (four subtypes).
4 Quasi-stationary frontal pattern. Two subtypes according to the frontal orientation.
5 High pressure to the south (with the North Pacific subtropical anticyclone extending over Japan in summer), low pressure to the north (summer weather pattern).

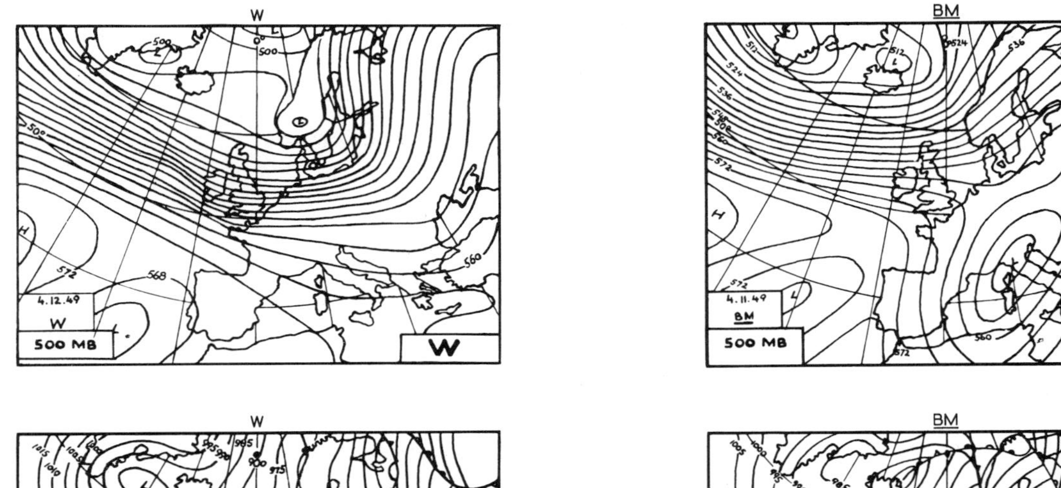

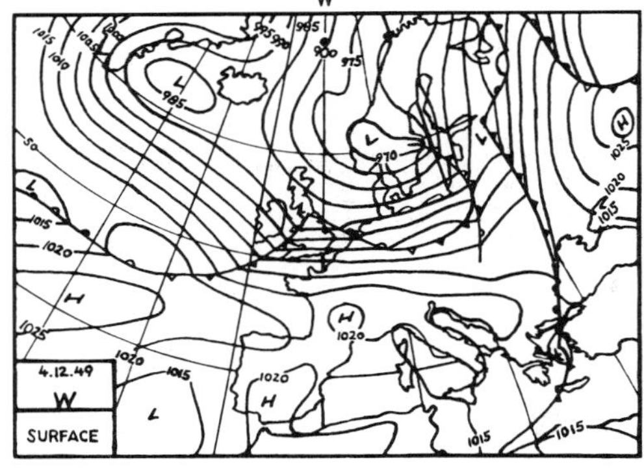

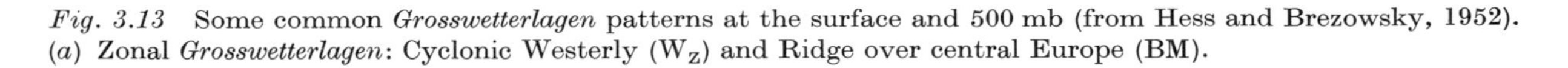

Fig. 3.13 Some common *Grosswetterlagen* patterns at the surface and 500 mb (from Hess and Brezowsky, 1952). (*a*) Zonal *Grosswetterlagen*: Cyclonic Westerly (W_Z) and Ridge over central Europe (BM).

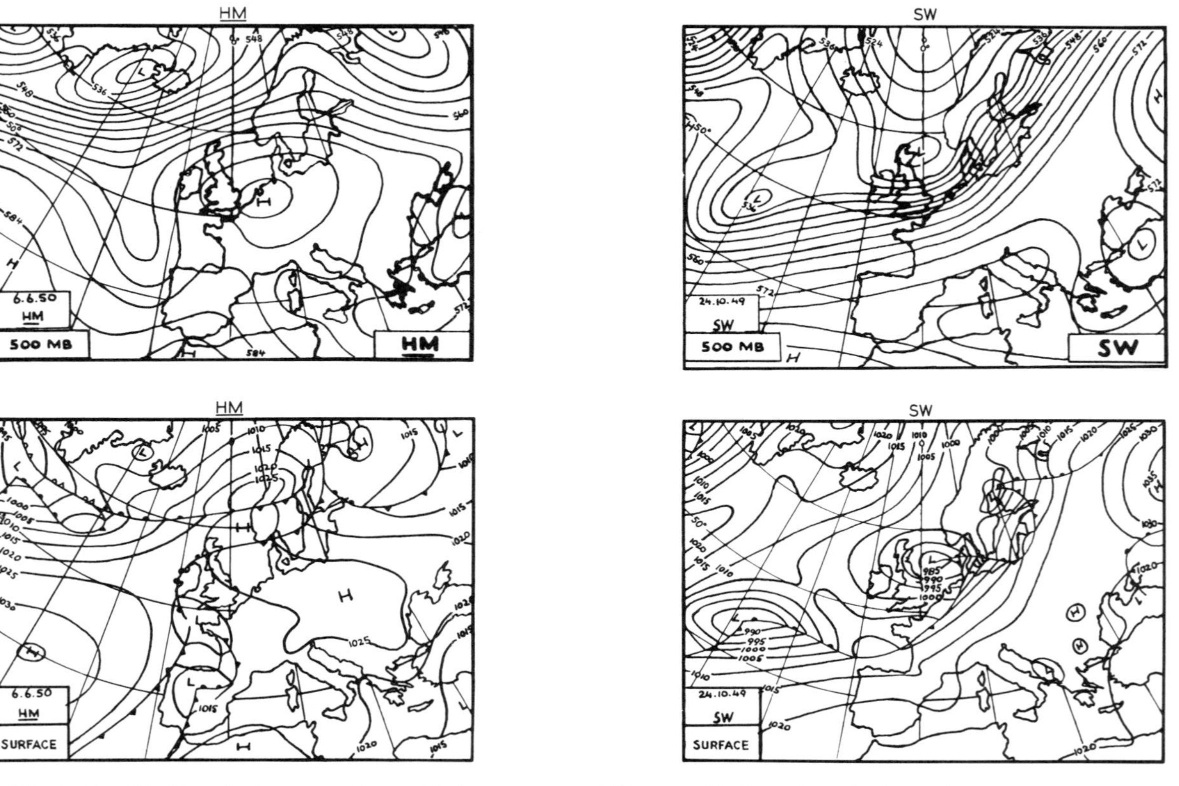

Fig. 3.13 (*b*) Mixed *Grosswetterlagen*: high over central Europe (HM) and cyclonic Southwesterly (SW_Z).

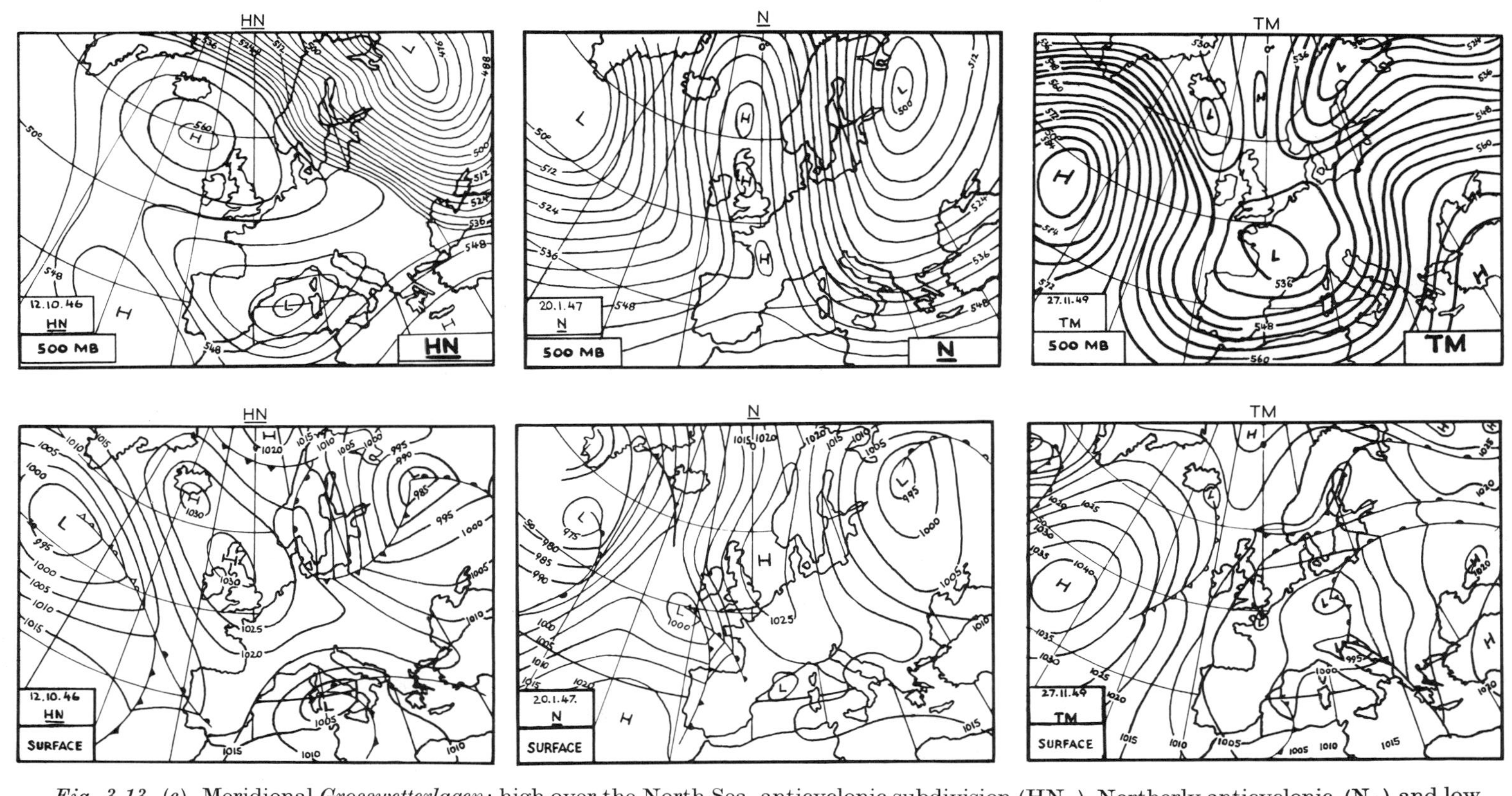

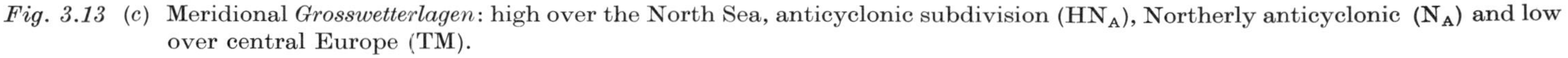

Fig. 3.13 (*c*) Meridional *Grosswetterlagen*: high over the North Sea, anticyclonic subdivision (HN_A), Northerly anticyclonic (N_A) and low over central Europe (TM).

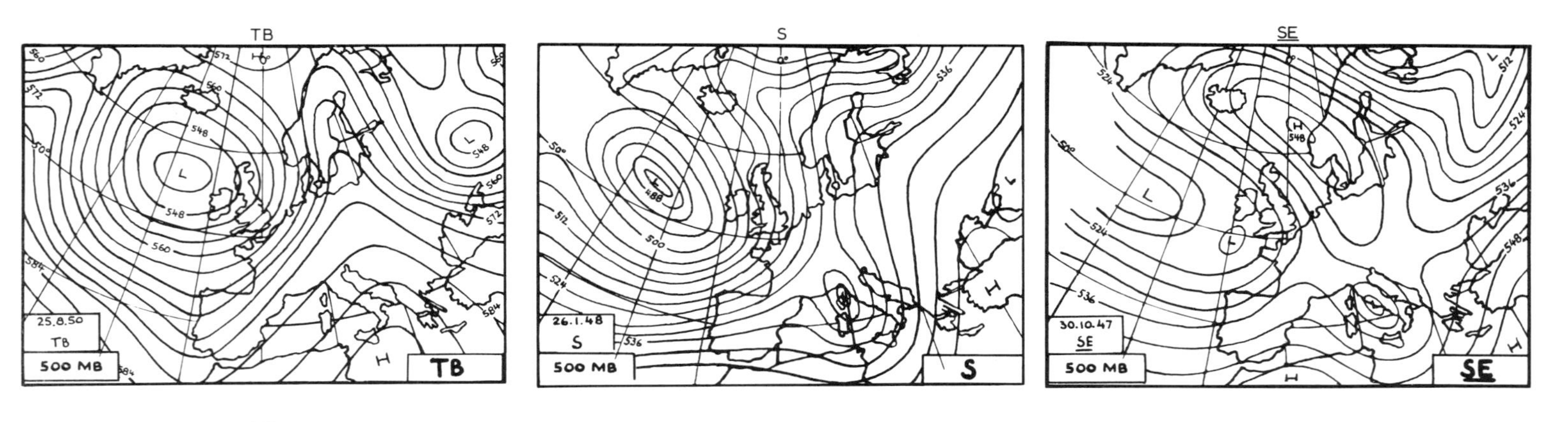

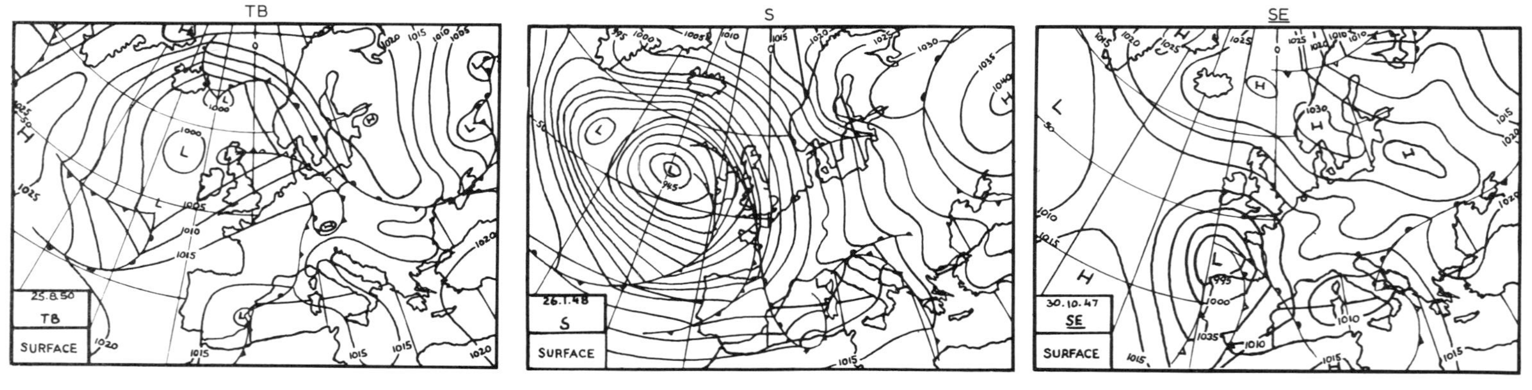

Fig. 3.13 (*d*) Meridional *Grosswetterlagen*: low over the British Isles (TB), Southerly cyclonic (S_Z) and Southeasterly anticyclonic (SE_A).

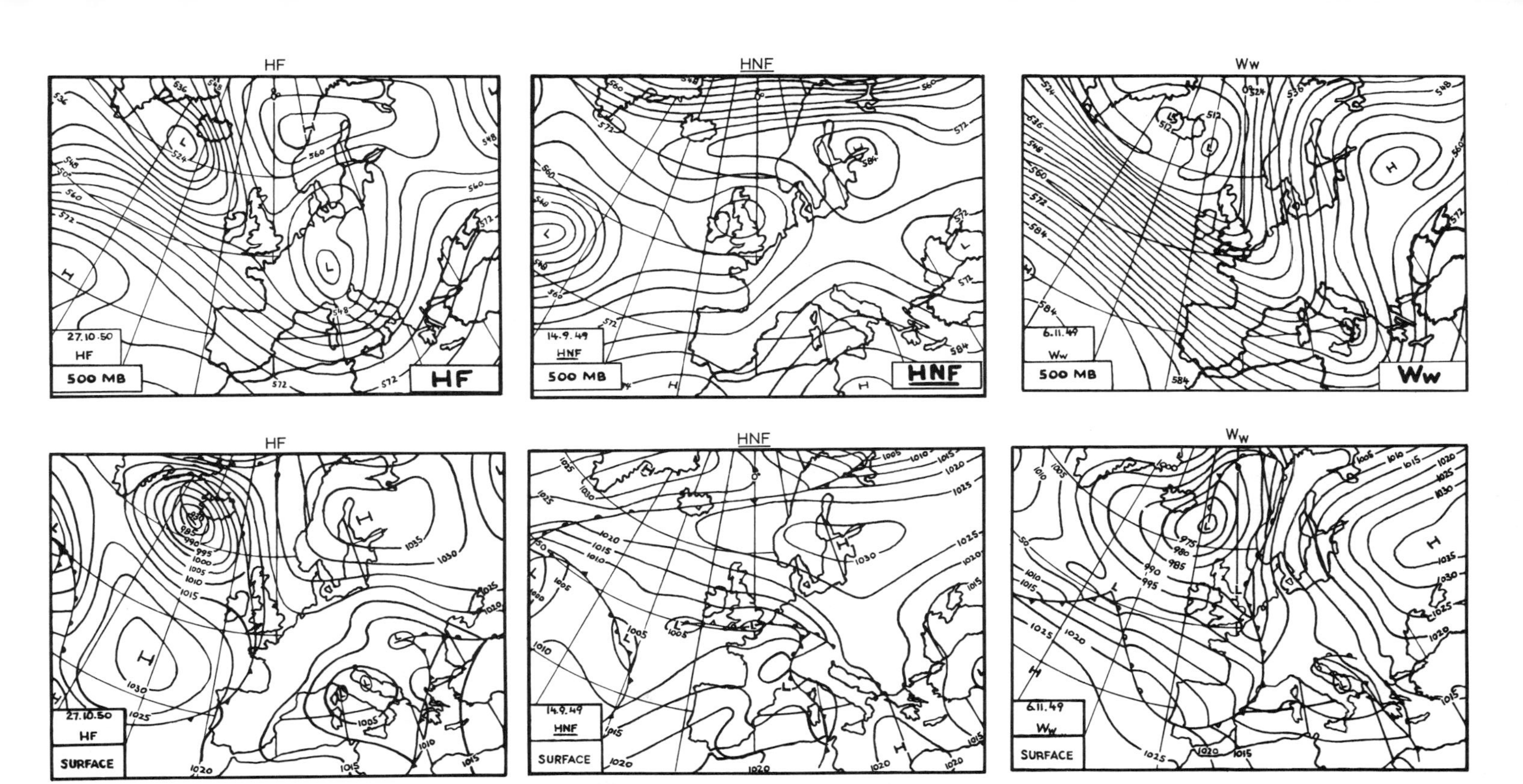

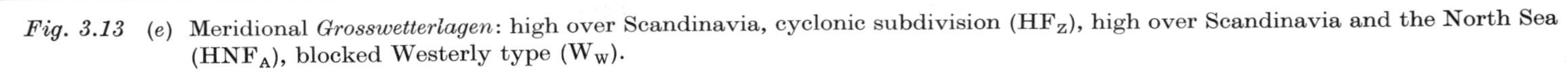

Fig. 3.13 (*e*) Meridional *Grosswetterlagen*: high over Scandinavia, cyclonic subdivision (HF_Z), high over Scandinavia and the North Sea (HNF_A), blocked Westerly type (W_W).

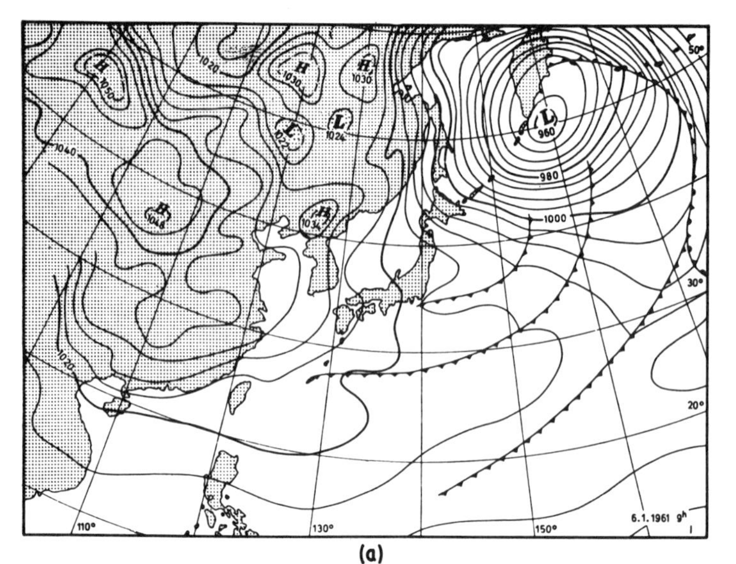

(a)

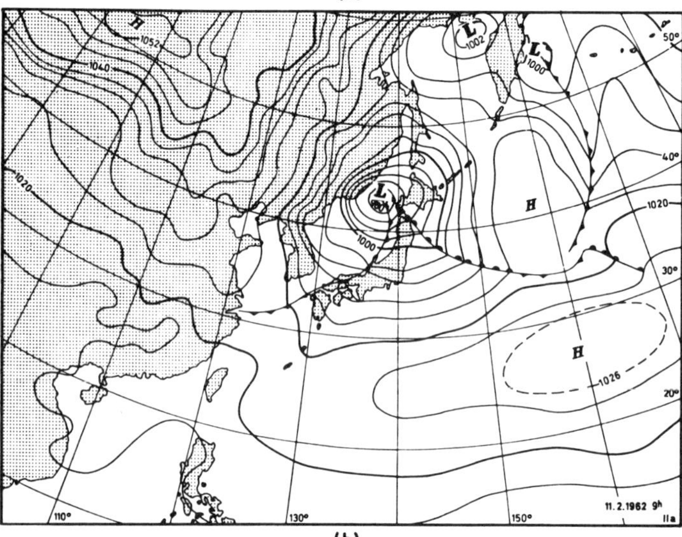

(b)

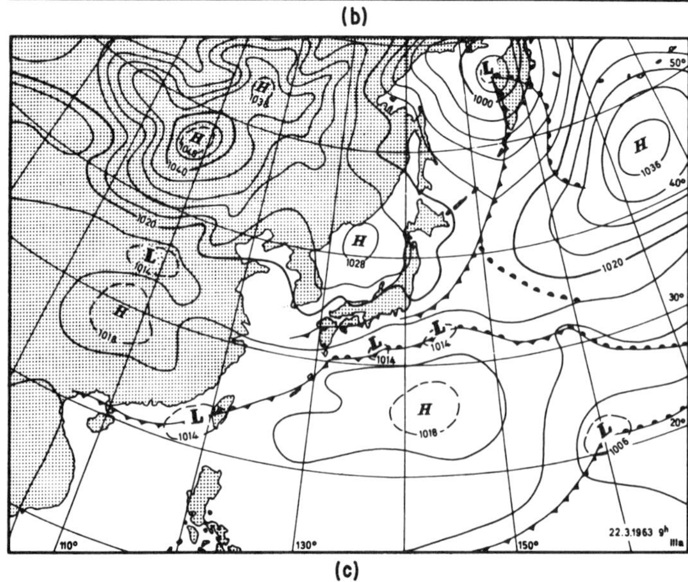

(c)

Fig. 3.14 Circulation types for east Asia (from Yoshino, 1968).

(*a*) An example of pattern I, west-high–east-low pattern.

(*b*) An example of pattern IIa. Patterns IIb, IIc and IId are variations of this pattern.

(*c*) An example of pattern IIIa. Patterns IIIb, IIIc and IIId are variations of this pattern.

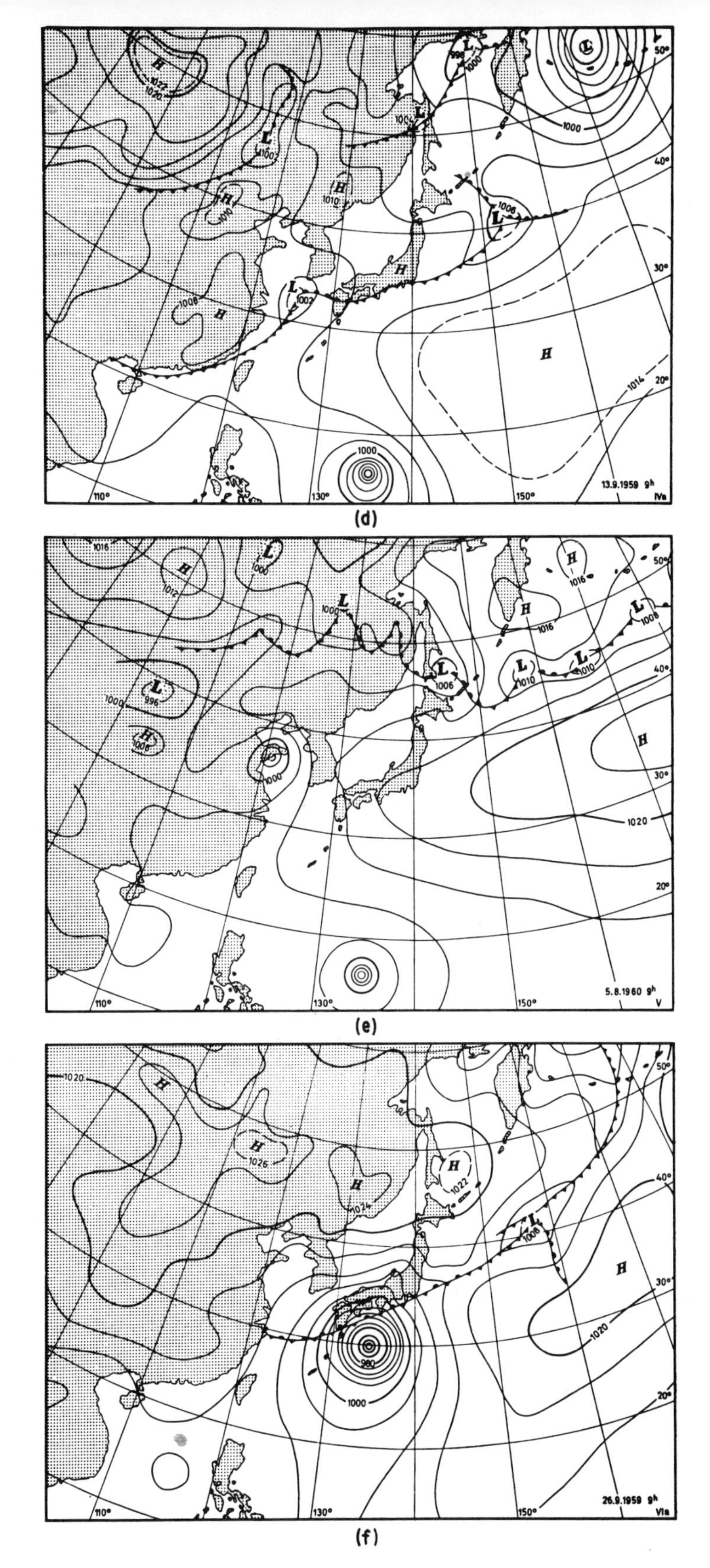

(*d*) An example of pattern IVa. Pattern IVb is a variation of this pattern.
(*e*) An example of pattern V, summer weather pattern.
(*f*) An example of pattern VIa.

6 Typhoon pattern. Three subtypes according to the location of the typhoon.
7 Hybrid patterns.
8 Double pattern (mainly 2*a* with 2*c*; lows over Hokkaido or Sakhalin and from Taiwan towards the Pacific coast of Japan).

The first six are illustrated in fig. 3.14. Yoshino gives frequency data by pentads (see fig. 5.7) and on this basis defines four 'natural seasons'. These, and associated singularities, are discussed in Chapter 5.A.

Most of the work on *Grosswetter* calendars for North America is not readily available. As Elliott (1951) reports, it was carried out just before and during the Second World War principally at the California Institute of Technology (1943). In the early studies a six-day interval was used (Blewitt and Paulhus, 1942), but it proved unrealistic to force the types into this interval. Subsequently (California Institute of Technology, 1945), a catalogue of three-day types was developed, relating the time interval to the average duration of a single depression family across the region 30–60°N, 135–90°W. Similar schemes were also evolved for other mid-latitude sectors – 135–180°E, 180–135°W, 98–45°W and 45°W–45°E and catalogues were prepared for each sector (Air Weather Service, 1944, 1955; Rempel and Stone, 1945).

The major circulation features used in the classifications were the Aleutian low, the North Pacific subtropical anticyclone, the trajectory of polar air outbreaks and the movement of depressions. The second scheme of three-day types also took upper air features into account. It was recognized that the major distinction is between zonal and meridional flow patterns. The main types for North America are illustrated in fig. 3.15. These are for winter although the same scheme was found to be essentially applicable in all seasons. The diagnostic characteristics are summarized in table 3.6. Elliott (1949, 1951) has described their general weather characteristics. The calendars were applied in analogue forecasting experiments but little use seems to have been made of them in analysing local or regional climate, apparently because they are not generally available, unlike the *Grosswetter* catalogue.

In discussing the observed sequences of surface and 500 mb features of circulation and their vertical coupling over the western United States, Sands (1966) concludes that *Grosswetter* are non-existent. This may be due to the method of classification and size of area used (see section B of this chapter, pp. 97–8) although the vertical structure of synoptic systems over the western United States is certainly inhomogeneous between the Pacific coast, the mountains and the Great Plains. On the other hand, the circulation types of Elliott are comparable to *Grosswetter*, and other studies for this area successfully identify synoptic regimes.

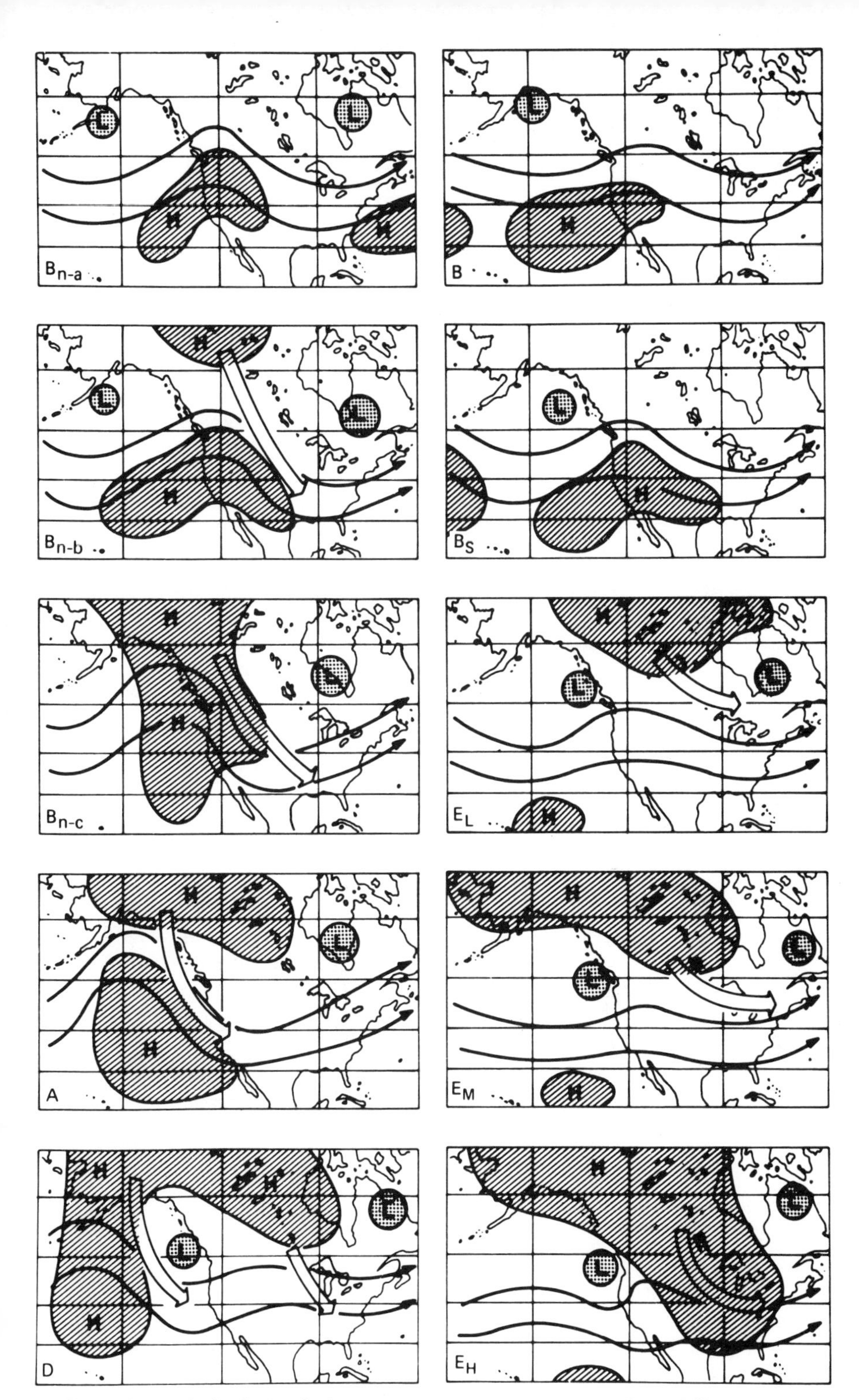

Fig. 3.15 Principal circulation types for North America (from Elliott, 1951). Heavy lines indicate upper level mean flow. Stippled areas = quasi-stationary low pressure centres. Hatched areas = persistent surface anticyclones. Open arrows = polar air outbreaks.

Table 3.6 North American circulation pattern types (after Elliot, 1949, 1951)

Type	*Characteristic features*		
Meridional:	Position of western ridge	Position of eastern trough	
B_{n-a}	115–120°W	90°W	Trough over central North America; strong Great Basin high.
B_{n-b}	115–120°W	90°W	Similar to B_{n-a} but stronger frontal contrasts.
B_{n-c}	135°W	100°W	Strong western ridge, deep eastern trough; intense polar outbreaks over mid-West.
A	145–150°W	110°W	Strong Pacific high extending well north; lows move SE'ward across Pacific northwest U.S.
D	160°W	130–135°W	Extreme meridional pattern; west coast trough.
C_L	125°W	85°W	Great Basin high displaced to British Columbia; very strong in C_H. (C_L, C_H)
C_H	130°W	75°W	
Zonal:	*Latitude of depression tracks across Rockies*		
B	59°N		Zonal; storm track displaced to N.
B_s	55°N		Similar to B_{n-a} but weaker western ridge.
E_L	47°N		Westerlies displaced to S; high over northern Canada.
E_M	40°N		Cyclones much farther S across continent.
E_H	34°N		Extreme type; strong high over Alaska–NW Canada extending SE'ward.
Others:			
Gulf Types			
G_a			Gulf cyclones develop and move northeastward; ridge over central U.S. in G_b and lows further E. (G_a, G_b)
G_b			
Hudson Bay Types			
H_a			High of polar air origin extending S.
H_b			Similar but with strong subtropical high over SE U.S. (summer).

3 REGIONAL AIRFLOW

Early studies

Attempts to relate weather conditions at a place to the general direction of air movement are probably the oldest form of synoptic climatological analysis. As soon as systematic weather observations were available over a period of years in the early nineteenth century, frequencies or averages of weather elements, singly or in combination, were determined for the winds from the eight points. Von Buch (1820) and Dove (1827),

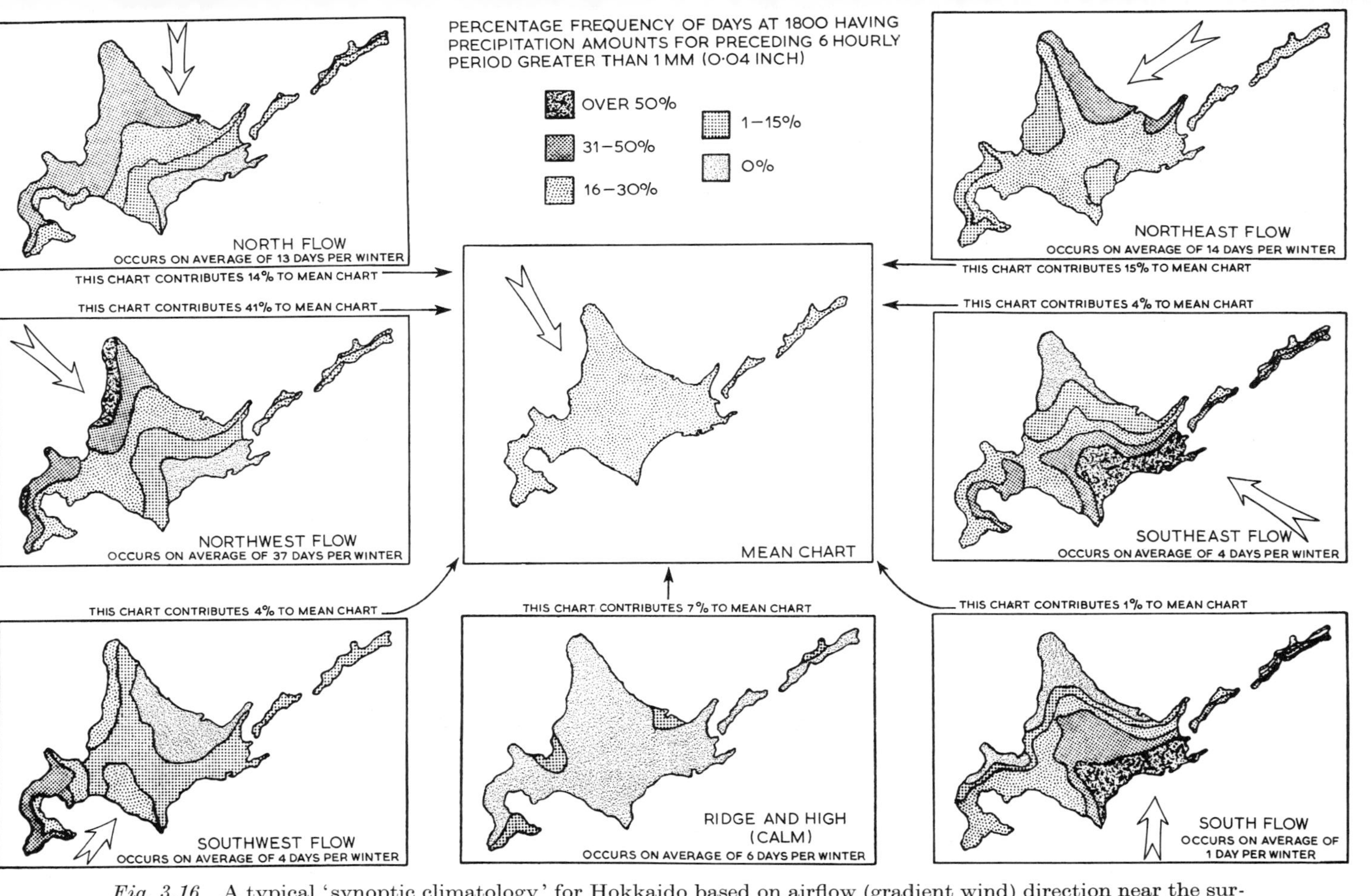

Fig. 3.16 A typical 'synoptic climatology' for Hokkaido based on airflow (gradient wind) direction near the surface for December–February. The mean chart shows no evidence of pattern (from Jacobs, 1947).

for example, prepared wind rose diagrams of average barometric pressure and other parameters for each wind direction.

The first comprehensive study of airflow types was made for the British Isles by Abercromby (1883), who described the weather associated with airflows from north, south, east and west A similar scheme was adopted some fifty years later (Reed, 1932) in a study for the northwest Pacific Ocean. He added SW and NW types to the four basic flow directions and described the pattern of pressure deviations and weather over the Pacific coast of North America. Although Reed gave no synoptic maps to illustrate these types they were subsequently provided by Vernon (1947) in a study applying Reed's classification to forecasting precipitation at San Francisco.

Modern developments

Interest in airflow climatology waned during the 1930s with the concentration on air masses and fronts, and it only revived during and after the Second World War. In a discussion of the methodology then adopted Jacobs (1947) indicated that the regional airflow pattern represented the lowest ranked element in features of primary synoptic classification (table 3.1). As such, regional airflow can be specified with respect to direction and speed of the wind, isobaric curvature and vertical air mass structure. Broadly speaking we are concerned with features persisting for at least a day and affecting an area with a scale of up to about 1000 km (10° latitude). Jacobs illustrated his review with results of a study for Hokkaido, Japan, based on eight directions of gradient airflow at the surface, together with four pressure patterns – ridge or high, trough, low centre and col. Calculations of the precipitation distribution occurring with each type (fig. 3.16) allows the contribution of each to the overall mean distribution to be assessed.

The first long-period catalogues of surface airflow types for the British Isles were drawn up by Levick (1949, 1950) for 1898–1947. Additional data were published later (Levick, 1955). The classification had five simple directional types (NW, W, S, E and N) although the last of these was designated northerly-cyclonic. The non-directional anticyclonic type was subdivided according to the origin of the airflow or recorded as indeterminate (Ai).

Lamb (1950, 1964) developed a rather more elaborate scheme based on Levick's work. There are eight directional types, the direction referring to the general airflow over the British Isles and to the overall movement of synoptic systems from each cardinal point (each subdivided into anticyclonic, cyclonic, hybrid and unspecified categories) and three non-directional types. Two of the latter are anticyclonic (A) and cyclonic (Z) types defined, respectively, as situations where high pressure dominates the map area of the British Isles, and where depressions stagnate over or cross the area. The third non-directional type includes unclassifiable cases (U). The hybrid types, such as anticyclonic westerly (AW) and cyclonic westerly (CW), are applied as follows: westerly type (W) refers to the eastward movement of synoptic features from the Atlantic; if the track of the depressions is displaced to the north and an anticyclone dominates the pressure field over the southern part of the map area the type will be AW, whereas if the track is such that the centres pass more or less over the British Isles the type will be CW. Table 3.7 shows the complete classification and fig. 3.17 illustrates some of the major types.

Table 3.7 The synoptic classification of H. H. Lamb

	Anticyclonic	*Unspecified*	*Cyclonic*
Northwesterly	ANW (A)	NW (B)	CNW (C)
Northerly	AN (M)	N (N)	CN (O)
Northeasterly	ANE (G)	NE (H)	CNE (I)
Easterly	AE (D)	E (E)	CE (F)
Southeasterly	ASE (J)	SE (K)	CSE (L)
Southerly	AS (R)	S (S)	CS (T)
Southwesterly	ASW (£)	SW (P)	CSW (Q)
Westerly	AW (V)	W (W)	CW (X)
Non-directional	A (Y)	U (U)	C (Z)

It is necessary to emphasize that each type is envisaged as comprising a pattern which may last for several days. The airflow direction is essentially that of the middle troposphere which in turn is related to the *steering* of depressions and anticyclones. Hence the actual surface wind direction and the weather characteristics will vary with the passage of these systems. The initial catalogue has recently been slightly modified and the revised 'homogenized' classification has just been published (Lamb, 1972). The catalogue was originally evaluated on a daily basis for 1883 onwards, but efforts are currently being made to extend it back in time. Daily synoptic charts have been reconstructed for 1780–85, for example, using weather records, ships' logbooks and weather diaries of the period (Kington, 1970). The application of such historical catalogues is discussed in Chapter 5.B and 5.C.

Lamb (1950) gave considerable attention to the persistence of the individual types and to the occurrence of spells at particular times of year (see Chapter 5.A, pp. 304–6), and more recently Perry (1970) has determined the mean and maximum durations of weather type groups in January, April, July and October 1910–30 and 1948–68 and also the

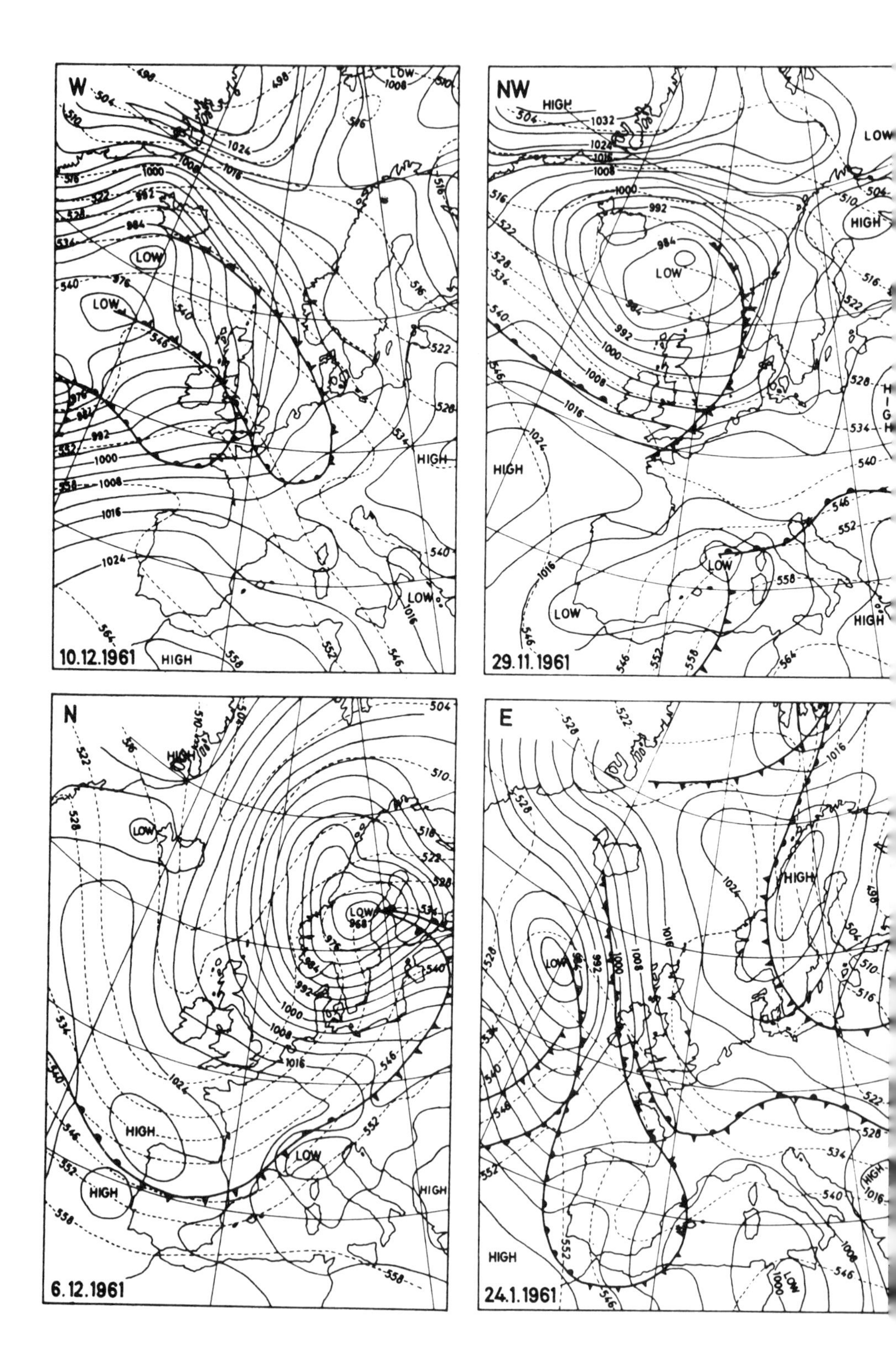

W
10.12.1961
NW
29.11.1961
N
6.12.1961
E
24.1.1961

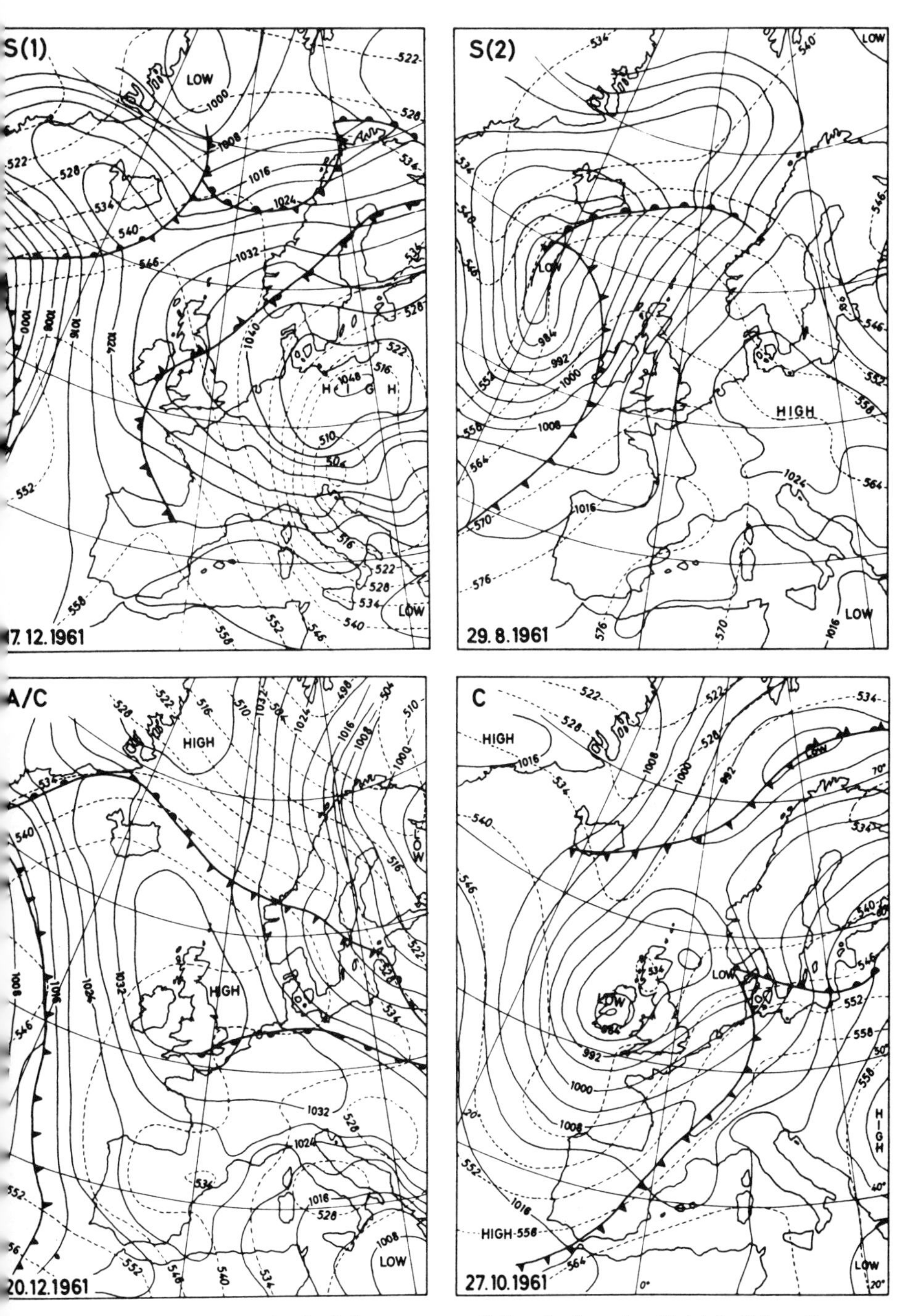

Fig. 3.17 The principal airflow types of Lamb for the British Isles (from Gregory, 1964). Solid lines = MSL isobars (mb) at 0600 hrs GMT and dashed lines = 1,000–500 mb thickness (geopotential decametres) at 0001 hrs GMT.

type transitions frequencies for the second period. The mean duration is between about one and two days although the westerly types had an average of 4·1 days for January 1910–30 (which decreased to 1·7 days for 1948–68!).

The Lamb catalogue provides considerable detail and for some purposes a more generalized summary is preferable. Simple numerical indices based on the catalogue have therefore been devised by Murray and Lewis (1966) as follows:

1 *Progression* (P index). Positive scores are given for each day of progressive synoptic types (NW, W and SW); easterly and meridional types are given negative scores. Non-directional types are scored according to the types on the preceding one to three days, ignoring unclassifiable days. The details of this may be found in Murray and Lewis' paper. The basic scoring is:

Type	*Score*
NW, W, SW	2
Unclassifiable	0
N, S	−1
NE, E, SE	−2

2 *Meridionality* (M and S indices). The basic scoring is:

Type	*Score*
N	−2
NW, NE	−1
W, E, A, Z, U	0
SW, SE	1
S	2

The meridionality (M) is the sum of the daily scores regardless of sign. The S index, which indicates the degree of northerly or southerly flow, is the algebraic sum of the same scores.

3 *Cyclonicity* (C Index). This measures the general cyclonic or anticyclonic character of any time period. The scores are allocated as follows:

Type	*Score*
Non-directional anticyclonic	−2
Directional anticyclonic	−1
All unspecified categories	0
Directional cyclonic	1
Non-directional cyclonic	2

Applications of these indices are discussed in Chapter 5.A and 5.B.

A classification of similar order to the schemes of Levick and Lamb has been developed by Van Loon (1958) for Tristan da Cunha. Ten types were identified based on isobaric curvature and the position of troughs and ridges:

1 Ridge to west	6 Anticyclonic zonal flow
2 Ridge to east	7 Low to south
3 Trough to east	8 Low to north
4 Trough to west	9 High to south
5 Cyclonic zonal flow	10 High

This scheme is intermediate between the static pressure pattern and the airflow approach. The study is of interest because, in addition to statistics of type frequency, type duration and weather characteristics, Van Loon determined the most common sequences of types in each season. Fig. 3.18 illustrates the summer sequences based on data for 1950–5.

More clearly in the regional airflow category is the classification developed for Labrador–Ungava by Barry (1959, 1960). The categories were based on the general trajectory (subjectively assessed) of the airflow near the surface for two or three days prior to its passage over the area. The types are:

1 *Anticyclonic* (*A*). MSL pressure $\geqslant$ 1016 mb with neutral or anticyclonic curvature, subdivided according to the origin of the air or termed indeterminate.
2 *Northwesterly* (*NW*). Northwesterly airflow from the Arctic Archipelago or Keewatin (example given in fig. 3.19(*a*)).
3 *Northwesterly modified* (*NWm*). Northwesterly flow over Keewatin becoming zonal over Hudson Bay and Labrador–Ungava.
4 *Westerly* (*W*) *zonal flow*. This may occur as warm sector air with a depression on the Arctic front or ahead of the warm front and/or to the rear of a depression on the Maritime front (example given in fig. 3.19(*b*)).
5 *Westerly modified* (*Wm*). The air swings southeastward from western into central North America and later recurves northeastward across the area (example given in fig. 3.19(*c*)).
6 *Southwesterly* (*SW*). Predominantly as warm sector air in lows moving northeastward from the Great Lakes area (example given in fig. 3.19(*c*)).
7 *Southerly* (*S*). Associated with depressions moving northward along the east coast of North America.
8 *Easterly* (*E*). East coast depressions often become slow-moving near Newfoundland and especially if they continue northward easterly flow affects much of Labrador–Ungava (example given in fig. 3.19(*d*)).

9 *Northeasterly* (*NE*). Sometimes similar to E, but the flow is dominated by a high over Greenland–Baffin Bay.

10 *Northerly* (*N*). With a trough over the Labrador Sea–Davis Strait area. It may be the final stage of a S–E–N sequence.

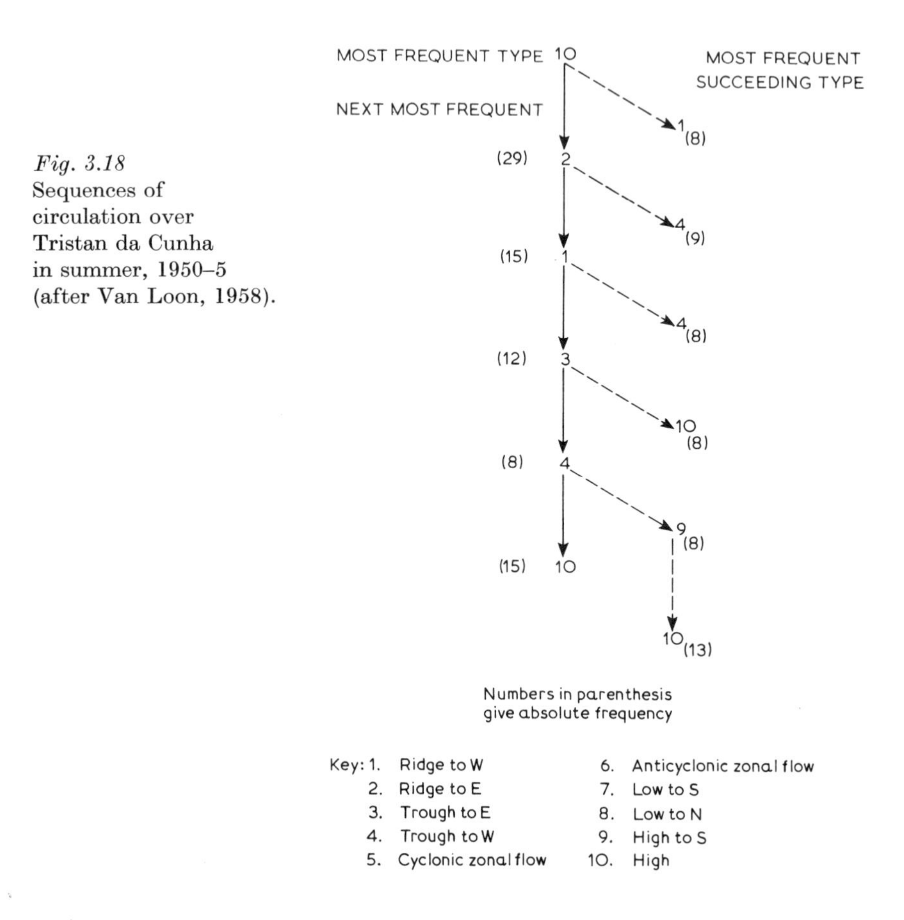

Fig. 3.18 Sequences of circulation over Tristan da Cunha in summer, 1950–5 (after Van Loon, 1958).

A synoptic classification for Poland prepared by Litynski (1970) has identical types to the Lamb scheme (table 3.7) but the definitions are objective. They are based on indices of zonal and meridional flow for the sector 40–65°N, 35°E–0° and on the MSL pressure over Poland. The eight directional classes are determined by dividing the frequency distributions of daily means of the three indices into three equal classes. Cyclonic, neutral and anticyclonic subtypes are also distinguished. This type of approach merits further investigation in other areas. The mean duration of type spells was 5·7 days for the period 1960–6 based on January, April, July and October which corresponds well with the *Grosswetter*. The weather associated with each circulation type in these months is described by Litynski in terms of tercile categories of daily precipitation and daily mean temperature at Warsaw.

Interestingly, it was found that sector indices of zonal and meridional flow for Labrador–Ungava (Barry, 1959) were not closely related to the subjectively determined airflow types except in the cases of NW and E types. Nevertheless, these subjective types did serve quite adequately in discriminating weather conditions. In this study the sector indices probably represented a larger scale of motion than the flow types.

Surface–upper air links

Apart from the *Grosswetter* system, few synoptic classifications have dealt satisfactorily with the problem of surface–upper air relationships. Many classifications consider only the surface pressure or airflow pattern. Mosiño (1964) uses a scheme of ten surface types and fifteen upper air (700 and 500 mb) types for Mexico, but the two sets are not directly related to one another. Gazzola and Montalto (1960) similarly used eleven patterns at the surface and at the 500 mb level in a study of winter precipitation in Italy. Subsequently, Gazzola (1969) identified twenty-two patterns at each level based on the period December 1956–November 1965. The types were in ten groups as follows:

A Low pressure over Italy and the Tyrrhenian Sea (four types).
B Low over the western Mediterranean (three types).
C Low over the central Mediterranean (three types).
D Secondary depression crossing Italy from the north (two types).
E Secondary depression crossing Italy from the east (two types).
F High pressure over Europe influencing Italy (two types).
G High pressure over the western Mediterranean influencing Italy (two types).
H High over Italy (two types).
I Straight isobars over Italy.
L Ill-defined situation.

After the classification had been applied to the surface and the 500 mb level separately, contingency tables (see Chapter 4.A, pp. 219–21) of their mutual occurrence were prepared on a seasonal basis. In all there are 484 possible intersections but it was found that 211 (43·6%) did not occur and a further 94 (19·4%) occurred three times or less. Sixty-nine per cent of the days in the eleven year period were accounted for by only 13·4% of the paired types. The most important paired types (500 mb first) were:

Aa – all year, summer minimum
Ba – especially winter
Bb – especially winter

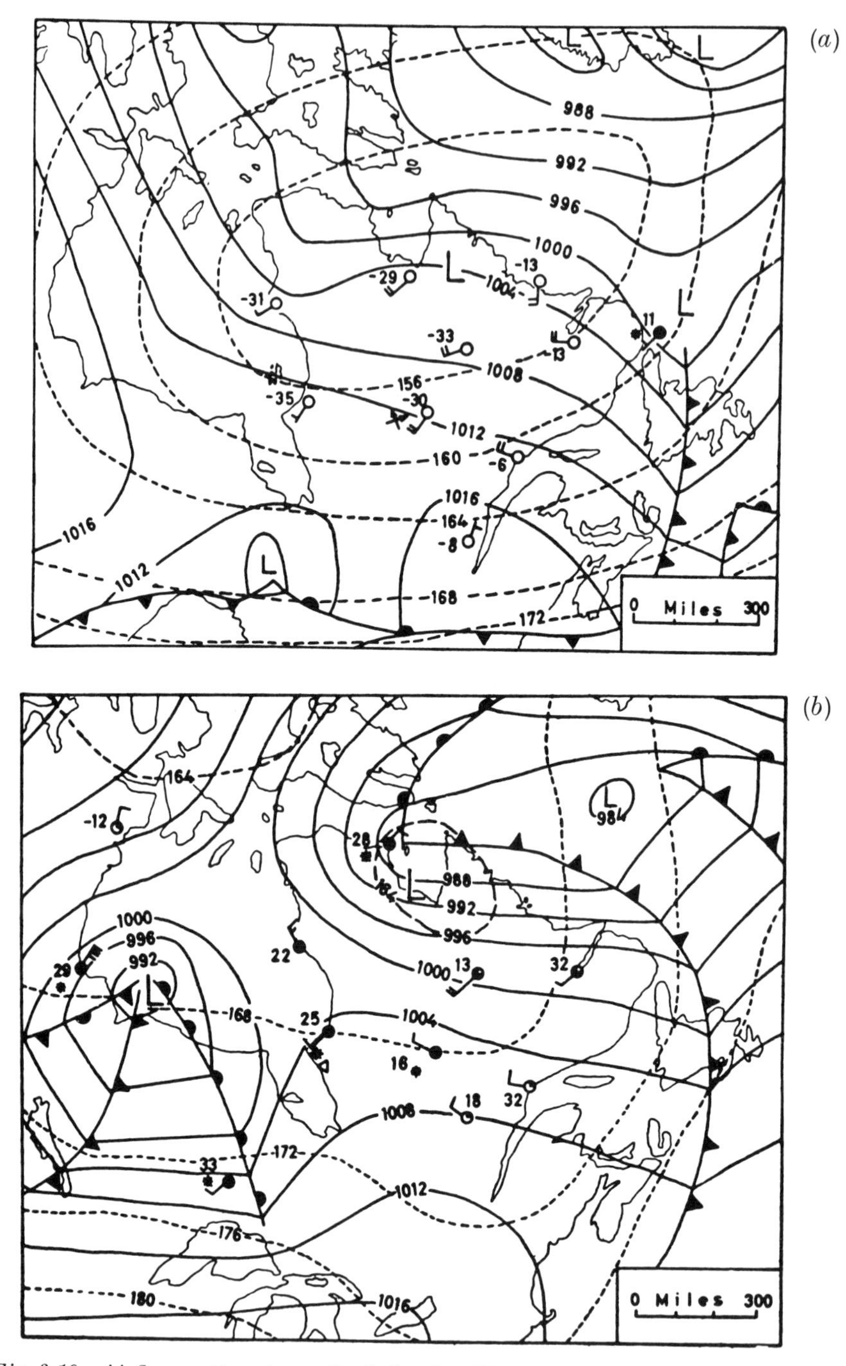

Fig. 3.19 Airflow pattern types for Labrador–Ungava (from Barry, 1960). Maps show surface isobars, 500 mb contours and weather data (temperature in °F).

(*a*) Northwesterly (NW) type, 28 January 1957; MSL 0000 GMT isobars (solid); 500 mb 0300 GMT contours (dashed) – hundreds of feet.

(*b*) Westerly (W) type, 6 November 1957; MSL and 500 mb 1200 GMT.

(*c*) Westerly modified (Wm) and Southwesterly (SW) types, 19 December 1957, and 500 mb 1200 GMT.

(*d*) Easterly (E) type, 18 February 1957; MSL 1200 GMT, 500 mb 1500 GMT.

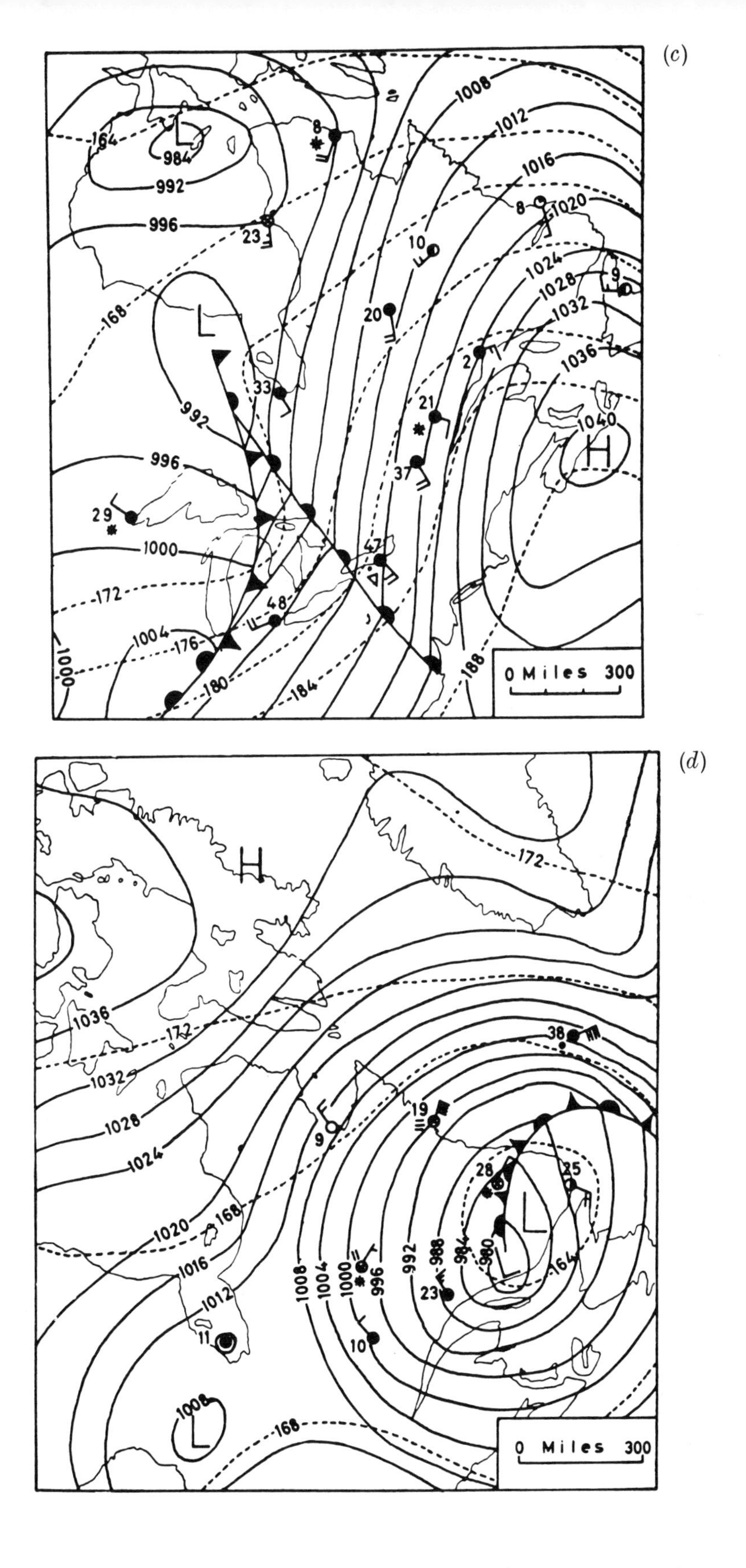
(c)
1008
1012
1016
1020
1024
1028
1032
1036
1040
984
992
996
1000
1004
164
168
172
176
180
184
188
L
H
0 Miles 300
(d)
H
172
1036
1032
1028
1024
1020
1016
1012
1008
168
1004
1000
996
992
988
984
980
164
L
0 Miles 300

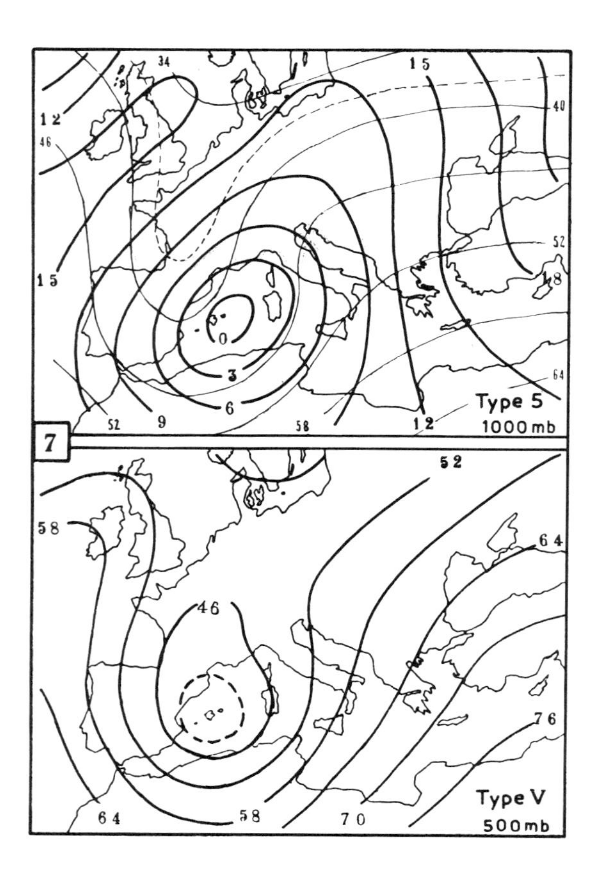
7
Type 5
1000 mb
Type V
500 mb

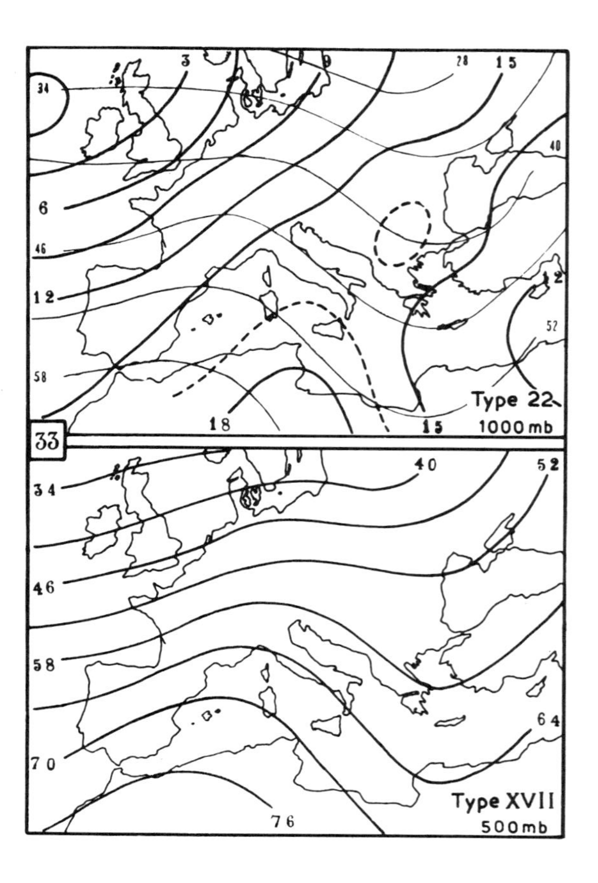
33
Type 22
1000 mb
Type XVII
500 mb

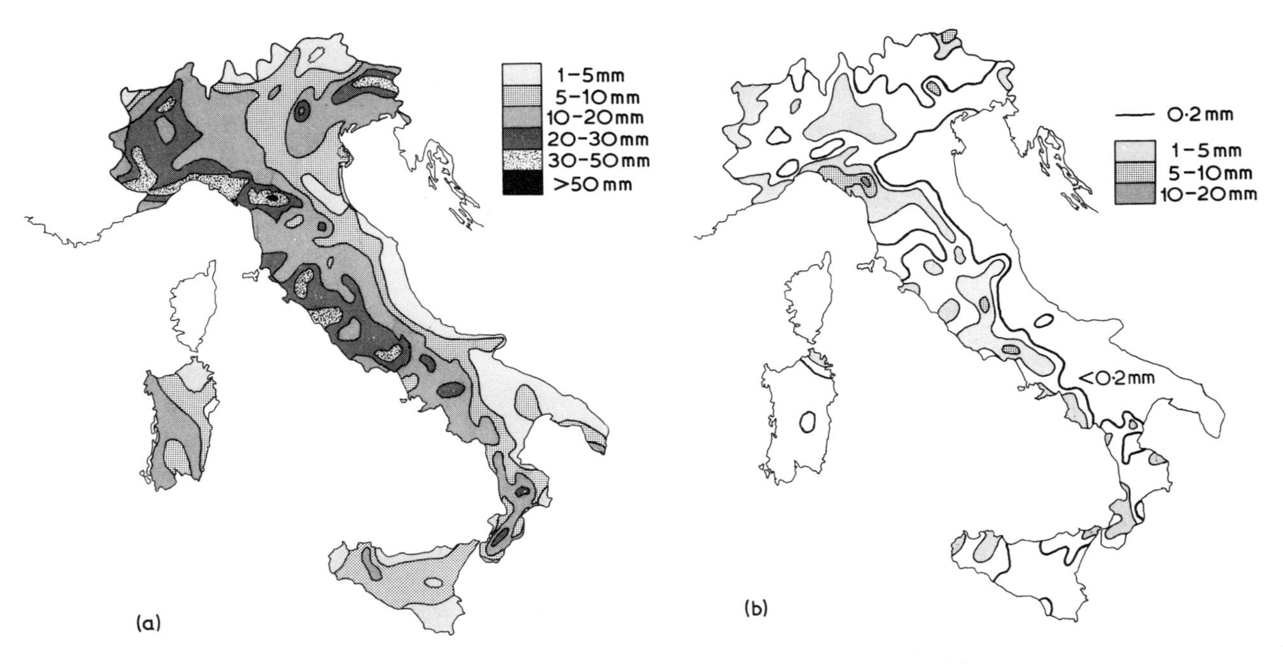

Fig. 3.20 Twenty-four hour precipitation for ten situations of two airflow types over Italy in winter (from Gazzola, 1969). (*a*) Bb, (*b*) Gl. Solid line = 1,000 mb contours and fine lines = 1,000–500 mb thickness (above), 500 mb contours (below).

Da – all year
Ge – all year
Gh – late summer especially
Gl – all year, summer maximum
Il – late summer especially

Fig. 3.20 illustrates the mean 1000 mb and 500 mb contours, 1000–500 mb thickness and twenty-four-hour precipitation (based on 450 stations) for the paired types Bb and Gl for ten situations in winter. This particular approach is instructive in several respects. First, the upper air–surface patterns are not unnecessarily constrained by an oversimplified scheme relating to both. Instead, the two-way frequency analysis serves to indicate which couplings are the most important. Secondly, Gazzola calculates the map patterns for the specimen ten situations of each paired type for the day of occurrence and for the following day. This evolutionary aspect of the synoptic climatology is rarely considered in such detail although it is extremely important in connection with forecasting applications.

The problem of a multiplicity of types resulting when the vertical structure is considered may be more apparent than real. Jenkinson (1957), in a discussion of the relationship between standard deviations of wind and contour height, states 'migratory pressure systems do pass considerable vertical coherency' and this is 'chiefly linked with major weather types'. This coherence is said to be evident up to the level of the tropopause. Jenkinson points out that in fact this vertical coherence of pressure systems is implicit in the characteristic disposition of the climatological mean fields in the troposphere.

One other approach to this question, that of using a unified system with surface and upper air fields, has been developed by Schüepp (1959) for the Alps. This is discussed below.

Evaluation of type classifications

Generally, it is not possible to make objective evaluations of the merit of particular classification schemes unless, for example, they are being tested with respect to forecasting success. It is useful to determine standard deviations, as well as means, for climatic parameters investigated with reference to classification categories. Belasco (1952) found, for example, that a number of the air mass types he distinguished were not statistically distinguishable from one another in terms of certain parameters at all levels in the troposphere. Barry (1963) has similarly shown that not all of the Lamb types are distinct from one another at individual stations in central southern England in every month of the year. For example, the mean daily maximum and minimum temperatures for 1921–50 at Southampton show no significant difference between the following types:

January	*July*
N–E	E–S
NW–S	
W–CW	

More complete analyses of the variance within and between the type categories for various parts of the country would be of interest as a measure of the homogeneity of these categories as 'weather types'.

Such considerations provide some evidence as to the merits of particular classifications schemes, but apart from perhaps suggesting a need to group certain types together, they are of limited usefulness. For the European Alps, however, there is a variety of classification catalogues available and this unique circumstance has enabled climatologists to

Table 3.8 Synoptic classifications developed for the Alps

Reference	Cadez (1957)	Mertz (1957)	Lauscher (1954, 1958)	Gressel (1959)	Schüepp (1957)	Schüepp (1959, 1968)
Area	Ljubljana	Alps	E. Alps	Alps	Alps	Alps
Basis	Airflow direction near surface	500 mb flow	Modified *Grosswetter*	*Wetterlagen*	*Wetterlagen*	*Witterungslagen*
No. of types	7	19	(17) 19	23	121	33
Catalogue period	1942–7	1950–2	1946→	1946–57		1955–67

compare the effectiveness of different proposed schemes. Table 3.8 summarizes the classifications which have been specially developed for this area. In addition, the *Grosswetterlagen* of Baur/Hess–Brezowsky have been applied to the Alps although, as pointed out by Willfarth (1959), there is the problem of the scale of these classification categories compared with the Alps and the fact that this central European based system takes little account of Mediterranean influences which are important in the Alps.

The classification proposed by Cadez (1957), and exemplified with reference to Ljubljana, is considered to be applicable to any point location. It is based on the direction of airflow in the lower troposphere (advective types – W, N, E and S) together with anticyclonic and cyclonic patterns and the addition of a convective type. Essentially, this compares with the earliest approaches to airflow classification and it need not be discussed further. The Mertz (1957) classification is somewhat similar except that it refers to the 500 mb level. He uses the eight compass point directions, subdivided according to dominantly cyclonic or anticyclonic curvature, plus cyclone, anticyclone and col situations, giving nineteen categories. This scheme has not been applied beyond the three-year period he investigated.

More extensive use has been made of the other four schemes listed in table 3.8 Schüepp provides considerable discussion of the rationale of his methodology. The 1957 classification relates to a regional division of the Alps (fig. 3.21). It is based firstly on the flow patterns at 1000 and

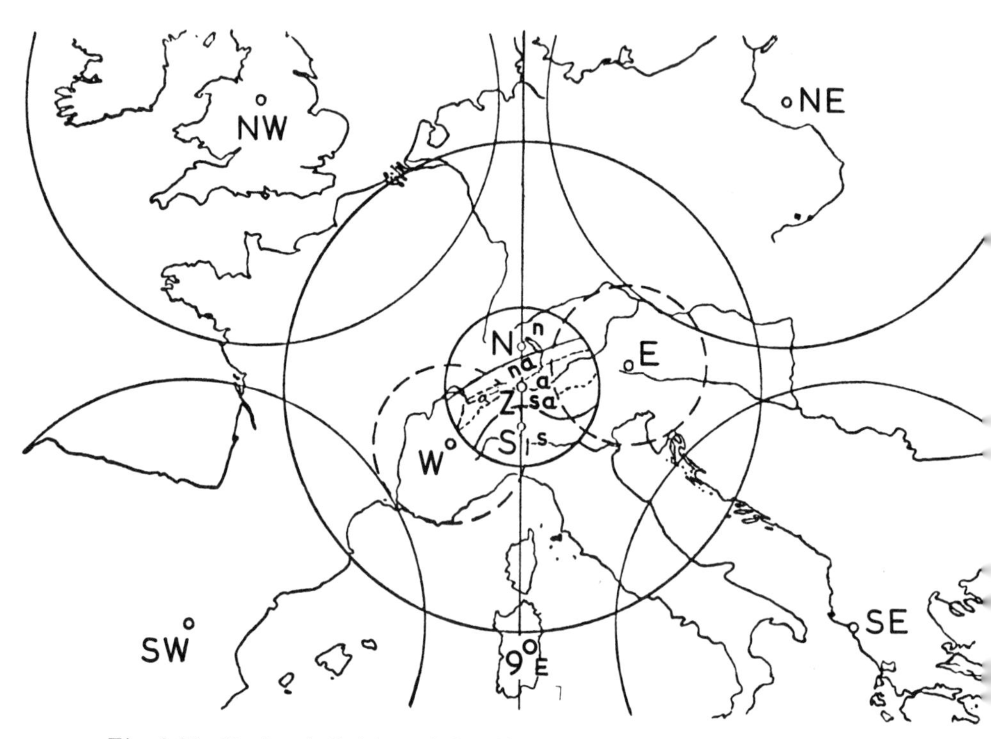

Fig. 3.21 Regional division of the Alps used in Schüepp's 1957 classification (after Schüepp, 1957 and reference sectors for airflow).

500 mb with eleven types at each level, giving 121 categories. The way in which these are combined into *Wetterlagen* is shown in fig. 3.22. These include types with more or less uniform flow at one level, five categories with strong winds – bise (a cold, dry northeast wind similar to the mistral), SE, föhn, W and stau (flow blocked by the mountains). Secondly, it was proposed that the 1000–500 mb thickness be considered with respect to each of the eight directional and three non-directional *Wetterlagen*. Ten categories of thickness were also to be differentiated according to the occurrence or non-occurrence of a frontal zone.

Gressel (1959) considered this scheme to be unnecessarily detailed and noted that it did not provide a perspective on the evolution of weather processes. In other words, it remained essentially a static picture of the weather map. His own approach will be discussed after we have looked at Schüepp's later work which largely meets these objections.

In 1959 Schüepp reported on a *Witterungslagen* system that was based on the principles set out by Lauscher (1954, 1958) with respect to the

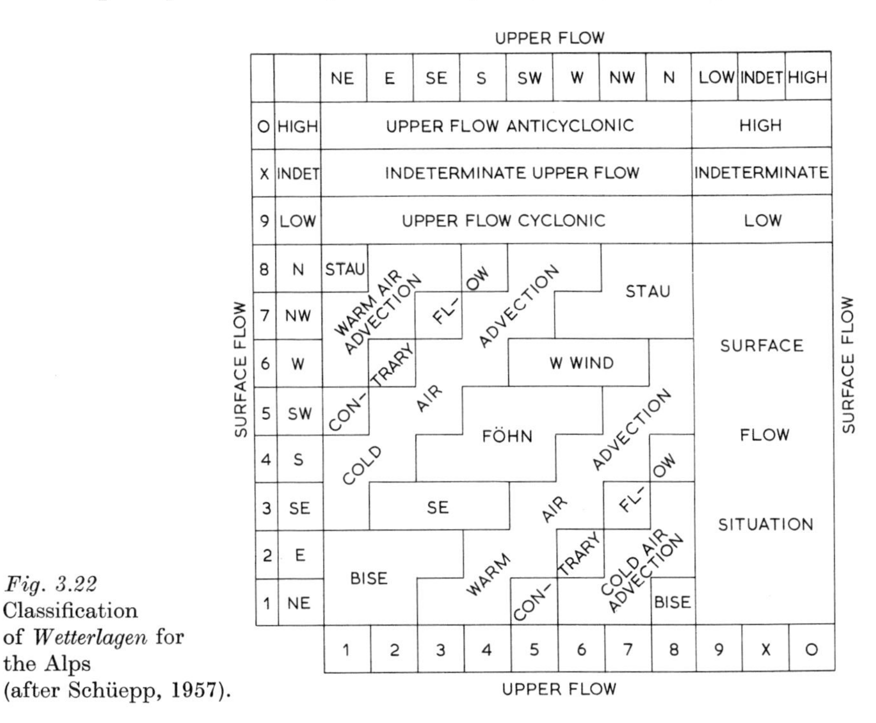

Fig. 3.22 Classification of *Wetterlagen* for the Alps (after Schüepp, 1957).

low-level airflow direction and the related upper-level circulation pattern. Lauscher's final scheme, as reported by Fliri (1962), is:

A High pressure situations
1 High (H)
2 Cold high between two low pressure systems ('*Zwischenhoch*') (h)
3 Zonal ridge from the Azores high to a cell over east Europe (Hz)

B High pressure margin situations
1 High in the east and southeast (HE)
2 High over Fennoscandia (HF)
3 High in the north (HN)
4 High in the northwest, over the British Isles (HNW)

C Extensive upper airflow situations
1 N (North and northeast)
2 NW

3 W (the *Grosswetter* differentiation of northern and southern W situations is not significant for the eastern Alps)
4 SW
5 S (south and southeast)

D Low pressure margin situations
1 Low over the British Isles (TB)
2 Low over the western Mediterranean (TwM)
3 Low over the central Mediterranean–Balkans area (TS)
4 Low over the southwest (TSW)

E Low pressure situations
1 Low over central Europe (TK)
2 Trough (TR)
3 Low northward from the Adriatic Sea to Poland (Vb)

Schüepp (1959) modified Lauscher's scheme principally with respect to the curvature of the airflow and by the inclusion of explicit reference to the vertical motion. Table 3.9 summarizes the basis of the classification and shows the coding of the thirty-three types. In accordance with his earlier classification he indicates the association between certain weather characteristics and the advective categories:

NE (01, 11, 91) E (02, 12, 92) SE (03, 33, 93)	Bise on the north side of the Alps
S (04, 44, 94) SW (05, 55, 95)	Föhn situations
NW (07, 77, 97) N (08, 88, 98)	Stau situations

This classification is now used routinely by the Swiss Meteorological Institute and a catalogue is available for 1955–67 (Schüepp, 1968).

Finally, Gressel (1954, 1959) has also considered the special problems of developing a classification system for the Alps. His approach is to eliminate the static aspects of the *Grosswetter* approach and to emphasize the airflow. The details of anticyclonic and cyclonic types, as in Schüepp's convective group, are regarded as being relatively unimportant. The twenty-three categories are:

1	H	All types of high pressure situation where there is no uniform large-scale airflow over the Alps.
2	T	All types of low pressure situation where there is no uniform large-scale airflow over the Alps.
3	Tsa	Low moving from the Gulf of Genoa area to the Balkans and influencing alpine weather.

Table 3.9 Witterungslagen *scheme developed for the Alps (Schüepp, 1959)*

A. BASIC WITTERUNGSLAGEN CLASSES

I Anticyclonic II Indeterminate pressure pattern III Cyclonic	*Witterungslagen* without a uniform basic flow (convective situation)
IV Low pressure margin V Airflow VI High pressure margin	*Witterungslagen* with a uniform basic flow at the surface (advective situations)

B. TYPES

	CONVECTIVE			ADVECTIVE							
	ANTICYCLONIC			ANTICYCLONE-MARGIN							
Anticyclonic subsidence	*High*	*Zonal ridge*	*Cold high*	*Anticyclonic advection situations*							
	H	Hr	h	Na	NWa	Wa	SWa	Sa	SEa	Ea	NEa
	00	0×	09	08	07	06	05	04	03	02	01
	INDETERMINATE			ADVECTION SITUATIONS							
Indeterminate or changing between cyclonic and anticyclonic motion	*Upper high*	*Indeter-minate*	*Upper low*	*Indeterminate or changing between cyclonic and anticyclonic*							
	Fa	F	Fz	N	NW	W	SW	S	SE	E	NE
	×0	××	×9	88	77	66	55	44	33	22	11
	CYCLONIC			CYCLONE-MARGIN							
Cyclonic ascending air	*Secon-dary low*	*Trough*	*Low*	*Cyclonic advection situations*							
	E	Tr	T	Nz	NWz	Wz	SWz	Sz	SEz	Ez	NEz
	90	9×	99	98	97	96	95	94	93	92	91

4 N **5 NE** **6 E** **7 SE** **8 S** **9 SW** **10 W** **11 NW**	**General airflow (cyclonic or anticyclonic) without any effective pressure centre in the alpine area.**
12 N + Tsa 13 NE + Tsa 14 E + Tsa 15 SE + Tsa 16 S + Tsa 17 W + Tsa 18 NW + Tsa	Large-scale **airflow** (cyclonic or anticyclonic) with a pressure centre to the south of the Alps affecting the weather of the region.

19 H+Tsa	High pressure influence from the north and a low pressure area south of the Alps also affecting the weather of the region.
20 W+Vb 21 NW+Vb	Large-scale airflow (cyclonic or anticyclonic) with a low on the Vb track affecting the weather of the Alps.
22 G	Slack pressure pattern with no significant centre or airflow affecting the Alps.
23 K	Complex weather patterns with either two large-scale airflows, or with high pressure influence affecting one part of the Alps and a large-scale airflow in another part.

The effectiveness of the classifications of Lauscher, Gressel and Schüepp have been compared by Fliri (1965) using records of cloudiness and temperature departures at Säntis, Switzerland for 1955–63. Only types represented by at least thirty days were examined. Fig. 3.23 shows the overall standard deviation for each classification system as a percentage of the standard deviation for the climatological series for both

Table 3.10 Standard deviation of temperature departures from normal (°C) at Säntis, 1955–63, for different synoptic classification systems (Fliri, 1965)

	Winter	*Spring*	*Summer*	*Autumn*
1955–63 series value	5·2	4·6	4·2	4·2
Schüepp 1957	4·1	3·2	3·0	3·1
Schüepp 1959	4·2	3·4	3·3	3·5
Gressel	4·1	3·5	3·1	3·1
Lauscher	4·5	3·8	3·4	3·3

parameters. The largest variability occurs with the Lauscher system which has the fewest categories and the smallest variability with the 121 type Schüepp system. However, there is no appreciable reduction beyond about thirty types. Table 3.10 shows the standard deviations of temperature departures in the four systems. As indicated here and in fig. 3.23 there are seasonal variations in the relative order of the four systems. The deviations are lowest for cloudiness in autumn and winter, but the temperature deviations in the Lauscher and 121-type Schüepp schemes are large in winter. More detailed consideration of the 121-type system showed that the standard deviation for cloudiness is reduced (and therefore this parameter is more closely specified) by the upper air types for high pressure and by the surface types for low pressure, except in the summer season. In summer both sets give similar deviations while smaller deviations occur for upper airflow from between east and south-

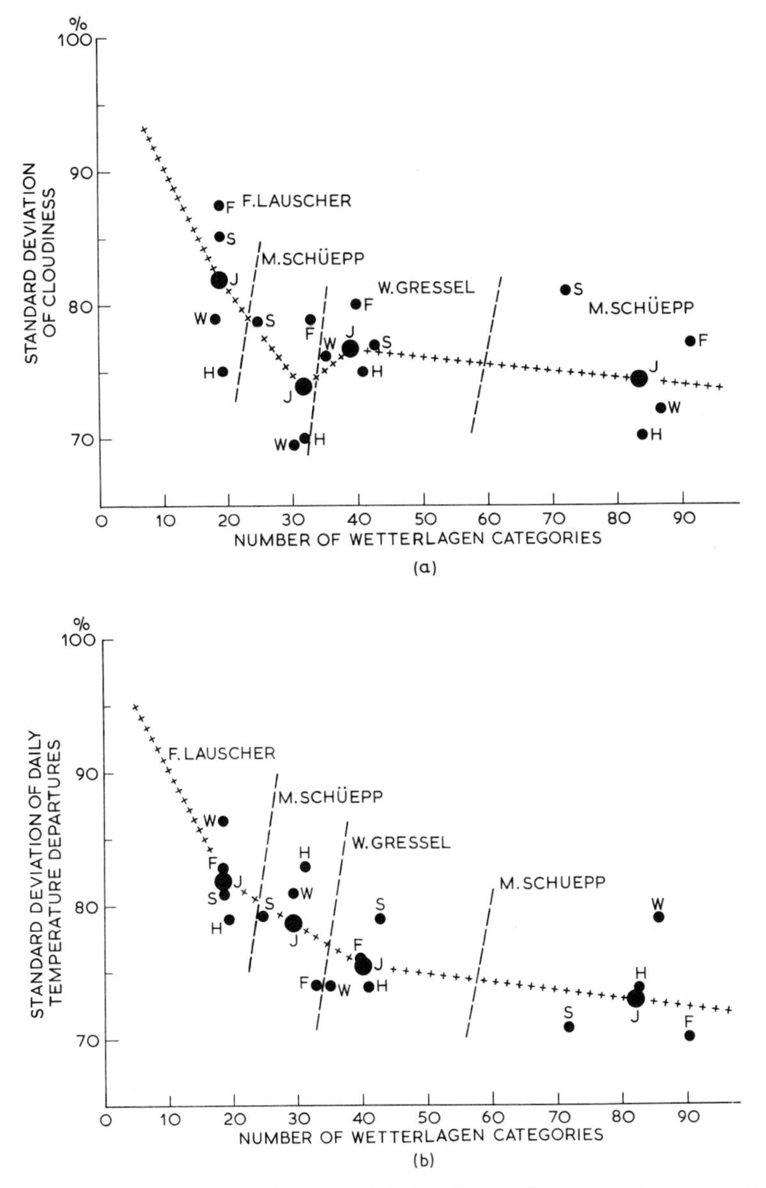

Fig. 3.23 The standard deviation of (*a*) cloudiness, (*b*) temperature departures, for each classification system for the Alps, in each season and for the annual mean, as a percentage of the standard deviation for the 1955–63 series. W = winter, S = summer, F = spring, H = autumn, J (line) = annual mean (from Fliri, 1965).

west and for surface airflows between west and northeast. In the other seasons the surface types of airflow are invariably better.

A further consideration is the mean duration of weather situations. Fliri notes that for 1955–63 this value is 1·2 days for the Schüepp 121-type scheme, 1·4 days for Gressel's, 1·7 for Lauscher's and 4·6 for Schüepp's thirty-three *Witterungslagen*. The effect of different time and space scale emphases is striking. Grunow (1965) has made a more limited comparison of the Hess–Brezowsky, Lauscher, Gressel and Schüepp schemes with respect to sunshine data at Hohenpeissenberg, Bavaria. By calculating frequency distributions of daily sunshine duration for each weather category on a seasonal basis it is possible to examine the occurrence of heterogeneous distributions. He concludes that the Schüepp thirty-three-type scheme is the most satisfactory.

4 HEMISPHERIC CIRCULATION PATTERNS

At the hemispheric scale it is clear that the classification scheme must be very generalized. This applies not only to the space dimension; meridional or zonal motion of the atmosphere, and perturbations within the flow, need to be considered over a time span of weeks rather than days. Subjective and objective procedures have been used to develop classifications and although, surprisingly, some of the former types are the more recent they provide a convenient starting point.

Subjective studies

As an illustration of the difficulties, it is worth looking first at an unsuccessful attempt. Lamb (1951), in an unpublished study of five years (1899, 1911, 1923, 1935 and 1948), found that only about half of the days could be grouped into nine hemispheric types. Six types were geometrical patterns and three geographical:

	Frequency (%)
1 Zonal pattern (similar to the climatological mean). 2 Modified zonal – winter (subtropical high sloping ENE from the Atlantic towards northeast Asia).	8·8
3 Double-structure zonal (subtropical and subpolar high pressure belts; mid-latitude and polar lows).	4·3
4 Slanting axis circulation (tracks of highs and lows orientated SW–NE).	11·2
5 Northerly and southerly meridional patterns (in surface winds) predominant.	9·4
6 Slanting inverted pattern (high about 60°N with axes NW–SE; zonal depression track in low latitudes and minor lows moving from SE to NW in mid-latitudes).	5·1

7 Inverted pattern (zonal easterlies in mid-latitudes). 4·2
8 Monsoonal winter pattern (continental highs, oceanic lows). 6·4
9 Monsoonal summer pattern (continental lows, oceanic highs). 4·7

It was found that 44% of the remaining patterns were 'chaotic' and 2% 'weak'. Consequently, the project was not continued.

A scheme relating surface and 500 mb patterns was proposed about the same time by Baur (1963). The schematic types, illustrated in fig. 3.24, refer to a sector of approximately 45° longitude and are in two groups – zonal and meridional.

A Zonal

1 Zonal flow, west–east movement of highs and lows.
2 Frontal zones and jet stream south of normal position. Cold polar anticyclones to the north. This is mainly a winter type.
3 Frontal zones and jet stream north of normal position. Extensive subtropical anticyclone. This is mainly a summer type.
4 Warm anticyclone in the north, cold low in the south. At the surface the high is warm in summer, cold in winter.

B Meridional

5 Subtropical category. Warm high to east. Depressions moving northward.
6 Polar category. Warm high to west. Frontal depressions less developed than in 5.
7 Meridional blocking anticyclone.
8 Meridional cold polar trough.

The idea is significant in the present context in that a comparable scheme was developed independently by Kletter (1959, 1962), as discussed in the following section.

Russian meteorologists have been interested in large-scale classifications of the circulation since the formulation of Multanovskii's views (see pp. 119–20) on natural synoptic periods. Dzerdzeevski (1968), for example, points out that the study of atmospheric processes over a sector with artificial boundaries weakens the genetic element in a classification since there will be differences in the recent synoptic history, outside the selected area, of the pressure patterns included in any particular type category. Kletter (1962) noted earlier that a hemispheric approach is necessary from the point of view of extended range forecasting since the 'influence field' (Charney, 1949) for twenty-four-hour forecasts at a point in middle latitudes covers an area of 80° longitude by 55° latitude. This problem is graphically demonstrated in fig. 3.25. Dzerdzeevski advocates the recognition of *elementary circulation*

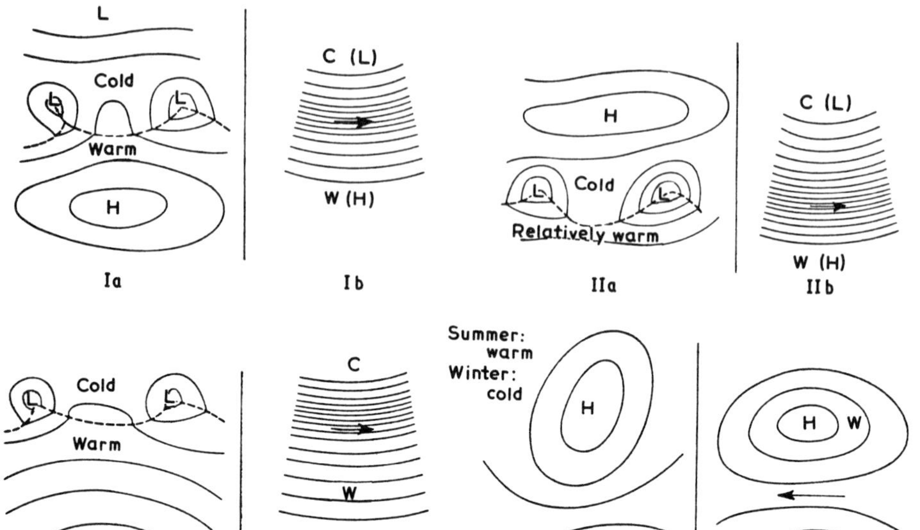

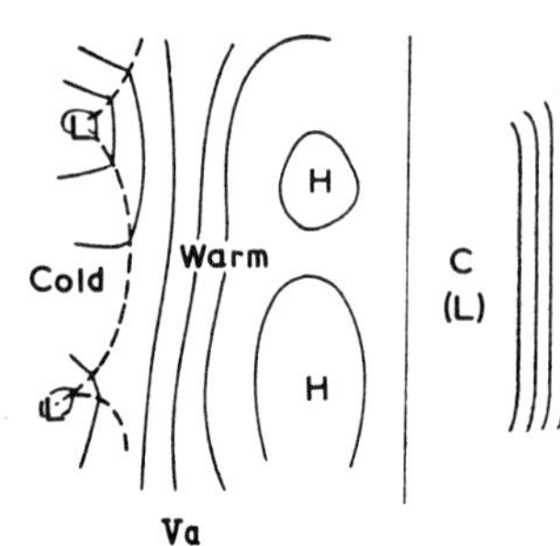

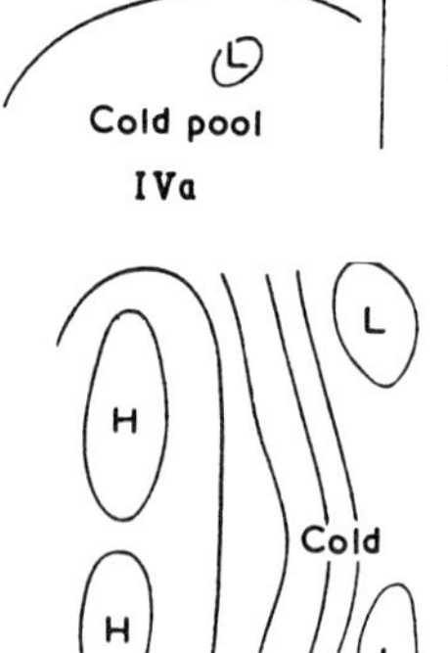

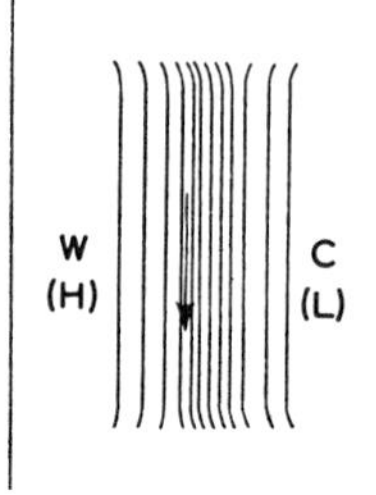

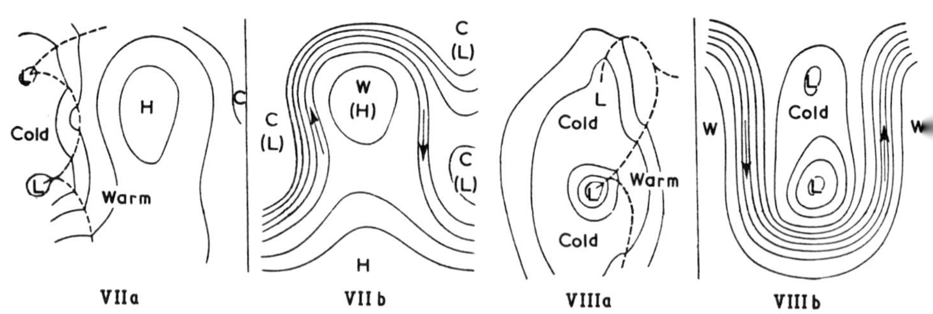

Fig. 3.24 Eight large-scale circulation types with schematic surface pressure and temperature distribution and 500 mb contours (after Baur, 1963). (*a*) zonal group, (*b*) meridional group.

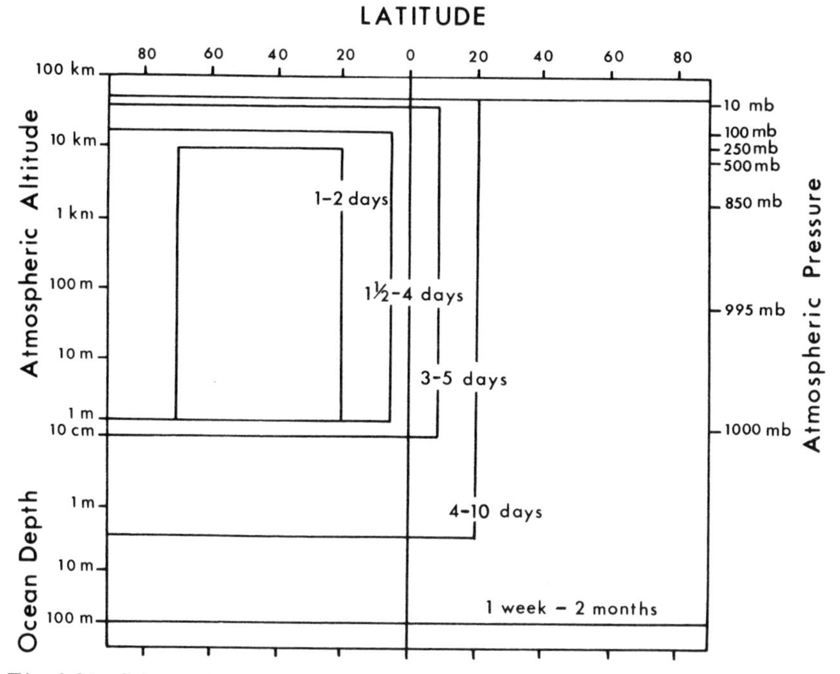

Fig. 3.25 Schematic of the data required for forecasts on different time scales in mid-latitudes (after Barber, 1970).

mechanisms (ECM) which operate over a short time interval and govern the circulation pattern over an entire hemisphere. The basis for distinguishing discrete categories of atmospheric process is derived from the following two postulates:

1 The hemispheric circulation is determined by a finite number of characteristic circulation mechanisms. The number of these mechanisms is small over short time periods when the insolation and the properties of the earth's surface are constant, but their characteristics differ greatly with season.
2 The features of each circulation mechanism (including its spatial organization) persist longer than the time scale of synoptic processes. Thus, the hemispheric circulation is a real macroprocess, not a chance combination of independent synoptic processes. Individual disturbances and fronts are regarded as 'noise'.

Although not formulated as precisely as this, the basic approach is to be found in Vangengeim's classification of elementary synoptic processes (1935, 1946). He distinguished three basic types of circulation in the northern hemisphere – westerly (W), easterly (E), and meridional (C).[1]

[1] The meridional type is C in Russian which is S in English. However, in English publications Russian authors themselves refer to the C type.

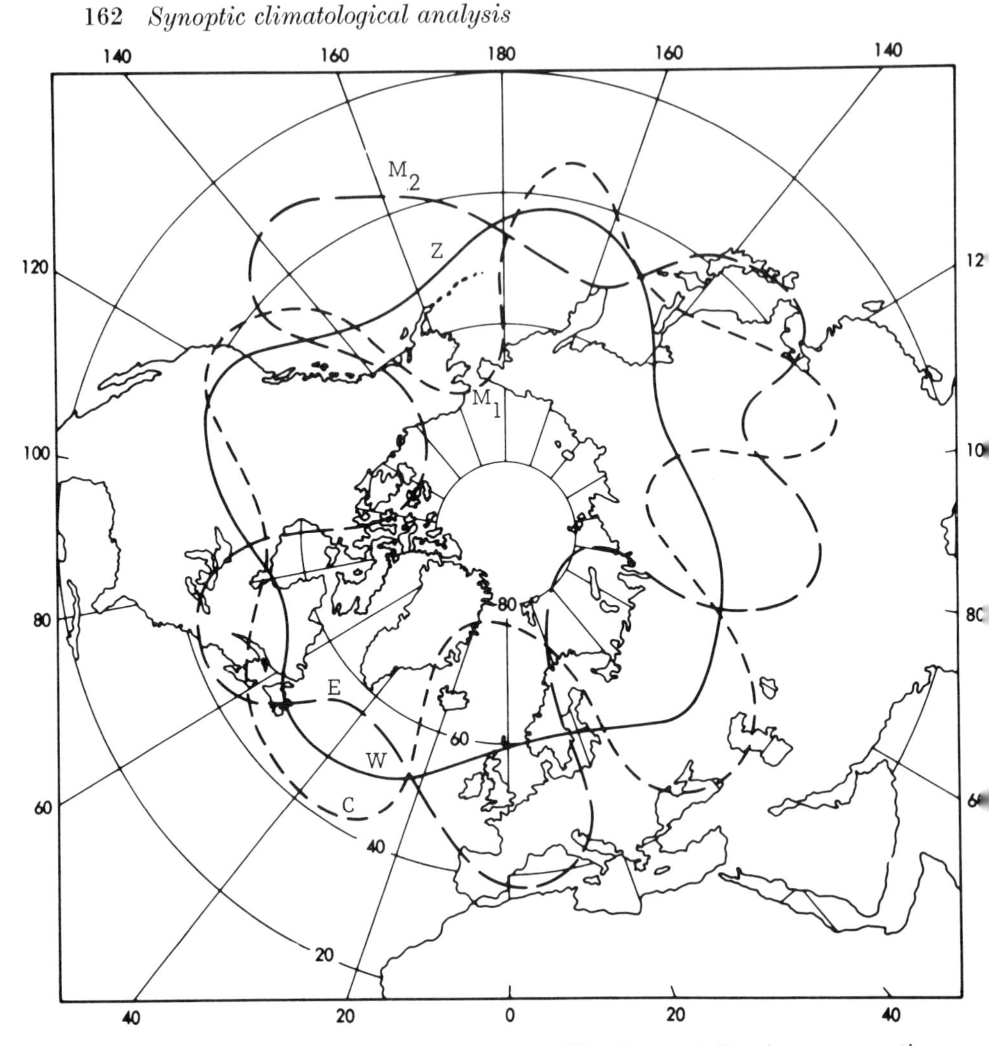

Fig. 3.26 500 mb trough positions for the W_z, C_{M1} and E_{M2} 'macrosynoptic processes' of Vangengeim (from Girs, 1966).

As illustrated in fig. 3.26, each type is characterized by a particular distribution of depressions and anticyclones at the surface and by an organization of the major long wave pattern. W type refers to patterns with essentially zonal movement of small-amplitude waves, nine subtypes are distinguished on the basis of the latitude of the subtropical anticyclone cells. With C type (seven subtypes) there are large-amplitude, stationary waves. The subpolar lows are shallow, there is a well-developed high, and the subtropical anticyclone cells are split and displaced northward. Pressure over Europe and western North America is low. E type (ten subtypes) is comparable with C, but the troughs are in different locations. The subpolar lows are well developed, the Siberian high is weaker and further west than with C, the Azores and Pacific anticyclone cells are also west of their normal position, and there are stationary highs over Europe and western North America. Subsequently, Girs (1948, 1960) showed that in the North America–Pacific Ocean sector two meridional patterns (M_1 and M_2) and one zonal type (Z) are the most significant, and he indicated that these may combine with W, E and C to give nine basic types. Table 3.11 indicates these and the annual

Table 3.11 Frequency of Vangengeim's types, 1900–57 (%)

	TYPE								
	W_Z	W_{M1}	W_{M2}	E_Z	E_{M1}	E_{M2}	C_Z	C_{M1}	C_{M2}
Annual frequency	9·9	7·5	9·1	14·3	11·1	9·0	10·1	8·9	10·1
		26·5			44·4			29·1	

Z = Zonal in the North Pacific sector.
M_1 = Surface anticyclone near the Aleutian Islands, lows to the north.
M_2 = Ridge from the Pacific high extending to western North America.

frequencies for 1900–57. Much attention has been given to the dominance of particular types during 'circulation epochs' since 1891. There has, for instance, been a marked decline in the frequency of W types and an increase of E and C since the 1930s. This aspect is discussed further in Chapter 5.C.

The parallel studies of Dzerdzeevski (1945, 1962, 1966, 1968; Dzerdzeevski *et al.*, 1946) are concerned with the degree of organization of the hemispheric flow. Cyclone and anticyclone tracks at the 700 or 500 mb level are used as an indicator of the main mid-tropospheric steering currents (Dzerdzeevski and Monin, 1954). It is emphasized that charts averaged over several days provide the best view of the various types. Special attention is given to polar intrusions and associated blocking in the westerlies in six hemispheric sectors of 50–60°. The four major patterns are shown in fig. 3.27. They are:

1 A zonal ring of cyclone tracks in high latitudes; two or three breakthroughs of southern cyclones (two types, five subtypes).
2 Interruption of zonality with a single polar intrusion; one to three breakthroughs of southern cyclones (five types, according to the sector of the intrusion, thirteen subtypes).
3 Northerly meridional motion with two, three or four polar intrusions (five types according to their location, twenty-one subtypes).
4 Southerly meridional; no polar intrusions and poleward movement into the Arctic in two to four locations (one type, two subtypes).

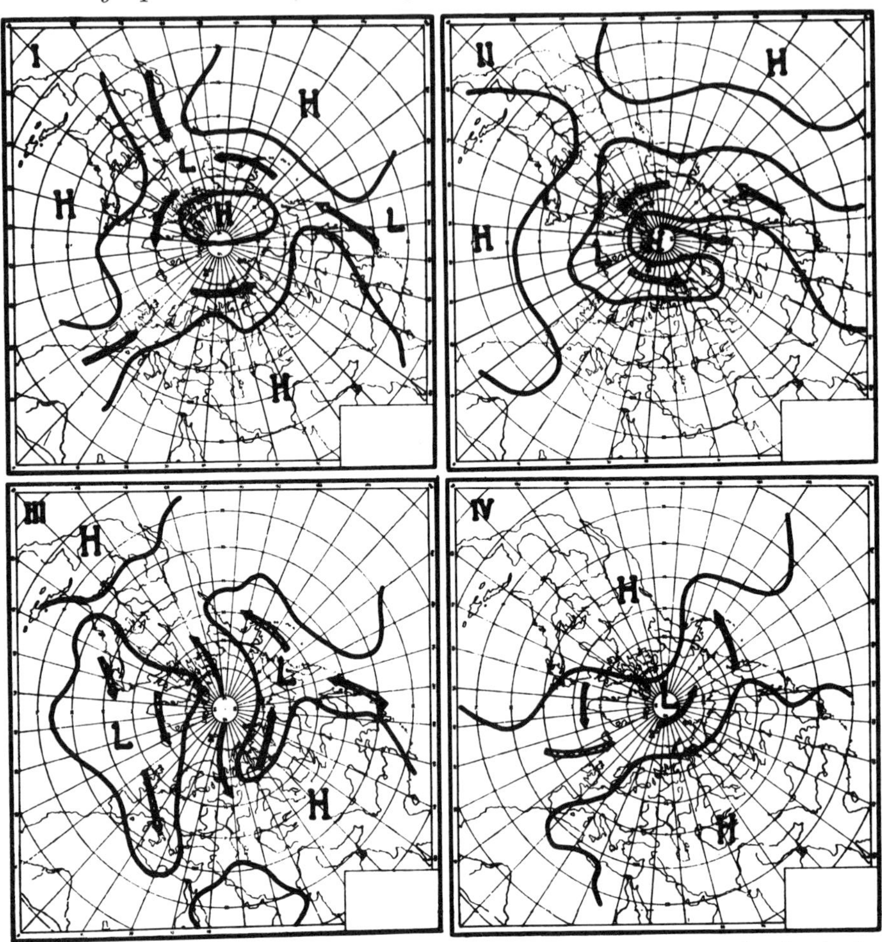

Fig. 3.27 The four basic types of 'elementary circulation mechanism' of Dzerdzeevski (1962). Solid arrows = cyclone tracks, open arrows = anticyclone tracks.

The patterns for the forty-one subtypes are illustrated in Dzerdzeevski (1970).

A similar approach is evident in recent Soviet studies of the southern hemisphere (Davidova, 1967). The term 'synoptic process' is used to refer to the movement of pressure systems over a two to three day period. The classification refers to the three southern oceans from 20°S to the coast of Antarctica. The types, between five and seven for each ocean, refer primarily to the zonal or meridional character of the circulation at MSL. Davidova shows that meridional patterns of circulation are dominant in all three oceans; in the South Pacific meridional forms have an 81% annual frequency, with 94–8% in the winter months. Fig. 3.28 shows the most common patterns in the South Indian Ocean. In the South Pacific there is a fairly uniform distribution of type frequencies. The analysis refers to a six-year period (1956–68, January–June 1959, July–December 1962, and 1963–4).

The essential feature of all of this work is the recognition that there are world-wide interactions of the atmospheric circulation. As noted by Namias (1951), this derives from the interrelated behaviour of the centres of action, demonstrated by Rossby *et al.* (1939), and the fact that energy in the planetary waves propagates from one system to another at a speed which often exceeds the wind speed (Namias and Clapp, 1944). However, the Russian work is almost entirely empirical in its approach to circulation interactions. A sound physical basis for relating the *actual* mechanisms to Dzerdzeevski's ECM circulation types has yet to be developed, although analytical studies of vertical motion and kinetic energy distributions related to preselected type patterns have been carried out (Dzerdzeevski, 1968).

Objective approaches

The need for objective means of analysing and classifying circulations patterns on the hemispheric scale is obvious from the foregoing discussion. Attempts to devise representative indices for large-scale atmospheric processes have a long history. The best-known one – the *zonal index* – represents the strength of the circulation in a specified band around the hemisphere, or a sector of it. The concept, originally outlined by Clayton (1923), was applied to point values by Walker (1924) in studying the 'North Atlantic Oscillation' – the variations in strength of the Icelandic low and the Azores high pressure and, consequently, of the North Atlantic westerlies. Long-period values of this particular index, for pressure differences between Ponta Delgada in the Azores and Stykkisholm in Iceland, are given by Rudloff (1967).

Rossby *et al.* (1939; Rossby, 1940) extended the application of the hemispheric zonal index in his formulation of long-wave theory, where they related wave speed, c, to geostrophic zonal wind speed at the mean level of non-divergence (about 600 mb), U, wavelength, L, and to the poleward variation of the Coriolis parameter, β ($=\partial f/\partial y$). U and c are positive eastward:

$$c = U - \frac{\beta L^2}{4\pi^2}$$

The idea was quickly adopted in connection with extended-range forecasting experiments, and a study of sector and hemispheric indices was carried out by Allen *et al.* (1940). They found that the North American (130–80°W) and Atlantic (70°W–10°E) sectors were significantly correlated with one another and with the hemispheric index whereas the Asian and Pacific sector indices were more independent of one another

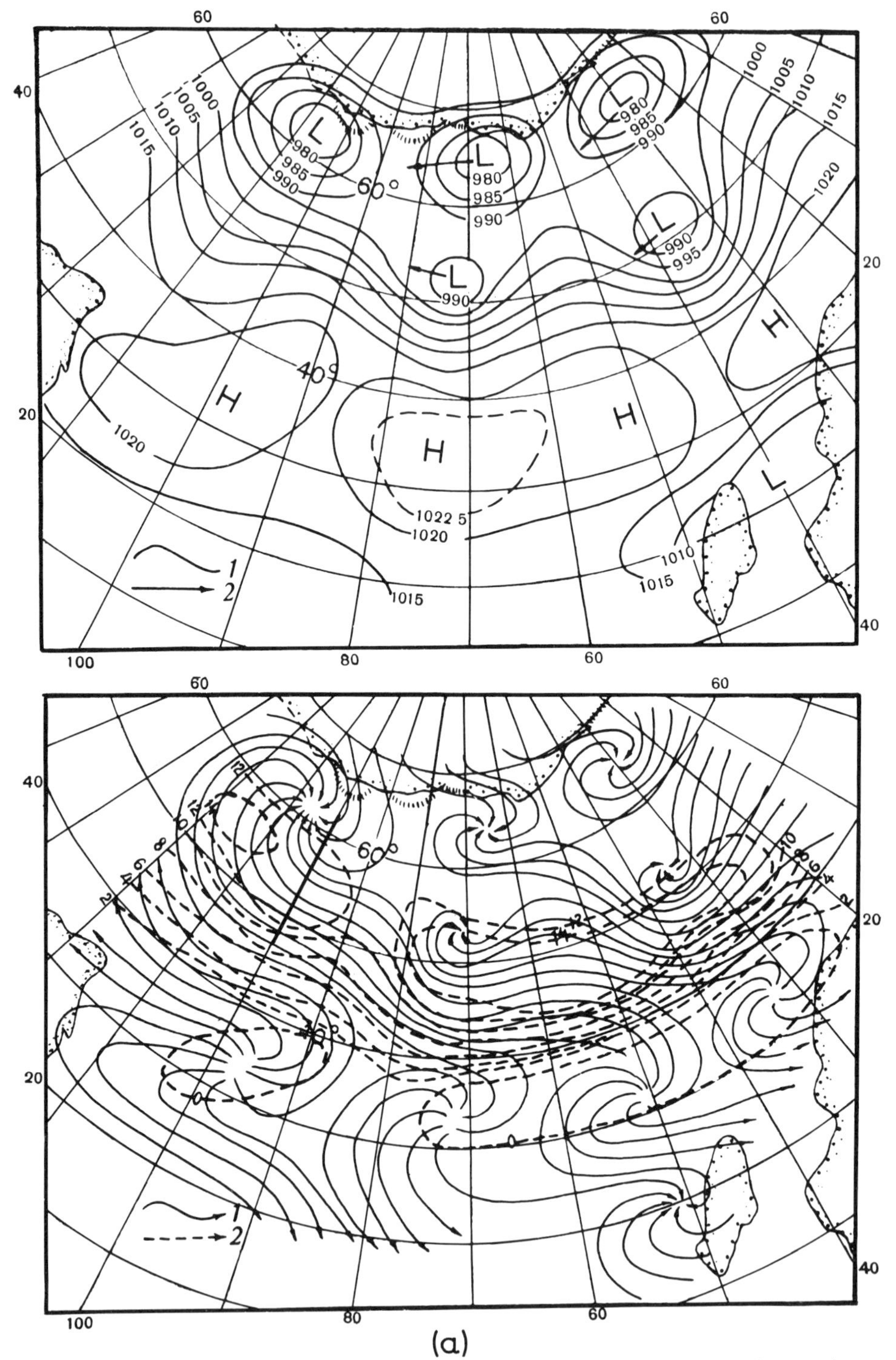

(a)

Fig. 3.28 The most frequent large-scale circulation patterns over the south Indian Ocean (from Davidova, 1967).
(*a*) Type 1, zonal with mid-latitude travelling depressions; 42% frequency. Above 1 = MSL isobars (mb), 2 = cyclone tracks. Below 1 = streamlines, 2 = isotachs (m s^{-1}).

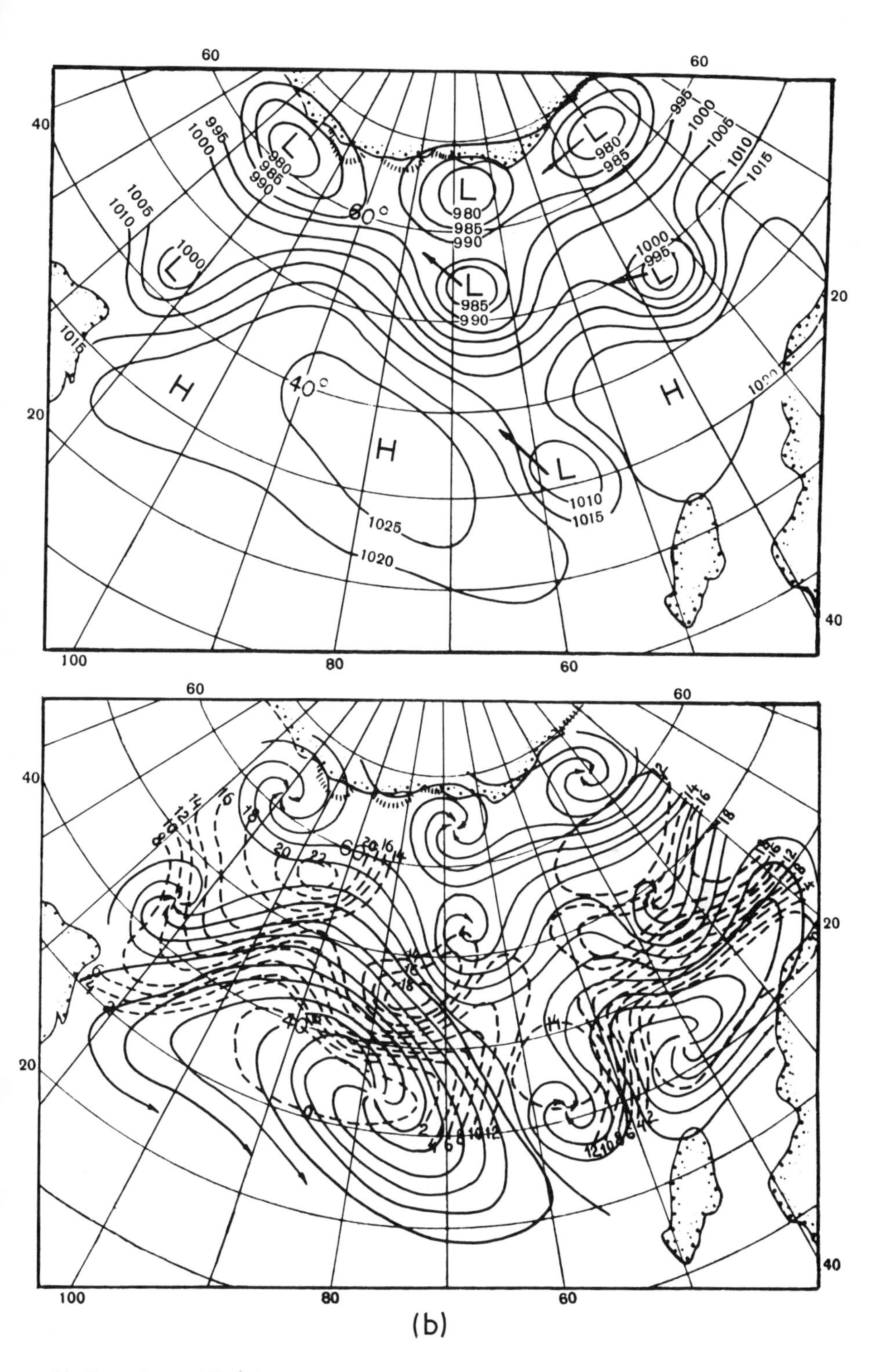

(*b*) Type 2, meridional with break-up of the subtropical high to the southeast of Madagascar; 26%.

and of the hemispheric value. Other standard indices in common use (Namias, 1947) are those for the 'polar easterlies' (55–70°N) and the subtropical easterlies (20–35°N). It is usual to calculate these, and the mid-latitude westerly index, for MSL and 700 mb between 5°E and 175°W only. Dinies (1968) has recently published monthly values of the index for 1899–1967 and also monthly and annual means.

It is a basic premise that the selected zone (35–55°N) centres on the position of the maximum westerlies. The MSL hemispheric westerlies are a maximum at about 50°N for the greater part of the year, but the speed of the westerlies varies greatly during the year, as shown in fig. 3.29. This seasonal trend is often removed from the index values. More

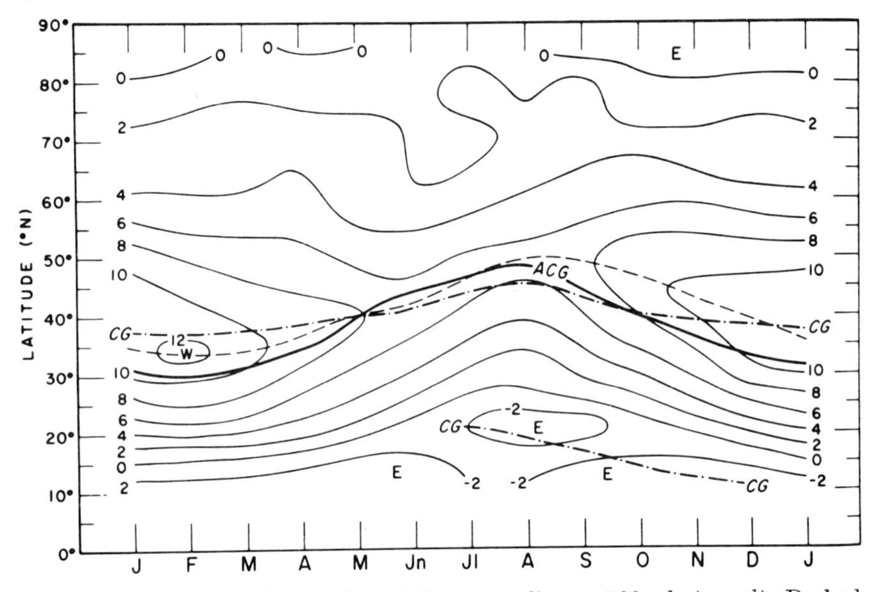

Fig. 3.29 The annual variation of the westerlies at 700 mb (m s^{-1}). Dashed line = maximum velocity; CG = latitude of maximum cyclogenesis; ACG = latitude of maximum anticyclogenesis (after Klein, 1958; from Palmen and Newton, 1969).

serious is the effect of longitudinal variations due to the asymmetry of the long-wave patterns in the zonal flow. Namias (1950) states that 'the total hemispheric index is not a primary parameter. . . . It is a derived quantity representing the degree of latitudinal organization of certain large-scale energy producing mechanisms in the middle and upper troposphere.' For example, a moderate index may represent a medium amplitude wave pattern or intense zonal flow in one sector and cellular patterns elsewhere. Another consideration is the occurrence of eccentricity of the circumpolar (Ferrel) vortex, especially in winter. Displacements of the circulation pole from the geographical pole amount to 10° latitude in the sector 160–170°W (La Seur, 1954). These shifts introduce large inconsistencies into a fixed-latitude index. To overcome this problem in studying the Ferrel vortex, La Seur advocated the use of a 'moving co-ordinate' system related to the position of the *circulation pole*, and showed that the corrected indices obtained in this way vary directly with the degree of displacement of the circumpolar vortex.

An alternative solution to this problem is to calculate the geostrophic west wind for 5° latitude belts. Such data, plotted against time, form a 'zonal profile'. Their application in determining northward or southward trends of relative west wind maxima and minima is illustrated by Riehl *et al.* (1952). However, these 'core' zones are not always easily identified, and during blocking spells especially trends in the profile may be absent for prolonged periods.

A different procedure, with a different end in view, is to determine indices for limited sectors. This was proposed by Flohn (1954, p. 223),

Table 3.12 Circulation characteristics of high and low index

Synoptic feature	*High index*	*Low index*
Icelandic and Aleutian lows	Deep, east of normal position	West of normal position
Subtropical highs	Strong, elongated east–west	Cellular, with north–south elongations
Movement of extra-tropical cyclones	Rapid, west–east	Large meridional components
Temperature gradients	Strong meridional gradient	Large east–west gradients

apparently overlooking Allen's investigations (Allen *et al.*, 1940). Examples may be found in the work of Putnins (1966, 1968) for the Greenland and Alaskan areas. Such indices provide a better measure of the fluctuations of individual centres of action, or of blocking activity, than the hemispheric index, but the movement of troughs and ridges past the sector boundaries can introduce 'false' oscillations into the index (Riehl *et al.*, 1954; Barry, 1959).

It has become usual to refer to the extremes of zonal flow as 'high index' and 'low index', and relationships between the intensity of the zonal circulation and large-scale synoptic patterns have been studied by Allen *et al.* (1940), Rossby and Willett (1948), Namias (1950) and Bradbury (1958). Table 3.12 summarizes the basic model proposed by Allen and his associates.

Bradbury examined the distribution of anticyclone and cyclone centres during high and low index circulations (fig. 3.30(*a*), (*b*)). It is evident from her analysis that, although there are fairly consistent changes

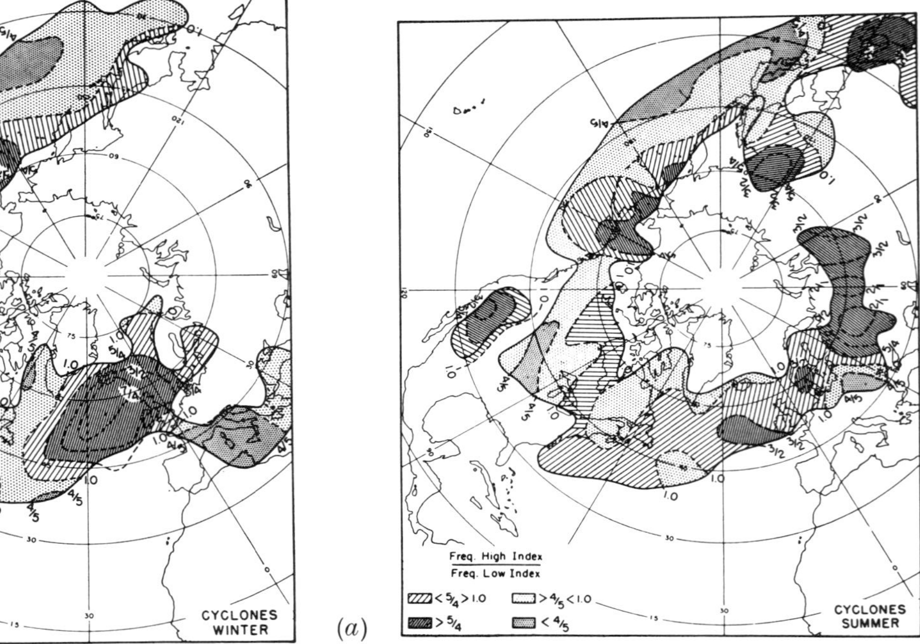
Freq. High Index
Freq. Low Index
< 5/4 > 1.0
> 4/5 < 1.0
> 5/4
< 4/5
CYCLONES
WINTER

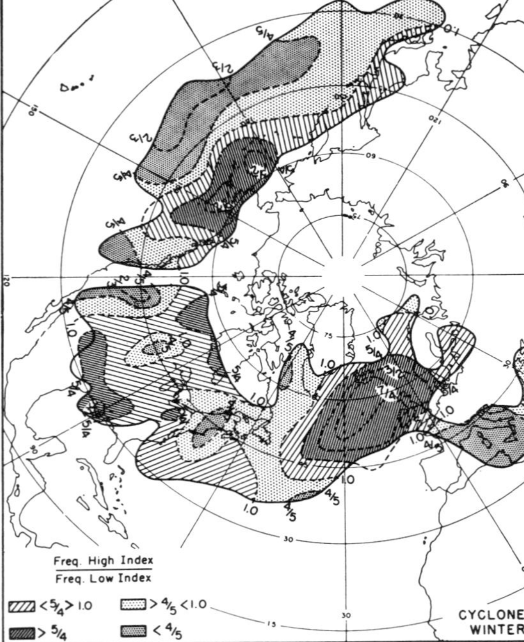
Freq. High Index
Freq. Low Index
< 5/4 > 1.0
> 4/5 < 1.0
> 5/4
< 4/5
CYCLONES
SUMMER

(a)

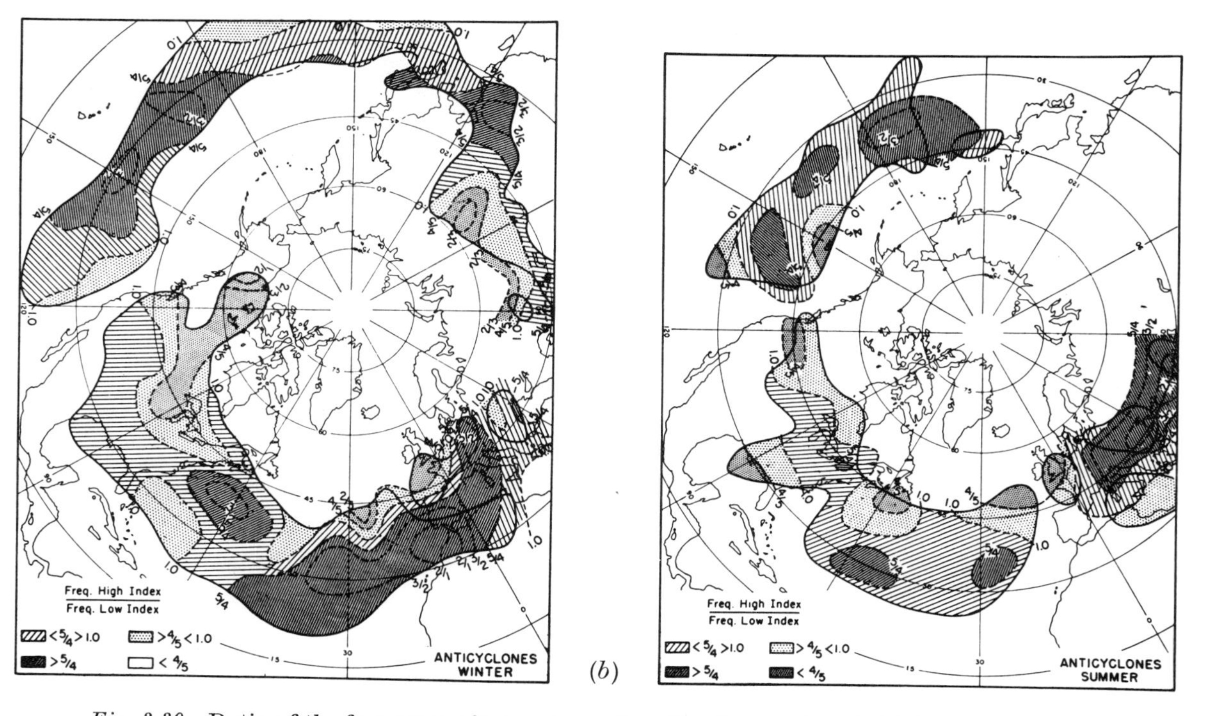

Fig. 3.30 Ratio of the frequency of (*a*) cyclone, (*b*) anticyclone centres, in high index/low index months. Line encloses area with $\geqslant 50$ centres/5° square (from Bradbury, 1958).

over the oceans, orographic and solenoidal effects exert an overriding influence on the patterns over continental areas. The contrast between the two stages in summer are generally weaker due to the lower mean zonal index.

A hemispheric classification of 500 mb pressure patterns related to the zonal index at 500 mb has been developed by Arai (1964). He takes into account La Seur's findings by incorporating the degree of eccentricity of the circulation into the classification. The nine types he distinguishes are:

Eccentricity	Index: *High*	*Normal*	*Low*
Large	H_1	N_1	L_1
Normal	H_2	N_2	L_2
Small	H_3	N_3	L_3

Subtypes are determined with reference to the amplitude of the third harmonic (i.e. wave number 3) for five-day mean heights of the 500 mb surface at 50°N. Combining large, normal and small anomalies of this amplitude with the basic types gives twenty-seven types in all.[1] The classification was used by Arai in an analysis of severe cold spells in northern Japan.

Recently workers in the Societ Union have tried to identify hemispheric circulation types (ECM) by indices for the 500 mb level and the 1000–500 mb thickness,[2] but Altykis (1966) considers that the criteria are not yet sufficiently discriminatory to obtain satisfactory results by this means.

Rossby and Willett (1948) identified fluctuations in the index with a time scale of about four to six weeks, during which the mid-latitude westerlies increase in strength, associated with an expansion of the circumpolar vortex, and then decrease. This feature is termed an 'index cycle' although Kletter (1962) showed that the 'characteristic' pattern is, in fact, rare. Moreover, statistical analysis by Julian (1966) has demonstrated that there is no recurring periodicity of this time scale in the circulation. Riehl *et al.* (1952) provided a more complex model by distinguishing three stages in both southward and northward drifts of relative wind maxima and minima in the zonal profile. Details of the associated synoptic patterns may be found in their report. The major importance of this scheme is its demonstration of *four* low index states, each characterized by distinct synoptic patterns at the surface and in the middle troposphere. It is worth noting that Namias (1950) had pointed out earlier that the terms high and low are misnomers. A high surface index tends to be associated with weak zonal flow aloft, and vice

[1] There is evidence (Nemoto and Kuboki, 1968) that the summer 500 mb field may be described by a smaller number of types.

[2] This is a 'solenoidal' index relating to the thermal wind.

versa. The level to which the terms refer always needs to be clearly stated.

In connection with an attempted classification of hemispheric circulation patterns, Kletter (1959, 1962) produced additional insight into the evolution of large-scale circulation characteristics. The classification, which was based on the Rossby wave formula, distinguished three possible solutions for C:

$$C > 0$$
$$C = 0 \quad \text{i.e.} \quad U \gtreqless \beta L^2/4\pi^2$$
$$C < 0$$

corresponding respectively to the following types:

1 Drift – a zonal, or westerly, flow
2 Wave – a meandering westerly flow
3 Vortex – cellular patterns.

Time-latitude profiles of the 850 mb zonal wind in the northern hemisphere were used to classify the groups. Subtypes were recognized of which the major ones are:

Type	Description	*Frequency Feb. 1955– Jan. 1958* (%)
1a '*Ringstromlage*'	– zonal flow between 40–65° latitude over most of the hemisphere. Main low centres north of 60°.	2
b '*Zonale Lage*'	– zonal flow; subdivided according to the degree of uniformity in sectors North American and Europe–North Atlantic, and to the latitude of the maximum flow.	27
c '*Trendlage*'	– a northward or southward trend of the wind maximum in one or both sectors.	20
2a Stationary, b Travelling, c Retrogressive	– planetary waves (at least three troughs or ridges).	23
3a Blocking pattern	– blocking anticyclone, with closed circulation and split westerlies. Subdivided according to location of the block.	11

		Frequency Feb. 1955– Jan. 1958 (%)
b '*Omega-lage*'	– blocking anticyclone linked to the subtropical high pressure belt.	9
c Cellular	– numerous weak circulations with no uniform flow over most of the hemisphere.	8

The mean duration of the subtypes ranges between four to eight days with an overall mean of 6·4 days. The figures correspond well with Baur's estimate (1936b) of a 5·5 day average duration for *Grosswetter* situations in central Europe.

Kletter analysed the successions of these patterns and determined the most frequent sequences for the four-year period. As fig. 3.31 shows, the

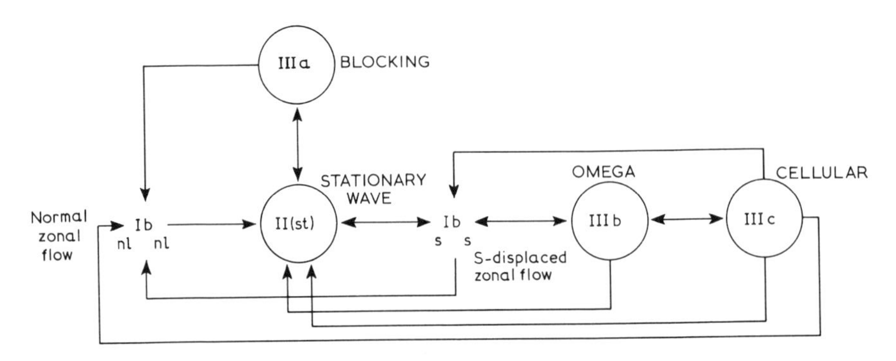

Fig. 3.31 A model of the most frequent circulation sequences in the northern hemisphere westerlies, 1955–9 (see text) (after Kletter, 1962).

evolution is much more complex than a simple index cycle. His results provide a valuable addition to the findings of Riehl *et al.* (1952).

The major shortcomings of Kletter's approach for climatological application is that the geographical aspects of the patterns are neglected. An ideal scheme would perhaps combine the methods of Kletter and Dzerdzeevski, thereby linking a circulation index to a classification incorporating a spatial description of the features. The significance of location in the case of blocking and a split jet-stream, for example, has been amply demonstrated by Wallén's study of summer conditions in Sweden (1953) and Butzer's work on Mediterranean weather (1961). Both show the need to distinguish between southerly and northerly meridional circulation types and their associated steering patterns.

We have already noted (p. 106) that the use of specification techniques provides a basis for describing succinctly the hemispheric or global circulation but that to proceed from this stage to classification presents difficulties which do not yet appear to have been overcome. Nevertheless, the eventual possibility of using one of these approaches in large-scale synoptic climatological studies should not be dismissed too quickly.

D Weather elements

1 COMPLEX CLIMATOLOGY

The idea of studying the most common combinations of the individual weather elements was originated in the Soviet Union in the 1920s by Federov (1927). He referred to the weather occurring at one location during a twenty-four-hour period (though daily mean values are used for the most part!) as a *weather case* and tried to group a number of cases into *weather types* by considering the joint occurrence of several weather elements. In general terms, these might be 'cold, snowy weather', 'hot humid weather', 'overcast, rainy weather', and so on. Since these constitute weather complexes, Federov termed his approach *complex climatology*. Climate was likened to a living organism – a complex of natural phenomena. Moreover, it was noted that weather elements, acting in combination, determine physiological processes in plants, whereas mean values of individual climatic parameters are of little significance. Thus, the practical application of climate data, especially in agricultural planning, was a major consideration.

The most detailed account of the method in English is given by Lydolph (1959) although early studies of human comfort zones in central America (Nichols, 1925; Switzer, 1925) were based on Federov's methods. For quantitative evaluations of the weather case, arbitrary limits must be selected for the various classes of each element. Inevitably a very large number of possible combinations results. For example, a system with ten classes for mean temperature, six for relative humidity, four for wind speed, and three for cloudiness, has 720 possible types. Lydolph notes that one scheme had 10^7 possible combinations! Some conditions may in reality be mutually exclusive; nevertheless, the method clearly has severe practical limitations. The usual procedure, apparently, is to combine the detailed types into a much smaller number of groups for specific purposes.

During the period 1921–69 Federov devised numerous classification schemes, some giving more weight to a set of primary parameters and distinguishing, for example, the diurnal occurrence as a significant item. Table 3.13 illustrates one such scheme. Many different tables and diagrams were also developed in order to present the results as effectively as possible. Fig. 3.32 illustrates a diagram for Moscow showing the annual regime. An elaborate study of this type has also been carried out for the northern Great Plains in winter months by Calef *et al.* (1957). 'Standard weather days' are defined with reference to maximum and minimum temperatures, precipitation, wind speed, relative humidity and cloudiness. The focus of interest here related to the suitability of conditions for military operations.

The general method has several obvious shortcomings. First, the selection of the significant weather elements and their class boundaries is quite arbitrary. Precipitation data, for example, are frequently ignored; second, frequency rather than magnitude of individual elements is emphasized; third, the time sequence of particular weather types is not considered. In the past the complexity of the classifications meant that the analysis was very laborious although computer procedures have overcome this practical difficulty.

Table 3.13 An example of a complex climatological classification scheme proposed by E. E. Federov

Primary	Wind, cloud
Secondary	Mean temperature, relative humidity
Tertiary	Diurnal temperature range, interdiurnal temperature change, time of precipitation

The method is still extensively used in the Soviet Union and eastern Europe (Pelzl, 1955; Petrovic, 1967; Wos, 1970). Climatic graphs for seventy-four stations in the U.S.S.R. are included in the *Physico-Geographic Atlas of the World* (Chubukov and Shvareva, 1964). In this case weather types are divided into three groups: frost-free days, days with air temperatures crossing the 0°C threshold, and frosty days. A further breakdown into sixteen categories is based on mean temperature, relative humidity, cloudiness, precipitation, and wind data. Such information on certain specified contingencies (see also Chapter 4A, pp. 219–22) may have practical value – Petrovic has studied spa towns in Czechoslovakia, for example – but it is doubtful that a graphical presentation is the most convenient means of summarization. Most of the work of this type concerns only point data. However, Quitt (1968) illustrates its application to an objective regionalization of climatic conditions in Czechoslovakia using fourteen parameters.

Complex climatology in Lydolph's view (1957) is not a research method. Certainly it provides little basis for further study along synoptic lines in its basic approach. Chubukov (1949) to some extent answered this criticism by incorporating synoptic data in the specification

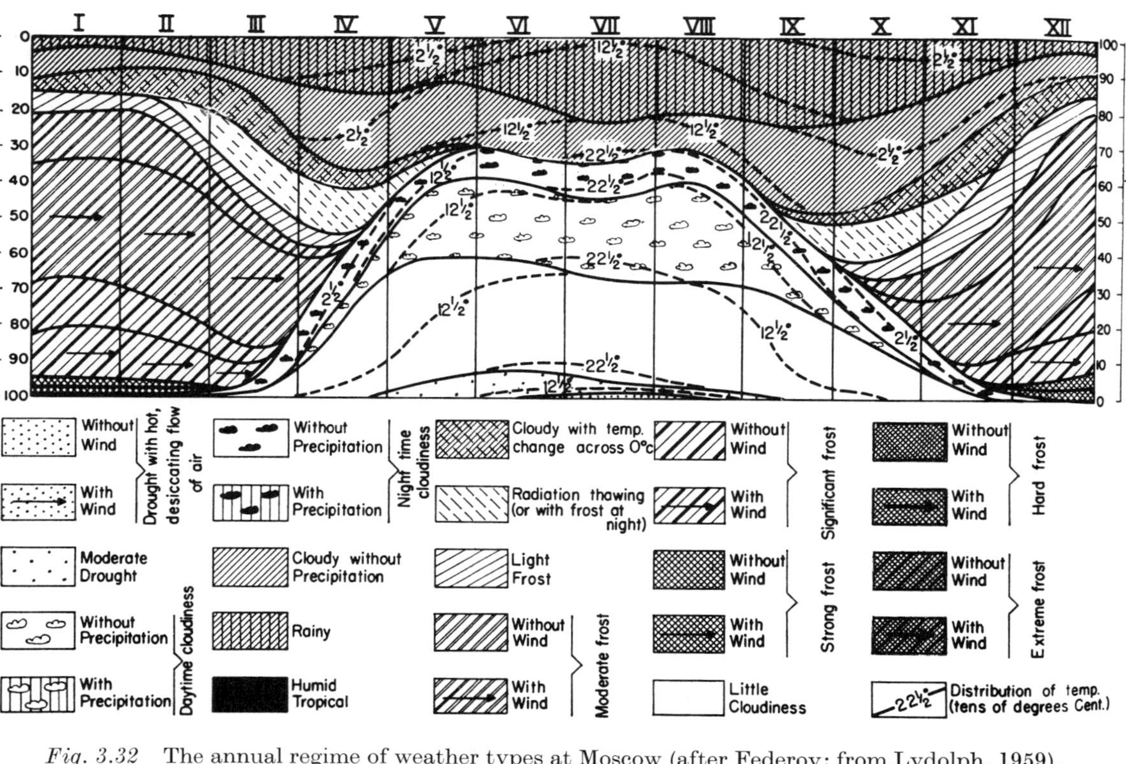

Fig. 3.32 The annual regime of weather types at Moscow (after Federov; from Lydolph, 1959).

of weather case, although this practice does not seem to have been emulated. Petrovic (1969) compares weather conditions at Strba Lake, Slovakia, as specified by a complex climatological classification and a synoptic one, but the latter scheme is evidently a Czech classification on a regional scale. In view of this, the detailed results that he presents are of very limited generality. Potapova (1968), on the other hand, uses the larger scale ECM catalogue of Dzerdzeevski to analyse January weather data at Moscow for 1899–1954 organized in the Federov–Chubukov type of schemes, but since only the major ECM types, rather than the subtypes, are used the results are too general. Overall it can safely be stated that, unless our interest is concerned simply with the absolute frequency of certain conditions, there are more fruitful approaches to climatological data than complex climatology.

2 AIR MASS CLIMATOLOGY

The literature on air masses is immense and it is only intended here to highlight a few of the more important studies in the field. By contrast, relatively little attention has been given to the underlying basic concepts and it is these that we wish to emphasize.

Definition

An ideal air mass is a *barotropic* fluid in which isobaric and isosteric (constant specific volume) surfaces do not intersect. Such a structure implies that the density (or temperature) field is a unique function of pressure and that the geostrophic wind remains constant with height. Air masses in this sense are largely confined to high and low latitudes. Less restrictively we can regard an air mass as a *quasi-barotropic fluid* (James, 1969). This implies the inclusion of cases where the isothermal and isobaric surfaces intersect, but isotherms are parallel to the isobars at a given pressure level so that the geostrophic speed (but not the direction) varies with height. This concept is examined further below. An air mass can be defined then as an extensive body of air with more or less uniform conditions of temperature, moisture content, and lapse rate, in a horizontal plane.

Classification

The earliest investigation of air mass properties appears to be that of Goldie (1923) although the classification and detailed characterization of air masses is due to Bergeron (1928, 1930), swiftly followed by Moese and Schinze (1929), Schinze (1932a, 1932b) and Willett (1933). Bergeron indicates two procedures that may be used in air mass classification. The first is to select the most conservative properties which reflect the net effect of the source area and subsequent modifications. Table 3.14

summarizes some of the main physical tracers of air mass movement and their properties.[1] Dew point temperature is commonly used in the case of surface data and wet-bulb potential temperature in the free atmosphere. This approach requires knowledge of the trajectory of the air and its vertical structure. Even so it does not provide a generic basic for climatological classification because air masses defined in this way are determined by local and regional conditions. In contrast, the second method emphasizes the most variable air mass characteristics such as

Table 3.14 Parameters used as air mass tracers (after Petterssen)

	Conservative with respect to temperature changes caused by:		
Element	*Dry adiabatic process*	*Saturated adiabatic process*	*Non-adiabatic process*
Dew point	Quasi	No	Yes
Specific humidity	Yes	No	Yes
Relative humidity	No	Yes	No
Potential temperature	Yes	No	No
Wet-bulb potential temperature	Quasi	Quasi	No

lapse rate. This in fact provides part of the framework for the usual subdivision of air mass types identified on a global scale.

This classification has the following bases:

A primary division relating to the geographical source area – arctic (A), polar (P), tropical (T).

A secondary division designating the maritime (m) or continental (c) nature of the source area.

A tertiary subdivision indicating the relative temperature and stability conditions. k denotes air moving toward a warmer surface and w toward a colder surface.

The major types are:

	Continental	*Maritime*
Arctic	cA	mA
Polar	cP	mP
Tropical	cT	mT

mA and mP represent modifications of cA and cP air, respectively, due to passage over ocean areas. cA and cP air masses originate in polar

[1] Gaseous tracers, such as ozone, and radioactive particles provide useful tracers in the upper atmosphere (Christie, 1965).

anticyclones (and not all workers distinguish between these groups) while cT and mT originate in the subtropical anticyclone cells, although Ozorai (1963) has shown that in March 1961 only twelve of forty-six air mass cases over Budapest derived from an anticyclonic source area in the preceding seven days! The term 'polar' is quite misleading in view of the frequent non-polar origin of such air, but this usage has become firmly established in the literature. Nevertheless, we should note that Russian meteorologists are more logical in preferring the designation 'temperate' (Borisov, 1965), while James (1970) suggests 'subpolar'. Certainly, one or other is preferable. Regarding the distinction made by Bergeron between arctic and polar air, it should be noted that this is often non-existent in winter over northern Canada and Eurasia. Moreover, samples of cA and cP air may comprise air of quite variable properties in the middle troposphere since polar anticyclones are normally shallow systems (see figs. 2.34 and 2.35). Flohn (1952) found that at Yakutsk, for example, air mass changes affected only the lowest 1500 metres in winter.

The most frequently encountered tertiary subtypes are cPk and mPw air. Occasionally the stability (s) or instability (u) of the air mass is also denoted in the codes – for example, mPws. The suffix will in general apply over limited regions, although stability changes may be produced in the free air as well as at the surface.

It is worth recording that the bioclimatologists Linke and Dinies (1930; Dinies, 1932) modified the Bergeron concept by stressing surface observations; they used the term *Luftkörper* rather than *Luftmassen*. Meteorologists rejected this approach in view of the great local variability within the friction layer (Scherhag, 1948, p. 137). However, even in the free atmosphere, air mass properties are far from conservative.

An air column does not move as a unit, despite statements to this effect in some textbooks. The instantaneous vertical structure reflects the disparate trajectories followed by the air parcels at each pressure level. For example, a speed difference of only 3 m s^{-1} between two levels, with no change of wind direction, will result in a 100 km separation between air parcels moving at these levels after four days. Fig. 3.33, after Ozorai (1963), illustrates the effect of vertical wind shear on air reaching Budapest.

In the twenty years following Bergeron's formulation of the concept there were numerous investigations of air mass frequencies and properties. Notable among these are the studies for North America by Willett (1933) and Showalter (1939), China (Tu, 1939; Lu, 1945), Japan (Arakawa, 1937), the British Isles (Belasco, 1952), the Mediterranean (Schamp, 1939; Serra, 1949), India (Roy, 1949), and the southern hemisphere (Gentilli, 1949; Taljaard, 1969). Petterssen was able to present a synthesis for the northern hemisphere in 1940, and Brunnschweiler (1957) subsequently tried to use this and other information as a basis for a genetic classification of climate. Nevertheless, global schemes of this type encounter serious difficulties when attempts are made to apply the air mass designations regionally.

In the 1930s to 1940s air mass and frontal analysis were inseparably linked in synoptic forecasting practice. The delimitation of different air masses was an essential preliminary to the location of frontal zones. This approach quite often led to practical difficulties in interpreting the synoptic pattern correctly and Godson (1950) tried to overcome some of these problems by providing a definitive analytical procedure for the identification of four air masses and three frontal zones over North America (see Chapter 2.A, pp. 58–9). Wet-bulb potential

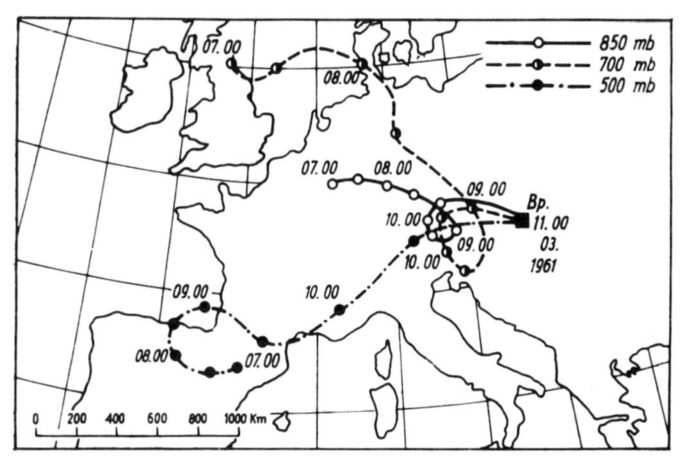

Fig. 3.33 The trajectories of air over four days at 850, 700 and 500 mb arriving at Budapest on 11 March 1961 (from Ozorai, 1963).

temperature is the primary working criterion (Harley, 1962). The system is logical, but there are various practical difficulties in the recognition of the baroclinic zones – as, for example, when the air masses are undergoing modification which weakens the concentration of the baroclinicity. Problems are also raised in attempting interpretation of the average location of the frontal zones determined from synoptic analyses which use the model. It has been shown (Barry, 1967) that neither the median continental Arctic front, nor the median Maritime Arctic front, for July over eastern Canada, accord with the 'Arctic front' position as defined by surface streamlines and collectives of daily maximum temperature (Bryson, 1966), although there is a reasonable correspondence between Bryson's analysis and the continental Arctic front over the Northwest Territories. Bryson emphasizes that he adopts 'regional' air mass names in preference to the hemispheric terminology. As noted in

Chapter 2.A, p. 62, modern frontal analysis techniques take less account of air mass identification. However, this is not to say that the characterization of air mass properties is no longer of climatological interest, as Taljaard (1969) points out in his study for the southern hemisphere.

In view of the numerous investigations of air mass it is remarkable that the question of their homogeneity has been relatively neglected. Belasco (1952) selected soundings 'well within' each air mass, but still found that not all categories were significantly different at some levels in the troposphere. James (1969) suggested that the spatial homogeneity of air masses need not be absolute and for temperature proposed the following categories:

	Standard deviation
Very homogeneous	< 1°C
Homogeneous	1·0–1·2°C
Weakly homogeneous	> 1·2°C

Polar air masses show greater spatial variability (a higher 'noise level') than tropical air masses as a result of the strong local contrasts in surface conditions in high latitude source areas and the rapid modifications of cP air by diabatic heating over the oceans. In this context the studies of air mass modification by Burke (1945), Craddock (1951), Manabe (1957), and numerous others listed in the bibliography, should be mentioned. It is also worth emphasizing that such modifications are not limited to low level effects due to heat and moisture sources. Air moving towards lower latitudes and conserving its absolute vorticity has to increase in relative vorticity since the Coriolis parameter is decreasing. If the actual curvature is cyclonic, which is commonly the case on the west side of an upper trough, then the air column shrinks vertically causing adiabatic heating. It is not uncommon for subsidence to create temperature and moisture discontinuities in the middle and upper troposphere unconnected with any frontal zone. As fig. 3.34 shows, polar air outbreaks only reach low latitudes with any considerable depth if the trajectory is cyclonic and the air is shallower on the west side of the outbreak than on the east.

Ozorai (1963) investigated temperature frequencies in the free air over Budapest for January and July 1951–60 and found many more maxima in the histograms than categories in the basic classification. More important, he showed the number of maxima to be different at each level (fig. 3.35). Longley (1959) and Creswick (1967) report similar findings with respect to the Canadian three-front model. Nevertheless, Bryson (1966) demonstrates that workable air mass classifications can be developed by applying the method of partial collectives (see Chapter

4.A, p. 217) to frequencies of daily maximum temperature at stations in North America. He has also suggested (personal communication) that the relative frequencies provide a measure of air mass dominance in terms of 'information content' (I) or relative entropy. This is given by

$$I = \frac{\sum_{i=1}^{k} f_i \ln_e f_i}{N \ln_e N}$$

where f = frequency of the ith air mass
N = total frequency (100%).

Consider three combinations, where the frequencies are in percentages

(*a*) $f_A = 80\%$, $f_B = 20\%$; $I = 0{\cdot}896$
(*b*) $f_A = 50\%$, $f_B = 50\%$; $I = 0{\cdot}850$
(*c*) $f_A = 33{\cdot}3\%$, $f_B = 33{\cdot}3\%$, $f_C = 33{\cdot}3\%$; $I = 0{\cdot}761$.

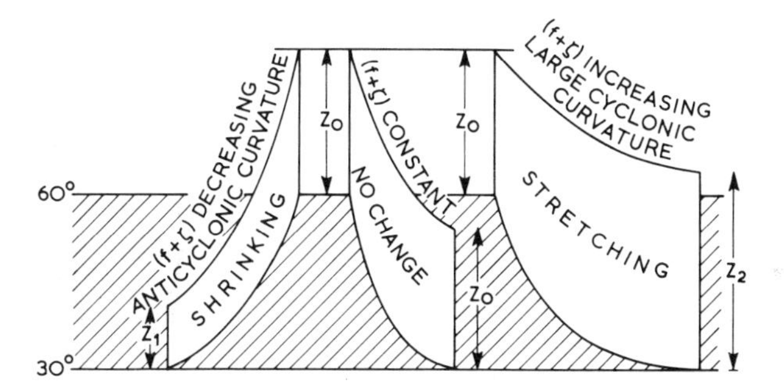

Fig. 3.34 The effect of the trajectory on the depth of polar air outbreaks in lower latitudes.

Thus, this index decreases as the diversity of air mass conditions increases.

An alternative approach which takes account of the non-homogeneity of air masses considers the flow in terms of *air streams*. An air stream has a distinctive origin but is recognized as being slightly baroclinic. Outbreaks of arctic air in North America fit well into this system as it is rare that they are quasi-barotropic. The mid-latitude westerlies can properly be regarded as comprising a number of slightly baroclinic air streams separated by hyperbaroclinic frontal zones. In low latitudes too the air stream concept is more useful than that of air masses, especially with respect to the identification of frontal zones (John, 1949; Watts, 1955). Moreover, it is a concept that can be married with the more recent perturbation view of tropical weather systems.

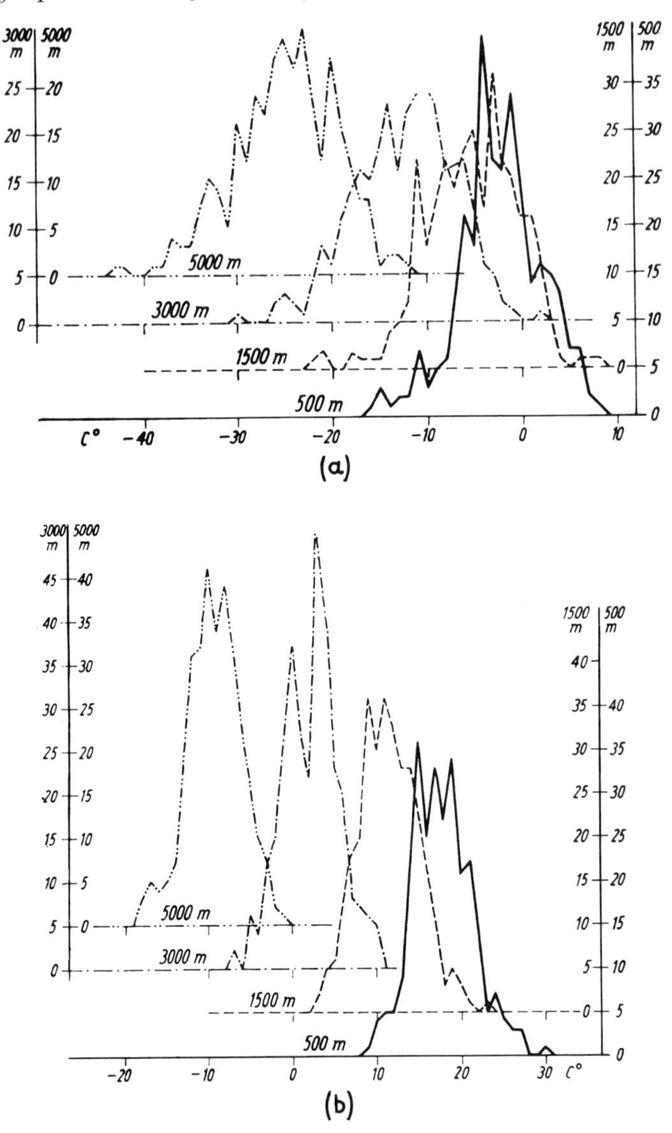

Fig. 3.35 Frequency curves of temperature at four levels over Budapest, 1951–60. (*a*) January, (*b*) July (from Ozorai, 1963).

Air masses in relation to the general circulation

The classical view of tropical and polar air, separated by the polar front, does not accord well with modern knowledge of the general circulation. Indeed figs. 2.40 and 3.36 show that it is more realistic to distinguish between polar, middle latitude, and tropical air, related to the three segments of the tropopause (Defant and Taba, 1957). This pattern is especially clear in the middle and upper troposphere where there are two baroclinic zones and the associated jet streams, although at low levels the divergence in the sub-tropical anticyclones obscures the distinction between tropical and mid-latitude air. Studies by Green, Ludlam and McIlveen (1966) demonstrate that low level tropical air ascends around the western margins of the subtropical anticyclone cells and enters the middle latitude upper westerlies where it may circulate for weeks (fig. 2.20). Within the simplified framework of fig. 3.36 it is

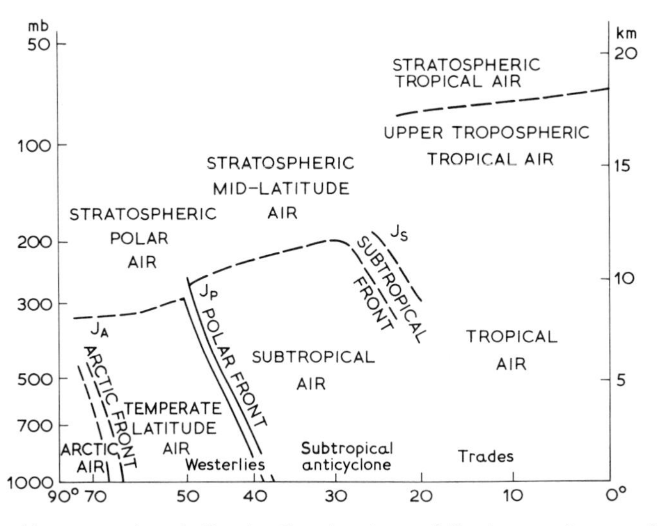

Fig. 3.36 Air masses in relation to the structure of the troposphere and stratosphere (partly after Palmen and Newton, 1969). J = Jet stream.

evident that a vertical profile may show polar or tropical air at low levels, related to cold outbreaks or warm air incursions, respectively, with relatively homogeneous subtropical air above.

A very different, but fundamental, view of air mass climatology has recently been stated by James (1970). He suggests that a revised treatment of air mass analysis can provide important data for general circulation studies. He shows (1969, 1970) that the quasi-barotropic relationship $[T = f(p)]$ can be expanded in a Taylor series to

$$T = T_0 + \frac{\partial T}{\partial p}(p - p_0) + \cdots$$

where T_0 = the air mass temperature at a conventional pressure level p_0 (say, 1012 mb for surface air).[1] In practice the relationship is linear so

[1] Similarly, for a constant pressure surface at height Z.

$$T = T_0 + \frac{\partial T}{\partial Z}(Z - Z_0) + \cdots$$

that higher order terms can be safely neglected. Two distinct regimes occur. Where $\partial T/\partial p$ is positive ('hyperthermal') high temperature is associated with high pressures and vice versa. This is the case in the mid-latitude westerlies where the wind speeds increase with height, whereas in the trades $\partial T/\partial p$ is negative ('hypothermal') and winds decrease with height. This reflects the thermal wind relationship (see pp. 12 and 47 and also fig. 2.41). Seasonal variations of T and p are considered in Chapter 5.A, pp. 311–14.

The method is illustrated by a study of the structure of maritime trade wind air over tropical Australia. Trade wind parameters are calculated from observations at Darwin, on the northern fringe, and Brisbane, on the southern fringe of the trades (table 3.15). The zonal

Table 3.15 Trade wind parameters over northern Australia (James, 1970)

Month	*Zonal temp. gradient* (°C)	$\partial T/\partial z$ (°C/100 m)	*Air mass temp.*, T_0(°C)	*Total KE (arbitrary units)*	*H (km)*
Jan.	3·7	8	31·9	54	3·6
Feb.	4·1	8	32·1	74	3·7
March	5·7	9	32·1	100	3·3
April	8·3	13	32·8	72	2·3
May	10·5	17	33·1	55	1·8
June	10·8	19	32·1	36	1·5
July	10·8	20	30·7	32	1·5
Aug.	10·8	19	33·6	36	1·6
Sep.	9·9	20	34·3	31	1·5
Oct.	8·6	18	33·3	29	1·7
Nov.	7·4	14	32·2	80	2·1
Dec.	5·4	12	31·6	41	2·6

temperature gradient increases threefold from summer to winter, but the wind strength, as reflected by the pressure gradient varies much less. More striking is the variation in depth of the easterlies represented by $H = T/-(\partial T/\partial z)$ (in the trades $\partial T/\partial z$ is negative). This is the thickness of a layer having the same momentum as the actual air mass but with constant velocity (the effective depth of the easterlies is about $2H$ where the wind speed is 13% of its surface value). The air mass temperature, T_0, shows a half-year cycle although the amplitude is small. A measure of total kinetic energy is given by the square of the pressure gradient multiplied by the thickness, H. This shows (fig. 3.37) a major build-up from October to March and sharp decrease thereafter to mid-winter. The steady level of KE from July to November occurs in spite of increasing solar radiation input, which probably reflects the storage of energy as sensible and latent heat.

This type of approach offers a novel link between the familiar ground of air mass climatology and dynamic concepts of the atmospheric circulation. It remains to be seen how much new information it will afford, but the ideas are sufficiently intriguing to merit further exploration in other quasi-barotropic regions.

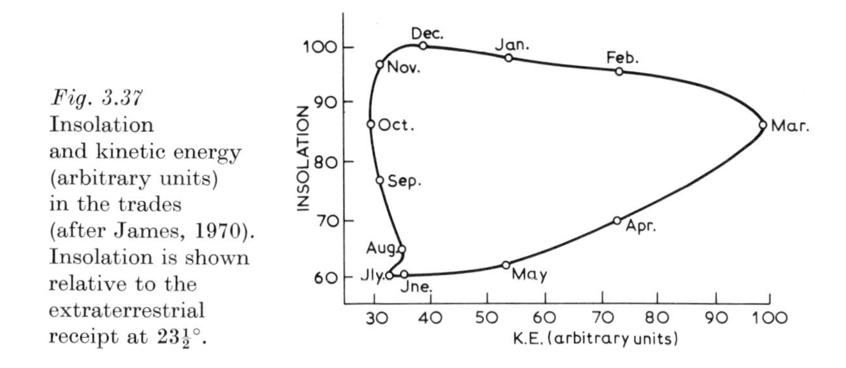

Fig. 3.37 Insolation and kinetic energy (arbitrary units) in the trades (after James, 1970). Insolation is shown relative to the extraterrestrial receipt at $23\frac{1}{2}°$.

3 RELATIONSHIPS BETWEEN WEATHER CONDITIONS AND SYNOPTIC FEATURES

There are several methods of examining the relationships between general weather characteristics and selected synoptic features which provide simple, but effective, exploratory tools. Although the methods are rather diverse it is convenient to discuss them together at this point.

Composite maps

One procedure is to construct a *composite map* by averaging the pressure field, or deviations from the mean pressure field, for individual cases of selected weather conditions. In order to emphasize any special features of the pattern it is common to select extreme weather conditions. The method has been applied, for example, in determining the synoptic situations likely to favour tornado outbreaks (Beebe, 1956; Lowe and McKay, 1962), summer rain in Alberta (Mokosch, 1962), precipitation in Hawaii (Solot, 1948) and snowfall at different stations in Baffin Island (Barry and Fogarasi, 1968), apart from the work of Multanovski discussed in section B of this chapter (p. 121). Surface pressure, geopotential, or other fields, corresponding to the time of the selected event and perhaps for twelve or twenty-four hours earlier, may be used.

This method is suitable in so far as a single type of pattern is primarily responsible for the conditions under study. However, Namias (1960) demonstrates that a variety of different patterns may give rise to years with light snowfall over the eastern United States and in this instance an average map will tend to show a weak and possibly quite misleading pressure field. A further point to note is that the 'normal' value of the precipitation (or other condition) may in some areas of the map be quite different from the mean of the sample which includes the wet (or dry) cases selected for study (Stidd, 1954).

Instead of using composites for an extreme condition it is sometimes helpful to construct *composite difference* charts (Solot, 1948) such as the pressure field for wet conditions minus that for dry conditions. Solot was able to demonstrate significant differences in the meridional pressure gradient over the North Pacific during years with heavy and light winter rains in Hawaii by this method.

Stidd shows that correlation fields (see Chapter 4.C, p. 251) relating average climatic conditions at a point (or for an area) to mean pressure or height values are analogous to the methods of composite charts or composite difference charts, but they represent a more precise technique.

Moving coordinates

Rather than analysing weather phenomena for a particular geographical area, it may be desirable to relate them to the synoptic systems themselves. Thus, the reference frame may be a depression, frontal zone or jet stream. The early cyclone models, which have been fully reviewed by Ludlam (1966), were developed intuitively on the basis of the results of case studies although the Norwegian polar frontal model was only evolved after several years of detailed synoptic analysis. The possibility of averaging weather conditions with reference to a system of moving coordinates was referred to by Jacobs (1947) but this method was seldom applied until computer facilities became available. Murray and Daniels (1953) used this approach to show that the occurrence of precipitation with respect to jet stream core zones reflects organized patterns of transverse and vertical motion, but it remained to Jorgensen (1963) to develop synoptic climatologies of the precipitation characteristics of winter depressions. Lows over the central and intermontane western United States were grouped into broad intensity classes according to pressure or height departures from normal, and precipitation amounts were determined for grid points with an origin related to the low centre (Jorgensen, Klein and Korte, 1967; Klein, Jorgensen and Korte, 1968). Fig. 3.38 illustrates the twelve-hour precipitation in relation to intense 850 mb lows over the western plateau states. In this case orographic effects have been eliminated by expressing the totals as percentages of seven-day station normals. Goree and Younkin (1966) apply the method to analyse heavy snowfalls over the central and eastern United States. Moving coordinates have also been used by Orgill (1967) in studying tropical disturbances over southeast Asia. Analogous studies of cloud distributions in relation to synoptic features are now possible from satellite photographs. Models of cloud patterns for mid-latitude and tropical systems have already been put forward (Boucher and Newcomb, 1962; Fett, 1964; for example) but so far no use has been made of moving coordinates as the framework for averaging.

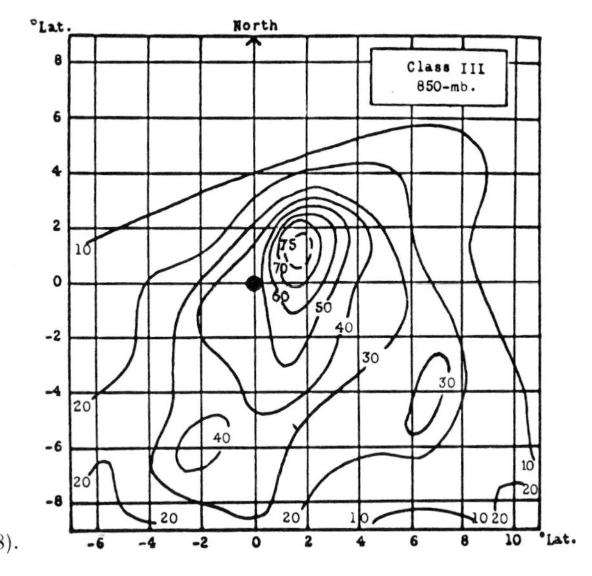

Fig. 3.38
Twelve hour precipitation related to intense 850 mb lows (in a moving coordinate system) over the western plateau states. Amounts are percentages of seven day normals (from Klein, Jorgensen and Korte, 1968).

'Weather types' – a new approach

A novel approach to the problem of linking the space–time distributions of weather phenomena to synoptic situations has been offered by Christensen and Bryson (1966). They use empirical orthogonal functions (eigenvectors) to remove redundancy amongst interrelated weather parameters. The method is illustrated for four years of observations in January, and five years in July, at Madison, Wisconsin. Fifteen weather variables recorded twice-daily, giving thirty variables for each set, were reduced by principal component analysis to nine new components accounting for about 80% of the original variance. The physical significance of the new components was interpreted on the basis of a regression analysis between the original data and the components. Air mass properties accounted for one fifth and the synoptic situation for four fifths of the 'explained' variance in January, whereas in July the air mass contribution increased to a quarter. This result, if generally true in inland areas, is of great significance. It implies that disturbances are the most important control of weather and that an air mass approach to synoptic climatology will tend to provide inadequate explanation and discrimination of conditions.

An objective grouping of the days into weather types was then made using Lund's correlation method (see p. 103) on regression coefficients for the nine components for each day. In January twenty-five groups represented 107/124 days and conditions on the remaining days were clearly transitional. In July 134/154 days were grouped into thirty classes. Fig. 3.39 shows the spatial relationship between the January types and a schematic synoptic weather situation patterned after observed cases within the sample for 1200 GMT charts. The shape and orientation of the pressure system as well as the stage of cyclonic development are generalized and the distances are relative only. The labelled areas indicate the core areas of the weather types occurring at Madison under different synoptic regimes. The diagram illustrates many more weather characteristics than in the classical schemes for

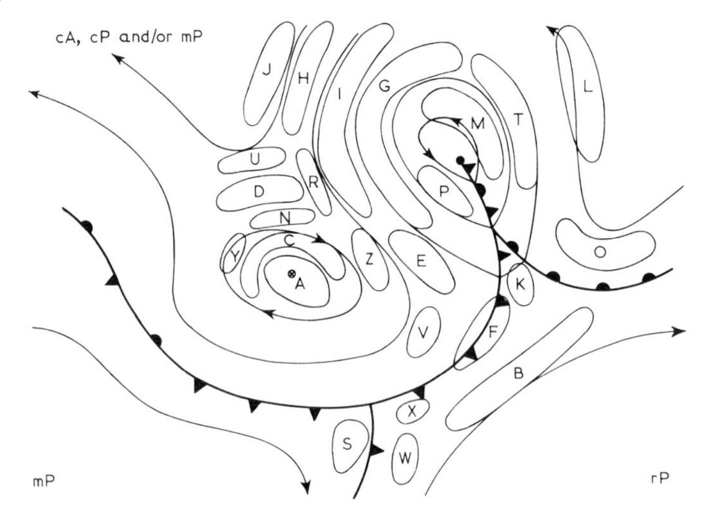

Fig. 3.39 Objective weather types in relation to the generalized synoptic situation for Madison, Wisconsin, in January (after Christensen and Bryson, 1966).

depression models. The actual weather sequences occurring on particular days at Madison will of course depend on the maturity of the systems, their dimensions and movement. Tests of the results against conditions at Minneapolis–St Paul demonstrate that the weather types are broadly representative of conditions in the upper Midwest.

This method, although involving complex computations, undoubtedly yields much more information than the conventional approaches discussed earlier. It will be interesting to see the results of other studies of this type.

The problem of frontal discontinuities

The question of the relationship of weather conditions to discontinuities such as fronts raises many problems, notably one of scale. As we have seen, broad synoptic classifications incorporate depression passages into the various type categories, while most air mass studies disregard occasions when fronts are in the vicinity. This problem led Frisby and Green (1949) to exclude 60% of days from their analysis of air masses over the British Isles. Several studies of frontal air mass precipitation have been made for individual stations. Shaw (1962) and Smithson (1969) use the following categories:

1 Frontal cases (subdivided into warm front, warm sector, cold front, occluded front).
2 Air mass type (mP, cP, Arctic).
3 Non-frontal depressions.
4 Non-frontal thunderstorms.

Their investigations make use of recording precipitation-gauge records. Table 3.16, illustrating results for Scotland and northern England, shows a noticeable increase in the contribution of occlusions in Scotland, with warm sector rainfall undergoing a corresponding decrease. There is nevertheless evidence of a marked orographic effect in warm sector situations at upland stations in both areas. A similar analysis has also been carried by Hiser (1956) for Illinois. More specific studies examining orographic effects on frontal precipitation in Scandinavia have been made by Spinnangr (1942) and Rossi (1948).

The results of such studies are necessarily complicated by sub-synoptic scale patterns which, it is usually assumed, are more or less eliminated by averaging a sufficiently large sample. Efforts are now being made to delineate mesoscale patterns within synoptic systems (Wallington, 1963). Browning and Harrold's work (1969) is the most detailed so far attempted. Autographic data were used to prepare hourly maps of rainfall intensity over England and Wales and these were converted into sequential maps showing the movement of individual rain bands. They identified three precipitation regimes:

Type A uniform rain (ahead of the warm front).

Type B_1 bands of rain areas aligned parallel to the warm front.

Type B_2 bands of rain areas aligned parallel to the winds in the warm sector.

Fig. 3.40 depicts the schematic distribution of these types with respect to the depression of their cast study. It was found that orographic effects were negligible for type A, but very pronounced for the B_2 bands in the warm sector. More analyses of this type are necessary to establish the generality of these findings.

We should not leave this topic without pointing out that other research findings have demonstrated that precipitation is often more

Table 3.16 Synoptic origin of total rainfall, (%) (after Smithson, 1969; Shaw, 1962; Matthews, 1972)

1959–1963 Scotland	*Warm front*	*Warm sector*	*Cold front*	*Occluded front*	*mP*	*Arctic*	*cP*	*Non-frontal depression*	*Thunderstorm*
Tiree (W Scotland) 10 m	21·2	6·4	16·5	21·8	18·4	3·8	0·4	10·9	0·4
Inveruglas (Dumbartonshire) 9 m	12·3	15·8	19·6	19·0	22·6	1·4	0·2	8·5	0·5
Leuchars (Fife) 11 m	15·1	1·9	15·6	29·8	6·1	2·8	6·8	19·1	2·9
1956–1960 Wales and Northern England	*Warm front*	*Warm sector*	*Cold front*	*Occluded front*	*mP*	*Arctic*	*cP*	*Non-frontal depression*	*Thunderstorm*
Cwn Dyli (Snowdonia) 99 m	18	30	13	10	22	0·1	0·8	5	0·8
Squires Gate (Lancashire) 10 m	23	16	14	15	22	0·2	0·7	7	3
Rotherham (Yorkshire) 21 m	26	9	11	20	15	1·5	1·1	14	3

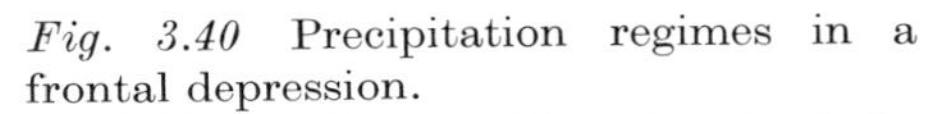

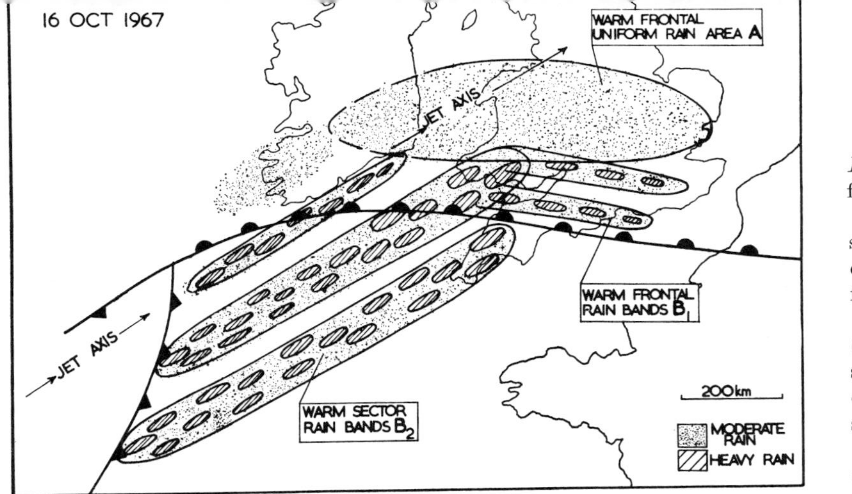

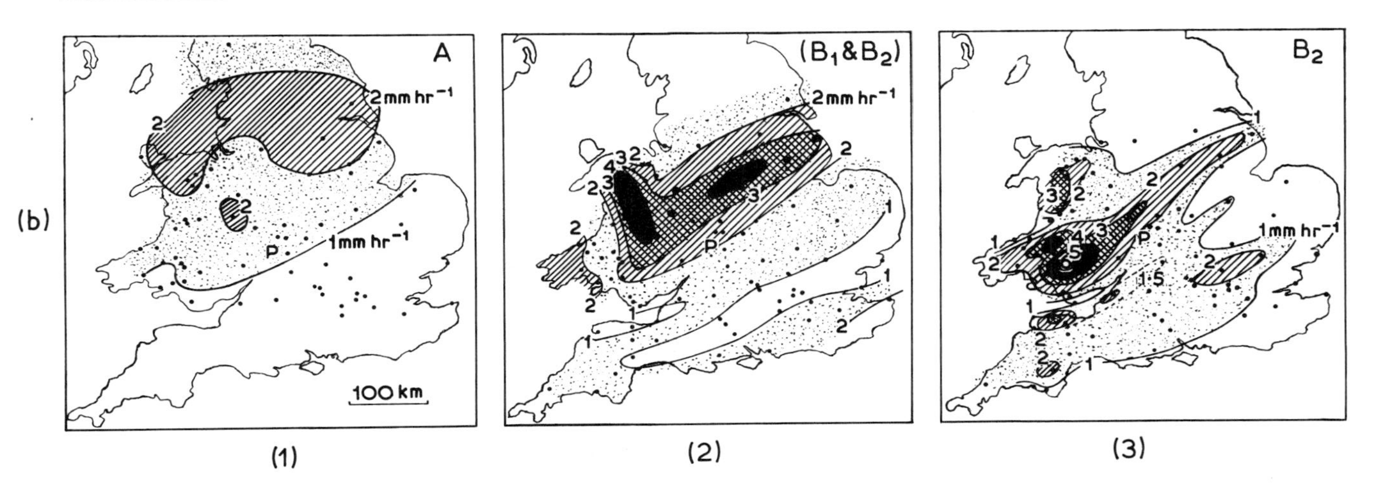

Fig. 3.40 Precipitation regimes in a frontal depression.

(*a*) Composite model based on land observation (the B2 bands were orographically enhanced and may have been lesser features over the sea).

(*b*) Distribution of precipitation rate (mm hr^{-1}): (1) 150–350 km ahead of the surface warm front; (2) 0–150 km ahead of the surface warm front; (3) in the warm sector.

Dots show autographic rain gauge records (from Browning and Harrold, 1969).

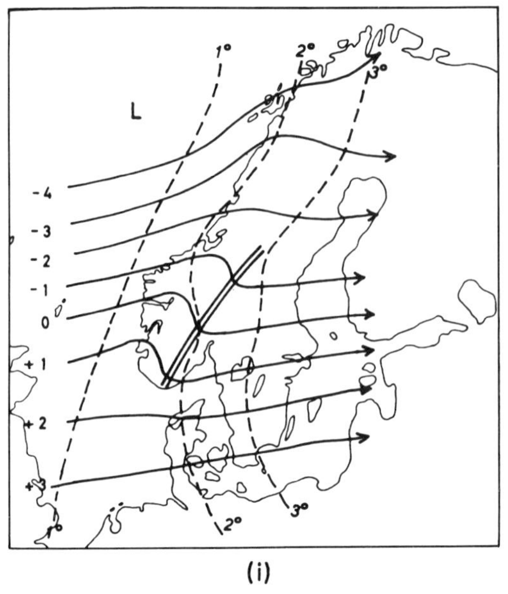

L
1°
2°
3°
-4
-3
-2
-1
0
+1
+2
+3

(i)

(a)

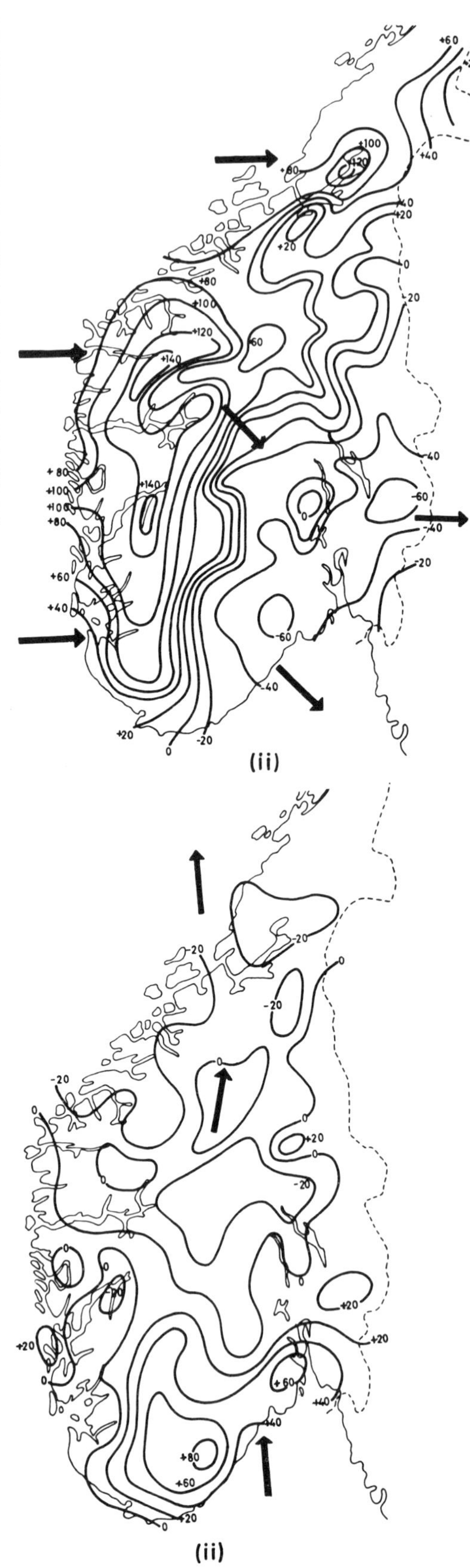

+60
+100
+120
+80
+40
+20
0
-20
+140
-60
-40
+100
(ii)

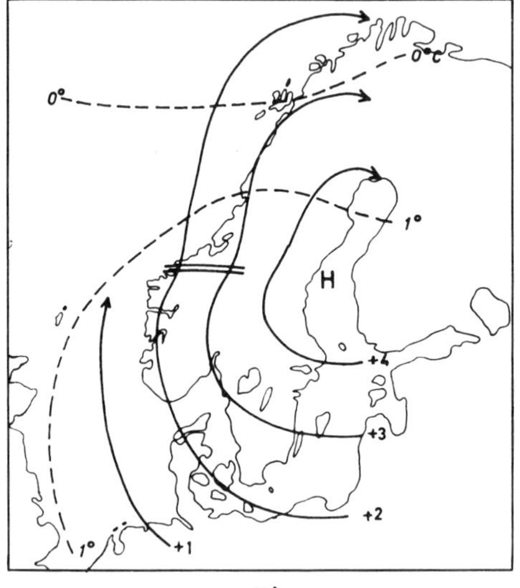

0°
0°c
1°
H
+4
+3
+2
+1

(i)

(b)

(ii)

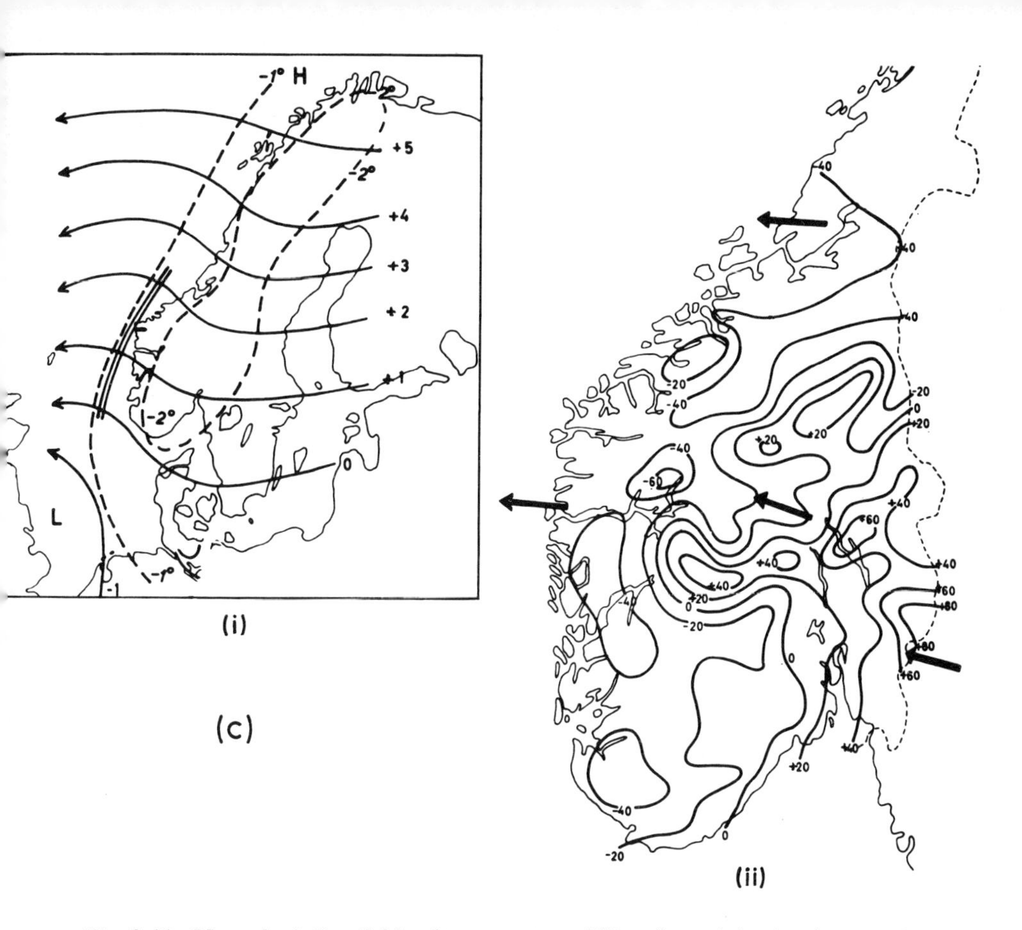

Fig. 3.41 Mean deviation fields of temperature (°C) and precipitation (per cent) over Norway related to 'additional gradient wind components' from west, south and east (from Hesselberg, 1962). (*a*) spring 1921, (*b*) summer 1917, (*c*) spring 1909. (i) Shows isallobars (mb) and anomalous flow with respect to the seasonal normals (solid lines with arrows), and temperature deviation – °C (dashed line). The double line indicates zone of mountain effects on the isallobars, (ii) shows precipitation anomalies (per cent) and anomalous flow (arrows).

closely related to the position of the jet stream than to surface fronts (Murray and Daniels, 1953; Richter and Dahl, 1958). An extension of such investigations (in space and time) is desirable, although it is to be anticipated that neither jet stream nor front locations alone will serve as a reliable discriminator of surface conditions.

4 ANOMALY PATTERNS

Significance

The spatial distributions of departures of weather elements from their long-period mean values are of considerable interest in synoptic climatology. These anomaly maps, which may refer to observed weather elements, such as precipitation, or field variables, such as the height of isobaric surfaces, provide a means of clarifying our understanding of interactions between surface and atmosphere and within the circulation itself. They also form an important element in the analogue and kinematic approaches to long-range forecasting (see Chapter 5.D, pp. 382–90). It must be noted, however, that anomaly patterns are different in kind from the synoptic map patterns we have so far considered. They show the average deviation from the mean, which is to say that the synoptic variability has been filtered out and only its net effect for some time interval (usually a month) is shown. For example, Kutzbach (1970) determined eigenvectors of monthly mean pressure anomaly maps but emphasized that the component fields 'are not hemispheric weather types', they are abstractions. If, however, pentad anomalies are being considered then effects with a time and space scale of planetary waves can be detected.

Since 1949 much use has been made in the United States of 700 mb height departures, particularly with respect to forecasting applications (Klein, 1956). Hawkins (1956) showed that monthly mean 700 mb height anomaly fields reflect the anomalous component of lower tropospheric wind flow for that particular month. This is demonstrated by high correlations between anomalies of 700 mb height and MSL pressure. Moreover, the 700 mb anomalies also correlate closely with mean 700 mb geostrophic relative vorticity. These relationships are less satisfactory, however, over continental areas and whenever the anomaly fields are weak. Practical verification of this approach is provided by the established relationships between surface temperature anomalies and 700 mb height departures (Martin and Hawkins, 1950). Another application of this method has been made for seasonal mean fields over Norway by Hesselberg (1962). In this case mean departures of temperature and precipitation were related to the 'additional gradient wind' indicated by the fields of surface pressure anomaly. Fig. 3.41 illustrates the mean deviations related to three such component wind directions.

Analysis and classification

It is customary to plot the absolute departure from normal in the case of temperature, geopotential, pressure, thickness, etc., but in the case of precipitation or sunshine the figures can be converted into a percentage of normal. It should nevertheless be noted that discontinuous elements, such as precipitation, are not as satisfactory for analysis as temperature fields although when long periods are considered this problem diminishes. Absolute values of temperature anomaly have a more or less normal frequency distribution in all months, though with a larger standard deviation in the winter half-year. Accordingly, it has been proposed (Klimenko, 1962) that easier comparisons could be effected if temperature anomalies were plotted as a percentage of the amplitude of the variation of monthly mean temperature. A base for such computations has been provided by Girskaia and Klebaner (1969) in maps of the standard deviation of mean monthly air temperature for the northern hemisphere relating to the period 1881–1960.

The earliest attempt to classify anomaly patterns appears to be that of Defant (1924). He identified four types of monthly pressure anomaly pattern over the North Atlantic and prepared a catalogue of their occurrence for 1881–1905. A more ambitious study was made by Brooks and Quennell (1926) for monthly pressure fields over large segments of the northern hemisphere. Using above or below average pressure in the Faroes as the first criterion, they identified two groups each with five categories:

Group I (Excess)	*Group II (Deficit)*	*Location of the pressure anomaly*
A	A	Over or near Scandinavia
B	B	From the British Isles across Europe
C	C	Over the British Isles
D	D	Over Iceland or southern Greenland
E		Generally over the Arctic, with deficit 40–50°N over the Atlantic and southern Europe
	E	Reverse of I.E.

The percentage frequency of these major types for 1873–1918 was as follows:

	IA	IB	IC	ID	IE	IIA	IIB	IIC	IID	IIE
Summer	10	4	17	15	4	8	6	12	20	4%
Winter	10	8	10	20	2	12	6	9	18	5%

The high frequency of ID and IID represents the natural tendency for pressure departures over Iceland and the middle latitudes of the North

Atlantic to be of opposite sign. A deep Icelandic low reflects more northerly depression tracks while any blocking in this region leads to the reverse. The extreme of this latter pattern was well illustrated in January–February 1963. The related phenomenon of a winter temperature 'see-saw' between Greenland and Europe (warm in one area and cold in the other) was first recognized 200 years ago (Loewe, 1966). It reflects the year-to-year variations in pressure and wind fields over the North Atlantic. The perhaps unexpectedly high summer frequency of excess pressure over the British Isles points to blocking anticyclones.

Brooks and Quennell also compared their categories with Abercromby's 'weather types'. They equated them as follows:

Abercromby	*Brooks and Quennell*
S type	Subtype of II.D
W type	Subtype of II.A
N type	Subtype of ID
E type	Subtype of I.D

Obviously they identified many more distinct patterns than Abercromby!

For the United States, Smith (1947) classified patterns of monthly temperature anomalies into twenty-four types (fig. 3.42) based on thirty-five years of data. A catalogue was then prepared for 1886–1952 (see Namias, 1953, p. 71) as a background for thirty-day forecasts. Certain types clearly reflect the position of long-wave troughs and ridges. For example, a ridge over the western United States is indicated by type 13, and over the eastern United States by type 19. Types 14 and 20 are suggestive of wintertime patterns due to radiative cooling in the surface high over the Great Basin, while type 1 points to chinook effects over the northern Great Plains (in winter and spring) associated with zonal westerly flow.

It is only recently that research into the temporal and spatial characteristics of anomaly patterns has been actively resumed and a systematic attempt made to determine their origin. The literature is nevertheless replete with numerous case studies, apart from the monthly analyses of 700 mb height anomalies in the *Monthly Weather Review* since 1950. Attention has focused particularly on notable hemispheric patterns of winter temperature departure, such as those of 1946–7, 1949–50 (Namias, 1947, 1951) and 1962–3 (Murray, 1966b), and on drought patterns, especially those which affected the northeastern United States during 1962–5 (Mitchell, 1968; Namias, 1966).

Much recent work on anomaly patterns has been carried out in the Soviet Union. Gedeonov (1967) and Zorina (1967) have examined the spatial distribution of major air temperature anomalies over the

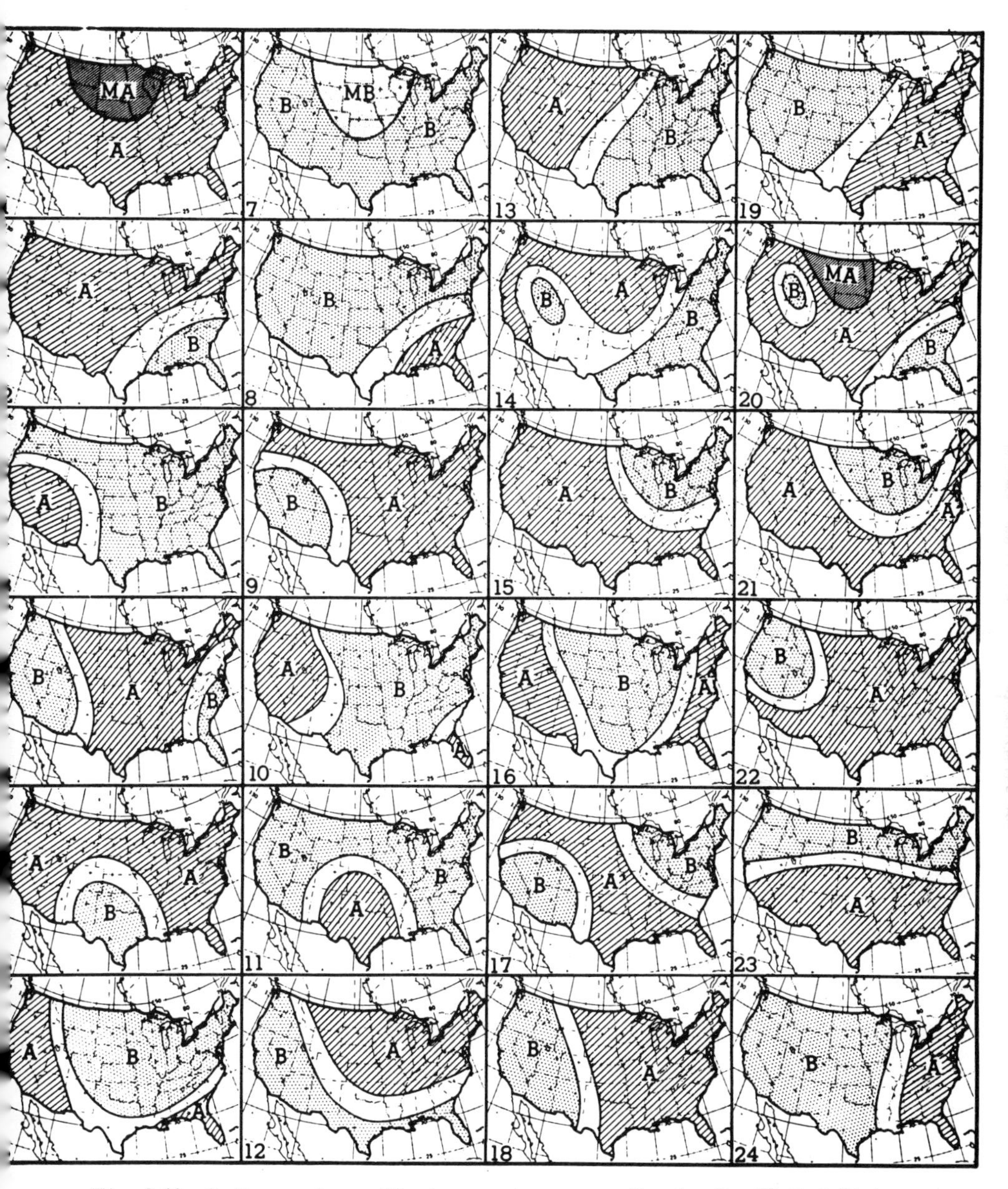

Fig. 3.42 Patterns of monthly temperature anomalies for the United States (from K. Smith, 1947; in Namias, 1953). (M)A = (much) above, (M)B = (much) below.

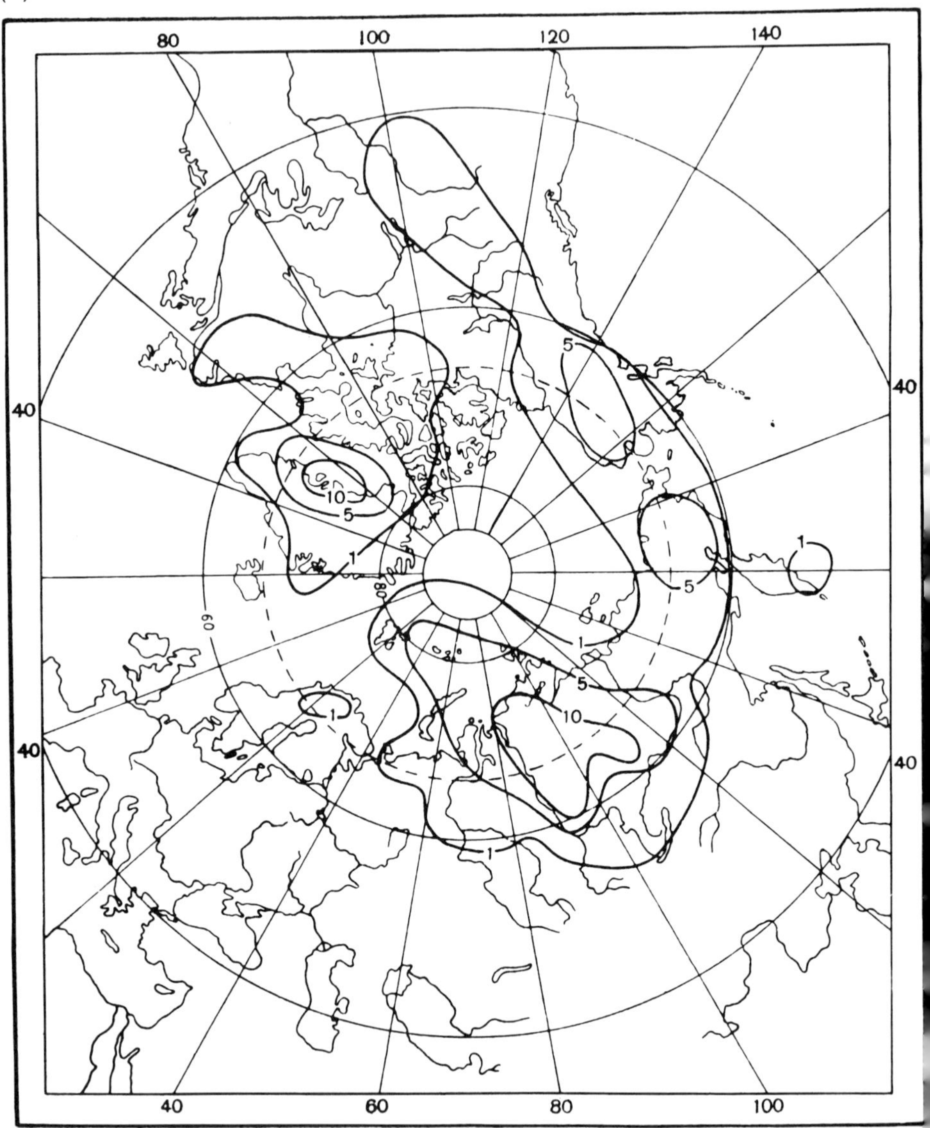

80
100
120
140
40
40
40
40
60
10
5
1
5
1
5
1
1
5
10
1
40
60
80
100

(*b*)

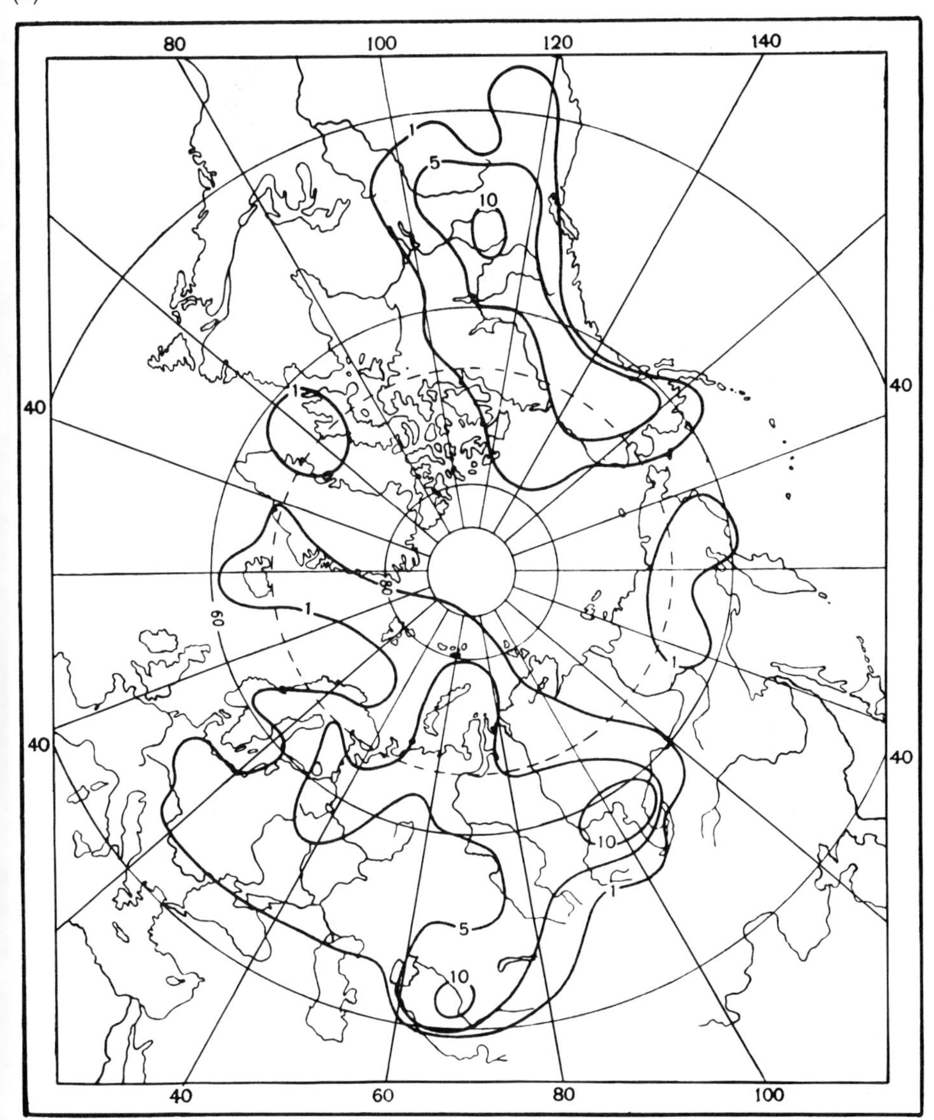

Fig. 3.43 The distribution of (*a*) positive and (*b*) negative departures of monthly mean air temperature of 10°C or more from the long-term mean (from Gedeonov, 1967). Figures are percentage frequency for a seventy-year period.

northern hemisphere and Zavialova (1965) performed a similar analysis for the Arctic. The most intense anomalies occur in winter in an area extending from north-central Siberia and the Kara Sea across the Polar Basin to northern Canada. In the area of Baffin Bay–west Greenland anomalies of +16°C are not infrequent between November and April. As fig. 3.43 shows, positive departures are most frequent over north-central Siberia and west Greenland. Negative departures are

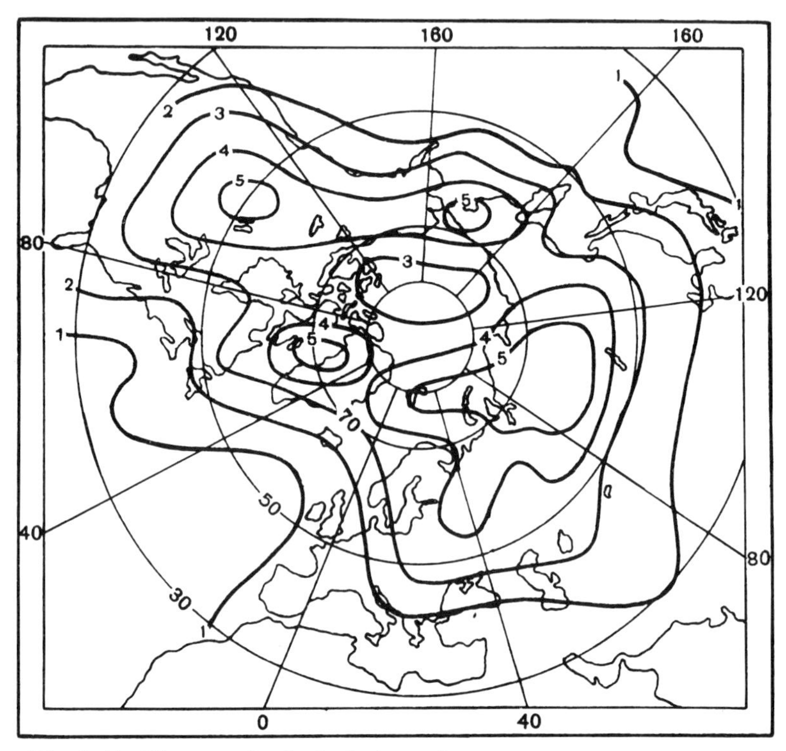

Fig. 3.44 The standard deviation of mean monthly temperature anomalies over the northern hemisphere in January 1903–60 (from Mescherskaia and Kliukvin, 1968).

more extensive occurring particularly over middle latitudes in central Asia and the North American high plains.

The distribution of standard deviation of mean monthly temperature anomalies in January (fig. 3.44) shows a basic land–sea contrast with maxima in northern Siberia and the high plains of North America, but there is also evident variability in the types of persistent pattern occurring over Baffin Bay and the Bering Strait. Factors determining such patterns are discussed below. The range of anomalies is much less in summer but again the continental areas experience greater variability than the oceans.

A number of regional analyses of types of anomaly patterns have begun to be published. For example, with reference to Rumania, Rahau *et al.* (1969) identified five types of January temperature anomaly over Europe. Bauman and Kanaeva (1967) classified monthly precipitation anomalies in four sectors of the northern hemisphere for October–March 1891–1940. Table 3.17 shows the frequencies, for

Table 3.17 *Percentage frequency of types of anomaly of mean monthly precipitation total over Europe, 1891–1940 (Bauman and Kanaeva, 1967)*

	Jan.	*March*	*Oct.*	*Winter season*
1 Positive anomalies over $\geqslant 75\%$ of the area	32	18	16	24
2 Negative anomalies over $\geqslant 75\%$ of the area	40	25	23	28
3 Positive anomaly in the north, negative in the south	8	8	16	10
4 Reverse of type 3	6	8	14	9
5 Positive anomaly in the west, negative in the east	2	8	8	8
6 Reverse of type 5	6	6	6	6
7 Irregular distribution	6	24	16	15

Europe, of the seven categories which they distinguished. While types 1 and 2 are dominant overall there is an interesting variability in the frequency of irregular patterns. They also note that type 1 months show a decreasing frequency over the period of record. Types 1 and 2 also form more than 50% of all types in the winter season over western and eastern Siberia, but not in North America. In the latter region the irregular pattern is common.

The areal framework used for classifying anomaly patterns over a limited hemispheric sector is usually arbitrary. Practical experience in connection with the swift determination of analogues for long-range forecasting purposes is usually the key consideration (Hay, 1960). A one-month time scale has been used in most studies, mainly because of the problem of greatly increased computation for shorter intervals. Miles (1963) argues that, although this may be appropriate for temperature anomaly patterns over the major continental areas, it is less appropriate for areas such as the British Isles where Craddock (1957) has demonstrated, by means of a band-pass filter analysis (see p. 238), the occurrence of fluctuations with a ten-day and a thirty-day duration in temperatures at Kew (London). The dominance of submonthly intervals (twenty to twenty-four day duration), which results in part from

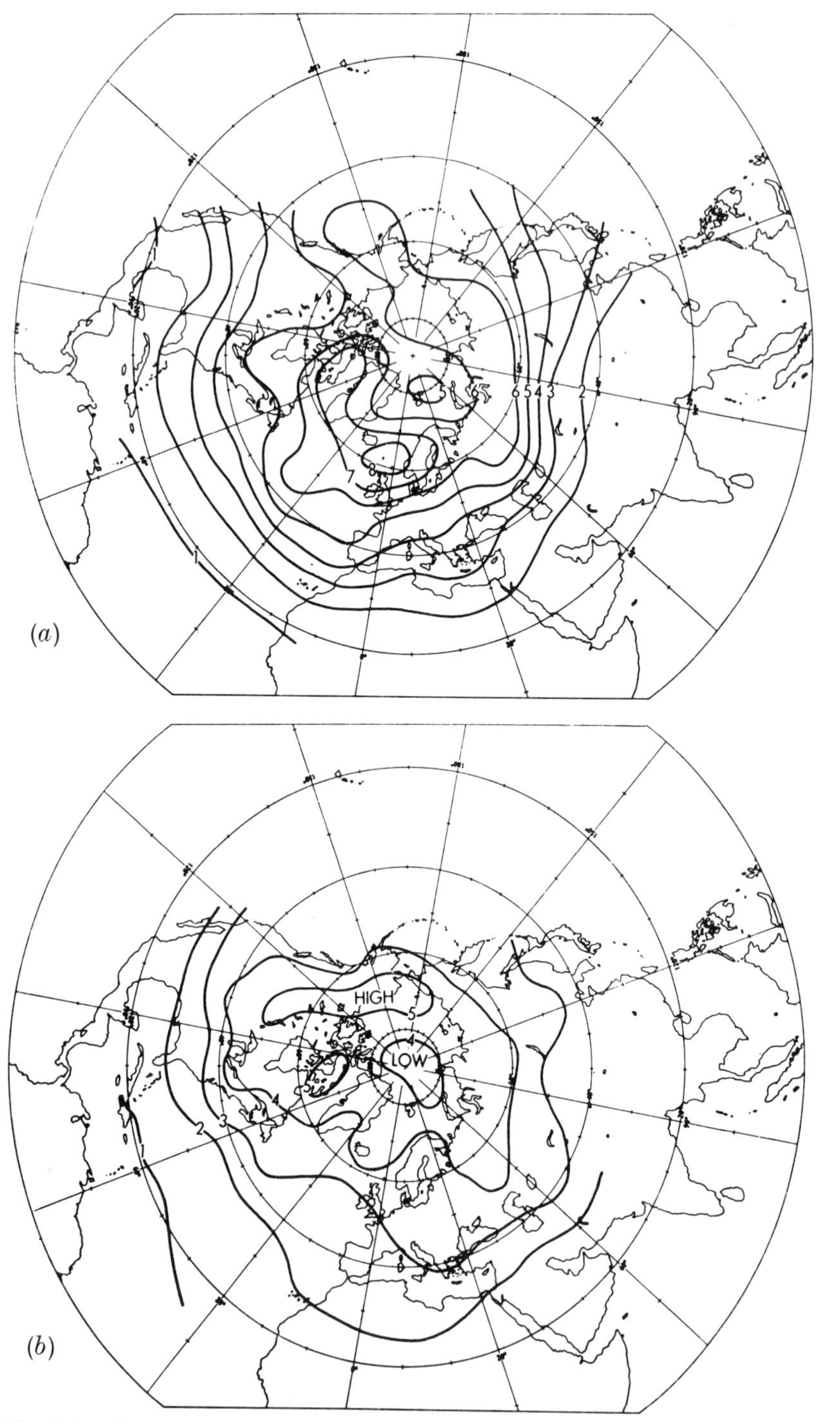

Fig. 3.45 The distribution of the standard deviation of (*a*) 500 mb height and (*b*) 1000–500 mb thickness, filtered to show anomalies of twenty to sixty days. Based on data for 1961–3 (from Sawyer, 1970).

blocking activity, is also apparent from similar analysis of zonal indices (*Drift Zahlen*, published by the Deutsche Wetterdienst) for 1948–57 (Miles, 1963). This question has been dealt with more recently in the spatial context by Sawyer (1970). Fig. 3.45, showing the standard deviation of the time-filtered 500 mb height and 1000–500 mb thickness, is a useful indicator of weather anomalies on the time scale of twenty to sixty days. The role of fluctuations < 15 days has also been

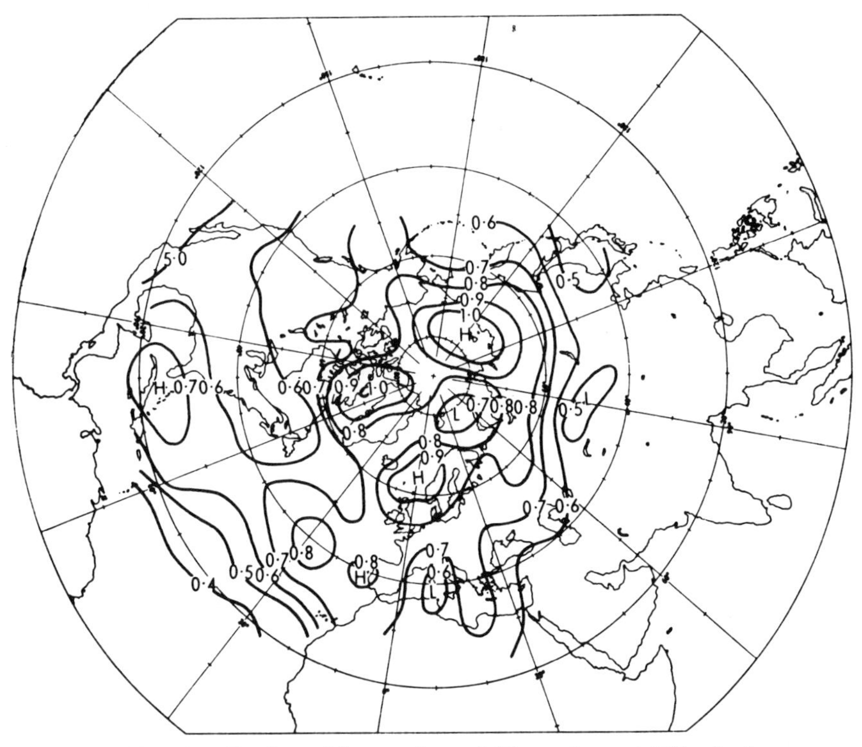

Fig. 3.46 The distribution of fluctuations of fifteen days at 500 mb shown as a ratio of the standard deviations of the series filtered to retain long-term components to that retaining short-term components (from Sawyer, 1970). Low values indicate dominance of short-term components.

examined by taking height departures from an eleven-day running mean. Fig. 3.46 illustrates their relative importance at the 500 mb level. It would seem in fact that the longer variations are most significant about the Arctic Circle and in the subtropical anticyclone areas, and are not unimportant over the British Isles.

The persistence of anomaly patterns is of particular interest with respect to the possible improvement of long-range forecasts (see p. 390). With this in mind, Craddock and Ward (1962) investigated the persistence of monthly temperature anomalies over Europe and Siberia for

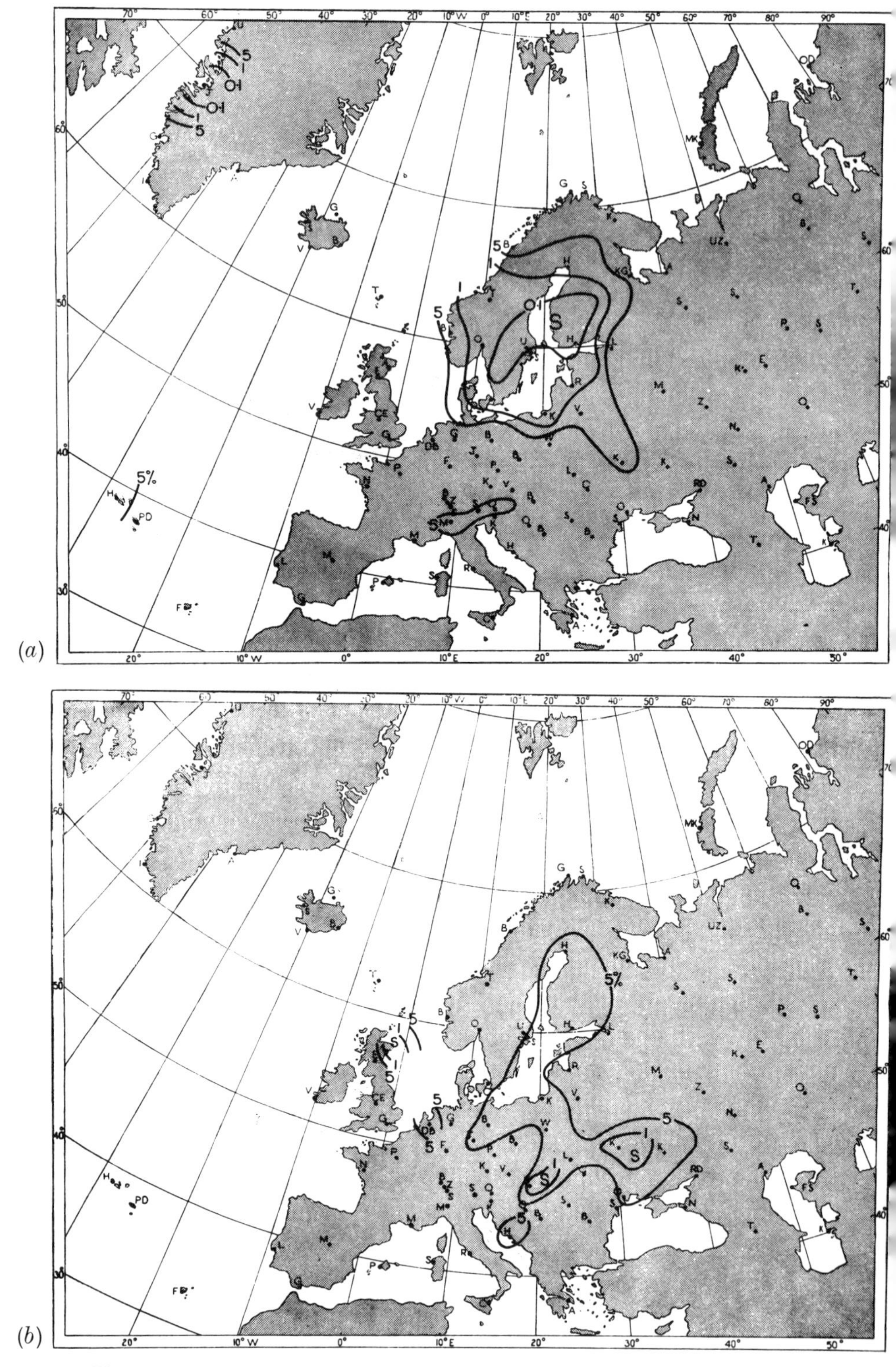

Fig. 3.47 The association of temperature anomalies over Europe and Siberia (from Craddock and Ward, 1962; Crown Copyright reserved). Significance levels for (*a*) February–April, (*b*) January–April.

lags of up to six months. Of the longer intervals only December–March and January–April showed significant persistence (fig. 3.47) but the month-to-month relationships were important for Scandinavia and oceanic areas. It is interesting that the month-to-month persistence is small over continental areas, contrary to popular opinion. Indeed, Blair (1933) stressed that the persistence of pressure anomalies appeared to be greatest in Asia, less in North America and least in the area of the British Isles. Osuchowska (1970) has examined the persistence of 700 mb height departures using correlation methods and he, too, found strong month-to-month persistence in the Scandinavian area. This may reflect the physical characteristics of the area and perhaps also the planetary wave structure (see fig. 2.36, for example, showing the distribution of ridges in the monthly mean 700 mb field).

In spite of the present limited knowledge of controls, attempts have been made to forecast monthly temperature departures statistically. Drogaitsev (1966) found evidence of periods of about five months in the monthly anomalies in European Russia and western Siberia, although this would not appear to be supported by the results of Craddock and Ward (1962). Rakipova (1959) used the month-to-month persistence of the current anomaly together with the anomalies of some of the radiation balance components in European Russia. Only 61% of the variance was accounted for, however.

The interrelation of temperature anomalies in the northern hemisphere has also received attention (Girskaia, 1968a, 1968b). The correlation coefficients between points at meridians along eleven latitude circles were calculated and the results used to indicate the longitudinal interactions. Girskaia found that in winter there are temperature 'see-saws' with negative correlation (*r* values exceeding the three standard deviations level) of anomaly patterns between the following:

Western Europe–west Atlantic/eastern Canada (35–60–60°N).

Chukotski peninsula/west Pacific–western North America (50–70°N).

Western North America–eastern North America (30–45°N).

In summer the most pronounced oscillations are the American (30–55°N), the Urals (45–70°N) and the Siberian (50–55°N). Fig. 3.48 shows that the western limit of the Siberian one coincides with the eastern limit of the Urals one. Synchronous coupling around the hemisphere has also been recognized in patterns of 1000–500 mb thickness anomaly (Rafailova, 1968). Further discussion of such global teleconnections is given in Chapter 5.D (pp. 414–7).

The studies discussed so far treat individual parameters. Obviously, however, there is close interaction between the anomaly fields of pressure, temperature and precipitation (Namias, 1952). In an attempt

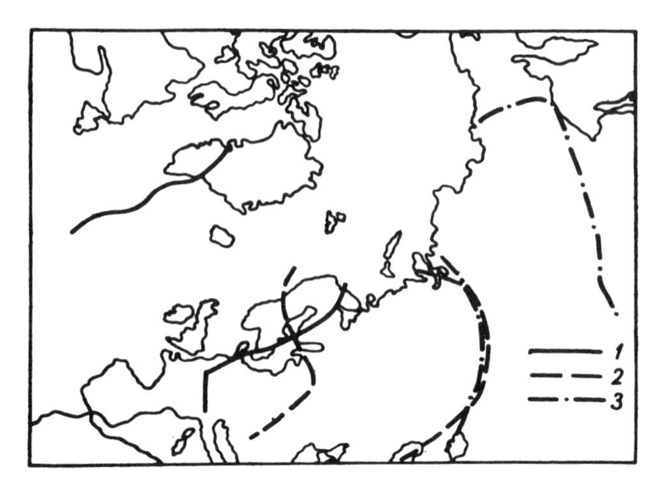

Fig. 3.48 Mean axes of inverse correlation of temperature anomalies, June–August 1881–1960 (from Girskaia, 1969). 1 = Atlantic, 2 = Urals, 3 = Siberian.

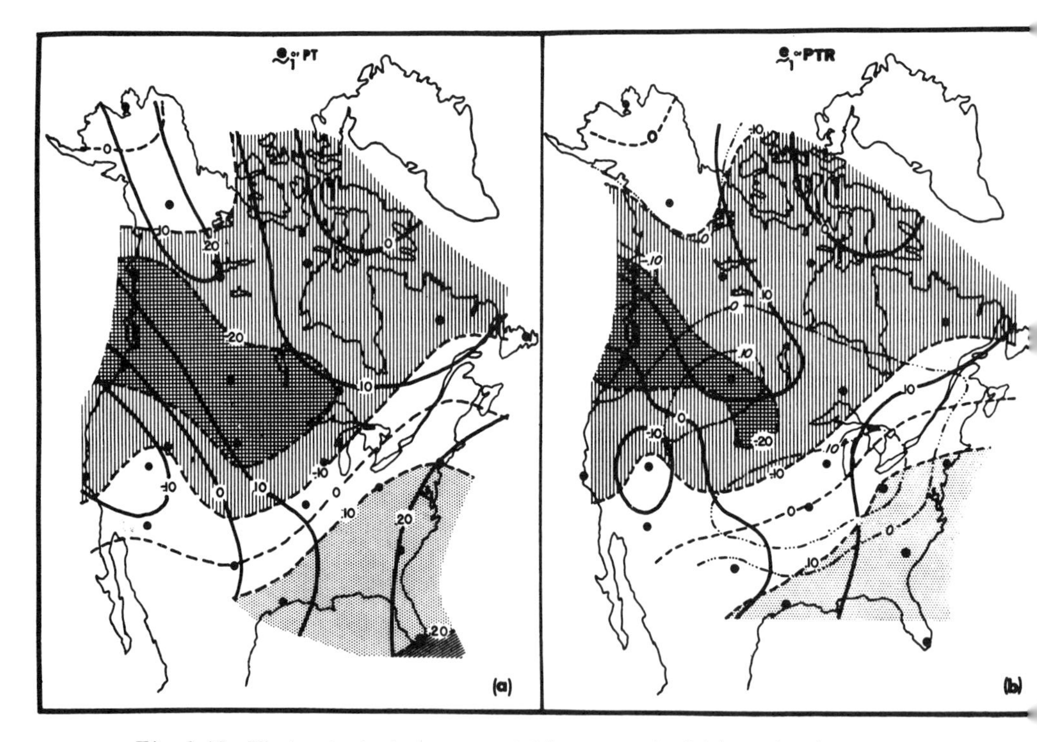

Fig. 3.49 First principal component (eigenvector) of (*a*) sea level pressure and surface temperature, and (*b*) sea level pressure, surface temperature and precipitation over North America. Solid lines = pressure, dashed lines = temperature, dash dotted lines = precipitation. Regions of maxima/minima in the temperature patterns are hatched/stippled. The component in (*a*) accounts for 28·6% of the total variance, in (*b*) for 24·3% (from Kutzbach, 1967).

to assess these relationships Kutzbach (1967) has used empirical orthogonal functions (eigenvectors) to simpilify the physical structure. The analysis was performed for departures from the three mean fields for January 1941–65 over North America. Principal components were calculated for each anomaly field individually and for combinations in twos and threes in order to facilitate the interpretation of the patterns. Fig. 3.49(*a*) showing the first component for MSL pressure and temperature departures delineates cooler areas in the northwest related to more northerly wind components than normal (as indicated by the pressure departures) and warmer areas with more southerly wind components in the southeast. This same pattern appears in fig. 3.49(*b*) for the first component on the three variables. Also, there is a positive precipitation departure in the area of increased meridional temperature gradient in the central United States. The nature of this analysis is discussed further in Chapter 4.C (p. 268) and 4.D (p. 282).

Basic controls

The associations between temperature or other anomalies and the characteristics of the circulations are complex, varying with time of year and with place. A large negative anomaly of air temperature is generally associated with persistent trough or cold pool in the 1000–500 mb thickness pattern. A positive anomaly over the continents in winter is normally caused by persistent advection of maritime air (Kats, Morskoi and Semenov, 1957; Siskov, 1962). More specifically, Ragozina (1967) shows that the sign of the anomaly can be related to the basic types of circulation according to Vangengeim's classification (see fig. 3.26). This approach has been elaborated using Dzerdzeevski's hemispheric circulation types (see fig. 3.27) by Kurshinova (1970). Negative temperature anomalies over the European U.S.S.R. are related to incursions of northerly airflow in winter, whereas in summer they are due to more frequent cyclones from the west. However, as Caplygina (1968) points out, the distribution of temperature anomalies is determined not only by the circulation, but also by the influence of the underlying surface. In low latitudes radiation rather than circulation conditions is the primary control of the sign of the anomaly. Table 3.18 illustrates the relative importance of radiative and advective factors in the formation of temperature anomalies and also provides a broad classification of their characteristics. Grimmer (1963) attempted to isolate some of the determinants of temperature anomaly patterns 1881–1960 over Europe and the eastern North Atlantic by principal components analysis. He showed that in each month of the year continentality is the dominant pattern with a centre over north-central Europe and a strong gradient over the European coastline (although in mid-summer the position of the pattern has shifted). The second pattern appears to be related to thermal advection in the westerly depression tracks. These two patterns account for an average of 34% and 24%, respectively, of the total variance throughout the year. A third pattern which appears to reflect blocking in the vicinity of the British Isles is also of some importance in the period December–April. This type of investigation can usefully complement more conventional analyses. A similar study has been performed recently by Perry (1970) for the British Isles. Rafailova (1968) suggests that patterns of thickness anomaly, which are closely related to surface temperature anomalies, are influenced by events in the stratosphere. Important changes in the patterns take place during winter stratospheric warmings in high latitudes and during the evolution of the summer easterly flow in the middle and upper stratosphere. It remains to be established, however, that these changes are antecedent and causally linked.

Table 3.18 General characteristics of types of temperature anomaly (Sazonov and Girskaia, 1969)

Type of anomaly	*Main zone of development*	*Range of monthly anomalies*	*Character of circulation*	*Predominant character of pressure field*	*Role of radiation in formation of anomalies*	*Role of advection in formation of anomalies*
A	Polar (over continents)	Large	Unstable	None	Variable	Variable
B	Temperate	Medium	Slightly unstable	Depressions	Small	Large
C	Subtropical	Medium	Slightly unstable	Depressions	Large	Small
D	Subtropical	Small	Stable	Anticyclone	Large	Small

Synoptic studies of anomalies of surface air temperature and precipitation for periods of five days were started in Britain in 1955 (Craddock and Lowndes, 1958) and are now produced as standard procedure by the Meteorological Office using a network of about 250 stations covering the Atlantic and European sectors of the hemisphere. Charts of the MSL pressure and heights of the standard isobaric levels have also been prepared for much of the northern hemisphere, for the internationally agreed pentad calendar. Following this work it is now possible to trace sequences of anomaly patterns and to relate them to changes in the circulation. For example, Murray (1966a) has illustrated the value of tracing five-day mean 1000–500 mb thickness anomaly centres in a study for autumn 1965 over the British Isles. Fig. 3.50 shows the relationships between five-day centres and the monthly mean 500 mb patterns for September and October 1965. Cyclonicity over Britain in September was replaced by blocking in October while

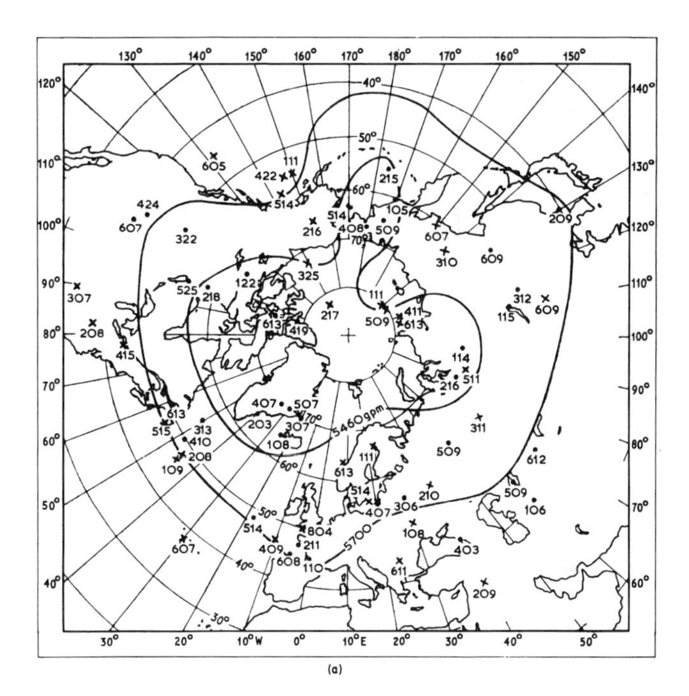

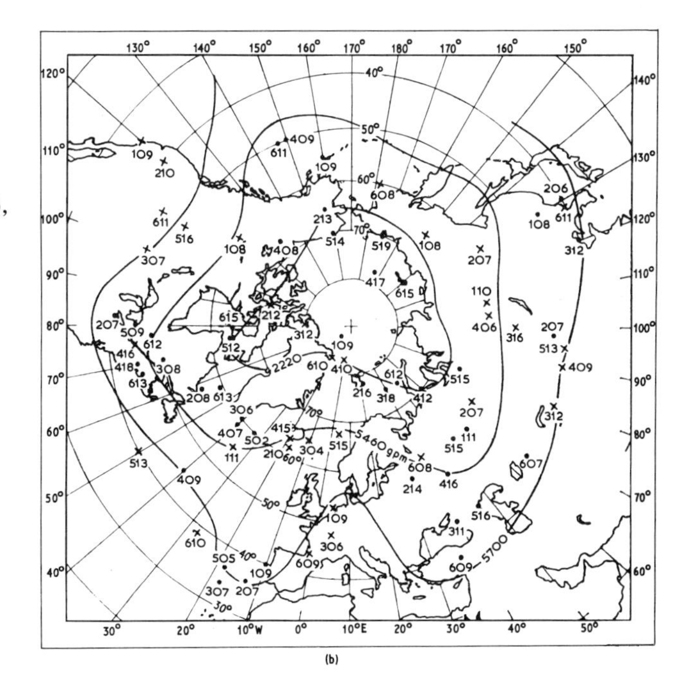

Fig. 3.50
The relationship between five-day centres of 1000–500 mb thickness anomaly and the monthly mean 500 mb pattern in (*a*) September and (*b*) October 1965 (from Murray, 1966a; Crown copyright reserved). Solid circles = negative centres (NAC), crosses = positive centres (PAC), line = selected **monthly** mean 500 mb contours in geopotential metres. In the three-figure groups the first figure is the pentad number and the remaining figures give the anomaly in geopotential decametres.

Five-day periods:

No.	*Begins*
1	3rd
2	8th
3	13th
4	18th
5	23rd
6	28th

opposite changes took place over the Gulf of Alaska. These changes can also be seen in the trough-and-ridge diagram of fig. 3.51.[1]

Much work still remains to be done on the nature and causes of anomaly fields. In the case of pressure or geopotential anomalies, such questions are a basic part of the problem of understanding the general circulation. Iakovleva and Turikov (1968), for example, found a good correlation between eigenvectors of anomalies of geopotential and circulation indices over the northern hemisphere in winter. Likewise,

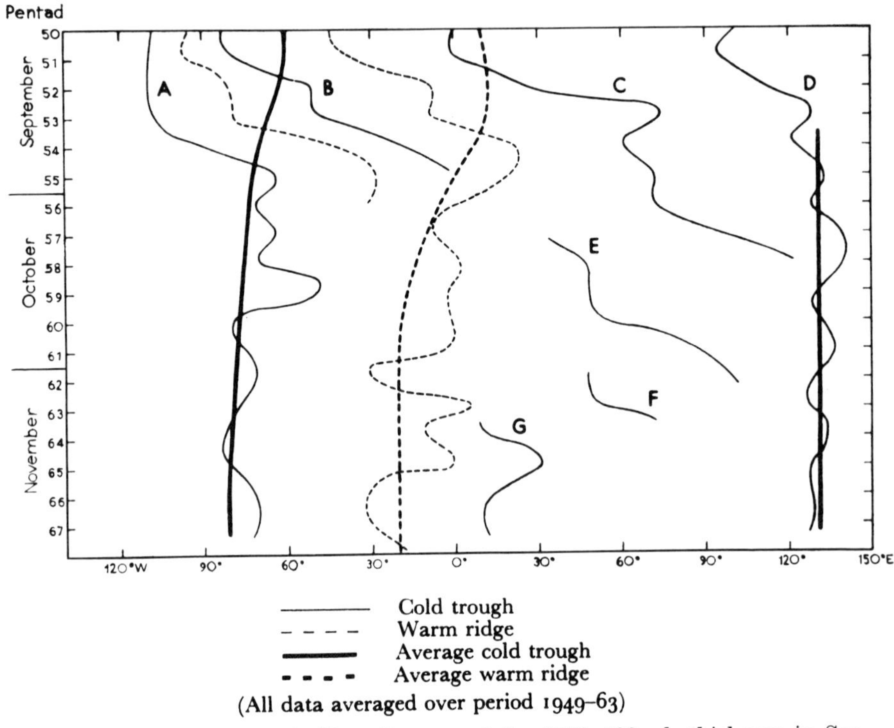

Fig. 3.51 A trough and ridge diagram of the 1000–500 mb thickness in September–November 1965 (pentads 50–67) (from Murray, 1966a; Crown Copyright reserved).

Kutzbach (1970) shows that for January and July monthly mean MSL pressure departures are related to the major centres of action. In addition, time changes can be seen, especially comparing pre-1920 and post-1950. However, such results still leave unanswered the question of fundamental causes. Work on abnormal heat sources and sinks is vital here, and while this has been begun for the oceans (Bjerknes, 1962; Namias, 1962; Perry, 1969) it will ultimately be required over the

[1] This diagram was devised by Hovmöller (1949) for studying the time–space displacements of upper air troughs and ridges.

various land surfaces. Anomaly patterns of climatic elements, are, if anything, more difficult to deal with. As we have seen, they reflect the complex interaction of atmospheric circulation, underlying surface and energy transfers. We shall examine some aspects of these problems further in Chapter 5.D and 5.E (pp. 418–25).

4 Statistical methods

A wide variety of statistical techniques is to be found in the synoptic climatological literature. Some of these are well known, whereas others have only recently been applied to meteorological problems. Our aim here is to illustrate the present or potential climatological application of existing techniques in four major areas – frequency and probability analysis, time series, spatial analysis, and classificatory methods. The formal basis of these techniques is not considered; in most cases they can be found in standard references on mathematical statistics (see the general references to this chapter). The essential properties of various statistical functions are, however, summarized.

A Frequency and probability analysis

> 'Distributions which fit a normal curve with sufficient accuracy are the exception' (*E. Czuber, 1938, quoted by Conrad and Pollak, 1950*).

1 FREQUENCY DISTRIBUTIONS AND RELATED STATISTICS

We begin by reviewing some basic aspects of frequency distributions and the statistics of samples. It is important to define a climatological series. It is 'a sample series of data consisting of one climatological value for each year of the record' (Thom, 1966a). For example, in the case of a thirty-year record, typical series are:

the thirty hourly temperatures for 12 GMT, 1 January,

the thirty daily mean temperatures for 1 January,

the thirty mean January temperatures.

The use of mixed populations commonly produces a multimodal distribution. Unless the population is first clearly defined subsequent statistical analyses may be improperly applied.

Each weather element tends to have a distinctive frequency distribution in different climatic regimes and for each month, or season, of the year. Many distributions are not Normal (see p. 224) and therefore the types of statistical analysis that are applied to them must be carefully chosen. Some examples of frequency distributions for different weather elements at Bergen are given in fig. 4.1. Those for fog and cloudiness are particularly striking. We should also note that the distributions of short- and long-period characteristics are often markedly different, as in the case of daily and annual precipitation totals. It is well known that the distributions of annual totals in arid and humid climates respectively show an analogous pattern to those for daily and annual totals and we need not examine this further. The statistical properties of common distributions are reviewed on pp. 221–4.

The statistics of a frequency distribution are:

$$\text{the mean } \bar{x} = \frac{1}{n} \sum_{i=1}^{n} x_i$$

and the moments, which describe the relative frequency of values about the mean. The second moment, or variance, is

$$s^2 = \frac{1}{n} \sum_{i=1}^{n} (x_i - \bar{x})^2.$$

The standard deviation is given by s. The rth moment is

$$\mu_r = \frac{1}{n} \sum_{i=1}^{n} (x_i - \bar{x})^r.$$

Skewness is usually determined from μ_3/s^3 and the kurtosis from μ_4/s^4. Skewness is a measure of the symmetry of the distribution; the skewness is positive (negative) when the distribution has a long tail towards high (low) values of the variate. Kurtosis is often interpreted as peakedness of the distribution although Kaplansky (1945) shows that there is no reliable relationship between them.

Bennetts (1967) nevertheless uses these statistics to illustrate differences between months of anomalies of monthly mean temperature for central England. The annual variation of the mean, standard deviation, skewness and kurtosis are given in fig. 4.2. The distributions are negatively skewed in the winter half-year, but the variation of kurtosis is less regular being mainly leptokurtic (peaked) in autumn and winter and platykurtic (flattened) in spring and mid-summer. Another detailed example of month-to-month differences may be found in the work of Wachter (1968). He uses cumulative percentage frequency graphs to compare the distributions of maximum, mean and minimum

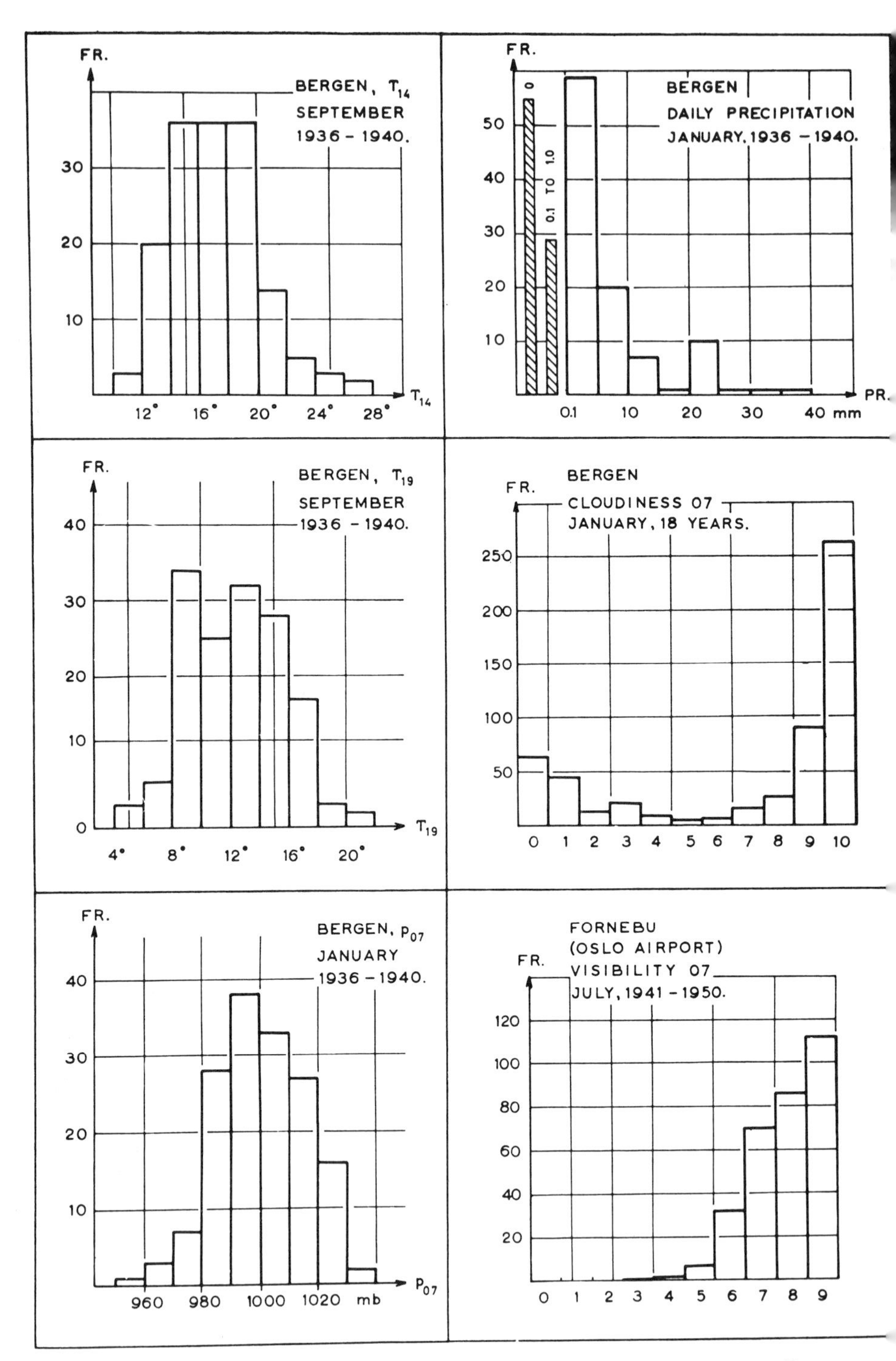

Fig. 4.1 Frequency distributions for selected climatic parameters at Bergen, Norway (from Godske, 1966). P = pressure, T = temperature (°C). Subscripts refer to hour of observation.

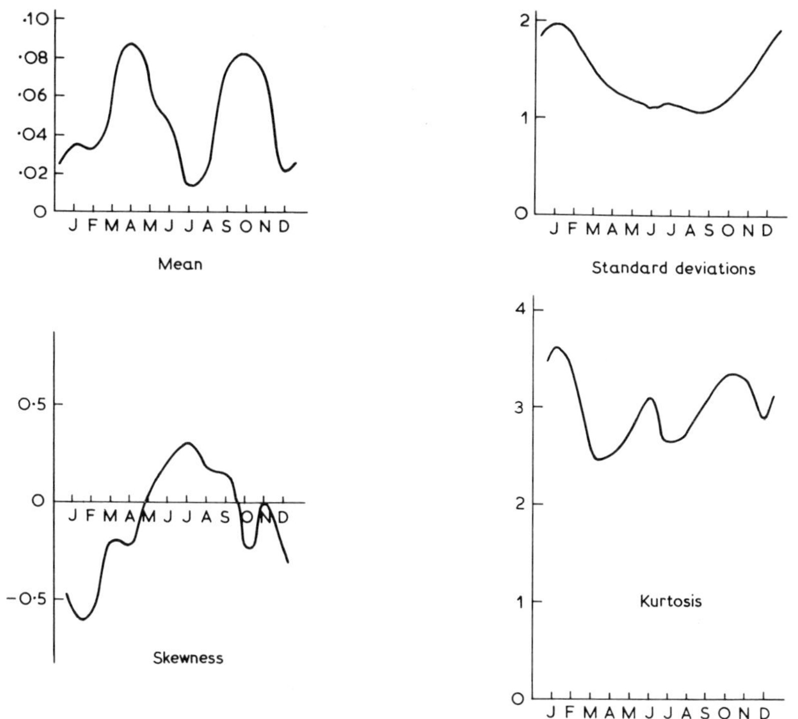

Fig. 4.2 The annual variation of statistics of anomalies of monthly mean temperature for central England, 1698–1952 (after Bennetts, 1967). The anomalies refer to departures from moving averages taken over the immediately preceding 25-year period.

temperatures and their deviations from long-term means at Frankfurt-am-Main (fig. 4.3).

2 PARTIAL FREQUENCY ANALYSIS

The distribution of daily mean temperatures occurring with a specific air mass or airflow type will, in general, be approximately normal. If we graph all daily means, however, the distribution is likely to indicate several peaks associated with the various air masses affecting the locality (Fiedler, 1965). A numerical method for decomposing such distributions into several collectives or partial frequency distributions has been developed by Essenwanger (1954, 1955), and further work at the University of Wisconsin has shown that sufficiently accurate results for many purposes can be obtained by direct graphical analysis (Bryson, 1966). Fig. 4.4 illustrates the principle and compares the collectives obtained by Essenwanger for vapour pressure at Karlsruhe with air mass frequencies determined independently.

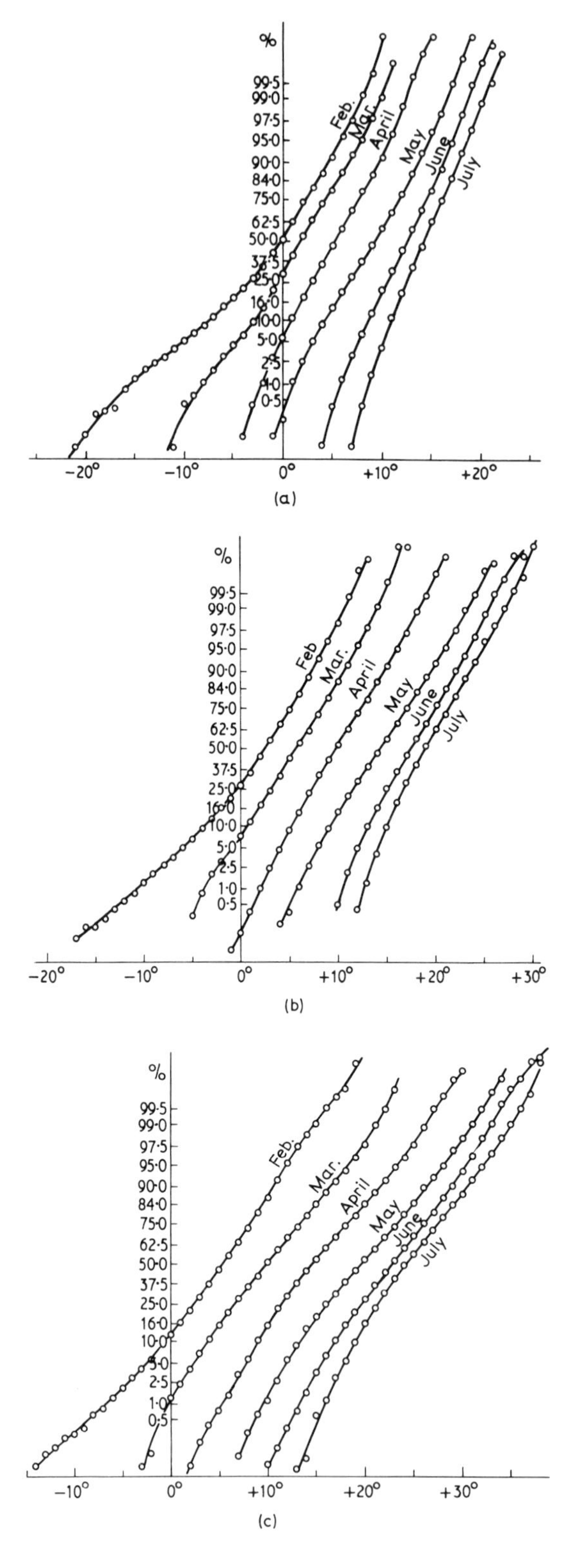

Fig. 4.3
Cumulative percentage frequency of (*a*) daily maximum, (*b*) mean and (*c*) minimum temperature in February to July at Frankfurt, 1870–1960 (from Wachter, 1968).

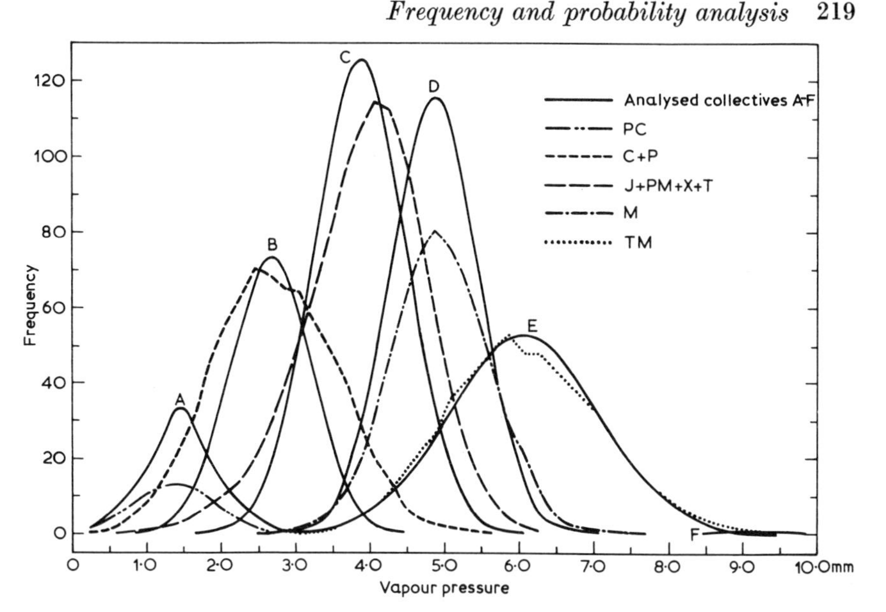

Fig. 4.4 Partial frequency analysis of vapour pressure at Karlsruhe, winter 1887–1941 (after Essenwanger, 1954). The broken lines refer to various air masses.

Essenwanger (1960a) has also shown that precipitation frequency distributions for short time intervals can be analysed into collectives as a means of determining the physical processes that are operating. At Asheville, North Carolina, for example, a distinction can be made by this method between excessive daily rains in autumn due to extra-tropical cyclones and hurricanes. With hourly data there is adequate expression of convective and advective precipitation (Essenwanger, 1960b).

3 CONTINGENCY ANALYSIS

Next we consider joint frequencies of several variables. The mutual occurrences of specified conditions of two or more variables are commonly arranged in the form of a contingency table. Each cell refers to a chosen subset of the variables. Table 4.1 illustrates a contingency table for the cyclonicity index (see Chapter 3.C, p. 142) and tercile categories of rainfall over England and Wales. Other examples of their synoptic climatological application may be found in studies of associations between months or seasons (Hay, 1968; Murray, 1968) or shorter time intervals such as pentads (Craddock, 1963).

The construction of such a table is a necessary preliminary to tests of association. Unless the observations are very numerous it is desirable to have a few cells. A 3×3 or 2×2 table is often the most satisfactory.

Table 4.1 Contingency table for cyclonicity index and average rainfall over England and Wales in January, 1873–1964 (Murray and Lewis, 1966)

Rainfall (terciles)	Cyclonicity (quintiles) Anticyclonic 1	2	3	4	Cyclonic 5
Dry 1	13	11	3	3	0
2	4	7	12	7	2
Wet 3	1	1	4	9	15

Table 4.2 Observed and theoretical contingencies of the cyclonicity index and tercile categories of rainfall over England and Wales, January 1873–1964 (after Murray and Lewis, 1966)

	OBSERVED *Cyclonicity index*				THEORETICAL *Cyclonicity index*			
Rainfall	*Anticyclonic*		*Cyclonic*		*Anticyclonic*		*Cyclonic*	
Dry	24	3	3	30	12·05	6·2	11·75	30·0
	11	12	9	32	12·9	6·6	12·5	32·0
Wet	3	4	24	30	12·05	6·2	11·75	30·0
	37	19	36	92	37·0	19·0	36·0	92

Note: these values are approximate to avoid calculations to several decimal places.

$$\chi^2 = \frac{11 \cdot 95^2}{12 \cdot 05} + \frac{3 \cdot 2^2}{6 \cdot 2} + \frac{8 \cdot 75^2}{11 \cdot 75} + \frac{1 \cdot 9^2}{12 \cdot 9} + \frac{5 \cdot 4^2}{6 \cdot 2} + \frac{3 \cdot 5^2}{12 \cdot 5} + \frac{10 \cdot 05^2}{12 \cdot 05} + \frac{2 \cdot 2^2}{6 \cdot 2} + \frac{12 \cdot 25^2}{11 \cdot 7}$$

$$= 48 \cdot 2$$

For four degrees of freedom this value is significant at the 0·1% level.

The usual significance test of association is chi-square (χ^2) which is non-parametric, so that the type of frequency distribution is unimportant.

$$\chi^2 = \sum^{n} \frac{(O-E)^2}{E}$$

where O = observed frequency in a cell

E = expected frequency in the same cell

and $\sum^n$ denotes summation of all cells in the table.

The expected, or theoretical, contingency table is determined by calculating jk/N for each cell

where j = the row total
k = the column total
N = the grand total of frequencies.

As a rule of thumb E should be $\geqslant 5$ in any cell to ensure the statistical validity of the test; in table 4.1 the original quintiles of the cyclonicity index have been grouped into three classes to meet this condition. For r rows and s columns, χ^2 has $(r-1)(s-1)$ degrees of freedom[1]. Table 4.2 demonstrates a highly significant relationship between the cyclonicity index and rainfall over England and Wales in January.

The table illustrates the use of the χ^2 test as a measure of association. It is also used as a test of goodness of fit of an observed distribution to a theoretical one (see p. 225).

4 PROBABILITY

Probability theory concerns the likelihood of specific change events occurring. Probabilities (p) are expressed on the scale $p = 0$ to $1{\cdot}0$. The total probability in a situation where there are several possible outcomes is always equal to one. For example, if the average frequency of rainy days at a particular station in April is six, then the probability of any day in the month having rain is $p = 0{\cdot}2$ and the probability of it being dry is $1 - p = 0{\cdot}8$ (assuming that the rainy days occur at random). Suppose that wind speed at the same station is unrelated to precipitation and that the probability of a day with winds below 5 m s^{-1} is $0{\cdot}05$ in April, then the joint probability of a rainy day with light wind is $p = 0{\cdot}2 \times 0{\cdot}05 = 0{\cdot}01$. Some implications of probability concepts in synoptic climatology were noted in 3A.3.

Four statistical distributions which are of particular importance in determining probabilities are now discussed. They are the Normal, Binomial, Poisson and Gamma distributions.

Normal distribution

The well-known Normal (Gaussian) distribution, symmetrical and bell-shaped, is the basis of many methods in parametric statistics. Only a brief summary of its properties need be given. The Normal probability density function is

$$f(x) = \frac{1}{\sigma(2\pi)^{1/2}} e - \frac{(x-\mu)^2}{2\sigma^2}$$

[1] Degrees of freedom = the number of measurements of a variable that can be assigned arbitrarily before the remainder are automatically determined, e.g. if $(s-1)$ cells of a row are assigned, the sth cell is determined by the row total.

where μ = the population mean, estimated by $\sum_{i=1}^{n} x_i/n$ $(=\bar{x})$
and σ = the population standard deviation, estimated by $\hat{\sigma}$,

$$\hat{\sigma} = \left[\sum_{i=1}^{n} (x_i - \bar{x})^2/n-1 \right]^{1/2}.$$

The use of $n-1$ in the denominator relates to the degrees of freedom.

This function expresses the proportion of the distribution under a specified portion of the Normal curve when the total area beneath the curve is equal to 1. The required portion of the distribution is specified in terms of σ. For example,

68·26% of the distribution is within $\pm 1\sigma$ of the mean
95·46% of the distribution is within $\pm 2\sigma$ of the mean
99·73% of the distribution is within $\pm 3\sigma$ of the mean

Binomial distribution

This is an approximation to the Normal curve for data in discrete classes. It is determined from the expansion $(p+q)^n$ where $q=1-p$, when p and q are mutually exclusive events. The Binomial probability law (probability mass function because discrete values are involved) is

$$f(x) = \binom{n}{x} p^x q^{n-x}, \quad \text{for } x = 0, 1, 2, \ldots, n$$

where $\binom{n}{x}$ is the number of combination of x items out of a total of n,

$$\binom{n}{x} = \frac{n!}{x!\,(n-x)!}$$

$x!$ (x factorial) denotes the expression $x(x-1)(x-2)\ldots 3.2.1.$

In a simple case we might be interested in the probability of days being wet or dry. If $p=0{\cdot}5$ for a wet/dry day, and assuming no interdependence, then the probabilities for two days picked at random are:

State:	2 wet	1 wet, 1 dry	2 dry	Total
Probability:	$p^2=0{\cdot}25$	$2pq=0{\cdot}5$	$q^2=0{\cdot}25$	$(p+q)^2=1$

For probabilities over three days we use the expansion $(p+q)^3=p^3+3p^2q+3pq^2+q^3$ and so on. It will be noted that the distribution assumes constant probability, a situation referred to as Bernoulli trials.

The mean of a Binomial function is np and the variance is npq.

Poisson distribution

For many types of meteorological event the frequency of non-occurrence cannot be specified. This is true of storms, floods and droughts, all of which are rather rare events occurring 'at random'. The Poisson distribution, a limiting form of the Binomial, is applicable to many of these situations. The frequency distribution follows an exponential form (based on the constant $e = 2{\cdot}7183$).

If z is the average number of events during the total time interval,[1]

$$e^z = 1 + z + \frac{z^2}{2!} + \frac{z^3}{3!} + \cdots + \frac{z^r}{r!} + \cdots = \sum_{x=0}^{\infty} \frac{z^x}{x!}.$$

This infinite series converges to e^z for all values of z. Now, for positive integers of a random variable x,

$$f(x) = \frac{z^x e^{-z}}{x!}.$$

This function satisfies the conditions of a probability density function since

$$f(x) > 0$$

and

$$\sum_{x=0}^{\infty} f(x) = \sum_{x=0}^{\infty} \frac{z^x e^{-z}}{x!} = e^{-z}e^z = 1$$

(i.e. the total area beneath the frequency curve is 1).

A variable satisfying a function of this form is said to be Poisson distributed. A characteristic of the Poisson distribution is that the mean and variance are both equal to z. The application of Poisson probability is illustrated in table 4.3 relating to annual frequencies of hurricanes (see also Thom, 1966b). This illustrates the physical meaning of a Poisson distributed variable. The more extreme an event (e.g., four hurricanes in one season), the much less likely it is to occur in a given interval. The model is useful for computing the probability that exactly k events occur in a specific interval, given that the average occurrence in that interval is z.

Gamma distribution

The Gamma distribution, like the Poisson, is positively skewed, but it is a continuous distribution for x between 0 and ∞ (Thom, 1958). Here the assumption is that the events constitute a 'renewal process' where the time intervals between these events are distributed independently and identically (Cox, 1962).

The gamma function

$$\Gamma(\alpha) = (\alpha - 1)!$$

[1] It is assumed that the average remains constant from trial to trial, so that there is no time trend, and that the probability of an event is unaffected by the time elapsed since the preceeding one.

where α is any positive integer. It can be shown that

$$\Gamma(\alpha) = \int_0^\infty x^{\alpha-1} e^{-x} dz.$$

α is a shape parameter estimated from

$$\alpha = \frac{1}{4A}\left\{1+\left(\frac{1+4A}{3}\right)^{1/2}\right\}$$

$$\text{where } A = \ln \bar{x} - \frac{(\sum^n \ln x)}{n}$$

$$\bar{x} = \alpha\beta$$

and $\sigma^2 = \alpha\beta^2$

where β is a scale parameter.

The probability density function[1] is

$$f(x) = \frac{x^{\alpha-1} e^{-x/\beta}}{\beta^\alpha \Gamma(\alpha)}.$$

To determine cumulative probabilities, for example, of precipitation $\leqslant x_i$ we use

$$t(F) = x_i/\hat{\beta}.$$

Tables of this incomplete gamma function are available (Pearson *et al.* 1951) for F against $\hat{\alpha}$ and $t(F)$.

The distribution is particularly useful for zero-bounded variables such as short-period precipitation totals (Suzuki, 1964, 1967). Decker (1952) has also applied it to evaluate the probabilities of damaging hail occurrences in Iowa.

Transformations

Many meteorological data series are so skewed that the more common distribution curves cannot be fitted. As it is often desirable to use parametric statistical methods based on the Normal distribution an appropriate transformation of the original data must be made. For example, distributions of short-period precipitation data are invariably truncated on the 'dry side' and a log-normal curve is usually appropriate[2] (Essenwanger, 1960b). Thus, for each x_i we use the logarithm of x_i. A cube root transformation has also been advocated (Stidd, 1953; Kendall,

[1] It is worth noting that the χ^2 function is a special case of the Gamma distribution where $\beta = 2$ and $\alpha = n/2$, where n = degrees of freedom.

[2] Chow (1955) deals with some theoretical and practical aspects of log-normal probabilities.

1960; see also Huff and Neill, 1959). At the same time we should note the results of careful analyses at Hohenpeissenberg Observatory (Grunow, 1956) which indicate that the number of completely dry days diminishes as greater care is taken with the observations. Collectives for very small precipitation amounts also appear to follow a log-normal distribution.

The adequacy of an assumed distribution function, after transformation of the original data, can readily be checked by using χ^2 as a test of goodness of fit. The expected frequencies according to the assumed model are compared with the observed frequencies (see table 4.3, for example) and the χ^2 statistic is evaluated as shown in table 4.2. For a satisfactory fit the χ^2 value should exceed the 95% probability level (i.e., the *opposite* end of the χ^2 tables from that used in table 4.2 to assess the significance of the association).

5 SIMULATION

A common problem in applied studies is the inadequacy of available data series. The advent of computers has greatly facilitated the use of mathematical simulation – specifically, that category known as *Monte Carlo methods* – to overcome such difficulties by generating a set of synthetic data. The procedure has three basic steps:

1 Identify a probability model appropriate to the sample series.
2 Sample from the probability distribution by using a sequence of random numbers.
3 Repeat this sampling to obtain approximate solutions of the model. A succession of such trials tends to converge towards an average solution.

Alternatively, a function may be converted into a cumulative probability function and the latter used for the Monte Carlo sampling.

In applied statistics the method has been used to determine parameters for functions with complex distributions such as Gamma and, also, as a means of establishing criteria for significance tests (Lund, 1970). The main interest from our point of view, however, is the simulation of some random process. The probability model may be based on three types of trial: Bernoulli trials where the probability remains constant throughout; Poisson trials where the probability varies from trial to trial in an individual series but remains constant in each series and also in the *corresponding* trials of different series (for example, values of temperature on a set of dates in a series of Januaries); and Lexian trials where the probability is constant within a series but varies between series.

As a simple illustration we shall use the hurricane frequencies of table 4.3 and postulate that they follow a Poisson frequency distribution. This may be simulated by assigning 100 numbers in accordance with the Poisson distribution as shown in table 4.4(*a*) and then tabulating the frequency of sets of 100 random numbers read sequentially from appropriate tables. Table 4.4(*a*) shows that the average of only two trials provides a close approximation to the Poisson distribution. In the second example (table 4.4(*b*)) the distribution function is not assumed. Instead, the observed cumulative distribution is used as the basis for random sampling proportionate to the strata. This method provides a long-term estimate of hurricane frequency and, where the mean is not

Table 4.3 Frequency of formation of hurricanes in the Gulf of Mexico, 1886–1958 (data from Jordan and Ho, 1962)

Number per year (x)	*Number of years frequency occurred*	*Probability*	*Expected frequency (years/73)*
0	36	0·464	33·9
1	23	0·357	26·1
2	10	0·137	10·0
3	3	0·035	2·6
4	1	0·007	0·5

The probabilities are determined as follows, given a mean frequency over the 73 years of 0·77:

$$P(x) \approx e^{-0\cdot 77}\left(1+0\cdot 77+\frac{(0\cdot 77)^2}{2}+\frac{(0\cdot 77)^3}{6}+\frac{(0\cdot 77)^4}{24}+\cdots\right)$$

$$= 0\cdot 464+0\cdot 357+0\cdot 137+0\cdot 035+0\cdot 007 = 1\cdot 00.$$

The probabilities are converted to expected frequencies by multiplying them by 73.

accurately known, it represents a useful application of Monte Carlo simulation. Many refinements of this basic technique have been developed to make the procedures more rapid and more efficient for complex problems (see, for example, Hiltier and Lieberman, 1967), but in climatology such techniques have only recently been adopted. Gringorten (1966) has derived simulated probability distributions of various climatological conditions such as the duration of cloudy skies at Minneapolis and of freezing conditions at Boston. Here, successive hourly values of a variate are generated by a simple Markov chain model (see section B of this chapter, p. 239) with a constant hour-to-hour correlation between the successive values. Sufficiently stable results were obtained with 100 samples of 768 random normal numbers (768 hours = 32 days).

A further interesting application of randomized methods has been described by Palmieri *et al.* (1969). A computer method of using random line segments has been developed for recognizing meteorological pressure patterns. Illustrations are given relating to cyclogenesis in the Gulf of Genoa and to precipation occurrence over Italy. Where a synoptic classification scheme has been devised this technique will evidently greatly facilitate its application to a weather map series.

Table 4.4 Illustration of a Monte Carlo Experiment

(a)

Hurricanes/ year	*Poisson model frequency (%)*	*Random numbers*	*Frequency of first 100*	*Frequency of second 100*	*Mean*
0	46	0–45	41	51	46·0
1	36	46–81	38	29	33·5
2	14	82–95	16	14	15·0
3	3	96–98	4	4	4·0
4	1	99	1	2	1·5

(b)

Hurricanes/ year	*Observed frequency (%)*	*Random numbers*	*Frequency of first 100*	*Frequency of second 100*	*Mean*
0	49	0–48	50	46	48·0
1	32	49–80	31	39	35·0
2	14	81–94	12	11	11·5
3	4	95–98	6	3	4·5
4	1	99	1	1	1·0

B Time series

1 BASIC CONSIDERATIONS

Almost any set of climatological data is ordered chronologically, which is to say that it constitutes a time series.[1] A variety of causal factors may interact to generate a particular series and in order to try and distinguish these many elaborate techniques have been developed. Formally, we can characterize a time series as consisting of a *deterministic element* – this may be strictly periodic (annual and diurnal cycles) or a trend of indefinite period – and a *random element*. In practice the latter is the unaccounted for residual.

[1] General references include Matalas (1967), Quenouille (1957), Rosenblatt (1963) and Tyson (1969).

A number of fundamental problems arise in time series analysis. First, the series is invariably truncated owing to limitations of the available data. This greatly hinders the analysis of long-term fluctuations because they are inadequately defined by data points. Short-term fluctuations and the occurrence of singularities (see Chapter 5.A, p. 292) create particular problems when one is interested in, for example, means of daily or pentad values and only a limited record is available (Takahashi, 1965). This problem is common in the case of upper air and sea temperature data. Second, the observations represent a sample and the relevant population to which this refers cannot generally be defined because meteorological phenomena do not occur randomly (Panofsky, 1949). Moreover, it has usually to be assumed that a unique set of long-term statistical properties of the atmosphere do exist although we cannot be sure that this is the case. Third, most techniques assume *stationarity* of the time series, which is to say that a shift in the time origin does not affect the distribution function. This implies that the mean is constant and independent of time, that the probability distribution is the same for all time intervals, and that the variance and higher moments are independent of time. Most climatological series are probably non-stationary and the interpretation of certain analyses will therefore be uncertain.

In addition to these intrinsic problems other difficulties may arise as a result of heterogeneity in the data due, for example, to change in the site or its surroundings, to instrument modifications or to revised observing practices. Often a simple *double-mass curve analysis* will detect such changes (Bruce and Clark, 1966; Weiss and Wilson, 1953). Cumulative values of the station under test are plotted on the ordinate against cumulative mean values for six to ten neighbouring stations on the abscissa. If the series is not homogeneous the plot will show a break in the slope. For statistical evaluation of homogeneity a non-parametric test, such as the 'runs test', with respect to values above and below the median, is preferable (Thom, 1966) although, if it is known from the station history that some discontinuity is present, the two parts of the record may be compared using Student's t test (see p. 289).

2 PERIODIC ELEMENTS

It is convenient to discuss first the analysis of periodic components of a time series. If there is good reason to believe that a data series comprises dominant periodic elements, we can represent these components by a series of sine and cosine functions. We refer to this procedure as *harmonic or Fourier analysis.*

The form of the sine and cosine functions is illustrated in fig. 4.5(a) for the range 0 to 2π. This range may be equated to a time interval such as twenty-four hours or a year, or to a spatial interval (see section C of this chapter, p. 263). The series for a finite function $f(x)$ is

$$f(x) = a_0 + a_1 \cos x + a_2 \cos 2x + \cdots + a_n \cos nx + b_1 \sin x + b_2 \sin 2x + \cdots + b_n \sin nx +$$

$$= a_0 + \frac{2}{n}\left[\sum_{k=1}^{n} a_k \cos kx + \sum_{k=1}^{n} b_k \sin kx\right]$$

where a_0 = the mean, n = the total length of the time period. Note that in total there are $n/2$ harmonics. The expression can be rewritten

$$f(x) = a_0 + \sum_{k=1}^{n/2} [A_k \cos (kx - \phi_k)]$$

where $A_k = (a_k{}^2 + b_k{}^2)^{1/2}$, the amplitude of the harmonic and
$\phi_k = \arctan (b_k/a_k)$
$= \arcsin (b_k/A_k)$

which avoids the ambiguity of two solutions for the tangent between 0 and 2π. ϕ_k represents the phase difference (time interval) between each harmonic or wave. It follows that

$$a_k = A_k \cos \phi_k \quad \text{and} \quad b_k = A_k \sin \phi_k.$$

Computer programs for determining the amplitude and phase angle are now widely available so that the previous laborious methods are no longer necessary (Brooks and Mirrlees, 1930). Craddock (1965), for example, was able to analyse the annual temperature variation at forty-two stations in Europe in order to isolate possible singularities (see Chapter 5.A, p. 293). The most recent computer programs incorporate an algorithm known as the Fast Fourier transform which has significantly reduced the computation time and cost.

A number of studies have been made of seasonal precipitation types using harmonic analysis (Horn and Bryson, 1960; Sabbagh and Bryson, 1962; Walker, 1964; Lettau and White, 1964; McGee and Hastenrath, 1966). Here the first harmonic denotes an annual cycle with a single maximum and minimum, the second a semi-annual cycle and so on (fig. 4.5(*b*)). The contribution of each harmonic to the total variance is calculated from

$$\frac{A_k{}^2}{2s^2} \quad \text{(except for } k = n/2 \text{ where the contribution is twice this value)}$$

where $2s^2 = \sum_{k=1}^{n/2} A_k{}^2$. In cases where the first harmonic accounts for a large proportion of the variance it is worthwhile mapping this value (the amplitude) and the phase angle. Fig. 4.6 shows the phase angle of the

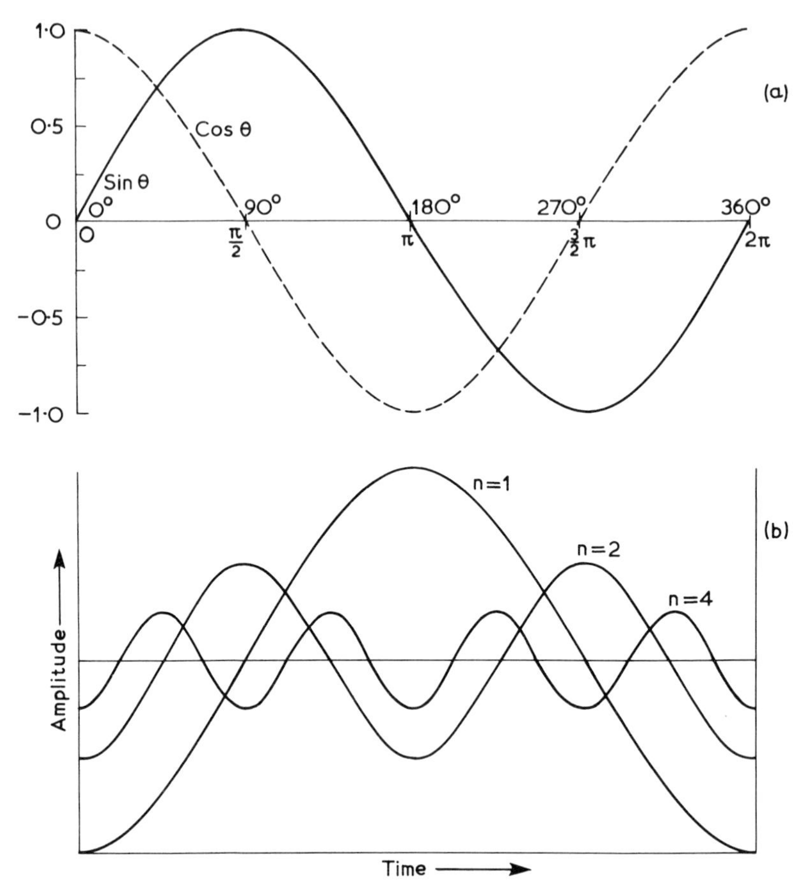

Fig. 4.5 (*a*) Sine and cosine functions between 0 and 2π (0–360°). (*b*) Simple harmonic curves for $n=1$, 2 and 4. The amplitude is shown as being reduced by 50% from $n=1$ to $n=2$ to $n=4$.

first harmonic over the United States and the ratio of the second to the first harmonic. The latter indicates the relative strength of the semi-annual component over the intermontane west. Fig. 4.6(*a*) indicates that the mid-winter maximum is later going southward along the west coast, although the amplitude (not shown) also decreases sharply, while east of the Rockies the spring maximum (300–315°) in Wyoming is delayed until mid-July (270°) in Wisconsin. In Tennessee the 60° line demarcates a February maximum. The map indicates some marked changes across New England but, in spite of the importance of the first harmonic evidenced by the ratio map, it should be noted that the amplitude of the first harmonic is small. Seasonal contrasts in this region are weak. As pointed out by Rayner (1971, pp. 30–2) the method is appropriate only in so far as periodic processes *are* operating. While this assumption finds some support in physical arguments for the annual and semi-annual periods (see p. 400), shorter periods do not. Moreover, harmonics of the annual cycle with frequencies of 2, 3, 4, 5 and 6 cycles/year are superimposed on the computations when monthly averages are used. The annual cycle is not affected, however.

In spite of the limitations the technique provides some guide to the delineation of regions with similar types of seasonal precipitation distribution.

3 IRREGULAR FLUCTUATIONS

Frequently the fluctuations in a time series consist of irregular, superimposed oscillations and random elements. The most powerful tool for sorting out these various components is *variance spectrum analysis*. The starting point is to examine the degree of correlation in the time sequenced data. We compute the coefficient of *serial correlation* (or autocorrelation), given approximately by

$$r_L = \frac{\sum_{i=1}^{N} x_i x_{i+L}}{N \sum_{i=1}^{N} x_i^2}$$

or more nearly by

$$r_L = \frac{\sum_{i=1}^{N-L} x_i x_{i+L} - \frac{1}{N-L} \sum_{i=1}^{N-L} x_i \sum_{i=L+1}^{N} x_i}{\left[\sum_{i=1}^{N-L} x_i^2 \sum_{i=L+1}^{N} x_i^2 \right]^{1/2}}$$

where L = the lag, or interval between the pair of observations being correlated and x_i are the values in the series for $i = 1, \ldots, N$. Most meteorological time series show large correlations for short lags (Schumann and Hofmeyr, 1942) and this affects the accuracy of arithmetic means. The standard deviation of the mean of n consecutive values, each with variance s^2 is $s/\sqrt{n_1}$, where n_1 is the number of independent days. For various values of r_L for $L = 1$ and pentad or monthly means, n_1 is much smaller than n as shown by Godske (1966):

r_L =	0·0	0·4	0·6	0·8	1·0
n = 5	5	2·64	1·91	1·38	1
30	30	13·28	8·00	3·91	1

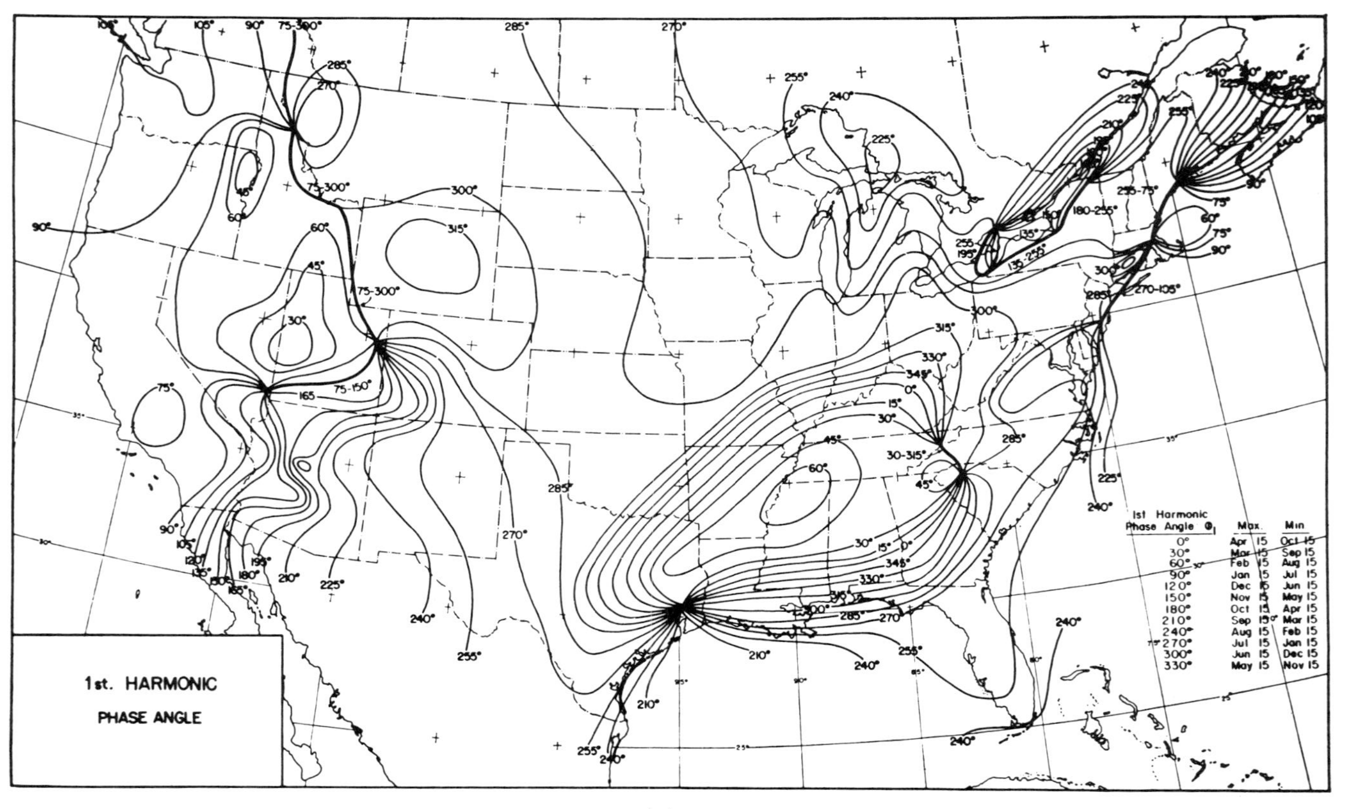
1st. HARMONIC
PHASE ANGLE
1st Harmonic Phase Angle Φ_1 | Max. | Min
0° | Apr 15 | Oct 15
30° | Mar 15 | Sep 15
60° | Feb 15 | Aug 15
90° | Jan 15 | Jul 15
120° | Dec 15 | Jun 15
150° | Nov 15 | May 15
180° | Oct 15 | Apr 15
210° | Sep 15 | Mar 15
240° | Aug 15 | Feb 15
270° | Jul 15 | Jan 15
300° | Jun 15 | Dec 15
330° | May 15 | Nov 15

(a)

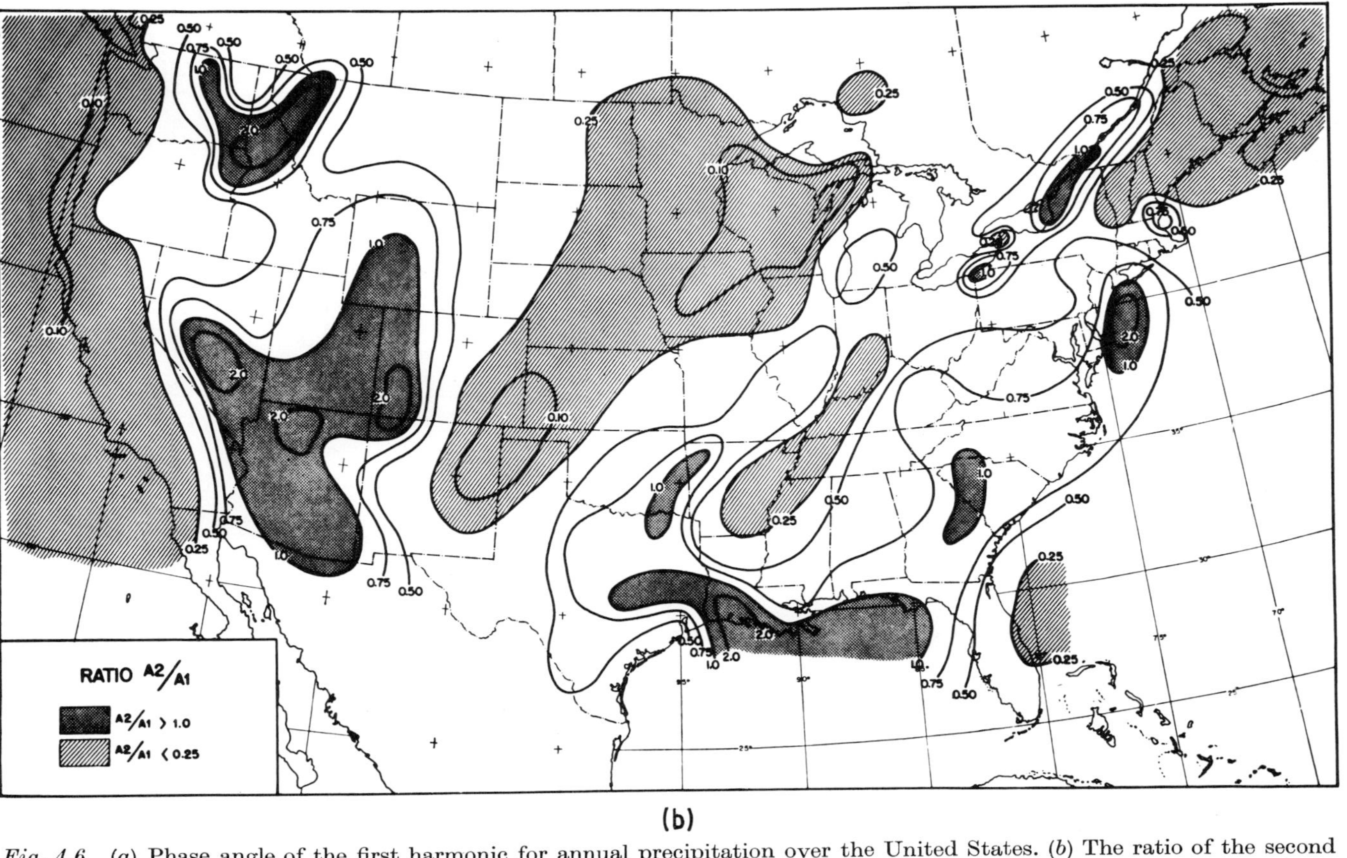

Fig. 4.6 (*a*) Phase angle of the first harmonic for annual precipitation over the United States. (*b*) The ratio of the second to the first harmonic (from Horn and Bryson, 1960).

For typical serial correlation values there may, for example, be only five to ten independent daily values contributing to a monthly mean.

This serial correlation, reflecting the persistence of cyclone and anticyclone systems over several days, is known as *coherence*. Plots of r_L against the lag, termed *correlograms*, as shown in fig. 4.7, illustrate this characteristic (Craddock, 1968). Pronounced peaks in a correlogram represent periodicities at about the lag indicated by L. In a study of British rainfall data for 1727–1929, Alter (1933) found no evidence for long-term cycles by this method. Unfortunately, this approach is severely limited. The statistical significance of the autocorrelation is not easily determined and the correlogram cannot be readily interpreted for

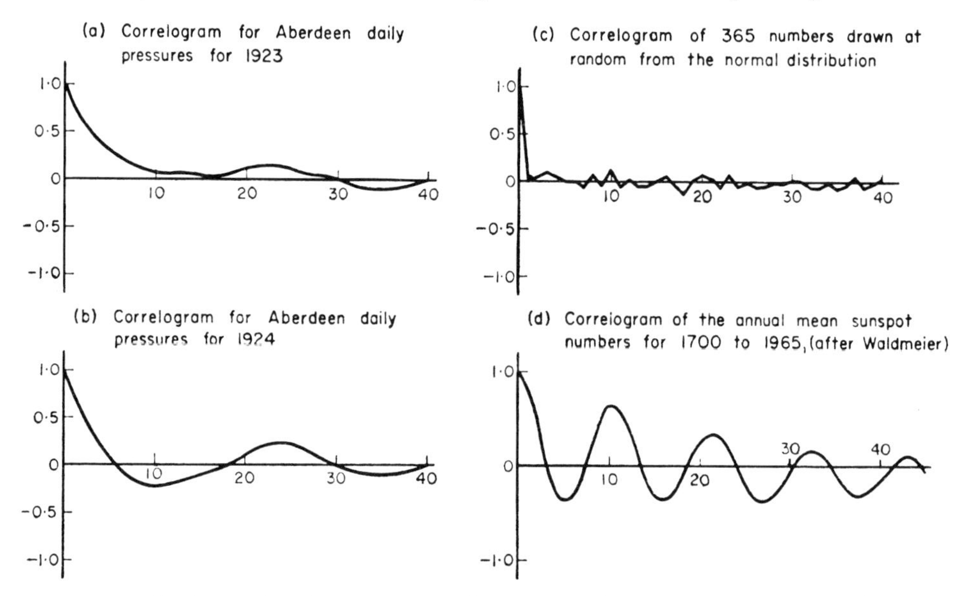

Fig. 4.7 Examples of correlograms showing varying degrees of coherence (from Craddock, 1968).

the longer lags. Little reliance can be placed on autocorrelation estimates when the sample series is short.

Instead of examining the correlogram, therefore, attention is focused on the *variance spectrum* which is more tractable with respect to evaluation of statistical significance and physical interpretation. This spectrum provides information on the contribution of the various frequency components to the total variance. Essentially, the method is to plot the variance of the quantity examined ($A_k{}^2/2$) against the frequency (cycles per unit time interval). Frequency is the reciprocal of the period, so that 0·14 cycles/day corresponds to a period of approximately one week and 0·033 cycles/day to approximately one month. This direct approach is unworkable in practice due to sampling variations. Instead, the serial correlation coefficients (up to $L \sim n/5$, where n is the total number of observations in the series) are computed and subjected to an harmonic analysis. The coefficients, a_k (see p. 229), are then smoothed by a weighted moving average and plotted against frequency. The use of the serial correlation coefficients (which have $s_x{}^2$ in the denominator) ensures that the spectrum is normalized, such that there is unit area under the curve. The basic procedures are detailed by Blackman and Tukey (1958), Jenkins (1961), Cox and Lewis (1966), Jenkins and Watts (1968) and Rayner (1971). Concise reviews are given by Panofsky (1955), Cavadias (1967) and Julian (1967). Landsberg *et al.* (1959) and Bryson and Dutton (1962) provide illustrations of some climatological applications.

At this point it is worthwhile identifying a number of basic types of spectrum. Fig. 4.8(a) shows, schematically, spectra for 'white noise' (random uncorrelated data), quasi-periodicities, and 'red noise', where long periods (i.e. low frequencies) contribute most of the variance.[1] While frequency is used for the abscissa in fig. 4.8(a) it is often preferable to plot log frequency. Mandelbrot and Wallis (1969b) suggest that simple frequency plots may indicate randomness where none exists. There is equal need for care in interpreting semi-logarithmic plots, where a white-noise spectrum has a maximum at 0·5 cycles per sampling interval if the ordinate shows frequency x variance (Julian, 1966).

An important point to note in spectral analysis is the occurrence of *aliasing* between different frequencies. The shortest period (highest frequency) about which a data series gives information is $\frac{1}{2}\,\Delta t$, where Δt is the time interval between observations. This is known as the Nyquist frequency, F_0. For example, at least two observations per day are necessary to estimate the diurnal variation. For any frequency, f,

$$f,\ 2F_0 - f,\ 2F_0 + f,\ 4F_0 - f,\ 4F_0 + f,\ \ldots,$$

are aliases such that variance is added to them by unresolvable frequencies which exceed F_0. Fig. 4.8(b) illustrates aliasing between frequencies. The shorter fluctuations cannot be resolved by the available data. The low frequency (long period) limit is between one cycle per $n/5$ to $n/10$, where n = number of observations.

The computational procedures of spectrum analysis are not detailed here because the whole subject has been radically transformed in the last few years by the introduction of the Fast Fourier transform in harmonic analysis (Tukey, 1967; Hinich and Clay, 1968; Rayment, 1970).[2] It is still desirable to eliminate sampling variations and the usual

[1] The term 'red noise' was proposed by E. N. Lorenz (Gilman, Fuglister and Mitchell, 1963).

[2] Many computer programs for spectral analysis which incorporate this approach are now available.

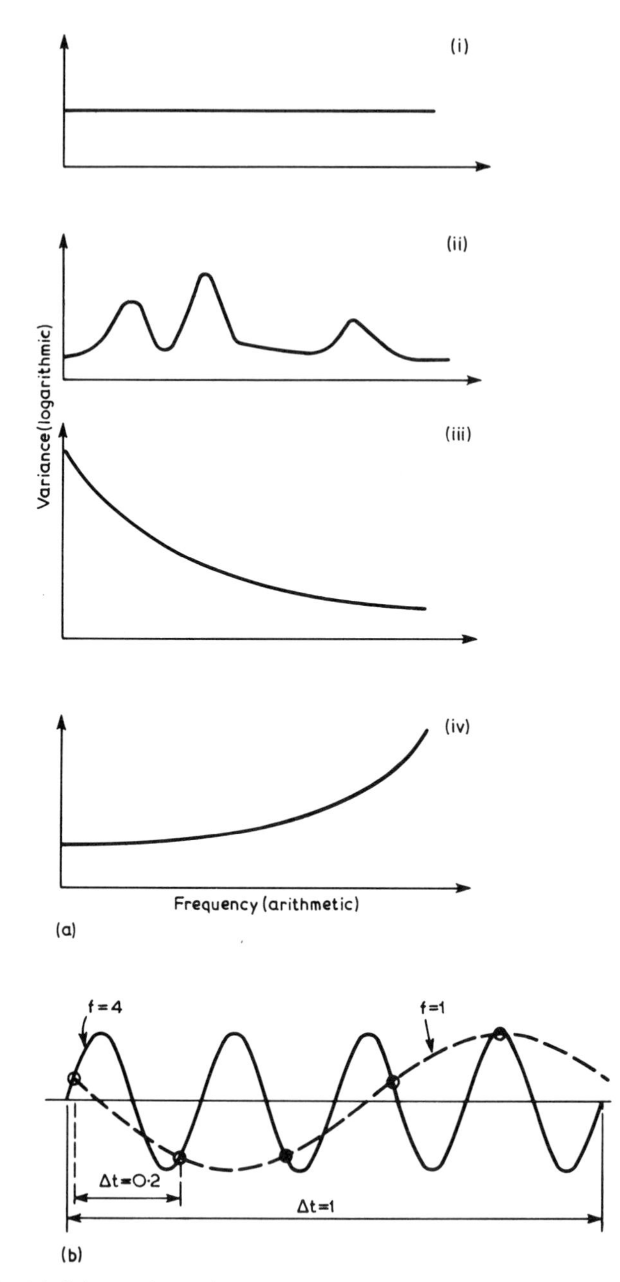

Fig. 4.8 (*a*) Schematic variance spectra: (i) random (white noise); (ii) quasi-periodic oscillations at three frequencies; (iii) persistence (red noise), an autoregressive series with low frequencies predominating; (iv) a rapidly oscillating autoregressive series with high frequencies predominating.

(*b*) Aliasing of waves (after Blackman and Tukey, 1958). The sampling causes the sinusoidal wave with a frequency of 4 to appear as a wave with a frequency of 1. For $\Delta t = 1$ the Nyquist frequency is 0·5. For $\Delta t = 0{\cdot}2$ the Nyquist frequency is 2·5.

practice is to divide the total period into several non-overlapping sections and compute the variance spectrum for each. The spectra are then averaged, which reduces the resolution of the frequencies but improves the stability of the estimates.

Spectra of daily temperatures have been used as a climatic indicator in North America by Polowchak and Panofsky (1968). At some stations in the continental interior in winter the variance is mainly concentrated in periods of about two weeks, whereas at east coast stations the major period is one week or less (see also section C of this chapter,

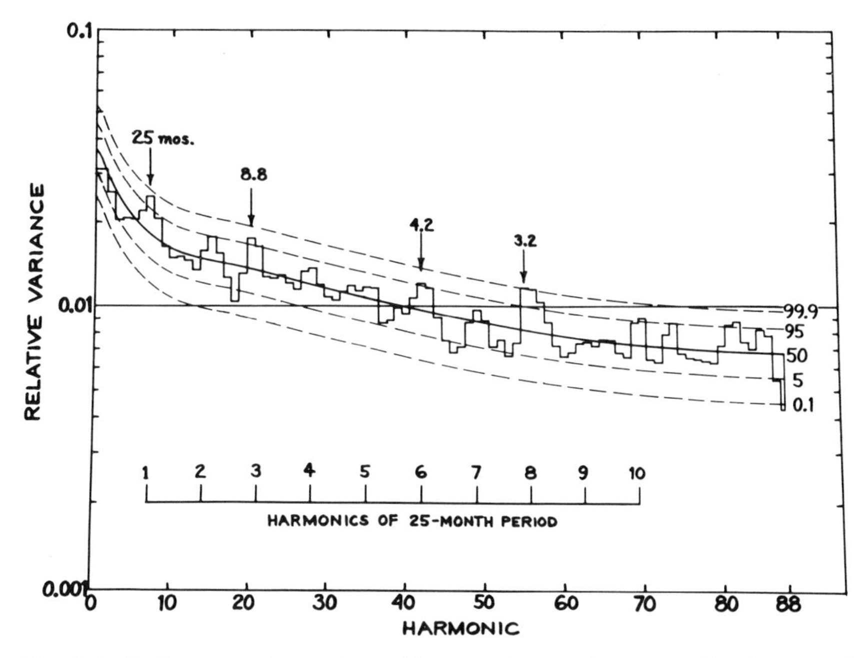

Fig. 4.9 Variance spectrum of monthly mean temperature anomalies for central England, 1698–1952, with a maximum lag of eighty-eight months. Smooth curves show 'null' spectrum and associated confidence limits (0·1–99·9). Arrows denote statistically significant peaks believed to be related to the biennial oscillation (from Mitchell, 1966).

p. 254). Spectrum analysis has also been applied to the problem of distinguishing different scales of tropospheric wave disturbance in the tropical Pacific (Wallace and Chang, 1969). Extensive work by Ward and Shapiro (1961) is of particular synoptic climatological interest. They have prepared spectra for the following: meridional and zonal indices in the northern hemisphere, persistence measures and also polynomial coefficients of the surface pressure field over North America and Europe, and polynomials of the western half of the northern hemisphere 500 mb height field. Their conclusion is that all of these series are characterized by persistence (red noise) and by diurnal and annual cycles. Fig. 4.9 shows persistence in a 269-year temperature series together with the biennial oscillation (see p. 398) and harmonics of it.

This characteristic of persistence creates serious problems for significance testing of the peaks in a variance spectrum. The spectral estimates are usually based on random samples, which is rarely true of meteorological data series. One approach, for the case where peaks are superimposed on a background of simple persistence, is discussed by Gilman *et al.* (1963), and Julian (1966) has applied their method in testing spectra for zonal index data.

For some purposes it is desirable to exclude frequencies other than those about some selected frequency. This general procedure is known as *filtering*. The removal of the mean from a series is the simplest example. Equally weighted filters, such as the simple running mean, have the effect of reducing the amplitude of oscillations with a period which is an integer multiple of the length of the moving average to zero; i.e., if the length of the moving average is t, the response function is 0 for periods of length t, $2t$, $3t$, etc. However, the response function is negative for periods between t and $2t$, $3t$ and $4t$, etc. which means that the *phase* of such periodicities in the smoothed series is inverted by comparison with the original data. Special filters can be designed to eliminate or suppress unwanted frequencies. For example, in studies of the biennial atmospheric oscillation (see p. 398) Landsberg *et al.* (1963) use 'band-pass' filters to allow only variations with a period close to twenty-four months. The same technique was employed by Craddock (1957) using a 'low-pass' filter to analyse long-term temperature variations at Kew. However, the design of filters for special purposes is a complex operation (Holloway, 1958; Craddock, 1968).

4 PERSISTENCE IN SPELLS

The atmosphere reflects the importance of persistence effects on many time and space scales. We shall look now at the occurrence of sequences of designated weather conditions, in other words *runs* or *spells* of a given type over a few days.

Efforts to develop appropriate theoretical probability models have a long history (Newnham, 1916; Gold, 1929; Cochran, 1938; Williams, 1952). Basically, these studies attempt to relate the length of runs of wet and dry days in particular, to simple probability models. Gold, for example, considered two-state events that occur independently and with equal probability – a geometric distribution – which was also found to be satisfactory for wet and dry spells in Canada (Longley, 1953) and Israel (Gabriel and Neumann, 1957), while Cochran extended analysis to conditions with unequal probability. Williams found that the distribution of the duration of spells of wet and dry days ($\geqslant$ or $<0{\cdot}01$ in) at Rothamsted, Hertfordshire, during 1938–47 approximates a logarithmic series (fig. 4.10), and Srinivasan (1964) applied a similar model to the occurrence of monsoon rain spells in India.

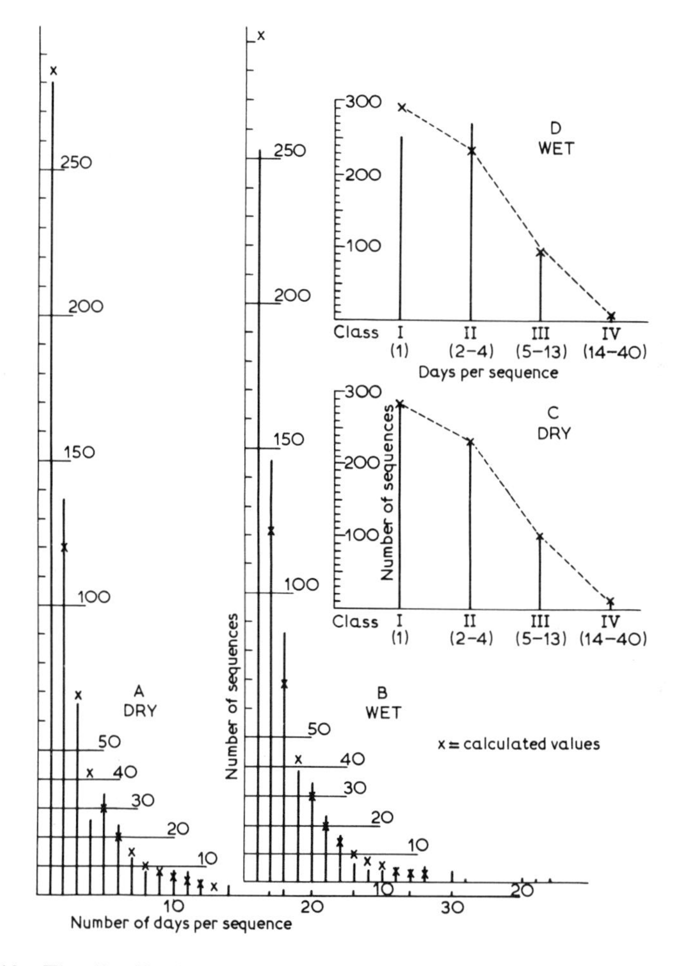

Fig. 4.10 The distribution of the duration of wet and dry spells, daily precipitation $\geqslant$ or $<0{\cdot}1$ inches, at Rothamsted, 1938–47 (from Williams, 1952). A = dry days, B = wet days, arithmetic scale; C = dry days, D = wet days, logarithmic scale; histogram = observed; crosses and dashed line = calculated for a log series.

Recent studies of spells have shown that often the probabilities can be well represented by a simple (first-order) *Markov chain* model, where the conditions on the preceding day only are considered (Gabriel and Neumann, 1962). If P_0 is the (conditional) probability of a wet day following a dry day, and $(1-P_1)$ is that of a dry day following a wet day, so that

$$P_1 = P(\mathrm{W} \mid \mathrm{W}) \qquad 1-P_1 = P(\mathrm{D} \mid \mathrm{W})$$
$$P_0 = P(\mathrm{W} \mid \mathrm{D}) \qquad 1-P_0 = P(\mathrm{D} \mid \mathrm{D})$$

where | denotes time succession, then the probability of a dry spell of length $n(pn)$ can be shown to be

$$P(n) = P_0(1-P_0)^{n-1} \quad n = 1, 2, 3, \ldots$$

and of a wet spell of length m, (Pm), is

$$Pm = (1-P_1)P_1{}^{m-1} \quad m = 1, 2, 3, \ldots$$

The usefulness of this geometric distribution approach has been demonstrated in connection with warm and cold spells (Caskey, 1964) and for wet and dry spells in many parts of the world (Caskey, 1963; Hopkins and Robillard, 1964; Weiss, 1964; Feyerherm and Bark, 1965; Fitzpatrick and Krishnan, 1967; Watterson and Legg, 1967). Some results have raised doubts as to its universal applicability to wet and dry spells (Williams, 1952; Green, 1965) but detailed tests by Feyerherm and Bark (1967) encourage confidence. They examined several stations in the north-central United States with average annual precipitation totals from 445 to 1062 mm (17·5–41·8 in). The fit of a first-order Markov chain

$$P(x_t, x_{t+1}, \ldots, x_{t+k}) = P(x_t)P(x_{t+1} \mid x_t)\ldots P(x_{t+k} \mid x_{t+k-1}),$$

where x may represent either a dry day or a wet day and t = the first day in the series, was compared with that of a second-order chain, where the state on day t depends on the two preceding days,

$$P(x_t, x_{t+1}, \ldots, x_{t+k}) = P(x_t)P(x_{t+1} \mid x_t)P(x_{t+2} \mid x_{t+1}, x_t)\ldots P(x_{t+k} \mid x_{t+k-1}, x_{t+k-2}).$$

They found that, on the one hand, the relative frequency for a wet day preceded by two wet days is less than that where the preceding two-day sequence was dry. On the other hand, the relative frequency for a wet day (t) preceded by a dry day is unaffected by the state on day $(t-2)$. They concluded that in spite of this difference, which is most marked in spring, there is no serious overall departure from a first-order model. Wiser's result (1965) also indicates that this simple model is usually adequate for daily data although probably not for information at hourly intervals. Analysis of precipitation data by Lowry and Guthrie (1968) suggests that higher order models may be required for regions where there is greater diversity of seasonal weather.

One practical problem is that conditional probabilities are seldom available. Direct evaluation may not always be essential, however.

Hershfield (1970) shows that at least for dry day sequences the unconditional probabilities (actual frequencies) of days receiving less than a stated threshold amount can be used to estimate the conditional probabilities.

The good fit of the Markov chain model indicates a widespread tendency for a rapid loss of dependence between weather events, although it should be emphasized that the model provides only a statistical description of the observations. It is possible to interpret the dependence on the previous day as being due only to the tendency of many meteorological systems to cause, or inhibit, precipitation across two or more adjoining time intervals. Alternatively, there may be physical mechanisms which cause one occurrence of precipitation to ehnance the possibility of subsequent precipitation in certain situations. There remains much work to be done in extending this type of analysis to other time intervals and for various precipitation thresholds in order to clarify the physical processes in operation.

While much attention has been given to the Markov chain in the literature, other studies should not be overlooked. An intensive examination of various persistence models was carried out by Lawrence (1957) for spells of dry days occurring during summer months in southern England. Frequency distributions at many stations were compared with four series: geometric (with the probability remaining constant with time, i.e. zero persistence), logarithmic (persistence increasing with time, i.e. positive persistence), 'Jenkinson probability' type (smoothed persistence effect), and 'natural persistence' type (actual persistence values were introduced). The last series gave the best results for area–mean frequency data, which indicates that mathematical representations are invariably inexact to a greater or lesser degree! Lawrence shows that at least in southern England in summer the persistence factor varies with time as follows:

Length of run of dry days	*Persistence*
Up to 8–10 days	Positive
8–10 to 18–20	Approximately zero or slightly negative
18–20 to 25	Slightly positive
25 to 30	Slightly negative
Over 30	Negative

Negative or anti-persistence implies an increased likelihood of change. Belasco (1948) noted a tendency for anti-persistence in sequences of anticyclonic weather after twenty days at Kew. These statistical results are therefore open to synoptic climatological interpretation. The wider question of the difference in persistence between different areas is one that has received little attention.

The problem of persistence has recently been treated in a very different manner by Mandelbrot and Wallis (1968, 1969a,b,c,d). Their starting point is the known tendency for many hydrometeorological records to show very extreme high (and low) values – the 'Noah effect', and intervals of persistent high (or low) values – the 'Joseph effect'.

The 'Joseph effect' is more significant in the context of synoptic climatology than the question of extremes. Mandelbrot and Wallis (1969c) demonstrate that a wide range of geophysical series exhibit interdependence over the longest available records, and they have succeeded in developing an appropriate mathematical model (1969a, 1969b) for such data. They show (1969b, 1969d) that when non-cyclic long-run effects are present in the data, and especially when the distribution is highly non-Gaussian (for example, with Noah effects), an analysis technique related to their model is much preferable to spectral analysis. They developed a statistic, the 'rescaled range', which is shown to be very robust, for testing for the presence of (non-cyclic) long-run dependence in records and, where it is present, for estimating its intensity. Although these methods cannot be detailed here, they provide valuable insight into persistence effects and new analytical tools for study of climatological time series. The concepts appear to be of particular importance for research into climatic change.

5 RELATIONS BETWEEN TWO TIME SERIES

Correlations between two time series can seldom be investigated adequately by simple methods (Quenouille, 1952; 1957; Matalas and Benson, 1961; Hamon and Hannan, 1963). The general relationship between two time series may be examined by cross-correlation but the existence of a time lag in one series relative to the other, or of positive correlations in one frequency range and negative correlations in another, for example, will usually be detected only by sophisticated techniques of analysis. The appropriate tool for this purpose is cross spectrum analysis (Cavadias, 1967; Rayner, 1971). This technique (based on the Fourier transform of the cross-correlation function) determines the relationship between two series in the frequency domain by evaluating the contribution of a specific frequency in both series to the total cross-covariance. Four items are calculated:

1 the *cospectrum* $C_{xy}{}^2(F)$ which specifies the contribution of each frequency, F, to the total cross-covariance of the two series x and y for zero lag.

Essentially, the cross-covariance of the two series for positive and negative lags are averaged and plotted against the lag. The covariances are then subjected to a Fourier analysis and the computed coefficients give the cospectrum.

2 The *quadrature spectrum* $Q_{xy}{}^2(F)$ which measures the contribution to the covariance components which are $\pi/2$ out of phase.

Differences between the cross-covariance of lag t and lag $(-t)$ are calculated and averaged; the coefficients determined from a Fourier analysis of these results then give the quadrature spectrum. The co-spectrum is the cosine component and the quadrature spectrum the sine component of the total cross variance.

3 The *phase difference* $\phi_{xy}(F)$ between the components of the two series at each frequency.

$$\phi_{xy}(F) \text{ in radians} = \arctan \left(\frac{Q_{xy}(F)}{C_{xy}(F)}\right)$$

4 the *coherence* $W_{xy}(F)$ which is a measure of the correlation between the two series at each frequency.

$$W_{xy}(F) = \left\{\frac{C_{sy}(F)^2 + Q_{sy}(F)^2}{S_x(F)S_y(F)}\right\}^{1/2}$$

where S_x and S_y denote the variance spectra of the two series. Again the use of the Fast Fourier transform has revolutionized the computational procedures. Computer library programs are widely available for cross-spectrum analysis and it is unnecessary to go into those details here.

Interpretation of cross spectrum analysis involves the study of the characteristics of the coherence spectrum and the phase spectrum.

Among the earliest meteorological applications of this technique was an analysis of the relationships between zonal indices at 25°, 40° and 60°N by Panofsky and Wolff (1957). Subsequently Julian (1966) examined this question further and concluded that neither the cospectra nor the coherence data indicate a preferred period in the range three to eight weeks, as is implied by the 'index cycle' concept. More recently, Doberitz (1967, 1968) has demonstrated the use of cospectrum analysis to delineate areas with homogeneous precipitation and sea temperature regimes in the tropical Pacific and Atlantic Oceans. Fig. 4.11 illustrates the distribution of four types of spectrum. Rayner (1965, 1967) uses spectrum and cross-spectrum analysis applied to time and space series to characterize weather regimes in the New Zealand area. This is discussed below in section C of this chapter (pp. 266–8).

A point to emphasize is the need for care in interpreting the results of complex analyses of this type. This is well illustrated in a study of the spectra of temperature, precipitation and pressure over the western

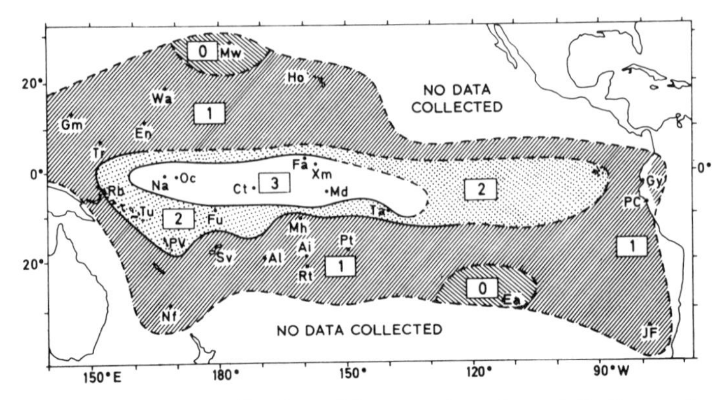

Fig. 4.11 The spatial distribution of four spectra of precipitation regime in the tropical Pacific (from Doberitz, 1968). 0 = no obvious high frequencies; 1 = well-marked annual cycle; 2 = mixture of 1 and 3; 3 = periodicities mainly > 1 year. Dots show stations.

United States by Rodriguez-Iturbé and Nordin (1969). Fig. 4.12 shows spectra and coherence graphs for temperature, precipitation and pressure at Eureka, California. The coherence graph (fig. 4.12(*b*)) suggests that temperature and precipitation are each strongly correlated with pressure, but the computation of *partial coherence* values (fig. 4.12(*c*)), analogous to partial correlation coefficients, demonstrates that the annual cycle in precipitation is related to that in pressure only through the effects of the annual temperature cycle.

C Spatial series

The statistical properties of data distributed in space rather than time have been much neglected by statisticians and meteorologists until very recently. Quite commonly statistical methods which rest upon the assumption of a normal distribution are applied in situations where this is untrue or where the actual characteristics are unknown. Fortunately many parametric methods are robust but this does not exclude the risk of misinterpretations. It is appropriate to review here, therefore, the limited information on spatial frequency data and joint space–time variability before proceeding to consider various analytical techniques.

1 SPATIAL COHERENCE

Most attention to the question of spatial coherence of meteorological data has focused on precipitation values. The work of Trischler (1956) on precipitation at thirty-six stations in northeastern Austria is a pioneer study in this respect. He showed that the proportion of days

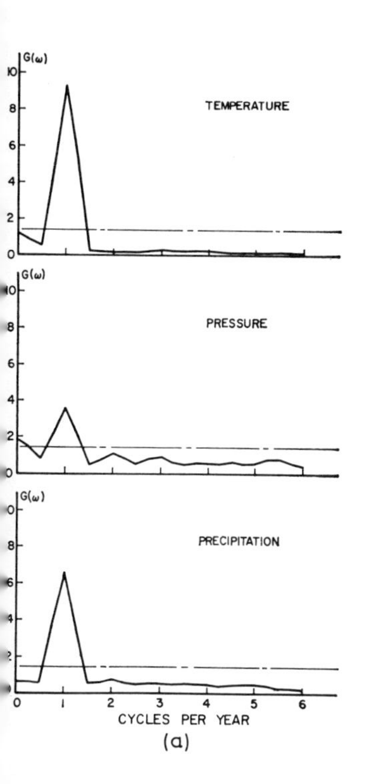

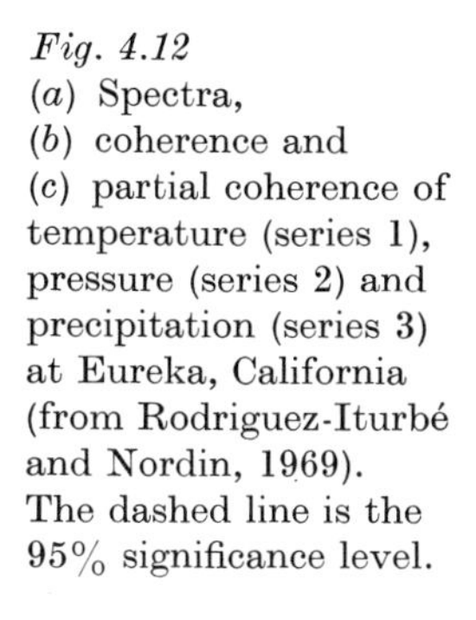

Fig. 4.12
(*a*) Spectra,
(*b*) coherence and
(*c*) partial coherence of temperature (series 1), pressure (series 2) and precipitation (series 3) at Eureka, California (from Rodriguez-Iturbé and Nordin, 1969).
The dashed line is the 95% significance level.

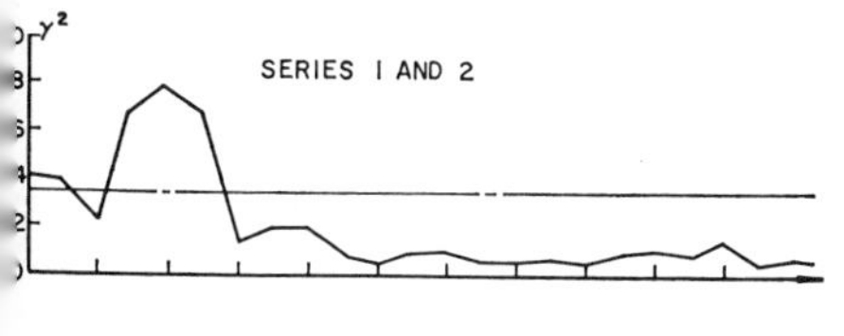

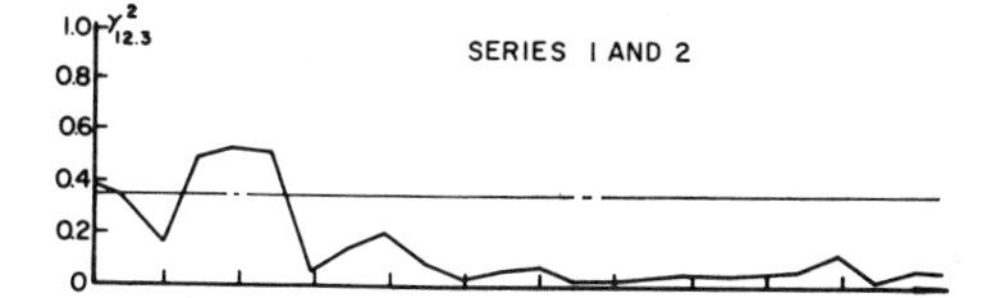

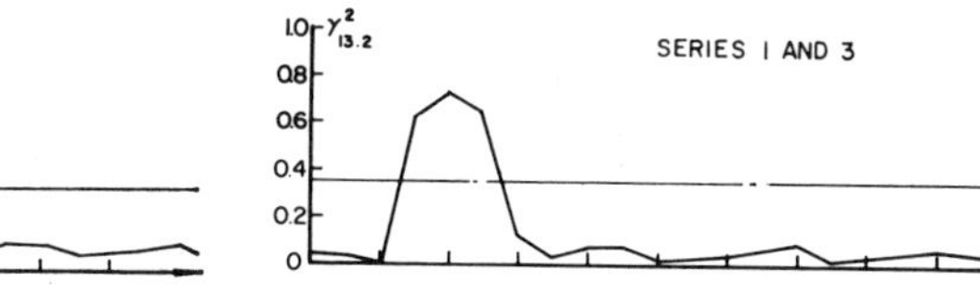

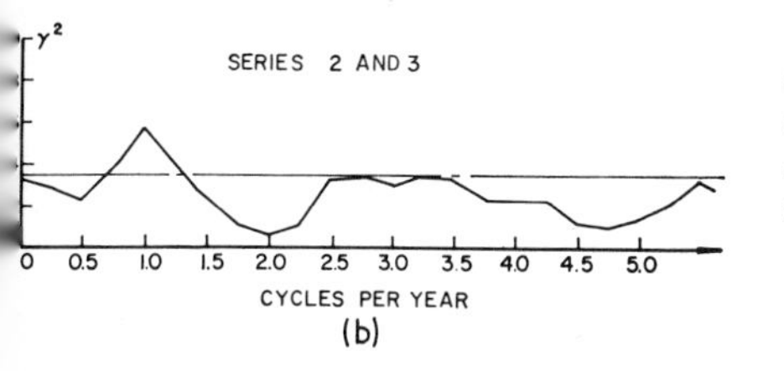

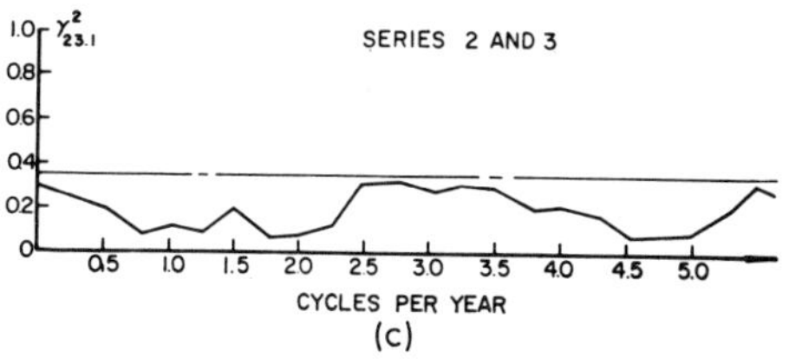

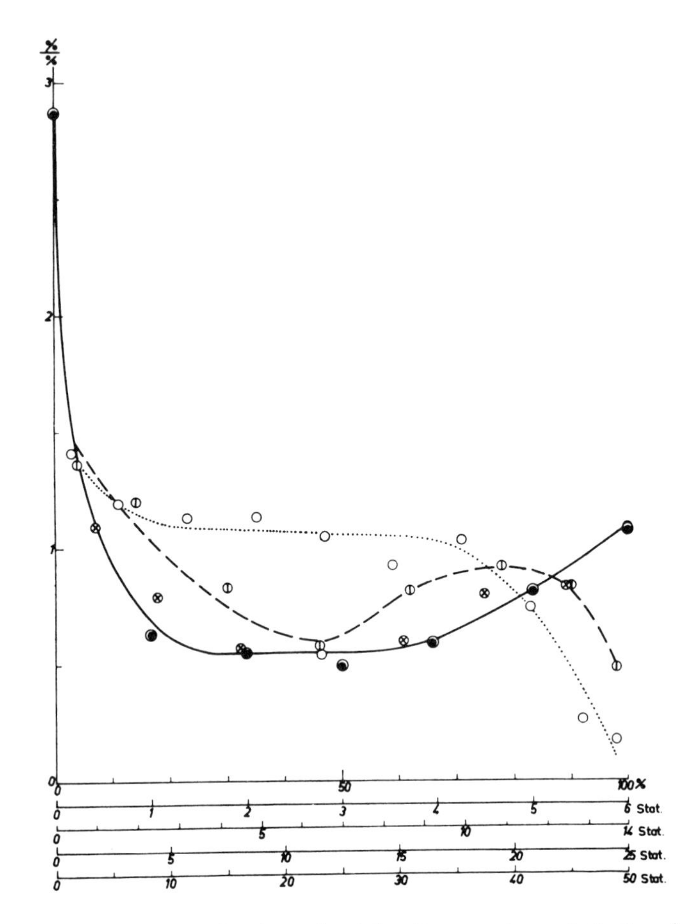

Fig. 4.13 Frequency distributions for 'simultaneous' precipitation between 0600 and 1200 hours in winter over different networks around Frankfurt-am-Main, 1951–60 (from Wachter, 1968). Continuous line = 6 stations, dashed line = 14 and 25 stations, dotted line = 50 stations. Ordinate shows frequency, abscissa shows percentage of stations with precipitation.

with 'coherent precipitation' (i.e. occurring at all of the representative stations in a specific region) decreased from about $\frac{1}{3}$ to $\frac{1}{10}$ as the area under consideration was increased from 3000 to 36,000 km^2. The percentage of the precipitation occurring in this category decreased correspondingly from 85% to 25%. More fundamental analyses have been carried out subsequently by Wachter (1963, 1965), who demonstrated that simultaneous occurrences of precipitation over a large area (1000 km radius around Frankfurt) can be treated as a random distribution if allowance is made for the probability at individual stations

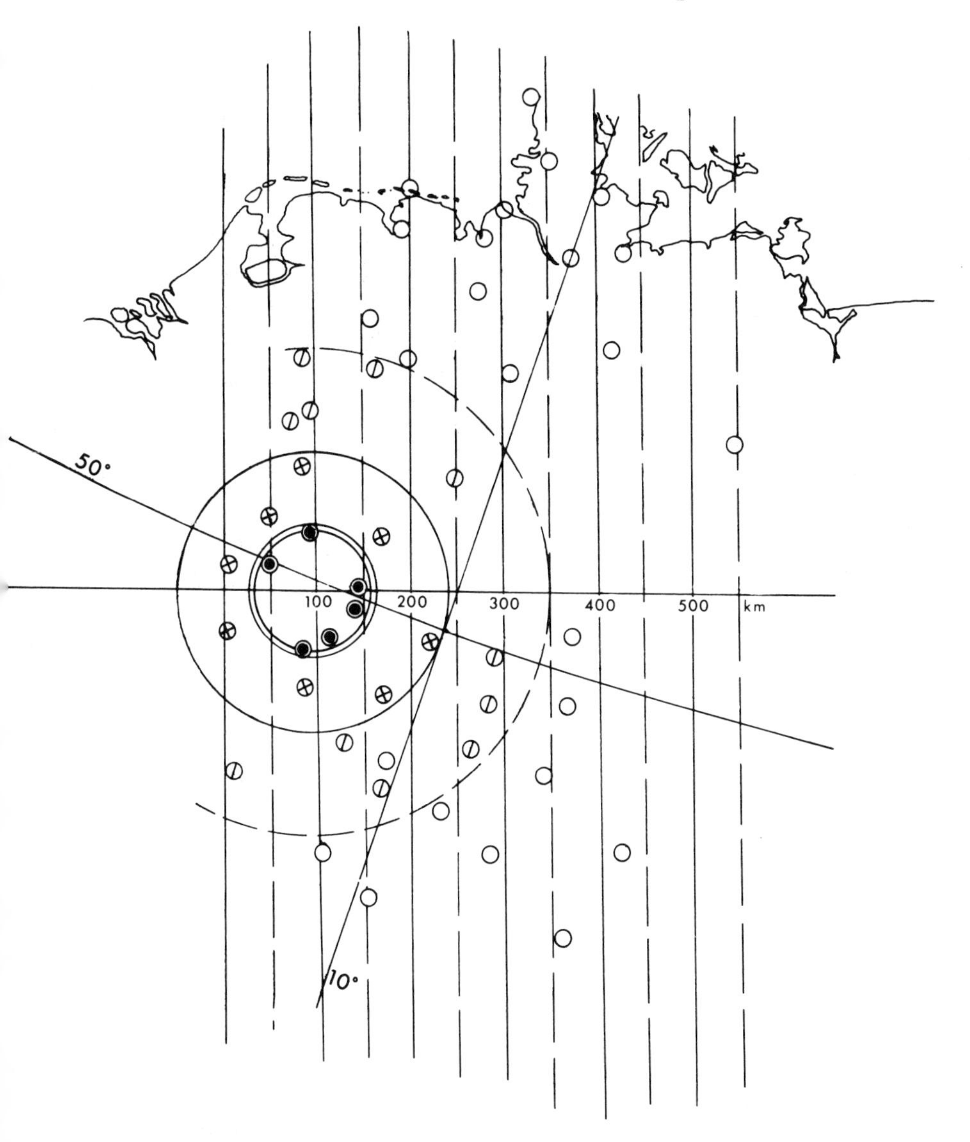

and for interdependence between stations. Elaboration of this approach by Wachter (1968) using synoptic 'present weather' codes for all forms of drizzle, rain, snow and showers ($ww = 50 - 99$) gives further insight into the question of spatial coherence. Fig. 4.13 illustrates the different frequency distributions which may occur with 'simultaneous' precipitation (during the period 06–12 hours) at stations around Frankfurt in winter over a ten-year period. The distribution is bimodal if the station area is smaller than the average-sized precipitation area.

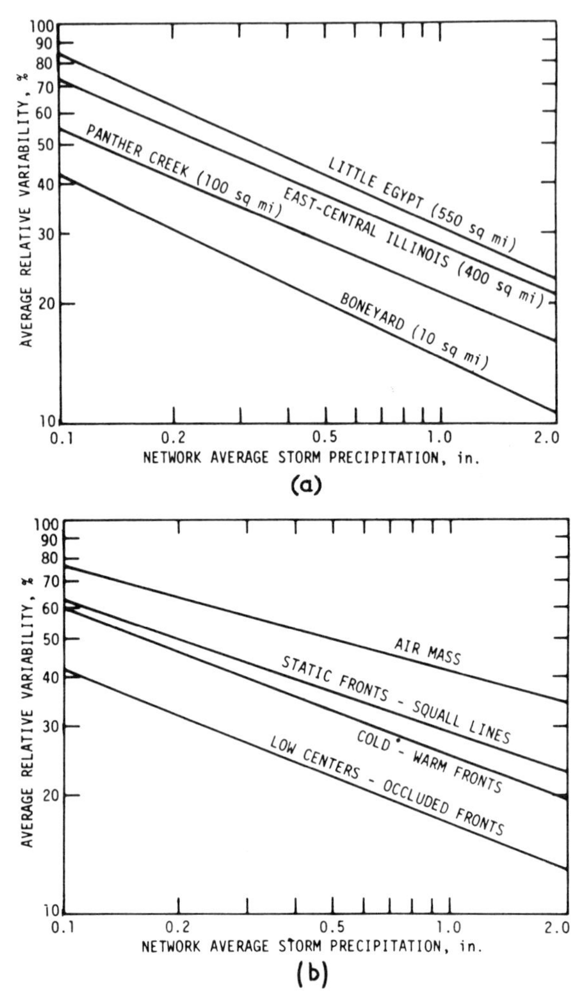

Fig. 4.14 Relationships between (*a*) network size, relative variability and mean storm precipitation, and (*b*) type of precipitation, relative variability and mean storm precipitation in Illinois (from Huff and Ship, 1968).

Huff and Shipp (1968) have adopted a different approach in studies of the spatial variability of storm rainfall over Illinois. They show that the average relative variability

$$\left(\frac{\sum |x_i - \bar{x}|}{\bar{x}} \times 100\right)$$

increases with increasing area and decreases exponentially with increasing network mean precipitation. This is illustrated in fig. 4.14(*a*) for different networks. They also show that the spatial variability is greater with unstable types of precipitation and less with frontal precipitation (fig. 4.14(*b*)). This effect of scattered shower activity is reflected in the increased variability from summer to winter, particularly for low monthly precipitation averages.

The problem of joint space–time variability has been even less explored. A fundamental analysis of homoscedasticity, or the constancy of variance, has recently been carried out, however, by Julian (1970). The study, which uses rank-order (distribution-free) statistics, demonstrates that for both streamflow and November–May precipitation over the conterminous United States there is a greater spatial coherence in dry years than in wet years. Fig. 4.15 illustrates the distribution of the rank sums for precipitation at eighteen stations for 1908–66 plotted on Gaussian probability paper. The spatial coherence varies considerably

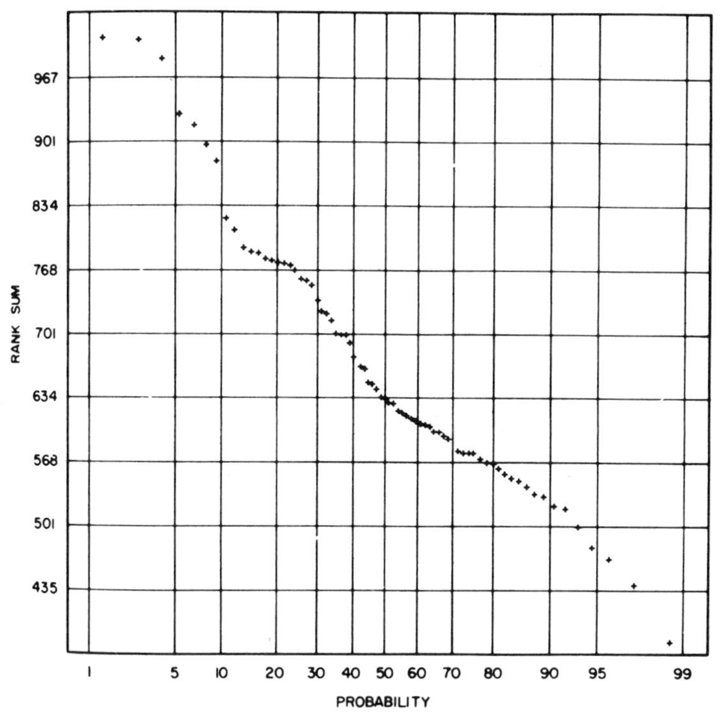

Fig. 4.15 The distribution of rank sums of total precipitation for November–May 1908–1966 at eighteen stations over the United States (from Julian, 1970).

with season. There is an autumn maximum, but even for summer rainfall, which is greatly affected by convective activity, the degree of coherence is still statistically significant. Further analysis of time variation of the driest and wettest years (arbitrarily selected by reference to the quartiles) suggests that for precipitation data there may be no heteroscedasticity (inconstancy of variance), so that dry and wet periods are distributed more or less at random. In the case of streamflow this appears to be true for high values whereas years of low streamflow indicate persistence. These results have considerable importance for other analytical studies because the assumption of homoscedasticity (and stationarity) is basic to the application of principal components and empirical orthogonal functions. When these conditions are not satisfied, the statistical efficiency of the technique is reduced and the physical interpretation of the computed functions may prove to be invalid.

2 SPATIAL CORRELATION AND REGRESSION ANALYSIS

Much attention has been devoted to the spatial variation of weather elements using correlation techniques. Beginning with the simplest problem, the spatial variation of a single element at a given time (see table 1.1), we may note the investigations of the space correlation fields of monthly or seasonal precipitation by Fliri (1967) for the Alps, and Nordø and Hjortnaes (1967) for Norway, among others. Fliri's work is discussed further in section D of this chapter (pp. 275–7). The method is a useful one for illustrating the varying effects of orography.

Similar analyses have been made by Godske (1952, 1965) and Huff and Shipp (1969) to determine the optimum spacing of weather stations. At the simplest level the association coefficient

$$Q = \frac{ad - bc}{ad + bc}$$

can be used for occurrence or non-occurrence of a selected weather condition at a pair of stations. a = occurrence at both stations, d = occurrence at neither station, b and c denote occurrence at one or other station. Godske suggests that a station may be characterized as highly locally representative ($Q > 0{\cdot}99$), fairly locally representative ($0{\cdot}98 < Q < 0{\cdot}99$), or slightly locally representative ($0{\cdot}95 < Q < 0{\cdot}98$).[1]

With reference to the product–moment correlation coefficient, r, Godske (1965) proposes that a desirable lower limit of correlation between any two stations might be when $r > +0{\cdot}80$ in 90% of all synoptic situations. At the upper limit, a station would be considered redundant if the correlation with observations at a neighbouring station exceeded $+0{\cdot}95$. In the foreseeable future, however, only special purpose mesoscale networks are likely to have a density sufficient for this degree of relationship. In the case of storm precipitation in Illinois, Huff and Shipp show that the correlation decay from a selected reference station is greatest for thunderstorms and air mass showers and least for cyclonic precipitation. It follows from this that the decay is also most rapid in the warm season.

[1] The χ^2 statistic of association has also been used as a measure of spatial similarity by selecting particular significance levels (see p. 206)

Since all climatological analysis depends on the available data, their reliability, and their power of resolution with respect to the various scales of weather system that may be of interest, further information regarding suitable station spacings for other climatic parameters in different synoptic regimes would be a considerable asset. In terms of the climatology of the free atmosphere it is worth noting that the decay of spatial correlation in fields of contour height for the major standard isobaric surfaces has been evaluated with respect to stations in the North Atlantic and Europe (Bertoni and Lund, 1963). Similar work has been performed by Gandin (1965). He examines the correlation structure of fields of MSL pressure, dew point temperature at 850 mb and geopotential at 500 mb, and include a separate analysis for the three basic circulation types of Vangengeim at the 500 mb level.

Australian studies (Cornish, Hill and Evans, 1961) have examined in some detail the correlation fields for six-day and monthly precipitation totals and show that they are usually elliptical in shape. This anisotropy is a complex function of moisture sources, precipitation types and topography (Caffey, 1965). Nevertheless it should not be overlooked that the spatial decay of correlation may also be affected by purely statistical factors. For example, for short-period precipitation, in particular, it is usually necessary to transform the values to obtain an approximately Gaussian distribution.[1] However, when this is done the effect is to increase the calculated r, particularly for log and cube root transforms. Izawa (1965) illustrates this graphically with precipitation data from Japan, and further shows that the spatial pattern of *isocorrelates* (lines of equal correlation) is distorted (fig. 4.16). He demonstrates that this can be explained if it is assumed that the transformed precipitation amounts have a Gamma type of distribution.

Many applications of correlation methods in synoptic climatology have been concerned primarily with relationships between weather elements and salient features of the atmospheric circulation pattern (Martin and Leight, 1949; Stidd, 1954; Klein, 1963, 1965). Stidd develops the mathematical relationship for the correlation of the geostrophic wind with the 'gradient' of isocorrelates of precipitation anomaly and he demonstrates that the simple elements of wind velocity and geopotential height in the geostrophic wind equation can be replaced by the regressions of precipitation anomaly on these two terms. The procedure used in these studies has been to correlate precipitation or temperature at a selected station with concurrent MSL pressure values, or 700 mb height anomalies in four parts of the United States (fig. 4.17). Such isocorrelate maps provide information on, for example,

[1] McDonald (1960) has argued that this is often unnecessary in view of the shortcomings of the basic data.

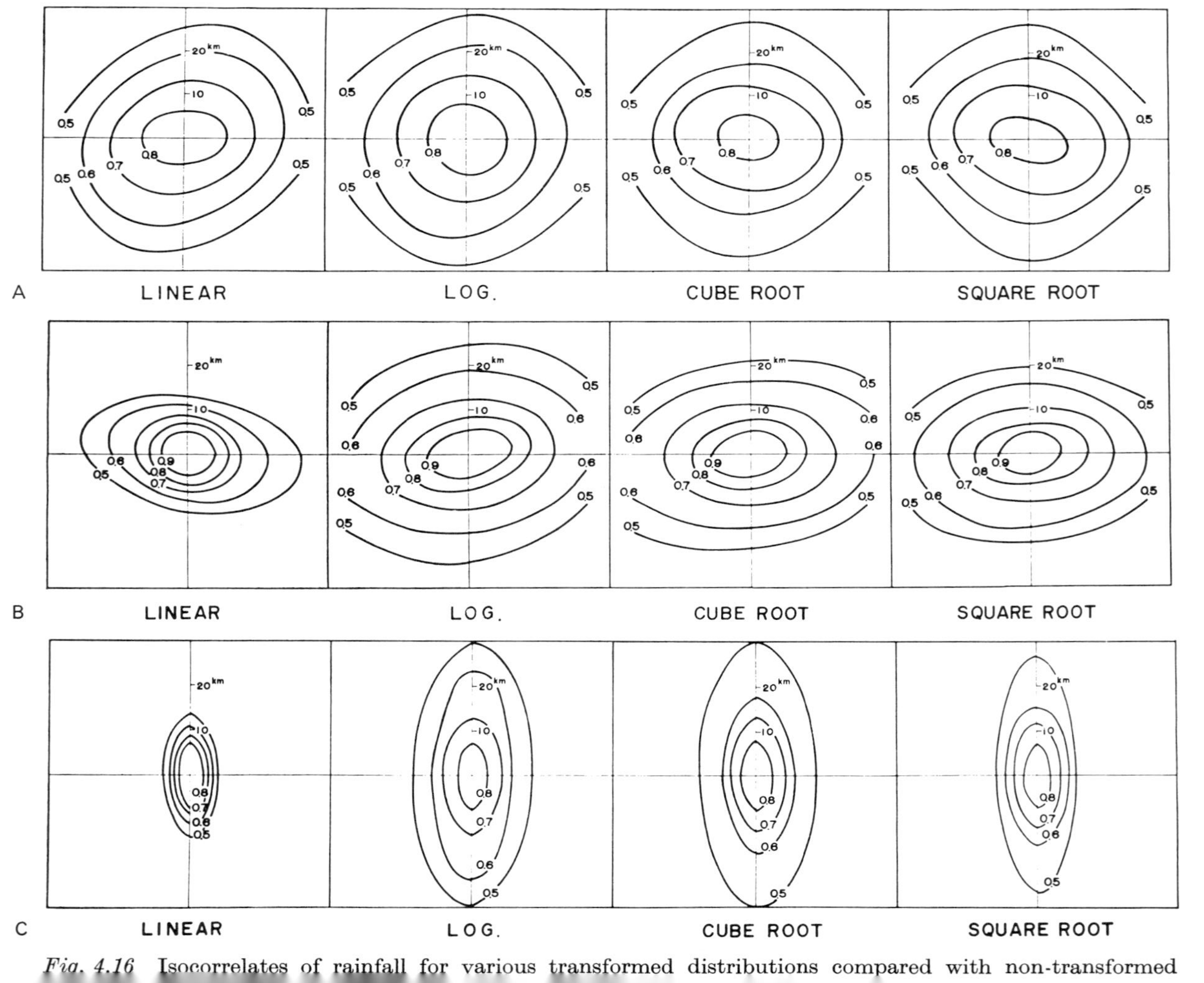

Fig. 4.16 Isocorrelates of rainfall for various transformed distributions compared with non-transformed

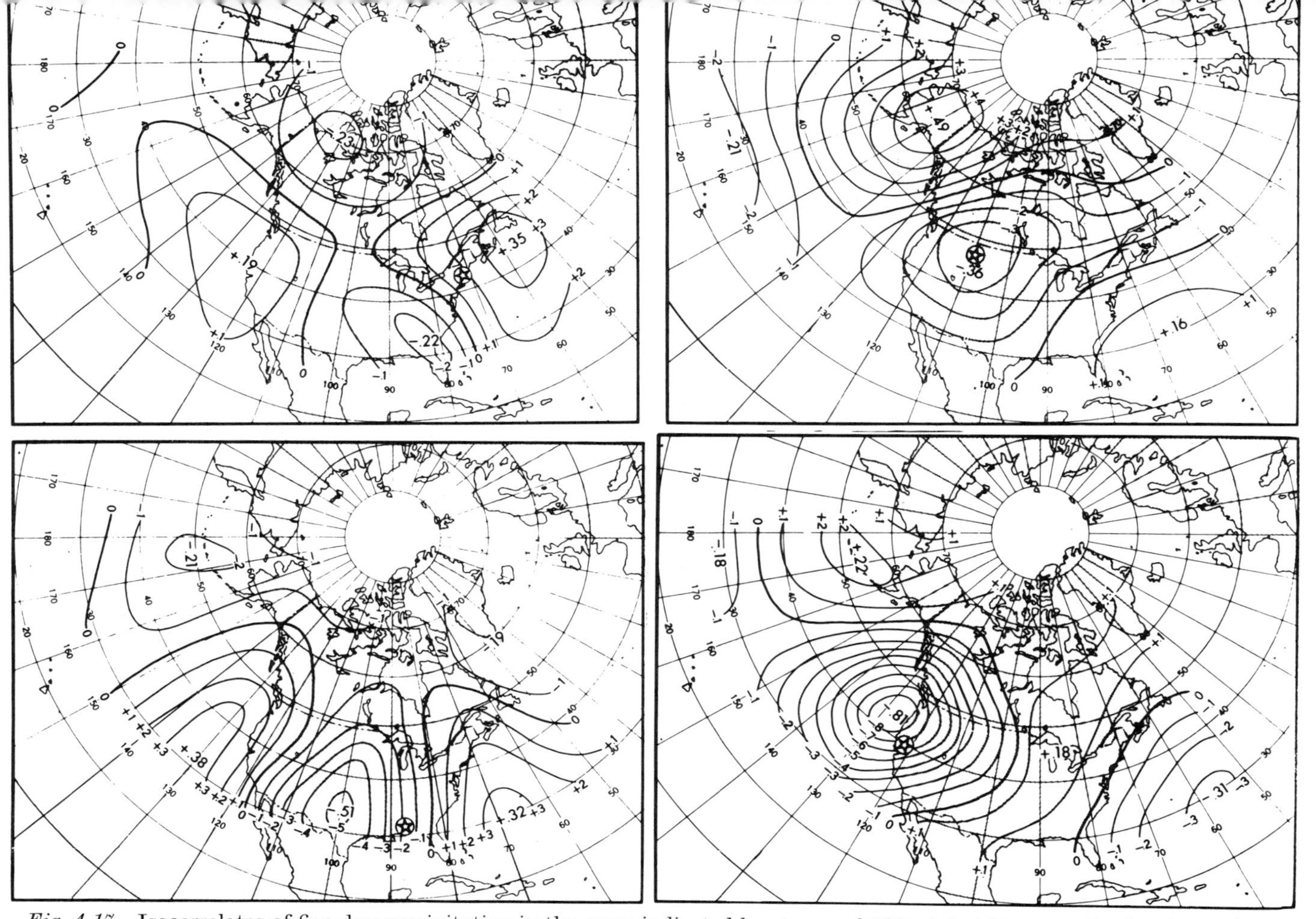

Fig. 4.17 Isocorrelates of five day precipitation in the areas indicated by stars and 700 mb height anomalies in winter (from Klein, 1963).

the direction, curvature and origin of anomalous flow components conducive to particular precipitation or temperature conditions (see fig. 5.41).

Correlation fields have been used on more complex data by Dickson (1971). He correlated the variance of daily mean temperature in selected spectral bands with seasonal mean heights of the 700 mb surface over the western hemisphere. The spectra of temperatures for twenty-one years at eight stations in the United States were partitioned into intervals of less than 3 days, 3–10 days (the synoptic range), 11–22·5 days, and greater than 22·5 days. The results (fig. 4.18) indicate that the mean circulation in winter shows increasing wave amplitude as the variance

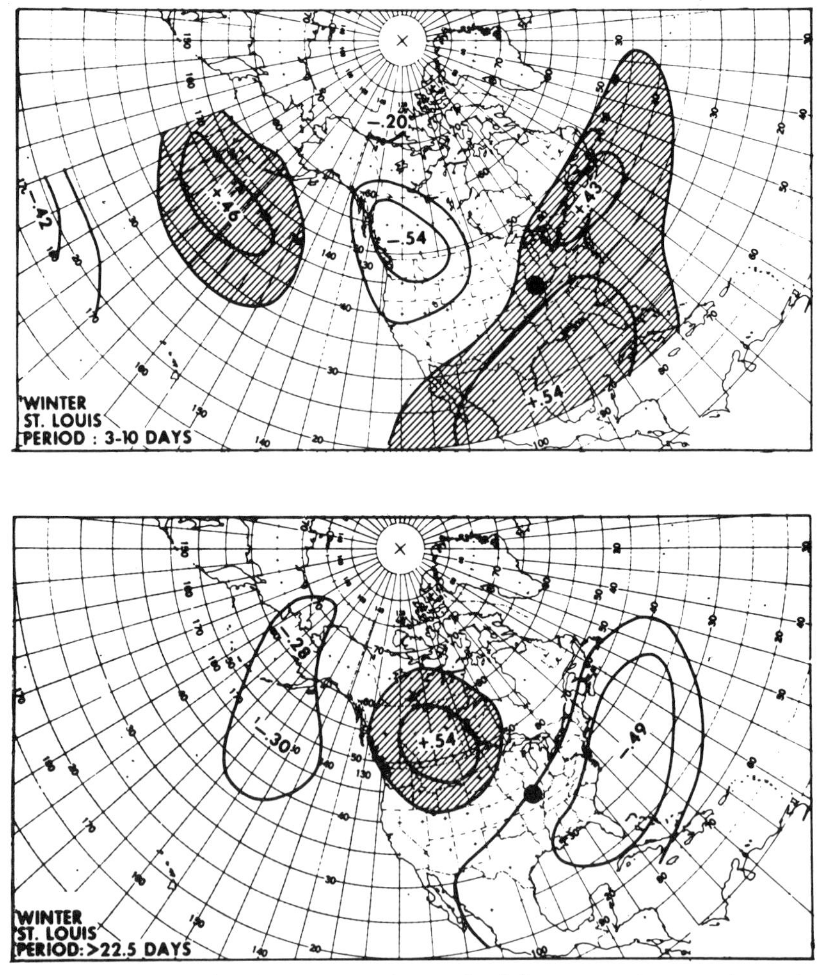

Fig. 4.18 Isocorrelates between two intervals of the temperature spectrum at St Louis and 700 mb heights in winter (from Dickson, 1971).

spectrum shifts to longer period temperature oscillations. The summer pattern for periods exceeding eighteen days is very similar to that for extended drought conditions which are associated with a well-developed upper level anticyclone over the United States.

It is worth emphasizing that the isocorrelate fields relate of course to a single reference station. Any direct attempt to correlate two fields runs immediately into the problem of *spatial* autocorrelation (King, 1969, p. 157; Curry, 1966). If, for example, we were to correlate point values for two elements, T and P, so that the correlation coefficient is given by $r(Tij, Pij)$, then we must recognize that Tij is almost certainly related to $Ti+1, j$; $Ti, j+1$; $Ti+2, j$; and so on. Moreover, information with respect to the correlations $r(Tij, Pi+1, j)$, $r(Tij, Pi, j+1)$, $r(Tij, Pi+2, j)$, etc., is not taken into consideration. The autocorrelation between observations at neighbouring and more distant points renders invalid tests of significance although such correlation coefficients can be used as a general guide to the relationships between pairs of variables for different time periods. Alternative, and generally preferable, procedures are discussed below.

3 MULTIPLE REGRESSION

Most of the spatial distributions of interest to climatologists are complex in their structure and interrelationships necessitating the use of *multiple correlation and regression analysis* to examine their properties. Such evaluations have become routine with the advent of computers and comprehensive program libraries although this development calls for increased attention to the assumptions of the regression model. The essentials of multiple regression analysis will be summarized briefly. For more details the reader may refer to Draper and Smith (1966, p. 104) or King (1969, pp. 135–49).

Multiple regression involves fitting a linear surface to a set of data for a chosen variable (X_0) such that the sum of squares of the deviations is a minimum. An equation of the form

$$X_0 = a + b_1 X_1 + b_2 X_2 + \cdots + \varepsilon$$

is required where $X_1, X_2, \ldots$ are independent variables, ε is the error term and the b's are partial regression coefficients. The latter express the rate of change of X_0 for unit change in their respective independent variable with the effect of the other independent variables held constant.

In the case of two independent variables we have

$$Y = a + b_{01\cdot 2} X_1 + b_{02\cdot 1} X_2$$

where Y = predicted value of the dependent variable X_0
$b_{01\cdot2}$ = regression coefficient for X_0 and X_1, with the effect of X_2 held constant.

The error term is dropped here since it represents the residual $X_0 - Y$.

The regression coefficients are estimated by minimizing the sum of squares of the deviation of the observed values (X_1, X_2) from the fitted linear surface. $b_{01\cdot2}$ and $b_{02\cdot1}$ are determined from the pair of simultaneous equations:

$$\sum x_0x_1 = b_{01\cdot2}(\sum x_1^2) + b_{02\cdot1}(\sum x_1x_2)$$
$$\sum x_0x_2 = b_{02\cdot1}(\sum x_2^2) + b_{01\cdot2}(\sum x_1x_2)$$

where x represents the deviation of the X values from their respective $\bar{X}$. Thus, $x_0 = X_0 - \bar{X}_0$, $x_1 = X_1 - \bar{X}_1$, etc., and

$$\sum x_1^2 = \sum X_1^2 - (\sum X_1)^2/N$$
$$\sum x_0x_1 = \sum X_0X_1 - [(\sum X_0)(\sum X_1)]/N, \text{ etc.}$$

The constant is obtained from

$$a = \bar{X}_0 - b_{01\cdot2}\bar{X}_1 - b_{02\cdot1}\bar{X}_2.$$

Critical assumptions of regression are that the relationships between X_0 and the independent variables are linear, that the independent variables are not themselves linearly correlated, and that the error terms are not autocorrelated and have constant distributions (i.e. they possess homoscedasticity). An appropriate data transformation (see p. 224) may be sufficient to ensure linearity between the variables and homoscedasticity of the error terms, as well as possibly meeting the further (less critical) assumption that each variable be normally distributed. These and other assumptions are fully reviewed by Poole and O'Farrell (1971) who also discuss ways of overcoming the problems that arise when certain assumptions are violated. Apart from these considerations, careful thought must be given to the interpretation of regression coefficients. Wallis (1965) illustrates this problem with generated data for multiple and 'stepwise' regression, as well as other multivariate techniques.

Examples of regression analysis, as applied to spatial patterns of precipitation, include studies for Norway (Nordø and Hjortnaes, 1967), Sierra Leone (Gregory, 1968), northeast England (Jackson, 1969) and the Sudan (El Thom, 1969). Gregory (fig. 4.19) uses five independent variables – latitude (as a measure of the role of the intertropical convergence zone), distance from the coast (as a measure of the monsoonal influence), altitude weighted with respect to distance inland, longitude (as a measure of east–west moving disturbance lines), and a southwest–northeast relief factor. Of particular interest is the varying importance

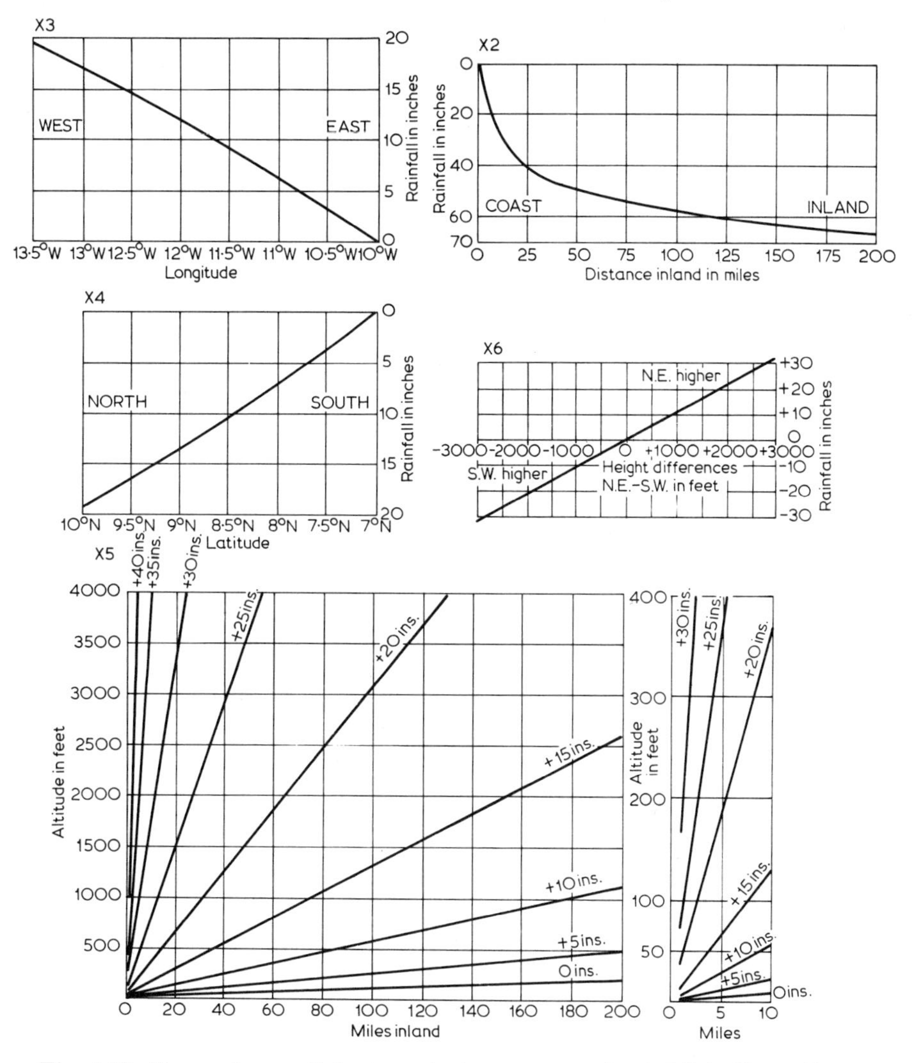

Fig. 4.19 Regression coefficients against mean annual rainfall in Sierra Leone (from Gregory, 1968).

of these variables with time of year allowing a quantitative assessment of their seasonal effectiveness.

The variance which the regression equations account for (the total 'explained variance') is determined from the regression sum of squares divided by the total sum of squares.

$$[b_1 \sum x_0 x_1 + b_2 \sum x_0 x_2 + \cdots +]/ \sum {x_0}^2$$

where $\sum x_0 x_1 = \sum X_0 X_1 - (\sum X_0 \sum X_1)/N$ (the sum of the cross-product of the deviations of X_0 and X_1 from their respective means), and so on. The significance of the individual contributions to the regression, and to the total variance, may be tested using the F ratio calculated from the ratio of the individual (or total) variance estimate (=sum of squares divided by the appropriate number of degrees of freedom) to the residual (=total − regression) variance estimate.

The pattern of the residuals from regression, $(X_0 - Y) = \varepsilon$, in spatial studies of the type referred to may be of considerable diagnostic value in suggesting new independent variables to be incorporated in the regression.

A common computational procedure for regression analysis is the 'stepwise' solution. This incorporates the independent variables one at a time, generating regression equations after each step. The first independent variable is incorporated on the basis of the maximum r value from a matrix of simple correlations between the independent variables and X_0. At each successive step the variable which is added is the one that has the largest partial correlation with X_0. The termination point of the analysis is set arbitrarily with reference to the F ratios, although synoptic experience can be used in conjunction with the statistical criterion (Klein, Lewis and Enger, 1959; Miller, 1962). Klein (1963) used stepwise regression procedures in developing equations that relate climatic conditions to circulation characteristics. The work of Klein and his associates has been summarized in the form of synoptic models showing, for different regions of the United States, those sectors of the 700 mb troughs and ridges that are most conducive to light and heavy precipitation (Klein, 1965). Fig. 5.41 illustrates these models.

A special form of multiple regression analysis is *trend surface analysis* in which a spatially distributed variable is described by the best-fitting polynomial equation involving as many terms as desired. These trend surfaces are obtained with increasing complexity from linear to quadratic, cubic, and possibly quartic, terms. The percentage of the pattern accounted for ('explained') at each step is expressed by the reduction in the total sum of squares.

The main function of trend surface analysis is to abstract the most simple large-scale patterns from irregular local and random variations. The technique has been widely used in the geological sciences and more recently by geographers (Chorley and Haggett, 1965; Norcliffe, 1969). Climatological applications are only just beginning (Page, 1968; Smithson, 1969; Unwin, 1969), but the techniques seem particularly pertinent. Unwin, for instance, finds that the method is more efficient than conventional regression as a predictor of rainfall totals in North Wales. Many of the computer programs now accept irregularly spaced data and provide the results in analysed map form. Apart from the basic trends, the local deviations of observed from calculated values (residuals) are of interest. Analysis of the residual pattern may indicate the influence of particular local factors so that the technique provides an analytical tool as well as a means of objective description. However, the problem of distinguishing the effects of the clustering of sample points from that of random spatial variations in the residuals presents difficulties (Robinson, 1970).

Smithson (1969) shows how trend surfaces can be used to generalize the distribution of rainfall and its frequency over Scotland during selected synoptic situations. Fig. 4.20(*a*) illustrates the pattern of rain-days with mP airstreams given by the cubic surface which accounts for 91% of the variance. There is a noticeable ridge over the Cairngorms. Nevertheless, Smithson notes that the available samples present difficulties in the interpretation of the analysis. For example, the trend surfaces to some extent reflect the frequency of the various wind directions within each sample. Fig. 4.20(*b*), showing the cubic surface for rainfall with mP air (which accounts for 82% of the variance), gives no indication of orographic increase over the Cairngorms. This may be caused by the low frequency of mP airstreams approaching from the southeast in the available sample. Further interesting results are provided by a comparison of the residuals from the cubic surfaces for mP, cP and Arctic air. The maximum residuals are shown below in cm:

	mP	*cP*	*Arctic*
Negative residual	−20·2	−6·7	−4·7
Positive residual	+30·0	+9·0	+5·2

The figures suggest that local instability effects are greatest with mP airstreams.

A similar application of trend surface analysis is in urban pollution studies (Page, 1968; Anderson, 1970). The problem is to assess the radial distance which needs to be considered in terms of pollution concentrations under light winds and to develop simplified diffusion models for different synoptic patterns in winter 1964–5. Days were selected with reference to the Lamb catalogue, using two consecutive days of the same type because the pollution period ends in the morning hours, and trend surfaces of sulphur dioxide concentration were prepared using forty-eight stations in the west Midlands. Attempts can be made to infer the contributions of topographic channelling and of local emission sources to the overall pattern by analysis of residuals from the trends. Such scrutiny may also indicate stations where the instrumental record is dubious. The method is a flexible one since comparisons are readily effected between the patterns on different days and the analysis may be extended to a wider area or applied to another conurbation for the same day.

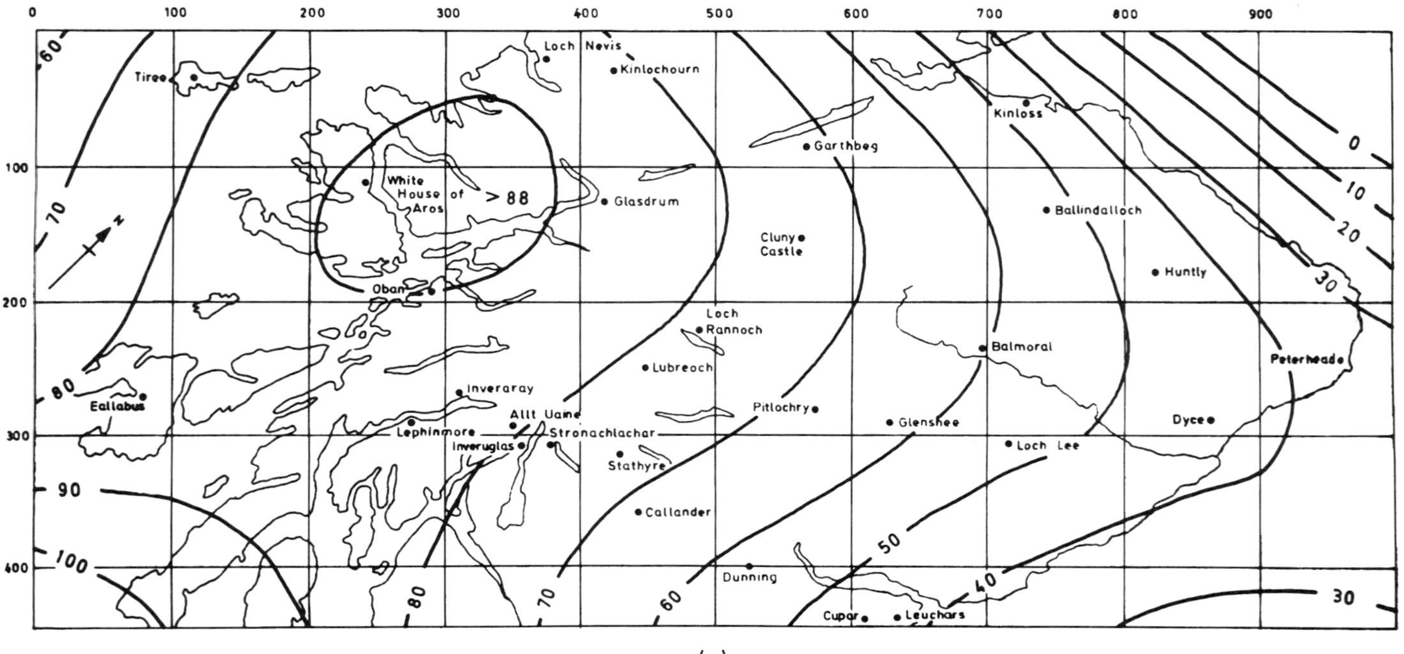

Loch Nevis
Kinlochourn
Tiree
Kinloss
Garthbeg
White House of Aros
> 88
Glasdrum
Cluny Castle
Ballindalloch
Huntly
Oban
Loch Rannoch
Balmoral
Peterhead
Lubreoch
Inveraray
Eallabus
Allt Uaine
Pitlochry
Glenshee
Dyce
Lephinmore
Stronachlachar
Inveruglas
Stathyre
Loch Lee
Callander
Dunning
Cupar
Leuchars
N
0
10
20
30
40
50
60
70
80
90
100

(a)

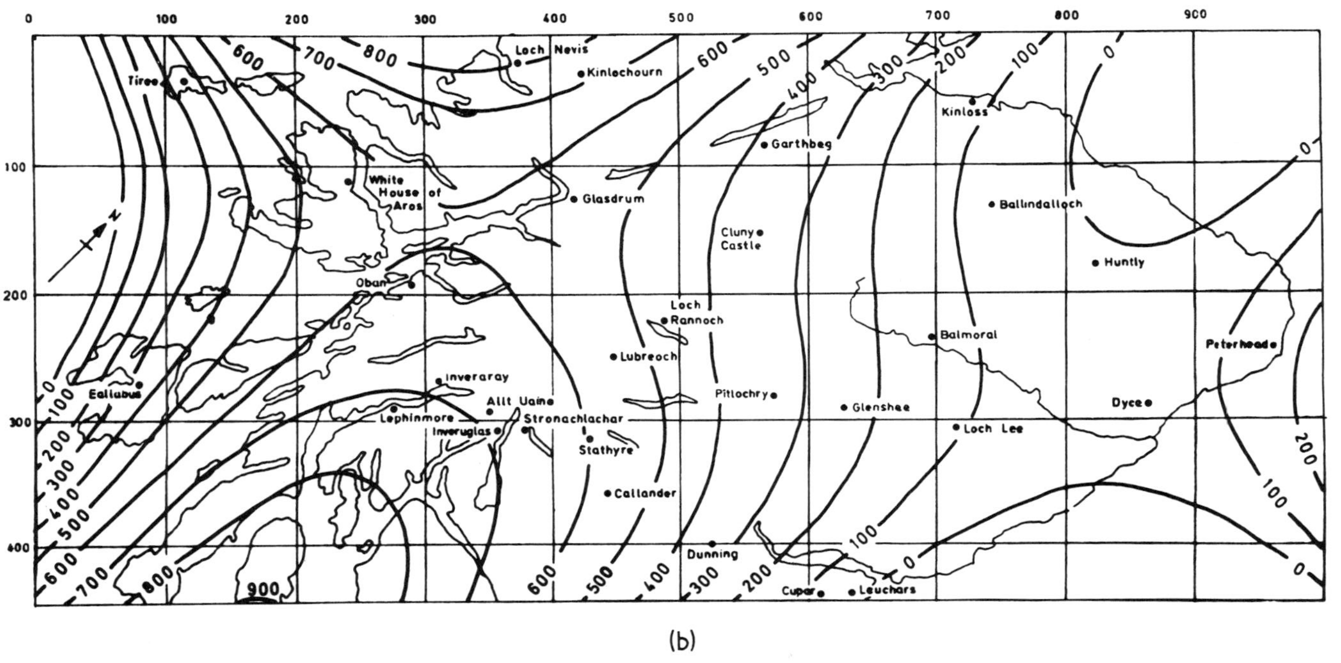

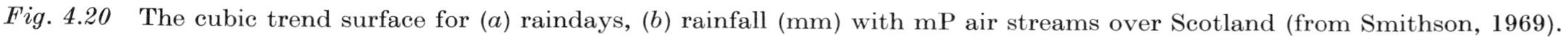

Fig. 4.20 The cubic trend surface for (*a*) raindays, (*b*) rainfall (mm) with mP air streams over Scotland (from Smithson, 1969).

4 ORTHOGONAL POLYNOMIAL SURFACES

Reference has been made to the problems created by spatial autocorrelation. One solution is to correlate polynomial surfaces which account for most of the variance of a given field, rather than the original fields themselves. The use of polynomials was described briefly in Chapter 3.B (p. 105) and we can now examine it further.

The field (pressure or height, for example) is described by a series of independent polynomial equations. These are orthogonal, which is to say that their coefficients are uncorrelated. If $P(x^r)$ is a polynomial of degree s, then orthogonality of these equations implies that

$$\sum_{x=0}^{n-1} P(x^r)P(y^s) = 0, \quad \text{for } r \neq s.$$

Three parameters are used to specify the field variable – the spatial mean, a measure of the variability, and a set of correlation coefficients between the field and the polynomial surfaces. There are three categories of correlation coefficient:

1 Column terms (10, 20, 30, . . .). The surfaces are parallel to the y axis of the grid and the coefficients specify the variability of the x direction.
2 Row terms (01, 02, 03, . . .). These surfaces are perpendicular to (1) and measure variability in the y direction.
3 Cross-product or interaction terms expressing variability in the x and y directions (11, 12, 13 represent asymmetrical; 21, 22, 23 symmetrical centres).

These are illustrated in fig. 3.3.

In simplified mathematical terms (Hare *et al.*, 1957) the problem is to represent the field variable, $H(x, y)$, by a polynomial of the form

$$H(x, y) = \bar{H} + a_1F_1(x, y) + a_2F_2(x, y) + \cdots$$

where $a_k = \overline{H(x, y)F_k(x, y)}/\overline{F_k^2(x, y)}$, the bar denoting a mean. If the orthogonal polynomials are normalized (orthonormal) then

$$\overline{F_k^2(x, y)} = 1$$

so that

$$a_k = \overline{H(x, y)F_k(x, y)}.$$

The coefficients can be determined independently without affecting the values of those already calculated which is a great computational asset. The reduction of variance due to the polynomials is given by

$$\overline{(H-\bar{H})^2} = \sum_{k=1}^{m} (a_k^2).$$

The degree of the polynomials in x and y (i.e. the number of terms employed, m) may not be the same; the appropriate number of terms has to be selected in relation to the particular problem.

The use of orthogonal polynomials to analyse pressure or height fields was in vogue during the 1950s especially (White *et al.*, 1957, 1958), although most detailed references to the subject appear in reports with limited circulation (see references to Chapter 3.B). A paper by Cehak (1962) exemplified the representation of height fields, in this case for 850 mb over Europe and the North Atlantic, by Tschebyscheff polynomials determined for a 45-point grid. Friedman (1955) used fourteen polynomial surfaces to specify fields over North America. For January 1948–52 it was found that these surfaces gave an average reduction of variance of 96·7% at 700 mb and 79·4% for the surface pressure field. The technique provides a useful approach to specifying surface weather conditions in relation to linear combinations of the polynomials. As an analytical tool, however, the method has to a large extent been superseded by empirical orthogonal functions, which are discussed on pp. 268–74 (but see Dixon, 1969), while as a basis for developing synoptic classifications it raises similar problems to most other objective procedures. This consideration is developed in Chapter 3.B (pp. 106–9).

5 SPATIAL FOURIER AND SPECTRAL ANALYSIS

The techniques of harmonic and Fourier analysis, which we have examined in the context of time series (p. 228), can also be applied to describe spatial variations in one and two dimensions as discussed in Chapter 3.B (p. 106). Little more needs to be added here concerning the actual computations. In the case of analysis of the wave patterns along a latitude circle the 0–360° interval is directly equated with 0 to 2π. The phase angles specify the wave spacing and the amplitudes indicate the overall degree of development of the waves. Most difficulty of interpretation probably occurs when an intense wave trough forms in isolation in one sector of the hemisphere. This may be reflected in added variance to more than one wave number.

For global studies spherical-surface harmonics can be used (Haurwitz and Craig, 1952). These are Legendre polynomial functions (comprising a set of orthogonal functions) which specify maxima and minima of wave forms in both the latitude and longitude directions. Fig. 4.21 illustrates a pattern of four maxima and minima around the hemisphere with maxima at 55°N, minima at 25°N, superimposed on a zonal distribution. The procedure for evaluating the harmonic coefficients, which is beyond the scope of our treatment here, is elaborated by Kubote *et al.* (1960) and Elsaesser (1966). Apart from a trial analysis by Deland (1965) of 500 mb patterns in the northern hemisphere for April

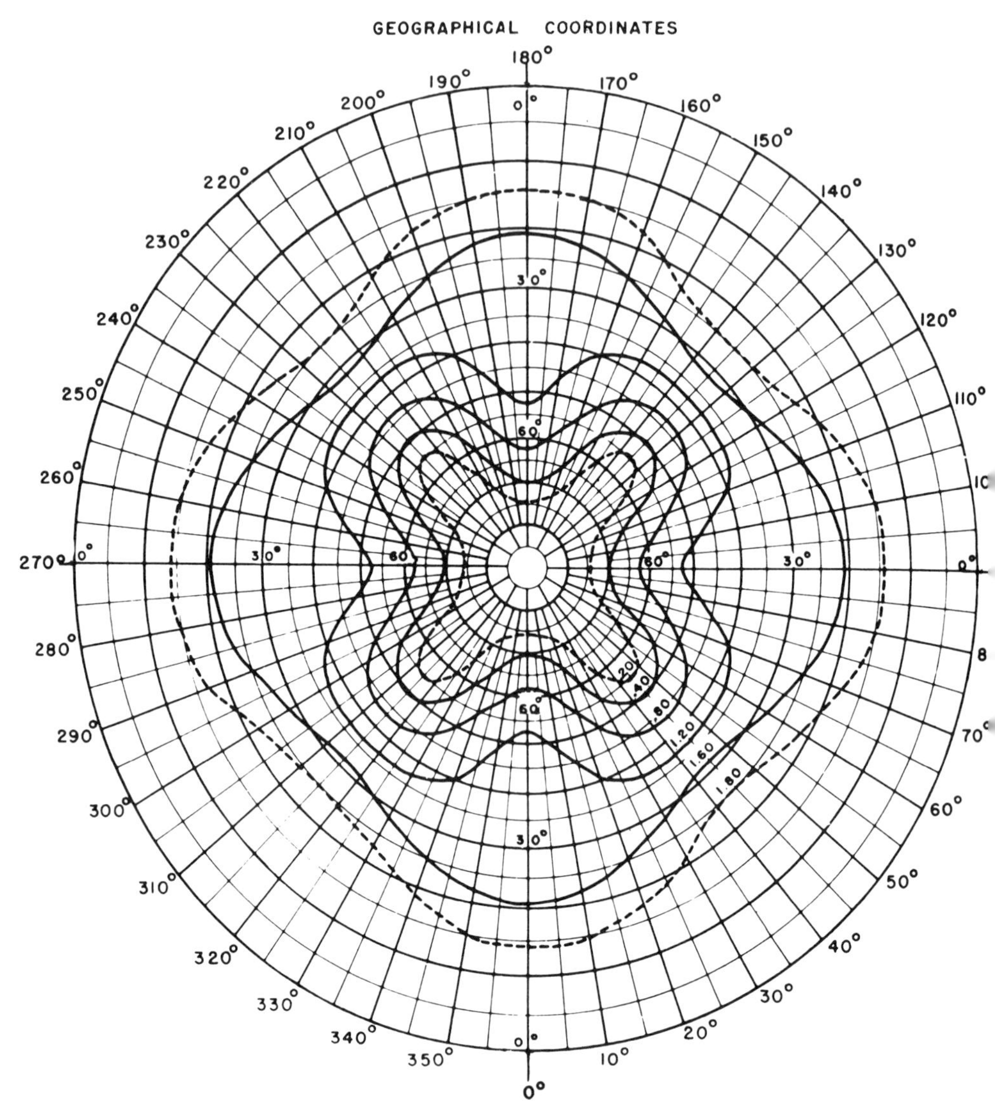

Fig. 4.21 The spherical harmonic cos $4\lambda E_8^4$ (cos θ) superimposed on a zonal distribution (2 $\sin^2 \theta$) (from Haurwitz and Craig, 1952).

1960 and the work of Eliasen and Machenhauer (1965), this type of representation of surfaces has been little used. Godson (Hare *et al.* 1957, p. 23) pointed out that the technique is not suitable for the polar regions and devised a special cylindrical set of Fourier functions weighted to correct for the convergence of the meridians (see also Hare, 1958).

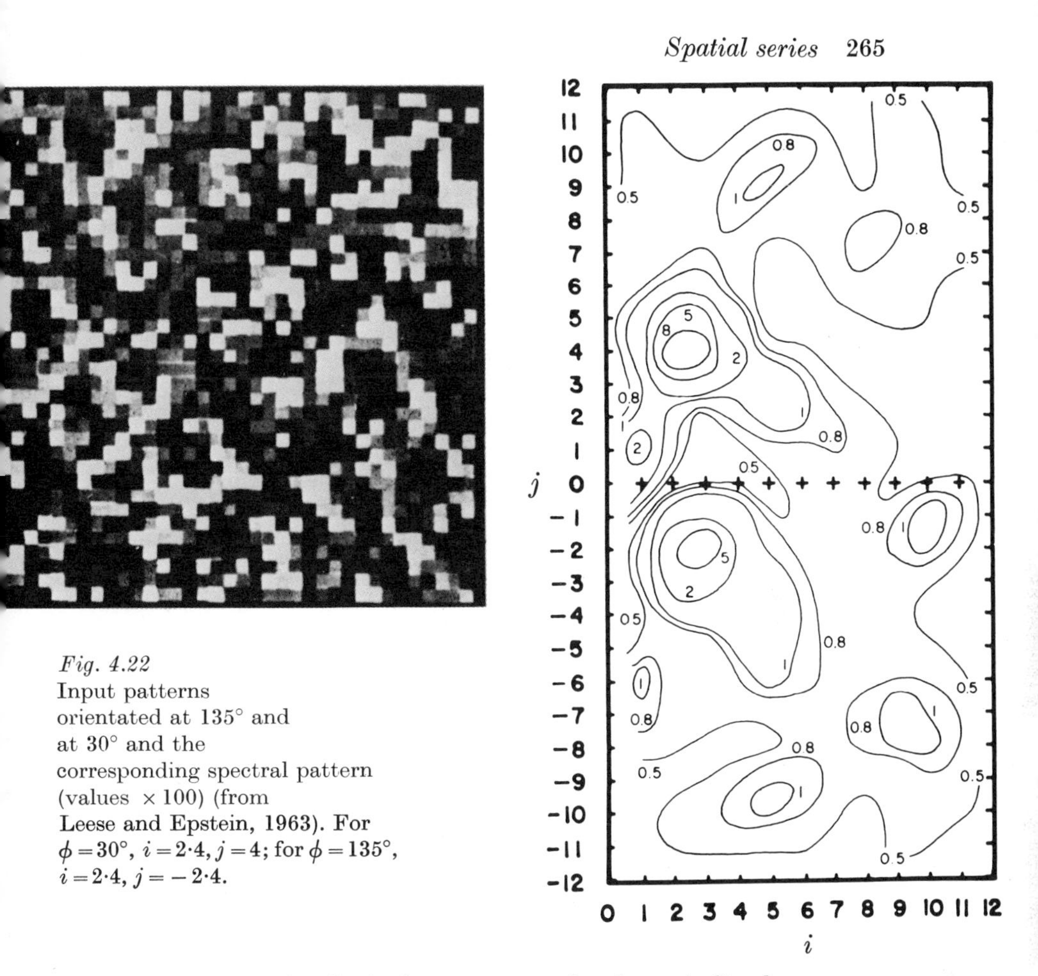

Fig. 4.22
Input patterns orientated at 135° and at 30° and the corresponding spectral pattern (values × 100) (from Leese and Epstein, 1963). For $\phi = 30°$, $i = 2{\cdot}4$, $j = 4$; for $\phi = 135°$, $i = 2{\cdot}4$, $j = -2{\cdot}4$.

In view of the rather limited occurrence of truly periodic phenomena, the use of harmonic analysis, where wave number has to be an integer, is somewhat restrictive. Nevertheless, the spherical harmonic approach is being studied as a basis for future numerical circulation models.

The frequency characteristics of two-dimensional spectra have been examined by extension of the simple one-dimensional formulations. Leese and Epstein (1963) used this method in studying satellite cloud photographs. For a wave number pair (i, j) the wave angle orientation with respect to the abscissa is given by

$$\phi = \arctan (i/j)$$

and the wavelength $\lambda = 2M\Delta(i^2 + j^2)^{-1/2}$ where M lags are used in the calculated autocovariances in both the x and y direction, and Δ denotes a uniform spacing between data points in these directions. Cloud streets are displayed as a single spectral maximum, while two maxima indicate a simple interference pattern representing cellular clouds (fig. 4.22).

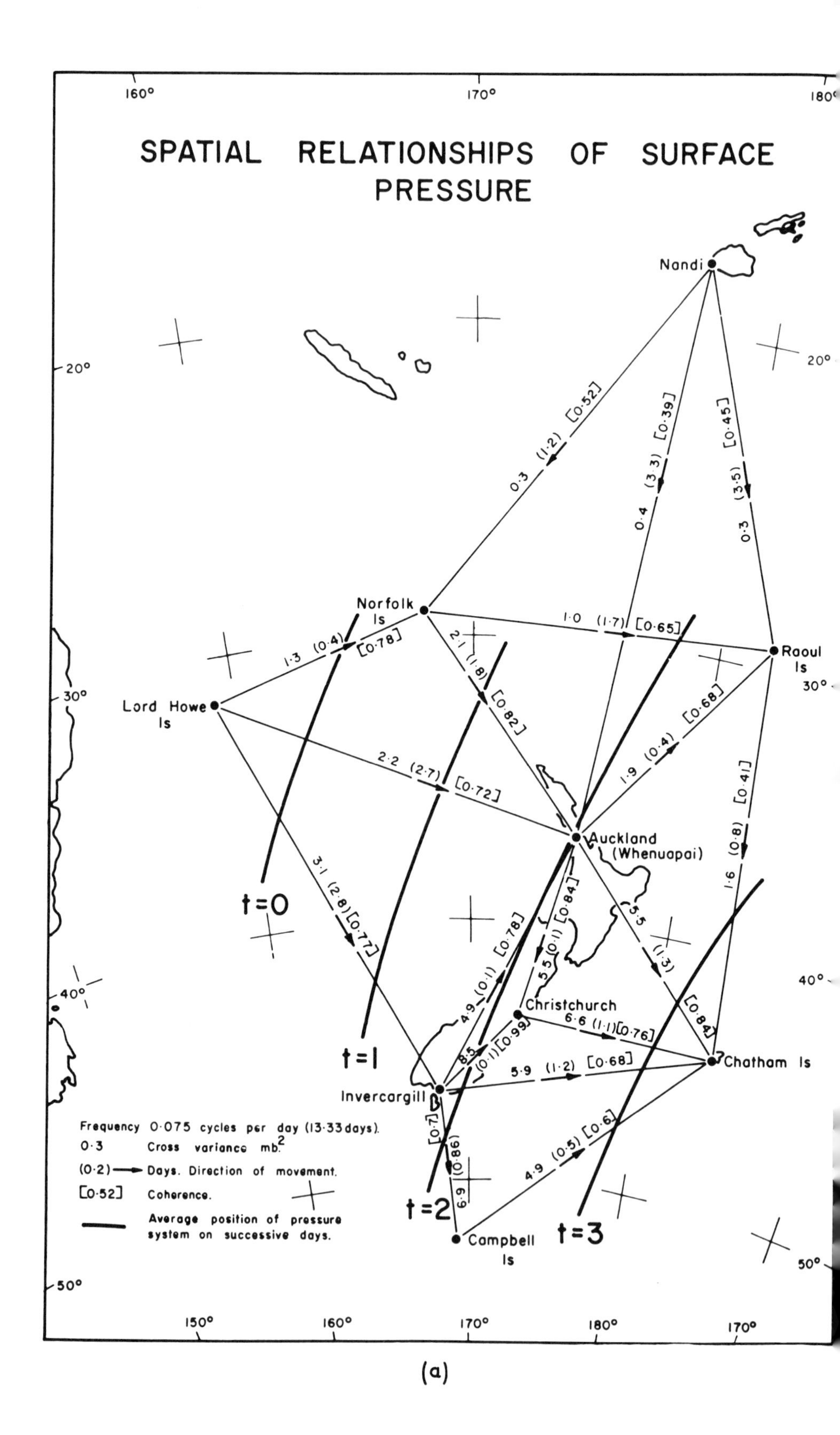
SPATIAL RELATIONSHIPS OF SURFACE PRESSURE
Nandi
Norfolk Is
Lord Howe Is
Raoul Is
Auckland (Whenuapai)
Christchurch
Chatham Is
Invercargill
Campbell Is
t=0
t=1
t=2
t=3
Frequency 0·075 cycles per day (13·33 days).
0·3 Cross variance mb.²
(0·2) Days. Direction of movement.
[0·52] Coherence.
Average position of pressure system on successive days.

(a)

Another aspect of this same approach is discussed by Rayner (1967a, 1971). Two-dimensional spectral analysis provides a means of correlating two surfaces using the coherence statistic (see p. 243). The virtue of this method is that it allows elements of different scale (frequency) in the spatial patterns to be distinguished. Evidently the potential of this procedure has so far been little appreciated. It is, however, particularly suitable for meteorological applications where data are available in cartesian coordinates.

It was mentioned earlier that Rayner (1967a) has applied cross-spectral analysis to temporal and spatial meteorological data in the New Zealand area. It is worth noting here that some conclusions can also be reached by his methods regarding space variations in two dimensions. Thus, the horizontal and vertical arrangements of one or more parameters can be analysed for selected frequencies at different stations and different levels at a single station, respectively. Fig. 4.23(*a*) illustrates the spatial relationships of surface pressure for waves with an approximate period of 13·3 days during 1962–3. South of 25°S the waves show coherence $>0{\cdot}5$ between most pairs of stations and consistency of lag times. The time–space transformation assumes that the waves move eastward and are in general relatively stable within the area. Fig. 4.23(*b*) shows the vertical organization of temperature and pressure

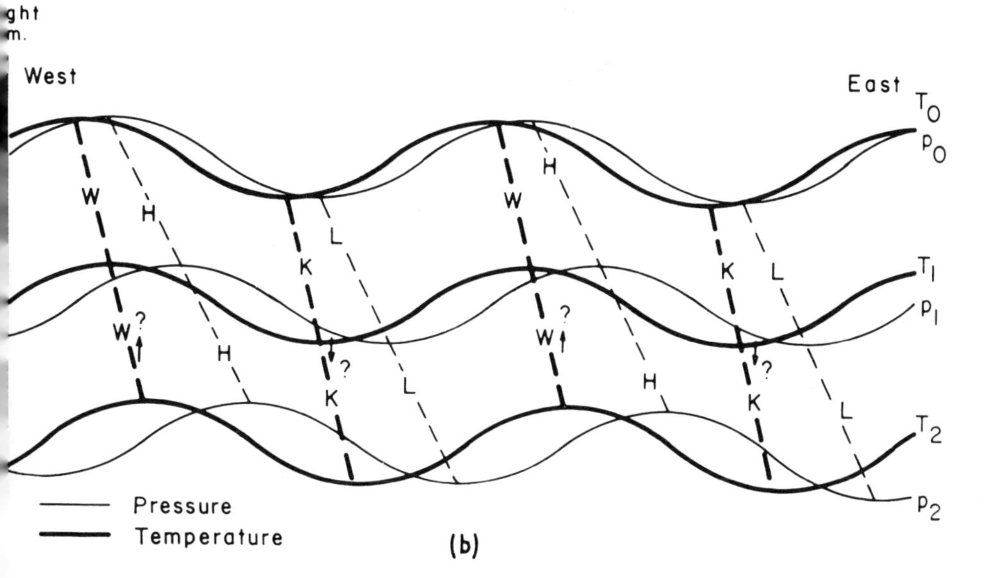

Fig. 4.23 (*a*) The spatial pattern of surface pressure waves with a period of 13·3 days over New Zealand (from Rayner, 1967a).

(*b*) The vertical relationship of temperature (*t*) and pressure (*p*) waves with a period of 7·3 days (from Rayner, 1967a). W = warm, K = cold, H = high, L = low. Inferred vertical motions are indicated by arrows.

for waves with a period of about 7·3 days. The normal westward tilt of pressure systems with height is evident.

Spectrum and cross-spectrum analyses along similar lines are now being used extensively in tropical meteorology. The sparse station network has led researchers to carry out detailed studies of time series of meteorological parameters at the available stations. The techniques can supply information on the frequency and wavelength as well as the vertical structure of disturbances and their energy transformation mechanisms. Wallace (1971) provides a thorough survey of the methods and the results so far obtained in the tropical Pacific.

6 EMPIRICAL ORTHOGONAL FUNCTIONS

The techniques described in the two preceding sections prescribe the nature of the result to some extent. It is possible, however, to generate from spatial or temporal data functions whose form is *not* predetermined and which represent a new combination of one or more physical characteristic of the anomaly patterns of some selected variable or variables. These are known as *empirical orthogonal functions* or alternatively as principal components, or eigenvectors.

These functions possess several important attributes. They are orthogonal (mutually uncorrelated) in space and the coefficients of different functions are also orthogonal in time. Moreover, if certain assumptions are met (see p. 250, for example), the functions provide the most 'efficient' possible representation of the variance of a set of data. That is to say, there is no set of fewer functions that will specify the pattern as precisely. A practical convenience is that the data are not required in grid coordinates, although it is not advisable to have a markedly uneven distribution of observation points.

The method was first developed for meteorological purposes by Lorenz (1956) and Bagrov (1959) and the earliest applications were made by Gilman (1957) and Grimmer (1963). Since about 1966 a quite extensive literature has developed and this is discussed below and in section D of this chapter (p. 282).

The detailed procedures cannot be discussed here. They are described by Cattell (1965), Jeffers (1967), Craddock (1969), and more generally by Gould (1967). General reference texts such as Cattell (1952) may be noted. The steps, in outline, are as follows:

1 Measurements of the variable(s) selected for analysis are organized in a data matrix. Qualitative information can be dichotomized and scored 1, 0, according to the presence or absence of some specified condition. Careful attention needs to be given to the relative magnitude of the variables under consideration. Standardized

values are normally used, where each variable is standardized with respect to a mean of 0 and variance of 1. Nevertheless, it must be noted that such changes in scaling affect the computed components.

2 The data matrix (for instance of grid point pressure values, Pij) is converted into a square matrix of *either* covariances *or* correlation coefficients between each of the n variables over N cases. If different variables are involved their scales of measurement must be weighted relative to one another. This is particularly important where highly correlated variables are concerned.

3 The covariance or correlation matrix $[X]$ is transformed into a $n \times k$ factor matrix $[F]$ where the new factors $k \ll n$.

$$[X] = [F][F]'$$

where $[F]'$ is the transpose of $[F]$: that is, rows and columns are interchanged. Diagrammatically,

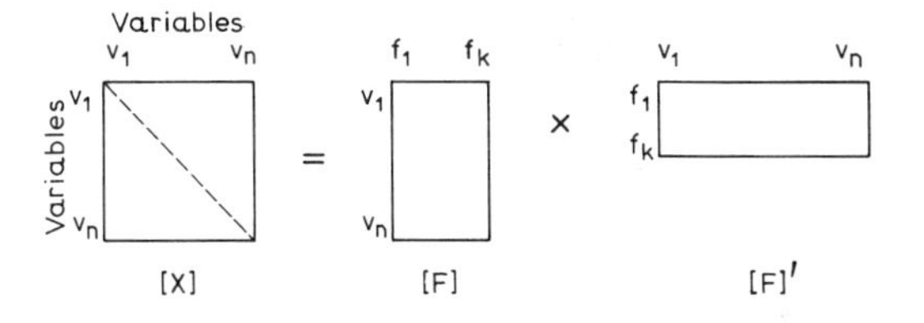

The dotted line in the correlation matrix is the principal diagonal or trace. To transform $[X]$ into $[F][F]'$ involves finding the eigenvalues (latent roots) and eigenvectors of the matrix $[X]$.

4 Eigenvalues are the roots of the polynomial equation corresponding to any particular square matrix. They represent the length dimensions of the principal axes. Gould (1967) provides a clear geometrical demonstration of their meaning. For example, given the matrix

$$A = \begin{bmatrix} 4 & 8 \\ 8 & 4 \end{bmatrix}$$

the characteristic equation is

$$\det |A - \lambda I| = 0$$

where $I =$ the identity matrix $\begin{bmatrix} 1 & 0 \\ 0 & 1 \end{bmatrix}$ with unity in the principal diagonal and zero elsewhere, and the eigenvalue a scalar matrix.

For matrix A, $\det|A-\lambda I| = 0 \equiv \det\begin{vmatrix}4-\lambda & 8\\ 8 & 4-\lambda\end{vmatrix} = 0.$

From matrix algebra, this is

$$(4-\lambda)(4-\lambda)-(8\times 8) = 0$$
$$\lambda^2-8\lambda-48 = 0.$$

Thus, the characteristic roots of the above quadratic equation are

$$\frac{8\pm(64+192)^{1/2}}{2}$$

so $$\lambda_1 = \frac{8+\sqrt{256}}{2} = 12 \quad \text{and} \quad \lambda_2 = \frac{8-\sqrt{256}}{2} = -4.$$

For a $n\times n$ matrix there are n eigenvalues. Each eigenvalue has a corresponding eigenvector which in a geometrical analogue (fig. 4.24) represents the slope of the lines with respect to the origin. Solving the two pairs of simultaneous equations

$$\left.\begin{aligned}4v_1+8v_2 &= 12v_1\\ 8v_1+4v_2 &= 12v_2\end{aligned}\right\}\text{ when } \lambda = 12$$

and $$v_1/v_2 = 1$$

$$\left.\begin{aligned}4v_1+8v_2 &= -4v_1\\ 8v_1+4v_2 &= -4v_2\end{aligned}\right\}\text{ when } \lambda = -4$$

$$v_1/v_2 = -1.$$

The $+1$ and -1 solutions show that the eigenvectors are normal to one another, or orthogonal. This means that they are completely uncorrelated.

For some purposes the principal axes are 'rotated' to obtain the simplest possible structure. Thus, each component is made to reflect

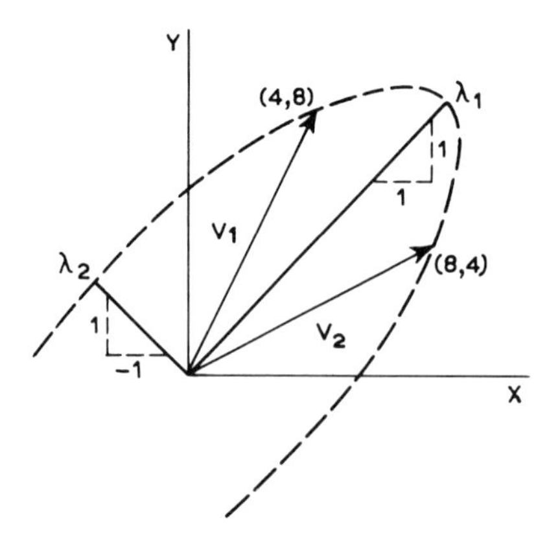

Fig. 4.24 Graphical analogue of eigenvectors (after Gould, 1967). λ_1 represents the primary axis and λ_2 the secondary axis of an ellipse fitting the row vectors v_1 and v_2 (see text). The slopes of the eigenvectors (1, 1), (1, -1) are shown by dotted lines.

only a few variables and each variable is constrained to correlate with only a few of the new components. It is also possible to obtain oblique instead of orthogonal factors, although the interpretation may prove more difficult.

5 The eigenvectors are examined and if possible interpreted in physical terms. To do this we examine the factor weightings which express the representation of the original variables in the new components. The role of factor weightings has already been discussed (p. 109 and fig. 3.6). The average sum of squares of all factor weightings on an eigenvector is equal to its eigenvalue. Also, the sum of all the eigenvalues of a symmetrical (square) matrix corresponds to the total variance of the raw data. The eigenvectors associated with the largest eigenvalues are those which are of most interest. That is to say, these *principal components* are new 'variables' which reduce the dimensions of the original data to the maximum degree possible. Beyond a certain level the least significant eigenvectors represent 'noise'. Appropriate procedures for determining this level have to be followed.

It must of course be emphasized that the new 'variables' are constrained to be mutually orthogonal and this may not adequately represent nature. Moreover, individual eigenvectors normally represent a combination of interacting physical processes which may not be readily recognized in terms of conventional parameters. Nevertheless, in a study of height and wind fields at several levels over northern Italy and adjacent areas, Palmieri (1968) was able to relate deformations in these fields relating to orographic effects to an individual eigenvector. It is also instructive to look at Stidd's study (1967) of rainfall in Nevada. Fig. 4.25 shows the seasonal variations of the first three eigenvectors he computed and compares them with three 'natural components' determined subjectively by J. G. Houghton. The first eigenvector is identified as a winter pattern associated with Pacific cyclones and the second as small-scale convective precipitation in summer derived from moisture originating over the Gulf of Mexico. The physical interpretation of the third component with a spring and autumn maximum is less certain. Nevertheless the three eigenvectors account for 93% of the total mean square of the original data.

Various analytical studies of hemispheric or continental scale have been carried out using the methods in the last few years. The eigenvectors representing the daily northern hemispheric fields of 500 mb height, 1000 mb height and 1000–500 mb thickness for 1965–67 have been determined (Craddock and Flood, 1969; Craddock and Flintoff, 1970) and it was shown that the first fifty patterns for the 500 mb fields give the most accurate representation (89–97% reduction of variance)

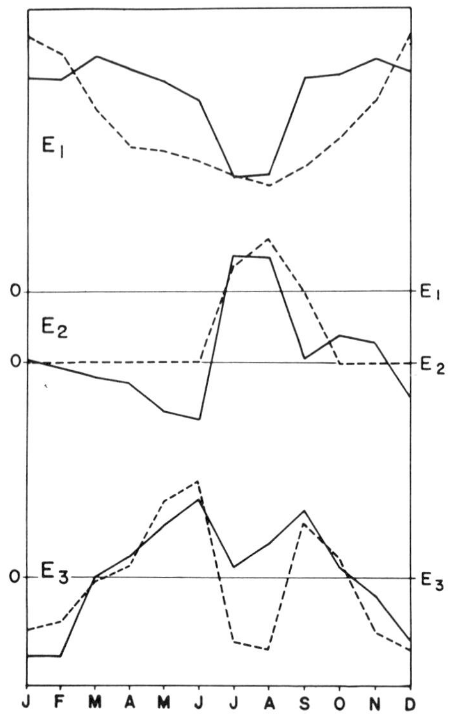

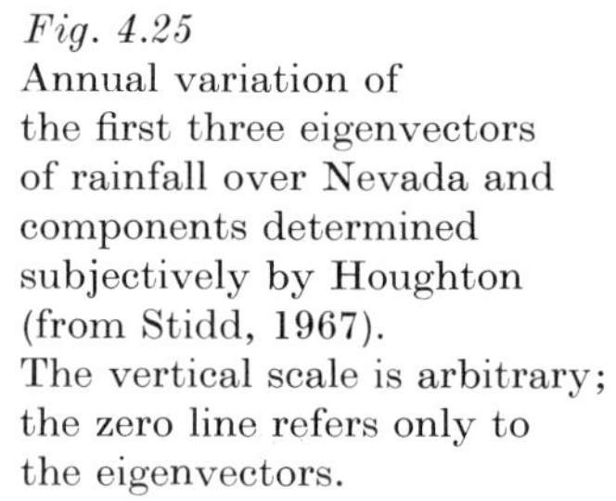
Fig. 4.25
Annual variation of the first three eigenvectors of rainfall over Nevada and components determined subjectively by Houghton (from Stidd, 1967).
The vertical scale is arbitrary; the zero line refers only to the eigenvectors.

of all three fields. On the continental scale of Eurasia the daily 500 mb fields in winter can be represented by only fifteen components accounting for 95% of the variance (Mdinaradze, 1968). Craddock and Flood used 130 data points for hemispheric geopotential fields, but Iakovleva *et al.* (1968) found that the main hemispheric 500 mb patterns can be described adequately by eigenvectors determined from as few as twenty-six points.

A major benefit of such procedures is in eliminating redundancy in raw climatological data and thereby economizing on computer storage in research or operational undertakings involving heavy computational requirements. Grimmer (1963) showed that the volume of data could be reduced to one quarter, whilst retaining 80% of the original variance, in an eigenvector representation of monthly fields of temperature anomalies for 1881–1960 over Europe and the North Atlantic. Space filtering can also serve as a basis for regional classification, but discussion of this is reserved for section D of this chapter (pp. 285–6).

Many studies of eigenvector representation have been avowedly exploratory. Investigations of the major fields of sea and air temperatures over the North Atlantic and of monthly anomaly patterns of pressure and precipitation over North America and the British Isles are of this type (Kutzbach, 1967; Perry, 1970; see also Craddock, 1968;

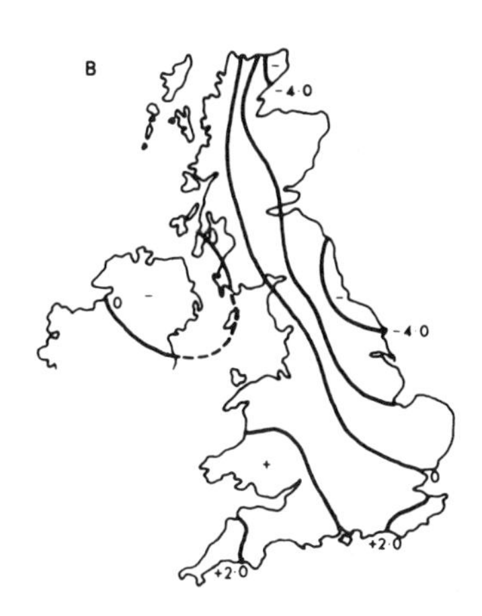

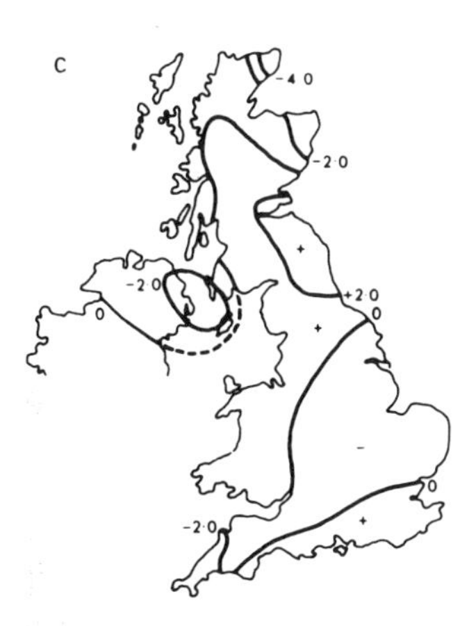

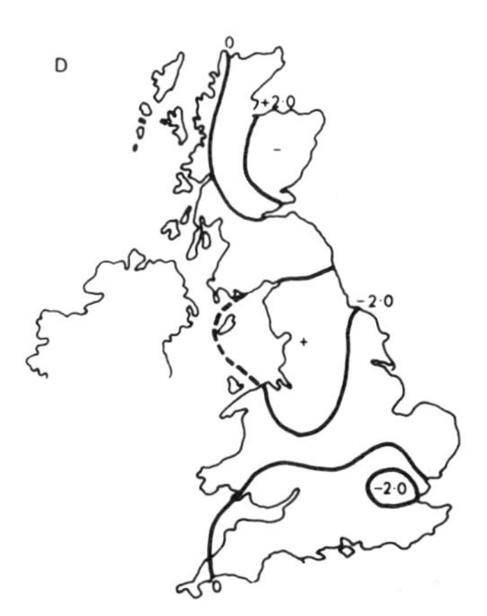

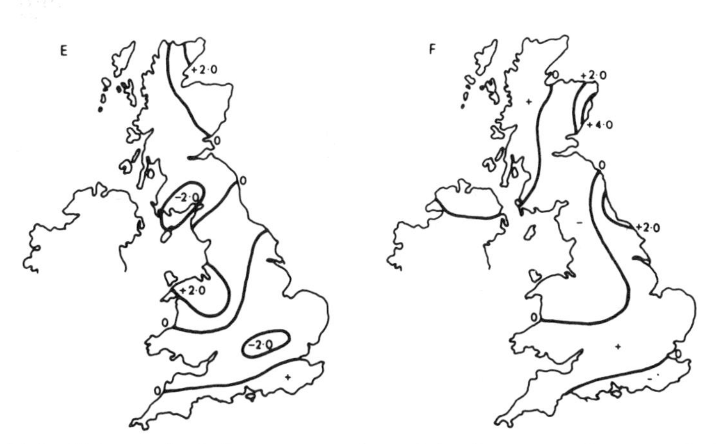

Fig. 4.26 The principal eigenvectors of precipitation anomalies over the British Isles, January 1920–60 (from Perry, 1970).

Sellers, 1968; Iudin, 1969). Fig. 4.26 illustrates the principal eigenvectors of precipitation anomalies over Britain for January 1920–60. Note that the sign of the eigenvectors is arbitrary. The first four eigenvectors contribute 24·7, 17·2, 10·4 and 8·6% of the variance. The first pattern is common in months of high zonal index, while the second has occurred in months with a low index of 'progression' (see p. 142). It is through such analyses of relatively familiar data that we gain insight into the operation of the method in unknown situations. A major collection of Russian studies using eigenvectors is worth noting (Iudin, 1969). The investigations deal with fields of pressure (including the southern hemisphere), temperature, precipitation and snow cover. On a smaller scale, Peterson (1970) determined the horizontal pattern of sulphur dioxide over St Louis in terms of eigenvectors. The first three components accounted for two thirds of the variance of mean twenty-four hour amounts on forty-nine winter days. Correlations between these components and meteorological variables were then determined.

Finally, a rather different application of eigenvectors occurs in canonical trend surface analysis. This technique is used to construct a polynomial surface describing the spatial variation of several variables considered together. This surface does not provide absolute values but it is useful in depicting the general trend and in terms of the data simplification achieved. Preferably, only three or four variables should be incorporated to facilitate interpretation. Lee (1969) gives a clear account of the details and geological illustrations. Apparently meteorological applications of such trend surfaces have not so far been made.

D Classificatory methods

Classification may serve the purposes of naming sets of things, and of grouping things by resemblance, by relationship, or by both. Properly, it is a means by which understanding of phenomena is increased, leading to models and theory, but in climatology, as in some other fields, classification has tended to become an end in itself. This emphasis is all the more unfortunate since climates can in any case only be distinguished on arbitrary criteria. It is essential, therefore, that the purpose of any putative classification be specified. As we saw in Chapter 3, there are various features that may be of interest in synoptic climatological classification. Here, the discussion is concerned with procedures for obtaining efficient classifications using modern numerical techniques. This review will be rather general since many procedures have been developed in other disciplines and have not yet been thoroughly investigated in climatology.

The fundamental basis of all systems of classification is to obtain the least variability within the groups and the maximum differences between them. Groups (classes) can be obtained by two basic methods – by *division* of a sample, or of a population; by *agglomeration* of similar individuals. The former has been applied particularly in taxonomy and ecology. Most commonly a hierarchical approach is used so that groups and subgroups are obtained and frequently the subdivision is made with respect to the occurrence or non-occurrence of a single specific attribute. This would be a monothetic-divisive classification. Agglomerative, or clustering, methods are generally polythetic, such that all measured attributes are taken into consideration. The groups may be arranged hierarchically or in a reticulate system. In general, divisive computations are quicker and the groups are little affected by anomalous data. Such anomalies can result in misleading groups at the lower levels in an agglomerative process and thereby affect all subsequent higher order groups.

We can recognize three aspects of classification. The first concerns the delimitation of groups in terms of the various possible measures of similarity between individuals. The second involves the assignment of additional individuals to established groups with minimum likelihood of error, and the third relates to tests of significance for established groups.

1 THE DELIMITATION OF GROUPS AND AREAS

The delimitation of synoptic types or synoptic regions is an essential part of much synoptic climatological work. However, the spatial and temporal relationships of meteorological phenomena present some special difficulties. Certain techniques are applicable to both spatial and temporal data whereas others are of more limited applicability.

Three principal types of procedure have been used and we begin with the simplest which relates only to the question of spatial similarity, or homogeneity.

Correlation analysis

An early application of correlation to the problem of assessing the homogeneity of a region for a single climatic parameter may be found in the work of Glasspoole (1925). By correlating annual precipitation totals at Oxford, and then at Glenquoich in west Scotland, with other stations in the British Isles he demonstrated with isocorrelate maps that the country behaves as at least two units, although no boundary line was proposed. The two key stations were selected on the grounds that they have a correlation of virtually zero for annual precipitation amounts. The same method has been used for precipitation over the Alps by Fliri (1967). His seasonal maps based on Hohenpeissenberg to the north and Colle Venda to the south as key stations are given in fig. 4.27. Specific

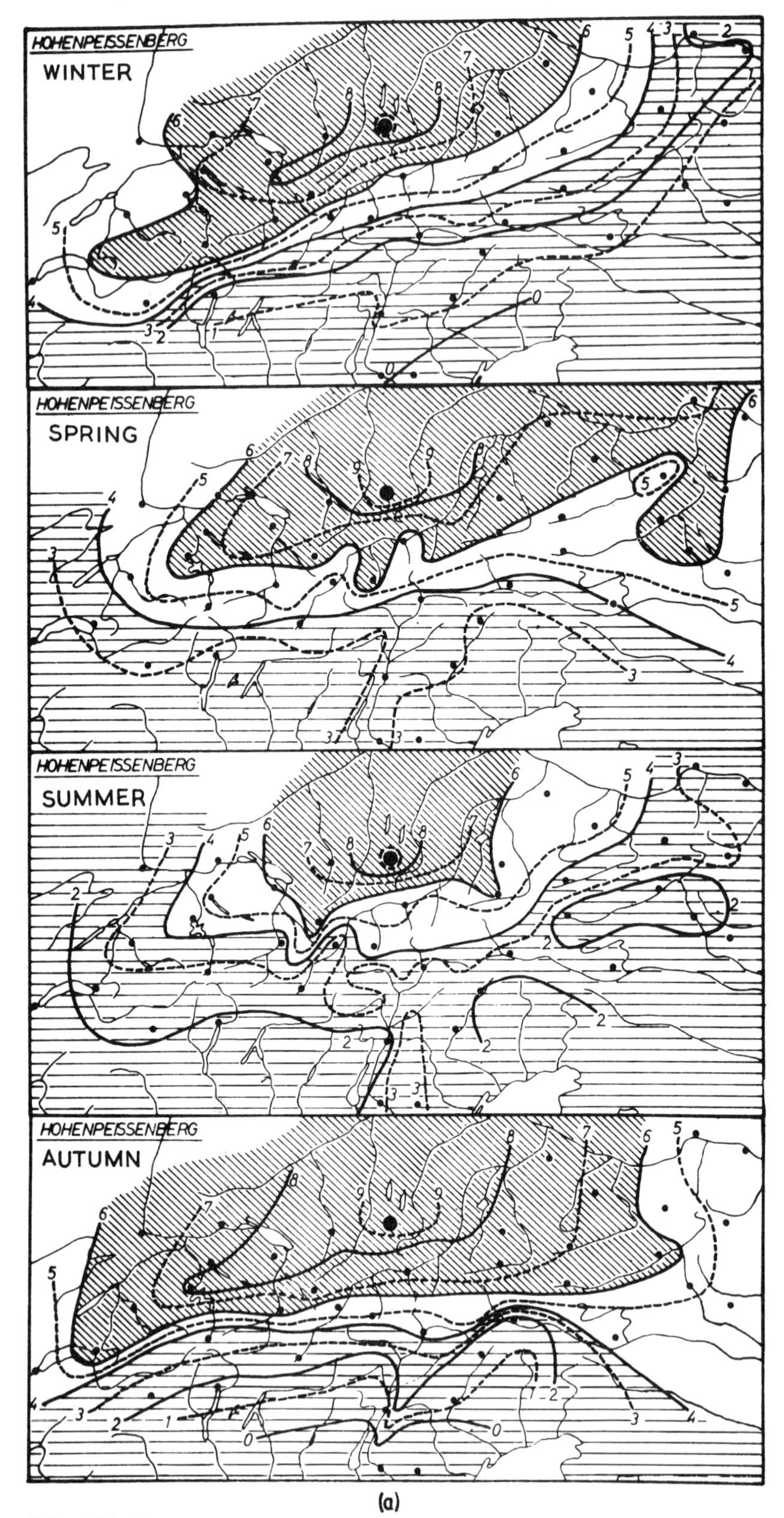

Fig. 4.27 Isocorrelates of seasonal precipitation over the Alps based on (*a*) Hohenpeissenberg, (*b*) Colle Venda (from Fliri, 1967).

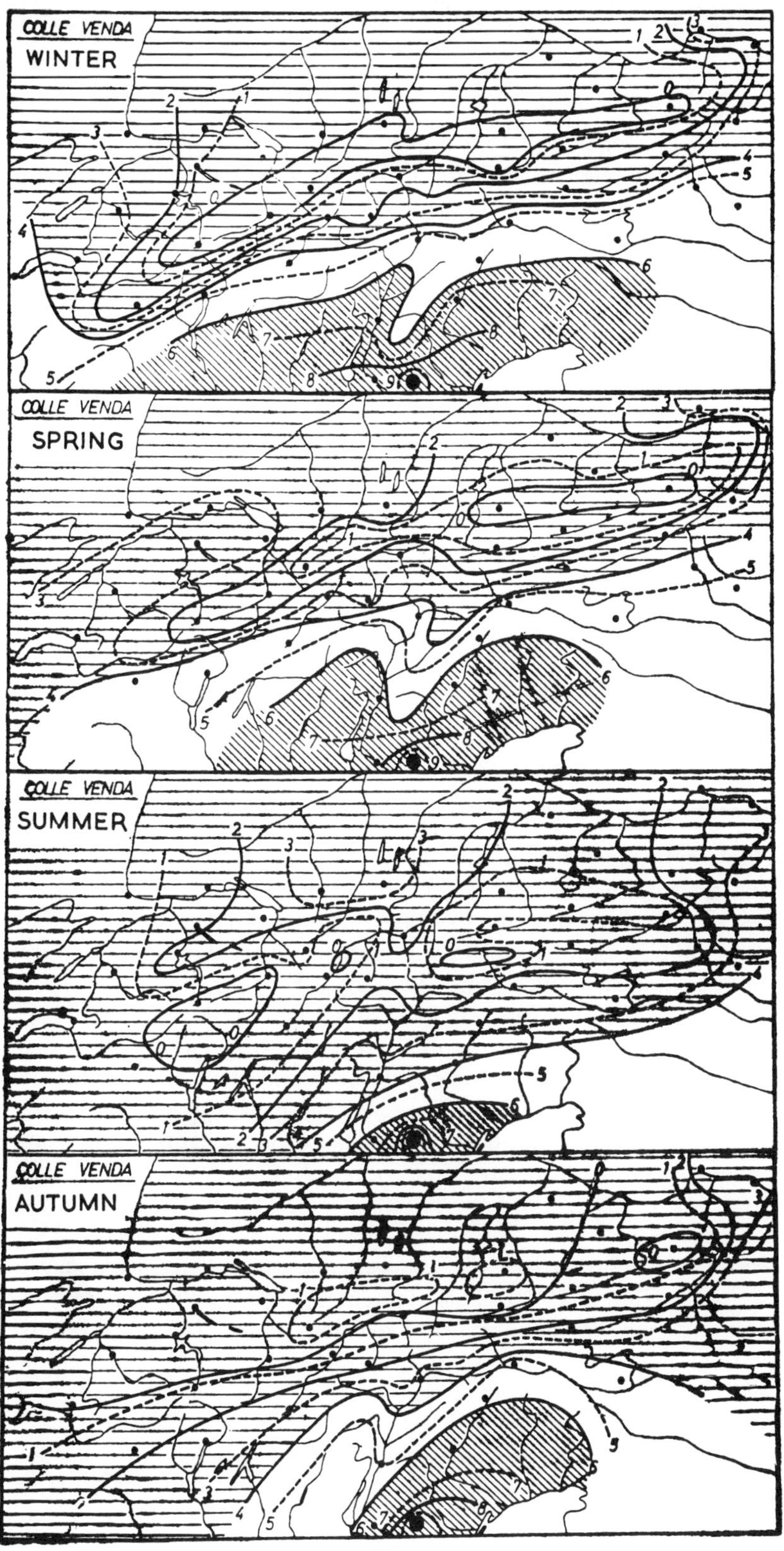
COLLE VENDA
WINTER
COLLE VENDA
SPRING
COLLE VENDA
SUMMER
COLLE VENDA
AUTUMN

(b)

use of this approach in delimiting climatic regions appears to have been made first by Sekiguti (1952) for Japan. A basic objection to this simple isocorrelate method is that the initial choice of station is arbitrary. Moreover, in practice the area of statistically significant correlations is often small and, more important, the area of useful correlation – where say at least 50% of the variance is accounted for (corresponding to $r \geqslant 0{\cdot}7$) – is even more restricted. A further drawback is that an arbitrary, subjectively determined value of r has to be used in order to delimit particular regions.

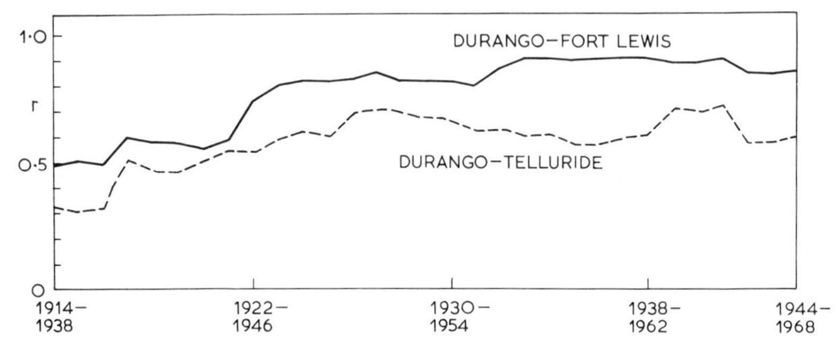

Fig. 4.28 Time trend (twenty-five year running means) of correlation of annual precipitation between pairs of stations in southwest Colorado.

It is worth digressing at this point to remark on the question of the statistical significance of correlation coefficients. Panofsky (1949) has drawn attention to the fact that meteorological data cannot in general be regarded as random selections from a particular finite or infinite population because it is manifestly evident that regimes of weather and climate exist. Nevertheless, either through ignorance of the problem, or due to the extra labour involved in collecting (wherever possible) additional samples, few workers concern themselves with the 'stability' of the calculated coefficients. Fig. 4.28 illustrates the changing level of correlation for annual rainfall at pairs of stations in southwest Colorado. Such changes may of course reflect a different response to climatic fluctuations or 'errors' due to inhomogeneities in the record at one or other station.

The first attempt to classify pressure fields by correlation methods was that of Lund (1963). This straightforward application and some associated problems are discussed in Chapter 3.A (p. 102). From the statistical point of view two major questions concern the effects of spatial autocorrelation and of non-linearity (see also section C of this chapter, p. 256). It is assumed that the data used in calculation of the product-moment correlation coefficient are not serially correlated, but the fact that one is spatially sampling a continuous field variable (pressure or height) puts the validity of this assumption in jeopardy. Since the whole question of spatial correlation is still hedged with uncertainties (see p. 255) caution must be used in interpreting such correlations, certainly in terms of their significance level. The second difficulty is that fields of pressure or geopotential height are generally non-linear and considerable information may therefore be lost in correlating fields on the basis of a linear technique.

It should also be mentioned at this point that the correlations between two sets of variables (or between two eigenvector patterns determined on different data sets) can be evaluated by canonical correlations. The canonical correlation coefficients are the eigenvalues of the variance–covariance matrices, or the correlation matrices, for the sets of variables (Cooley and Lohnes, 1962; Glahn, 1968). A recent analytic application by Fritts *et al.* (1971) used canonical correlation to compare anomaly patterns of tree growth with those of seasonal pressure fields for the period since 1700 (see Chapter 5.C, pp. 353–6).

Linkage analysis

Linkage analysis refers to a set of procedures that are employed to group individuals into clusters. The term graph clustering[1] is also used (Hararay, Norman and Cartwright, 1965; Haggett and Chorley, 1969, p. 244). Only a brief summary of the principal types of approach can be given here. Lance and Williams (1967b) and Spence and Taylor (1970) have provided reviews and evaluations of recent computer methods of clustering, and Johnston (1968) illustrated the resultant classifications obtained using a variety of procedures.

The basic operations in forming groups are: to initiate clusters, to assign additional individuals to the initial clusters, to fuse groups when appropriate, and to terminate the last stage at some suitable point. The initial step may involve taking an item at random and determining its similarity with all other items in order to obtain the first cluster. Individuals outside this group are then examined in the same way. Alternatively, each item treated in random order can be considered as a potential nucleus for a group. The actual sorting strategies are as follows (McQuitty, 1957, 1960, 1966; Sokal and Sneath, 1963, Chapter 7):

1 *Single linkage* – the clustering is based on relationships between pairs of individuals. The successive links may be based on 'nearest neighbours' (the 'elementary linkage analysis' of McQuitty (1957)), where the individual to be admitted is closer to one member of the group than to any member of some other group, or on 'farthest neighbours' (Macnaughton-Smith *et al.* 1964) where the fusions are

[1] The term 'cluster analysis' is commonly used in this context, although statisticians limit its usage strictly to *R*-mode factor analysis (see p. 283).

based on the least of the maximum distances between pairs of individuals and then pairs of groups.

2 *Multiple linkage* – the clustering is based on the relationships between all the items of data. Sørensen (1948) and McQuitty (1966) have used 'complete linkage' where the individual to be admitted is closer to every member of the group than to any member of some other group. More commonly the linkage is based on some reference item of the group, such as the group mean, or centroid (Sokal and Michener, 1958; Mather, 1968a). Group average sorting (Lance and Williams, 1967a), where groups are fused so as to minimize the average of the similarities between all pairs of individuals, seems to be a superior method. It takes account of the variance within each group in addition to their average separation.

Multiple linkage methods, in which the aim is to minimize the within-group sum of squares, involve more computation than single linkage because the group reference item has to be recalculated at each step. The particular constraints which are employed to minimize the within-group scatter are summarized and compared by Wishart (1969). He also emphasizes the problem of 'chaining effects' which occur with linkage methods based on reciprocal pairs of data. This means that separate clusters may not always be distinguished if there is considerable noise in the observations. Wishart shows how single linkage procedures can be modified so as to overcome this objection. Nevertheless, multiple linkage methods are better where the clusters are more or less 'spherical'.

In climatology only single linkage methods appear to have been used. Gregory (1964) used McQuitty's method with correlation coefficients between precipitation totals at pairs of stations in Sierra Leone, showing that rainfall types vary markedly in areal extent from season to season and between months in the wet season. Perry (1968) adopted the same method to examine the grouping of areas of the British Isles (fig. 4.29) on the basis of correlations between climatic parameters and the synoptic indices derived from Lamb's catalogue, and subsequently the grouping of Atlantic weather ships according to correlations between computed turbulent heat fluxes (Perry, 1969).

So far we have ignored the measure of similarity that may be used in forming clusters or groups. The three general categories are

1 Coefficients of association.
2 Coefficients of correlation.
3 Distance measures.

Sokal and Sneath (1963, Chapter 6) give a detailed review of their characteristics, in so far as numerical taxonomy is concerned, and Harbaugh and Merriam (1968, p. 161) do the same for geology.

Coefficients of association have occasionally been used as a simple measure of similarity in climatology (see p. 250) but not apparently in specifically classificatory studies. For most purposes the product-moment correlation coefficient can be used, although in some cases the non-parametric rank correlation coefficient may be more suitable. It is

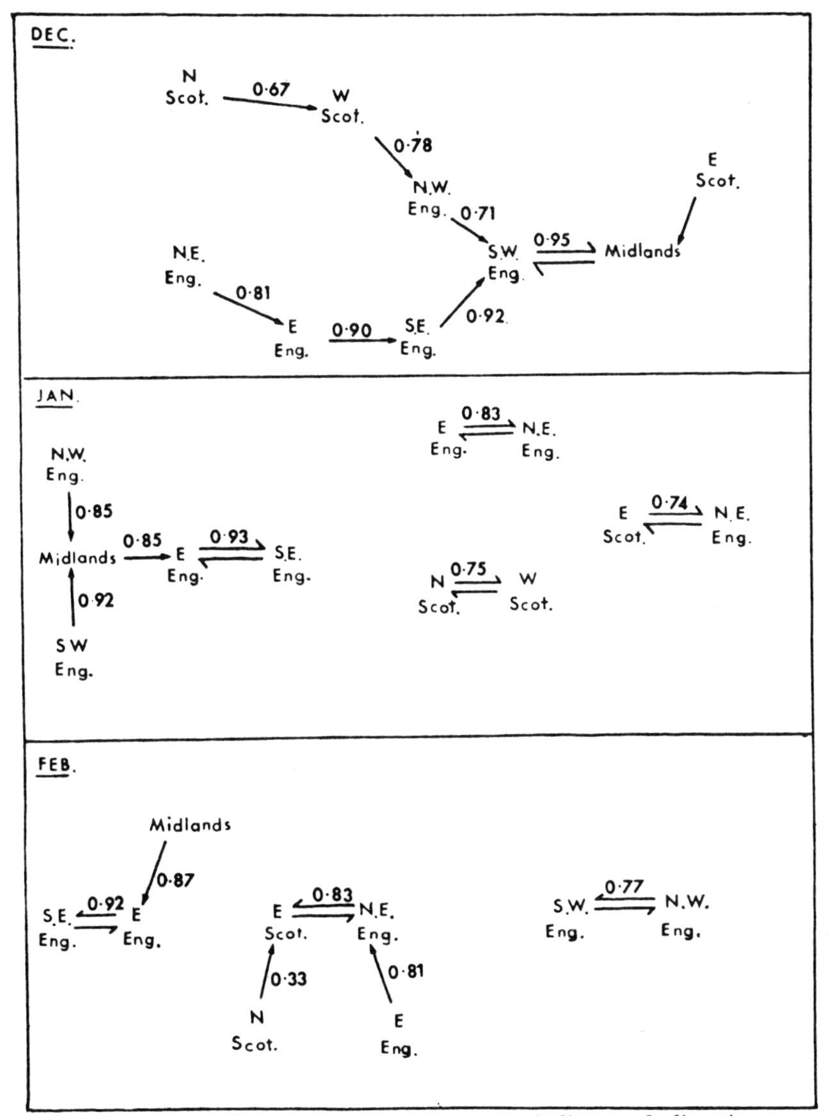

Fig. 4.29 Simple linkage relating to synoptic indices and climatic parameter for areas of the British Isles in winter months (from Perry, 1968).

also possible that one of the distance measures may be appropriate. Since these are much less familiar they will be briefly described.

The general form of the distance measure between two elements j and k in r-dimensional space with m $(=1, 2, \ldots, r)$ orthogonal axes is

$$d_{jk} = \left[\sum_{m=1}^{r} (|x_{km} - x_{jm}|)^a \right]^{1/a}$$

where a is any variable, not necessarily an integer. For $a = 2$ we have the well-known Euclidean (or taxonomic) distance. More specialized measures include the 'generalized distance' (D^2) of Mahalanobis (Rao, 1952) which takes into account the variance of individual characters and the correlations between them, the 'information content' (Rajski, 1961; Estabrook, 1967) which expresses the shared information, and angular separation

$$\cos\theta = \frac{\sum_{i=1}^{n} X_i Y_i}{\left(\sum_{i=1}^{n} X_i^2 \sum_{i=1}^{n} Y_i^2 \right)^{1/2}}$$

These measures can be transformed so as to relate them to the product-moment correlation coefficient. Additional measures are listed by Boyce (1969).

The appropriateness of a particular similarity coefficient is dependent on the type of data. The use of the 'information content' and of association coefficients for nominal or ordinal scale data should be especially noted. Biological and geological examples of hierarchical and reticulate clustering using information coefficients are to be found in Estabrook (1966), Wirth, Estabrook and Rogers (1966), and Andrews and Estabrook (1971; see also Williams and Lambert, 1966). Experimentation with some typical sets of meteorological data would be a useful contribution to our evaluation of the potential of some of these methods.

Component and factor analysis

The essential principles of principal component analysis (eigenvectors) have already been discussed in section C of this chapter (pp. 268–74). The application of these methods to classification problems in climatology is very recent, although they have been developed for non-quantitative and quantitative biological descriptors (Williams and Lambert, 1959; Lambert and Williams, 1962; Sokal and Sneath, 1963) for over a decade.

Two basic approaches are used. In 'classification analysis' (known commonly as 'Q mode' analysis) individuals are grouped into classes on the basis of their measured characteristics (attributes). For example, locations might be classified with reference to such parameters as pressure, thickness, temperature, etc. In 'classification analysis' (R mode) groups are determined amongst the measured variables on the basis of the different individuals. Indices of air mass humidity characteristics such as cloudiness, precipitation and visibility could be grouped using observations on different days, or at different locations, for instance.

Cattell (1965) also refers to T mode analysis in which pairs of individuals are correlated with respect to different variables over time, and P mode analysis where pairs of variables measured for an individual are correlated over time. The transpose of this latter, O mode analysis, examines correlations between pairs of time intervals in terms of different weather variables. Thus, one can classify days in terms of weather parameters at a given location. The distinction between time and space modes has apparently not been made in the meteorological literature, where the 'individual' may refer to a location or a time interval. Kutzbach (1967), for example, examines the eigenvector patterns derived from one, two or three meteorological parameters for twenty-three locations over twenty-five Januaries (see p. 368). This is in the category of Cattell's T mode. Interest is focused on the spatial distribution of time deviations for pressure, temperature and precipitation considered individually and in two-way and three-way interaction.

A modified form of component analysis, referred to as principal coordinate analysis (Gower, 1966, 1967), makes use of distance measures instead of correlation coefficients or covariances. This has recently been applied by Russell and Moore (1970) in a search for homoclimates of the Brigalow region of eastern Australia. The procedure produces an *ordination* (or ranking) rather than a classification. The 'distance' between a pair of stations in terms of weather characteristics is specified proportionally by the similarity coefficient (in this case the Euclidean distance). Russell and Moore also investigate the monthly sequence of three groups of weather parameters based on an R mode analysis of data at twenty-two stations. The seasonal structure is therefore taken into account.

It should at this point be noted that principal components analysis is a particular factor analytical model. In the principal components model the principal diagonal in the matrix of correlation coefficients between the variables (see section C of this chapter, p. 269) is assumed to be unity. The assumption is that all of the variance of the variables is accounted for within the particular sample. Alternatively, in factor analysis, we enter estimated values (each less than one) in the principal diagonal. These values, known as *communalities*, indicate the proportion of the variance of each variable accounted for by the variables included in the analysis. Some variance is attributed to errors and unrepresented factors. Cattell (1965) discusses estimation procedures in general terms.

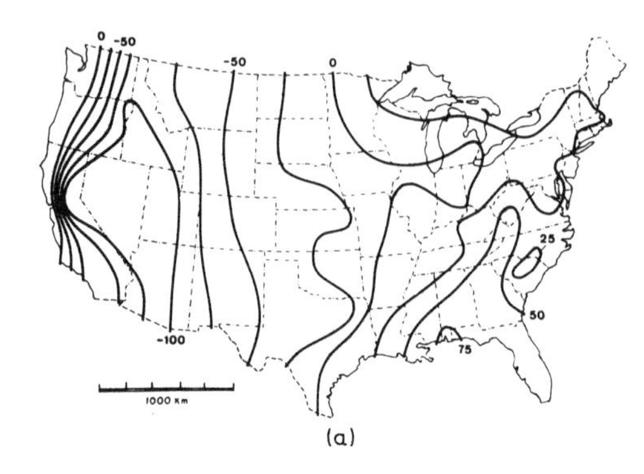

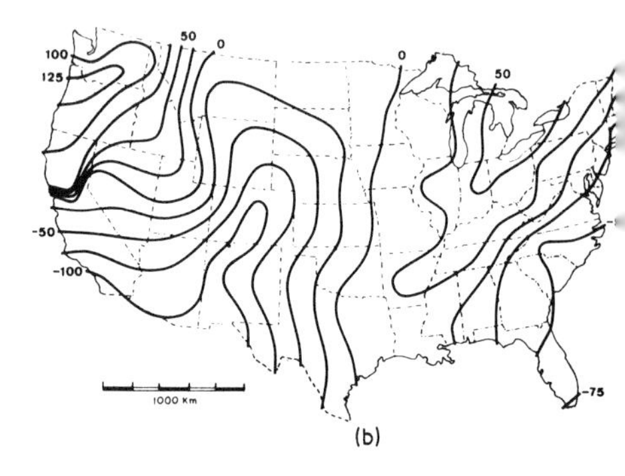

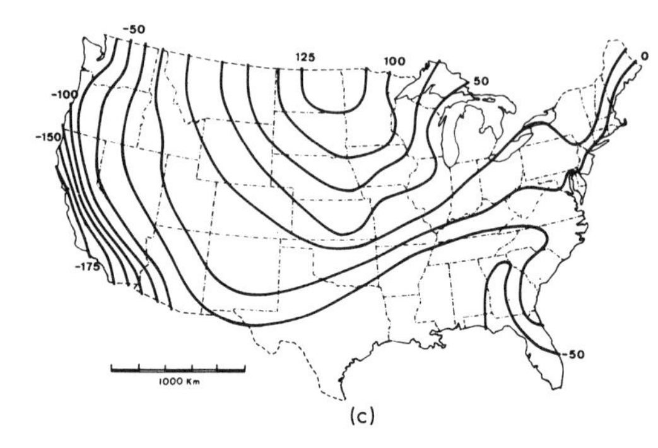

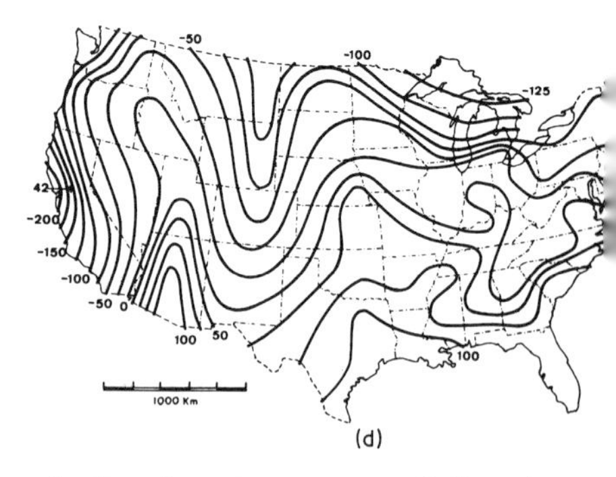

Fig. 4.30 Distribution of scores (× 100) on the first four eigenvectors of climatic characteristics over the United States (from Steiner, 1965). (*a*) 'Humidity', (*b*) 'Atmospheric turbidity', (*c*) 'Continentality', (*d*) 'Thermality'.

Component analysis may be regarded as an internal structuring technique, whereas factor analysis is used to generate hypotheses. However, Armstrong (1967) draws attention to the risks involved in this latter process if prior analysis is not performed in order to provide a conceptual model against which to judge the results of the factor analysis.

While empirical orthogonal functions are being used increasingly in the analysis of spatial patterns (see section C of this chapter, p. 272) there have been few attempts to apply the methods to classification problems in climatology. Iakovleva and Gurleva (1969) use eigenvectors as a basis for climatic regionalization of the U.S.S.R. The first eigenvector which shows a zonal pattern accounts for 40% of the variance of eighteen parameters (one annual, eight for January and nine for July) and the second, with a meridional pattern, accounts for 23%. However, their regionalization is limited to the selection of completely arbitrary boundaries on the maps of these two components.

A thorough illustration of the potentiality of the method is provided by Steiner (1965). Sixteen climatic parameters were chosen for sixty-seven stations in the United States. The frequency distributions were normalized where necessary and then subjected to a component analysis (with rotation of the four principal axes). The four orthogonal factors accounted for 88·6% of the total variance. By comparison with the original variables, in terms of the factor weightings, it was considered that the eigenvectors represented measures of 'humidity', 'atmospheric turbidity', 'continentality' and 'thermality'. Fig. 4.30 shows the spatial distribution of the scores of each eigenvector.

In the second phase of Steiner's investigation, the sixty-seven stations were grouped according to their factor weightings using as a distance measure the individual distances to group means. After obtaining a single group the process was reversed and groups were formed using minimum sums of squares of the within-group distance as the criterion. The ten groups so obtained (accounting for 80% of the within-group distance), which are shown in fig. 4.31(*a*), were then tested by discriminant analysis (see below) and only minor corrections were found to be necessary. Fig. 4.31(*b*) shows the final map and the hierarchical order of the boundaries.

The work of Kuipers (1970), discussed in Chapter 3.B (p. 106), illustrates another approach to the classification problem. The basis of the division is the minimization of the sum of the subset variances. The class limits and the mean values of the classes are obtained by integration of the normal distribution function. He uses the percentage of the total variance accounted for by a given number of divisions of the factor weightings on the major eigenvectors as a measure of the information provided by the classification. As noted in section C of this chapter (p. 271), the average sum of squares of all the factor weightings on an

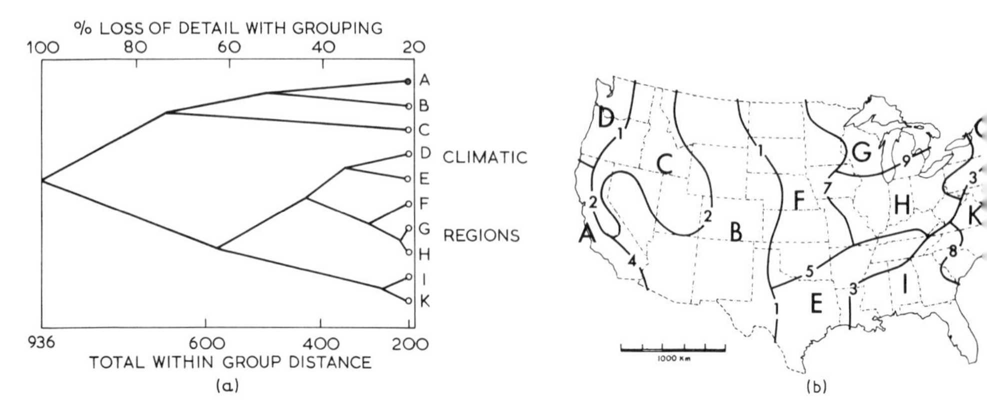

Fig. 4.31 (*a*) Linkage tree of the grouping of climatic regions. (*b*) Climatic regions of the United States and hierarchical order (1–9) of the boundaries (after Steiner, 1965).

eigenvector is equal to its eigenvalue and the sum of all eigenvalues corresponds to the total variance. Kuipers shows that for his results a two-way division of the factor weightings on the first eigenvector yields 64% of the information provided by the first eigenvalue, whereas a four-way division increases this to 88% (equivalent to 26% of the total variance). However, a two-way division of the factor weightings on the first and second eigenvectors gives 27% of the total variance. It is concluded by extension of this method that the optimal solution with twenty-four classes is for four, three and two divisions respectively on the factor weightings for the first, second and third eigenvectors. Fig. 3.6(*b*) shows examples of pattern types with specific weightings corresponding to the mean of each class.

It is worth noting that twenty-four classes still only account for 44% of the original variance. Additional eigenvectors and classes are required if more precise specification is desired. Clearly, this method demonstrates the fundamental problem of synoptic variability to which we have already referred, but at least it provides an objective measure of the variance unaccounted for and the dispersion within classes (if we can assume that the factor weightings are normally distributed).

While the work of Kuipers and also that of Christensen and Bryson (1966), discussed on p. 189, represent major contributions to climatological methodology there is still room for new and improved procedures. For example, no attempt has yet been made in climatology to combine Q and R mode component analysis to define clusters – a method developed successfully for ecological purposes by Lambert and Williams (1962). An approach such as this would at least produce useful operational classifications.

In summary, it should be re-emphasized that factor analytical techniques do not in themselves provide absolute classifications. The researcher still has to make decisions concerning the criteria used to locate class boundaries, the amount of variability that can be permitted within a group, and the size of cluster preferred in a particular investigation. In certain respects discriminant analysis can be a useful ancillary tool and this is discussed next.

2 DISCRIMINANT ANALYSIS

The primary purpose of discriminant analysis is to set up the best combination of variables to differentiate classes of some object. However, it can also be used to facilitate the assignment of new observational data to established classification categories with a minimum of error. This has great practical significance in the context of the present discussion because eigenvectors have to be completely recalculated for additional information.

The procedures, which were first developed by R. A. Fisher, have been elaborated by Rao (1952). Meteorological applications are detailed by Miller (1962) and Suzuki (1969). Simple accounts of the fundamentals are given by Panofsky and Brier (1958, p. 118) and Cooley and Lohnes (1962, Chapter 6). The method is based on multiple regression and is applicable to normal and moderately skewed data, provided the group variances are similar. Suzuki has extended it to deal with heterogeneous (discrete and continuous) variables.

For discrimination with two variables (predictors), x_1 and x_2, we define a linear discriminant function, D, with D_1 greater than a critical value for occurrence and D_2 less than the critical value for non-occurrence of some specified condition of interest,

$$D = a_0 + a_1 x_1 + a_2 x_2$$

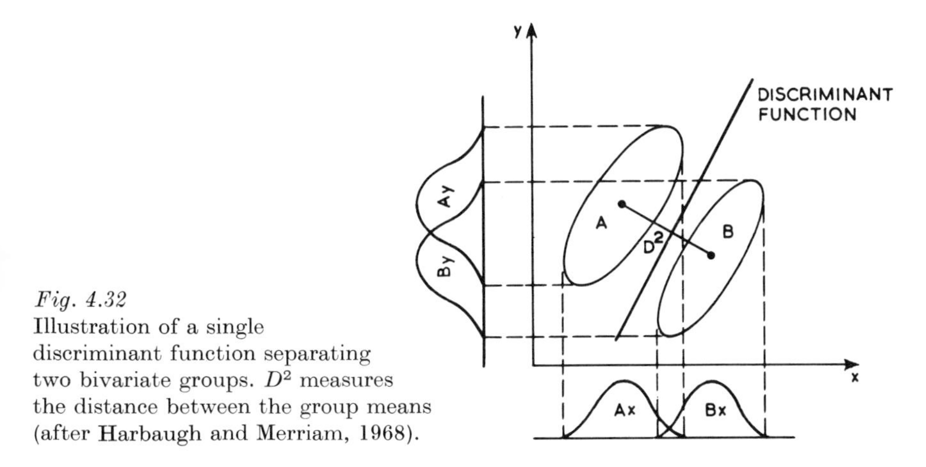

Fig. 4.32
Illustration of a single discriminant function separating two bivariate groups. D^2 measures the distance between the group means (after Harbaugh and Merriam, 1968).

a_1 and a_2 are then chosen to maximize $(\bar{D}_1 - \bar{D}_2)/s_D$ where the bar denotes the mean of the respective classes, s_D = the standard deviation of the pooled D and D_2 classes. Fig. 4.32 illustrates a simple discriminant function between two groups.

The method has been used by Suzuki (1964), for example, to discriminate between categories of light and heavy precipitation occurring with fronts and typhoons, and by Brinkmann (1970) in an attempt to distinguish chinook and non-chinook wind characteristics at Calgary. Casetti (1964) applied the technique to the determination of spatial boundaries in an investigation of climatic regions in North America. He took twenty-four climatic parameters (average precipitation and mean temperature for each month of the year) for seventy stations and first reduced the variables by principal component analysis. The first three components, which represented the levels of temperature and precipitation taken together, aridity, and seasonality, accounted for 93% of the original variance, and a total of six provided an optimum description. Multiple discriminant functions (see for example, Mather, 1969b) were then used to determine the 'coordinates' of the stations in discriminant space. From these, Euclidean distances of each station to the nearest class centroid are calculated and the stations allocated to the nearest class. The data are reordered with respect to the new classes and the same steps repeated. Fig. 4.33(*a*) shows the 'core' areas of the classification and the limits of the areas within which stations remained in the same class after the analysis, and fig. 4.33(*b*) gives the final boundaries determined after several iterations of the discriminant analysis. Further tests showed that, even when substantial noise was present, discriminant analysis distinguished the core areas of the classification categories.

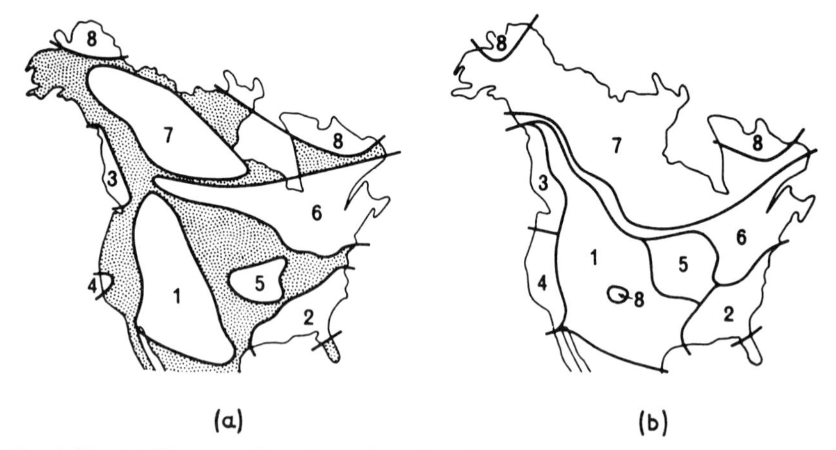

Fig. 4.33 (*a*) Köppen climatic regions based on discriminant analysis and the original 'core' areas. (*b*) Final regions following seven iterations (after Casetti, 1964; from Chorley and Haggett, 1970).

Mather and Doornkamp (1970) illustrates a further application of multiple discriminant analysis in a geomorphological study. After determining five groups of drainage basins by a centroid linkage method they show that 98% of the difference between each of the groups can be specified by two discriminant functions. These functions are then interpreted in terms of six eigenvectors used to describe the basin characteristics (in place of eighteen original variables). This analysis allowed a model relating the original variables and the basin groups to be constructed. The development of exploratory climatological investigations along similar lines would appear worthwhile.

3 TESTING GROUP DIFFERENCES

After developing a scheme of synoptic types (or climatological regions) by subjective means, a key question is the degree to which the types are different from one another with respect to their climatic characteristics.

Simple tests based on the mean and standard deviation may be applied, if the assumption of normality is applicable to the given data. Thus, considering the sample means $\bar{a}$ and $\bar{b}$ for two types (or regions), we can compare

$$\bar{a} \pm 2s_a/\sqrt{n_a} \quad \text{and} \quad \bar{b} \pm 2s_b/\sqrt{n_b}$$

where s = the sample standard deviation
n = the number of observations in each sample.

Alternatively, for small samples where skewness is likely, the t estimate is

$$\bar{a} \pm t_{05} \frac{s_a}{\sqrt{n_a}}$$

where t_{05} is the t statistic for the 5% significance level.

$$t = \frac{x_i}{(x^2/n)^{1/2}}$$

where x_i is a random value drawn from a standardized normal distribution.

In comparing two sample means $\bar{a}$ and $\bar{b}$ using the t statistic

$$t = \frac{|\bar{a} - \bar{b}|}{\{(s_a{}^2/n_a) + (s_b{}^2/n_b)\}^{1/2}}$$

with $(n_a + n_b - 2)$ degrees of freedom.

A related problem concerns the homogeneity within a particular group of data. For Gaussian distributions one measure of this is Bartlett's (M) for the homogeneity of the variance.

$$M = \frac{1}{C} \log_{10} \left[B - \sum_{1}^{m} (n_i - 1) \log_{10} s_i^2 \right]$$

where $B = \log_{10} s^2 \sum_1^m (n-1)$

and $$C = 1 + \frac{1}{3(m-1)} \left[\sum_{1}^{m} \frac{1}{(n-1)} - \frac{1}{\sum_{1}^{m} n-1} \right]$$

m = the number of types
s^2 = variance.

M is tested by χ^2 for $(m-1)$ degrees of freedom. The null hypothesis that the type samples come from a group with homogeneous variance is accepted if M is less than the appropriate χ^2 value.

A complete analysis of the variance within and between types (or regions) is often appropriate. The use of this technique was advocated by Godske (1959) although, apart from a case study by Nosek (1967) for temperature characteristics of weather types at Brno, Czechoslovakia, in June 1950–9, its application in synoptic climatology has apparently been very limited. The method is discussed in nearly all introductory statistical texts and is sufficiently familiar not to be repeated here.

In concluding this section we may note that many of the classificatory procedures which have been briefly outlined have not yet been applied in climatology. Our concern here is twofold. First, there is a need for greater experimentation with many of the analytical statistical tools now available. Second, in trying out new methods it is well to be aware of the experience developed with some of them in other disciplines. There is often a risk, for instance, that certain procedures will be adopted simply because they are available in a computer program package rather than for their intrinsic characteristics in relation to the problem at hand.

5 Applications

In this chapter the main types of study in which synoptic climatological methods have been used are discussed. Obviously there is some overlap between the themes in so far as many investigations have more than one objective, but the topics reflect the general lines of past and present research.

The first two sections deal with the description and analysis of climatic regimes in both temporal and spatial contexts. Then follow considerations of the role of synoptic climatology in climatic change studies and long-range forecasting. Finally, illustrations are given of the contribution of synoptic climatological investigations to more specialized problems concerned with air–sea interaction and bioclimatology.

A Description and analysis of climatic regimes

The seasons alter: . . . the spring, the summer,
The childing autumn, angry winter, change
Their wonted liveries; and the mazed world,
By their increase, now knows not which is which.
(*W. Shakespeare*)

The climatic regime of a given place, or area, refers to the seasonal pattern of weather elements which are observed in a majority – perhaps two out of three years. It is well known that differences between years may be considerable, although conventional climatic averages conceal such variability. In this respect the use of a synoptic reference frame is of considerable value. In studying the temporal and spatial patterns displayed by weather phenomena, three topics are of special interest – singularities, weather spells, and natural seasons.

1 SINGULARITIES

The weather lore of most countries reflects the belief that certain types of weather are associated with particular calendar dates (Inwards, 1950;

Yarham, 1966; Zimmer, 1941). Some of these adages have been successfully related to meteorological principles, but others have proved to be quite false. Observations in Belgium, France and central Germany (Lancaster, 1886–7), for instance, show that 11 November ('St Martin's summer') tends to be cold, on average, rather than warm![1]

A *singularity* may be defined as the recurrence tendency of some weather characteristic about a specified date in the year. The term was used by Schmauss (1928) in the mathematical sense of a singular point, where $dy/dt=0$, on a time plot of some weather element. It was then applied to specific calendar events. Three approaches have been used to isolate and study singularities:

1 Examination of mean daily values of individual climatic parameters for periods of years in order to identify irregularities in the seasonal trends.
2 Analysis of catalogues of circulation types to find periods when particular types are unusually prevalent.
3 Synoptic and spatial analyses of selected singularities.

Each of these techniques will now be examined in more detail.

The idea of finding singularities from records of mean temperature for each day, or five-day intervals (pentads), dates back to the late eighteenth century (Pilgrim, 1788) or before. It has been suggested by Brooks (1954) that some writers of seventeenth and eighteenth century weather almanacks had begun to keep careful note of weather phenomena, but for business reasons kept their discoveries secret! By 1919, Talman could cite 144 references on the subject. European scientists (Brandes, 1820, 1826; Dove, 1857) were particularly interested in cold spells during winter and spring. Dove wrote a classic memoir on May frosts, which were thought to be most frequent in central Europe on 11–14 May. These dates are referred to as the *Ice Saints* after the Christian saints who are commemorated at that time. Dove concluded that the outbreaks of cold polar air which cause the frosts were too irregular in their occurrence to be connected reliably with the specified dates, whereas Köppen (1884) not only accepted their reality but postulated possible cosmic effects on the circulation (Bayer, 1959). About the same period Buchan (1867–9), using records for 1857–66 for five stations in Scotland, identified a number of cold and warm spells which bear his name. Unfortunately, later work (McIntosh, 1953) did not substantiate their reality as persistent features, although Flohn (1941) recognized them in the late nineteenth century observations on Ben Nevis.

As early as 1884, Bilwiller denied the existence of temperature singularities and later Marvin (1919) showed that most of the variance in

[1] *Some* of the apparent anomalies in ancient weather lore may relate to the change to the Gregorian calendar (1752) in England.

mean weekly temperatures is attributable to the annual and semi-annual harmonic components, and also that departures from the smooth annual curves are poorly correlated between different periods. Flohn and Hess (1949) note that the Ice Saints occurred with 77% frequency from 1881 to 1910, but only 58% from 1911 to 1947, for instance. However, Flohn (1947) has also emphasized the persistent nature of some central European singularities over several centuries. In spite of Marvin's work the statistical significance of singularities was rarely considered until relatively recently (Bartels, 1948; Baur, 1948). Bartels selected random dates to construct a fictitious calendar for each year, displacing a few days with respect to the true calendar, to demonstrate that departures, similar in manner and magnitude to supposedly genuine singularities, occur in the fictitious series. Baur showed that false singularities are generated in any series of random numbers with an introduced persistence tendency.

Statistical tests of singularities have been very salutary. Brier, Shapiro and MacDonald (1963) were able to show, from an analysis of independent series of daily precipitation data in the United States, that there is no tendency for rainfall anomalies near specific dates, contradicting the findings of Bowen (1956, 1957) and others (Bigg, 1957; Kline and Brier, 1958; Nagao, 1960). Bowen had postulated that dust from meteor showers might stimulate precipitation by providing additional freezing nuclei.

Computers have made sophisticated analyses a straightforward task. For example, Craddock (1956) was able to carry out an harmonic analysis of long-term five-day mean temperatures for forty-two stations in

Table 5.1 Temperature singularities in northern and central Europe (Craddock, 1956)

Cold pentads	*Warm pentads*
	31 January–4 February
12–16 March	
	1–5 April
	21–25 May
	26–30 May
	31 May–4 June
20–24 June	
	28 September–2 October
12–16 November	
17–21 November	

central and northern Europe. He found that the first two harmonics account for approximately 99·6% of the annual variance. The departures from the two-term harmonic can, however, be shown to contain a systematic component. The anomalies at any of the stations show serial correlations of 0·5 to 0·7 for consecutive five-day periods (lag one) and there are also spatial correlations. The most pronounced singularities are given in table 5.1. Of particular interest is the warmer than average period in late May–early June, followed by a cool spell in mid-June (temperatures differ by up to ± 1°C). Since this feature is best developed in central Europe it may represent a monsoonal effect induced by summer heating of the European land mass. It is interesting to note that Mädler (1843) found in a 100-year record at Berlin that there was a sharp temperature rise 31 May–3 June and that 16–20 June was usually a cool spell. However, he also reported a sudden decrease between 25 and 30 September, which is contrary to Craddock's results.

Many European pressure series show evidence of a 30-day wave or pulse related to the strength of anticyclones over central Europe and Scandinavia (Lamb, 1964, p. 155). This pulse appears to have greater amplitude in the stratosphere although the physical mechanisms involved are uncertain. In the North Atlantic area, however, cyclonic activity fluctuates in intensity over 15–20 day periods in winter and 20–25 days in summer. Apparently the necessary energy build-up takes longer in summer when the overall circulation intensity is weak. Obviously the relative degree of control of continental or oceanic influences in the British Isles during a particular season will determine the phase and amplitude of its circulation anomalies. In Antarctica Sir George Simpson detected pressure surges with a period of about forty-five days. These do not appear to have been further examined.

Before leaving this type of approach it is worth noting that most workers deal only with a single weather element.[1] In Germany numerous detailed investigations of a variety of parameters – temperature, precipitation frequency, pressure, wind frequency – have been made for each day of the year by Schmauss (1932, 1938, 1941). He was also the first person to prepare a calendar of singularities, although he erred in attributing 'significance', without statistical checks, to all irregularities in his graphs.

The identification of singularities in terms of circulation types is a relatively recent idea, although Talman (1919) noted that the Indian Summer in North America must be a type of weather, rather than a recurring temperature singularity, in view of its irregular occurrence. This

[1] An example of a rather more complex variant of a single element is the use of the minimum range in the frequency distribution of daily mean temperatures for the period by Nagao (1957).

Table 5.2 Major singularities in the British Isles (after Brooks, 1946)

Mean dates	*Type*	*Frequency % (1889–1940)*
5–17 January	Stormy	87
18–24 January	Anticyclonic	87
24 January–1 February	Stormy	85
24 February–9 March	Stormy	89
1–17 September	Anticyclonic	83
24 November–14 December	Stormy	98
25 December–1 January	Stormy	83

approach was adopted by Brooks (1946), who classed the pressure patterns over the British Isles for 1889–1940 into stormy, anticyclonic, and other. Table 5.2 lists those singularities with a frequency greater than 80%. It will be noticed that the two anticyclonic spells compare well with European singularities given in table 5.3. Ehrlich (1954) calculated the mean pressure field for selected five-day periods in winter when troughs or ridges occurred over the North Atlantic in at least twenty-seven of the forty years 1899–1938. The patterns for the periods of singularity noted by Brooks accord well with what would be expected. Fig. 5.1 illustrates the maps for 9 and 19 January, which according to Brooks give, respectively, stormy and anticyclonic conditions.

Some of the most extensive studies have been based on the European

Table 5.3 Singularities in central Europe, 1818–1947 (Flohn and Hess, 1949)

Period	*Circulation type*	*Characteristics*	*Frequency %*
15–26 January	Continental anticyclones	Dry, night frosts	78
22 May–2 June	Northern and central European highs	Dry	80
9–18 June	Northwesterly	Summer monsoon, thundery rains, cool	89
21–30 July	Westerly	Summer monsoon, thundery rains	89
1–10 August	Westerly	Summer monsoon, thundery rains	84
3–12 September	Central European highs	Dry	79
21 September–2 October	Central and south-eastern European highs	'Old Wives summer', dry	76
1–10 December	Westerly	Mild	81

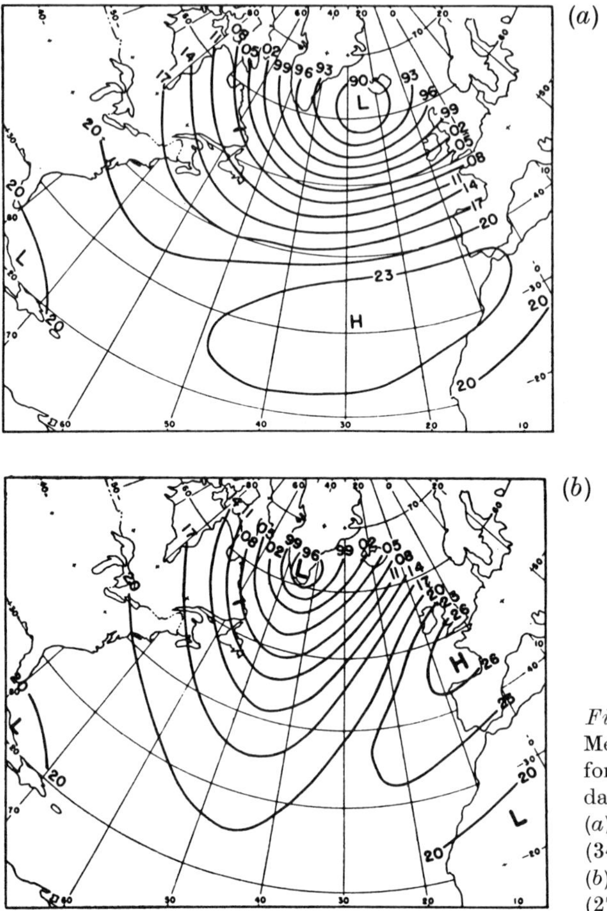

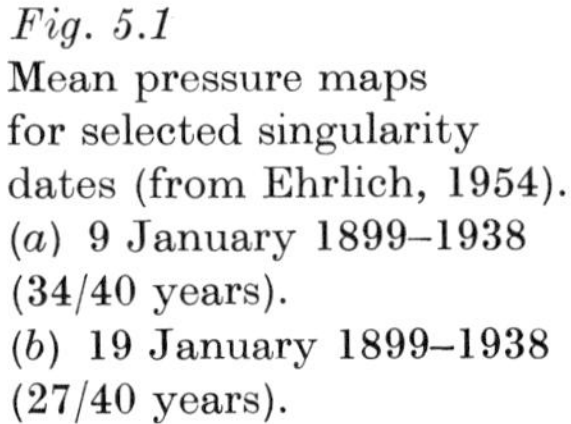

Fig. 5.1
Mean pressure maps for selected singularity dates (from Ehrlich, 1954).
(*a*) 9 January 1899–1938 (34/40 years).
(*b*) 19 January 1899–1938 (27/40 years).

Grosswetter catalogue (Baur, 1947, 1958). Flohn and Hess (1949) used the modified Hess–Brezowsky catalogue for 1881–1947. They identify periods of ten to twelve days during which a particular *Grosswetterlage* occurred on three or more consecutive days in at least two thirds of the period of record. The definitions allow a variation of ± 5 days about the middle date for each singularity. Table 5.3 summarizes those periods with a frequency of 75% or more over the sixty-seven years. Particularly striking are the summer events. Singularities in the winter half-year, not shown in the table as they have a frequency of only about 70%, are mainly anticyclonic in character. From September to March these occur with an interval of about thirty days, reflecting the pressure pulses referred to above.

Lamb (1950) has carried out a similar analysis for the British Isles using his own catalogue for 1889–1947. These have subsequently (Lamb,

1964, pp. 175–89) been extended to cover the years 1873–1961 with separate consideration of the periods 1873–97, 1898–1937 and 1938–61. Table 5.4 summarizes the primary features.

It will be noted that several of the features correspond closely to the dates shown for central Europe (table 5.3). Other correspondences may be found in Lamb's more detailed tables. However, precise agreement is not to be expected since the circulation controls in the two regions are often quite separate, although not necessarily independent of one another. Similarly, but on a smaller scale, there is normally a different response in the weather characteristics over the northern and western parts of the British Isles, compared with southeastern England.

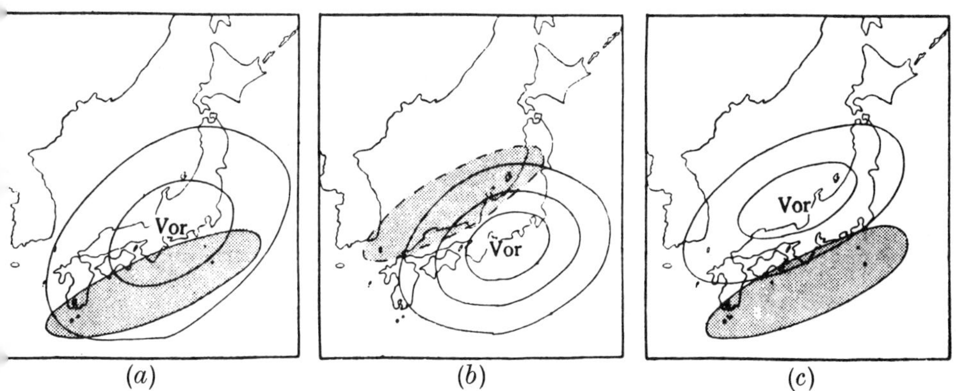

Fig. 5.2 Models for temperature singularities in winter and spring in Japan (from Nagao, 1960). (*a*) High temperature singularity. (*b*) High temperature singularity with maximum range of frequency distribution of daily mean pressure. (*c*) Low temperature singularity. Shaded area = singularity occurrence. Vor = anticyclonic vorticity.

Synoptic analyses of particular singularities have a surprisingly long history. For example, a detailed study for 1890–1908 of the 'Old Wives Summer' (*Altweibersommer*) in Europe[1] (comparable to the Indian Summer of North America) showed that high pressure over eastern Europe leads to southeasterly airflow, particularly around 28 September–2 October (Lehmann, 1911). Nevertheless, it was pointed out also that this weather regime may last anything from a few days to three or four weeks. Recently, Nagao (1960) in a study of singularities in Japan took into account the patterns of horizontal divergence, relative vorticity and the solenoidal field as well as the pressure field characteristics. A simple model for high and low temperature singularites in winter and spring is shown in fig. 5.2. When an anticyclone forms over Japan the supply of cold Asian air is cut off producing a high temperature

[1] Flohn (1947) provides a map of the spatial extent of this phenomenon according to Schmauss and Koncek.

Table 5.4 Singularities in the British Isles (Lamb, 1964)

Period	*Circulation type*	*Characteristics*	*Type frequency % and significance level*	*Period*
20–23 January	AC, S, and E together	Generally dry and sunny in central and southern England.	50	D
		Year's lowest frequency of C type (10–12%) 24–26 January.	5% level	D
12–23 March	AC, N and E together	Notable rainfall minimum in central and southern England.	70	D
		12–14 March peak of AC.	35 (1% level)	D
12–18 May	N type	Annual maximum about these dates; 14–20 May is sunniest week of the year in Ireland.	30	A, B, C
21 May–10 June	AC type	Annual maximum frequency, 40% or more on some days during most of this period; driest weeks of year in Scotland, Ireland; more year-to-year variations in southern half of England.	5% level	A, B, C
18–22 June	W, NW and AC together	Generally dry and sunny in southern England; cloudy and wet in Scotland and Ireland.	70	D
		W type frequency 52% on 20 June.	1% level	D
31 July–4 August	C type	Sharp peak (replaced by twin maxima around 20 July and mid-August).	35% + (5% level)	B, A & C
17 August–2 September	W and NW together	Wet in most areas.	70	D
	C type	Peaks 19 and 28 August.	30 (5% level)	D
6–19 September	AC, N and NW together	Dry, especially east and central England.	80	A, B, C,
		C type frequency, < 20% between 6–12 September.	5% level	A, B, C
5–7 October	AC type	Slight check to seasonal cooling.	40 (5% level)	D

Table 5.4 Singularities in the British Isles (Lamb, 1964) – Cont.

Period	*Circulation type*	*Characteristics*	*Type frequency % and significance level*	*Period*
24–31 October	C, E and N types	Great decline to year's minimum frequency of AC type ($\leqslant$10%) about 28–31 October.	1% level 5% level	A, B, C
		Stormy, wet weather.		A, B, C
17–20 November	AC type	Dry, foggy period in central and southern England.	30 (1% level)	A, B, C
3–11 December	W and NW together	Wet and stormy in most areas with 3–9 December generally wettest week of year on average.	70	A, B, C
17–21 December	AC type	Generally dry, foggy weather.	25	A, B, C

Period A = 1873–97, B = 1898–1937, C = 1938–61, D = 1890–1950 ± about 10 years.

singularity with a small range of daily mean pressures (fig. 5.2(a)), although with the development of an intense high there is a large range of daily mean pressures during the spell (fig. 5.2(*b*)). If the centre of anticyclonic vorticity is over the Japan Sea then cold northerly flow creates a low temperature singularity (fig. 5.2(*c*)). It is doubtful, however, whether the amount of effort in determining is justified by the small gain in information over that provided by a straightforward examination of the pressure fields.

A more general trend has been the consideration of regional phenomena in a global context. This reflects the view of Wahl (1953) that a differentiation should be made between *primary* singularities affecting the general circulation, and *secondary* ones which are important regional manifestations of the former. Subsequently, formal definitions have been proposed by Bayer (1959, p. 620). He regards a primary singularity as an increased tendency to intensive meridional exchange of air masses in a specified period and specific regions of the northern hemisphere. This tendency to meridional exchange manifests itself through an increased frequency of development of long waves in the circumpolar 500 mb flow. A secondary singularity represents the increased tendency of appearance of particular *Wetterlage* in a specified period and in a specific region of the northern hemisphere due to the occurrence of a primary singularity. The secondary singularity manifests itself in the region as a tendency for an anticyclonic *Wetterlage* to develop, or for

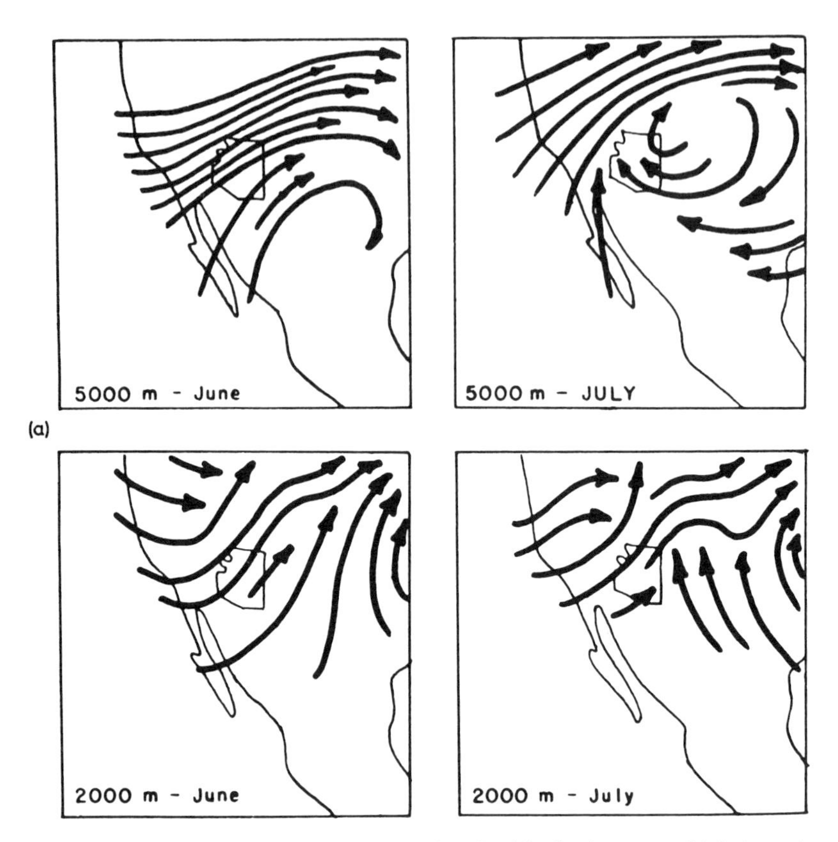

Fig. 5.3 Circulation changes associated with the 'summer high jump' over the southwest United States (from Bryson and Lowry, 1955).
(*a*) Mean flow at 2 km and 5 km.
(*b*) Five-year mean contours at 500 mb on selcted dates.

the advection of particular air masses, and causes the occurrence of singular points in the annual curve of meteorological elements at a given location.

The January thaw in New England (Wahl, 1952) and the 'summer high jump' in the southwest United States (Bryson and Lowry, 1955) are examples of regional features linked to larger scale events. In late June–early July the anticyclone over the eastern North Pacific shifts suddenly northward and air from the Gulf of Mexico is able to move into Arizona on the southwestern margin of the Atlantic (Bermuda) subtropical anticyclone. This southwesterly airflow gives a sharp increase in rainfall about 1 July in Arizona–New Mexico. Fig. 5.3 illustrates the change in the pressure field and flow pattern. Wahl shows that in New England the tendency for mild weather about 20–23 January, followed by a colder than average period, is most pronounced in years of low (westerly) index. It would seem that when the zonal flow is strong

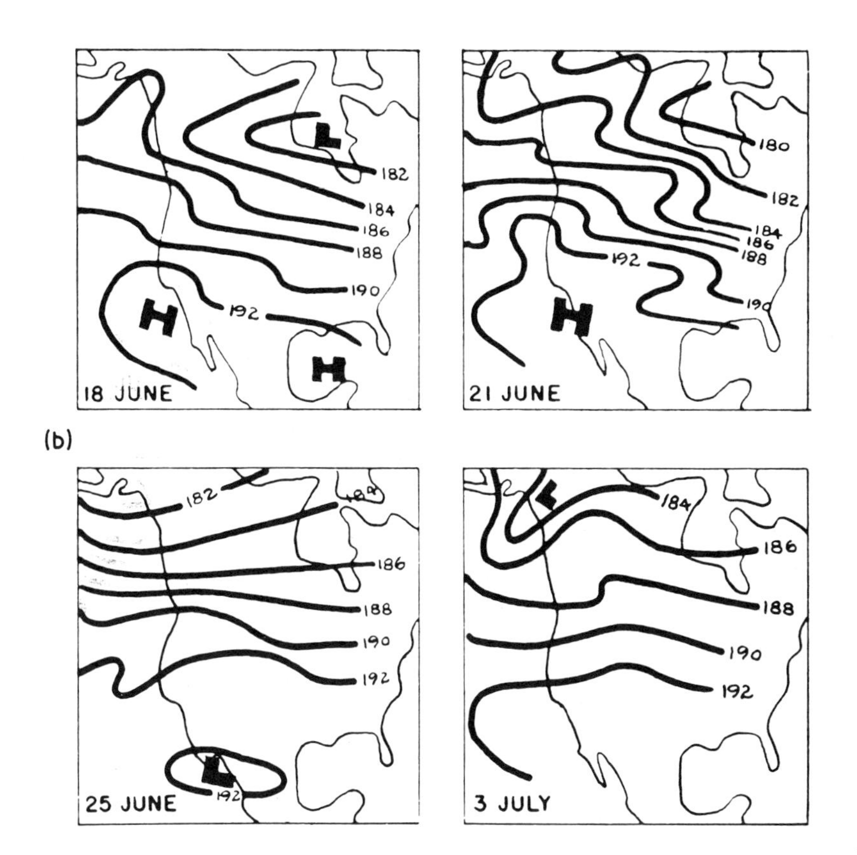

in middle latitudes regional anomalies tend to be dampened out. This approach reflects Lettau's suggestion (1947) of treating 'specific singularities' by considering a group of years selected according to some particular criterion. The significance of the January thaw has also been studied by Duquet (1963). He re-examined Marvin's weekly temperature data for 1872–1916 and found a positive correlation between the latitude of the Azores high and the amplitude of the warm spell at Easton, Maryland (fig. 5.4), in accord with Wahl's findings for low index conditions. Duquet suggests that the warmth in mid-January results from Gulf Coast cyclones moving along the Appalachians and he postulates that a primary singularity occurs as the planetary circulation shifts from an early- to a late-winter mode. Frederick (1966) has reconsidered the 'January Warm Spell' from a geographical standpoint. Fig. 5.5 shows his map of the mean southeastward progression of the warming during 1927–56. These dates, it should be noted, are not in fact in accord with those cited by Wahl. The progression is too slow to be related to synoptic scale troughs and, moreover, the subsequent cooler weather is not traceable spatially. Frederick did confirm that the warm spell comprises almost entirely above average temperature conditions

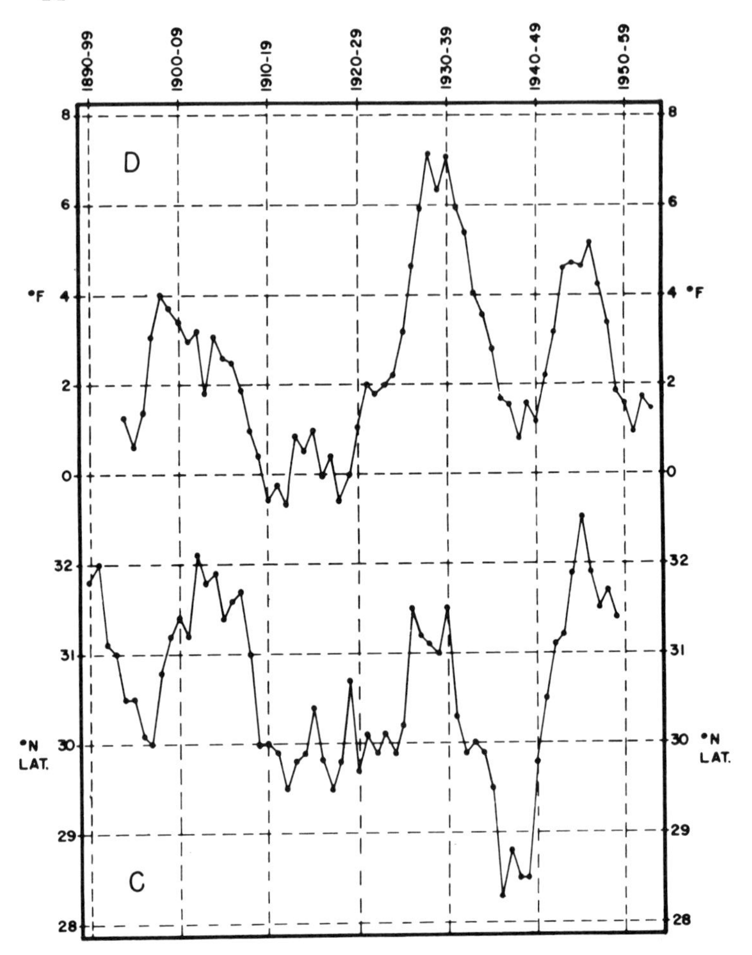

Fig. 5.4 Relationship between the average latitude of the Azores high and January temperature at Easton, Maryland (from Duquet, 1963).

and it is not, therefore, a statistical phenomenon, although clearly its occurrence and explanation still pose many questions.

In Asia there is additional evidence for regional singularities linked with seasonal readjustments of the general circulation. To refer only to two well-known features, the onset of the Bai-u rains in Japan seems to be related to circulation patterns, determined at least in part by the heat source over the southeast Tibetan Plateau (Asakura, 1968), as well as to a more local northward shift of the polar front (Maejima, 1967), while the Shurin rainy season in late August is associated with large-scale blocking (see also fig. 5.7).

Other global links undoubtedly remain to be found. Karapiperis

(1953) identified cold spells over the eastern Mediterranean especially about 24 September, 16 October and 20–25 November, and showed that these were related to the presence of cold anticyclones over northwest Europe. These last dates coincide with a singularity noted by Namias (1968) at Washington, D.C., where the mean daily maximum temperature falls by about 3°C from 21 to 25 November, and at Denver, Colorado, where the mean maximum *rises* by 5°C (8°F) for the same dates. The circulation patterns indicate that the east coast trough at 700 mb amplifies greatly and an upper ridge enters the western United States during this period. Likewise, Wahl (1954) shows that snowfall likelihood increases greatly over the whole Midwestern United States about 15 October in response to a switch from southeasterly to northwesterly flow. It is quite conceivable that the synchronous events in the Mediterranean are part of a global readjustment of the hemispheric circulation, although the details of these changes and possible reasons for them remain to be determined. The occurrence of a minimum in Hawaiian precipitation in February can be traced to an expansion of the circumpolar westerly vortex and the consequent readjustment of the wavelength between the upper trough over the Pacific downstream of the major east Asian one (Namias and Mordy, 1952). Subtropical

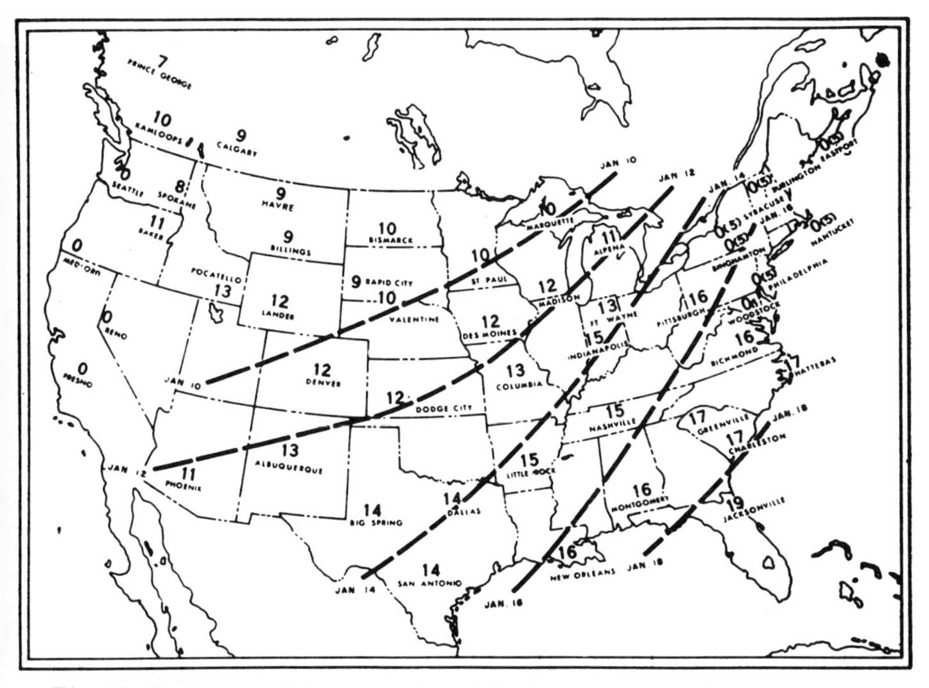

Fig. 5.5 Isochrones of the progression of the January warm spell 1927–56 (from Frederick, 1966).

cyclones (Kona storms), which bring precipitation in January, are excluded from the area and the trade wind flow is also displaced. The pattern is seen as part of a (primary) index cycle which frequently develops in late winter. Precipitation increases again in March when the trades are restored.

In the context of global studies it is important to recognize that singularities will be more readily detectable in some areas than others. This is determined in part, at least, by regional differences in the intensity of the annual and more particularly the semi-annual pressure oscillation (see p. 398). Interestingly enough, these oscillations are weak in central Europe and eastern North America (among other areas) where many singularities have indeed been recognized.

Under the general heading of singularity we have, in fact, treated three types of event:

1 Broad maxima and minima of particular circulation patterns during the year.
2 Shorter period peaks and troughs of certain types of weather regime about a particular date.
3 The abrupt onset or decline of given types about a particular date.

The broad maxima and minima can to some extent be used to characterize seasons, and the abrupt changes their beginning or termination. Bryson and Lahey (1958) have demonstrated that many seasonal transitions are abrupt rather than gradual and it is in this connection, therefore, that we shall consider the question of singularities further.

2 SPELLS AND NATURAL SEASONS

In many climates the character of an individual season is quite frequently marked by the persistence of some recognizable type of weather, known as a *spell*. Early studies of weather spells were limited to individual elements such as rainfall or temperature conditions. For example, *British Rainfall* lists spells of absolute drought (fifteen or more successive days each with at least 0·25 mm rainfall). Statistical studies of runs of dry and wet days have been made by Longley (1953) and Lawrence (1954, 1957). For dry spells in summer in southern England, Lawrence finds positive persistence effects for runs of eight to ten days, and for runs of about eighteen to twenty-five days. Anti-persistence is marked for runs of more than thirty days (see also Chapter 4.B, p. 241).

Subsequent workers have examined the occurrence of pressure pattern or weather types. As noted earlier, Brooks (1946) studied the occurrence of stormy and anticyclonic conditions lasting at least three days in the British Isles. More detailed work has been carried out by Lamb (1950). He examined the occurrence of spells of twenty-five or more days

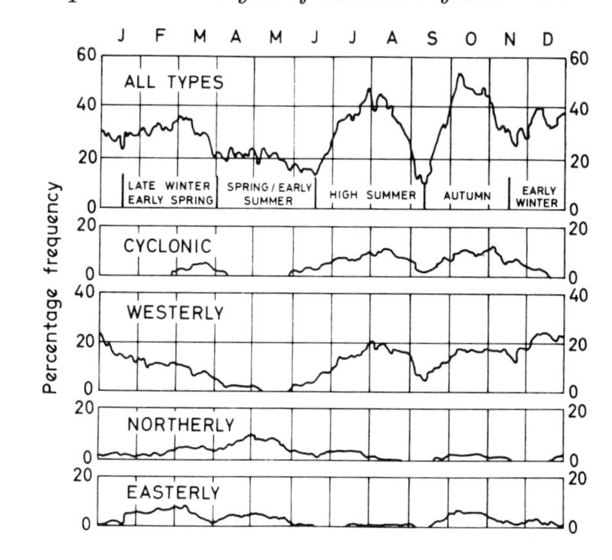

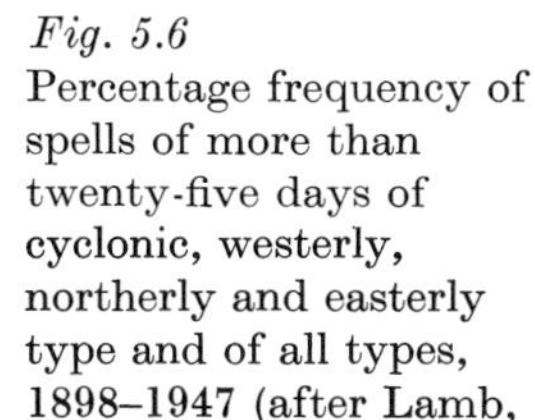
Fig. 5.6
Percentage frequency of spells of more than twenty-five days of cyclonic, westerly, northerly and easterly type and of all types, 1898–1947 (after Lamb, 1950; from Barry and Chorley, 1971).

of a particular type, or groups of related types (such as Westerly, Cyclonic Westerly, and Cyclonic – see Chapter 3.C, pp. 138–9) disregarding brief interruptions up to three days in length. 158 such spells occurred during 1898–1947; 79 involved Westerly type (without Anticyclonic hybrids) and 56 involved Anticyclonic and a hybrid type. Seasonal analysis of all spells indicated a minimum in April–June and maxima in October and in mid-summer (fig. 5.6). This is reflected in Craddock's analysis (1963) of pentad temperatures at Kew, London. There is a deficiency of runs (persistent anomaly regimes) in spring whereas they are frequent in late summer, especially 14 August–7 September, and during the period 25 June–19 July. Lamb also found that the duration of his spells showed a peak at twenty-nine days, which agrees with Lawrence's result for dry days, and another at forty-five days.[1]

Lamb used the occurrence of spells, in conjunction with the singularity table (table 5.4) to define *natural seasons*; that is to say, calendar intervals characterized by particular kinds of weather regime. A similar seasonal calendar has been developed in Germany (Baur, 1958) using the *Grosswetter* catalogue and the two are compared in table 5.5. British holidaymakers will undoubtedly be surprised to learn that 'High Summer' lasts longer in Britain than in central Europe!

Natural calendars have been developed for other parts of the world.

[1] Although not as specific, we should note Veitch's (1965) finding of a quasi-periodic source of pressure variability in the range ten to forty days, which accounted for about 18% of the total variation over Australia for the period 22 January–21 April 1950–9. Madden and Julian (1971) also report a forty to fifty day oscillation in the zonal wind over the tropical Pacific Ocean.

Table 5.5 Natural seasons in the British Isles and central Europe (Baur, 1958; Lamb, 1950)

Season	*Calendar period in central Europe*	*Calendar period in the British Isles*
High winter	1 Jan.–14 Feb.	20 Jan.–31 March
Fore-spring	15 Feb.–31 March	
Spring	1 April–16 May	1 April–17 June
Fore or early summer	17 May–30 June	
High summer	1 July–15 Aug.	18 June–9 Sept.
Late summer or early autumn	16 Aug.–30 Sept.	
Autumn	1 Oct.–15 Nov.	10 Sept.–19 Nov.
Fore-winter	16 Nov.–31 Dec.	20 Nov.–19 Jan.

In Japan we may note the work of Sakata (1950, 1953), Takahashi (1955), Maejima (1967) and Yoshino (1968). Maejima determines the duration of six seasons – winter monsoon, spring, the Bai-u, mid-summer, the Shurin, and late autumn – from curves of five-day means and daily means of sunshine duration, cloudiness, and precipitation. Temperature and vapour pressure are secondary criteria for the winter monsoon season. The scheme of Yoshino, based on five-day periods of pressure patterns, is more elaborate (table 5.6). Fig. 5.7 illustrates the

Table 5.6 Natural seasons in Japan (Yoshino, 1968)

Winter	19 December
Early spring	7 February
Spring	17 March
Late spring	6 May
Early summer	18 May
Bai-u rains	7 June
Summer	20 July
Late summer	6 August
Early autumn	21 August
Shurin rains	10 September
Autumn	10 October
Early winter	26 November

Dates refer to beginning of season.

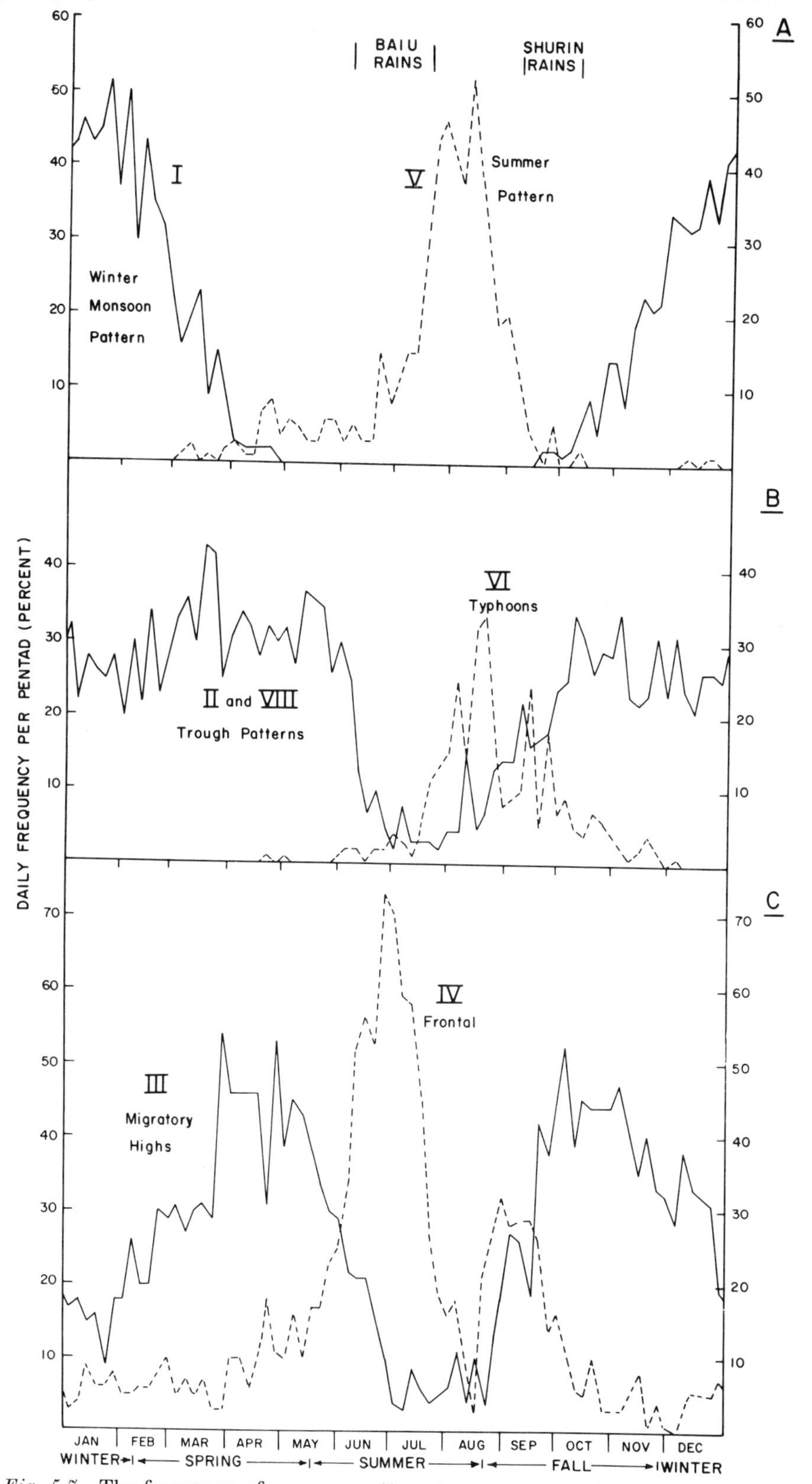

Fig. 5.7 The frequency of pressure pattern types over east Asia, 1946–55, the two rainy seasons, and a seasonal calendar (based on Yoshino, 1968).

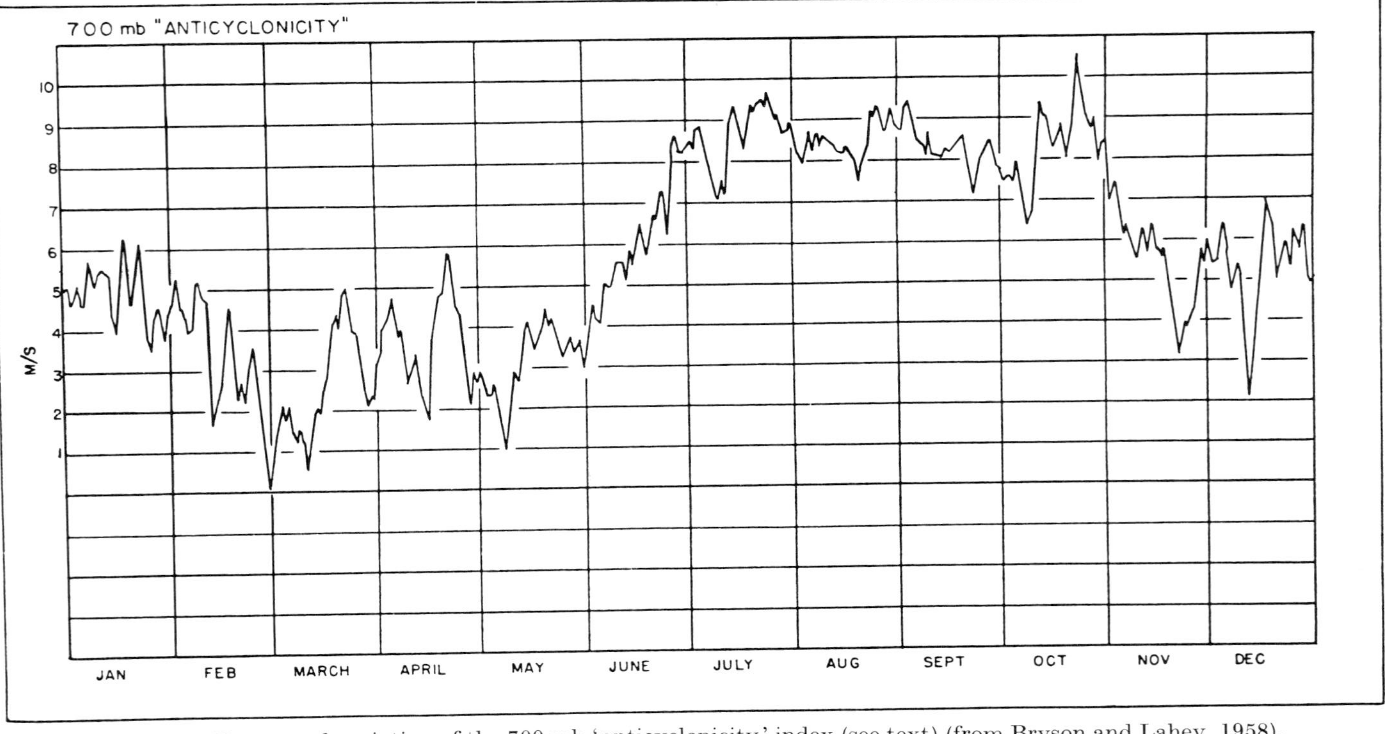

Fig. 5.8 The annual variation of the 700 mb 'anticyclonicity' index (see text) (from Bryson and Lahey, 1958).

four major seasons in relation to the daily frequency for 1946–65 of Yoshino's seven main types of pressure pattern over east Asia (see also fig. 3.15).

The idea that seasonal changes of the general circulation, rather than being gradual, are quite abrupt has been emphasized by a number of workers (Liu and Wu, 1956; Yeh *et al.*, 1959; and Bryson and Lahey, 1958). The wave structure of the tropospheric westerlies undoubtedly contributes to such switches since the wave number must be an integer. De la Mothe (1968) shows, for instance, that the 500 mb European trough is situated farther east in summer than in winter and a subsidiary upstream trough often develops in summer. It remains for the future to investigate heat budgets for circulation types and singularities, although a start has been made on such synoptic studies (see section E of this chapter (pp. 422–5).

Bryson and Lahey provide a synthesis of various circulation indices in an attempt to define primary singularities and hence the seasonal structure of the year. They extended Winston's analysis (1954) of the annual variations in the 700 mb zonal index with several derived indices. For example, fig. 5.8 illustrates the variation of 700 mb 'anticyclonicity' – difference between 700 mb zonal indices for 0° westward to 180° longitude for the zones 20–35°N and 35–55°N. It shows major changes in late June, following a transition period, and at the beginning of November. Another descriptor is the annual variation of persistence correlation at 500 mb. Fig. 5.9 shows the one-day lag correlation for the western hemisphere between 10° and 70°N for 1945–53 obtained from orthogonal polynomial specification. The upper curve shows that persistence drops sharply in late March. The lower curve, which excludes the zonal gradient, demonstrates that in spring most of the persistence is due not to zonality, but to higher order terms.[1] The same is true in the autumn. The index contribution to persistence is very high in July–August, and again, though to a lesser degree, in December–February. In terms of climatic anomaly patterns of temperature, precipitation and 700 mb heights over the United States, however, the major breaks in month-to-month persistence occur in April–May and October–November (Namias, 1952). The latter, at least, fits the autumn/winter changeover. Fig. 5.10 summarizes the numerous parameters studied by Bryson and Lahey and shows the seasons and subseasons which they identified. It is interesting to note that the decrease in the variability of the zonal indices that they show in early June is corroborated in a recent study by Zaharova (1969) with respect to surface indices and also 500 mb heights.

An attempt to provide a more fundamental approach is evident in the

[1] The blocking maximum occurs at this time of year, for instance (Rex, 1950).

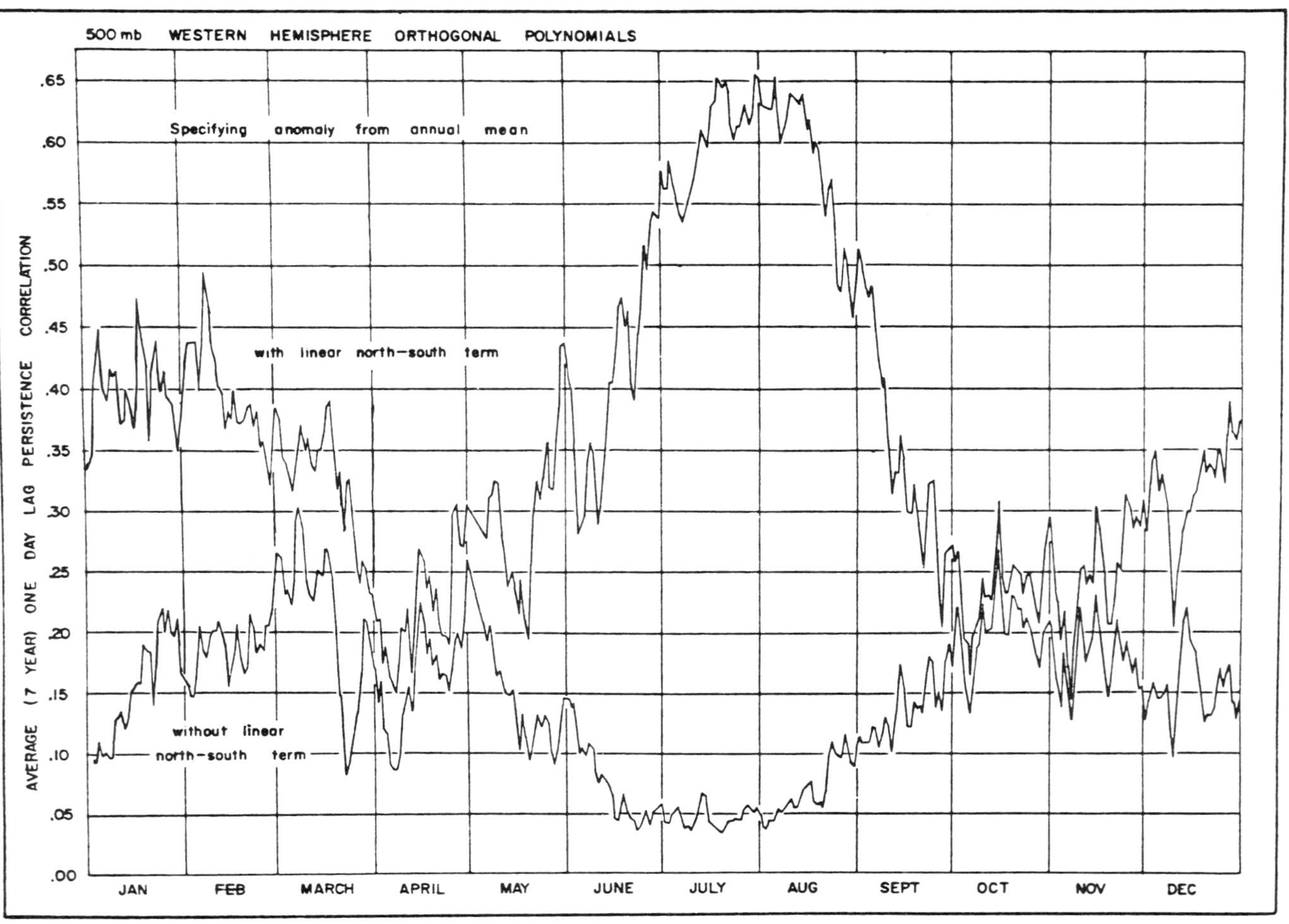
500 mb WESTERN HEMISPHERE ORTHOGONAL POLYNOMIALS
Specifying anomaly from annual mean
with linear north–south term
without linear
north–south term
AVERAGE (7 YEAR) ONE DAY LAG PERSISTENCE CORRELATION
.65
.60
.55
.50
.45
.40
.35
.30
.25
.20
.15
.10
.05
.00
JAN
FEB
MARCH
APRIL
MAY
JUNE
JULY
AUG
SEPT
OCT
NOV
DEC

work of Bradka (1966). He defines a natural season as an interval of time in which the basic features of a heat budget are reflected in the type of circulation for this time interval in a characteristic manner. The application of the heat budget concept is, however, rather generalized. Bradka notes that there are two factors affecting the northern hemisphere circulation. First, the latitudinal heat budget gradient leads to basically zonal flow in the middle and upper troposphere. Second, meridional components of circulation develop during the periods of extreme solar insolation under the influence of warm and cold air centres which are related to the distribution of land and sea. In addition, circulation patterns occur which are not directly dependent on solar radiation or surface characteristics. Thus, the accumulation of cold air in high latitudes intensifies the meridional temperature gradient and encourages blocking situations.

Another more specific use of energy concepts to delimit seasonal patterns at individual locations is suggested by James (1970). He shows first that a plot of mean daily maximum temperature for each month against mean pressure reflects the air mass climate as determined by the quasibarotropic relation (see Chapter 3.D, p. 185). Fig. 5.11 illustrates such diagrams for Hobart (Tasmania), Brisbane and Darwin (Queensland). Air traversing the seasonal loop performs work or has work done on it, in *approximate* proportion (because p is used rather than the height of an isobaric surface) to the area enclosed; neglecting net latitudinal and vertical transfers, the analogy of a simple heat engine (*in situ*) is applicable. Fig. 5.11(*a*) shows an isothermal change in summer (the heat source) and in winter (the heat sink). During the spring build-up the surface air does work against the 'environment' and the reverse is true in autumn. Since the latter phase is the greater in magnitude it is concluded that, over the year, work is done to sustain the surface airflow against friction by the 'environment' – implying the higher level circulation. In other words, the flow in middle latitudes is thermally indirect. At Brisbane (fig. 5.11(*b*)), which is primarily under trade wind influence, the autumn part of the loop is only a little larger than the spring part and the total area is very small. The higher level flow is expending little energy to maintain the low level circulation. Note that the summer season – January, February – is short as defined here. For Darwin (fig. 5.11 (*c*)) virtual temperature is plotted because of the moist atmosphere. Here the seasonal vector operates in the reverse sense (spring is warmer than autumn) indicating that the surface air imparts net energy to the higher level flow – a thermally direct circulation as assumed by general circulation models. The seasonal energy total is not large due to the relatively small range in insolation at 12°S.

In spite of the simplifications, therefore, this thermodynamic approach indicates a definite relationship between temperature, pressure

NATURAL CALENDAR

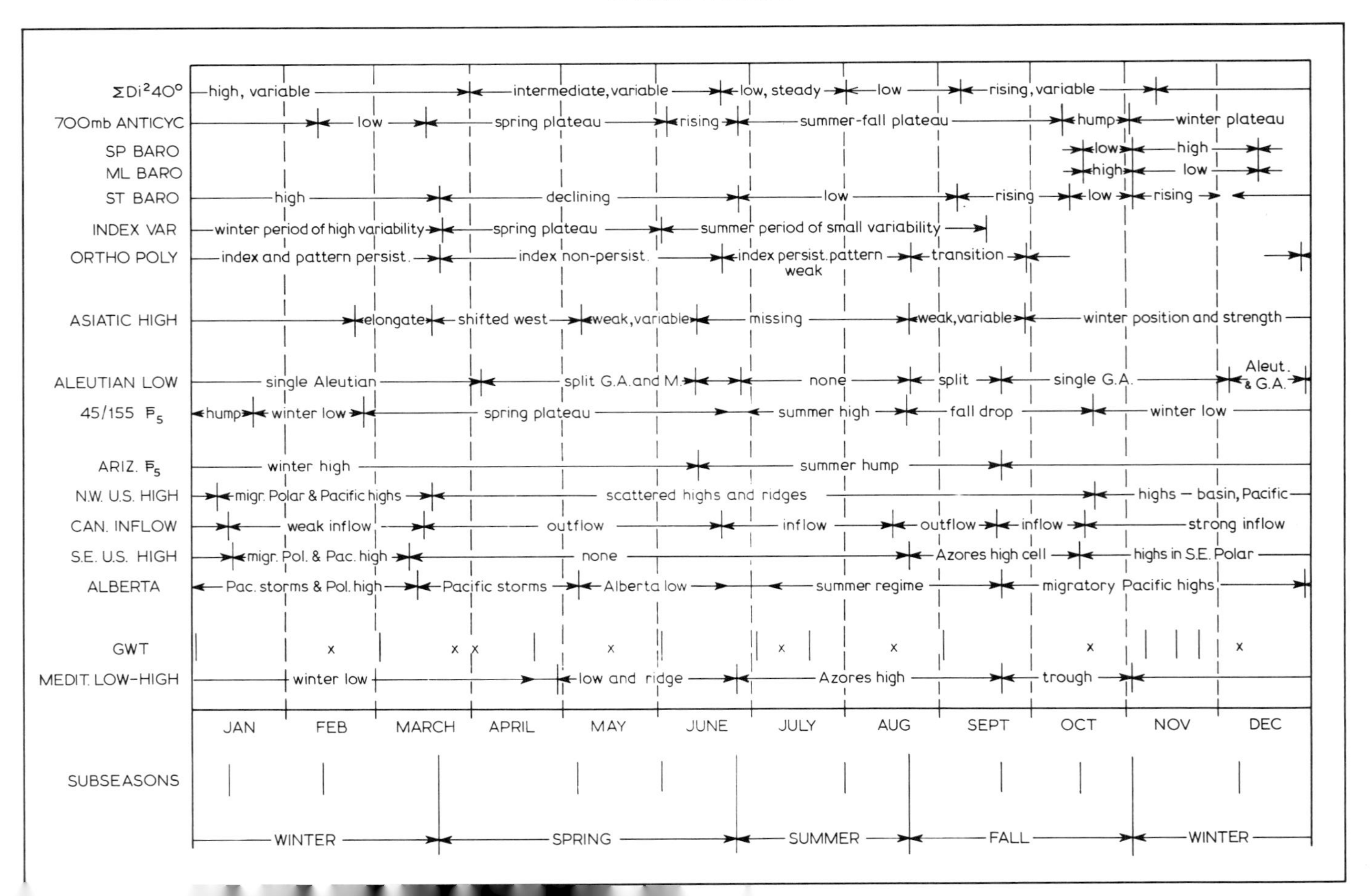

Fig. 5.10 A summary of criteria used to establish a natural calendar (from Bryson and Lahey, 1958).

Abbreviations:

$\sum D_i^2$ 40°	Wadsworth 'storminess' index; a measure of meridional circulation at MSL. Pressure differences are taken for each 5° around 40°N then their squares are summed.
Anticyc	Anticyclonicity; the difference between the mid-latitude westerlies and the subtropical easterlies.
Baro	Baroclinicity; the difference between the MSL and 700 mb zonal indices. SP = subpolar, ML = mid-latitude, ST = sub-tropical.
Index Var	Standard deviation of circulation indices.
Ortho Poly	Orthogonal polynomials for the western hemisphere at 500 mb.
G.A. and M.	Gulf of Alaska and Manchuria occurrences of the Aleutian low.
$\bar{P}_5$	Normal five day mean pressure. 45/155 = 45°N, 155°W.
N.W. U.S. High	Anticyclones, location and origin, in the northwestern U.S.
Can. Inflow	Sea level geostrophic net flow from Canada into the U.S. between Alberta and Ontario.
Alberta	Synoptic pattern dominating the southern Alberta area.
GWT	*Grosswetter* type $\mid = 3\sigma$ changes, $\times = 2\sigma$ changes.
Medit. Low–High	Cyclonic or anticyclonic regime in the Mediterranean area.
migr.	Migratory. Pac. = Pacific; Pol. = Polar.
basin highs	Great Basin high, over Utah, Nevada, Idaho.

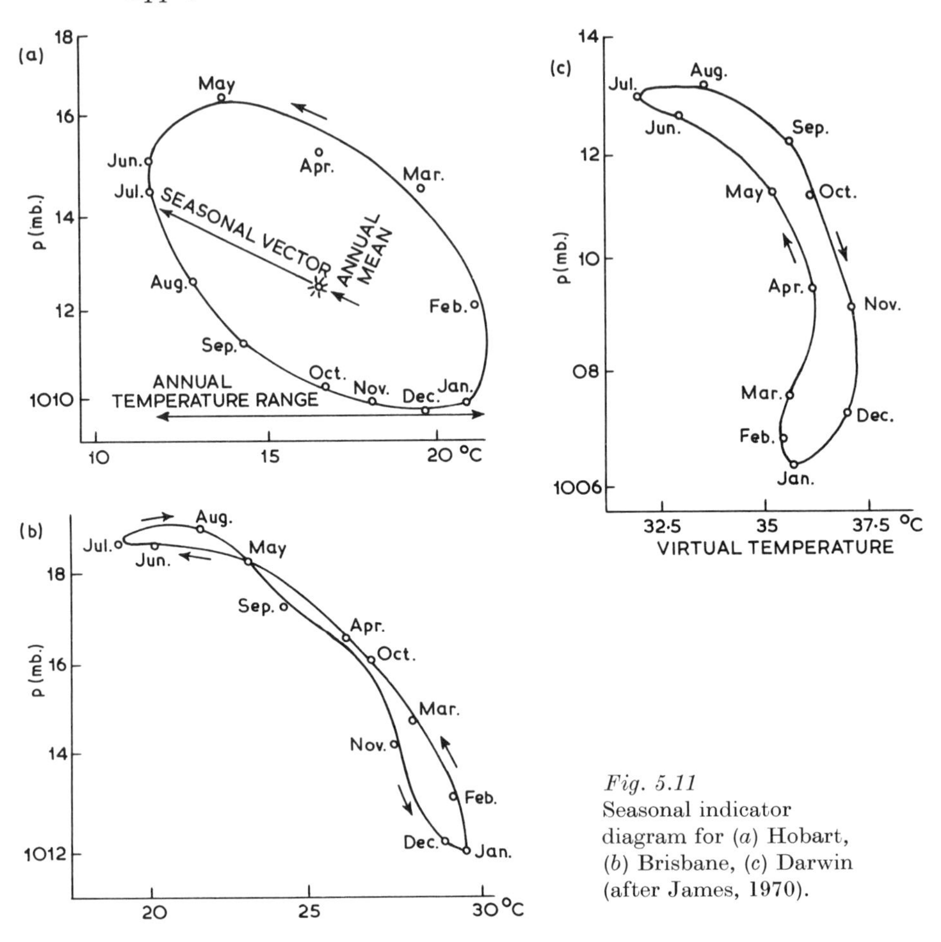

Fig. 5.11
Seasonal indicator diagram for (*a*) Hobart, (*b*) Brisbane, (*c*) Darwin (after James, 1970).

and humidity according to season and latitude. This relationship can be developed further in terms of the variation of insolation with latitude and season. James also points out that this approach can be extended with regard to the variance of pressure and temperature which reflect fluctuations in air mass properties as well as synoptic variability. The method at present provides only a crude seasonal classification, but there is no reason why the method should not be applied to shorter time periods and comparisons made with the other dynamic approaches that have been discussed.

The question as to what determines the seasonal transitions remains problematical. In order to provide some perspective on circulation dynamics it is helpful to consider what Lamb (1964) has referred to as *master seasonal trends*. These are the broad patterns of development of the circulation in the northern hemisphere during the year. Following Lamb's and Bradka's discussions we can recognize seven major factors:

1 The mid-winter maximum of the zonal index, related to the maximum gradient of the heat balance.
2 The southernmost position of the main belt of depressions about February–March, related in part to the limit of snow/ice cover. Meridional patterns now prevail with greater north–south interchange of air masses.
3 The development of the main centre of the polar anticyclone over the Canadian Arctic Archipelago in March–April, related to a persistence in the region of negative net radiation values.
4 Minimum intensity of the general circulation in May over the Atlantic sector, related to the minimum gradient of the heat balance from low to high latitudes. Over the Pacific the circulation remains predominantly zonal.

5a Northward shift of the subpolar depression tracks over western North America in July and August.

5b Displacement of the depression tracks southward over the Atlantic–European sector together with a shift in the cold pole towards this sector.

The contrasts in surface heating between the northeast Asia–Alaska sector, where there is mainly land surface, and the ice and cold seas of the sector from northeast Canada to Iceland are probably responsible for these trends.

6 Accentuation of the upper trough over northeast North America in September related to early autumnal cooling probably effected by the cold sea surface. The circulation intensifies in the Atlantic frontal zone with deepening of the Icelandic low. Depressions now move northeastward towards the Barents Sea.
7 Expansion and intensification of the circumpolar westerly vortex, and the associated southward displacement of depression tracks, particularly from October onward.

Seasonal changes in circulation intensity are less pronounced in the southern hemisphere. The circulation intensifies in the autumn months but the maximum development of surface winds is delayed until late winter.

In this global context it is apparent that the results of synoptic and dynamic climatology begin to converge and provide fuller understanding of the determinants of the seasonal structure of global weather and climate.

To fix the beginning and end of seasons to a specific date for the whole hemisphere is of course out of the question, since the circulation regimes

Table 5.7 Natural calendars according to various studies (partly after Bradka, 1966)

	Average beginning dates of the seasons		
	Autumn	*Winter I*	*Winter II*
Baur (1958)	16 Aug.	16 Nov.	1 Jan.
Bradka (1966)	30 Aug.	1 Nov.	19 Jan.
Bryson and Lahey (1958)	21 Aug.	1 Nov.	
Chu (1962)*	8 Aug.	7 Nov.	
Dzerdzeevski (1957)	26 Aug.	7 Oct.	6 Dec.
Multanovski (1920, in Bradka, 1966)	13 Aug.	5 Oct.	27 Dec.
Lamb (1950)	10 Sept.	20 Nov.	20 Jan.
Sakata (1950)	29 Aug.	27 Nov.	27 Dec.
Yoshino (1968)	21 Aug.	26 Nov.	19 Dec.
Craddock (1957)	15 Sept.		4 Dec.
Zaharova (1969)	24 Aug.	16 Nov.	25 Dec.
	Early spring	*Spring*	*Summer*
Baur (1958)	15 Feb.	1 Apr.	17 May
Bradka (1966)	9 March	21 Apr.	7 June
Bryson and Lahey (1958)		21 March	25 June
Chu (1962)*	4 Feb.		6 May
Dzerdzeevski (1957)	13 March	19 Apr.	22 May
Multanovski (1920, in Bradka, 1966)	13 March		18 May
Lamb (1950)		1 Apr.	18 June
Sakata (1950)	10 Feb.	22 March	10 June
Yoshino (1968)	7 Feb.	17 March	18 May
Craddock (1957)	13 March		16 June
Zaharova (1969)		10 March	7 May

* These refer to some of the twenty-four 'solar terms' of fifteen-day duration recognized in China *c.* 247 B.C.

Note: Some of the definitions used by different workers are not identical. Baur and Yoshino apparently include 'early summer' in summer and 'early autumn' in autumn.

in the Pacific and Asian sectors are to a large degree independent of those in the Atlantic and North American sectors (Allen *et al.*, 1940). These differences are reflected in table 5.7 showing the start of the main seasons according to various workers in different parts of the world. It is recognized that there is a wide range of dates on which the seasonal changes may be identified. If we consider Bradka's dates, for instance, there is a large spread about the mean, as illustrated in fig. 5.12. In particular, the second half of winter may start as early as Christmas or as late as the beginning of February.

It is argued by some meteorologists that these differences between individual years, and also between different periods, vitiate the concept of a 'natural calendar'. Yet in the case of singularities we have seen that important differences do arise when the zonal index is high or low. It seems that the occurrence of long spells of a particular circulation type may suppress the development of particular singularities. Lamb (1964, p. 146) aptly likens the singularity calendar to a steeplechase: 'If a particular long spell succeeds in getting over a hurdle – i.e. in surviving dates which the singularity calendar indicates as unfavourable for it – the spell may be firmly enough established to continue a good while longer.' In the long term there are trends in the spacing of the upper troughs and in the strength of the zonal flow (Lamb and Johnson, 1959, 1961) which makes the occurrence of calendar shifts not at all unlikely. Nevertheless, it is worth noting that the evidence presented by Bryson and Lahey incorporates parameters which refer to several different periods.

The jumps which commonly accentuate the seasonal transitions have been attributed to a variety of factors. Perhaps inevitably the role of sunspots as a trigger in determining singularities has been proposed (Nagao, 1960), although many other more likely terrestrial factors may supply the necessary stimulus. Swift changes in albedo with the formation or disappearance of a snow cover may be critical (Lamb, 1955; Houghton, 1958). The delay imposed on the seasonal retreat poleward of the subtropical jet stream over northern India in early summer, as a result of the influence of the Himalayan–Tibetan massif, is probably important over much of southern Asia. Changes in the external environment may not, however, be required. It is conceivable that the atmospheric circulation can possess more than one state for a given set of boundary conditions. Lorenz (1967, pp. 124–6) refers to the possibility of this situation which he terms the 'almost intransitive' nature of the circulation. Rotating annulus experiments (Fultz *et al.*, 1959, 1964), for instance, show that there are heating and rotation rates for which either of two Rossby regimes may occur. Intransivity is also present for the transition from the Hadley to the Rossby regime in such experiments. If intransivity were a property of the atmospheric circulation there

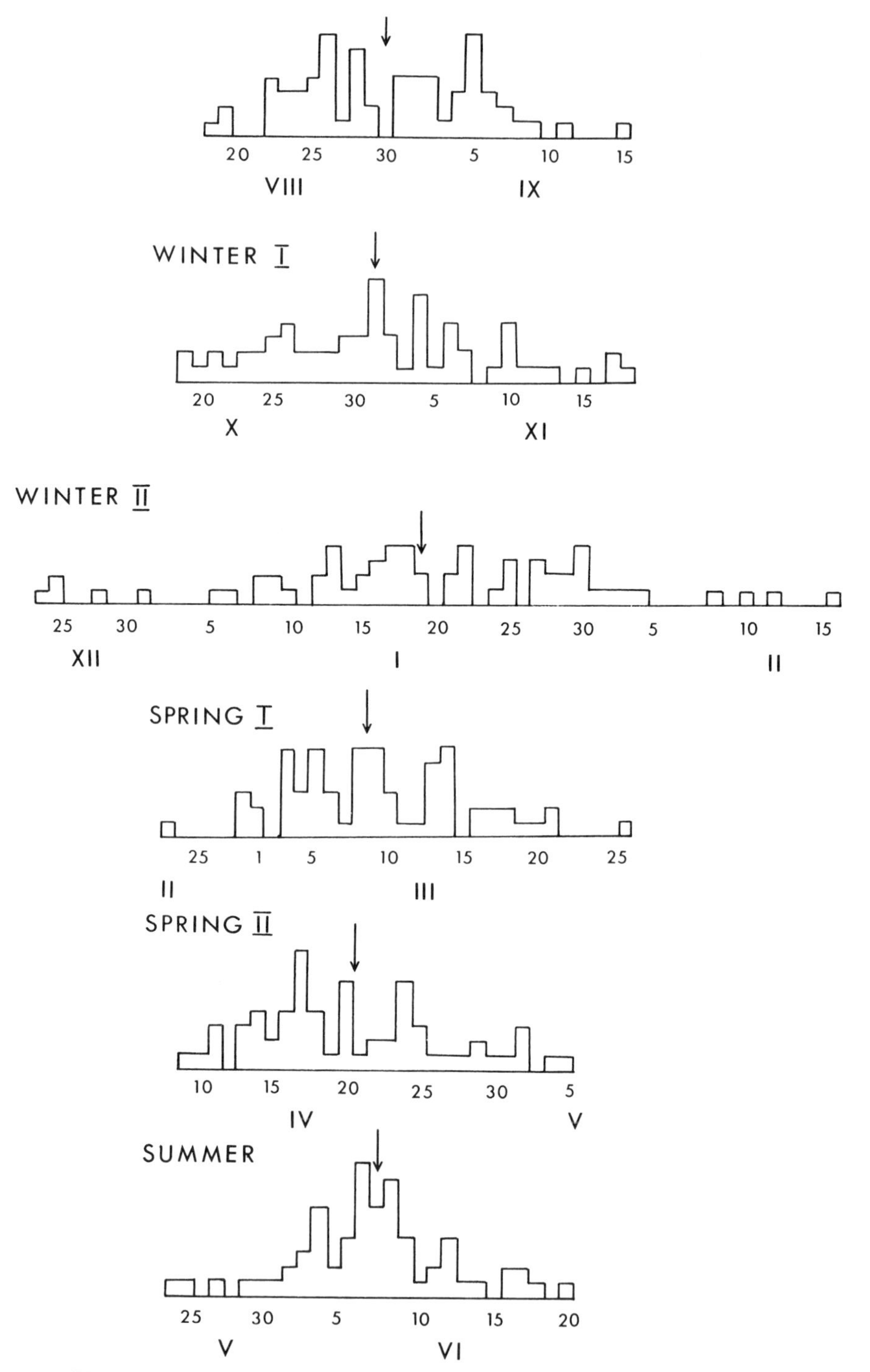

Fig. 5.12 Frequency distribution of beginning dates for natural seasons 1899–1965 (after Bradka, 1966). Heavy arrow shows mean date.

would be no necessity to seek the explanation of circulation changes in terms of external forcing. No final answer to this question can yet be given. It is one of considerable interest, of course, in connection with climatic change.

B Description and analysis of spatial characteristics of climate

The various approaches which have been adopted in synoptic climatology have been discussed in depth in Chapter 3. Here we are concerned with the appropriateness of particular synoptic climatological methods in terms of regional problems being considered. In addition to presenting this emphasis in a new framework of discussion, additional examples of the different types of synoptic classification schemes are included. In the first section studies which deal with the regional macroclimate on a scale 5×10^2–10^3 km are considered. Then local analyses where topoclimatic effects become important are examined.

1 REGIONAL CLIMATIC CHARACTERISTICS

For a variety of reasons research seldom proceeds in a regular, systematic manner. It is convenient for purposes of presentation, however, to picture the synoptic climatological study of particular regional climates as evolving through three stages of increasing complexity. The first involves investigations of the effect of mean features of the atmospheric circulation on the area in question. Next the major components of the more common weather regimes of the area are studied. Finally synoptic calendars are prepared with a variety of applications becoming possible and new problems at various levels, being generated. We shall follow this putative scheme now in more detail, although the actual examples will, of necessity, not refer to the same area in most instances.

Basic characteristics

When little detailed knowledge of an area exists it is usually helpful to begin by examining the relationship of mean features of the atmospheric circulation to conditions in the area. The reliability of this procedure is, of course, critically dependent on the representativeness of the mean in the particular climatic region under study (see Chapter 2.B, p. 64). In a study of this type for rainfall in the central Sudan, Osman and Hastenrath (1969) examine the mean monthly pressure, surface streamlines, relative humidity and 300 mb contours for May–October. The Intertropical Front is identified as a discontinuity in the surface moisture and wind fields. Its significance for the average precipitation conditions is underlined by case studies of anomalously dry and wet years when the ITF is respectively further south or north than usual.

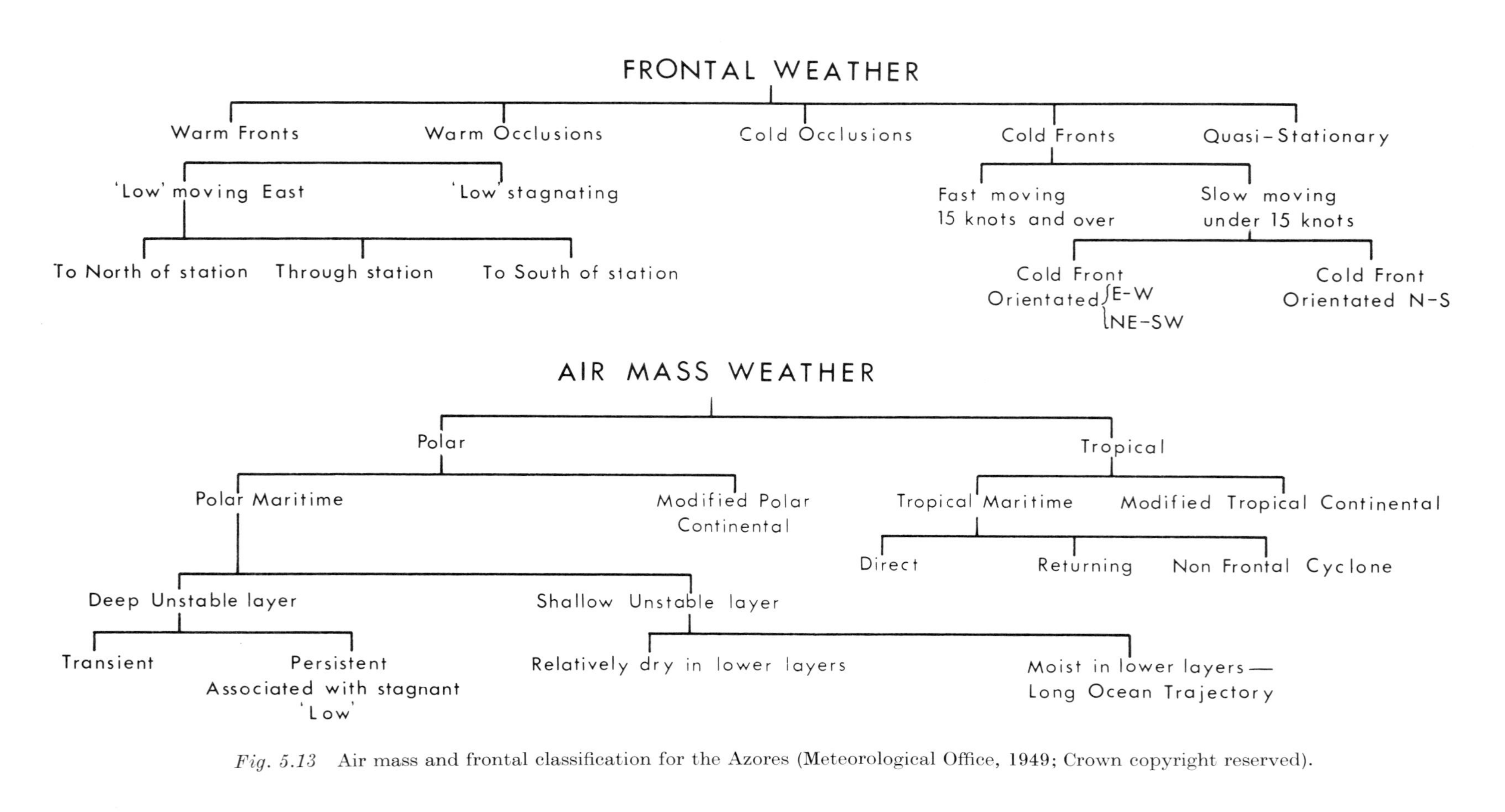

Fig. 5.13 Air mass and frontal classification for the Azores (Meteorological Office, 1949; Crown copyright reserved).

For the somewhat similar area, in terms of data availability, of Amazonia, Vulquin (1971) examines the occurrence of a summer monsoonal flow situation. By analysis of the wind flow, moisture characteristics, and the configuration of the ITCZ as evidenced by satellite cloud photography, he shows that a 'drift pattern' (in the terminology of Johnson and Mörth, 1960; see fig. 6.2) occurs with interhemispheric airflow from the north to south. This is not a *seasonal* characteristic but takes place on a synoptic time scale with an, as yet, undetermined frequency. This particular study illustrates the value of exploratory synoptic climatological research in the tropics in highlighting specific problems for further investigation, and leads us on to the next, more complete, stage of analysis.

The principal weather regimes

The second stage in the development of regional synoptic climatological knowledge involves the delimitation of the principal types of weather regimes in a particular area. This may be in terms of air mass characteristics, as illustrated by the numerous regional studies referred to in Chapter 3.D (pp. 178–87), or generalized *weather types* (*sensu stricto*). The analysis is not necessarily exhaustive, but at least the dominant weather characteristics are accounted for. For example, Kossowski (1968) notes four major synoptic patterns occurring on about 50% of days over Poland. They are: cyclonic weather of North Atlantic origin (103–132 days/year, according to region of the country), anticyclone (55–64 days), cyclone of Arctic origin (6–9 days), and cyclones of Mediterranean origin (3–5 days).

Regional air mass studies such as those of Belasco (1952) for the British Isles are sufficiently well known not to require discussion in this context. However, it is worth noting the type of air mass and frontal classification (fig. 5.13) used in an analysis of 'synoptic weather types'

Table 5.8 Frequencies of air masses and weather types in the Paris Basin (Pédélaborde, 1957)

	Air masses (%)				*Weather types (%)*	
	1899–1939	*1946–1950*			*1926–1950*	*1946–1950*
cA	0·4	1·1	Continental Ac	L	21	21·1
mA	3·0	7·7		A	8·0	10·3
cP	27·0	33·7	Oceanic Ac		13·0	10·0
mP	69·2	53·0	Cyclonic	W sector	31·0	36·5
cT	0·1	0·8		N sector	17·5	15·0
mT	0·3	3·7		S sector	9·5	7·0

L = local origin, A = distant origin.

affecting the Azores (Meteorological Office, 1949). In more recent regional handbooks, such as that for the Mediterranean area (Meteorological Office, 1962), a combination of air mass and circulation types description is adopted and in the one for the Black Sea (Meteorological Office, 1963) simple circulation types alone are used. In an earlier but very comparable analysis for New Zealand, Watts (1947) determined the annual frequency of weather influences from various directions for July 1943–June 1946 as follows:

Influences	*N. of 39°S*	*Central Area*	*South Island*
	%	%	%
NW, W, SW	39	41	37
S, SE, E	14	21	21
Anticyclonic	28	19	20
Moist northerly	5	4	3
Depressions, fronts	14	15	19

Garnier (1958) illustrates some of these synoptic situations and describes their regional significance in his book on the climate of New Zeland.

There are many studies, particularly in the French climatological literature, which adopt a combination of a 'weather type' (*type de temps*)[1] and a circulation type methodology. These include investigations for the Mediterranean (Clerget, 1937), Australia (Fénelon, 1951), the Rhône valley (Blanchet, 1959), Morocco (Noin, 1963) and the Adriatic Sea (Penzar, 1967). The method that is used in several of these studies corresponds with that referred to in the German literature as *Witterungsklimatologie*. The term *Witterungslage*, which has no precise equivalent in English or French, implies the time duration of a weather type. An illustration is provided by the *Witterungsklimatologie* of the High Veldt in South Africa (Vowinckel, 1955). His seven types are a mixture of descriptive and genetic categories:

1 Fine weather type
2 Warm air type
3 Cold air type
4 Monsoon type
5 Shower type
6 Equatorial type
7 Bad weather type

For the Paris Basin, Pédélaborde (1957) prepared a detailed synoptic climatology using both air mass categories and weather types (*types de temps*). The latter, which were defined specifically for this area, had a duration of one to three days. It is of some interest to compare the annual frequency of the major categories (table 5.8). It is apparent that the air mass classification provides no information as to the degree of

[1] The term has also been used, however, by French meteorologists (Mézin, 1945) to refer to large-scale circulation types of one to two weeks duration.

anticyclonic or cyclonic control while there is no indication of the relative warmth of the designated weather type categories.

Under the general heading of 'weather type' studies can also be placed investigations of selected synoptic phenomena such as outbreaks of cold air and warm 'waves'. Blüthgen (1942, 1966, p. 353), for example, classified cold outbreaks over northwest Europe in winter into six categories according to their origin and extent and then analysed their frequency at seventeen stations. The climatological characteristics of cold waves in India have recently been examined by Raghavan (1967). Information on the synoptic aspects of such phenomena helps to complete the synoptic climatological picture of a region, although whether or not the study should be approached in terms of an initial classification of circulation types to which weather phenomena are then related, rather than the reverse, depends largely on the objectives of the research.

Next come studies that employ a classification of circulation patterns (although the authors not infrequently use the term weather types). In a study of rainfall over Algeria, Pédélaborde and Delannoy (1958) used the following categories, placing emphasis on the 500 mb circulation:

1 Cyclonic northerly.
2 Cyclonic westerly and southwesterly.
3 Cyclonic southerly with the Saharan front moving northward.
4 Anticyclonic types.
5 Thundery weather in summer.

Fig. 5.14 illustrates the significance of northerly type for precipitation at Algiers during 1951–3. The analysis also shows that the west and southwest type is important during March–April.

In dry climates it is the storm characteristics that are of major interest. Aelion (1958), for example, examined storms over Israel during November–April 1947–52 which contributed 30% or more of the mean monthly rainfall. Eight types of pattern were identified and characterized by surface and 500 mb maps. The analysis showed that the major storms occurred with cold air drawn into the eastern Mediterranean from Russia with only a short sea track.

A more complete study by Snead (1968) for southern West Pakistan illustrates the types of analysis that can be regarded as a logical step beyond the basic consideration of the regional significance of mean circulation features. Snead identified six weather patterns:

1 The predominant subtropical anticyclone.
2 Winter cyclonic pattern.
3 Arabian Sea cyclones, in the hot pre-monsoon season mainly.
4 Local convective storms (duststorms or squalls).
5 Modified (summer) monsoon pattern bringing thick cloud.
6 Monsoon lows from the east.

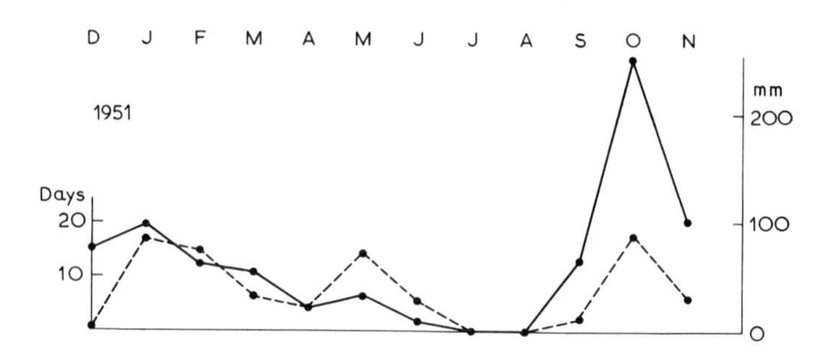

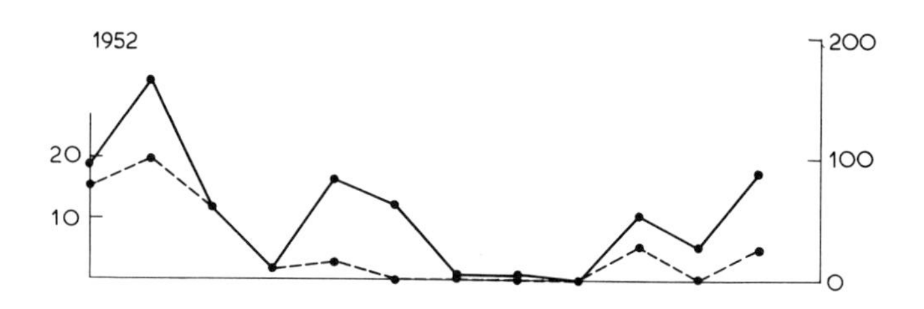

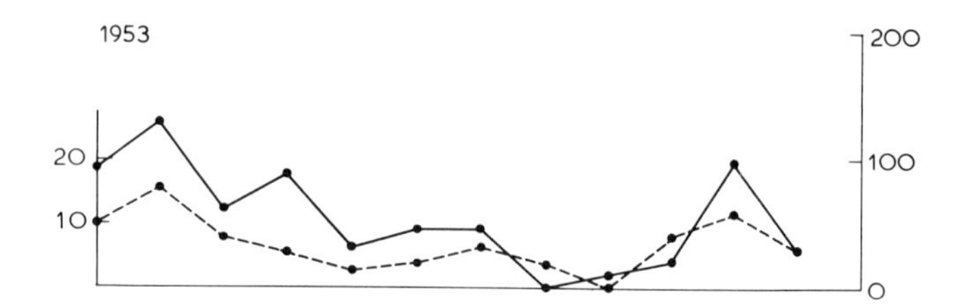

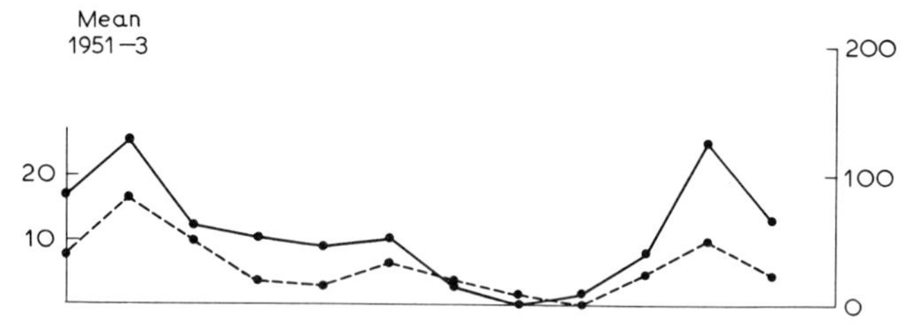

Fig. 5.14 Precipitation associated with northerly types at Algiers 1951–3 (after Pédélaborde and Delannoy, 1958).

Since the climate of the region is transitional between that of monsoon India and the Middle East region of winter cyclone activity, Snead was able to demonstrate the changing influence of the storm types from east to west across the region.

A somewhat similar but more detailed investigation was carried out much earlier for Oahu (Hawaii) by Yeh, Wallén and Carson (1951), again with reference to rainfall. They identified eight basic synoptic patterns:

1 Anticyclonic situations (two types).
2 Cyclonic situations: cold front or polar trough (three types).
3 Warm front from the southwest.
4 Closed cyclonic circulation (two types).
5 Mixed pattern with polar front orientated west–east, north of the islands.

Fig. 5.15 shows these patterns and a graph of the seasonal variation of the average daily rainfall occurring with three groups of related types. Recently, Worthley (1967) updated their work with reference to the frequency of cyclone types and frontal passages near Hawaii for the period 1957–66. A comparable approach using nine large-scale patterns was adopted by Howell (1953) in a study of rainfall in central Cuba. In this study the winds at 1000 feet and at the 700 mb level were found to be of value as additional indicators of the airflow.

Another example, which uses the 700 mb flow direction as the synoptic descriptor, is the work of Walker (1961) for British Columbia. He examined temperature, dew point and wind speed at 500, 700, 850 mb and MSL, as well as synoptic scale vertical motion (computed from numerical models) and precipitable moisture, in relation to the 700 mb flow. The results showed that this synoptic parameter differentiated the weather characteristics satisfactorily. They were also used to verify an orographic precipitation model in terms of precipitation occurring with flow from westerly, southwesterly and southerly directions.

Regional synoptic catalogues

The final stage in the development of a regional synoptic climatology requires the preparation of a catalogue of synoptic types. The most notable examples are those of Hess and Brezowsky (1969) for central Europe, Lamb (1971) for the British Isles and Vangengeim (Bolotinskaia and Ryzakov, 1964) and Dzerdzeevski (1968) for the northern hemisphere. Others, for shorter periods, include eastern Asia (Yoshino, 1968), Alaska (Putnins, 1966) and the Alps[1] (Schüepp, 1968).

These circulation or flow classifications have not all been examined in detail in terms of the associated weather types. The various schemes

[1] Data on Lauscher's classification (see Chapter 3.C, p. 153) are published in the *Jahrbücher, Zentralanstalt für Meteorologie und Geodynamik*, Wien, from 1948.

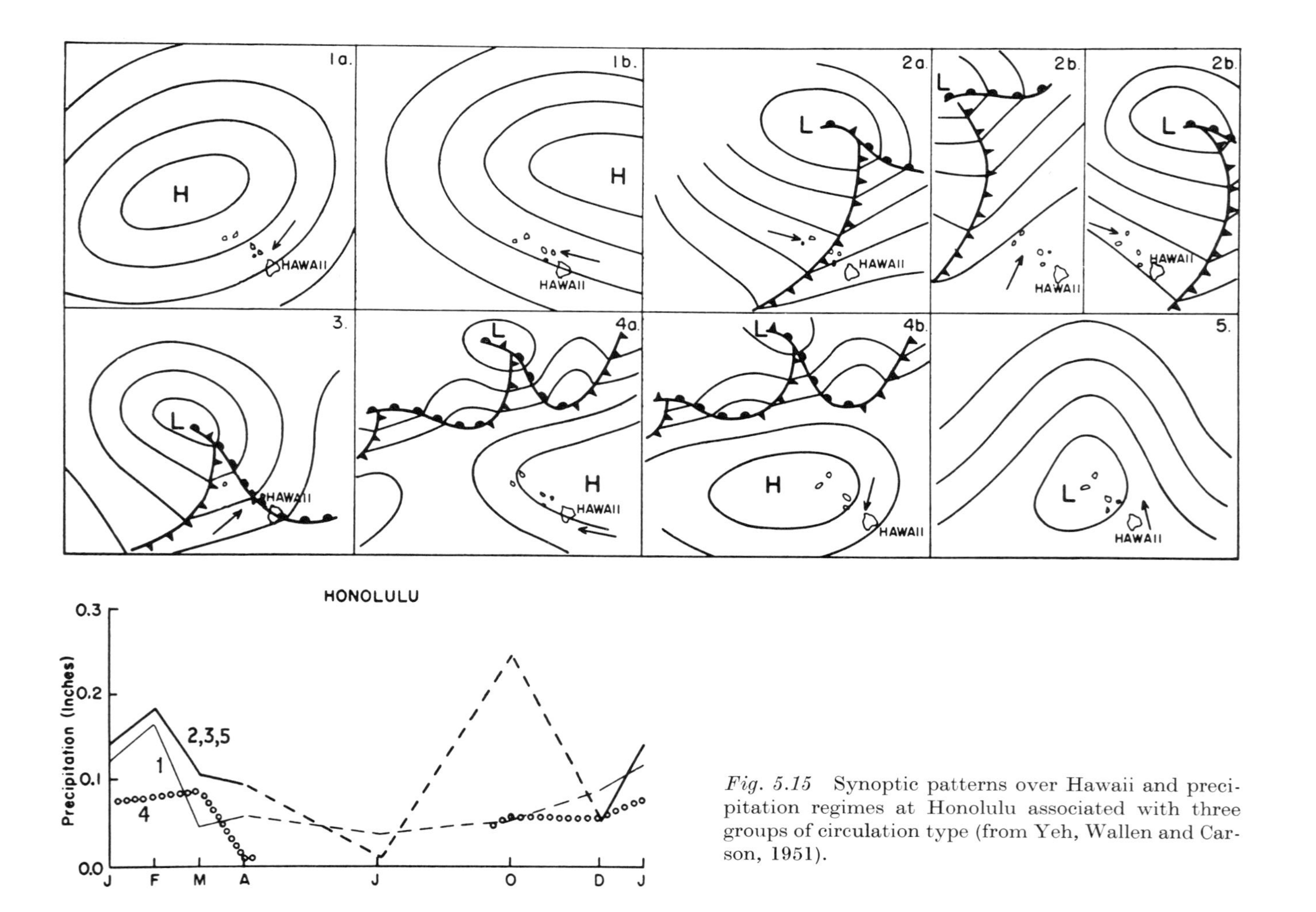

Fig. 5.15 Synoptic patterns over Hawaii and precipitation regimes at Honolulu associated with three groups of circulation type (from Yeh, Wallen and Carson, 1951).

proposed for the Alps have been quite thoroughly evaluated as we noted in Chapter 3.C (p. 156). Putnins and Langdon (1969) provide detailed weather tabulations for Alaska, but little explanatory discussion. The Lamb catalogue has so far only been analysed with respect to climatic conditions associated with the individual types for central–south England (Barry, 1963), although Lamb (1950) provided a general description of the weather types.[1] The generalized P.S.C.M. indices derived from the catalogue (see Chapter 3.C, p. 142) have also been applied to determinations of regional climatic conditions over the British Isles by Murray and Lewis (1966) and Perry (1968, 1969). The Vangengeim classification is obviously very large-scale and most applications have dealt with conditions over a large sector or all of the hemisphere. However, it has also been applied to some extent in regional studies.

Table 5.9 Mean daily precipitation (mm) at Berlin and Munich with Grosswetterlagen, *1890–1944, 1946–1950 (Bürger, 1958)*

Grosswetterlagen	*Berlin*	*Munich*
N_A	1·25	2·69
N_Z	2·35	3·66
NW_A	1·81	3·26
NW_Z	3·35	4·73
W_A	1·58	1·99
W_Z	2·56	3·24
SW_A	0·71	0·97
SW_Z	1·49	1·16
S_A	0·27	0·53
S_Z	1·21	1·85
SE_A	0·62	0·99
SE_Z	1·13	2·14
E_A	0·79	1·29
E_Z	1·64	2·15
NE	1·63	3·87
HM	0·48	1·01
BM	0·77	1·89
TM	2·77	3·59
W_W	1·75	2·89
Mean	1·61	2·53

[1] Average daily rainfall over England and Wales as a whole for 1920–59 has recently been determined for Lamb's synoptic types by Lawrence (1971). See also Lawrence (1972).

(a)

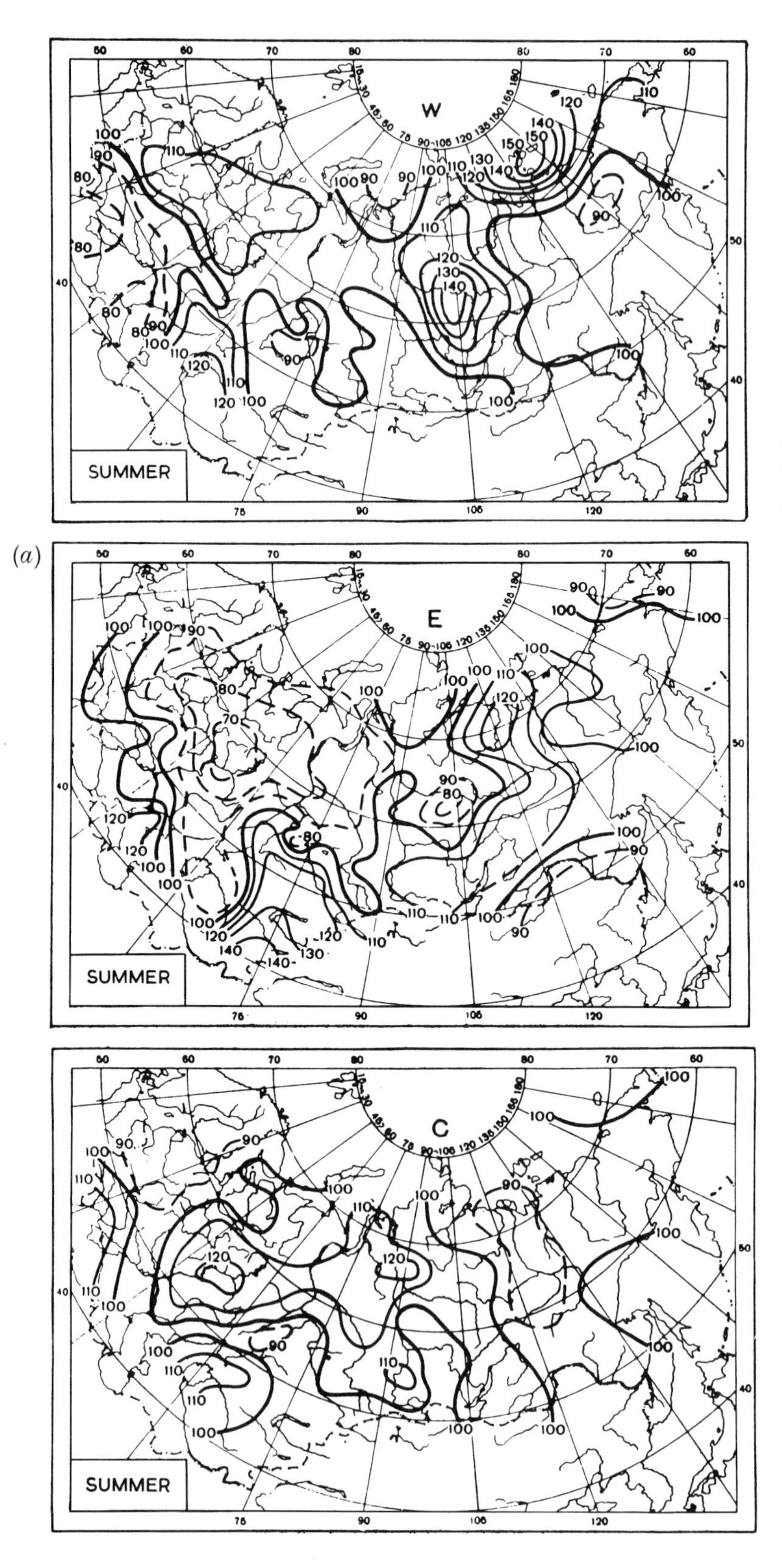

Fig. 5.16
Seasonal precipitati
departures from nor
(1891–1962) over th
U.S.S.R. for the thr

n circulation types
Vangengeim in
summer, (b) autumn
m Vorobieva, 1967).

(b)

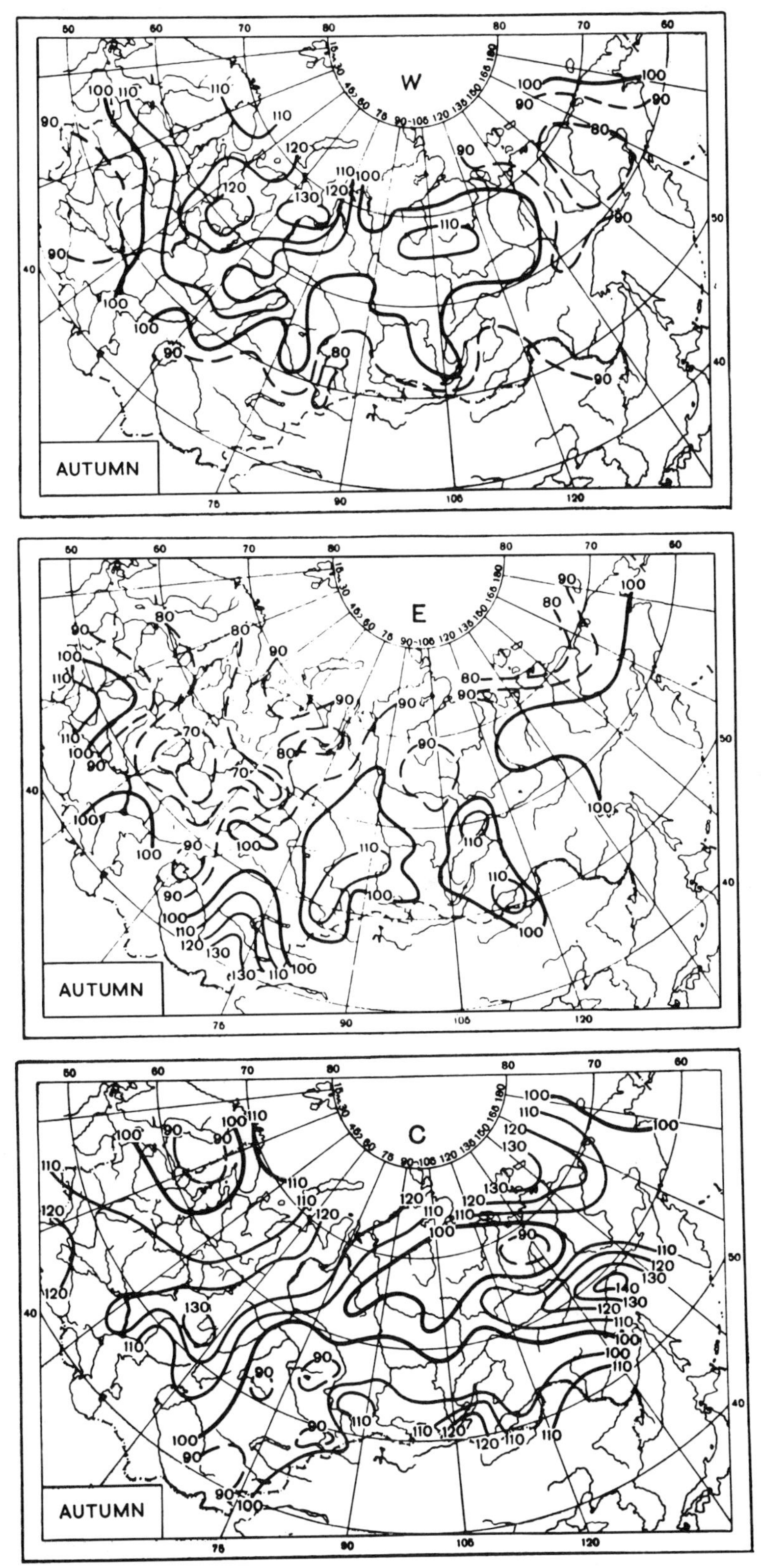

Shklyaev and Alikhina (1968), for example, calculated temperature departures in summer and winter in the central and southern Urals in realtion to the three main categories, W, E and C, and Vorobieva (1967) has mapped seasonal precipitation departures from normal for 1891–1962 over the U.S.S.R. for the three main forms in spring, summer and autumn (fig. 5.16).

Undoubtedly, the scheme that has been examined most intensively is the Baur/Hess–Brezowsky classification. Flohn (1954) characterized the climate of central Europe in terms of *Grosswetterlagen* in *Witterung und Klima in Mitteleuropa*, a book that has no equal in the English language. Some of the most detailed work has been carried out by Bürger (1952, 1958). He determined the properties of individual *Grosswetterlage*, the second study dealing in depth with conditions at Berlin, Bremen, Karlsruhe and Munich for 1890–1950. Type averages and frequency values were determined for various climatic parameters. Fig. 5.17 gives an example of the relationship between the frequency of precipitation days and the *Grosswetter*. An analysis of mean daily precipitation for each *Grosswetterlage* (table 5.9) illustrates the importance of convective activity over southern Germany with NE, N and NW types. The frequency of falls >5 mm is greatest with NW_z type at both Berlin (23·7%) and Munich (32·4%). Examination of the temperature data for the four German stations showed little spatial variations with respect to seasonal departures from normal (table 5.10). Over the year as a whole SW, W and TB types gave the warmest conditions, TM, HFA and NE types the coldest conditions.

Table 5.10 The occurence of temperature anomalies with Grosswetterlagen *in Germany (Bürger, 1958).*

	Berlin	*Bremen*	*Karlsruhe*	*Munich*
Winter				
$>+5$°C	—	SW_Z	SW_Z	SW_Z
$+3·0$ to $4·9$	W_A, W_Z, SW_Z	W_A, W_Z, SW_A	W_Z, W_S	W_Z, W_S, TB
$-3·0$ to $-4·9$	SE_Z	HN_A, SE_Z, NE	HN_A, SE_A, HF_A	HN_A, HB, SE_A
$>-5·0$	SE_A, HF_A	SE_A, HF_A	—	HF_A
Summer				
$+3·0$ to $4·9$	TrW, HF_A	HF_A, HM	—	—
$+1·0$ to $2·9$	TB, HM	TB, TrW, NE	TrW, HF_A, HM, BM	W_A, TrW, HF_A HM, BM
$-1·0$ to $-2·9$	N_A, HB, NW_A, NW_Z, W_S	N_A, HB, TrM, NW_A, NW_Z	N_A, HB, TrM, NW_Z, Ws, TM	N_A, HB, TrM, NW_A, NW_Z, T_M
$-3·0$ to $-4·9$	N_Z	N_Z	N_Z	N_Z

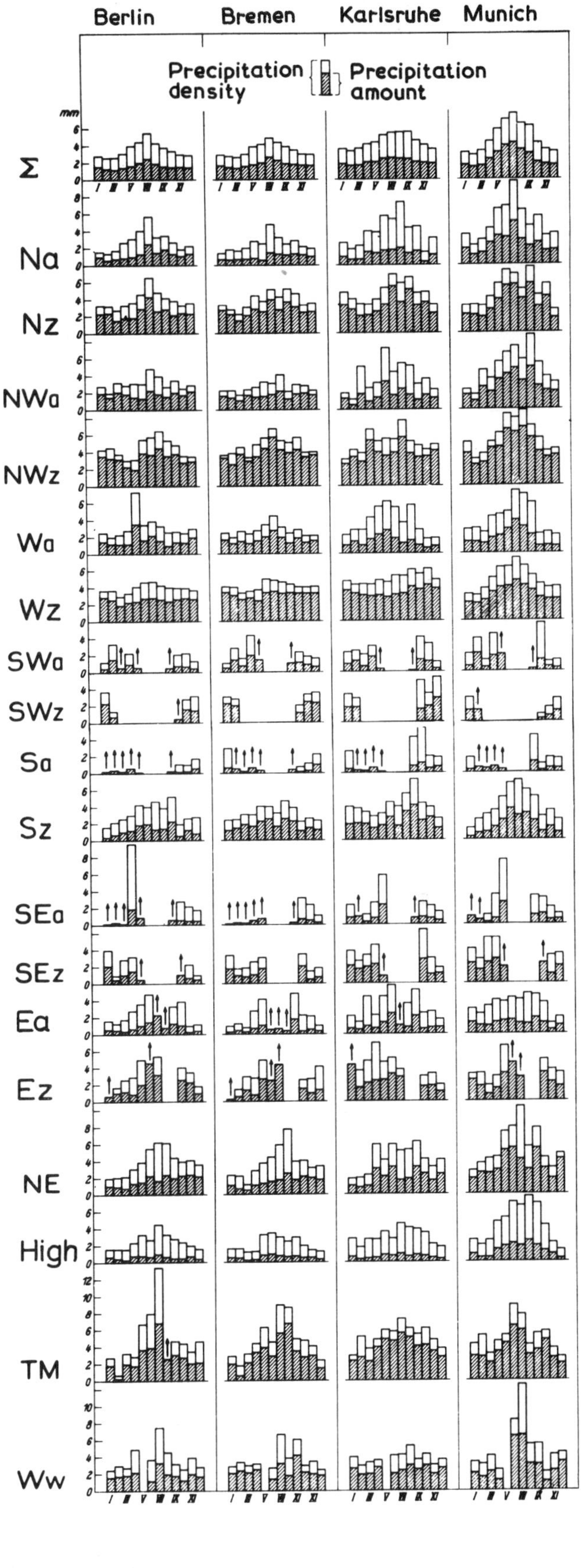

Berlin Bremen Karlsruhe Munich

Precipitation density | Precipitation amount

Σ, Na, Nz, NWa, NWz, Wa, Wz, SWa, SWz, Sa, Sz, SEa, SEz, Ea, Ez, NE, High, TM, Ww

Fig. 5.17
Mean daily precipitation and specific precipitation density (mm) for *Grosswetterlagen* at Berlin, Bremen, Karlsruhe and Munich, 1890–1950, except 1945 (from Bürger, 1958).
Months with < 15 cases are omitted. Precipitation density omitted for cases with < 15 precipitation days (denoted by arrow).

Other workers have also considered general temperature conditions in relation to *Grosswetter* types. Böer (1954) calculated the frequency of *Grosswetterlagen* for months when the temperature departed by more than $\pm 1\sigma$ from the mean value at stations in Germany. Based on the period 1881–1952 he postulated the following primary relationships:

	Grosswetterlagen		
Temperature pattern	*Zonal*	*Mixed*	*Meridional*
Cool summer, mild winter	W_S, W_Z, W_A	SW_A, SW_Z	N_A, S_Z
Warm summer, cool winter	BM		HN_A, TM, SE, HF, HNF, NE
Spring and autumn warm		NW_A, NW_Z	TB, TrW, S_A
Summer and winter warm			HB, W_W
Cold year			N_Z, HN_Z
Warm year		HM	

Certain of these associations look of potential predictive value if the *Grosswetter* pattern could be forecast successfully. On the other hand, a more specific study of winter conditions by Hesse (1953) gave much less homogeneous results. He classified the winters from 1852/3–1952/3 in northwest Saxony into categories such as early/late, short/long, mild/severe, dry/wet, and determined *Grosswetter* frequencies for each category. 'Pre-winter' is particularly associated with BM type, for example, whereas 'post-winter', late, long and severe winters frequently have HF type. Early, short and mild winters are often related to HM type but this gives rise to both wet and dry conditions. Perhaps more satisfactory is Heyer's (1955) analysis of eleven severe winters at Berlin (and surrounding stations). The percentage frequencies of the principal types that occurred in these years were as follows:

	Grosswetter type					
	W	*HM*	*N*	*E*	*S*	*SE*
November	26·8	12·4	10·9	5·1	10·9	
December	27·0	23·2	14·4	8·5		
January	10·0	11·4	12·9	20·8		11·7
February	14·1	19·3	6·7	23·2		
March	30·0	15·8	23·5	10·2		
Winter	21·6	16·4	13·8	13·5		

The variable importance of W and E types in early and mid-winter is especially striking.

There have been numerous applications of the *Grosswetter* classification on a regional level. Grutter (1966), for example, carried out a detailed study of precipitation patterns in Switzerland associated with each of the twenty-six *Grosswetterlagen*. Maximum twenty-four-hour precipitation amounts have been evaluated by Kern (1963) for each of the twenty-two years 1934–44, 1947–57, at 102 stations in Bavaria. The ratio of the percentage of days with high rainfall occurring with a specific *Grosswetterlage* to the percentage frequency of occurrence of that type was determined, with the following results, for the more extreme cases:

Maximum ratios		*Minimum ratios*	
NW	1·5 (8·4%)	S	0·2 (2·9%)
N, NE, Ww	1·4 (2·7, 5·8, 2·8%)	NHF	0·4 (3·1%)
W, TrW	1·3 (23·2, 2·9)	SE	0·5 (3·5%)

Frequency of *Grosswetter* occurrence is shown in parentheses (with respect to a total of 315 cases). When the data were subdivided into smaller subregions a distinct east–west difference was noted (fig. 5.18) except in the case of W type. The variation for N and NE types is almost opposite to that for NW type. Kern does not provide an explanation for these spatial contrasts, although they may be associated with the greater relief in the Ammer–Isar sector. For 170 stations in the Erzgebirg area (around Dresden), Fojt (1967) found that daily totals in excess of 50 mm, which occur most often in summer, were associated during 1934–59 particularly with TM, NW and W types, while in

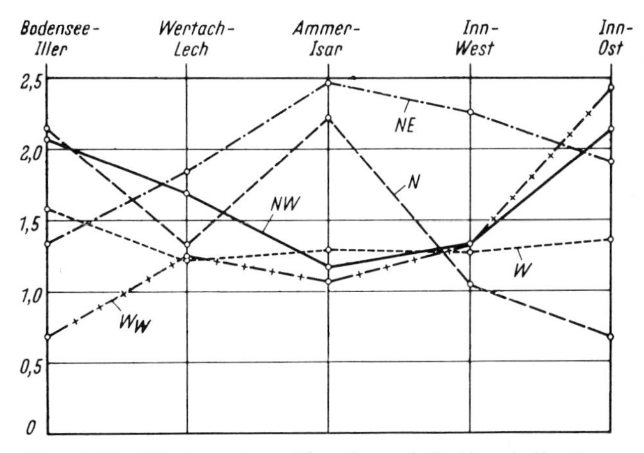

Fig. 5.18 West–east profile of precipitation 'effectiveness' with selected *Grosswetterlagen* in Bavaria (from Kern, 1963).

Thuringia heavy daily falls occur mainly with N_z and W_a types (Kirsten, 1960). This local and regional variability emphasizes the need for great care in attempting to generalize or extrapolate from synoptic climatological results obtained in one area. Conversely, these methods provide a useful basis precisely for the analysis of climatic differentiation.

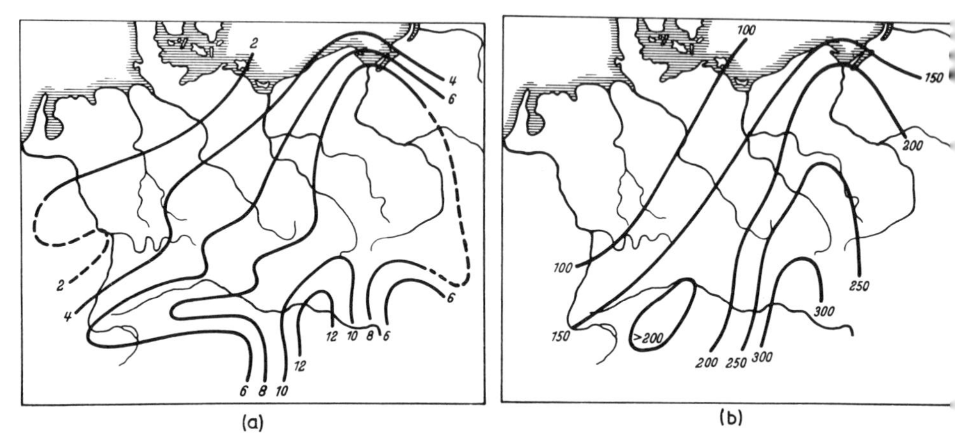

Fig. 5.19 The spatial extent of Vb type influence in Europe (from Flohn and Huttary, 1950). (*a*) Percentage contribution to annual precipitation. (*b*) Relative specific precipitation density (per cent).

This type of application of the *Grosswetterlage* catalogue is well illustrated by Flohn and Huttary (1950). They examined the spatial extent of the influence of the Vb situation with respect to precipitation over Central Europe. Fig. 5.19(*a*) shows the percentage contribution of this type to the annual precipitation while fig. 5.19(*b*) represents the *relative specific precipitation density* – the mean amount per precipitation day ($>0{\cdot}1$ mm) for the ninety-six Vb cases which occurred in a fourteen-year period expressed as a percentage of the overall mean precipitation day.[1] Flohn and Huttary suggested that the 100% line, corresponding to an annual contribution of about 3% of the total precipitation, be regarded in this case as the effective limit of this type. The 100% line of course implies that the amount per precipitation day with a particular type is the same as the overall mean. In a smaller scale study for Mecklenburg, Maede (1951) considered the seasonal and annual pattern of specific precipitation density for the Vb and W types although he was

[1] Maede (1951) gives the following convenient expression:
specific precipitation density for type *i*,

where *f* is the type frequency, *m* the precipitation amount, and $(n-i)$ denotes all types other than *i*.

only able to use days with precipitation > 1 mm. He argues that the limit might more appropriately be related to 25% of the percentage contribution of a particular type to the total precipitation in its 'core area'. Since the latter is 12% (fig. 5.19(*a*)) the limit would still be 3%. Maede emphasizes that this determination cannot in general, therefore, be based on a local study. In a discussion of Maede's paper, Flohn essentially concurs and certainly it would seem that in general this type of criterion, rather than a measure of the *relative* raininess of a type, is more appropriate in determining type effectiveness.

One of the most thorough regional synoptic climatologies is provided by Fliri (1962) for the Tirol, based on Lauscher's classification (see p. 153). Precipitation frequency and average amounts for each of the nineteen types are mapped on a seasonal basis using data for 1948–60 at 200 stations. Other parameters, tabulated for stations with the appropriate data, include temperature, relative humidity, height of condensation and freezing levels, cloudiness, and sunshine. An interesting feature of Fliri's study is his use of graphical presentations of the results. This topic is considered in some detail below. He also examines the effect of record length on average precipitation amounts with the Lauscher types. For annual type averages at Innsbruck the values are in most cases reasonably stable with about five years of data and satisfactorily so with twelve years.

The overwhelming majority of regional synoptic climatologies discuss only the basic climatic parameters and most often temperature and precipitation. A shift of emphasis is nonetheless discernible in keeping with other developments in climatological research. Considerable attention is now being given, for example, to the synoptic climatological aspects of energy transfers and this theme is discussed in some detail in section E of this chapter (pp. 422–5). Atmospheric vapour content and horizontal vapour transport have been examined in relation to the synoptic classification developed by Barry for Labrador–Ungava which was discussed in Chapter 3.C (p. 143) (Barry, 1967). But in spite of these examples we are far from utilizing the range of possible useful parameters which were discussed in Chapter 2.

Few of the studies we have discussed deal extensively with spatial patterns of weather conditions. The investigations of precipitation by Mosiño (1965) for Mexico and by Gazzola (1969) for Italy are good examples of what can be done in this respect. Mosiño found that the contribution of each circulation type to the total mean annual precipitation is a better indicator of circulation effects than the simple average per type day since the former largely eliminates topographic effects. Figs. 5.20(*a*) and 5.20(*b*) illustrate two of the dominant patterns of precipitation and the corresponding circulation types at 700 mb. Mosiño was also able to demonstrate that westerly circulation patterns are generally dry,

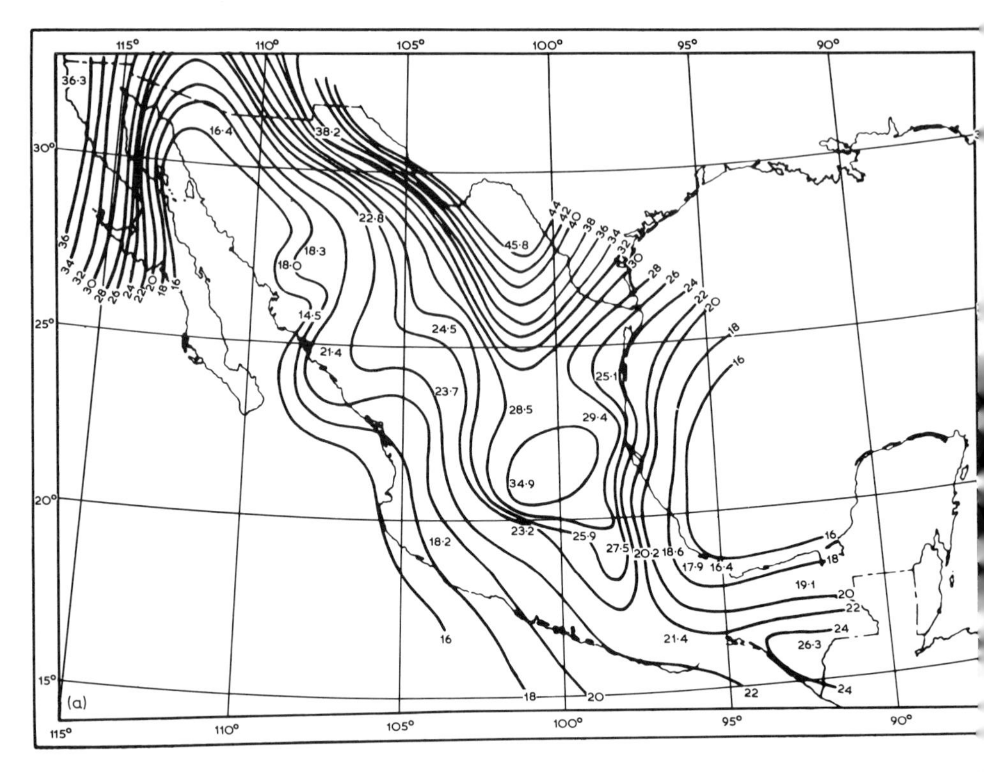

115°
110°
105°
100°
95°
90°
30°
25°
20°
15°
36·3
16·4
38·2
22·8
18·3
18·0
14·5
45·8
44
42
40
38
36
34
32
30
28
26
24
22
20
18
16
36
34
32
30
28
26
24
22
20
18
16
24·5
21·4
25·1
23·7
28·5
29·4
34·9
23·2
25·9
27·5
20·2
18·6
17·9
16·4
18·2
19·1
16
18
20
22
24
26·3
21·4
16
18
20
22
24
(a)

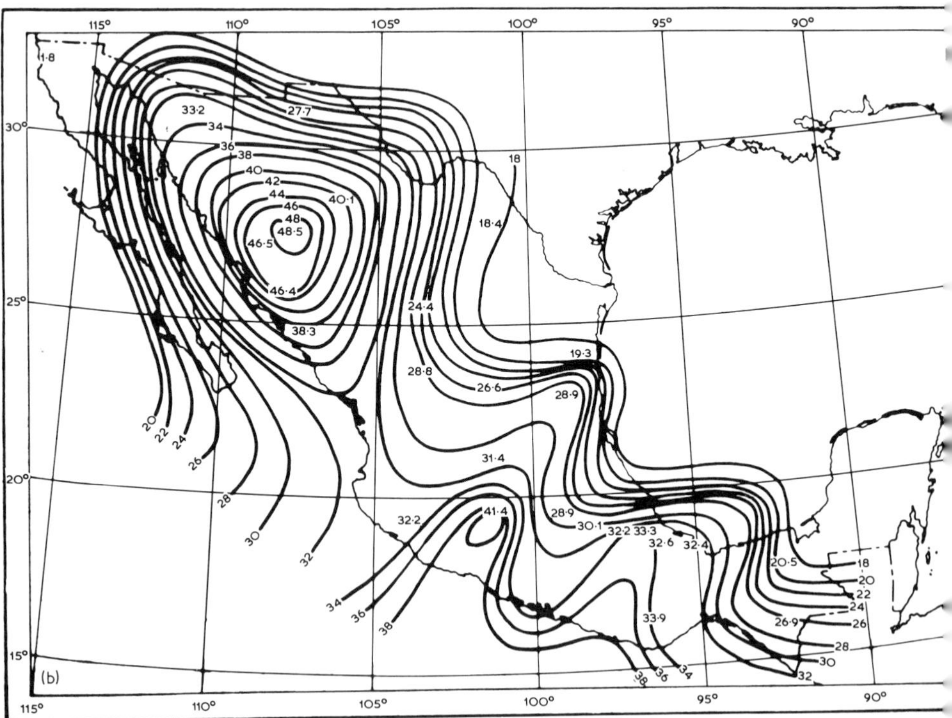

115°
110°
105°
100°
95°
90°
30°
25°
20°
15°
1·8
33·2
27·7
34
36
38
40
42
44
46
48
48·5
46·5
40·1
18
18·4
46·4
24·4
38·3
19·3
28·8
26·6
28·9
20
22
24
26
28
30
32
31·4
32·2
41·4
28·9
30·1
32·2
33·3
32·6
32·4
20·5
18
20
22
24
26
26·9
28
30
32
33·9
34
36
38
34
36
38
(b)

in spite of trajectories over the Pacific Ocean, whereas easterly types are invariably sources of precipitation regardless of the isobaric curvature.

The North American weather type catalogue (which has not been published) has been applied in considerable detail to analyses of weather parameters over the continent. Indeed there are seasonal maps for the United States and Canada showing the distribution of the 50% probability isoline of various precipitation totals for each day of the occurrence of a five to seven day synoptic sequence and also maps of the average departure from the six-day seasonal normals (California Institute of Technology, 1943). The data that have been collected could undoubtedly contribute greatly to regional synoptic climatology in North America if they were synthesized and integrated with other information. The limited availability of the appropriate reports appears to have restricted this work. The approach is nevertheless still used in connection with commercial analogue forecasting (based on at least post-1928 records of the classification).

Analysis of types on this continental scale obviously presents many problems of interpretation. In particular, the developments ahead and to the rear of the systems on which the specific type classification category is focused may be quite variable. However, information on this problem goes some way to providing knowledge of the simultaneous spatial coupling of the airflow patterns in adjacent areas which Jacobs regards as an important synoptic climatological parameter (see table 3.1). Essentially it is only the Soviet hemispheric classifications that shed light on this question. At the regional level there are only occasional, and usually incidental, data serving as pointers. Frequency figures for flow types in adjacent areas do not generally consider simultaneous occurrences.

Data presentation

It is readily apparent that the larger a classification scheme is the more difficult it is to get an overview of the numerous results that may be generated. This problem is becoming increasingly more severe with the analysis of records by computer. For example, if we are determining averages of three weather elements in the four seasons for the Lamb classification of twenty-seven types then we have 324 results to consider. This problem led Fliri (1962, 1965b) and Schüepp and Fliri, 1967)

Fig. 5.20 opposite Percentage contribution to annual precipitation over Mexico by two surface flow patterns (from Mosiño, 1964). (*a*) Type 5: trough in the lee of the Rocky Mountains and Sierra Madre Oriental. (*b*) Type 6: anticyclone cell over the north Gulf of Mexico, trough over northwest Mexico. This shows (*a*) horizontal convergence effects over the Mexican plateau, (*b*) a maximum over the Sierra Madre Occidental and another related to a moist tongue commonly over the southern Pacific coast area.

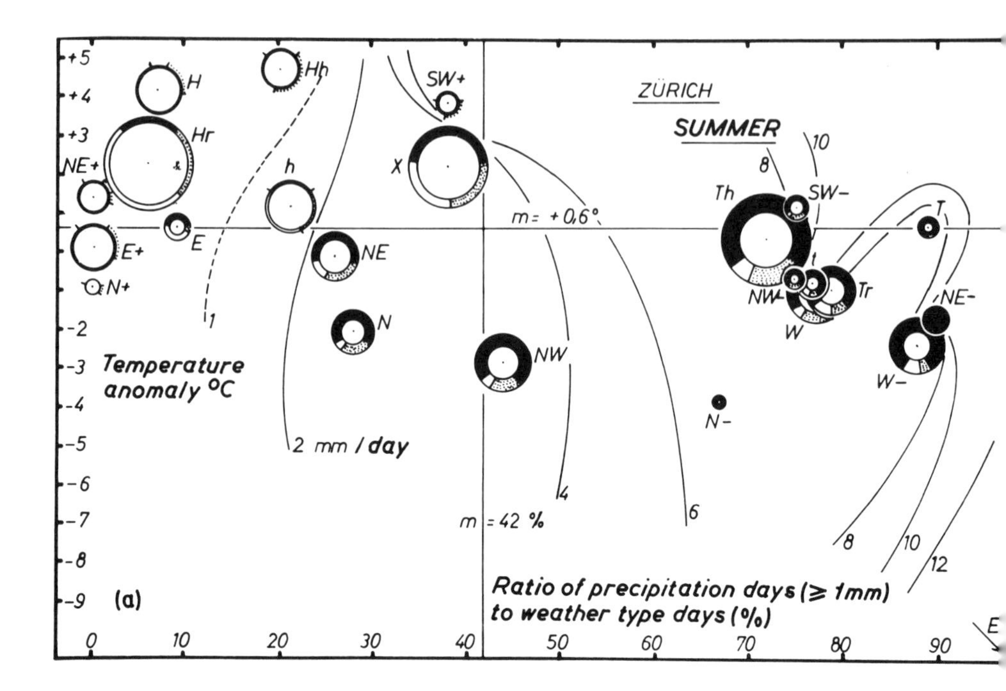
ZÜRICH
SUMMER
Temperature anomaly °C
Ratio of precipitation days (≥ 1mm) to weather type days (%)
m = +0,6°
m = 42 %
2 mm / day
(a)
H
Hh
Hr
NE+
E+
N+
E
h
NE
N
SW+
X
NW
N-
Th
SW-
NW
W
Tr
T
NE-
W-

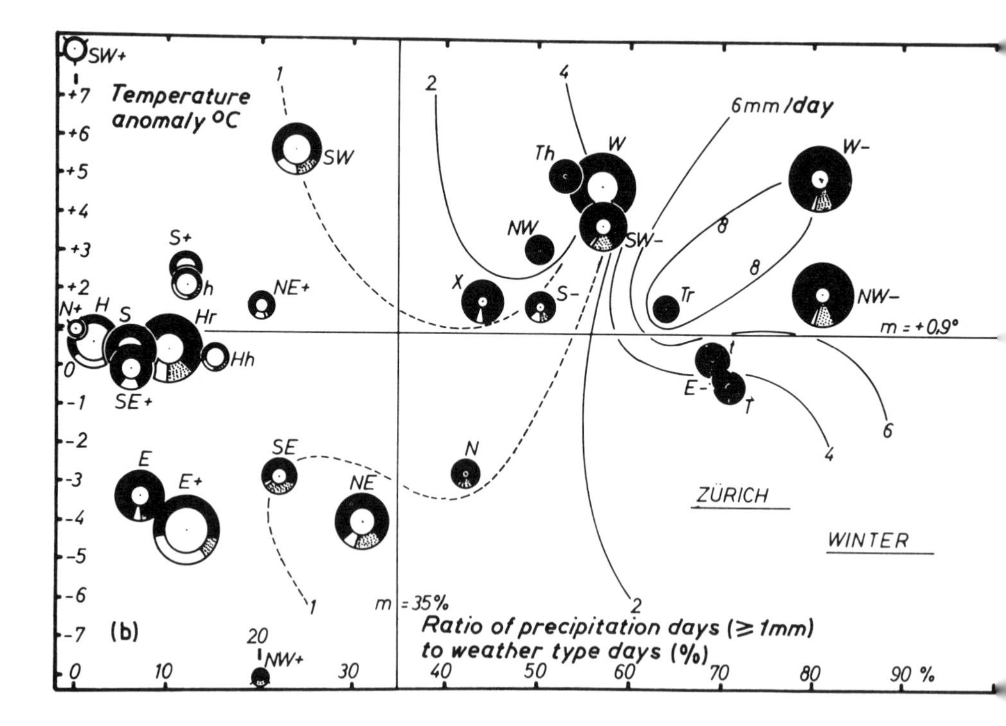
ZÜRICH
WINTER
Temperature anomaly °C
Ratio of precipitation days (≥ 1mm) to weather type days (%)
6mm /day
m = +0,9°
m = 35%
(b)
SW+
SW
S+
h
NE+
N+
H
S
Hr
Hh
SE+
E
E+
SE
NE
N
NW+
X
S-
NW
Th
W
SW-
Tr
W-
NW-
E-

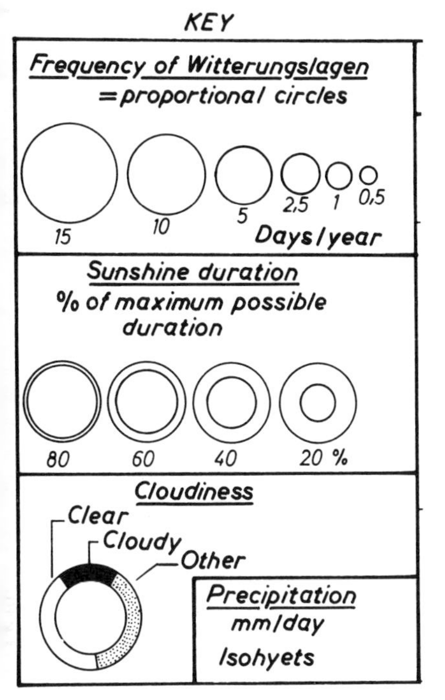

Fig. 5.21 Synoptic climatological diagrams for Zurich in summer and winter (from Schüepp and Fliri, 1967).
m = mean values
h = col
H = stable high
Hh = upper high
Hr = zonal ridge
t = secondary low
T = central low
Tr = meridional trough
Th = upper low
X = flat pressure field
+ = anticyclonic subtype
− = cyclonic subtype

to construct a special synoptic climatological diagram which provides a synopsis of circulation and weather type relationships. Fig. 5.21 illustrates the information that they select and an example for Zurich. In fact the detail here even seems excessive (Barry, 1970). A simpler plot of mean daily temperature departure against specific precipitation density (which indicates the relative raininess) for the various weather types would seem to be preferable. In this case a graph can also be used to compare seasonal trends, as shown in fig. 5.22 for Southampton. The degree of parallelism of the arrowed lines indicates the extent to which types have a similar response to seasonal trends, while the length of the line shows the amount of change involved. Neither system lends itself especially to the problem of spatial comparison, where the best that can be done at the present time basically involves isoline plots of individual climatic parameters for a particular circulation type. In the case of a comparison between only two stations, however, it is feasible to use a graphical analysis. Müller (1969), for example, plots the temperature deviations from the mean, for different *Grosswetterlagen*, at Munich against those for Obersiebenbrunn. The graph shows clearly which types produce similar anomalies at the two stations. It is evident that reduction of the dimensions of the weather variables through principal component analysis may be helpful in attempts to formulate a synthesis

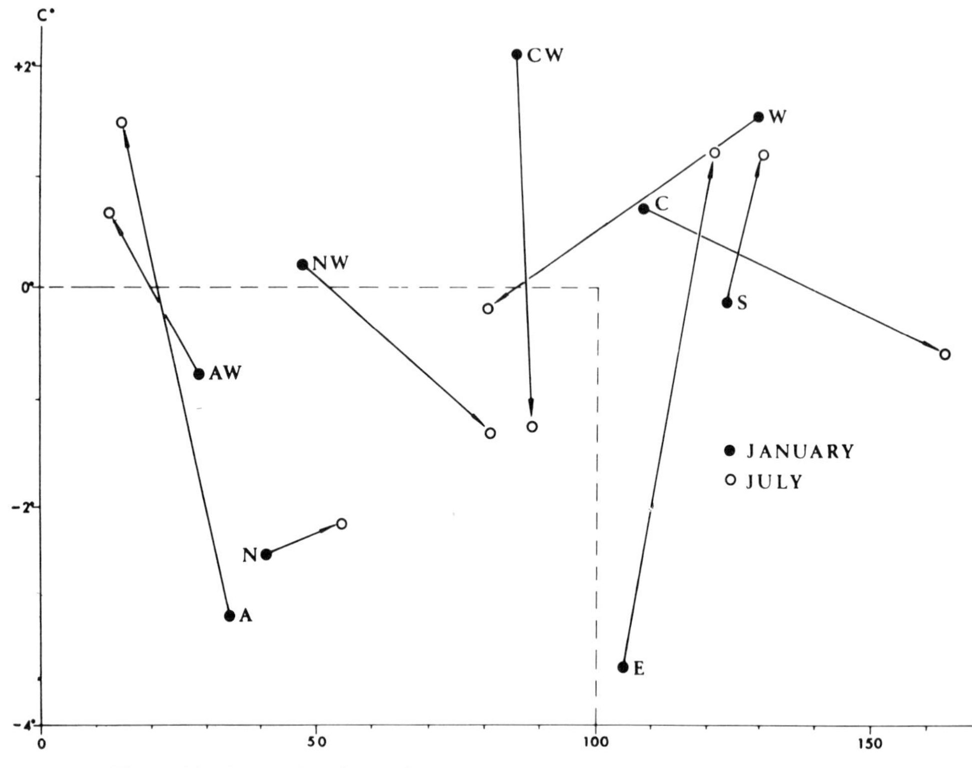

Fig. 5.22 Synoptic climatological diagram for Southampton, 1921–50 (from Barry, 1970). Letters refer to Lamb types. Ordinate = departure of temperature (°C) from monthly mean, abscissa = relative specific precipitation density (%). Arrows indicate January–July change in characteristics.

of climatic complexities. Nevertheless, further thought needs to be given to the most effective means of presenting results of spatial and temporal synoptic climatological studies.

2 LOCAL CLIMATIC CHARACTERISTICS

Early synoptic climatological studies of local climatic conditions were largely directed at short-term local forecasting (Pick and Wright, 1925–27, for example). This approach has obviously been superseded by more sophisticated techniques, although the value of synoptic climatology information as a supplementary aid in such prediction should not be overlooked. A recent example of a detailed local synoptic climatology is provided by Fliri and Dimai-Feucht (1970) for Innsbruck using Schüepp's 1968 calendar (see Chapter 3.C, p. 155). For the period 1955–67 tabulations are given on a seasonal basis of the average frequency and duration of each type and of the corresponding weather characteristics (mean departure of temperature, mean soil temperature, sunshine duration, cloudiness, wind speed, vapour pressure, relative humidity, potential evaporation, daily precipitation and also the frequency of snowfall, of selected amounts of precipitation and of föhn conditions). The authors conclude that the large standard deviations limit the usefulness of the results for forecasting purposes, but stress that the data usefully complement the conventional climatological averages.

The variability within synoptic airflow categories arises from sub-synoptic scale features and from local topographic effects, apart from the intrinsic variability of meteorological conditions on the synoptic scale. It is interesting that Hufty (1962) found wind sector classes provided better discrimination of temperature conditions than circulation types in winter at Florennes, Belgium, whereas in summer the latter were quite satisfactory. This question of the most suitable synoptic parameters for use in climatic analysis has received little detailed attention.

Most local climate studies during the 1930s were based on air masses. Landsberg's (1937) investigation for Pennsylvania is a classic example. More recently Richardson (1956) analysed temperature differences at three sites across the South Tyne valley, Cumberland, in relation to Belasco's categories. The absence of air mass catalogues precludes such studies, however, apart from the basic problems discussed in Chapter 3.D (pp. 178–84).

Following the attempts to use synoptic climatology for local forecasting which culminated in the Second World War (Jacobs, 1947), local studies have been rather neglected. This neglect stems in part from the unavailability of most catalogues of daily airflow or pressure types as well as the necessity for computer facilities to analyse adequate statistical samples. In Europe there have, however, been numerous studies of local conditions using the *Grosswetter* catalogue. For example, Maede and Matzke (1952) calculated mean radiation receipts at Greifswald (54°N, 13°E) for the *Grosswetterlagen* during 1931–50. Peczely (1961) determined mean temperature, precipitation amount, cloudiness, pollution concentration and fog frequency at six to eleven stations in Hungary for each of his thirteen categories of *Grosswetter*. For England the Lamb catalogue has been used by Barry (1964, 1967). Averages of temperature, precipitation and sunshine totals were calculated for the major flow types for 1921–50 at stations in central–south England.[1] Fig. 5.23 illustrates two ways of displaying seasonal variability at a given locality

[1] Thompson (1970) has recently performed a similar analysis in northeast New South Wales on both a regional and topoclimatic scale using a scheme of four surface airflow groups, eight surface pressure types and four types of 500 mb pattern.

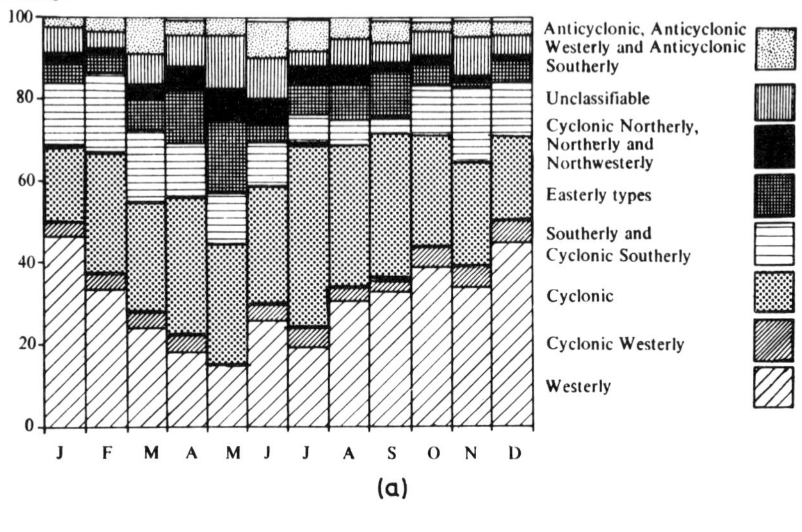

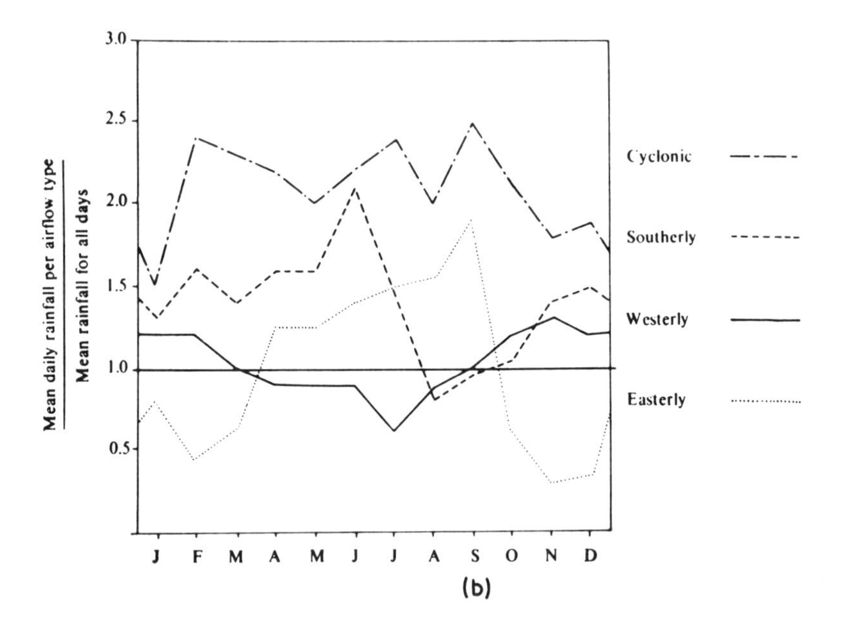

Fig. 5.23 Illustrations of the contribution of Lamb types to precipitation at Southampton, 1921–50 (from Barry, 1967). (*a*) Percentage contribution to the monthly mean. (*b*) Ratio of the mean daily rainfall for a type to the mean rainfall for all days.

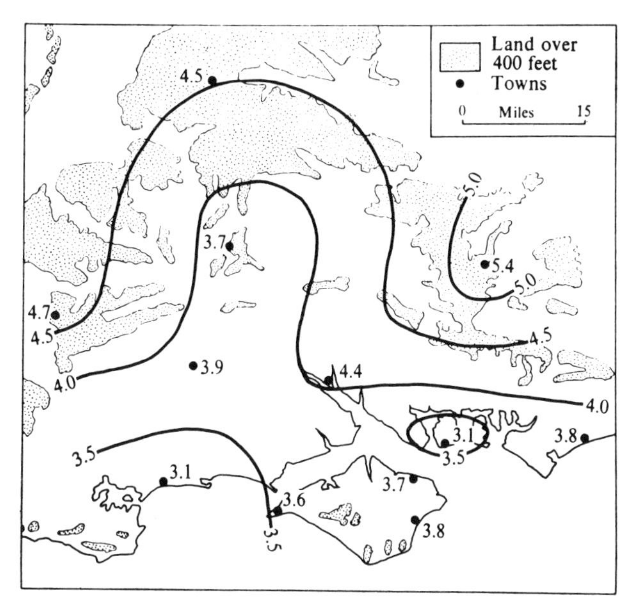

(a)

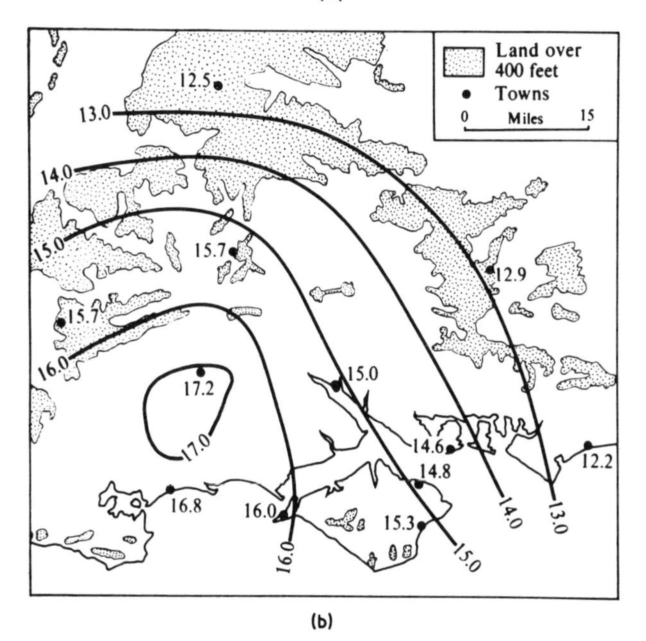

(b)

Fig. 5.24 Precipitation distribution over central south England with two of Lamb's types (from Barry, 1967). (*a*) Average daily rainfall (mm) with cyclonic type July 1921–50. (*b*) Percentage of the monthly precipitation with southerly type, January 1921–50.

and fig. 5.24 shows the possibilities of spatial analysis with synoptic climatological data. The use of the percentage contribution of a type to the total precipitation largely eliminates the effect of topographic factors.

Recently, there have been signs of renewed interest in the application of local synoptic climatology, particularly with respect to extreme conditions. Lowndes (1968, 1969), investigating heavy rainfalls in the Dee and Clwyd River basins of North Wales, found that nearly all of the sixteen occasions with amounts exceeding 50 mm (2 inches) during September–February 1911–68 were associated with Lamb's Westerly type and, in particular, within the warm sector of deepening waves or depressions. About half of the falls in this category during March–August 1911–68 occurred with the Cyclonic type. Fontaine (1963) made a similar hydrometeorological study of precipitation and floods in the Durance basin, France, employing a scheme of five zonal and two meridional types of circulation patterns. Earlier, he used a similar approach with six types to examine conditions associated with the initiation and removal of snow cover in the French Alps (Fontaine, 1959).

One aspect of local climatology to which airflow classification categories can usefully be applied concerns the nature of the climatic transition inland from a coastline. In many areas it is difficult to separate coastal–inland differences from the effects of elevation and topography, as noted by Howe (1953) for the Aberystwyth area of Wales. The Baltic coast of East Germany provides a very suitable region in which to examine the coastal transition zone. Maede (1952, 1953a, b, c) has made detailed studies of the nature of this transition with different *Grosswetterlagen*. Table 5.11 illustrates the type of results he obtained for maximum temperature data based on a rather variable number of case examples. Similar calculations were performed for minimum temperature, relative humidity and wind speed. Account was taken of sea temperature and other conditions in selecting the cases. The differences relate to an average of five inland stations.

The major factors contributing to the differences between the coast and inland are the different thermal and moisture properties of land and sea and the change in roughness. The effect of coastal friction in setting up patterns of convergence and divergence has been referred to in Chapter 2.A (p. 30). The change of sign of the coastal–inland temperature difference with NW flow (from early to late winter months) (table 5.11A) reflects the development of ice cover on the Baltic in late winter.[1]

[1] The incidence and duration of ice along the Baltic and North Sea coasts of Germany have themselves been related to the frequency of easterly *Grosswetterlagen* (Goedecke, 1955).

Table 5.11 Coastal–inland differences in maximum temperature (°C) with selected Grosswetterlagen *(Maede, 1953a, 1953b)*

A. Seasonal changes – NW type

	Jan.–March	*May–Sept.*	*Nov.–Dec.*
Baltic Sea	−1·4	−3·4	1·5
North coast	−0·9	−3·1	1·8
Northeast coast	−1·0	−2·3	0·9
North Mecklenburg	−0·5	−1·6	1·0
South Mecklenburg	0·0	−0·9	0·4
Potsdam	−0·1	0·1	−0·1

B. Type differences in spring

	Low types (March–May)	*Vb type (Apr.–May)*	*E types (May)*	*W types (March–Apr.)*
Baltic Sea	−4·3	−3·6	−7·8	−5·1
North coast	−2·8	−2·1	−3·0	−2·3
Northeast coast	−1·6	−1·6	−4·0	−0·8
North Mecklenburg	−0·8	−0·8	−2·4	−1·1
South Mecklenburg	−0·4	−0·4	−0·2	−0·5

The differences in minimum temperature for NW flow in November–December are even more striking (+3·7°C for the Baltic and +2·0°C for the North Coast, but declining very rapidly inland).

An exploratory study along similar lines for Britain has been carried out by V. Perry (1969). Table 5.12 summarizes some of her results relating to the Lamb catalogue for January 1952–61. The selected airflow directions are more or less perpendicular to the coastline, although the actual winds may depart considerably from these directions.

Another type of problem which is currently receiving more attention is the nature of urban climatic effects in a synoptic context. Eriksen (1964), for example, has noted the varying intensity of the urban heat island of Kiel under various *Grosswetter* types:

Type	*Town–country difference in minimum temperature in winter*
W	0·7°C
HF	0·7
NE	1·3

Munn, Hirt and Findlay (1969) investigated the interaction of the daytime heat island of Toronto in relation to the lake breeze circulation over Lake Ontario with reference to broad circulation types. Wilmers (1968), however, found that large-scale weather situations were not a satisfactory basis for study of city effects. For a study of Hanover he devised a cruder scheme with four classes: radiation weather type, cyclonic weather type, squall weather type, neutral weather type. These four categories were considered adequate to indicate the general significance of advection, net radiation and vertical motion.

Table 5.12 Mean temperature (°C) at coastal and inland stations in England and Wales at 0900 January 1952–61 with airflow types (V. Perry, 1969)

Station	*Elevation (m)*	*Distance inland (km)*	*Airflow type NW*	*S*	*E*
Dale Fort, Pembroke	26	0	6·4		
Haverfordwest, Pembroke	37	10	3·6		
Swanage, Dorset	4	0		5·1	
Ringwood, Hampshire	76	19		3·1	
Felixstowe, Suffolk	16	0			4·6
East Bergholt	4	24			3·7

The differences are all significant at $<5\%$ by Student's t test.

In an examination of the effect of urban areas on precipitation, Atkinson (1966a, 1966b, 1968) made use of a more detailed classification scheme. He differentiated first between thunder-rainfall in frontal and non-frontal situations. Air mass storms accounted for 62·5% of all thunderstorm outbreaks over southeast England during 1951–60 and five of the eighteen air stream categories showed clear urban effects in summer (May–September). Fig. 5.25 illustrating thunder-rainfall with NW cyclonic flow points to a general maximum on the dipslope of the Chiltern escarpment, related to orographic triggering or accentuation of instability, and a secondary maximum in southeast London induced by the city effect. Of the frontal situations, warm front thunder–rainfall also seems to be particularly sensitive to urban effects (fig. 5.26). Atkinson suggests three possible mechanisms which may contribute to this effect:

1 The increased roughness over the urban area.
2 The role of pollutants in producing additional condensation nuclei.
3 The higher vapour pressure in the city leading to more humid thermals which can therefore reach condensation level at a lower level than in the surrounding countryside.

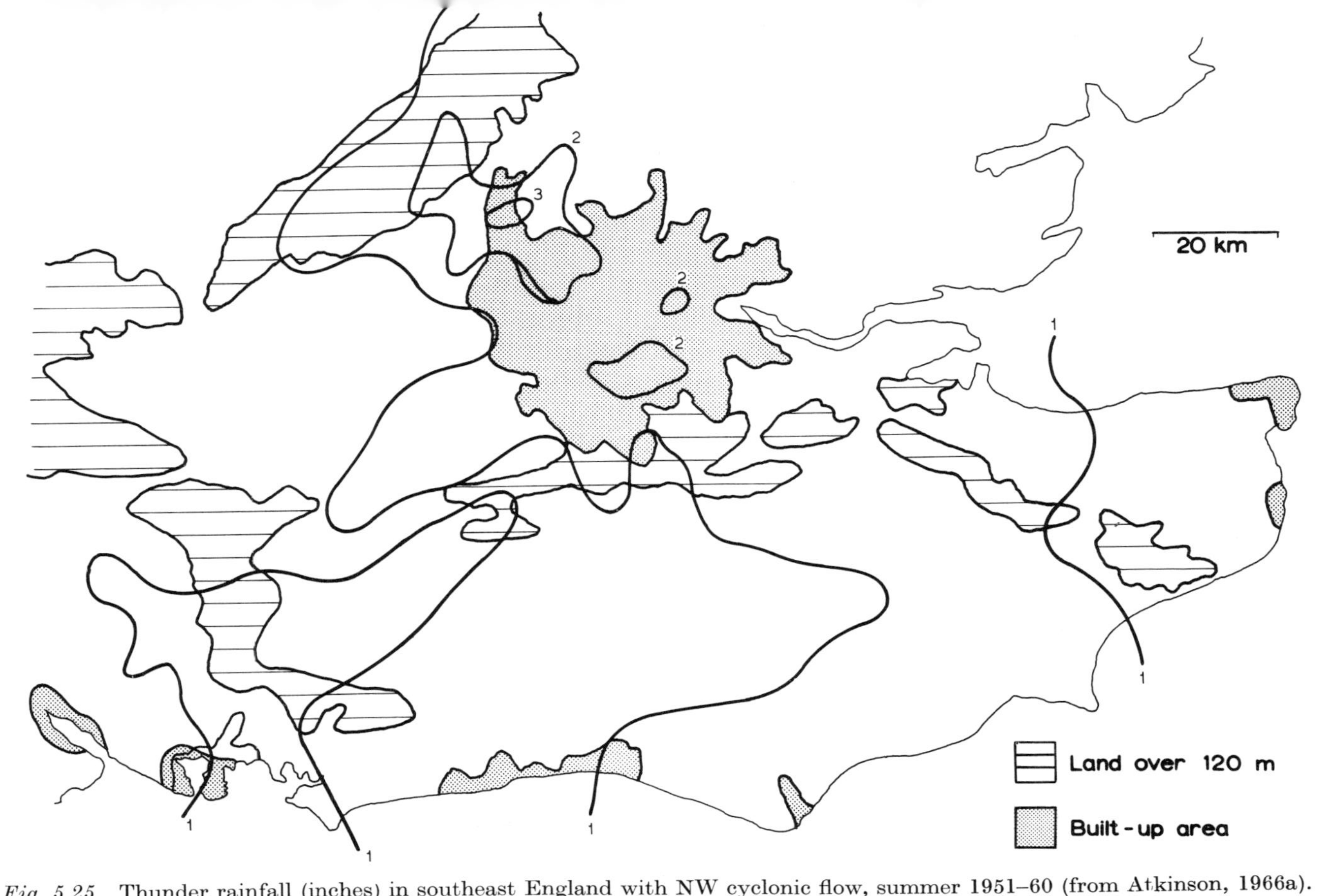

Fig. 5.25 Thunder rainfall (inches) in southeast England with NW cyclonic flow, summer 1951–60 (from Atkinson, 1966a).

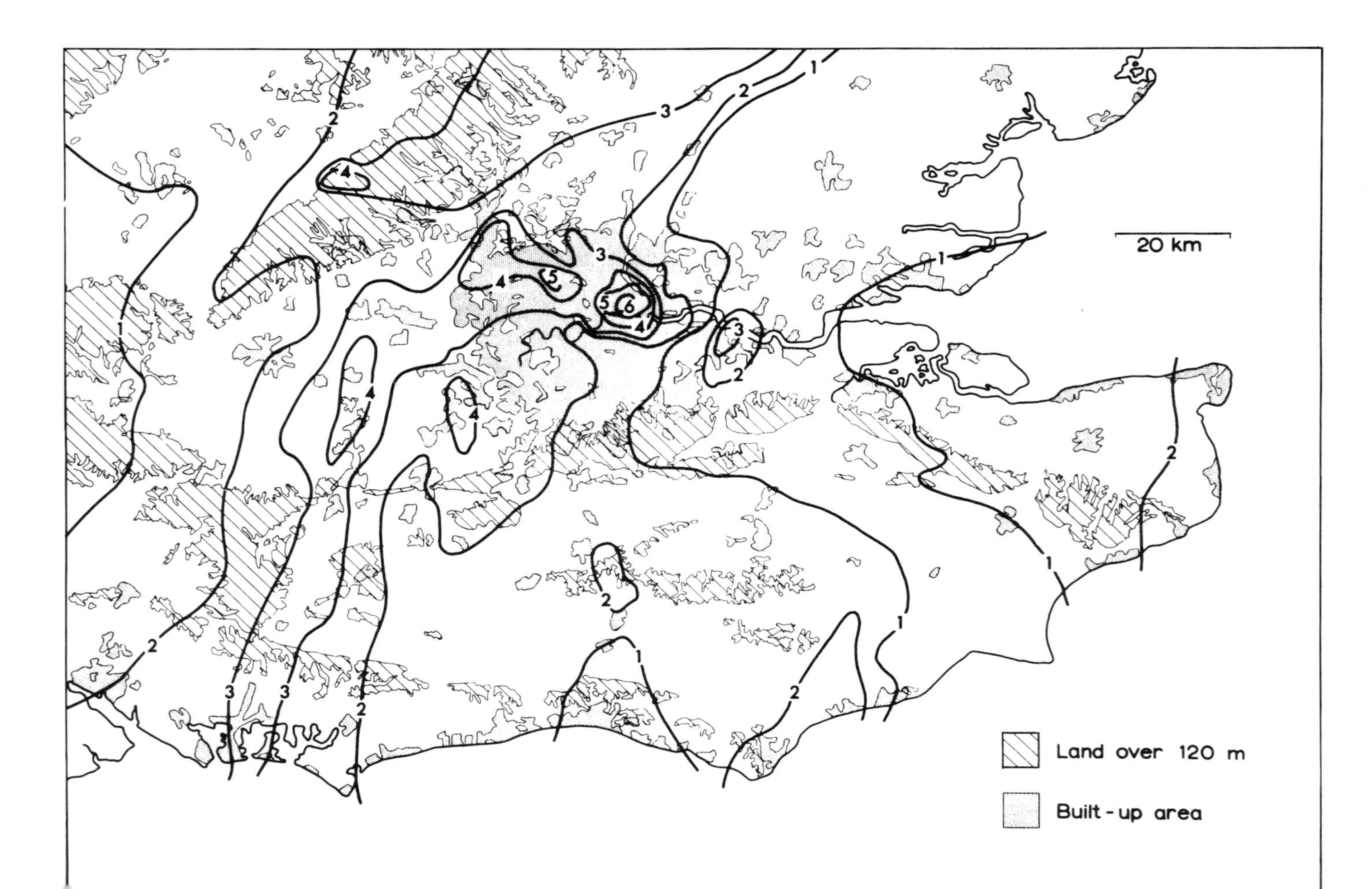
20 km
Land over 120 m
Built - up area

Analysis of the distribution of thunder-rainfall with respect to time of day would probably help to clarify the principal physical mechanism responsible for the observed distributions.

A synoptic climatological approach has also been found to be of value in studies of air pollution. In particular, analysis of pollutant levels over a period of years with respect to a synoptic catalogue facilitates the assessment of trends in relation to any fluctuations in frequency of particular circulation types. Velds (1970) found that daily concentrations of sulphur dioxide in Rotterdam may exceed 300 $\mu g\ m^{-3}$ with the HM, HN_a and S_z *Grosswetterlagen* types. The first type, which is associated with very stable conditions and light winds, is a particular hazard when it persists for five days or more. HN_a type gives rise to north or northeasterly flow over the Netherlands which in winter brings very low temperatures. The concomitant increase in domestic heating appears to be the cause of high SO_2 concentration. The association with S_z (southerly cyclonic) type seems to be 'fortuitous' in that a spell of this type occurred during the very cold month of February 1963. Velds notes that the occurrence of high SO_2 concentrations in winter has decreased over the period 1962–3 to 1967–8, but the frequency of the three types referred to has also declined. Careful analysis is required, therefore, to evaluate the effectiveness of control measures.

Schmidt and Velds (1969) argue that since the frequency of HM type in particular seems to be related to phases of the sunspot cycle it may be possible to predict the general trends of pollution. In Britain, Lawrence (1966) has demonstrated that smoke pollution in London reaches its highest average levels about two years prior to a minimum in the sunspot cycle, apparently in relation to an increased frequency of cold anticyclones with inversion conditions at those times. Further testing of these proposed relationships in other locations would be of interest.

C Synoptic climatology and climatic change

1 GENERAL POSTULATES

Before we look at the various applications of synoptic climatology in climatic change studies it will be helpful to consider the assumptions on which such studies are based. The most general concept is that the present is the key to the past (Mather, 1954, for example). The uniformitarian principle – that processes have operated in the same manner throughout geological time – encounters some difficulties in the case of investigations into atmospheric processes. Certain features of global geography such as ocean currents, vegetation zones, and particularly ice cover, have not remained constant even during the Quaternary era, and Tertiary events certainly involve continental drift. This has been incorporated into a synthetic model of ice ages by Flohn (1969), for example. Many recent modifications of surface conditions are known to have been consequent upon climatic change, although it is possible that some oceanic changes are antecedent. In any event the complexity of air–sea interaction makes it difficult to ascribe primary causes. Any differences in surface characteristics, whether of primary or secondary origin, undoubtedly make reconstructions of past climate more difficult and uncertain.

Even more fundamental is the problem already referred to (p. 317) concerning the existence of a unique set of atmospheric circulation statistics for a given set of boundary conditions. This as yet unsolved problem is obviously crucial for climatic change studies (Lorenz, 1970).

Finally, as far as Pleistocene studies are concerned, it is widely assumed that recent climatic fluctuations are similar in kind to those of the past. Faegri (1950) expressed this idea in the following terms: 'The shorter the duration of a climatic fluctuation, the smaller is the area similarly affected; the longer the cycle, the greater the area within which it is felt in the same way'. This viewpoint has not been universally accepted, however. Simpson (1957, p. 485), for example, held that the major Pleistocene changes reflected temperature trends of the order of 0·1°C per 1000 years, whereas Willett (1949) had earlier suggested that the amplitude of past fluctuations increased with the length of period involved. Faegri stresses that this is an apparent effect caused by the slow response of palaeoclimatological indicators. The lag may vary from 10–10^2 years in the case of vegetation to 10^3–10^5 years in the case of ice caps and ice sheets. Bryson and Wendland (1967) have formalized this concept, as shown in fig. 5.27. It is assumed that the atmosphere

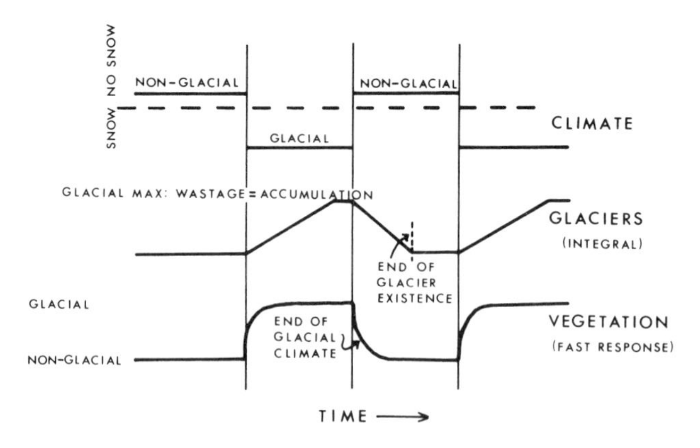

Fig. 5.27 The characteristic response of vegetation and ice sheets to hypothetical step changes in climate (after Bryson and Wendland, 1967).

behaves in a Markovian fashion. There is some evidence to support this contention. Kutzbach *et al.* (1966) have estimated variations in the thermal Rossby number (a measure of the effect of differential meridional heating and the earth's rotation) for the last 500,000 years on the basis of the computed incoming radiation, assuming Milankovitch's (1969) orbital oscillations. The present-day relationship between incoming radiation and thermal Rossby number was used to estimate Pleistocene variations in the thermal wind, and the results showed deviations of about $\pm 2\%$ from present winter values and $\pm 5\%$ from present summer values of the thermal wind. Examination of the shifts between planetary wave numbers and thermal Rossby numbers in the atmosphere produced analogous results to annulus experiments (see p. 317) with lag effects in the circulation response to a change of heating gradient and sudden switches at certain critical thermal Rossby numbers. These sharp transitions lend plausibility to the view that circulation changes represent a Markov process.

2 TYPES OF PROCEDURE

A number of different types of climatic change study can be identified and it is convenient for our purposes to distinguish between hemispheric (or global) and essentially regional investigations.

Hemispheric models

The first attempts to model the circulation patterns during glacial and interglacial periods were based on the index cycle concepts of Rossby and Willett (1948). In essence the interglacials in middle latitudes were regarded as being periods of high index, zonal flow giving mild, damp winters and relatively cool, moist summers, while glacial conditions were thought to be associated with low index patterns which give rise to meridional air movements in higher middle latitudes. Willett (1950) constructed a schematic mean pressure map for the northern hemisphere winter under full glacial conditions along these lines, and similar attempts have been made by other workers (Flohn, 1953).

Further work on the nature of index variations showed that they were considerably more complex than envisaged in the simple high–low model (see Chapter 3.C, p. 169) and these findings led Willett to modify his earlier conclusions. The correspondence of high index with interglacial conditions was maintained but glacial periods were now linked with an expanded circumpolar vortex rather than the extreme cellular type of low index pattern. The westerlies were considered to have remained strong but to have been shifted to lower latitudes (Willett and Sanders, 1959). This kind of approach has been elaborated by Lamb (1964) using hemispheric comparisons relating to the latitudinal extent of snow and ice during 1920–40 and during maximum glaciation. Table 5.13 summarizes his evidence. Lamb infers that the circulation in Ice Age summers would be much more vigorous than that in 'present' summers in the northern hemisphere. The winter Ice Age circulation was probably weaker than now in high latitudes of both hemispheres, but conditions in lower latitudes were probably more stormy than at present.

Table 5.13 *Estimated snow and ice (including sea ice) limits at present and at maximum glaciation and estimated circulation intensity (after Lamb, 1964).*

	SUMMER MINIMUM			WINTER MAXIMUM		
	% of hemisphere covered		*Equivalent latitudinal limit**	*% of hemisphere covered*		*Equivalent latitudinal limit**
1900–c. 1940						
Northern hemisphere	5·5		71°	25		49°
Southern hemisphere	7·7		68°	13·5		60°
Pleistocene maximum glaciation		Difference from now			Difference from now	
Northern hemisphere	16·5	11	57°	34·5	9·5	41°
Southern hemisphere	(21)	13·5	(53°)	24	10·5	50°
Probable circulation intensity at maximum glaciation						
Northern hemisphere	Intermediate between present N. hem. winter and S. hem. summer.			Highest latitudes weaker than now.		
Southern hemisphere	Similar to present N. hem. winter			Lower latitudes – stronger than now, particularly stormy.		

* Denotes latitude corresponding to a circular ice area centred on the pole.

Lamb (1965b) has also outlined five varieties of circulation regime, with particular reference to the British Isles, and suggested periods when they may have been predominant (table 5.14). The wide range of historical, archaeological, botanical and geological data which can provide evidence for such reconstructions is well illustrated for the contrasting conditions of the eleventh and sixteenth centuries in the Aspen Conference discussions (Lamb, 1962; Bryson and Julian, 1962), and also in a survey of the last 1000 years by Le Roy Ladurie (1967, 1971).

More recently new approaches to palaeoclimatic reconstruction have been developed. The first can properly be regarded as deriving from dynamic meteorology (Lamb, Lewis and Woodroffe, 1966; Lamb and Woodroffe, 1970). Nevertheless, these investigations cannot be overlooked here. The procedure is to use palaeobotanical and other estimates of temperature to compute the mean thickness field. From this, areas of cyclogenetic and anticyclogenetic tendencies are specified through Sutcliffe's 'development' equations and then the mean MSL pressure is derived from observed relationships between areas of cyclogenesis or anticyclogenesis and the mean pressure field. Figs. 5.28 and 5.29 illustrate the circulation patterns determined by this means for *c.* 20,000 and 6500 B.C. It is essential to recognize that such computations represent only the circulations occurring with given distributions of glacial ice and *not* the preceding conditions which gave rise to glacierization or deglaciation. Similarly Barry (1973) and Williams *et al* (1973) have recently used a global circulation model to simulate Ice Age conditions.

Table 5.14 Circulation epochs in the east Atlantic since the last Glacial Maximum (after Lamb, 1965b)

Period	*Regime*	*Characteristics*
1959	Anticyclonic	
1930s–1940s	High-latitude Westerly (approx.)	
1920s	Low-latitude Westerly (approx.)	
1550–1700 Neoboreal	Low-latitude Westerly (extreme)	Surface Westerlies S of British Isles; meridional patterns over Britain; dry cold winters, wet summers.
1300–1400	Low-latitude Westerly	
1150–1300	High-latitude Westerly (approx.)	
A.D. 900–450 B.C. Sub-Atlantic	Low-latitude Westerly	Main surface Westerlies across England; mild winters, cool summers, generally moist.
3,000 B.C. {Late Atlantic / Sub-Boreal}	High-latitude Westerly	Surface Westerlies N of British Isles; warm; periods of summer drought.
6,500 B.C. Boreal	Anticyclonic	Circumpolar vortex displaced towards North American sector where ice persisted; Europe warm and dry.
18,000–20,000 B.C. Glacial Maximum		Expanded circumpolar vortex and polar cap displaced towards Atlantic sector.

The second method relates to conditions over the last few centuries. La Marche and Fritts (1971) apply principal components analysis to forty-nine tree ring chronologies from the western United States for 1931–62, and find a useful agreement between the eigenvectors of anomaly patterns of tree growth and the eigenvectors obtained for monthly precipitation by Sellers (1968). Eigenvectors are then determined from the tree ring chronologies for 1700–1930; four eigenvectors

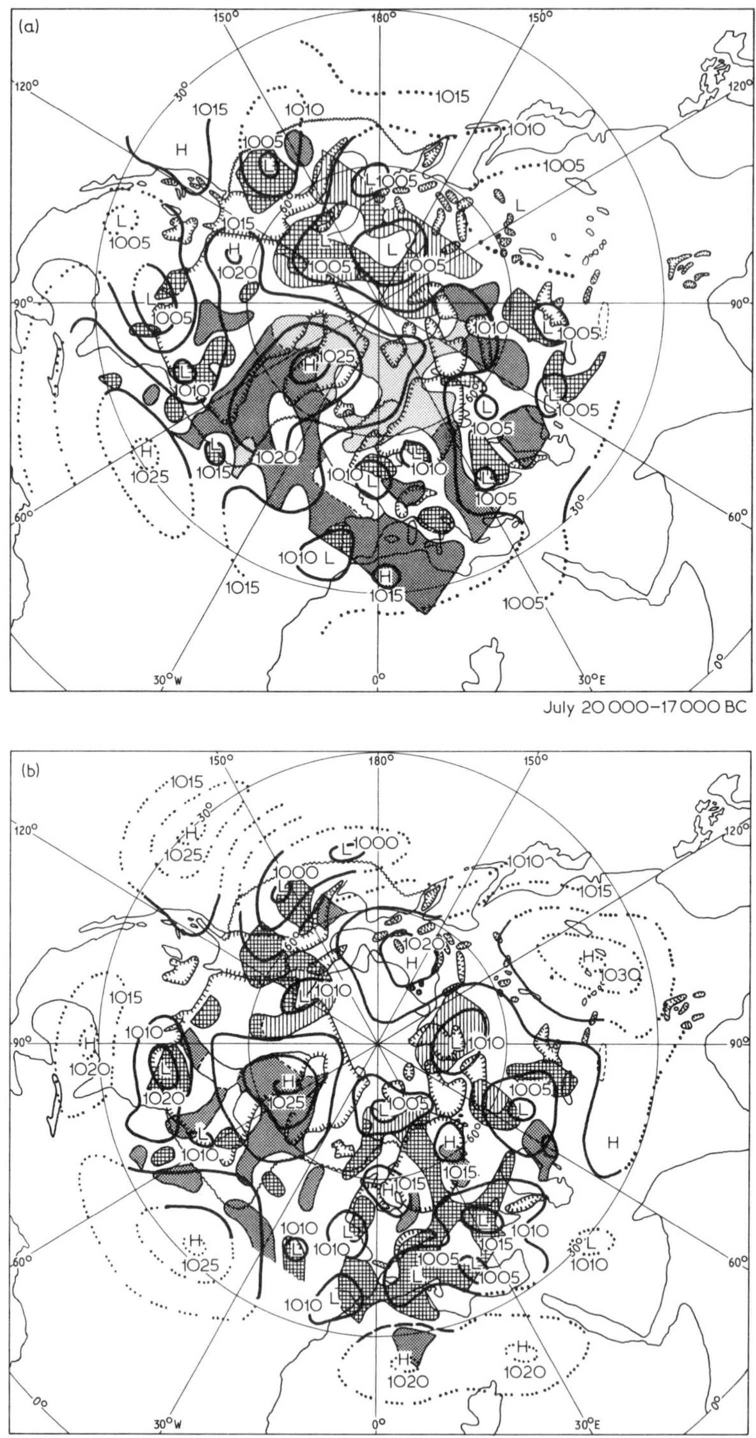
(a)
July 20 000–17 000 BC
(b)
January 20 000–17 000 BC

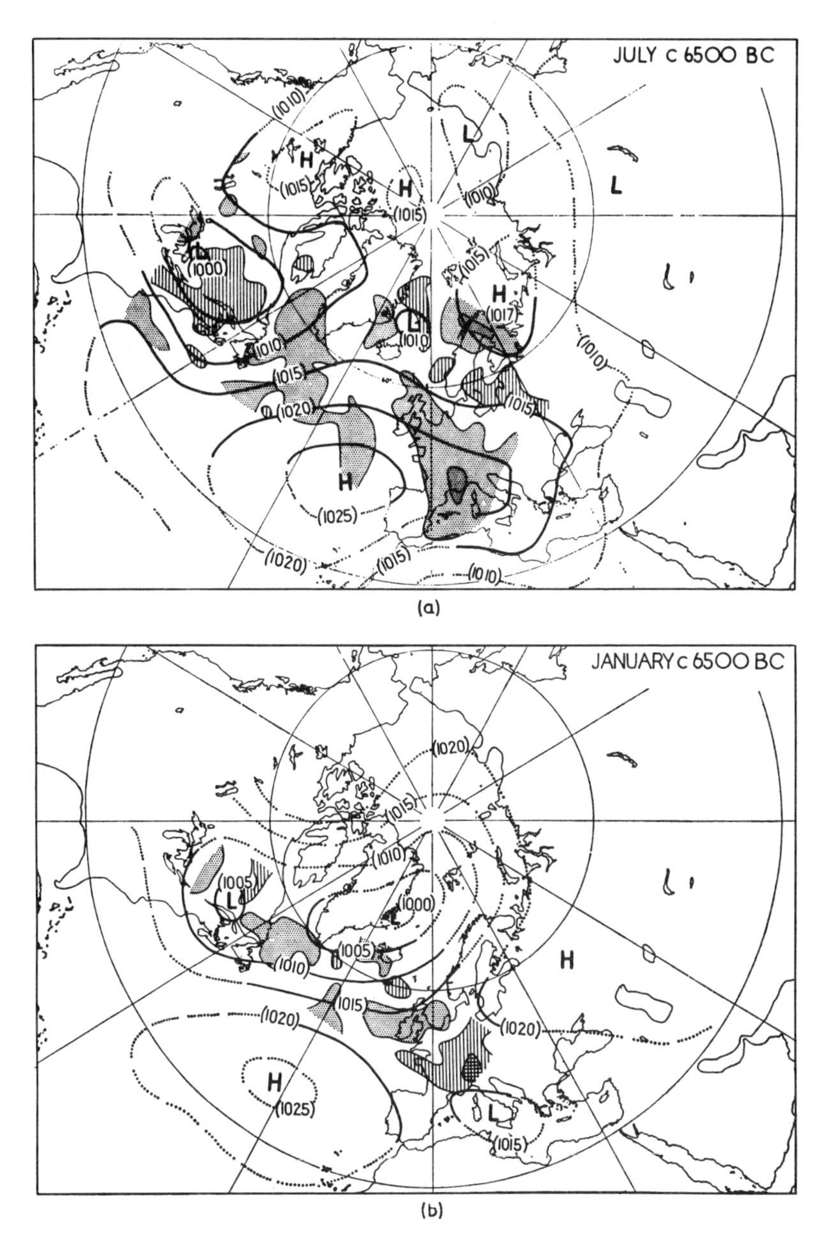

Fig. 5.29 The inferred MSL circulation in the northern hemisphere 6500 B.C. (from Lamb, Lewis and Woodroffe, 1966). (*a*) July, (*b*) January. Key as fig. 5.28.

Fig. 5.28 opposite The inferred MSL circulation in the northern hemisphere 20,000–17,000 B.C. (from Lamb and Woodroffe, 1970). (*a*) July, (*b*) January. Isobars schematic.

Key: Cyclonic development predominant at > 5 of the nearest 9 grid points heavy cross-hatching. Weak cyclonic development and zero divergence at > 5 of the nearest 9 grid points, vertical hatching. Anticyclonic development predominant at > 5 of the nearest 9 grid points bold stipple. Weak anticyclonic development and zero divergence at > 5 grid points, light shading. The grid point spacing was about 300 km.

account for 45% of the total variance of tree growth anomaly patterns and for 36% of that in the period 1931–62. They conclude therefore that the major anomaly patterns of precipitation represented in the last three decades have been occurring for at least 260 years. Fritts *et al.* (1971) have attempted to extend this method with the same chronologies to estimate anomalies of seasonal pressure over the western sector of the northern hemisphere for each year of the period 1700–1899. Rather variable results are obtained from tests on data for 1900–30, 1945–62, but the 'prediction' is best (> 30% of the variance is accounted for) over the tropical oceans in all seasons, over western North America and north-central Canada in summer, and over the Aleutian Islands and northern Labrador in winter. With refinement these techniques should provide useful long-term records of the annual variability of circulation patterns.[1]

A number of recent quasi-hemispheric studies have been concerned with the question of the role of the oceans in climatic change (Rasool and Hogan, 1969). Particular attention has been devoted to problems of large-scale interaction and teleconnections in the Pacific Ocean. Bjerknes (1966, 1969) shows that higher ocean temperatures in the central and western equatorial Pacific, resulting from a weakening of the wind-driven oceanic circulation, induce higher equatorial cloudiness and precipitation. The accompanying heat input to the ascending branch of the atmospheric Hadley circulation will, he argues, intensify that circulation and lead to an increase in the mid-latitude westerlies through the augmented poleward transport of angular momentum. Positive anomalies of equatorial ocean temperature during the winters of 1957–8, 1963–4 and 1965–6 were associated with stronger westerlies over the North Pacific, for example. It is also noteworthy that Charney (1969) has demonstrated the instability of the intertropical convergence zone. Small fluctuations in ocean temperature could, therefore, drastically modify the equatorial circulation and, in turn, that in middle latitudes of both hemispheres. Namias (1970) postulates that positive anomalies of sea surface temperature over much of the North Pacific Ocean in autumn and winter almost certainly help to amplify the atmospheric long-wave pattern. This seems to have occurred during the last decade. An intensified ridge and trough pattern downstream over North America led to frequent outbreaks of arctic air in the eastern half of the continent, resulting in average winter temperatures 0·5° to 2°C below the 1931–60 normals. The possible implications of such studies and of related work on the Southern Oscillation and the El Niño phenomenon (see section D of this chapter, p. 417), including cross-spectral analyses

[1] A similar approach to numerical estimates of palaeoclimate has been made from pollen spectra (Webb and Bryson, 1972) and from plankton records in deep sea cores (Imbrie and Kipp, 1971).

of precipitation and sea temperatures in the tropical Pacific (Doberitz, 1968), have so far been little considered in terms of past climatic fluctuations, although Quinn (1971) illustrates their possible significance in the Pleistocene. Geological evidence points to a persistent dry zone extending across the equatorial Pacific during the glacials. At the present this generally reaches only from South America to 180° longitude. Quinn argues that lower sea levels exposed more land in the Malayan–Indonesian area, excluding cool waters deriving from equatorial upwelling and the South Equatorial Current. This caused the Southern Oscillation to remain in its high index phase, with an Indonesian low and a well-developed subtropical anticyclone in the South Pacific, giving rise to a strong oceanic circulation with vigorous upwelling in the Peru Current and along the equatorial zone, thereby intensifying the dry zone. Development of many of these ideas will depend on global ocean–atmosphere models of the type being developed by Manabe and Bryan (1969). Already they have been able to demonstrate that an ice age would develop if oceanic poleward heat transport ceased. Refinement of such models will undoubtedly clarify the importance of oceanic conditions and anomalies for the atmospheric circulation.

On the shorter time scale Kraus (1956) finds evidence of a probable tropical–middle latitudes link during the period 1881–1940. He postulates that the strength of the standing meridional (monsoonal) circulation over Asia varies inversely with that of the zonal westerlies. Prior to 1850 the circulation in the northern hemisphere was more zonal. Subsequently this decreased with greater development of summer lows and winter highs over the continents. Kraus (1956, 1958) has also suggested a weakening of the trades and an associated reduction of evaporation and decrease in precipitation in low latitudes around the turn of the century, but this is not borne out by the circulation indices subsequently determined for the North Atlantic trades (Lamb and Johnson, 1966). On this evidence, changes in cloud amount or other factors must have been responsible.

It will be readily appreciated, following the discussion in Chapter 3.C (pp. 167–72), that longitudinal variations cannot be overlooked. Sawyer (1966) affirms that 'we should not look for interpretations (of climatic change) in terms of the poleward or equatorward shift of the subtropical highs and belts of westerlies, etc., but rather in terms of shifts in position and changes in amplitude of the favoured forms of the waves in the westerlies of the temperate belt'.

Thorough analyses of changes in the intensity of the zonal flow and of associated shifts in trough and ridge positions during the last one and two centuries, respectively, have been carried out for the southern and northern hemisphere by Lamb and Johnson (1959, 1961, 1966). Reconstructed decadal mean charts back to the 1760s have been prepared

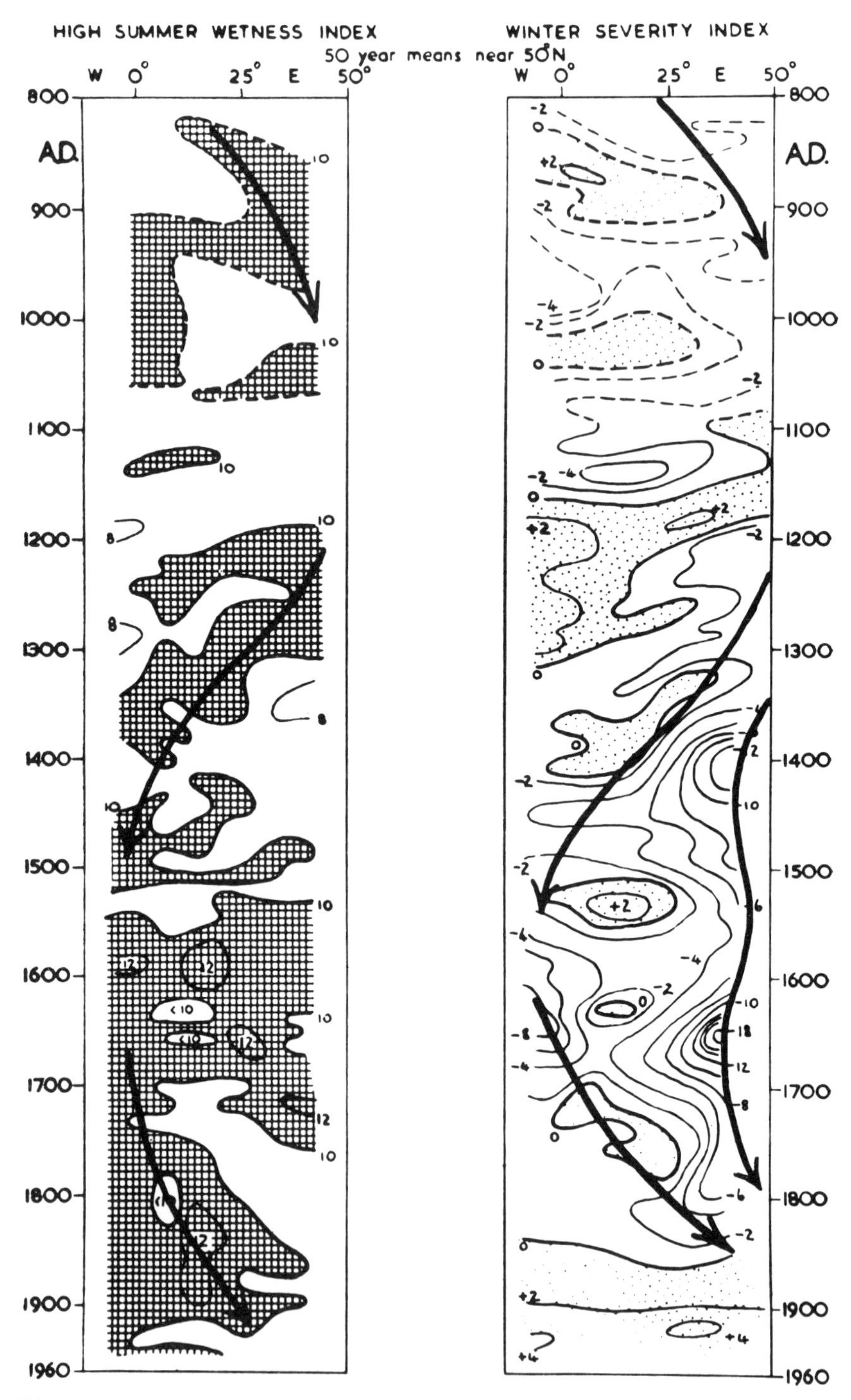

Fig. 5.30 Fifty year mean values of summer wetness and winter mildness/severity indices for different longitudes in Europe near 50°N since A.D. 800. Hatched (dark) areas had excess of wet over dry Julys and Augusts. Stippled (grey) areas had more very mild than very cold winter months (Decembers, Januarys and Februarys). Arrows have been drawn to make it easier for the eye to follow the movement of the maximum incidence of anomalies (from Lamb, 1963).

for the North Atlantic–Europe sector.[1] Historical records of summer wetness and winter severity have also been utilized to reconstruct longer term trends in Europe. The results (see fig. 5.30) indicate that the climatic deterioriation from about 1300–1600 reflects a reduction in wavelength with weaker westerlies and their displacement equatorward. The improvement in climate after 1700 was associated with opposite tendencies.

Regional models

Regional investigations seeking to use present synoptic patterns as a guide to past events generally make the assumption, at least tacitly, that a change in frequency of certain types of pattern would favour a particular climatic mode. We shall examine the validity of this in terms of recent climatic fluctuations on pp. 374–7. For the present we might note Dzerdzeevski's statement (1963, p. 293) that 'all changes in temperature and precipitation . . . are connected with changes in frequency and duration of large-scale circulation patterns', and use this as a working hypothesis.

A number of studies have focused on the synoptic climatology of present extreme climatic conditions (cold summers, snowy winters, etc.); others have looked at unusual pressure fields and then determined the related weather patterns. One of the first major contributions was of this second type. Rex (1950) analysed blocking anticyclones over Scandinavia and showed that precipitation is much below normal over most of Europe during such episodes in summer and in winter (fig. 5.31). He concluded that such patterns would inhibit glacierization. An investigation carried out by Namias (1957) of the circulation conditions accompanying warm summers and cold winters in Sweden generally supports Rex's views. Namias shows that warm summers and *interdependent* cold winters occur during weakened zonal flow. Such extreme continentality would certainly interdict a glacial regime. However, it is important to note that Rex found Scandinavian blocking to have little effect on the 500 mb zonal index, since the upstream westerlies are then stronger than normal. Scandinavian blocking must be characterized as an essentially high index type in contradistinction to the so-called 'jump type' which occurs in lower middle latitudes (Berggren, Bolin and Rossby, 1949). From a study of the winter climate of Japan, Tsuchiya (1963, 1964) suggests that a stationary hemispheric blocking pattern, with blocks over the North Atlantic and Alaska, would facilitate continental glaciation. This circulation produces below normal temperatures and above average precipitation over eastern North America, but Tsuchiya concedes that the evidence is not decisive for

[1] Attempts are also being made to prepare synoptic maps for the eighteenth century (Kington, 1970; Oliver and Kington, 1970).

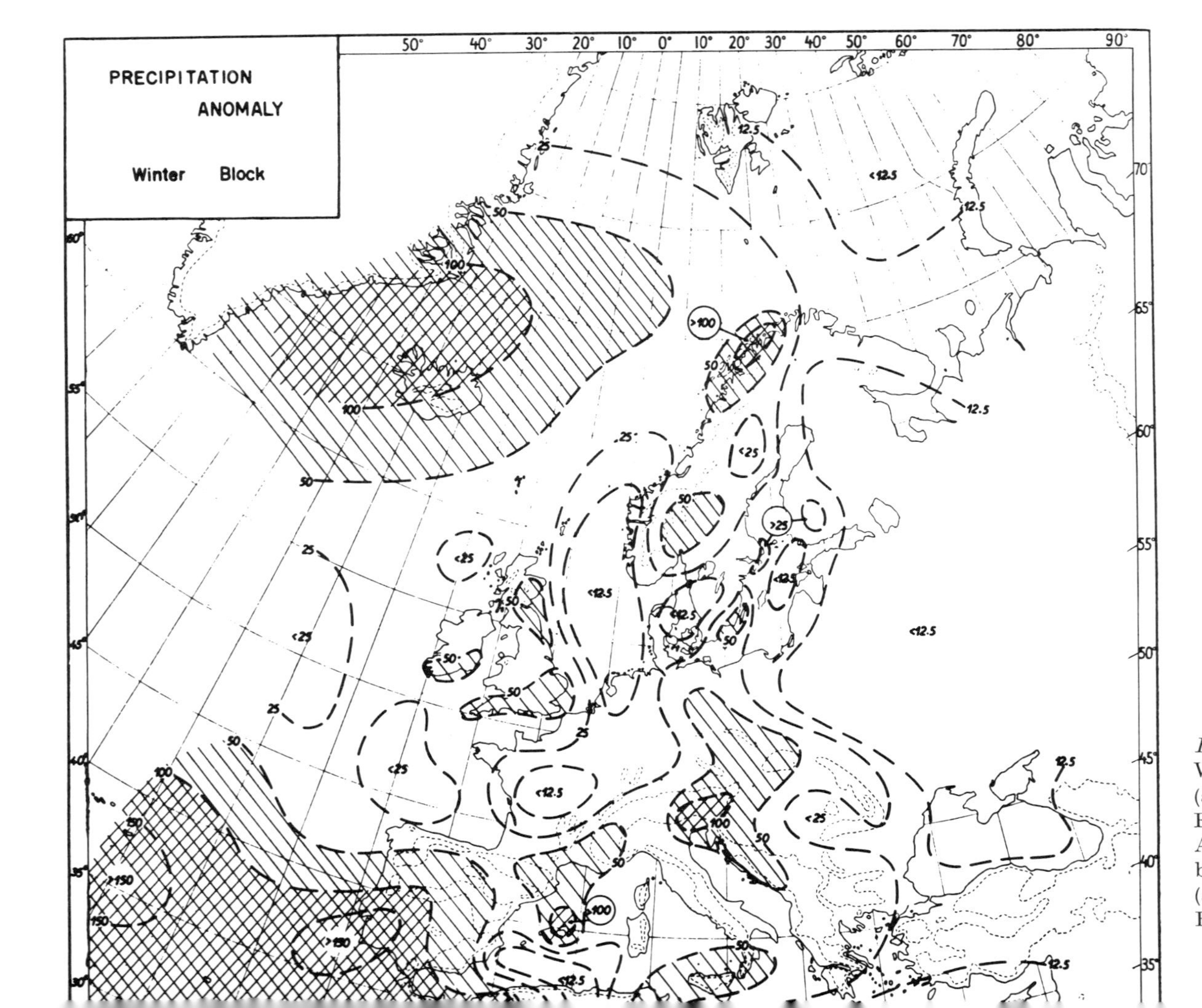

Fig. 5.31
Winter precipitation (as per cent of normal) over Europe and the eastern North Atlantic during Scandinavian blocking situations (after Rex, 1950; from Barry and Chorley, 1971).

Scandinavia. As Flohn (1952) has emphasized, the location of blocking patterns is all important.

The difficulty of reconciling reconstructed regional and hemispheric circulation models is a familiar one, as the work of Butzer (1957a, 1958) demonstrates. For the Mediterranean and Near East, Butzer has attempted to synthesize the geomorphological and archaeological evidence and to build a picture of the changes in regional climate which can be related to global circulation models. He suggested (Butzer, 1957b) that the zonal (high index) circulation regime of about 1920–50 over Europe and the North Atlantic may provide a model for inter-pluvial conditions in the subtropics, and subsequently (1961, 1971) elaborated his ideas on the types of climatic phase and circulation regime for the Upper Pleistocene as follows:

Phase	*Europe*	*Mediterranean*	*Postulated changes in circulation (see table 5.15)*
Dry Interglacial	As today	As today (Interpluvial)	As today
Moist Interglacial	Warmer	Moister, warmer (Subpluvial)	Moist S. meridional
Early Glacial	Cooler, moist	Moister, cooler (Pluvial)	N. and warm meridional
Full Glacial	Colder, drier	Drier, colder (Interpluvial)	Warm meridional Zonal (?)
Late Glacial	Cooler, drier	Drier, cooler (Interpluvial)	Zonal, S. meridional

Recognition of the need to distinguish south and north meridional patterns was made by Wallén (1953) in studying summer conditions in Sweden. The two types which are mainly associated, respectively, with southwesterly and northwesterly steering, reflect the location of a split in the main jet stream current. There is also a warm meridional pattern related to anticyclonic *Grosswetterlage* situations. The circulation–weather relationships that he envisages, together with those established for the Mediterranean by Butzer, are summarized in table 5.15.

In a more limited study for Labrador–Ungava, Barry (1960, 1966) looked at the circulation and weather conditions of winters 1956–7 and 1957–8 as a guide to the relative significance of westerly depressions and airflow, and of easterly airflow, with respect to vapour flux-convergence and snowfall amounts in winter. The overall conclusions were that no single circulation type is the sole determinant of large flux-convergence values in the south-east, but in the central and northern parts of the peninsula high (hemispheric) index situations result in little precipitation. In these areas spells of cyclonic easterly flow with generally low index patterns give rise to many of the heavier snow falls. Such patterns were postulated as being of considerable importance for glacierization. Derbyshire (1969, 1971) has adopted a somewhat similar approach in a study of the possible circulation patterns during the inception phase and maximum glaciation of Tasmania. Seven circulation patterns and their associated precipitation and temperature conditions at the present day are considered. He suggests that a disturbed westerly type, which gives abundant precipitation in the mountains of western Tasmania, would be important for glacierization.

Table 5.15 Circulation type – weather relationships in Europe and the Mediterranean (after Wallén, 1953, and Butzer, 1961, 1971).

Circulation type	Grosswetterlage	*Weather in* North and central Europe	*Weather in* Mediterranean (in winter)*
Zonal	W, NW, SW, BM	Moist; mild winters, cool summers	Dry, warm
South meridional	Ww, SE, S, TrM, TrW, TM, TB	Moist; mild winters	Moist, warm
Warm meridional	HF, HNF, NE	Clear, cool winters; warm summers	Cool
North meridional	N, HN, HB, HM	Cool; snowy winters	Relatively moist, cool

* Summer conditions in the Mediterranean are of less palaeoclimatological significance.

The feasibility of relating glacier variations to frequencies of particular circulation patterns has been demonstrated by Hoinkes (1968) for glaciers in Switzerland between 1891 and 1965. He evaluated frequencies of *Grosswetterlagen* patterns favourable and unfavourable to glacier mass budgets by considering the seasonal weather characteristics of each type. Favourable conditions were considered to be oceanic weather types in winter, cold spells in spring, and 'monsoon' type weather in summer; and unfavourable conditions were considered to be anticyclonic weather in summer, and fine weather in September. Fig. 5.32

Fig. 5.32 opposite (*a*) An index of favourable minus unfavourable *Grosswetterlagen* for glacierization.
(*b*) Glacier response in the Alps (from Hoinkes, 1968).
W1 = Ocean type winter weather
F4 = Cold spells in spring
S6 = Monsoon type summer weather
S5 = Anticyclonic summer weather
S7 = Fine weather in September.
The heavy line in (*a*) shows five-year running means, (*b*) shows the percentage frequency of advancing, retreating and stationary glaciers as well as the number of glaciers under observation.

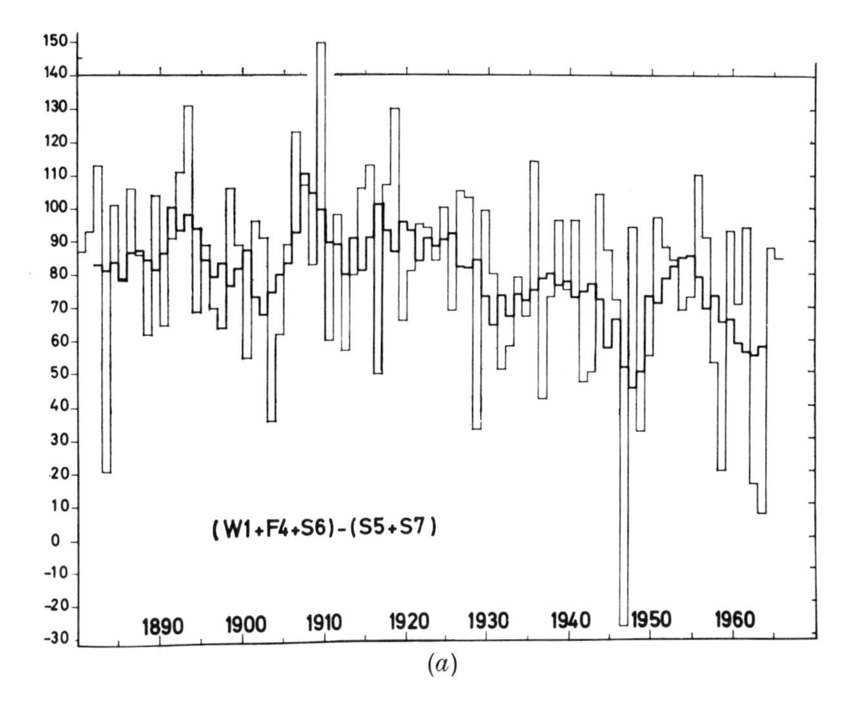

150
140
130
120
110
100
90
80
70
60
50
40
30
20
10
0
-10
-20
-30
(W1+F4+S6)-(S5+S7)
1890
1900
1910
1920
1930
1940
1950
1960

(a)

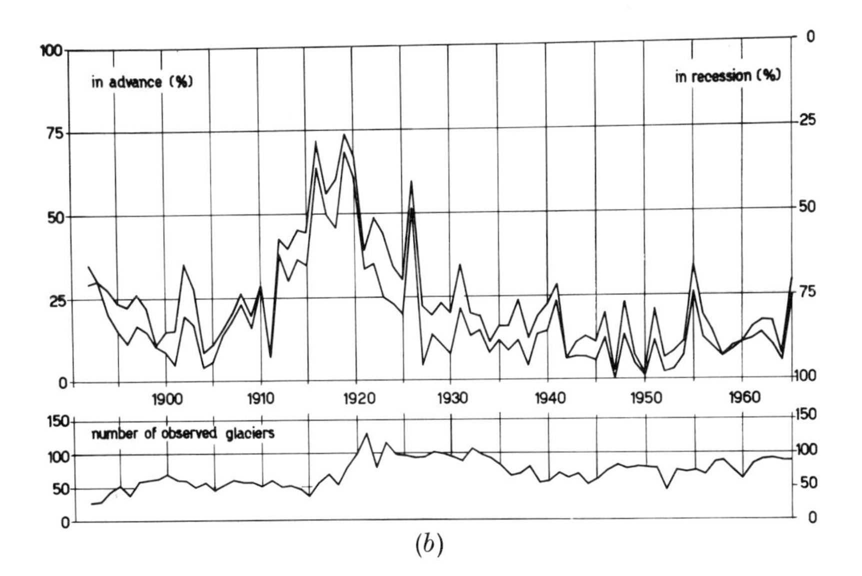

in advance (%)
in recession (%)
100
75
50
25
0
0
25
50
75
100
1900
1910
1920
1930
1940
1950
1960
number of observed glaciers
150
100
50
0
150
100
50
0

(b)

shows the simple combination of the number of favourable minus unfavourable days averaged by five-year running means and the relative frequency of advancing and retreating glaciers. The overall agreement is encouraging. Hoinkes suggests that the relationship might be improved if some weighting of different types were incorporated. The duration of spells of particular types – during the ablation season, for instance – will be important. The degree of agreement is also limited by the various possible combinations of ablation and accumulation conditions and the glacier volume. Increased winter precipitation from 1906 to 1920 resulted in advances, but maximum winter precipitation around 1950 was more than counterbalanced by sunny, dry summers and a lengthened ablation season.

The procedure of considering the circulation patterns related to extreme conditions at the present time has been used by Brinkmann and Barry (1971) to assess those that might have favoured glacierization in Keewatin, Northwest Territories – an area thought to have been an ice centre for Pleistocene glaciations. Composite maps were constructed of 700 mb height anomalies for six snowy winter months and six cool summer months in Keewatin and concurrently in Labrador–Ungava. These maps and consideration of synoptic patterns within the selected months suggest that, at least at present, high winter precipitation receipts occur with a rather weaker than normal 700 mb trough (over eastern Canada) displaced west of its mean position. This reflects a hemispheric low index type of circulation. More frequent southeasterly flow components are important in both areas, in line with the earlier suggestions of Barry referred to above. Cool summer conditions in both areas are associated with a deep 700 mb trough over eastern Canada although such a pattern gives generally dry conditions in Keewatin and wet in Labrador–Ungava. It must of course be recognized that other combinations of hemispheric and regional circulations which occur rarely, if at all, at the present time, *may* also give rise to similar climatic patterns.

A different approach to palaeoclimatic reconstruction based on air masses and frontal zones has been developed by Bryson and Wendland (1967). The apparent relationship between the summer and winter locations of the arctic front over central and western North America and the northern and southern limits, respectively, of the boreal forest at the present day (Bryson, 1966; cf. fig. 2.29) are assumed to hold for the last glaciation and postglacial time. Thus, it is postulated that the relationships between air mass characteristics and the flora remain essentially constant and that the air mass properties themselves are unchanged. If these assumptions hold, then from a reconstruction of the vegetation zones based on pollen data it is feasible to reconstruct the positions of the frontal boundary at different periods in the past.

3 STUDIES OF RECENT FLUCTUATIONS

If we accept the postulate, stemming from Faegri's ideas, that the Pleistocene climatic variations were similar in kind to those of the historical period, but simply persisted much longer, then it is essential to examine these more recent changes thoroughly. As we have already noted, many attempts to do this are continuing to be made. The period of instrumental records, covering about 100 years for much of the northern hemisphere, provides a somewhat more reliable source of evidence and, since it also spans a by no means negligible temperature fluctuation in middle and high latitudes, serves as a more direct guide to possible past conditions. This approach is well illustrated by Mather (1954) who examined changes in the northern hemisphere centres of action.

More specifically synoptic climatological investigations have used the daily catalogues of airflow or circulation types which are now available for several parts of the world. These span many decades and provide a useful basis for examining recent climatic fluctuations. The earliest date for which data exist for some of the major catalogues is as follows:

Hess and Brezowsky (1969):	Central Europe	1881
Lamb (1971):	British Isles	1861
Peczely (1957):	Hungary	1877
Vangengeim (Bolotinskaia and Ryzakov, 1964):	Hemispheric sectors	1891
Dzerdzeevski (1968b, 1970):	Hemispheric sectors	1899

Hess and Brezowsky have calculated decadal frequencies of the major circulation forms (zonal, meridional and mixed) and of the main *Grosswetter* categories for 1881–1966 (see table 3.5 and fig. 5.33). Earlier, Wege (1961) compared frequency changes of the *Grosswetter* types for each month between the periods 1901–30 and 1931–60. Such studies represent a useful start, but they may blur some important changes. Lamb (1965a) used ten-year running means to depict the variation of the westerly type over the British Isles thereby emphasizing the broad trends (see fig. 5.36) and also compared type frequencies for three successive twenty-five-year periods. His findings are discussed below. When arbitrary time intervals of this type are selected there is a risk that differences in different phases of the various sunspot cycles may influence the results.

Dzerdzeevski (1969) has used his circulation type catalogue to identify three circulation epochs during the first half of this century. Similar analysis has been attempted by other workers, and the results are compared in table 5.16. On the basis of a modification of Vangengeim's classification, Girs (1966a) distinguished several epochs of ten to thirty

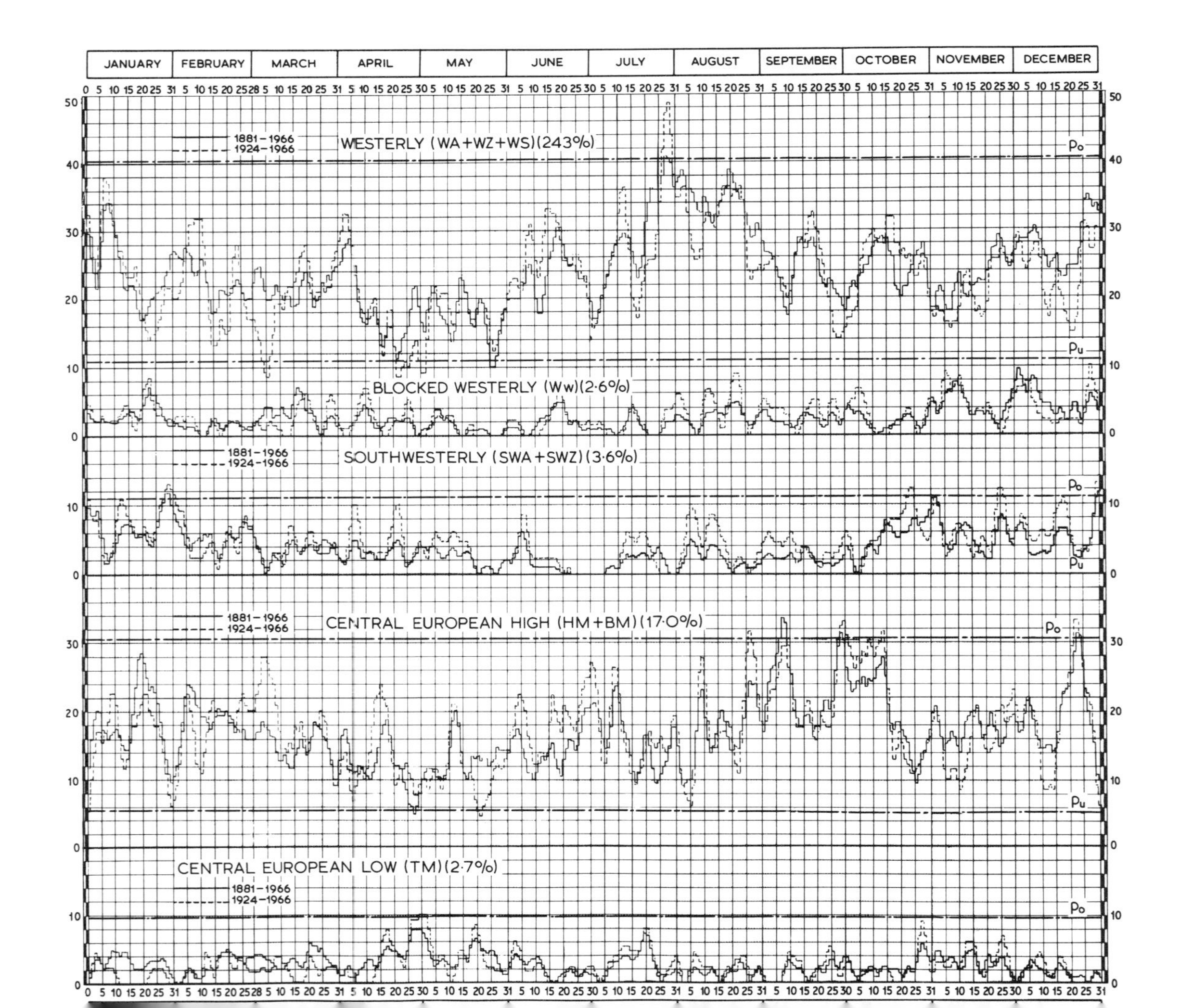
JANUARY
FEBRUARY
MARCH
APRIL
MAY
JUNE
JULY
AUGUST
SEPTEMBER
OCTOBER
NOVEMBER
DECEMBER
1881–1966
1924–1966
WESTERLY (WA+WZ+WS)(243%)
Po
Pu
BLOCKED WESTERLY (Ww)(2·6%)
SOUTHWESTERLY (SWA+SWZ)(3·6%)
CENTRAL EUROPEAN HIGH (HM+BM)(17·0%)
CENTRAL EUROPEAN LOW (TM)(2·7%)

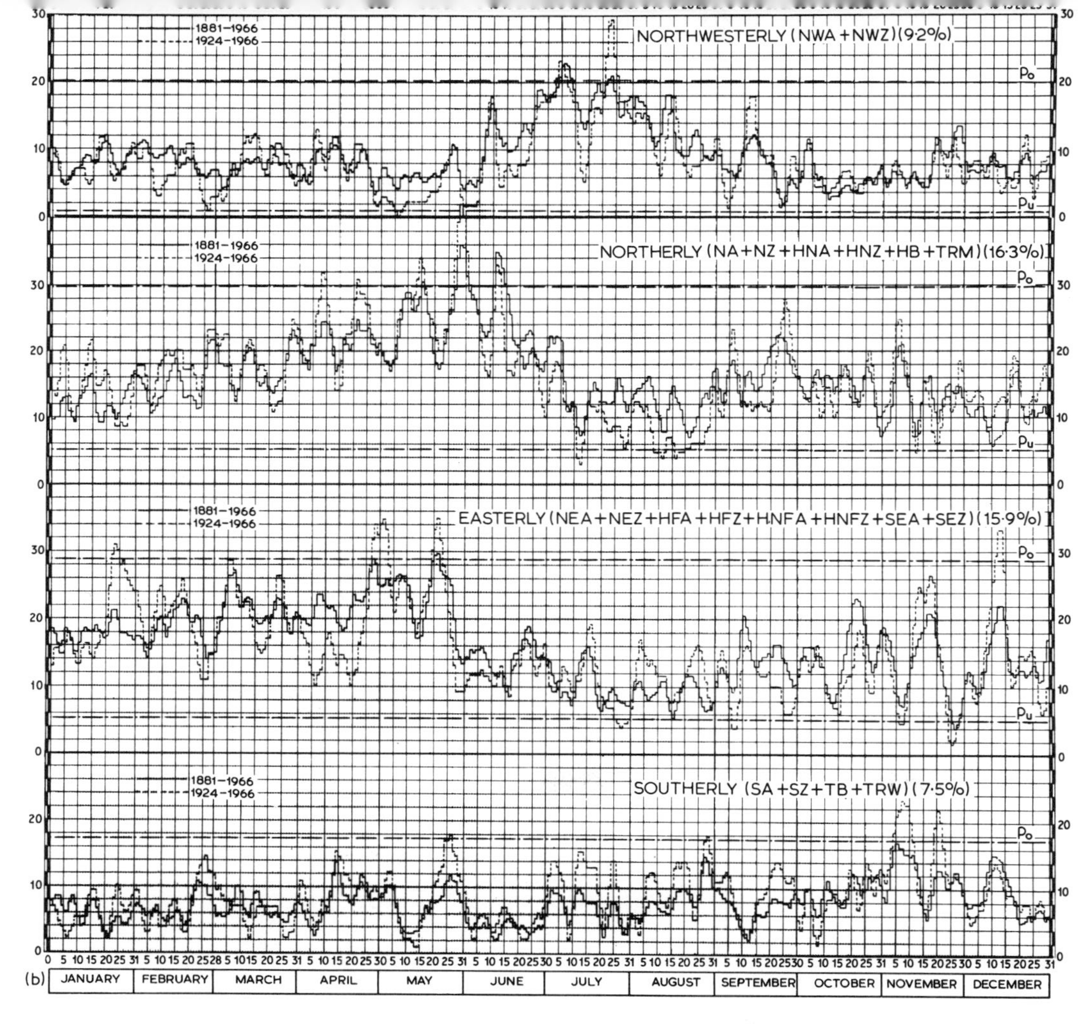

Fig. 5.33
Decadal percentage frequencies of major *Grosswetter* groups, 1881–1966
(from Hess and Brezowsky, 1969).

years duration which he considers may be linked to long-term variations of sunspot activity. Greater solar activity appears to be associated with an increased frequency of meridional types and blocking. According to Dzerdzeevski (1968a) shorter fluctuations occur within the broad trends of zonal and meridional types. The annual frequency curve (fig. 5.34) has a 'saw-tooth' character with the amplitude of the deviations increasing and decreasing several times, suggesting resonance or auto-oscillations of the basic macroprocesses. The maxima on the curves of meridional circulation are on average 2·3 years apart and the minima 2·7 years, while for the zonal circulation the corresponding figures are 3·3 and 3·0 years giving a mean value for a complete cycle of thirty-one months.

A more objective definition of circulation epochs has been provided by Kutzbach (1970) on the basis of time series analysis of eigenvectors of hemispheric pressure anomaly maps for January and July. Table 5.16 shows that the results are in line with other determinations, though less

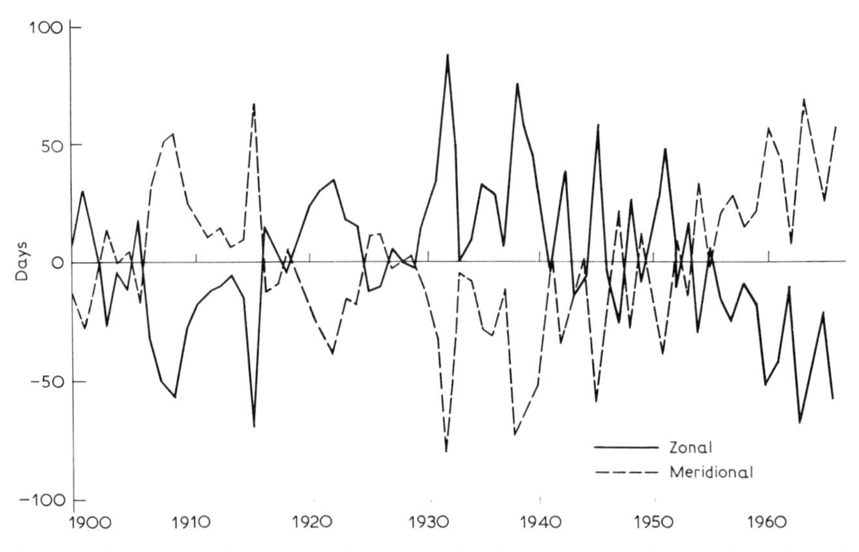

Fig. 5.34 Annual frequency of zonal (solid line) and meridional (dashed line) hemispheric circulation types expressed as departures from the mean, 1900–65 (after Dzerdzeevski, 1968a).

Fig. 5.35 opposite Time series of the coefficients associated with the first three eigenvectors (C_1, C_2, C_3) for (*a*) January, (*b*) July, of the anomalies of MSL pressure with respect to the mean (from Kutzbach, 1970). Units are in mb. The eigenvectors must be multiplied by the coefficients to obtain spatial departure patterns in mb. The dotted line represents the filtered series. These eigenvectors reflect primarily changes in the major centres of action.

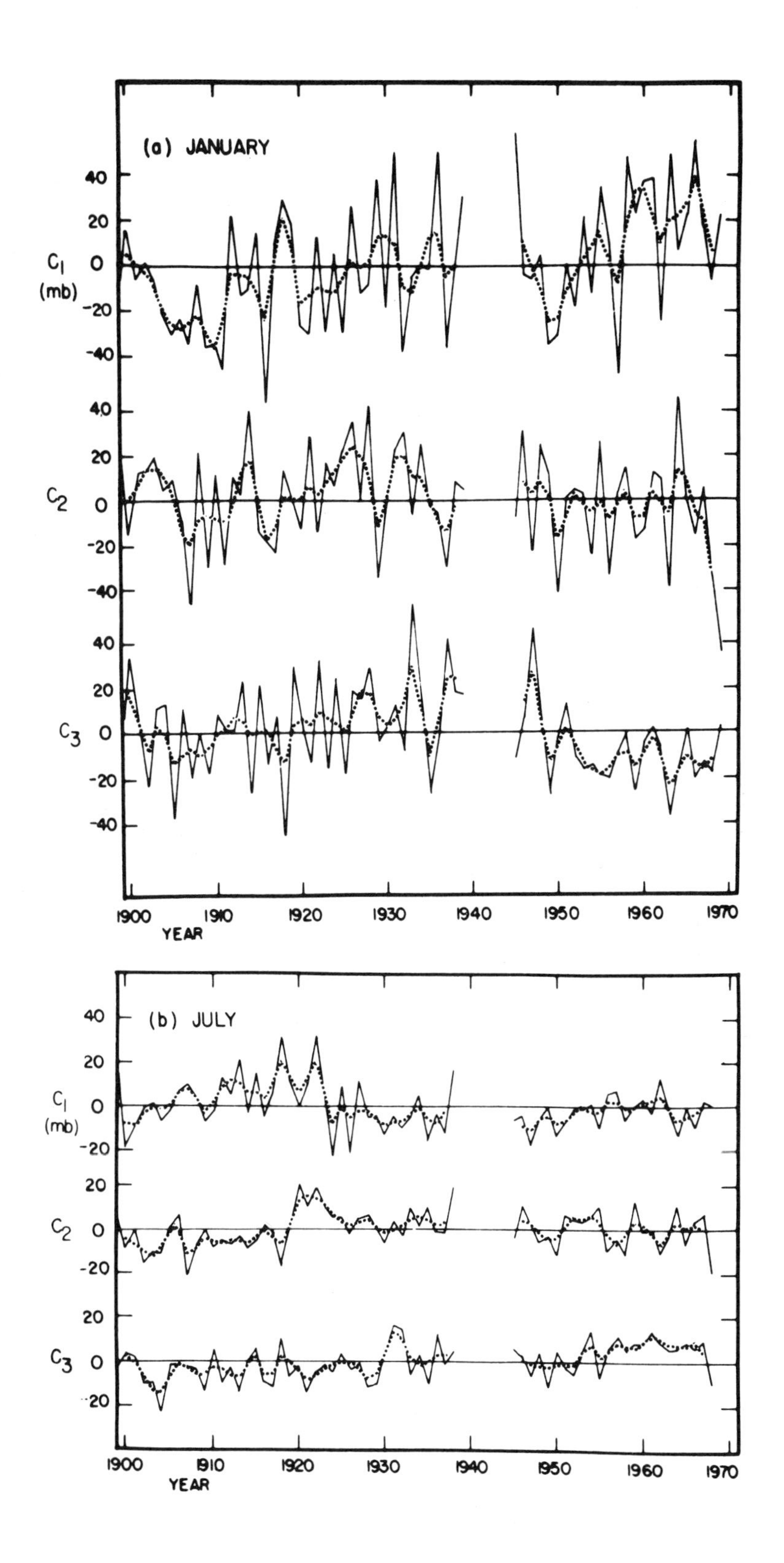

(a) JANUARY
C_1 (mb)
C_2
C_3
40
20
0
-20
-40
1900
1910
1920
1930
1940
1950
1960
1970
YEAR
(b) JULY

detailed than the Russian work. Undoubtedly the internal coherence of these epochs is rather variable, as is apparent from fig. 5.35.

Table 5.16 Circulation epochs since the late nineteenth century

Lamb and Johnson (1966)	*Dzerdzeevski (1963, 1969)*	*Girs (1966a, b)*	*Kutzbach (1970)*
Late 19th century		1891–99 W+C	
to *c.* 1940. Zonal	1900–late 1920s: meridional, continental and cool	1900–28 W	1899–early to mid 1920s
		1929–39 E	Mid-1920s–early to mid-1950s
	Late 1920s–late 1950s: zonal	1940–48 C	
Post-1940 low zonal index	Post-late 1950s: greater meridionality	1949–60 E+C	Post mid-1950s to present

The centres of atmospheric activity in the northern hemisphere undergo substantial changes of position and intensity upon changes of circulation epoch and these have been described for the Icelandic low (Abramov, 1966), the Azores high (Maksimov and Karklin, 1970) and the Siberian high (Maksimov and Karklin, 1969). The Azores high and the Icelandic low both weakened by 5–6 mb from the 1920s to the late 1960s, although the pressure gradient between them has also been affected by a southward shift of the Icelandic low. In addition the latter has been subject to a westward displacement whilst the Azores high has been displaced eastward. In the case of the Siberian winter anticyclone no long-term changes in position can be detected during the present century. However, changes in intensity appear to have a predominant two to three year period. In this time range the occurrence of any periodicity may be difficult to separate from the effects of the quasi-biennial oscillation (see p. 398) on the centres of action (Angell, Korshover and Cotten, 1969), and of the fourteen-month oscillation thought to reflect the influence of the earth's instantaneous pole of rotation (Maksimov, Sarukhanian and Smirnov, 1970). This fourteen month oscillation, or nutation, is believed to cause an elliptical migration of the Icelandic low and to generate a pressure wave of maximum amplitude (between 1 and 2·7 mb in the monthly mean pressure) at 70° latitude.

As fig. 5.36 shows, the annual trend over the British Isles since the 1930s has been for a reduction of westerly types of airflow and a corresponding increase of easterly and northerly types (Lamb, 1965a, 1969). These changes are also reflected in the peak of the progressive index (Murray and Benwell, 1970) over the British Isles in 1920–4 and its minimum in the recent period. The latter period shows more tendency to cyclonicity and increased northerly bias of the southerly index (fig. 5.37). Confirmation of these trends is provided by Sorkina (1965) for the North Atlantic. She gives statistics of frequency and duration for six basic fields of pressure and wind direction over this area for 1899–1939 and 1949–62. In the winter months (October–April) the frequency of type 4, shown in fig. 5.38(*b*), has increased progressively from less than twenty days in the period 1899–1908 to over seventy days during 1959–63, whereas type 1 (fig. 5.38(*a*)) increased to a peak of fifty days for the period 1919–28 and then decreased to only twenty days in 1959–63. This represents an increase in meridionality over western Europe and eastern North America.

Table 5.17 Changes in frequency and mean run length of weather type groups in the British Isles between 1910–30 and 1948–68 (Perry, 1970)

	January				*July*			
	% frequency		*Mean duration (days)*		*% frequency*		*Mean duration (days)*	
Type groups	*1910–30*	*1948–68*	*1910–30*	*1948–68*	*1910–30*	*1948–68*	*1910–30*	*1948–68*
Westerly	45·0	29·2	4·1	1·7	28·9	28·2	2·6	2·7
Northerly	5·6	9·6	1·5	1·8	10·0	11·4	2·0	1·8
Easterly	4·7	7·6	1·9	2·6	5·2	4·7	2·0	1·5
Southerly	10·0	7·8	2·0	1·8	4·6	2·6	1·7	1·3
Cyclonic	8·9	10·9	1·4	1·3	17·4	16·1	1·7	1·9
Anticyclonic	8·7	16·5	2·2	1·9	14·6	18·6	2·5	2·2

In an extension of Lamb's analysis, Perry (1970) determined to what extent shorter, or less frequent, spells of the westerly type group contributed to the recent decrease in their overall frequency. The results (see table 5.17) for the periods 1910–30 and 1948–68 show that, in general, a decrease in type frequency has been accompanied by a reduction in the mean run length of the type group (see Savina, 1970). An exception to this tendency is anticyclonic type, which in January has almost doubled in frequency, apparently occurring as brief interruptions in other spells. In winter anticyclonic type frequently follows days of the northerly and easterly groups and is itself most commonly succeeded by westerly types. Such changes in run length or in the structure of long spells may be related to the accentuation or suppression of singularities in particular circulation epochs, although this specific question does not so far seem to have been examined.

The changes in frequency of surface circulation patterns and in the positions of the centres of action have been accompanied by shifts in the preferred locations of the upper air long-wave troughs and ridges. For example, shortening of wavelength and a change in the longitudes most frequently subject to cold upper troughs seems to be responsible

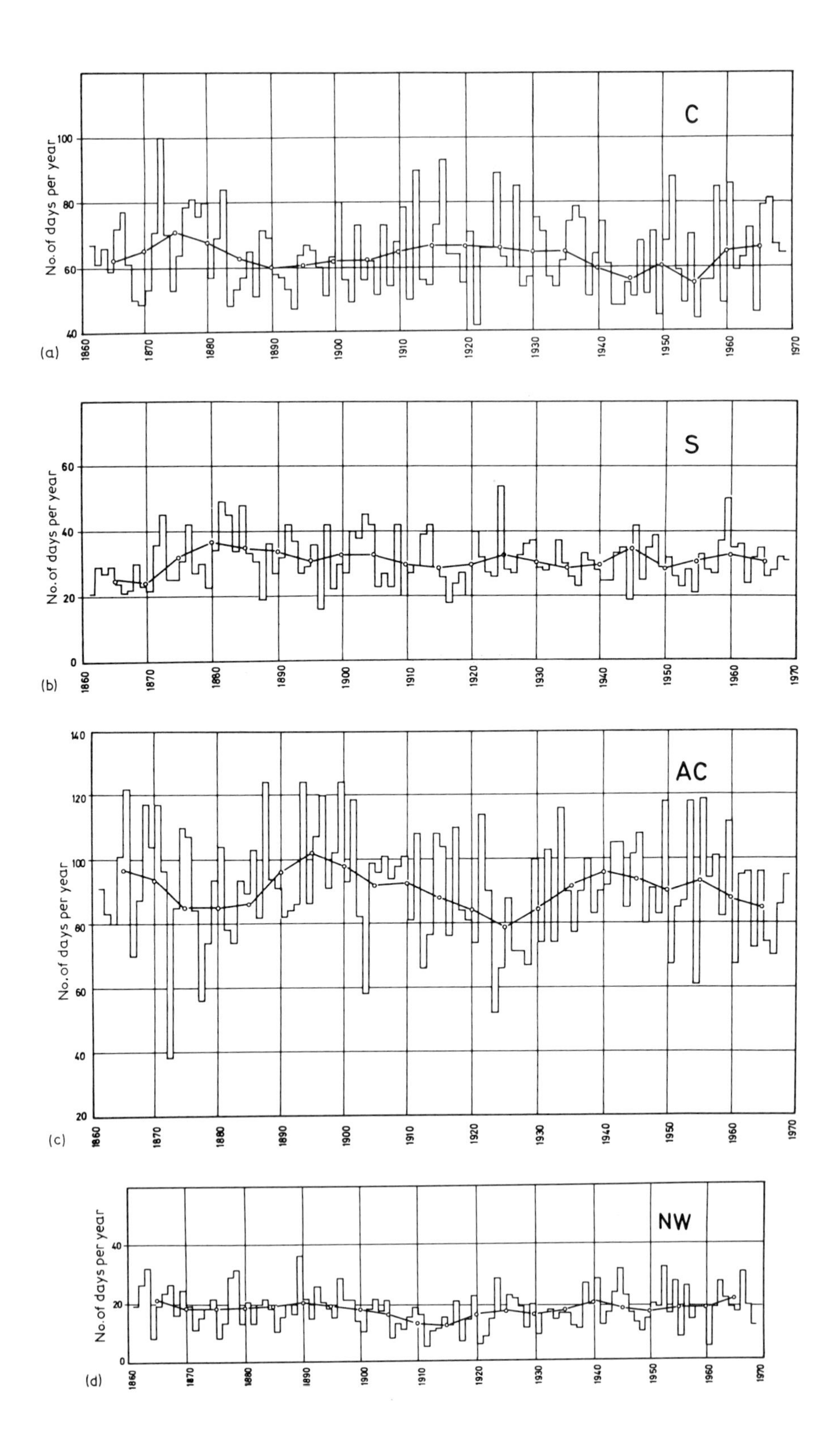
C
No. of days per year
100
80
60
40
(a)
1860
1870
1880
1890
1900
1910
1920
1930
1940
1950
1960
1970
S
No. of days per year
60
40
20
0
(b)
1860
1870
1880
1890
1900
1910
1920
1930
1940
1950
1960
1970
AC
No. of days per year
140
120
100
80
60
40
20
(c)
1860
1870
1880
1890
1900
1910
1920
1930
1940
1950
1960
1970
NW
No. of days per year
40
20
0
(d)
1860
1870
1880
1890
1900
1910
1920
1930
1940
1950
1960
1970

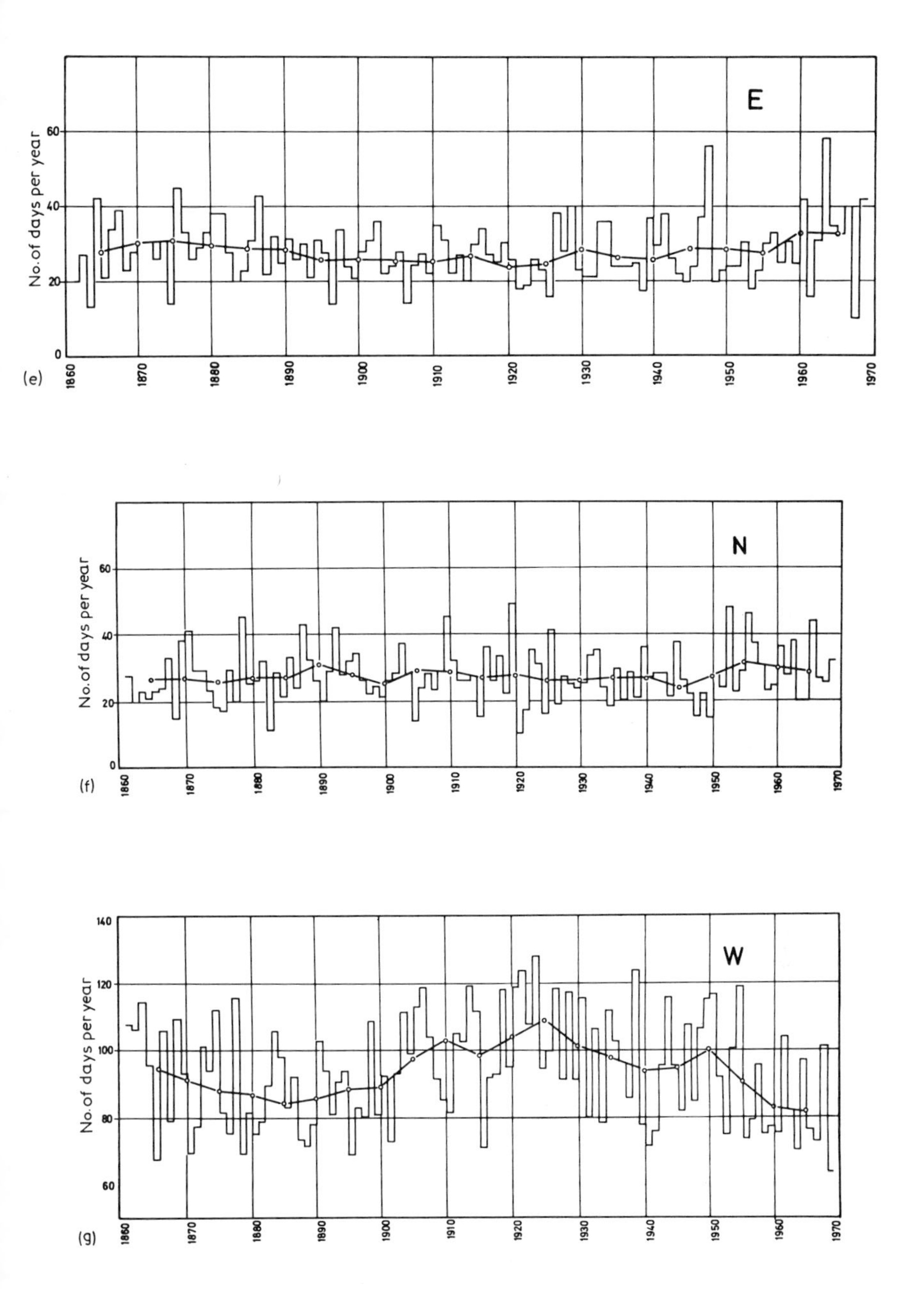

Fig. 5.36 Ten-year running means, plotted at five-year intervals, and annual frequencies of Lamb's airflow types for the British Isles since 1861 (from Weiss and Lamb, 1970).

(*a*) Cyclonic type
(*b*) Southerly type
(*c*) Anticyclonic type
(*d*) Northwesterly type
(*e*) Easterly type
(*f*) Northerly type
(*g*) Westerly type

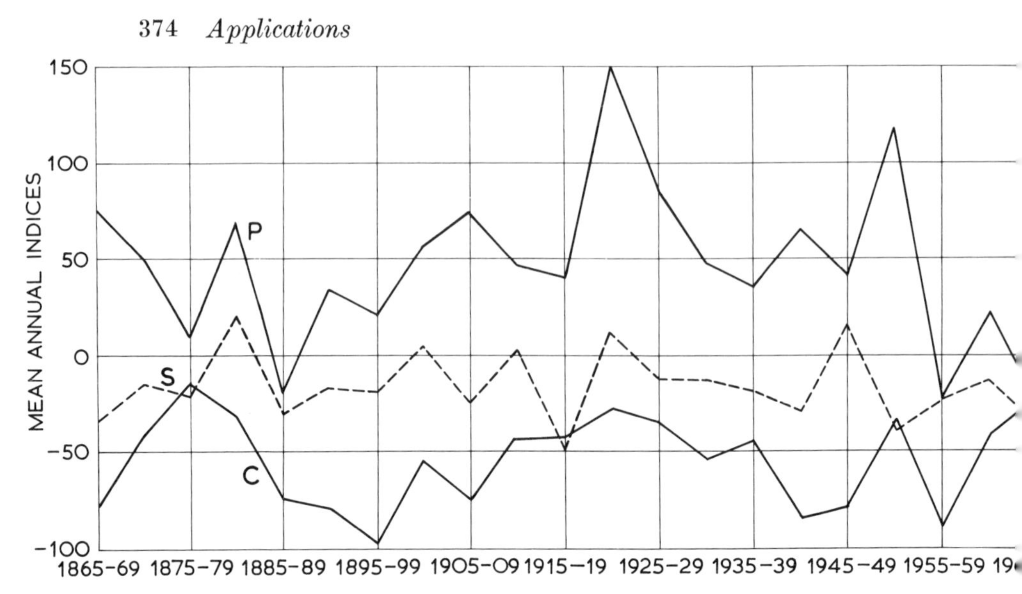

Fig. 5.37 Indices of progression (P), meridionality (S) and cyclonicity (C) for 1865–9 to 1965–9 (after Murray and Benwell, 1970).

for more frequent northerly outbreaks over the Norwegian Sea and, to a lesser degree, over the British Isles since the early 1950s (Lamb, 1963). However, the increase in more severe winters and cooler summers in Britain does not seem to be solely a result of these changes. Barry and Perry (1969, 1970) have drawn attention to the fact that changes have also taken place in the weather characteristics of individual circulation types.

The change in mean daily temperature for a given month between two periods can be regarded as comprising a component due to changes in circulation type frequency and another component due to changes internal to particular types.

The total change in mean temperature can be expressed

$$\Delta\bar{T} = \sum_{i=1}^{k} [(\Delta f_i(T_i + \Delta T_i)/n + f_i\,\Delta T_i/n]$$

where f_i = frequency of type i during the first time period
T_i = mean temperature of type i during the first time period

Fig. 5.38 (*a*) Synoptic type 1A over the North Atlantic. (*b*) Synoptic type 4 over the North Atlantic (from Sorkina, 1965).

Key:

1 = Tropical front　　4 = Isobars
2 = Polar front　　5 = Direction of cyclone movement
3 = Arctic front　　6 = Occasional trajectory of cyclones

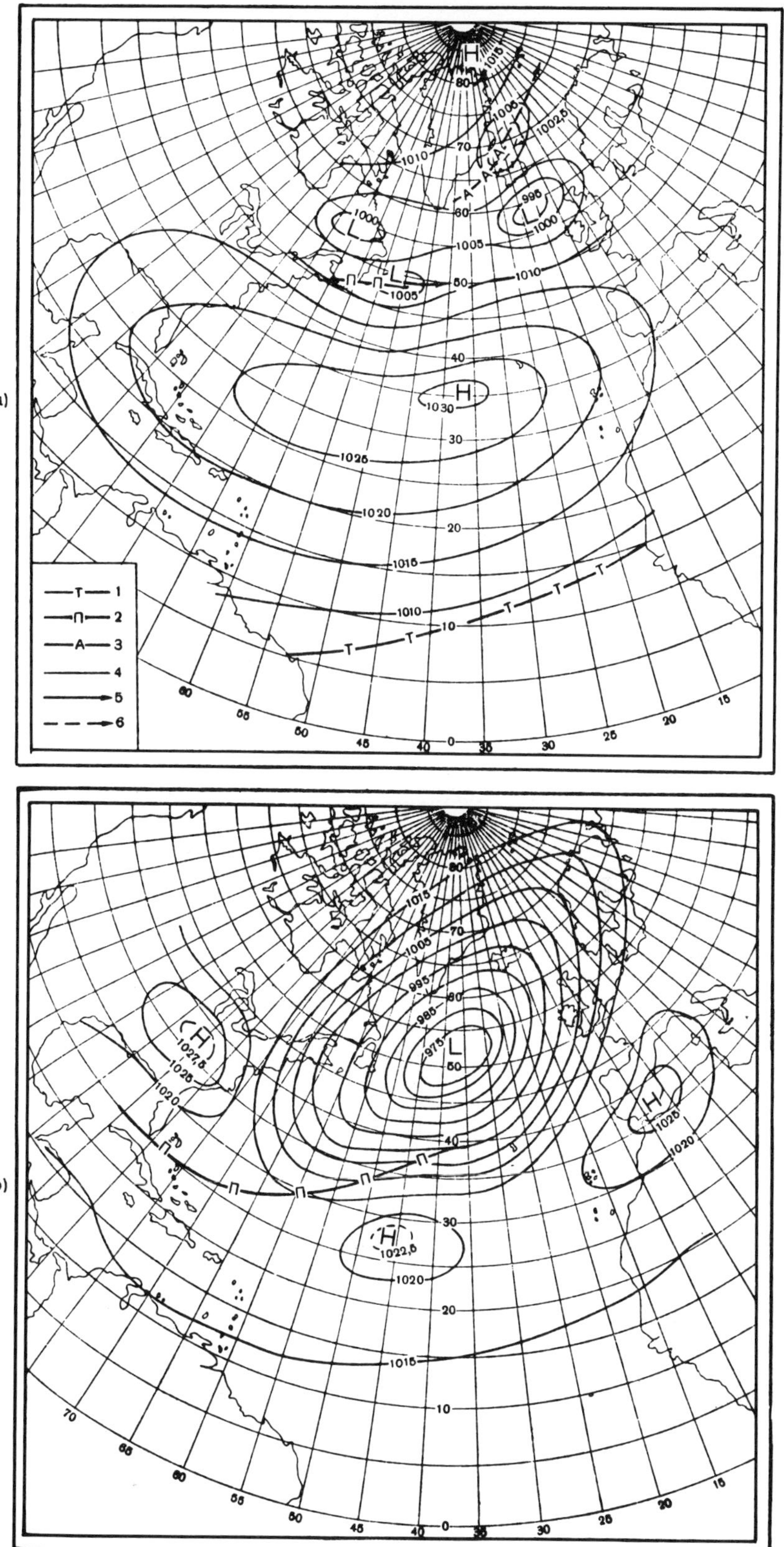

(a)
H
L
1000
1002,5
1005
1010
1015
1020
1025
1030
995
T 1
Π 2
A 3
4
5
6
80
70
60
50
40
30
20
10
0
60
55
50
45
40
35
30
25
20
15
(b)
H
L
1027,5
1025
1020
1022,5
1015
1005
995
985
975
70
65
60
55
50
45
40
35
30
25
20
15

n = total number of days in the first time period
$f_i + \Delta f_i$ = frequency of type i during the second time period
$T_i + \Delta T_i$ = mean temperature of type i during the second time period.

The term $f_i \, \Delta T_i/n$, which is independent of any change in frequency, represents a component due to within-type changes of temperature. The first term on the right-hand side represents the effect on ΔT of a change in type frequency when a change occurs in the temperature of type i in the second period. For some purposes it may be of interest to separate $\Delta f_i \, T_i/n$ and $\Delta f_i \, \Delta T_i/n$.

Table 5.18 Calculated changes in temperature between 1925–35 and 1957–67 due to (a) *changes in type frequency and* (b) *changes 'within type'* (°*C*)

	January				*April*			
	Max		*Min*		*Max*		*Min*	
	a	*b*	*a*	*b*	*a*	*b*	*a*	*b*
Valentia	(−0·8	+0·3)	−1·0	−0·1	+0·2	+0·8	(0·0	+0·5)
Eskdalemuir	−0·8	−0·6	(−0·8	+0·1)	+0·3	+0·4	−0·2	+1·1
Buxton	−1·2	0·0	−1·0	−0·3	(+0·2	+0·2)	(0·0	+0·4)
Gorleston	−1·0	−0·4	(−0·6	−0·3)	(−0·1	−0·1)	(0·0	+0·3)

	July				*October*			
	Max		*Min*		*Max*		*Min*	
	a	*b*	*a*	*b*	*a*	*b*	*a*	*b*
Valentia	(0·0	−0·3)	−0·1	−0·6	+0·3	+0·4	(+0·1	−0·3)
Eskdalemuir	(0·0	−1·0)	−0·2	−0·8	+0·4	+0·6	+0·1	+1·5
Buxton	−0·3	−1·1	−0·2	−0·6	(+0·2	+0·4)	−0·1	+0·8
Gorleston	−0·2	−0·9	(−0·2	−0·2)	(+0·1	+0·1)	+0·2	+1·0

Parentheses denote changes of mean temperature between the two periods which are not significant at the 10% level by Student's t test.

Application of this method to the daily maximum and minimum temperature at four stations in the British Isles for 1925–35 and 1957–67 (Perry and Barry, 1973), with reference to Lamb's (1972) catalogue indicates that, while the frequency change was apparently important in January, the significant temperature changes in April, July and October were 'within-type' (table 5.18). The types contributing changes of ±0·1–0·4°C most commonly to the 'within-type' component were anticyclonic, westerly, northerly and easterly types. The causes of the 'within-type' changes have not yet been isolated. It is possible that the types are not homogeneous between different periods but the catalogue is a revised version of Lamb's original classification so that this seems unlikely. There could, nevertheless, be subtle differences such as in wind speed or trajectory within a single type grouping between the two periods (Lawrence, 1970), although for the larger samples it is reasonable to assume a comparable range of wind speeds and trajectories within each type. One possibility is a change in sea surface temperature. However, Brown (1963) shows that sea surface temperatures were of the order of 1°C higher in the northeastern North Atlantic in the decade 1951–60 than in the 1920s, so that although Perry (1969) notes a cooling of −0·6° to −0·8°C at *Ocean Weather Stations* D, I, J and K for July between 1951–60 and 1961–67 there is little support for this argument. Further work is necessary on this interesting problem, since one implication of the significance of 'within-type' changes is that palaeoclimatic reconstructions, based simply on an assumed variation in type frequency, may be quite unreliable.

4 OUTSTANDING PROBLEMS

A topic of concern at the present time is the nature and cause of the present climatic fluctuation. Increasing awareness of the significant climatic shifts that have occurred in our own time and of the possible role of man in contributing to them has given rise to a demand for climatic forecasts (Lamb, 1969). These are essential in economic planning in a vast number of spheres of activity (see, for example, Maunder, 1970, Mason, 1970).

There is evidence of several periodicities in the climatic record which may provide a guide to super long-range forecasting. For example, a twenty-two to twenty-three year periodicity is apparent in the frequency of blocking highs over Europe according to Brezowsky, Flohn and Hess (1951), and Lamb (1963) notes periods of about 23, 45–60, 100 and 170 years in decade indices of winter severity and summer wetness since A.D. 1100 for Europe. Wagner (1971) performed a spectrum analysis of sixty-six years of seasonal MSL pressure data for the northern hemisphere. A peak at twenty-one years was pronounced in semi-arid and arid areas and there were less marked peaks at five to six years in winter and summer over the northern North Pacific. The quasi-biennial (26–8 months) oscillation was also well represented over most of the northern hemisphere, especially in winter (see p. 398). A 770 year record of 18O variations in an ice core from northern Greenland (Johnsen *et al.*, 1970) exhibits periodicities of 78 and 181 years, both of which are thought to represent solar effects. Supposed regularities of this type, which, it is presumed, will favour particular circulation patterns, have led to attempts to predict climatic trends over periods of thirty to fifty years ahead (Scherhag, 1939; Girs, 1966a; Dzerdzeevski, 1968a). However, apart from doubts as to the reality of some of the claimed periodicities, a physical reason has been tentatively advanced for only a few of them. Periods of twenty-two to twenty-three years coincide with sunspot cycles, but even so a theory satisfactorily accounting for their effect on the atmosphere has yet to be developed (see section D of this chapter, p. 405). Perhaps the best that can be done at present is to extrapolate recent trends, such as those evidenced by Lamb (1966a) and Rudloff (1969), over the next five to ten years. On the basis of the Greenland ice core record Johnsen *et al.* make a longer extrapolation to the beginning of the next century. They foresee a continued cooling for one or two decades followed by a warming trend culminating about A.D. 2015. Reversions such as that of 1957–61 in northwest Europe, referred to by Lamb as 'the last flicker of the warm period behind us', will nevertheless almost certainly recur from time to time to confute this type of prediction!

This topic cannot be concluded without reference to the fact that some scientists question the whole concept of climatic change. Rather, they regard the atmospheric circulation as a stochastic process which allows for the occurrence of irregular fluctuations (Curry, 1962; Yevjevich, 1968). Certainly, as Mandelbrot and Wallis (1969) have shown, long-run interdependence is characteristic of most geophysical series. Lorenz (1965) demonstrates that in any aperiodic process exhibiting persistence (such as a first-order linear Markov process) fluctuations with extended duration are to be expected. The persistence of large-scale atmospheric circulation patterns could result either from a basic sluggishness of the atmosphere, or from some additional physical controls. The latter, if they are critical, could be (*a*) inherent characteristics of the atmosphere which cause the circulation to switch abruptly from one regime to another at irregular intervals, (*b*) extraterrestrial impulses, (*c*) terrestrial effects modified by feedback processes, or some combination of all three. It is hard to see how these effects can be distinguished at the present time. Sophisticated numerical models will certainly be vital to this process but doubt may always remain as to the existence of a unique and correct solution to the problem.

D Long-range forecasting

Research work concerned with the slower changes in the atmosphere is at the stage typified by Kepler rather than Newton, in that its main task is that of assembling facts and recognising regularities and patterns rather than seeking explanations. (*J. M. Craddock, 1964*)

1 BACKGROUND

The history of long-range forecasting extends back almost a century. Fig. 5.39 summarizes the main approaches that have been adopted and their exponents since Teisserenc de Bort (1883) pointed out the existence of macroscale pressure patterns which he called *centres of action* and their relationship to prevailing weather conditions. It soon became evident that the weather abnormalities of a given period were linked to the positions and configurations of the centres of action and that it would be necessary to forecast these in order to predict the prevailing weather. Since a satisfactory physical theory for the generation and maintenance of abnormal positions of the centres did not exist, statistical techniques, and in particular simultaneous correlations between sea-level pressures in different parts of the world, were investigated. Such *teleconnections* will be considered again later. Considerable light was thrown on these early results by the work of Rossby (1939) and his collaborators on vorticity redistribution associated with long waves in the westerlies.

Schmauss (1937) attributed the occurrence of unusual seasonal weather to more or less random effects such as the existence or non-existence of a snow cover at a given instant of weather development, and from such reasoning it appeared that long-range forecasting was impossible in principle. More influential was the early work of Baur (1936) in Germany who recognized that certain macroscopic features of the general circulation were effective in steering the individual cyclones and anticyclones along determined paths. Rather earlier the Russian meteorologist Multanovski (1933) had discovered large-scale weather situations during which individual systems followed a particular track or 'axis'. Wartime conditions had the effect of stimulating atmospheric research and of necessitating forecasts for an extended period for military planning purposes. The impetus was maintained in the post-war period and by 1962 eight meteorological services in the middle and subtropical latitudes were issuing monthly forecasts, and a further seven countries were actively participating in scientific work concerned with long-range forecasting (Craddock *et al.*, 1962). In Britain the first experiments in the 1950s were concerned with the analysis of five-day mean charts of air temperature anomalies in the northern hemisphere. Since the anomalies are strongly correlated with those of the 500 mb airflow, it is possible to use the pattern of the anomalies of air temperature near the earth's surface as an indicator of the state of the circulation

Period	Centres of action and weather typing	Correlations and regressions	Tele-connections	Analogues	Cycles and harmonics	Singularities and symmetry points	Key periods	Cosmic influences	Kinematics and synoptics	Feed-back studies or surface influences	Thermodynamic and dynamic numerical models
					Howard 26 (1842)	Buchan 34 (1867)	Old Sayings like St. Swithin's			Blanford 10	
1880	Teisserenc de Bort 1 Clayton 2 Abercromby 3	Blanford 10			Brückner 66						J. J. Thomson 58
1890	Hildebrandsson 4									Hildebrandsson 4 Meinardus 48	
1900		Lockyers 11 Walker 12 Braak 13	Walker 12							Schott 49	
1910								Walker 40 Clayton 41	Arctowski 46	Petersen 50 Helland-Hansen & Nansen 51 C. F. Brooks 52 C. E. P. Brooks 53	
1920	Multanovsky 5 Baur 6	Groissmayr 14 Baur 15	Defant 19 Exner 20	C.E.P. Brooks 22 Mascart 23	Alter 27 Petitjean 28	Weickmann 35 Schmauss 36		Baur 42	Clayton 2 C.E.P. Brooks 45	Weise 54	
1930	Bjerknes 7 Rossby 8		Rossby 8		Berlage 29 Abbot 30		Baur 39	Abbot 43	Baur 6 Rodewald 47	Sandström 55	
1940	Namias 9	Wadsworth 16	Martin 21		Haurwitz 31	C.E.P. Brooks 37		Willett 24	Namias 9		Rossby 8 Stewart 59
1950		Klein 17 Gilman 18		Willett 24 Craddock 25	Takahashi 32	Grappe 38			Craddock 25	Namias 56	Phillips 60 Blinova 61 Smagorinsky 62 Mintz 63
1960					Van Mieghem 33			Brier & Bradley 44		Bjerknes 57	Charney 64 Adem 65
1970											

Fig. 5.39 Types of approach to long-range forecasting (from **Namias, 1968**). Numbers refer to references in Namias's paper.

in the upper troposphere. If the present state of the circulation in the upper air is any indication of the future, then cases that are similar in their large-scale patterns of temperature anomalies should tend to evolve in the same way. These first experimental steps have been described by Craddock (1958). To produce a reliable basis for prediction it was realized that other parameters than temperature must be considered, and the use of weather type sequences (Craddock, 1964) and pressure patterns was added to the circulation procedures, followed more recently by analysis of physical parameters (Murray, 1970).

The United States Air Weather Service (1954) indicated that the methods of long-range forecasting in general use fall into three broad groups:

1 Those based mainly on statistical procedures and largely unconcerned with map presentations or physical reasoning.
2 Methods based on types of map presentations which are primarily empirical and which make only secondary use of statistical and physical reasoning.
3 Methods in which physical reasoning plays an important part but in which empiricism and statistics are also used to a certain extent.

Wholly dynamical means of long-range forecasting based on physical principles and taking into account thermodynamic processes still remain to be developed (Adem, 1965), partly because of the incomplete meteorological coverage of the tropics and the oceans, and hence the difficulty of specifying the initial conditions accurately enough to start a model on the right lines, and also due to incomplete knowledge of the behaviour of the general circulation. The actual techniques employed by most meteorological services represent a combination of the following:

1 Analogues
2 Extrapolation and kinematic procedures
3 Synoptic persistence and development
4 Study of cycles and oscillations
5 Ocean–atmosphere interactions
6 Teleconnections

The first three are the main procedures used in conventional extended and long-range forecasts for five to thirty days and even in seasonal outlooks, although here ocean–atmosphere interactions assume great importance. The last three represent the major lines of evidence that are being studied with a view to assessing trends for up to a decade or more. We shall consider them under these broad headings in spite of the inevitable degree of overlap in their application.

It will become evident that consideration of the nature of atmospheric teleconnections and oscillations takes us somewhat beyond the strict realm of synoptic climatology as we have defined it (p. 7). However, the regional manifestations of these phenomena are such that they cannot be ignored in any thorough discussion of synoptic climatology in the context of long-range forecasting. No further apology is necessary therefore for their inclusion here.

2 EXTENDED AND LONG-RANGE FORECASTING

Analogues

The assumption basic to the analogue method is that the features in which the current case and selected analogues are similar comprise those which to some extent determine the subsequent weather development. If a 'better than chance' standard of success is achieved, and Ratcliffe (1970) has shown that this is the case in Britain, we know that among the parameters we are comparing there are some that are related by physical processes to whatever we predict. The analogue method then provides a direct solution to a problem which is so complex that we almost certainly do not fully understand all of the controlling processes that are involved.

The basis of the monthly long-range forecasts prepared regularly by the British Meteorological Office, and issued to the public since 1964, is the recognition of analogues to the present weather situation. Both subjective 'pattern matching' (Murray, 1966) and objective analysis are used to select past situations which in certain respects show agreements with the current situation. A forecast may then be made on the basis that what happened in the most similar case will be repeated in the present one. The procedure has probably attracted meteorologists since the time when a few years' file of weather maps first became available. Brooks and Glasspoole (1922), for example, searched for sea level pressure analogues to the drought year of 1921 in the British Isles. Girs (1956) suggests that besides consulting the patterns of the sequence of types the long-range forecaster must select analogues that will help him assess the total impact of independent factors. Analogues must be picked out according to genetic symptoms, not simply for outward signs of similarity.

Proponents of the analogue method cite two major advantages of its usage. First, it is objective and does not rely upon complex and subtle reasoning inherent in physical methods which are far from perfected. Second, analogues automatically contain the climatological peculiarities of every location for which a forecast is required.

The success of the method partly depends on the completeness of readily available processed data and the efficiency with which they can be sorted and retrieved. The assembly of a comprehensive library of past cases is a very large task entailing the collection and collation of data from many sources and their presentation in a form suitable for ready reference and easy assimilation. Flohn (1942) described an analogue selection method used in Germany using tiny pictures of European maps, about one inch square, mounted on large cards. Each card could contain several hundred maps representing many years of data, and analogues were found by visual inspection, the pictorial representation affording a rapid means of considering similarity of pattern, historical development and prognostic evolution. In France analogues have been classified as simple analogue, composite analogue and partial analogue (Namias, 1951). Somewhat similar techniques were suggested by Vitels (1948) using composite or averaged maps of all the best analogues to obtain those features inherent in all of them. Those features, common to all the analogues, would reflect the most important characteristics of the large-scale weather processes, and deviations on separate analogues would stand out when compared to the averaged maps. Grytoyr (1950) portrayed many elements of a weather sequence on a chart called a 'clue diagram', by taking into account the detailed dynamic development of the weather processes in a rather small geographical area. The movements of air masses, fronts and pressure systems were charted in accordance with their positions relative to Oslo and the diagrams then used in the visual selection of analogues.

Modern data storage is invariably on punched cards and/or magnetic tape. These facilities and the large capacity of modern computers allows numerous parameters to be recorded. Table 5.19 illustrates the constituents of a modern data bank for long-range forecasting established by the British Meteorological Office.

The computer plays an increasingly important role, not only in selecting analogues, but also in printing out time-averaged fields, e.g. mid-month to mid-month surface pressure charts, when required. As an example of the type of routine computation carried out fortnightly, the procedure for selection of analogues using Lamb's (1972) weather catalogue is as follows:

1 The series of daily weather types for the past thirty days is matched against those of the same dates in a preceding year, as described by Murray and Lewis (1966a).
2 The comparison is repeated, using a shift backwards or forwards of up to fourteen days.
3 Stages 1 and 2 are repeated for all past years from 1873 to the present.
4 The results are printed out in tables, ready for consideration by a panel of forecasters.

A major problem is to find an appropriate measure of the similarity between two or more charts. In general each chart is represented by grid point data, and an index of the discrepancy has to be computed from these. Root mean square, correlation coefficient matrices and eigenvectors have all been used experimentally. In a recent study to investigate the predictability of the atmosphere and the growth rate of errors Lorenz (1969) found numerous mediocre analogues but no very good ones. However, this study covered three quarters of the northern hemisphere (using 1003 grid points) at the 200, 500 and 850 mb levels so that the lack of very close analogues is not particularly surprising.

Table 5.19 Components of the Meteorological Office synoptic climatology data bank (after Craddock, 1970)

	Data	*Time base*	*Period*	*Grid points*	*Million decimal digits*
1	Hemispheric 500 mb heights	Daily	1949→	592	17
2	Hemispheric 1000–500 mb thickness	Daily	1949→	592	17
3	Hemispheric 1000 mb heights	Daily	1949→	592	17
4*a*	Hemispheric MSL pressure	Daily	1899→		45
4*b*	Hemispheric MSL pressure excluding most of North Pacific Ocean	Daily	1881–1899		10
5	Hemispheric MSL pressure	Monthly mean	1873→	592	0·5
6	Synoptic catalogues				
	(*a*) Lamb classification	Daily	1868→	—	0·08
	(*b*) Ward's classification for London	Daily	1873→	—	0·07
	(*c*) PSCM indices	Daily	1868→	—	0·08
	(*d*) *Grosswetterlagen* (Hess and Brezowsky)	Daily	1880→	—	0·13
7	(*a*) Temperature for 350 northern hemisphere stations	Monthly mean	Variable (some 1850→)	—	3·75
	(*b*) Rainfall for 100 northern hemisphere stations	Monthly totals		—	
8	North Atlantic sea surface temperature anomaly fields	Monthly	Variable (some 1876→)	—	7
9	Bathythermograph data for nine North Atlantic weather ships.	Daily	1966→	—	5
10	Weather elements (max. and min. temperature, rainfall) at seven stations in British Isles.	Daily	Variable (since 1853 at Oxford)	—	2

Descriptions of objective computer methods for selecting analogues developed in the U.S.S.R. using correlation and other empirical coefficients of the movement of pressure systems (Rafailova, 1968) and eigenvectors (Kudashkin and Iudin, 1968) indicate the wide interest in this problem.

Elliott (1942) and Krick (1942) recognized that many earlier attempts at weather typing were based upon single typical synoptic charts and therefore did not truly take into account weather processes occurring during an extended period. Accordingly they developed the six-day North American type scheme and subsequently the three-day scheme, which have already been mentioned in Chapter 3.B (p. 134). These are based on three basic features of the circulation–zonal flow, meridional flow with high pressure or low pressure dominating the weather (United States Air Weather Service, 1954). It may well be that this type of classificatory scheme, which has much in common with Baur's *Grosswetterlagen* (Baur, 1951) and with Multanovski's (1933) axes and natural synoptic periods, is more useful in analogue selection than the daily typing schemes. The simplification of Lamb's daily catalogue into the PSCM indices reflects a similar viewpoint (Murray and Lewis, 1966b). It is also interesting to note that Vangengeim's large-scale circulation types have been found useful in developing forecasts for up to eight to ten days in the Arctic (Dydina, 1967). All of these approaches help to eliminate the shorter term synoptic variability.

Extrapolation and kinematic procedures

As in short-range forecasting these methods depend on the assumption that the salient features of the circulation are shifting in a quasi-regular manner. The three principal working rules are:

1 Persistence in pressure values.
2 A tendency for pressure to return to normal.
3 Assessment of future weather conditions from the past and present behaviour of the atmosphere.

Early work along these lines is represented by Arctowski's (1910) study of the movement of annual centres of pressure excess (pleion) and deficit (antipleion) and by the similar studies of monthly centres by Brooks and Quennell (1926). The paths of deficit centres are less regular than centres of excess pressure but both tend to move eastward in all seasons and have an average life of about three months.

Experience suggests that this approach is most suitable when the upper air chart has well-marked troughs and ridges and strong tendency fields. In the United States and Germany some success has been achieved by noting the past and current rate of evolution of the mid-tropospheric flow patterns, in particular the major features (ridges, troughs and centres), and taking these developments into account in deriving a forecast 500 or 700 mb chart for time periods of up to a month. It must then be decided if these kinematic displacements are in harmony with vorticity principles. The method is unreliable when the flow patterns are weak, but the most fundamental criticism is that new circulation modes cannot be predicted by extrapolation procedures. Only existing trends can be taken into account.

After the mean 700 mb prognosis has been completed, anomalies of temperature and precipitation are estimated based on experience of the relationship between particular upper air patterns and the resulting surface anomalies. Martin and Hawkins (1950) and Klein (1948) have noted the types of contour pattern which produce abnormal monthly

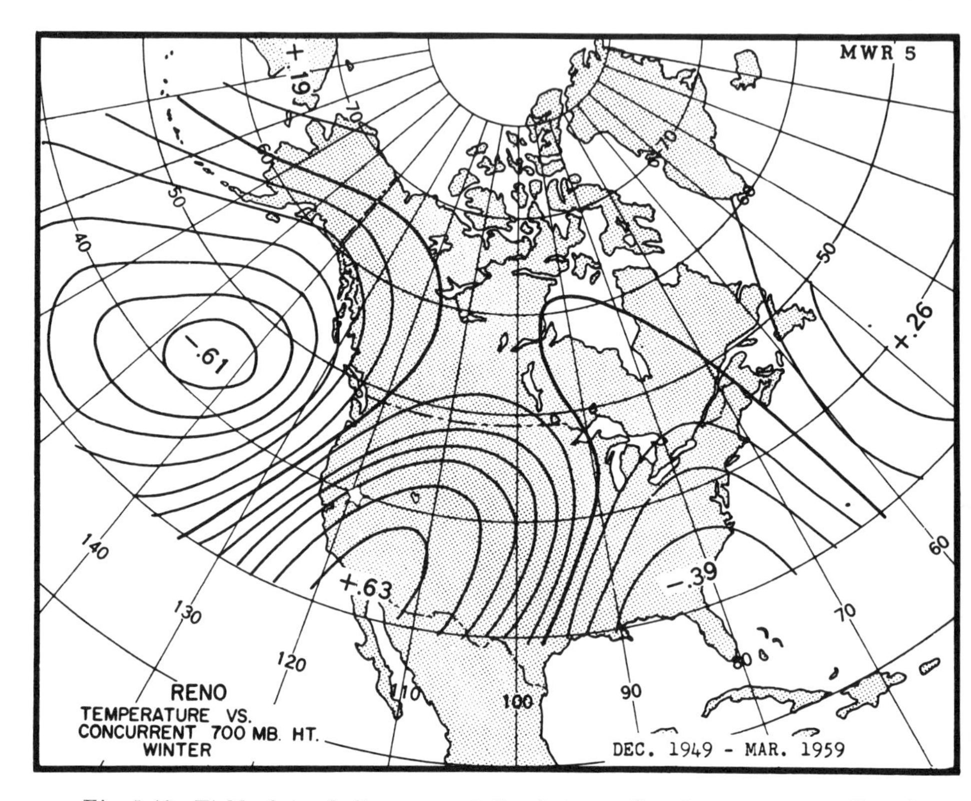

Fig. 5.40 Field of simple linear correlation between five day mean anomalies of surface temperature at Reno, Nevada (located by star), and 700 mb height two days earlier at 70 grid points for 140 winter cases. The lines of equal correlation coefficient are drawn at intervals of 0·10 with the zero isopleth heavier and centres labelled with highest interpolated value (from Klein, 1965c).

weather and they have developed models demonstrating the relation of precipitation areas and storm paths to the upper level wave pattern. Work of this type is reviewed by Namias (1953). For the United States Klein (1963, 1965b, 1968) has produced correlation coefficient values between the sea level (or 700 mb) circulation and precipitation amounts or temperature levels for points representing forty climatologically similar areas considered roughly homogeneous. In the example for Reno, Nevada (marked by a star in fig. 5.40), the gradient between the two centres of maximum correlation suggests that high temperatures are associated with anomalous southwesterly flow of Pacific air and cool temperatures with northeasterly flow. Similar correlation fields have now been constructed for many other cities and composite charts giving the direction of the anomalous flow for temperature and precipitation

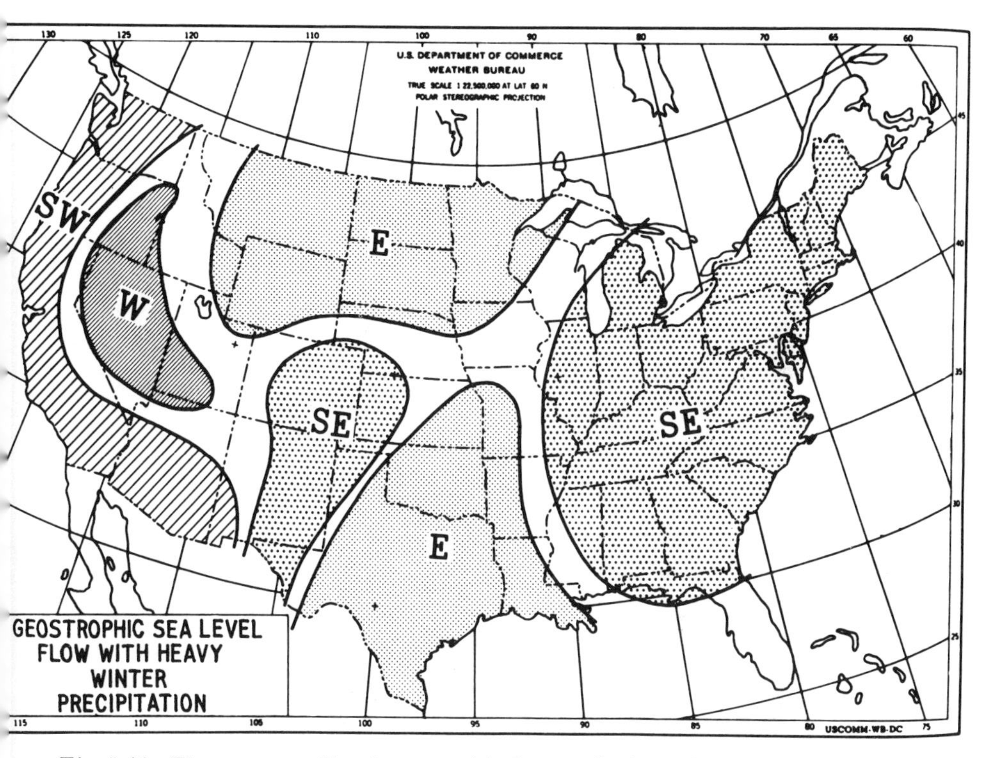

Fig. 5.41 The patterns of local geostrophic flow conducive to heavy precipitation in winter over the United States (from Klein, 1965b).

conditions in all parts of the United States (Klein, 1965a), together with charts of the zonal distance from each city to the nearby axis of maximum positive correlation. The isocorrelates indicate the direction, curvature and origin of anomalous flow. Fig. 5.41 illustrates the local geostrophic flow conducive to heavy winter precipitation determined from maps such as those in fig. 4.17. On the basis of the correlation values, schematic models such as that shown in fig. 5.42, illustrating the occurrence of light and heavy winter precipitation in relation to the simultaneous sea level pressure and 700 mb height distributions, have been prepared. The top of the figure is north; in fig. 5.42(*a*) the curves represent isobars at sea level and the numbers can be interpreted either as isobar labels or as distances in hundreds of miles from the low centres.

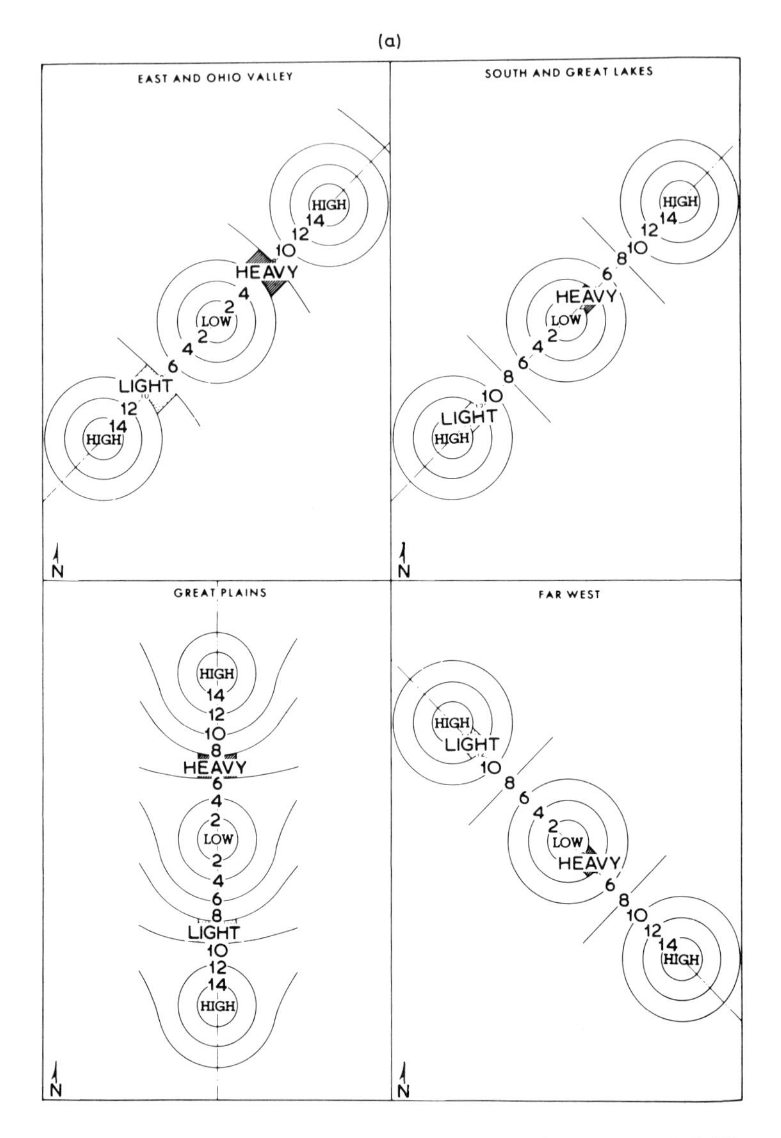

Fig. 5.42 Models of winter precipitation related to (*a*) MSL pressure distribution, (*b*) 700 mb contour regions of the United States. The figures are distances in hundreds of miles from the low centres (from Klein, 1965c).

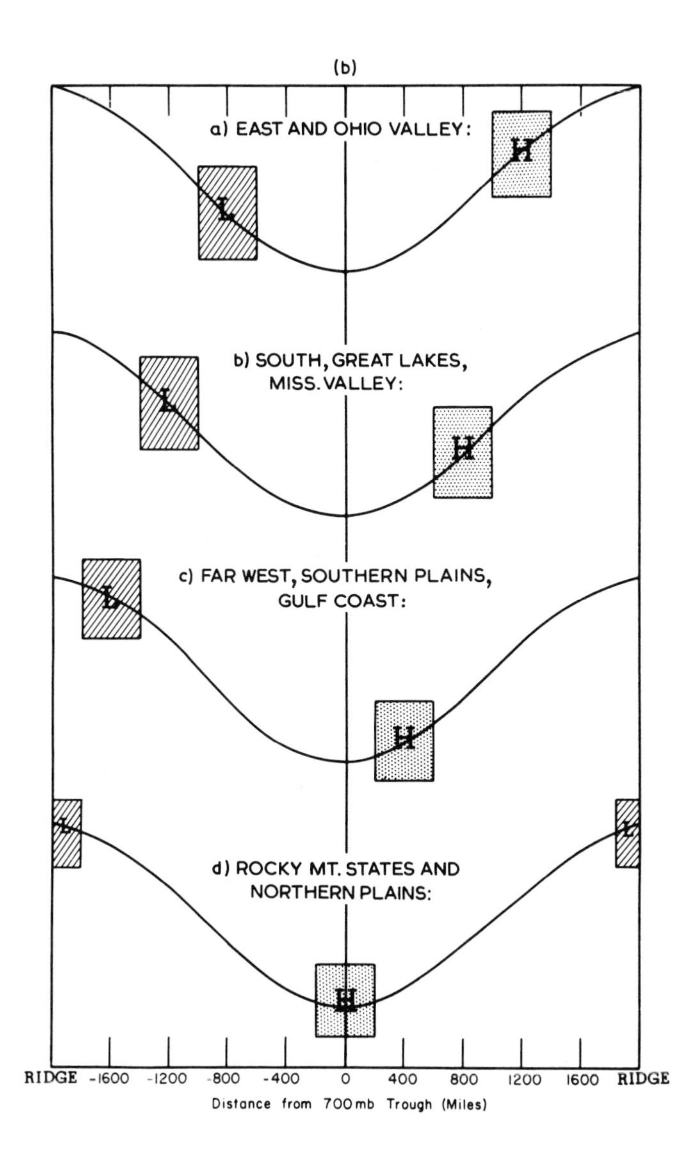
(b)
a) EAST AND OHIO VALLEY:
H
L
b) SOUTH, GREAT LAKES, MISS. VALLEY:
L
H
c) FAR WEST, SOUTHERN PLAINS, GULF COAST:
L
H
d) ROCKY MT. STATES AND NORTHERN PLAINS:
L
L
H
RIDGE
-1600
-1200
-800
-400
0
400
800
1200
1600
RIDGE
Distance from 700mb Trough (Miles)

Using a stepwise method of multiple regression called screening (Miller, 1958) objective specifications can quickly be calculated for both temperature and precipitation using a large number of possible predictors such as pressure, 700 mb height, relative humidity, height anomalies, etc. Numerous experiments have been conducted to determine which predictors will give the best objective forecasts, and complex parameters such as frontal positions and advection are being studied at present. Already equations of this type have been devised for sea level pressure (Klein, 1967), temperature (Klein, Lewis and Enger, 1959; Klein, Lewis and Casely, 1967) and clouds (Klein, Crockett and Andrews, 1965).

Synoptic persistence and development studies

It has long been known that considerable coherence exists in meteorological data both in space and time, with persistent recurrences of circulation patterns and their associated weather events over time intervals ranging from a few days to several weeks or even longer.[1] Indeed, it is dynamically reasonable that there are certain large-scale circulation patterns which have a built-in tendency to persist or to evolve in one way, in preference to another, at certain times of the year. Transition frequencies for the *Grosswetter* types of Hess–Brezowsky have been determined by Bryson *et al.* (1955) and Mertz (1959). Mertz used the nineteen individual types but did not differentiate between seasons, whereas Bryson *et al.* examined the ten basic groups on a seasonal basis, although they did not take into account the duration of a type prior to the change. Fig. 5.43 illustrates the modal transitions in summer and winter. The sequence HW–W–NW is common in all seasons. Thus, the winter values are 38%, 13%, 19%, and the corresponding summer ones are 53%, 22%, 24%. In both seasons, however, there is more likelihood of W succeeding NW than vice versa. Obviously, these frequencies are insufficient for forecasting purposes by themselves, particularly if the probability of change in relation to the duration of the type is not known, but such facts can serve as a supplementary aid in conjunction with other procedures.

Many sequential and recurrence tendencies were recognized in weather lore and, for Germany, Baur (1958, 1963) devised a set of physical–statistical rules based on such knowledge. Some popular beliefs proved less reliable guides than others. In the case of the 'Seven Sleepers', for example, it was popularly thought that rainfall on 27 June would be followed by stormy weather types with frequent rainy days. After an examination of rainfall statistics for the period during the years 1881–

[1] The earliest suggestion for using this approach for forecasts of up to four days ahead was made by Archibald (1898) based on the weather type studies of Abercromby and Köppen.

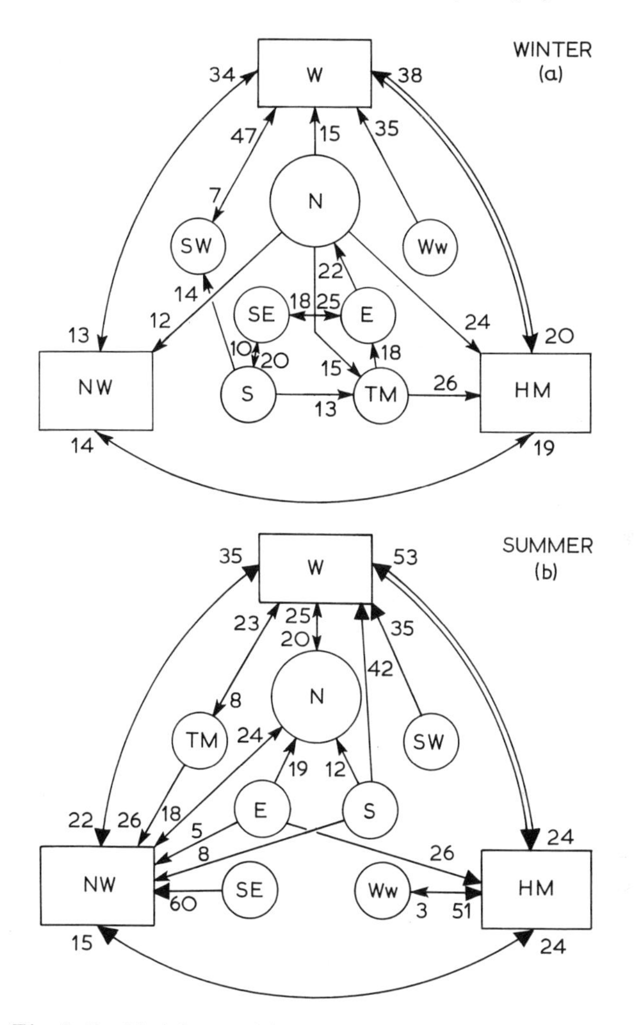

Fig. 5.43 Modal transitions for sequences of the *Grosswetter* types in (*a*) winter, (*b*) summer (after Bryson *et al.*, 1955).

1954 and an inspection of the prevailing *Grosswetter* for the same period, Baur concluded that warning signs in the large-scale weather of Europe at the end of June were not reliable enough to risk a conclusion as to what the weather might be like in the mid-summer period. By contrast, there is evidence that if the temperature in Berlin in the first ten days of December is more than 2·5°C above the normal due to westerly weather type, repetition and persistence tendencies which are prominent at this time of year make it probable that the winter as a whole will be mild in central Europe.

From numerous analyses of the long series of rainfall and temperature data available we are now well aware of the persistence and anti-persistence of runs of dry or wet months and cold or warm months at different times of the year (Keeley and Olien, 1961; Dickson, 1967). Monthly rainfall persistence (Bilham, 1934; Hawke, 1934) is generally less in evidence than temperature persistence, which is not surprising in view of the many different rainfall distribution and synoptic situations which make up the rainfall totals. Craddock and Ward (1962) have mapped areas where a strong statistical relationship exists between temperature anomalies in neighbouring months in Europe and western Siberia. Regions of high persistence nearly all appear to be associated with water, with freezing or breaking sea ice, or with the extension or decrease of snow cover (fig. 5.44).[1] Such features tend to have a conservative influence on the atmosphere, especially when the circulation is weak, and this probably accounts for the greater persistence of low index compared with high index patterns in the winter half-year. Obviously there is no one process leading to persistence, but there are a number perhaps depending on the local inertia resulting from the presence of water. Sharp gradients often occur between neighbouring areas, as for example between Helsinki and Leningrad (fig. 5.44). In the United States, Dickson (1967) finds maximum persistence in the central United States in summer and suggests that this may be linked with moisture transport and cloudiness. He also notes that the occurrence of persistence of warm and cold conditions is primarily limited to warm and cold epochs respectively, and postulates that 'the basic mechanisms responsible for secular temperature fluctuations and month-to-month temperature persistence are essentially the same'. If this is true it means that full understanding of persistence effects will be a major contribution to long-range forecasting and climatic change studies.

Recent studies have turned to general indices rather than individual elements. Murray (1967; Murray and Benwell, 1970) used the PSCM synoptic indices (see p. 142), 500 mb indices and other climatic pre-conditions to investigate monthly persistence (table 5.20). He found, for example that 'progressive', 'cyclonic' Junes tend to be followed by Septembers that are more cyclonic, or less anticyclonic, than usual over the British Isles. The relationships between adjacent months were in general not significant, however. Brier (1968) also reports little seasonal persistence in the zonal index with an almost total lack of correlation between month to month values, although Julian (1966) demonstrates a marked *seasonal* autocorrelation in 700 mb indices in all latitude zones, especially in winter and spring. For the Atlantic–European sector, Hay

[1] Sazonov and Girskaia (1969) have confirmed these findings and extended the analysis to the Pacific Ocean.

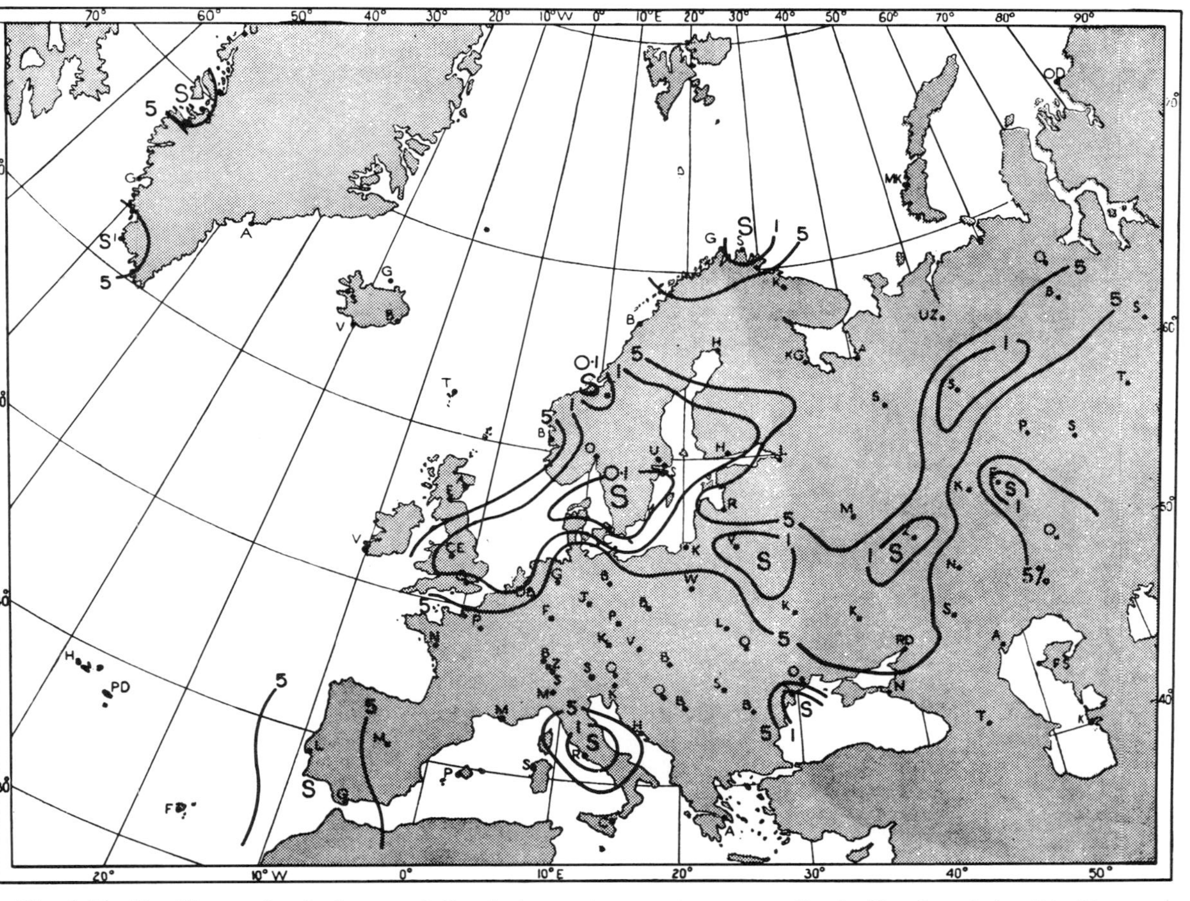

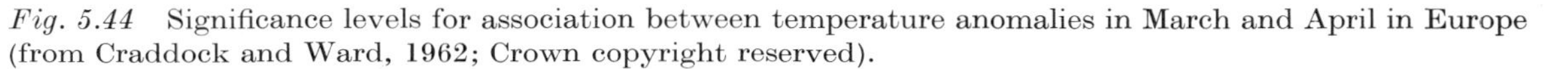

Fig. 5.44 Significance levels for association between temperature anomalies in March and April in Europe (from Craddock and Ward, 1962; Crown copyright reserved).

(1966) shows that an anomalous pressure distribution in autumn tends to precede winters having an abnormal distribution of temperature and hence pressure (see also Hay, 1969, 1970).

A tabulation of the seasonal association models developed for the British Isles has recently been published by Murray (1970), and table 5.21 taken from his paper summarizes work undertaken in the British Meteorological Office in the 1960s (Murray, 1967b, 1968a, b, c, 1969). As yet the synoptic significance of many of the sequences and associations shown in the tables have not been examined, but occasionally

Table 5.20 Some monthly relationships in Britain (Murray, 1970)

(a) Temperature and rainfall

Pre-condition	*Prediction*
1 Very cold (T_1) winter	R_3 *March* (2, 6, 9)
2 Very cold (T_1) winter	Cold March (5, 8, 2, 2, 1)
3 R_{12} spells of at least 5 months to October	R_{12} November (15/16)
4 R_{23} spells of at least 3 months to July	R_{23} August (24/26)
5 R_3 March and R_{12} April	R_{12} May (17/20)
6 T_{12} November and T_{345} spell December to January	T_{345} February (14/15)

N.B. Rainfall for England and Wales is in terciles R_1 (dry), R_2 (average) and R_3 (wet). Mean temperature over central England is in quintiles – from T_1 (very cold) to T_5 (very warm). R_{23} signifies that tercile 2 or tercile 3 applies, and so on. Frequency distributions in brackets refer to terciles or quintiles, with the lowest first. When fractions are used the numerator gives the cases which satisfy the prediction whilst the denominator indicates the total number of cases with the pre-condition shown.

(b) Monthly rainfall prediction (December–March) from 500 mb trough positions at 50°N

		A West of 75°W and 20°W to 5°E Forecast wet	Neither A nor B Forecast average	B 50° to 75°W and 5°E to 25°E Forecast dry
England and Wales rainfall (in terciles) in following month	Wet	4	3	4
	Average	2	7	4
	Dry	1	4	10

dynamical explanation of the evolving pressure field is possible. Hay (1968) notes the association between northern hemisphere pressure patterns in March and pressure values in Britain in April. Anticyclonic Aprils often follow an intensification of the North American upper trough, particularly when this lies farther west and extends more to the southwest than on the normal chart. This anomalous feature is often accompanied by an intensification of the ridge found on the mean March chart extending south-southeast in 120–100°W from an anticyclone over Alaska. This pattern in March should allow a stronger than normal flow of cold air to feed south over the frozen ground of North America and this in turn would disturb the west-southwest thermal gradient normally found over the western North Atlantic so that the 1000–500 mb thickness isopleths would assume a more south-southwest orientation. In such circumstances the favourable location for anticyclogenesis is downstream and on the right-hand side of the region of strongest baroclinicity, i.e. in the vicinity of the British Isles.

Table 5.21 Some seasonal weather relationships in Britain (Murray, 1970)

Pre-condition	*Prediction*
1 T_{12} winter and T_{12} February	T_{123} spring (10, 6, 6, 2, 1)
2 T_5 February	T_{345} spring (1, 2, 4, 4, 8)
3 T_1 April	R_{23} summer (2, 6, 11)
4 R_3 summer	R_{23} autumn (5, 11, 14)
5 R_3 Autumn	T_{23} winter (6, 8, 9, 4, 3)
6 R_3 September and R_3 October over both Scotland and England and Wales	T_2 winter (3, 10, 5, 0, 1)

N.B. See footnote to Table 5.20(*a*).

Similar work has been performed for Israel by Krown (1966). He shows that October precipitation correlates well with that of the following three months and that the major determinant appears to be the position and intensity of the 500 mb mean trough over the Mediterranean in October. A western position (10°E) leads to above normal precipitation, an eastern position (30°E) to below normal. The latter is essentially a low index pattern.

Intensive study has recently been given to the character and origin of various persistence patterns in the circulation. Particular attention has been devoted to the blocking pattern which occurs primarily when a surface anticyclone develops in comparatively high latitudes in association with an upper thermal ridge. This has the effect of producing an obstacle in the normal westerly flow. Sawyer (1970) used a filtering procedure to examine fluctuations in the range twenty to sixty days in 500 mb height and 1000–500 mb thickness (see pp. 204–5). He showed that their maximum amplitude occurs about 68–70°N over the north-eastern Atlantic and Pacific Oceans, areas well known with respect to blocking activity. Sumner (1954, 1959) has shown that the preferred longitude for such blocks lies just west of the British Isles and that they are most frequent in late spring (Mook, 1954),[1] when spells of westerly type are least common. Retrogression of the block westward is most common in the first six months of the year, but progression eastward occurs more often in autumn. Geb (1966) shows that there are in fact at least six different kinds of blocking which can be differentiated primarily according to the orientation and form assumed by the upper thermal ridge. Since the mean duration of blocking is sixteen days this atmospheric pattern is likely, once established, to dominate the circulation character over a large area for a considerable period of time. In the very cold winter of 1962–3 in Britain a very stable blocking pattern became established late in December 1962 and persisted with only small modifications until the following March, while repeated blocking action during the summer of 1968 (Murray and Ratcliffe, 1969) produced an inverted pattern of summer weather with dull wet conditions in south-east England and brilliantly fine warm weather in the northwestern part of the British Isles (Perry, 1970). Blocking has also been shown to be important over eastern Asia (Wada, 1962) and if the polar vortex is displaced towards this sector in April–May the summer in northern Japan tends to be cool.

3 VERY LONG-RANGE FORECASTS

The problem of assessing likely trends over years or decades introduces many new considerations including, of course, climatic fluctuations. Fundamentally, therefore, better understanding of the physical causes of these fluctuations is a prerequisite for prediction on such time scales (Sawyer, 1965). Some aspects of this problem were touched upon in section C of this chapter, but the possible approaches to prediction were not discussed. Five independent conditions involving relationships, past and future, between climatic and environmental variables have been stated by Mitchell (1968). Prediction is theoretically possible if one of the following requirements is met:

1 A climatic variable is autoregressive (so that it can be predicted from knowledge of its past history).

[1] For blocking situations $\leqslant 5$ days over Europe and the eastern North Atlantic during 1951–70 the maximum frequency is in February and October (Montalto, Conte and Urbani, 1971).

2 A climatic variable is *either* correlated with *or* physically determined by one or more environmental variables which are themselves autoregressive.
3 A climatic variable is *either* correlated with *or* physically determined by environmental variables which are predictable by physical laws.

In general our knowledge is inadequate to attempt very long-range prediction from physical laws. Rather, attempts have been made to make inferences based on statistical correlations. For example, it has been suggested that circulation patterns may exhibit marked persistence tendencies particularly with respect to zonality/meridionality and blocking (Girs, 1960b; Brezowsky, Flohn and Hess, 1951). However, in a study of decadal mean temperatures since 1775 at Bratislava, Koncek and Cehak (1969) demonstrate that only 'normal' decades show a tendency to develop runs. Moreover, combining four decades and examining the frequency of the various possible combinations (625 in all), while three combinations form 39% of the total, the remainder are accounted for by all other possible combinations. They conclude, therefore, that decade mean temperatures at least are unsuitable as a basis for long-term analogue/persistence forecasts. Other work has involved supposed links between weather and solar cycles, an approach particularly linked with the name of Baur (1956, 1958), and more recently weather–lunar relationships. These are discussed below. Several of the effects involve short-term interactions but it is convenient to treat them all together here, particularly as most of the results discussed have, for the most part, not yet been formally incorporated in forecast procedures.

Cycles and oscillations

Although events in the atmosphere appear to take place in an essentially random sequence, meteorologists have long held the view that there is some regularity with respect to time other than the diurnal and seasonal rhythms. Among the earliest documented references to cyclical weather changes is that of Howard (1842), but the reader has only to recall Joseph's prophecy to Pharoah of seven fat and seven lean years to realize the antiquity of the search for order in weather phenomena. Köppen (1881) was another early climatologist interested in the occurrence of weather periodicities. By 1936 Napier Shaw noted more than fifty studies claiming to have discovered periodicities, many related to movements of the heavenly bodies or variation in solar phenomena. One of the most systematic early attempts at finding cycles was Brückner's (1890) work with early European temperature and rainfall records. He deduced a thirty-five to thirty-six year cycle composed of a cool humid half and a warm dry half. The amplitude of the temperature oscillation did not exceed 2°F in any cycle and the mean variation in precipitation was of the order of 17% when considered for a large number of stations.

The statistical evidence for the existence for many of these cycles is weak or inconclusive, and it is only in the last decade or so that sophisticated filtering techniques have been available for the objective analysis of cycles. Many of these recent studies have supported the evidence for lunar and solar relationships and synoptic developments, with the result that the early studies have assumed a new interest and importance. An example to illustrate this is Bryson's (1948) finding, almost ignored at the time, of variations in the latitude of the eastern Pacific anticyclone related to variations in the equatorward component of the tide-producing force of the moon. The discovery and statistical confirmation of correlations between precipitation and lunar cycles (Bowen and Adderley, 1962; Brier and Bradley, 1964; Visvanthan, 1966) and the suggestion of other lunar–weather relationships has refocused interest on the whole subject of lunar influences and the weather.

The quasi-biennial oscillation. In scanning the meteorological literature the reader is soon struck by the fact that a few of the numerous alleged cycles reappear in many of the accounts. Most prominent among these is a period of slightly over two years, now referred to as the *quasi-biennial oscillation*. This was first noted by H. H. Clayton in 1885 on the basis of American temperature data taken at widely spaced localities for 1839–82. He placed the length at twenty-five to twenty-five and a half months, but in a later study revised this to two years. In Europe Voeikof (1895) called attention to a two-year cycle in ice conditions and snow cover. Useful bibliographies on the subject have been prepared by Landsberg (1962) and Craddock (1968). Perhaps the first work which attempted to explain the phenomenon was that of Helland-Hansen and Nansen (1920) who found a two-year wave in the sea surface temperatures off Norway and in the Stockholm temperatures for 1874–1911.

Recently, powerful analytical tools such as variance spectrum analysis have become available to confirm the reality of the biennial oscillation in many surface parameters (Landsberg *et al.*, 1963; Shapiro, 1964; Böhme, 1967; Angell and Korshover, 1968). Indeed, the evidence includes not only conventional meteorological parameters but features such as the Nile floods (Brooks, 1928b), lake levels, tree rings and varves. Mitchell (1966) reports a significant twenty-five-month peak (and associated shorter term harmonics at 8·8, 4·2 and 3·2 months) in Manley's 269 year series of monthly mean temperature anomalies for central England (see fig. 4.9). Wright (1968a) argued that the average period of the oscillation must be precisely, rather than approximately, two years since if it differed from this interval it would be in reverse phase as often as it was in direct phase. Subsequently, however, he detected a phase change at thirty-year intervals in wine-harvest records since the seventeenth century (Wright, 1968b). This would reflect an oscillation of 25·7 months, in good agreement with other evidence.

The world-wide extent of the phenomenon has now been verified although its amplitude decreases in high latitude. Here temperature oscillations are out of phase with those in the tropics and are most evident in the winter months (Landsberg *et al.*, 1963). From a review of observational evidence Wright (1971) suggests that the quasi-biennial features observed in Europe and the North Atlantic may reflect air–sea interaction and that several independently produced oscillations of about two years may be present in the atmosphere. These include fluctuations in blocking activity (Böhme, 1967; Geb, 1966) and the Southern Oscillation and North Atlantic Oscillation which are discussed on pp. 414–7.

After the discovery in the 1950s of east–west wind shifts with a period of twenty-six months in the equatorial stratosphere (Veryard and Ebdon, 1961) much interest was focused on the question of tropospheric–stratospheric links. The stratospheric oscillations have a maximum amplitude at about 25–30 km and propagate downwards at about 1 km/month. Ozone and associated temperature variations are closely linked to these wind changes, although their basic cause is still not known (Reed, 1965; Murgatroyd, 1970). Labitzke (1965) demonstrated that wintertime stratospheric warming began in the region of the Caspian Sea and moved westward in the even years from 1958 to 1964, but began in the vicinity of Bermuda and moved eastward in the odd years. A fundamental mechanism in the general circulation must be responsible for such phenomena and Landsberg (1962) has suggested the precession motions of the subtropical high pressure systems, possibly influenced by the asymmetry of circulation in the two hemispheres. The trade wind circulation and the meridional Hadley cell would be affected by these factors, and the Hadley circulation of course provides a stratospheric return current. Solar influences were proposed by Shapiro and Ward (1962) who found evidence of a twenty-five month peak (significant at the 5% level) in the spectrum of sunspot numbers between 1756 and 1955, but this hypothesis, and others involving external influences, remain in doubt (Berson and Kulkarni, 1968). Wallace and Newell (1966) show that eddy transport of heat and momentum at 30 mb and above is greater in winters of odd years in low and middle latitudes, although they could not determine whether the forcing was of tropospheric or stratospheric origin, or both. This question has been answered, at least in part, by Lindzen and Holton (1968). They demonstrate that momentum is transferred upwards from the equatorial tropopause to a critical layer below 40 km where the phase of oscillation is triggered by an existing semi-annual oscillation (see below). This momentum transfer to the mean flow causes the critical layer to move downward. The cycle repeats itself after some integer multiple of six months. In fact, five of the cycles in the stratospheric zonal wind during 1954–70 were approximately two years in length and two were close to three years. The extension of the oscillation into extratropical latitudes has been recognized since the work of Angell and Korshover (1962). Dartt and Belmont (1970) show that it is a global phenomenon but with irregular time variations in the high latitude stratosphere. Moreover, there are significant longitudinal differences in middle latitudes. Also there has been a curious reversal in the latitudinal phase progression. Prior to 1960 the oscillation showed poleward phase progression at 300 mb and equatorward at 30 mb, but these reversed directions in 1961. The whole topic remains a fascinating problem.

Work has started on the synoptic effects of the cycle. Shapiro (1964) suggested that 'the interest in such oscillations does not lie in their potential use as forecasting tools, as the amplitude of the spectral peaks are only about 1–2% higher than their respective continua', but Murray and Moffitt (1969) nevertheless believe that it is worth taking them into account as a second-order effect in long-range weather forecasting. Accordingly they have investigated the effect of the oscillation in terms of monthly mean pressures in different parts of the northern hemisphere using fifty-eight years of data. The mean difference in monthly pressure (odd–even years) is shown for the first six months of the year in fig. 5.45. There is a good deal of continuity from month to month in many of the main features and the results are also reflected in differences between odd and even years in the averages of mean surface temperature and monthly rainfall. These results are particularly interesting when it is remembered that the oscillation is not always equally pronounced. Also it may be masked by more localized effects such as variations in the strength of the North Atlantic circulation.

Attempts to predict forthcoming summer weather conditions over the British Isles using the evidence of the two-year cycle have been made by Poulter (1962) and Davis (1967) and it seems that if the ordinary man in the street took a double-length summer holidays only in alternate years – taking a month every odd year instead of a fortnight every year – he could expect to enjoy higher temperatures, more sunshine and less rain on his vacation! Many agricultural activities must be affected by the cycle and already the effects of honey production (Hurst, 1967) and wine quality (Wright, 1968b) have been considered.

Half-yearly oscillations. Half-yearly variations of surface pressure in middle and high latitudes of the southern hemisphere were first reported by Reuter (1936). A half-yearly component in the annual march of

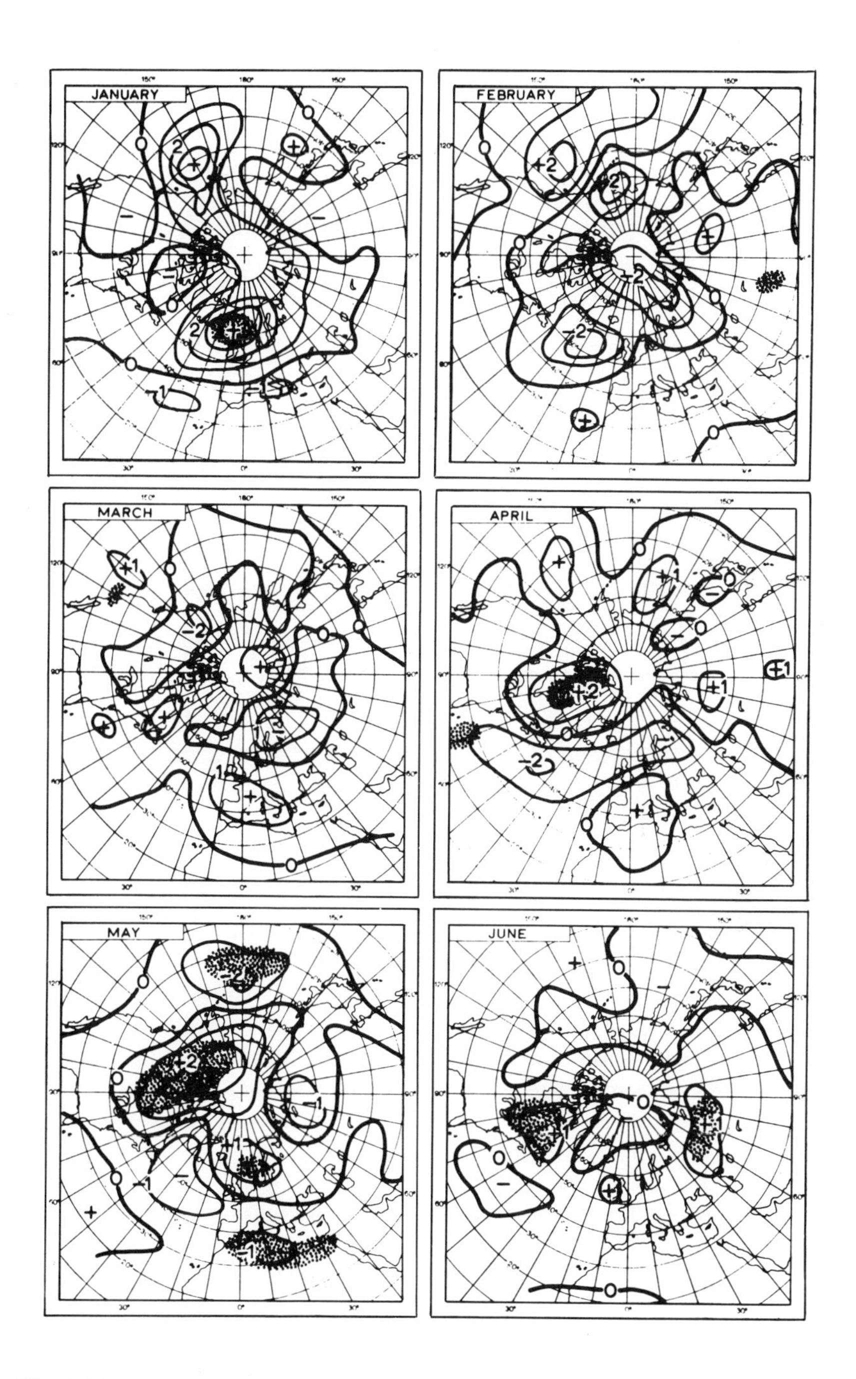

Fig. 5.45 Averages of monthly mean surface pressure differences (mb) between odd and even years for January–June. The stippled areas indicate differences significant at 5% level (from Murray and Moffitt, 1969).

temperature, geopotential heights and zonal wind in middle and high latitudes of both hemispheres, as well as another of different origin in the tropics, have recently been analysed by Van Loon and Jenne (1969, 1970). That in the tropics is most important between about 300 and 200 mb and is particularly pronounced between 60° and 120°E. Here, for example, the second harmonic accounts for 60% or more of the variance of mean temperature at latitudes 10°N and 25°S. Fig. 5.46 shows its contribution to the zonal geostrophic wind in the same sector.

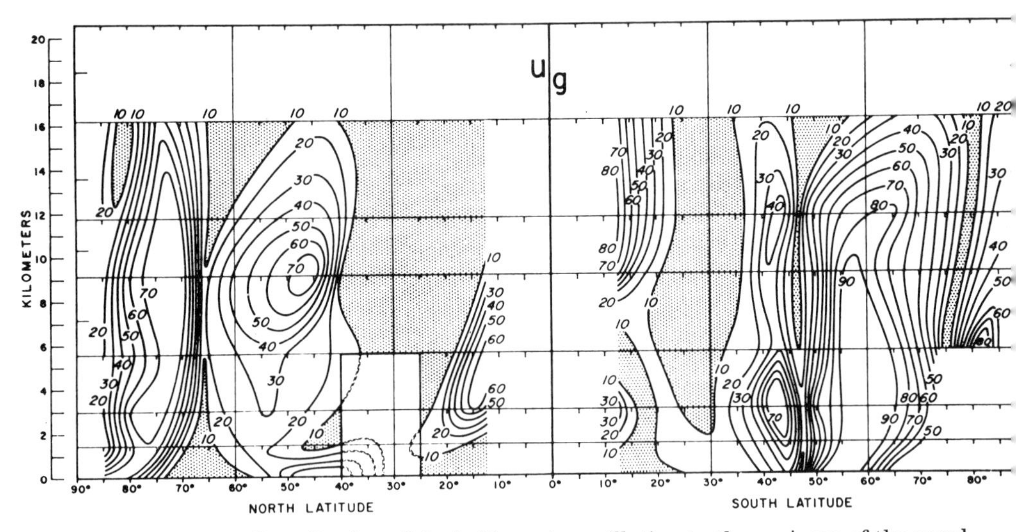

Fig. 5.46 Contribution of the half-yearly oscillation to the variance of the zonal geostrophic wind over 60–100°E (from van Loon and Jenne, 1970). Isopleths in per cent.

The maxima occur in May and November between 15–30°S and 20–35°N, but in February and August in the lower troposphere near the equator. It is suggested that these latter phase relationships reflect the effect of the large-scale monsoonal type of circulation moving from one summer hemisphere to the other. The maxima occur when the rising branch of the Hadley cell, which releases latent heat, is situated near the equator during the transition months. The phase relationship in the subtropics appears to be associated with latitudinal displacements of the subtropical jet stream and the associated baroclinic zone on the poleward side of the Hadley circulation.

In Antarctica Van Loon (1967) shows that the well-known *kernlose* (coreless) winter is related to the high latitude oscillations. This term refers to the reversal of the expected cooling trend at antarctic stations in winter and the lack of any well-defined seasonal minimum. The meridional temperature gradient in the subantarctic zone shows a dominant half-yearly variation in the middle troposphere as a result of the seasonal warming and cooling trends in middle and high latitudes, a hypothesis first suggested by Schwerdtfeger and Prohaska (1956). The temperature gradient between Antarctica and the oceans decreases in early winter and the subpolar low pressure belt moves equatorward (as it also does in spring). Simultaneous pressure increase over the continents, however, increases longitudinal pressure gradients amplifying the wave pattern. Meridional circulations are thereby enhanced with more frequent warm air advection into high latitudes offsetting the seasonal cooling trend, perhaps in part by wind disturbance of the near-surface temperature inversion.

The annual march of surface pressure in middle and high southern latitudes has a well marked second harmonic (of 1–3 mb amplitude) which is of opposite phase in these regions with maxima at the solstices poleward of 60°S and equinoctial maxima in middle latitudes (van Loon, 1971). This phase reversal leads to an augmented meridional pressure difference and consequently zonal wind components showing a like response, with spring and autumn maxima at 60–65°S. At latitude 60°S the second harmonic accounts for 87% of the total variance of the zonal geostrophic mean wind at MSL and 65% at 500 mb.

Atmospheric tides. The response of the atmosphere to the tidal potential is in the main less complicated than that of the oceans because the atmosphere has no lateral boundaries. However, the atmospheric tides must often be affected by the earth's surface topography and by the non-uniform and varying thermal properties of the surface. Naturally the difficulties in separating out these interactions are considerable. Brier (1968) has suggested that it may be possible to differentiate the following cycles in the atmosphere:

1 The solar atmospheric tidal oscillation called the S2 wave – 12·4 hours (discussed by Chapman (1951), Haurwitz and Cowley (1965) and Brier and Simpson (1969)).
2 Anomalistic month – 27·55 days (period from one perigee to the next). The effectiveness of the moon in raising tides may be expected to increase by some 22% as it moves to its position of closest approach to the earth (perigee).
3 Synodic month – 29·53 days (period from one new moon to the next).
4 Tidal year of twelve synodic cyles – 11·6 months.
5 Nodical cycle of 18·6 years (when the earth's rotation pole revolves elliptically about its 'mean' central position; a smaller 14-month oscillation is superposed on this cycle, see p. 370).

The role of the S2 tidal oscillation in producing a semi-diurnal frequency maximum of cloudiness and rainfall in the tropics around sunrise and sunset now seems to be confirmed (Brier and Simpson, 1969). The pressure changes of 2–3 mb created by the S2 wave are considered sufficient to account for the 5–15% increase in mean cloudiness and precipitation, although the causal links are still not known. A possible model involving convergence (Stolov, 1955) is shown in fig. 5.47. Brier and Carpenter (1967) suggest that the tropical atmosphere may be more susceptible to the development of synoptic disturbances when the amplitude of the S2 wave is large. However, it should also be noted that Simpson *et al.* (1967) show that travelling disturbances may not be the all-important determinant of tropical cloudiness and rainfall that they are usually considered to be. Many perturbations that are significant in these respects seem to develop and decay *in situ*.

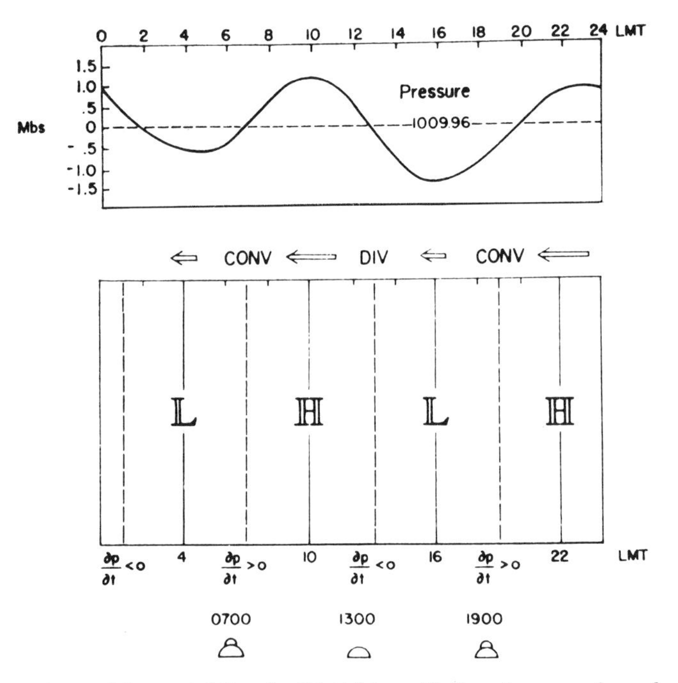

Fig. 5.47 A possible model for the S2 tidal oscillation (arrows above lower diagram) related to the diurnal pressure wave and cloudiness in the tropical ocean areas (from Brier and Simpson, 1969). Top: diurnal pressure wave, LMT = local mean time. Bottom: time (or space) pressure tendencies, open arrows show variations in tropical easterlies due to S2 wave.

The evidence of lunar tidal effects in precipitation data, already referred to, as well as in sunshine observations (Lund, 1965) and in temperature variability (Mills, 1966), relates primarily to the anomalistic and synodic months. The latter also shows up in a periodogram analysis of zonal index data (fig. 5.48) together with an unexplained peak at 162 months (thirteen and a half years). The causal linkage between synoptic disturbances and lunar influences is as yet unclear, but may include (Brier, 1966) factors such as air–sea interaction. If any predictable outside force can be shown to cause even a measurable fraction of the variation in atmospheric behaviour, then some aspects of weather particularly in the tropics may become more predictable. The need for careful analytical techniques, however, is shown by O'Mahony's (1965) study. He demonstrated that apparent correlations between rainfall and moon phase in Australia were spurious effects resulting from the use of running means in smoothing the data series.

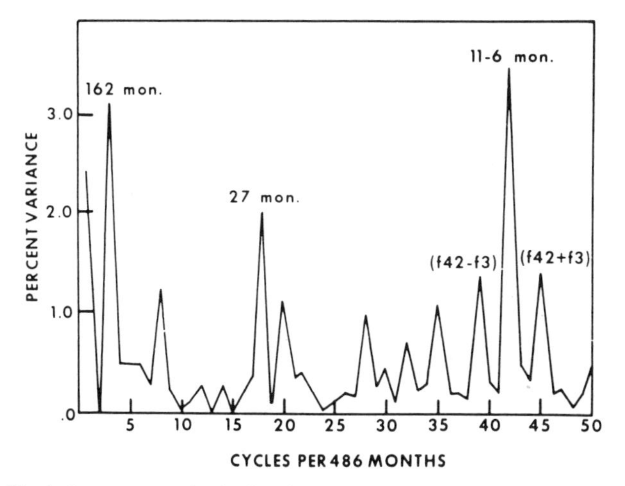

Fig. 5.48 Periodogram analysis for frequencies $f = 1$ to 50 of departures from monthly normal of the zonal index, 1899–1939 (from Brier, 1968).

Solar influences and weather cycles. Reviewing solar influences on the weather, Brooks (1951) wrote – 'the literature is extensive, complex and generally vague'. The eleven-year sunspot cycle has been known for centuries, but the relationship between the cycle and the frequency of such phenomena as solar flares has been known for a much shorter time. Solar–terrestrial events are on three time scales (Tucker, 1964) – the eleven-year period, the twenty-seven day period related to the occurrence of large magnetic storms, and the immediate period related to ejection from the sun of corpuscular streams. Various claims for effects on the short time scale have been advanced but none are convincing. For example, spectrum and cross-spectrum analysis of daily mean station pressure at polar stations shows no indication of an association with the emission of solar particles according to Shapiro and Stolov (1970), although MacDonald and Roberts (1960) detected such effects at the 300 mb level.

It is the eleven-year cycle that has attracted most attention from meteorologists. Brooks (1928a) could find no systematic variation of pressure type over western Europe during the eleven-year cycle, but Clayton's (1936) method of long-range forecasting was based on the belief that changes in temperature and pressure in all parts of the world have an intimate relation to changes in the solar condition. Increases in solar radiation were found to be accompanied by temperature rises and pressure falls in equatorial regions. However, Willett (1940) and Page (1940) pointed out that periodicities in Clayton's data are the result of his method of smoothing so that the physical basis of his work is insecure.

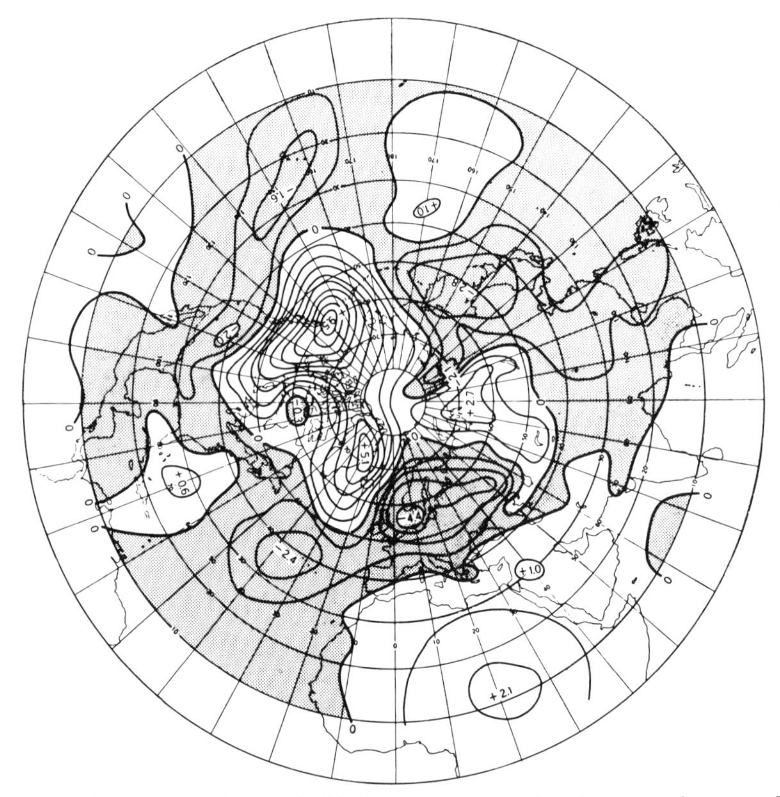

Fig. 5.49 Pressure difference (mb) for three consecutive Januarys between four sunspot maxima and four sunspot minima (from Wexler, 1956).

The eleven-year cycle varies over time in both amplitude and period. It may range in length from about eight to seventeen years and varies by a factor of approximately three in annual relative sunspot number between small and large maxima. Subperiods and beat frequencies between periods of different length may account for other intervals at 2·2,

2·8, 5·5, 22–23, 33–35 and 55–57 years. Longer term variations of 80–89, 178, 400 and 1700 years have also been postulated (see below and p. 378). Much fundamental work remains to be done, however, in order to establish the occurrence and magnitude of any variations in the solar constant (which cannot in any case exceed 1–2%) and the mechanisms by which such fluctuations could affect the troposphere.

A relationship between blocking pattern frequency and the solar cycle has been suggested by Wexler (1956). He considered the four solar cycles between 1900 and 1939 and took the average of three consecutive Januaries centred on each of the four sunspot maxima and minima. When the two are subtracted the average pressure difference (fig. 5.49) shows a coherent pattern with positive and negative centres of relatively large magnitude. Baur (1958) is critical of the significance of the results, since if a fifth cycle is substituted for the first, a completely different pattern is obtained. This suggests either that the patterns are not meaningful, or that any solar effect on the circulation is variable over time. Troup (1962) did in fact find a reversal of phase in the correlation values of tropical temperatures and sunspots. Great caution is needed if sunspot data are to be profitably employed by the long-range forecaster. Later work by Baur (1963) describes a double cycle of total solar radiation intensity, and particularly of ultraviolet radiation, reaching the atmosphere within the normal solar cycle. By carefully selecting limits and partitioning each sunspot cycle he has been able to draw up contingency tables relating precipitation anomalies in central Europe to the phase of the sunspot cycle. Dehsara and Cehak (1970) report on a more recent study of annual mean temperatures and precipitation totals at ninety-two stations. On average, temperature lags one quarter of a full cycle (nine to twelve years) behind the sunspot cycle and precipitation is up to a quarter of a cycle in advance.

With regard to pressure effects, Müller-Annen (1961)[1] finds evidence of more frequent westerly *Grosswetterlage* types in epochs with low sunspot cycles and vice versa. This theme has also been taken up by Schuurmans (1969) who considers that excess ultraviolet radiation may act as an intermediary between solar flares and pressure changes. The results of his investigation of the frequency of *Grosswetter* patterns following flares are shown in table 5.22. The results indicate that meridional *Grosswetterlagen* increase for the first three days following a flare, perhaps as a result of a northward displacement of the subtropical anticyclone, at the expense of zonal types. This pattern was evident in two subsamples of forty and forty-one flares, although Schuurmans

[1] Unpublished work by B. N. Parker of the British Meteorological Office using pressure data for all sunspot cycles between 1750–1958 shows increased westerly flow over the North Atlantic–European sector in winter for the descending phase of the cycles (see Lamb, 1972, pp. 496–8).

Table 5.22 *Deviations in the mean frequency of zonal, mixed and meridional* Grosswetterlagen *from the daily mean for the date of a flare and seven days thereafter* (*Schuurmans, 1969*)

Day	0	1	2	3	4	5	6	7
Meridional	−0·6	−1·8	+5·6	+5·6	−3·1	−6·7	+1·9	−0·6
Mixed	−3·2	+3·0	+0·6	+0·6	+3·0	+4·3	−4·4	−3·1
Zonal	+3·8	−1·2	−6·2	−5·0	+0·1	+2·4	+2·5	+3·7

emphasizes that it was not statistically significant and, since it does not follow every flare, is probably of limited forecasting value.

Willett (1962) suggested that the total amount of ozone varies during a sunspot cycle with a maximum of one and a half to two years before a sunspot minimum, accompanied by a tendency for more frequent blocking situations, although sunspot latitude appears to be an even more important variable than number of sunspots. London and Haurwitz (1963) have criticized Willett's interpretations, but the arguments have since been elaborated by Lawrence (1965). Willett (1964) has also drawn attention to an eighty to ninety year sunspot cycle (i.e. eight solar cycles). He postulated that in the first quarter with low sunspot numbers the circulation is zonal and in low latitudes, shifting to high-latitude zonal in the third quarter with increasing sunspots and to blocking patterns in the last quarter. A study by Eastwood (1965) finds agreement for July in the northern hemisphere for 1798–1877 and 1878–1962, but not for January, and no variations in the southern hemisphere. This model is, therefore, rather dubious, although Easton (1918) claimed an eighty-nine-year periodicity in winter temperatures since A.D. 760 over western Europe (see also Visser, 1959).

In spite of the considerable uncertainty about the significance of the various solar cycles for the atmospheric circulation, consideration is being given to their possible implications for climatological forecasts. A recent book by Pokrovskaia (1969) devotes considerable attention to this topic. Many more unequivocal results are required, however, before the proposed forecasting principles can be regarded as better than speculation.

Ocean–atmosphere interaction

It has long been recognized that the ocean surface circulation is primarily driven by wind stress. The spatial distribution of air–sea energy exchange, which in turn relates to the frequency and type of weather system (Hay, 1956), cannot be overlooked, however, since it can greatly modify the horizontal and vertical motion in the oceans. During the last decade it has also been shown that areas of anomalous ocean temperature can produce anomalous atmospheric circulation patterns by injecting latent and sensible heat into the atmosphere. The important influence of the ocean in storing and transporting heat has been demonstrated in numerical computations by Manabe and Bryan (1969), and a large number of empirical or semi-empirical studies, particularly by Namias (1964, 1968, 1969), have examined some of the forms that such interactions can take. Rigorous investigation of possible relationships between anomalies of sea surface temperature (SST) and atmospheric circulation requires a long historical series of anomaly charts, and fortunately these are now available for the North Atlantic. Riehl (1956) has produced such charts for areas south of 50°N for the period 1888–1936 and these data have been combined with anomaly data for the Atlantic north of 50°N (Smed, 1948) by Ratcliffe and Murray (1970). Since 1949 the ocean weather ships have provided considerable data, and more recently computer analysis of merchant shipping reports has been carried out (Wolff, 1966).

The role of areas of anomalous ocean temperature in causing persistence of particular synoptic types over a period of a few years is illustrated by the recurrence of blocking patterns over northern Europe during 1958–60 (Namias, 1964) associated with below normal sea temperatures in the western Atlantic and above normal in the eastern Atlantic. Probably the net effect of the anomalous ocean temperature pattern was to strengthen the east–west temperature gradient in the North Atlantic and consequently the southerly component of the thermal wind, thereby steering depressions persistently northward rather than eastward as is normal. Fig. 5.50 illustrates this schematically.

Persistence on a shorter time scale due to anomalous sea surface temperature (SST) is suggested by Namias (1969) in a study of autumnal anticyclogenesis in the North Pacific and North Atlantic. The cause seems to have been extensive pools of anomalously cold water in the Novembers of 1966 and 1967 (Petterssen (1950) has presented statistical evidence showing that bodies of cold water surrounded by warmer land favour anticyclogenesis). The abnormal pattern soon destroyed the cold water pool which originally caused it through increased insolation and less vertical stirring and upwelling. This negative feedback ensured that the circulation anomaly was only short lived.

Recent work has attempted to organize the body of information obtained from many of these case studies so that long-range forecasters might profit from an understanding of such feedback relationships. Semenov (1960b) related blocking activity in the Atlantic–European sector to sea temperature conditions in the North Atlantic, while Valerianova (1965) attempted to link on a monthly basis different types of circulation according to Vangengeim's classification with patterns of

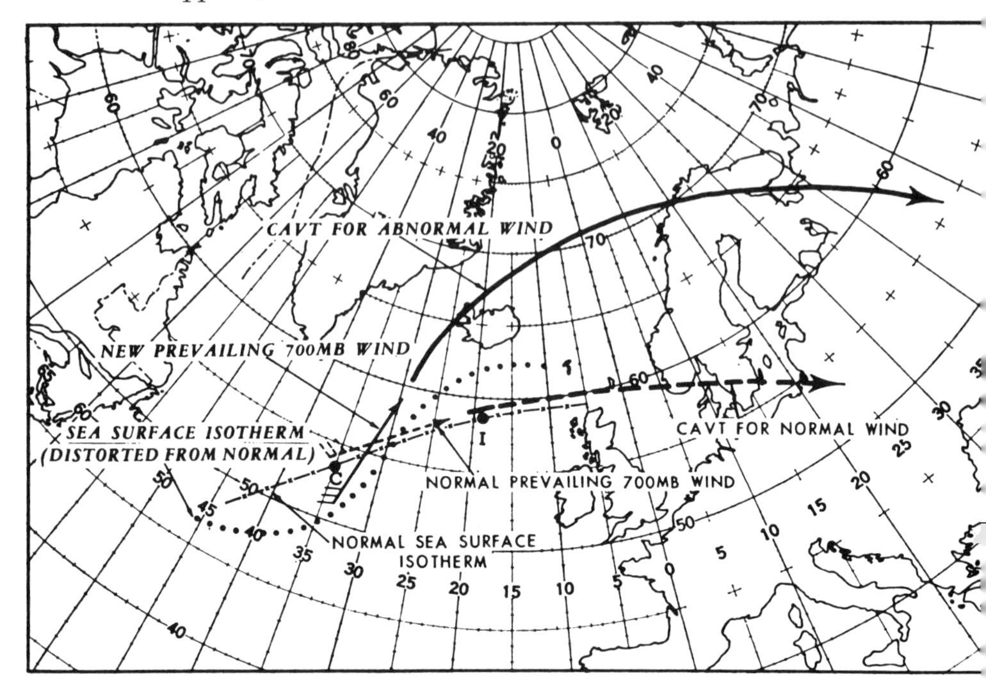

Fig. 5.50 A model of feedback effects due to sea surface temperature anomalies. CAVT – constant absolute vorticity trajectory (from Namias, 1964).

sea temperature anomaly in the North Atlantic. Perry (1969) has extended this analysis and the results suggest that anticyclone centres are one and a half to two times more frequent than normal over Scandinavia and the Norwegian Sea when positive water temperatures anomalies occur, and about half as frequent as normal when negative anomalies prevail.

Ratcliffe and Murray (1970) found from visual inspection of the SST anomaly maps for the North Atlantic that two broad-scale patterns are common. These are extensive anomalies of either positive or negative sign in the area south of Newfoundland. Surface pressure anomaly maps were produced for the month following the SST anomaly and the significance of the resulting pattern tested with Student's t test. The results for two sets of months are reproduced in figs. 5.51 and 5.52. There is little doubt that the results indicate a physically based lag relationship and hence have considerable forecasting potential. The physical mechanism in question may be similar to that described by Wolff and Laevastu (1968) with excess latent heat being carried many hundreds of miles downwind in the upper flow. Such a process tends to operate on a time scale of a few days but recurrences over the period of a month or so also appear possible. Our knowledge of the processes involved will

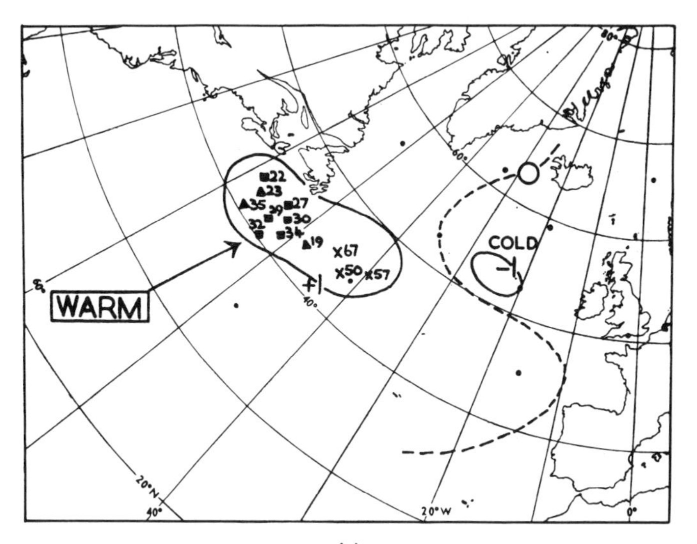

(a)

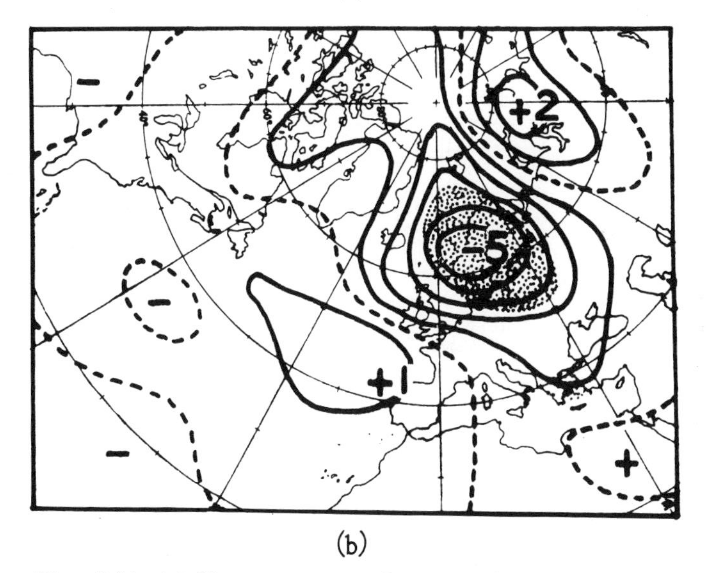

(b)

Fig. 5.51 (*a*) Mean pattern of sea surface temperature anomaly (°C) in September in years with a positive anomaly off Newfoundland. (*b*) Mean MSL pressure anomaly (mb) in October for years corresponding to (*a*). Stippled areas represent values significant at $\leqslant 5\%$ (from Ratcliffe and Murray, 1970).

Key:

■ positive anomaly > 2 degC

▲ positive anomaly between 1 and 2 degC

× position and intensity less certain.

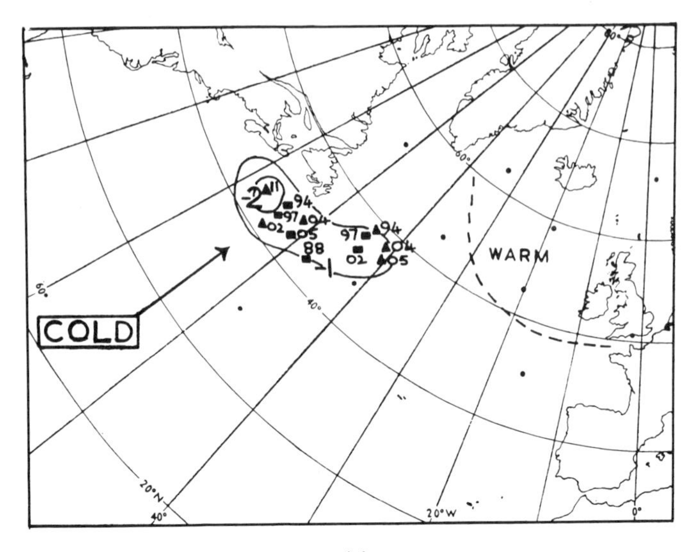

(a)

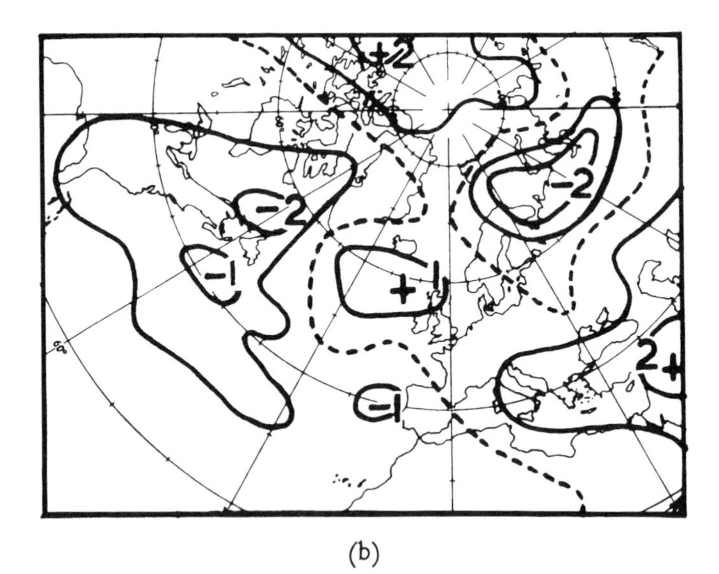

(b)

Fig. 5.52 (*a*) Mean pattern of sea surface temperature anomaly (°C) in December in years with a significant anomaly off Newfoundland. (*b*) Mean MSL pressure anomaly (mb) in January for years corresponding to (*a*). Stippled areas represent values significant at $\leqslant 5\%$ (from Ratcliffe and Murray, 1970).

Key:

■ negative anomaly > 2 degC

▲ negative anomaly between 1 and 2 degC

× position and intensity less certain.

increase as further synoptic climatological case studies of heat flux patterns during months with different circulation regimes are carried out (see section E of this chapter, p. 419). Particular importance is attached to the extension of such analyses because surface temperature anomalies of oceanic dimension have been found to persist over seven to twenty-four months (Nazarov, 1968). The mean duration of large-scale ocean anomalies in the North Atlantic, based on data for 1948–59, is about two years. A model of ocean–atmosphere interaction indicating negative feedback in the North Atlantic area has been proposed by Duvanin (1968). With positive temperature anomalies in cold currents (fig. 5.53(*b*)) the mid-latitude zonal air circulation is weakened. This favours an accumulation of warmer water in the tropical area which, after a time, is carried into the system of North Atlantic currents. As the sign of the anomalies reverses (fig. 5.53(*a*)) the zonal circulation strengthens.

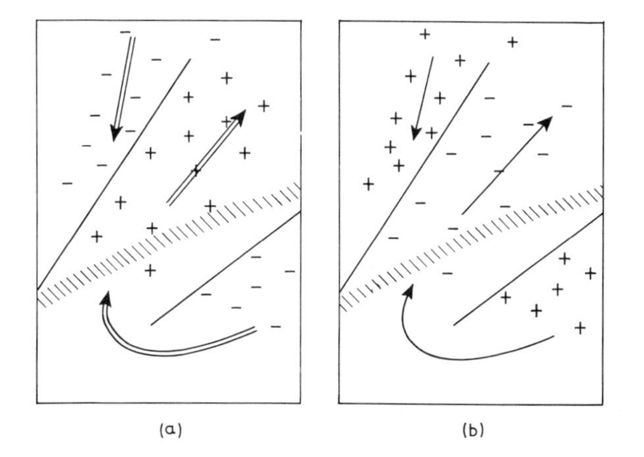

Fig. 5.53 Model of the interaction between water temperature anomalies and currents in the North Atlantic (after Duvanin, 1968). (*a*) Positive temperature anomalies in warm currents, negative in cold currents (leading to strong westerly circulation in the atmosphere). (*b*) Negative anomalies in warm currents, positive in cold currents (leading to weaker westerly circulation in the atmosphere).

Bjerknes (1966, 1969) demonstrates that air–sea interaction may operate on a hemispheric or even a global scale. Water temperature anomalies in the central and eastern Pacific occur as a result of changes in the strength of the trade winds in the southern hemisphere, probably themselves closely tied to the 'Southern Oscillation' (see next section). One effect is to alter the amount of heat supplied from the equatorial ocean to the ascending branch of the atmospheric Hadley circulation which, in turn, would increase the flux of angular momentum to the mid-latitude belt of westerly winds and result in stronger zonal westerlies. Krueger and Gray (1969), analysing meteorological records at Canton Island in the central Pacific, have in fact found low frequency oscillations in sea surface temperature with significant amplitude and periods of more than a year. These considerations therefore lead naturally on to the question of distant interactions.

Teleconnections

The existence of correlations between weather conditions in one part of the globe and those that are occurring or have occured elsewhere was first demonstrated by Hildebransson (1897). If the atmosphere is a unit, marked deviations from the normal over one large region must be compensated for elsewhere and such deviations should assume the form of slow pressure oscillations.

Walker's name (1923, 1924; Walker and Bliss, 1930; Bliss and Walker, 1932) is particularly linked with these teleconnection studies. By calculating contemporary and lag correlations between sea level pressure, rainfall and temperature in different parts of the world he discovered three large oscillations in which pressure in one region varies inversely with that in another:

1 North Atlantic Oscillation involving the Icelandic low and the Azores high pressure area.
2 North Pacific oscillation involving the Aleutian low and the North Pacific high.
3 Southern Oscillation involving pressure in the South Pacific Ocean and the equatorial Indian Ocean (fig. 5.54).

The persistence tendencies of these oscillations appear to be too small to make them significant for seasonal forecasting. Moreover, many of the correlation coefficients that Walker obtained were not statistically significant, and sampling fluctuations, trends and inhomogeneities in the data were largely overlooked. Walker (1938) also attempted to forecast monsoon rainfall in India by a study of the behaviour of the atmospheric centres of action. Several thousand correlation coefficients were calculated and regression formulae utilizing about thirty factors were drawn up but the results proved disappointing (Montgomery, 1940).

During the 1920s and 1930s teleconnection studies were particularly numerous, mainly adding to or modifying slightly the results obtained by Walker. Groissmayr (1930) found the mean winter temperature at Winnipeg to be an indicator for the temperature of the following spring in the Great Lakes region and he also correlated Argentine pressure with temperature six months later in the United States (Groissmayr, 1926).

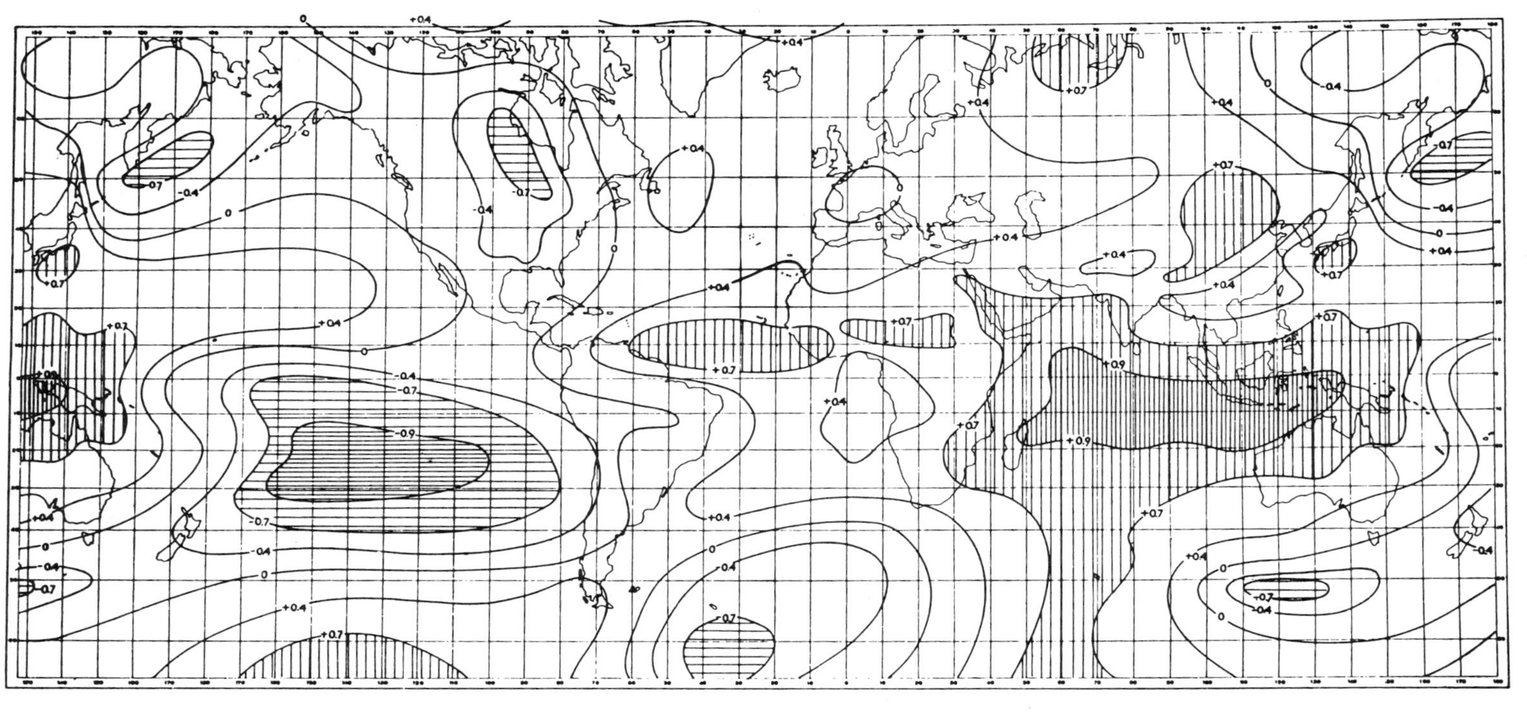

Fig. 5.54 Walker's 'Southern Oscillation'. Isocorrelates between local annual pressure anomalies and simultaneous ones at Djakarta, Indonesia, July 1947–July 1959 (from Berlage and de Boer, 1959).

Schell (1947, 1956b) proposed an extension of Walker's work using the concept of dynamic persistence. This is the persistence, in certain areas, of sea level pressure averages which affect subsequent weather in distant regions where no persistence occurs. For example, subnormal pressure values in tropical Australia were found to be followed by excessive storminess and precipitation in much of North America. Schell incorporated these and other relationships into simple regression equations for forecasting purposes and these had some limited success. Nevertheless, Weightman (1941) pointed out that physical explanation and synoptic evidence of the usefulness of many of these studies were lacking and this made the incorporation of many of the derived relationships into long-range forecasting of dubious value.

There has been a recent revival of interest in the Southern Oscillation using additional modern data (fig. 5.54) in order to test the reality of the phenomenon over time (Berlage and de Boer, 1959, 1960; Berlage, 1966; Troup, 1965). Although the signs of the earlier correlations can be confirmed there has been a decline in their magnitudes, supporting Berlage's (1957) suggestion that the Southern Oscillation probably undergoes secular changes. Table 5.23 indicates that this is the case. Probably fluctuations will only be successfully explained when more is

Table 5.23 The length and intensity of the Southern and North Atlantic Oscillations (after Berlage and de Boer, 1960)

	Southern Oscillation		*North Atlantic Oscillation*	
	1931–39	*1949–57*	*1931–39*	*1949–57*
Average length (months)	20·8	29·8	20·8	20·8
Amplitude/noise ratio in its own region	1·1	1·3	1·2	0·5

known about the causes of the phenomenon. Berlage and de Boer regard it as a stationary wave in its principal effect on the atmosphere, but the oscillation appears to originate in the equatorial low pressure over Malaya–Indonesia and is then sustained by coupling with the subtropical high in the South Pacific. Schell (1956c) believes that its existence is due to temperature differences between the relatively cool southeast Pacific and the relatively warm west Pacific–Indian Ocean region and that fluctuations in the strength of the oscillation may be sought in changes within the Humboldt–Peru Current complex. However, Quinn and

Burt (1970) demonstrate that the breakdown of the Peruvian coastal current associated with the El Niño phenomenon (when warm equatorial waters move southward along the coasts of Equador and Peru) is not apparently linked with the Southern Oscillation, as defined by MSL pressure at Darwin. At the same time they show that Darwin pressures in May–July are indicative of subsequent precipitation in the central equatorial Pacific dry zone. Further associations between the Southern Oscillation and droughts in Surinam and Indonesia have been postulated by Berlage (1966). Troup proposed a synoptic climatological model in which the cause of variations in the oscillation is linked to variation in the South Pacific trades and showed that energy exchange between atmosphere and ocean can be anomalous in amount for several months because of the tendency for persistence of sea surface temperature anomalies in this region. These anomalies have been seen to travel downwind in the trades (Berlage, 1957) probably as a result of evaporation effects.

The relationship between polar ice and the general circulation is a particular type of teleconnection study of some importance. A synthesis of the research of several investigators shows that heavy ice is associated with abnormally high pressure over the polar regions (Brooks and Quennell, 1928) and this in turn affects the general pressure distribution over large parts of the globe. In particular, cyclone tracks in mid-latitudes appear to be displaced southward in conjunction with the similar displacement of the zone of maximum thermal gradient. Schell (1940) constructed mean circulation maps for heavy and light ice years by averaging pressure throughout large portions of the hemisphere but he acknowledged that 'the role of polar ice in the atmospheric circulation is probably but a single phase of a more general and inclusive relationship and the variation of ice may also be the result of the character of the preceding state of the general circulation'. The generally recognized lowering of mean temperatures in much of the northern hemisphere in the 1940s occurred at a time when polar ice amounts were very small and it is only in quite recent years that heavy icing has been reported in the Barents and Greenland Seas (Marshall, 1958; Clarke, 1970). On the evidence of recent decades, therefore, the ice amount has *followed* rather than initiated the circulation changes, although on the shorter time scale Schell (1956a, 1970) has demonstrated that the summer limits of arctic ice are related to contemporary and subsequent temperature and pressure conditions to the east over Europe and Asia.

Much basic statistical work remains to be performed on teleconnections. The computation of the probabilities of the sign of 700 mb height anomalies over the northern hemisphere for1947–63 by O'Connor (1969) is a good illustration of the type of synoptic climatological data needed for further research.

E Other applications

In this section we shall outline a variety of areas in which synoptic climatological studies have found application. It is not intended to suggest that the synoptic climatological approach is the most useful or 'best' method in these particular fields, although at a certain stage or for some purposes this may be the case. Rather, we hope to illustrate the potential, as well as certain limitations, of the methods of synoptic climatology in some diverse topics in applied meteorology and climatology.

The topics considered basically involve problems in physical climatology. The rest concerns air–sea interaction where attention is focused on energy transfers and on temporal and spatial consequences which they create. The second involves several aspects of biometeorology and bioclimatology.

1 AIR–SEA INTERACTION

Basic problems

Hare (1966) observed that 'one of the outstanding recent changes in climatology has been the shift away from such parameters as temperature and relative humidity towards the measurement of fluxes'. A knowledge of the energy fluxes between ocean and atmosphere is essential to proper understanding of the atmospheric circulation and its prediction. The world distribution of the major terms in the energy budget for monthly and seasonal means has been known for some time; Budyko (1963) provided outstanding maps of this type. However, the magnitude of the individual terms on a short-term basis, and their range of variation, have only recently begun to be studied.

There are many theoretical and practical problems involved in attempting to determine transfers between sea and air. For example, since the heat transfer is effected via a very thin layer at the sea surface, and this is disturbed by wave motion, the proper theoretical formulation of the process is not determinable (Deacon and Webb, 1962). The accuracy of basic measurements, such as shipboard air temperature and sea surface temperature, required for the usual transfer equations, is often problematic (Edwards, 1970; Saur, 1963; Tauber, 1969). Moreover, as Laevastu (1965) points out, the formulae themselves have been determined from short-period observations, although they are frequently used for computing monthly or seasonal averages. In addition, non-linear interactions between the terms render the use of average values incorrect unless the computations are performed on a synoptic basis and then averaged. Diurnal variations of the elements are a further problem if a synoptic approach is adopted. Availability of sufficient synoptic data for reliable mapping of the spatial distribution of each oceanic and atmospheric parameter is a major practical difficulty. Laevastu shows that two analysts will produce somewhat different patterns of heat exchange with the same data, although the major features can usually be defined adequately. There can be little doubt, however, that significant detail is lost. Verploegh (1967), for example, draws attention to the local intensification of the wind field in the vicinity of fronts over the oceans. With these caveats in mind we can now look at the types of study that have been carried out and their results.

Synoptic case studies

In view of the data limitations and the other problems referred to, few synoptic studies of heat exchange over the oceans have been made. Case studies for the Gulf of Alaska (Winston, 1955) and the Sea of Japan (Manabe, 1957, 1958) have demonstrated the important role of latent heat in the short-term evolution of atmospheric flow patterns (Gambo, 1963). Indeed, the general evidence of synoptic charts indicates that most active cyclonic developments occur where the release of latent heat is pronounced (Sawyer, 1965). Computations of latent and sensible heat transfer in winter over the northern North Pacific by Laevastu (1965) show that cyclones predominate over areas where the transfer is to the atmosphere, anticyclones where it is in the reverse direction. Cyclones usually form where there are strong gradients of heat flux into the atmosphere.

A major contribution to the problem has been made by Petterssen, Bradbury and Pedersen (1962). Using the method of composite charts, they determined the distribution of sensible and latent heat exchange in five typical phases of the depression life cycle over the North Atlantic. Fig. 5.55(*a*) shows the pattern in a developing wave depression and fig. 5.55(*b*) that for a partly occluded system. It will be noted that, particularly in the former, the zones of marked convective activity are normal to the heating pattern. Simpson (1969) shows that there is, nevertheless, an important dynamic control on convective activity. Even with vigorous heat transfer to the atmosphere, cyclonic vorticity is essential for convective precipitation to develop over the oceans. She suggests that a mechanism to accomplish the necessary upward transfer of heat and thereby generate cyclonic vorticity is provided by cumulus convection. Heat release modifies the density field (the solenoidal term in the vorticity equation) although the convection may also exert a dynamic influence by its effect on the motion field. The role of cumulonimbus towers in maintaining the circulation in a mature hurricane by their high level warming effects is well known (Malkus, 1962; Garstang, 1967; Riehl, 1969). It is now becoming apparent that in middle latitudes

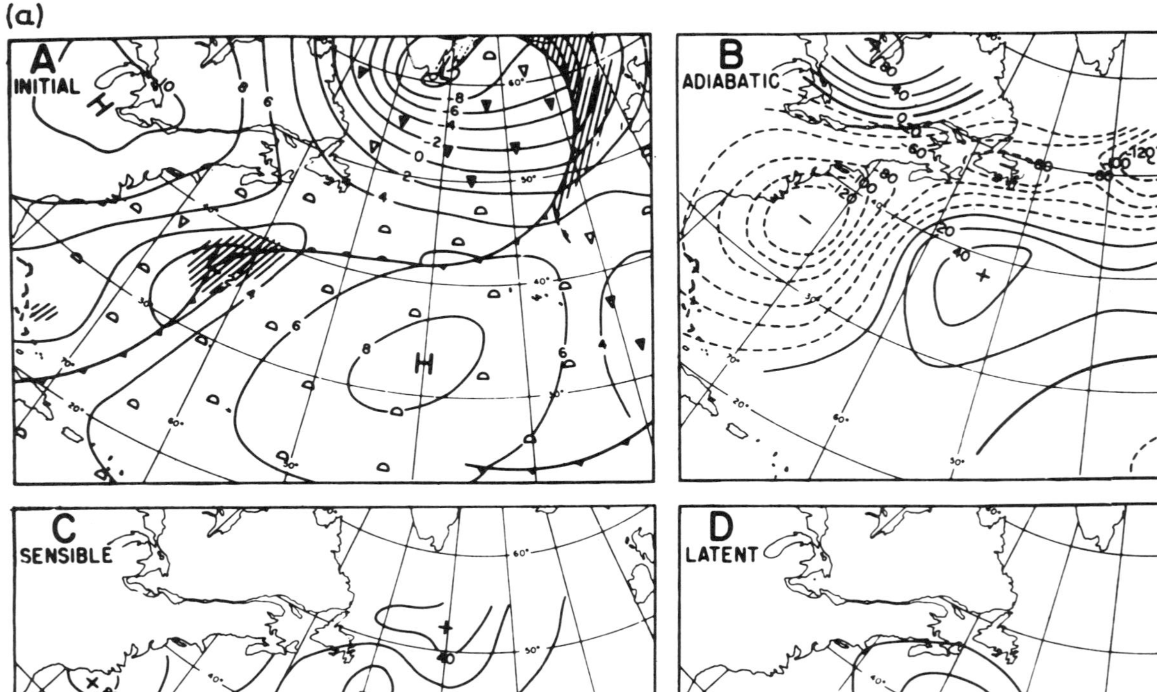

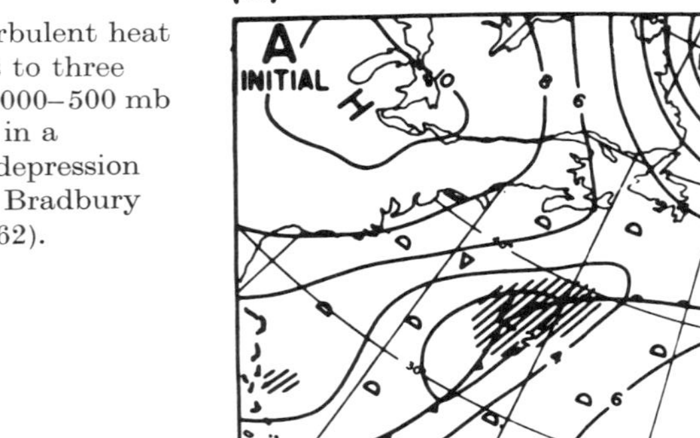

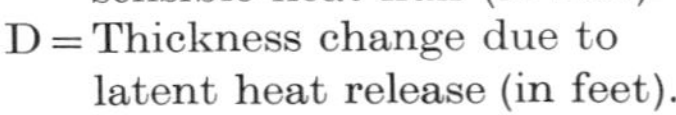

Fig. 5.55 (*a*) Turbulent heat flux contributions to three hour changes in 1000–500 mb thickness pattern in a developing wave depression (from Petterssen, Bradbury and Pedersen, 1962).

A = Synoptic map for 4 January 1958, 1200 GMT; 1000 mb contours (in hundreds of feet).
B = Thickness change due to adiabatic motion (in feet).
C = Thickness change due to sensible heat flux (in feet).
D = Thickness change due to latent heat release (in feet).

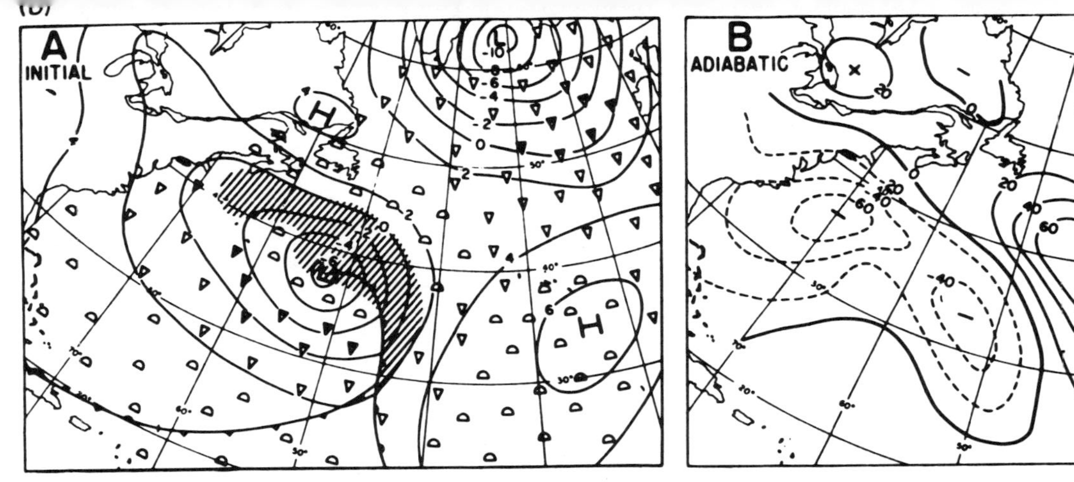

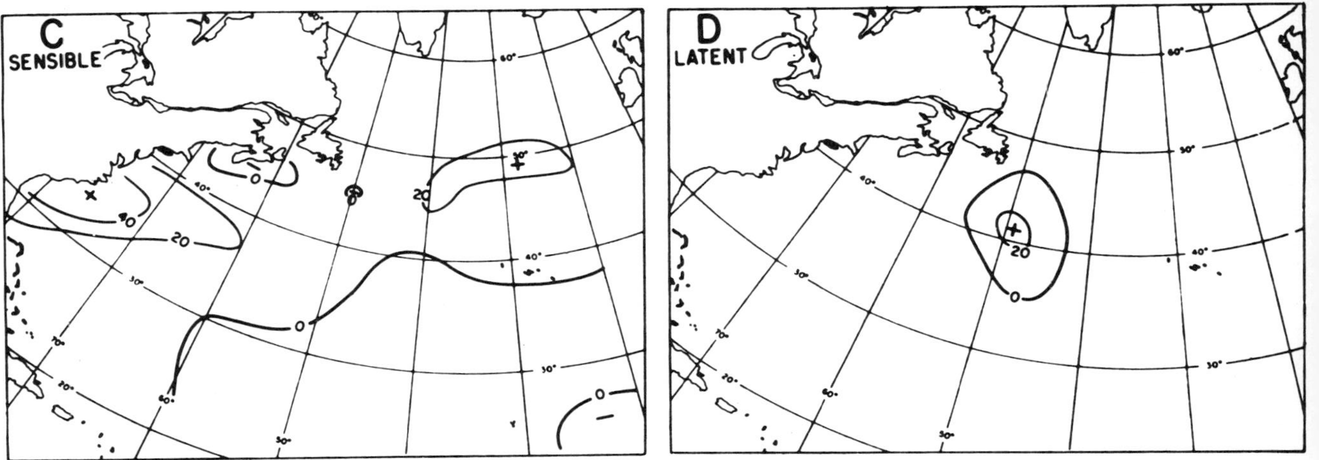

Fig. 5.55 (*b*) As (*a*) for a partly occluded depression on 28 January 1958, 1200 hrs GMT from Petterssen, Bradbury and Pedersen, 1962)

also convection can, in some circumstances, generate or augment cyclonic development.

Synoptic climatological studies

There is a growing body of literature dealing with air–sea interaction with respect to circulation patterns. It is assumed that different circulations will be related in a somewhat different manner with the oceanic conditions, in terms of their energy transfers, since each circulation type will have a different frequency distribution in at least some of the controlling parameters – wind speed, cloudiness, temperature, vapour pressure, etc. Hay (1956) investigated changes in air and sea surface temperature during cyclonic and anticyclonic regimes at Ocean Weather Stations (O.W.S.) I and J, while Manier and Möller (1961) examined turbulent fluxes at each North Atlantic O.W.S. in relation to wind speed and direction. Perry (1969) performed a somewhat similar analysis using a simple air mass typing scheme, based on origin and trajectory, for each North Atlantic O.W.S. Approximately 80% of the flux values calculated for each of the type categories showed statistically significant between-type differences at $\leqslant 5\%$ level.

For a more limited area, the Norwegian Sea, Gagnon (1964) determined energy budgets for five different isobaric patterns in winter. However, these categories accounted for only 50% of all days, since a great variety of circulation forms occur in this area. Gagnon showed that the radiative terms in the energy budget are relatively constant compared with the turbulent fluxes and of the same order of magnitude. Thus we can regard the energy budget as comprising a rather stable radiative component, dependent on latitude and on climatic fluctuations, and the unstable turbulent heat fluxes where irregular fluctuations created by the atmospheric circulation exceed the latitudinal variations.

Another detailed study for a small area – the Gulf of St Lawrence – has been carried out by Matheson (1967). Airflow types were determined based on daily surface and 500 mb charts. Year-to-year variations in ice severity were found to be related to the types of winter circulation. When types that facilitate the removal of large quantities of heat from the surface are prevalent early in the season, thereby creating extensive early ice cover, a severe ice year generally follows.

On a longer time scale, studies of anomalous months over the North Atlantic have been performed by Vowinckel (1965) and Perry (1968). During February 1965, for example, energy input to the atmosphere seems to have been below normal over much of the North Atlantic (fig. 5.56) and this was accompanied by reduced cyclonic activity and persistent blocking over the eastern region (see also Murray, 1966). Depressions in the western North Atlantic either turned northward into

Davis Strait or moved eastward in low latitudes. Fig. 5.56 clearly shows a tongue of relatively high values of energy transfer, corresponding to this latter movement, extending from the western Atlantic toward Spain. Sawyer (1965) considers that for sea temperature anomalies to be significant on a monthly time scale they must persist for this length of

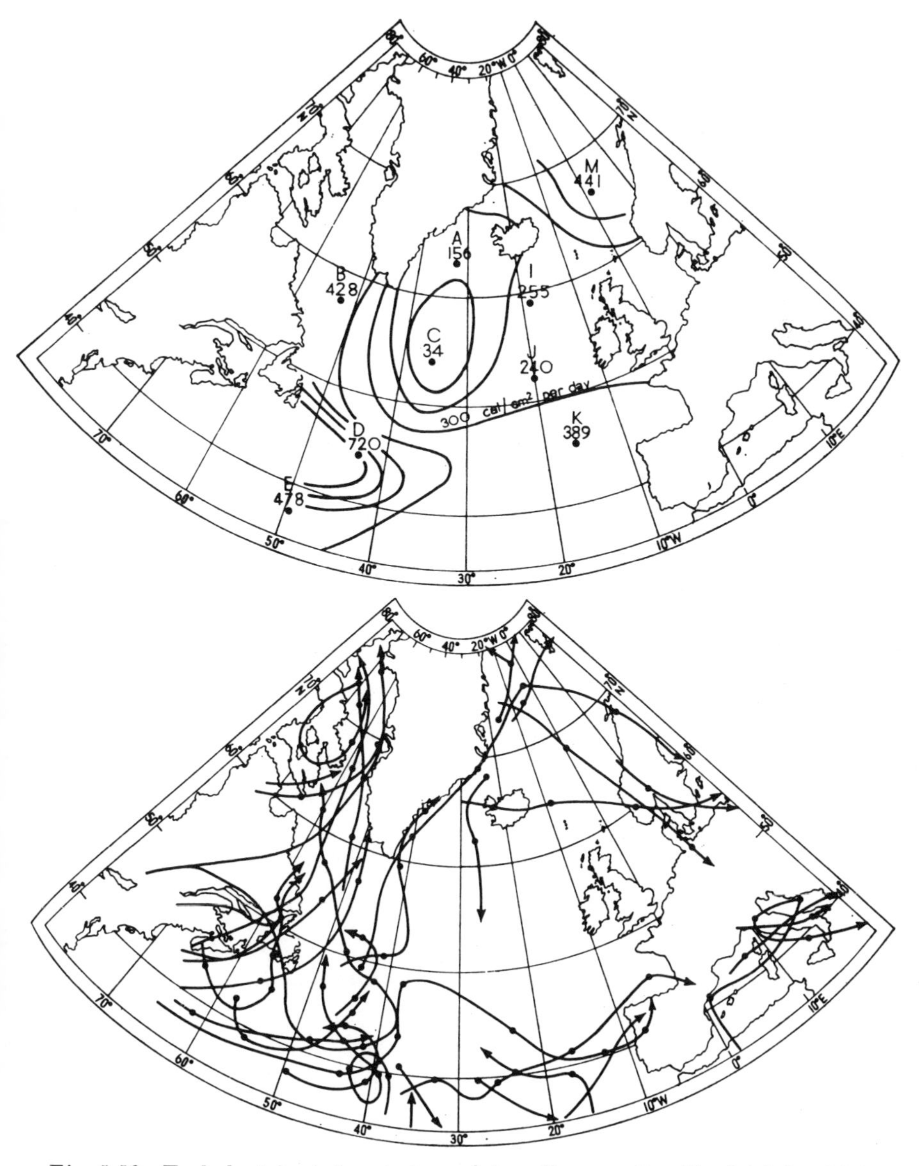

Fig. 5.56 Turbulent heat flux (cal cm^{-2} day^{-1}) over the North Atlantic in relation to depression tracks, February 1965 (from Perry, 1968).

time, extend over an area 1000 km or more across, and provide a sensible and latent heat input to the atmosphere of the order of at least 10% of the outgoing long-wave radiation (see also Kraus 1972, p. 202).

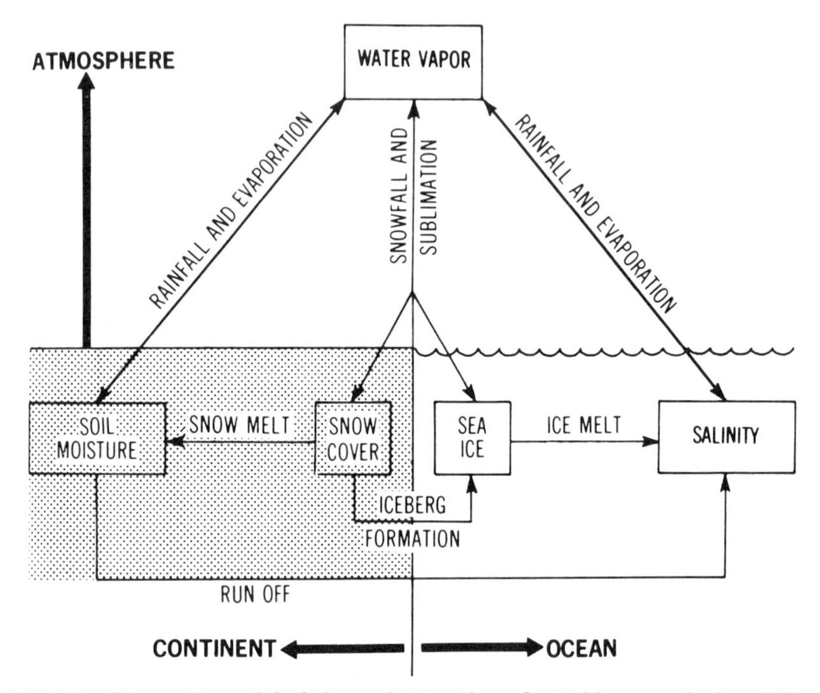

Fig. 5.57 Schematic model of air–sea interaction adopted in numerical modelling (from Manabe, 1969).

Such investigations have led to general acceptance of the view that the oceans affect the monthly and seasonal variations of climate as well as its long-term character. There is still considerable debate, however, over the relative importance of anomalies of heat flux from ocean to atmosphere as compared with anomalies in oceanic advection (Vinogradov, 1967). Adequate feedback models remain to be developed, although numeral experiments linking the general circulation with the ocean circulation have begun (Manabe and Bryan, 1969). Fig. 5.57 shows schematically their approach. The coupling is synchronized in a ratio of one atmospheric model year to 100 oceanic model years.

The work of Bjerknes (1963) illustrates the complexity of the problem in the long-term interactions. He shows that the relationship between the strength of the westerly winds and the surface temperature about 45°N in the North Atlantic varies according to the time scale. For variations of two to five years duration, strong zonal flow is associated with large turbulent heat fluxes and, since oceanic transport responds only

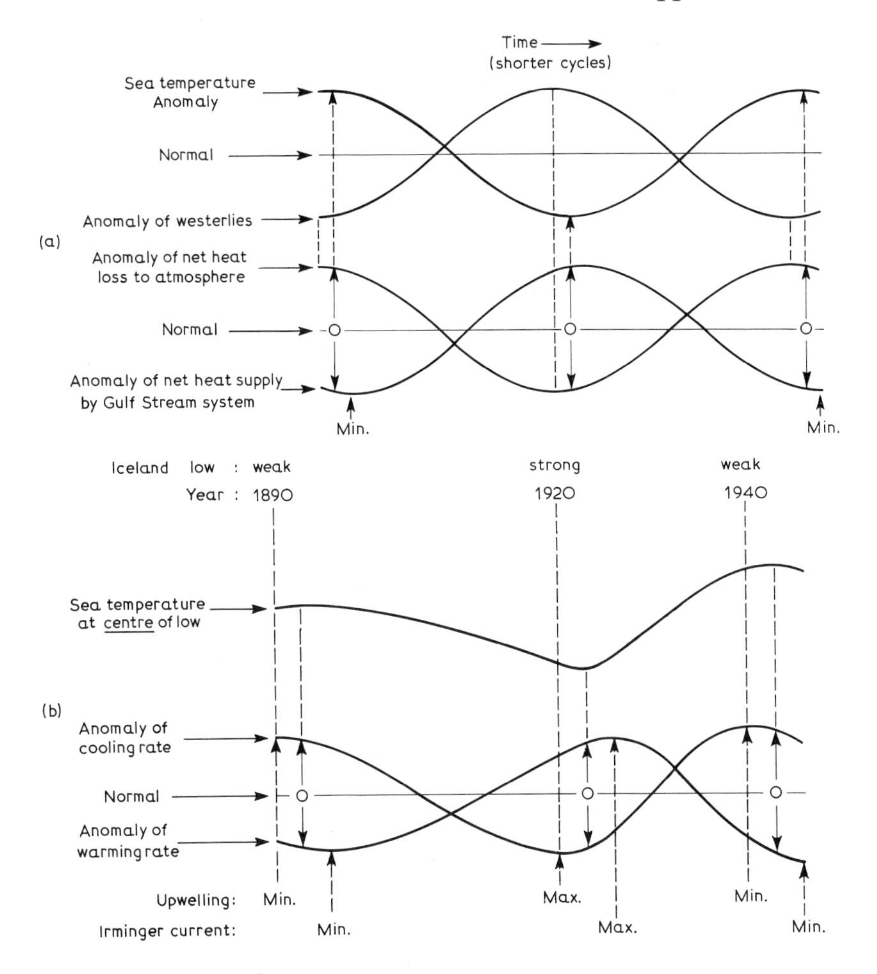

Fig. 5.58 Models of ocean–atmosphere interaction in the North Atlantic (after Bjerknes, 1963). (*a*) Oscillations of annual surface heat balance in the northernmost branch of the North Atlantic Drift at 35°W. (*b*) Variations of annual surface heat balance near the centre of the Icelandic low.

slowly, these are accompanied by surface cooling. Over fifty years, however, the wind speed and surface temperature are positively correlated. Fig. 5.58 summarizes these relationships and illustrates the need to take oceanic responses, such as upwelling, into account.

Further investigations

It is quite evident that the complexity and non-linearity of air–sea feedback processes are a hindrance to further progress in our understanding of them, especially in view of the limited availability of long-term records over the oceans. Even the present plans for new programmes to collect more extensive and precise data on oceanic and atmospheric parameters will not meet this need for some years to come. Until such time as adequate numerical models of air–sea interaction are available, empirical investigations can provide useful information.

A number of such attempts have been made to relate directly conditions in the North Atlantic or North Pacific to meteorological conditions either in the overlying atmosphere or over the adjacent continents. Pagava (1962), Semenov (1963) and Klimenko and Strokina (1969) have sought relationships between thermal conditons in the North Atlantic and the air temperature field over the European U.S.S.R., while Kryndin (1969) has examined links between the ocean and the 500 mb height field. Vinogradov (1967) dealt more specifically with circulation types and sea surface temperature in the northeastern North Atlantic. He concluded that blocking patterns are favoured by positive temperature anomalies, while westerly circulations are linked with negative anomalies of sea surface temperature. Ratcliffe and Murray (1970) have extended these analyses with reference to long-range forecasting (see p. 410) by demonstrating links between sea surface temperatures south of Newfoundland and the subsequent month's weather over northwest Europe. Comparable relationships have been indicated in the Pacific sector by Namias (1969). A firm basis for further studies of this type in the North Atlantic has recently been provided by Ratcliffe (1971). Anomalies of monthly mean sea surface temperature for 1877–1970 in the area 35–50°N, 40–60°W have been classified into six types according to the distribution of the sign of the anomaly. A study of secular trends has already been carried out by Valerianova (1965). She examined annual water temperature anomalies in the North Atlantic, north of 50°N, for 1876–1957 in relation to Vangengeim's circulation types. Cooler sea temperatures from 1902–25 corresponded to more frequent W type while the higher sea temperatures after 1929 corresponded to more frequent E type, and after 1939 to more frequent C type. It is likely that the more zonal circulation (W type) leads to a reduction in poleward oceanic heat transport which lowers sea temperatures north of 50°N. Conversely, E and C types result in enhanced ocean heat transport and higher sea temperatures.

Mitchell (1966) has proposed stochastic models to describe the behaviour of ocean temperatures and their interactions with the atmosphere. By prescribing the heat exchange as a Markov process he found that the results were compatible with the serial correlation function of observed monthly mean anomalies of sea surface temperatures. In a further example, the exchanges were treated as a random walk model. It was shown that such fluctuations can result in a cumulative heat storage in the ocean that could set up secular anomalies in air–sea parameters. Such studies, complemented by empirical investigations, have obvious significance for research into climatic change on the one hand and numerical forecasting on the other.

2 BIOCLIMATOLOGY AND BIOMETEOROLOGY

In many biometeorological research problems the limited availability of data and the state of the art often necessitate either a conventional climatological approach, averaging data over time intervals, or physical-numerical modelling of idealized situations. Nevertheless, biometeorology is a vast and expanding research area in which synoptic climatological methods should prove to be of considerable value. The discussion, which is limited by the present paucity of examples in the literature, is concentrated on, first, plant diseases and entomological problems and, second, human biometeorology.

Plant diseases and entomology

Plant pathologists and entomologists have long been aware of the importance of weather and climate for the development and spread of living organisms. In the case of spores, pollen, etc., the dispersion is very closely determined by the air motion and parameters such as temperature and humidity, whereas in the case of insects more complex interactions must be taken into account.

The warm, humid conditions favouring outbreaks of potato blight in the British Isles were identified specifically by Beaumont (1947). Subsequently, Bourke (1957) showed that these could be related to synoptic patterns. Favourable situations are: (1) the warm sector of a developing or mature cyclone with maritime tropical air, (2) slow-moving depressions, and (3) quasi-stationary fronts (provided that in the last two cases wet, overcast conditions result). Conversely, anticyclonic conditions and situations giving polar airflows with alternately bright and showery weather are unfavourable. Information of this type can be used to develop a warning system, but where spores are carried for long distances by the wind (black stem rust of wheat, for example) the forecasting problem is more difficult (Hogg, 1970). A single opportunity for the transference of spores from a source region is all that may be necessary for infection to spread. The same is true of wind-borne viruses such as foot-and-mouth disease in cattle and fowl pest (Smith, 1970). The importance of local meteorological conditions such as lee-wave occurrence has also been shown with respect to the initial spread of the 1967–8 foot-and-mouth epidemic in England (Tinline, 1970).

The distant travel of spores and pollen (Hirst and Hurst, 1967) depends greatly on their vertical concentration which is a function of their terminal velocity (determined by size essentially) and atmospheric

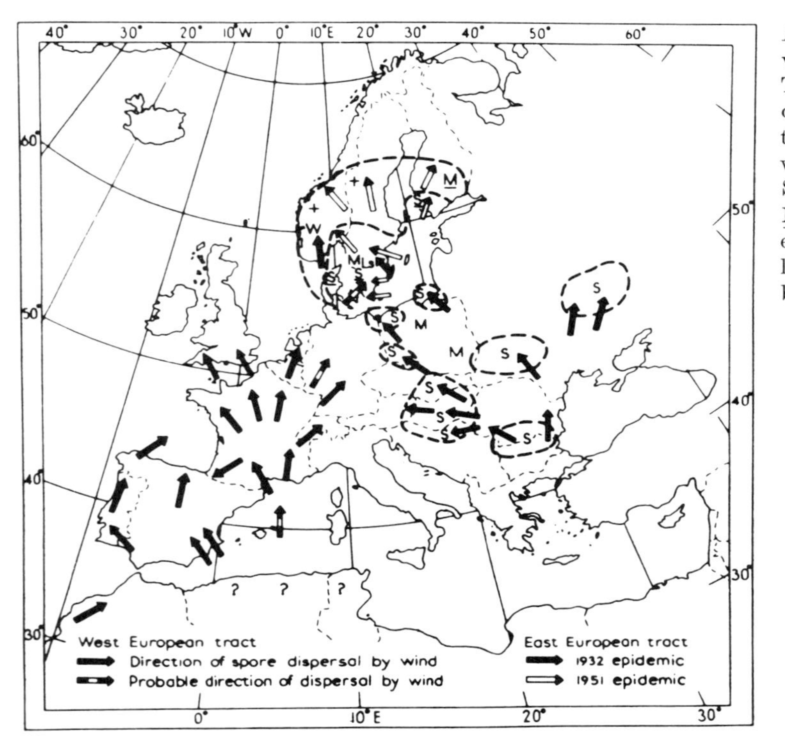

Fig. 5.59 'Dissemination tracts' for black rust of wheat in Europe (from Zadoks, in Hogg *et al.*, 1969). The arrow over south Norway (marked W) and that over the Danish islands refer to the west European tract. For the east European tract, letters refer to 1932 when not underlined and to 1951 when underlined. S, S̲ = several epidemics of national importance; M, M̲ = mild epidemics of regional importance or general epidemics with relatively small losses; Ls = severe local outbreaks; + = rust more abundant than usual but no yield reduction (1951 only).

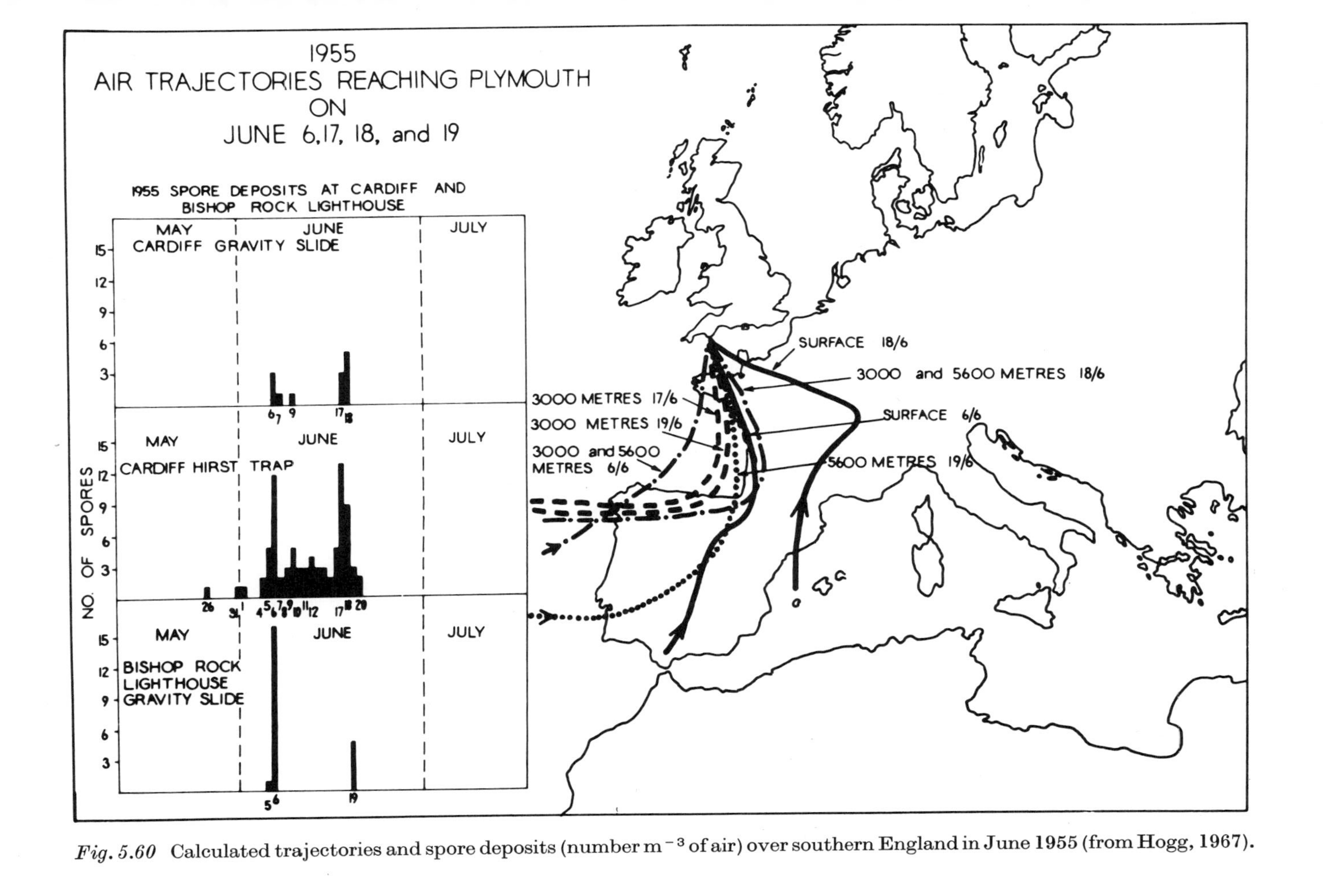

Fig. 5.60 Calculated trajectories and spore deposits (number m^{-3} of air) over southern England in June 1955 (from Hogg, 1967).

stability. It has been calculated that large spores with a terminal velocity of 1 cm s^{-1} or more have a probable flight range of the order of 10–20 km or less, whereas medium-sized ones with a terminal velocity of 0·1 cm s^{-1} may travel 1000 km. Wheat rust uredospores (summer spores) with a radius of the order of 5 μm may travel 1500 km in less than twenty-four hours with a horizontal wind velocity of 20 m s^{-1} (Hogg *et al.*, 1969). Even very large spores with terminal velocities comparable to the ascent rate in a frontal zones (10 cm s^{-1}) may be carried aloft in suitable convective situations and then transported horizontally. The 'red rain' which fell widely over Wales and southern England on 1 July 1969 (Stevenson, 1969) illustrates long-distance dispersal. Fine dust which apparently originated in the southern Sahara Desert was lifted by convective activity and then carried northward in the middle troposphere by a propitious southerly airflow. In the case of spores and viruses their viability must also be taken into account in assessing the risk of infection.

Observations of dispersal of spores have led to recognition of what are known as 'dissemination tracts'. Fig. 5.59 illustrates these for black rust of wheat in Europe. Clearly, determination of air mass trajectories is an important first step in investigating spore dispersal (Hurst, 1968). A number of case studies have been made by Hogg (1967) for black rust occurrences in southwest England during 1947–57. Fig. 5.60 shows specimen trajectories in June 1955 and spore deposits. On the basis of such agreement Hogg calculated trajectories at the surface, 700 and 500 mb levels relating to other occurrences during 1947–55 when spore catch data were not available. This approach cannot be used in all studies of aerial transport, however. Christie and Ritchie (1969) demonstrate in a study for Manitoba that pollen is not transferred along isentropic trajectories in the lower or middle troposphere. Rather, this transport seems to occur in the well-mixed layer within perhaps 1000 m of the surface.

Determining trajectory frequencies on a daily basis over a long period would involve considerable labour, and a short-cut approach has therefore been developed by Hurst, Cochrane and Rumney (1969). They evolved an approximation method using the Lamb airflow catalogue for the British Isles (see Chapter 3.C, p. 138). Each of five possible source areas – Norway, Denmark, the Low Countries, France and Spain – were examined separately and where it was considered that a particular airflow type was invariably associated with a trajectory which would affect the British Isles a weighting of one was assigned. A weighting of a quarter or half was given if the trajectory occurred on only about that proportion of days of the airflow category. They found good agreement between this method and the results of conventional trajectory analysis of daily weather maps on a seasonal basis for 1960. The combination of

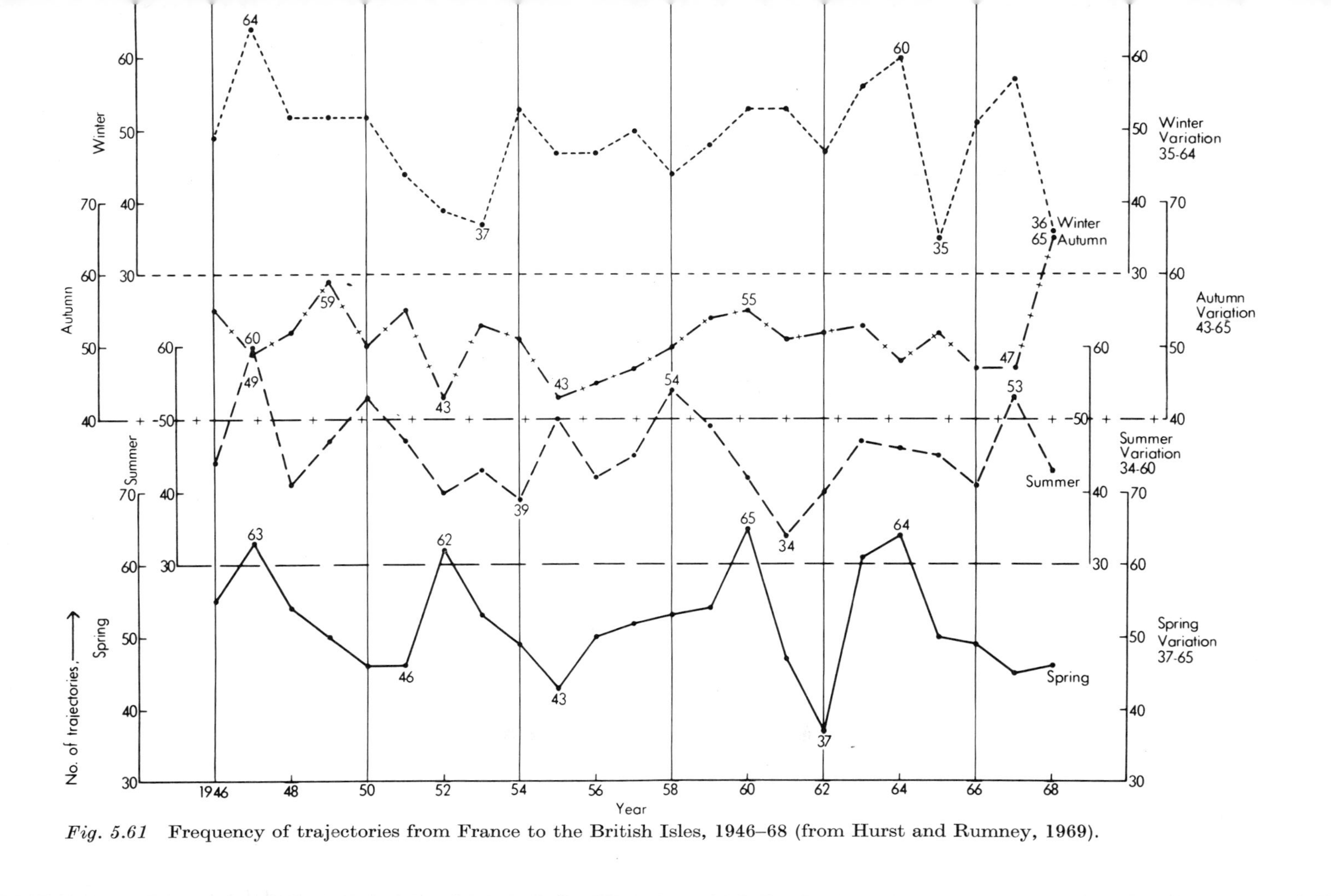

Fig. 5.61 Frequency of trajectories from France to the British Isles, 1946–68 (from Hurst and Rumney, 1969).

such results (fig. 5.61) together with pertinent environmental data should provide a suitable basis for analysis of past occurrences of airborne plant diseases and should eventually pave the way for provision of warning systems.

The trajectory information still leaves additional problems in plant pathological forecasting. First, the complex relationships between meteorological factors such as temperature, leaf moisture and light intensity on the one hand, and physiological phenomena like germination and penetration of the spores on the other, have to be evaluated. Second, the infection period is related to the weather conditions occurring at the time. This latter problem certainly lends itself to a synoptic climatological approach for specific diseases similar to that of Bourke, referred to above, for potato blight outbreaks.

In considering the influence of weather and climate on more complex organisms we have to recognize the interposition of second- and third-order effects which may be more critical in some instances than direct ones. To cite only two examples, climate affects the geographical distribution of an insect species in part by determining the spread of its food plants and also as a result of effect of climate on other, possibly competitive, species. In the direct sense 'climate is legislative but not governing in controlling insects', according to Richards (1961), although, near the geographical limits of its distribution, the numbers of a species may be determined by climatic factors (Harcourt and Le Roux, 1967).

Wellington (1957) has suggested that entomologists need to study actual crop climates at different stages of crop development and in different types of weather as a preliminary to an understanding of insect behaviour. Thus microsynoptic features like precipitation streaks (Schirmer, 1952) may be of some consequence for local vegetation, affecting in turn distribution of insects on a local scale. Standard observations may be of little use. For example, the wetting time of various kinds of vegetation with different types of air mass and the drying time in the different types of air that follow rain can be more significant parameters than precipitation totals or intensities.

Proper understanding of the effects of weather on insects requires not only an appreciation of the micro-environment. Haufe (1963) considers that synoptic correlations 'represent one of the important means of studying the relations between insects and weather'. Synoptic conditions control fluctuations of insect populations as well as expansion or contraction of the distribution area (Naya, 1967). Wellington (1965) has shown how differences in relief on Vancouver Island produce recurrent and consistent patterns of cloudiness during turbulent or convective synoptic types and these in turn affect the distribution of the western tent caterpillar (fig. 5.62). Neither enemies nor food were found to have as much influence on the distribution and abundance of the caterpillar as the local cloud conditions. Realistic classification of local climates can be developed easily and quite rapidly, even without meteorological instruments and used as the basis for population surveys by entomologists. Nevertheless, the increasing availability of satellite cloud data may prove to be of value in large-scale analyses of such populations. The need for comparability of results in assessing the synoptic aspects of local climatic differentiation makes it worthwhile considering large-scale weather systems in some suitable classification scheme (Brooks and Kelly, 1951). In a study of winter mortality of the lodge-pole needle miner (*Recurvaria* spp.) in Banff National Park, for example, Henson, Stark and Wellington (1954) found that the use of weather types yielded valuable information on the response in selected habitat extremes (and by this means they simplified environmental effects).

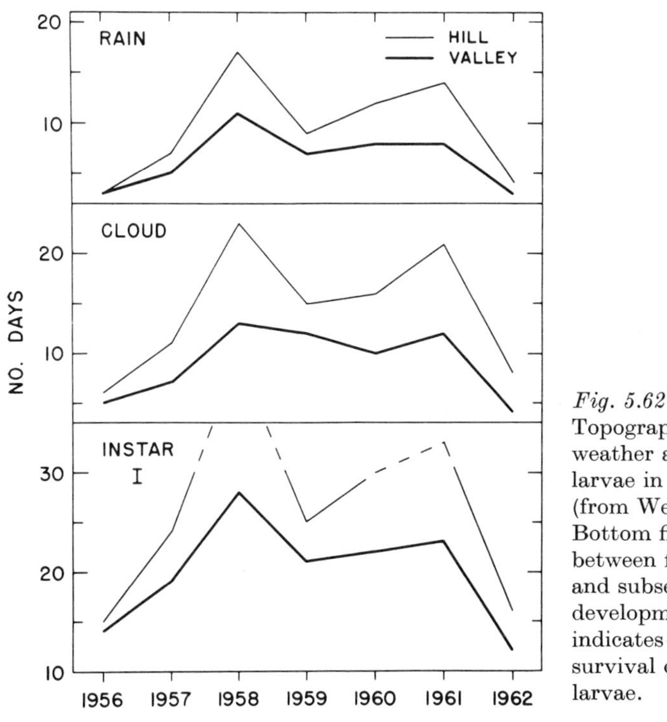

Fig. 5.62
Topographic effects on weather and first-instar larvae in spring (from Wellington, 1965). Bottom figure shows time between first emergence and subsequent larval development. Broken line indicates virtually no survival of first-instar larvae.

Weather may influence behaviour by evoking a specific response, such as overheating, that occurs only under specific conditions. Synoptic information may be usefully employed in investigations of such phenomena, as in the case of Green's (1955) study of the behaviour and activity of myrmeleontid larvae and their prey in relation to fluctuations in sand surface temperature. On the same short time scale frontal passages appear to influence behavioural response (Uvarov, 1931). Pressure fluctuations are thought to be more important than changes in temperature or humidity in augmenting the activity rates of many insects, and stimulation due to changes in electrical gradient, near cold fronts especially, may also be of significance. Obviously, however, more knowledge of the physiological responses will provide a better guide to the most appropriate grouping of weather characteristics.

Applied entomological studies sometimes overlook the great importance of favourable weather recurring, perhaps over several successive years, before major increases in population can take place. Wellington (1954a) makes particular note of this factor with reference to fluctuations in the spruce budworm (*Choristoneura fumiferana*, Clem.) and the forest tent caterpillar (*Malacosoma disstria*, Hbn.). These moth species both have a wide distribution in northeast North America. They feed on the foliage of several tree species, but whereas the forest tent caterpillar feeds exposed to the ambient air, the spruce budworm constructs its own feeding shelter, so that their microhabitats are quite distinct. Their larval development nevertheless occurs during the same spring–early summer period. The young larvae of spruce budworm need relatively dry weather and little cloud which is usually associated with cP and mP air masses. It has been found that the budworm population varies inversely with that of the forest tent caterpillar apparently in relation to air mass dominance (Wellington, 1954b). Favourable conditions for feeding during the larval stage of the forest tent caterpillar in Ontario seem to occur when mT air masses extend northward to the boreal forest zone. If one or other category of air mass is dominant in a succession of years, the particular population so favoured can increase with such rapidity that it cannot immediately be checked by any combination of adverse physical and biotic factors. Thus, the forest tent caterpillar population in northern Ontario shows a marked upsurge two to four years after a period of intensive cyclonic activity or a northward displacement of depression tracks. During 1946–8, for example (fig. 5.63), there were large increases in the population in the area of Lake Superior as a result of the passage of frequent cyclone centres in the early summer period (Wellington, 1954c). Conversely, below normal cyclonic activity along the north shore of the St Lawrence and in New Brunswick resulted in increased spruce budworm populations. These associations might be further improved if the source region of the cyclones was taken into account.

A number of case studies of insect migrations have been carried out, such as those for the diamond-back moth (Shaw, 1962; Shaw and Hurst, 1969) and the small mottled willow moth (*Laphygma exigua*) (Hurst, 1963). The diamond-back moth (*Plutella xylostella* L.), which has a nearly world-wide distribution, is indigenous to the British Isles. The caterpillar is very damaging to cruciferous crops. Sudden increases in

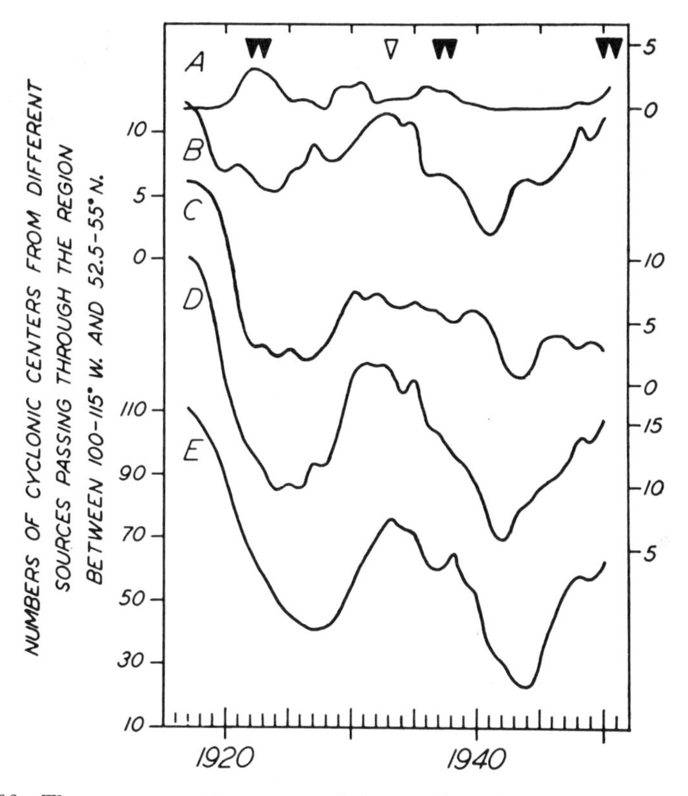

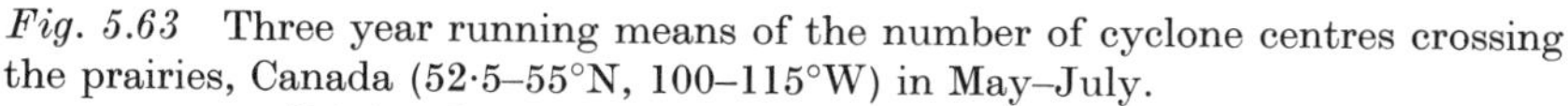

Fig. 5.63 Three year running means of the number of cyclone centres crossing the prairies, Canada (52·5–55°N, 100–115°W) in May–July.

Origin of centres:

A = Southwest

B = North and northwest

C = Pacific

D = Total number of all three types in May–July

E = Annual number of all three types

Triangles denote years in which outbreaks of forest tent caterpillar began in the region. The open triangles indicate an outbreak which died out the following year (from Wellington, 1954c).

numbers in Britain seem to be a result of immigration rather than population fluctuations, and fortunately such movements are relatively rare. The most serious upsurges prior to 1958 were in 1891 and 1914. The migration of early June 1958 can be traced by analysis of air trajectories to a source in the countries around the southeast margin of the Baltic Sea, where it is presumed there must have been a large over-wintering population. In general, migration of insects is known to be an adjustment to high population densities and consequent poor feeding conditions (Harcourt and Le Roux, 1967).

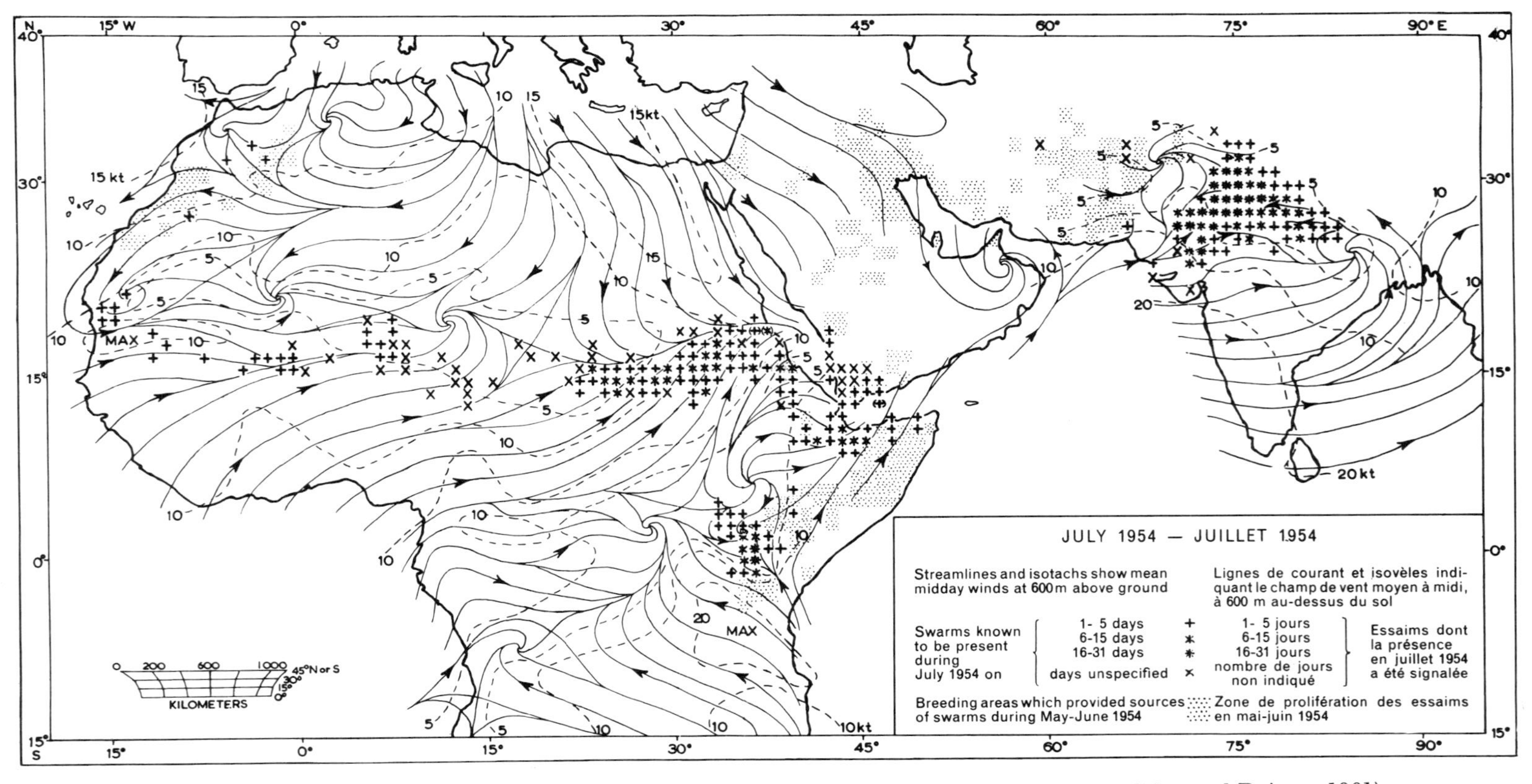

Fig. 5.64 Locust swarms in relation to the Intertropical Convergence Zone, July 1954 (from Aspliden and Rainey, 1961).

In economic terms, the most significant of such studies are undoubtedly those dealing with the migration of the desert locust (World Meteorological Organization, 1965). Pedgley and Symmons (1968) discuss the extent and effect of the 1967–8 upsurges which seem to have been associated with unusual rains in the Somali Republic, Saudi Arabia and the Yemen creating favourable breeding conditions. The subsequent movement and distribution of swarms is determined to a large extent by the low level wind field, since they fly mainly between 300–1000 m above the ground (Sayer, 1962). Rainey (1963) has reviewed this topic in detail. Swarms are essentially carried along by the wind at some fraction of its speed and usually arrive in a region of wind convergence. The Intertropical Convergence Zone (see fig. 5.64) is very important in this respect. The swarms then become slow-moving in this zone. Precipitation within such convergence zones also helps to provide the moist ground necessary for egg-laying as well as the vegetation growth required for feeding the adults and 'hoppers'.

The extent to which populations are affected by climatic fluctuations is still little known. Pschorn-Walcher (1954) suggests that outbreaks of insects in central Europe that were formerly destructive only in southern and southeastern parts of the continent may point to their northward movements during the warm years in the 1920–40 period. Further investigation of this question may, therefore, provide a basis for predicting the spread of a species. Wellington's (1964) work on the western tent caterpillar (*Malacasoma pluviale*) in Vancouver Island again gives some guide-lines. He noted that between 1953 and 1956 the regional circulation in spring was characterized by a decrease in frequency of westerly airflow with a corresponding reduction in cloudiness and precipitation. The caterpillars, favoured by these conditions, were able to establish themselves on otherwise marginal hillside forests. However, these locations were depopulated when conditions returned more to normal in the late 1950s (see fig. 5.62). If trends in large-scale flow patterns can be discerned with some reliability then the possibility of making useful estimates of population fluctuations will be enhanced.

Human biometeorology

The effects of weather and climate on human disabilities and mortality in particular have been the subject of increasing study during the last two decades. Tromp (1963) has provided a comprehensive survey of the field and here we shall restrict our consideration primarily to work of a synoptic climatological nature.

An inherent problem in medical climatology arises due to the different response of individuals to specific weather conditions as a result of their different genetic characteristics. The selection of proper samples for comparison is therefore vital. Even so there is always a risk that the medical diagnosis itself may be incorrect and hence the further need for samples of adequate size.

Investigation of relationships between weather conditions, working efficiency, and accident rates is an important area of practical research in which the work has been spearheaded by the Medical Meteorological Centre of the German Weather Service. Table 5.24 summarizes their principal findings. High pressure areas appear to be the most 'biologically favourable' conditions. The disturbing effects of weather begin in the transition to low pressure. This work, it should be noted, does not take into account the stimuli on the body created by synoptic diversity.

Table 5.24 Relationship between reaction time, working efficiency and work and traffic accidents, and biologically favourable and unfavourable weather phases (Sargent and Tromp, 1964)

Weather phase	*Anticyclone*	*Front part of depression*	*Passage of depression*	*Rear part of depression*
Temperature/humidity environment	*Cool–mild/ dry*	*Warm/ humid*	*Warm/ very dry*	*Cold*
Reaction time (msec)	239	254	256	254
*Working efficiency**				
(*a*) Reifferscheid (1954)	−5·0	1·6	2·0	1·4
(*b*) Brezowsky and Weisser (1961)	−35	19	16	−1
Accidents (%, difference from average)				
(*a*) Brezowsky and Weisser (1961)	−15	4	7	4
(*b*) Daubert (1956)	−8	6	2	6
(*c*) Spann (1956)	−18	8	6	3

* Mean number of complaints (*a*) and mean number of cases (*b*) per day, difference from average.

Accidents at work and in traffic are much more likely in 'biologically unfavourable' conditions. Kohn (1956) also found that 85–90% of traffic accidents in Hamburg occurred in such conditions and Reiter (1960) noted a 21% increase in the accident rate in south Germany during cold front weather. It is clearly difficult to separate the purely biological effects from such direct factors as more slippery road surfaces in bad weather, although most researchers claim that physical effects of this type play an appreciably lesser role in causing accidents than disturbances to the body.

Weather and climate factors also cannot be disregarded in studies of mortality rates. Indeed, in 1878 Buchan and Mitchell compared seasonal mortality patterns for different diseases in New York and London for 1871–7. Their results and those of recent studies show a basic seasonal trend in mortality with a maximum in the winter months in Europe and Japan particularly associated with respiratory diseases. This peak appears to be muted in the United States, perhaps due to the more widespread use of central heating systems (Momiyama and Katayama, 1967; Momiyama, 1968). Superimposed on the seasonal curve are peaks and troughs which, in part at least, can be ascribed to weather factors. For example, pronounced fronts producing a marked change of temperature lead to an increase in mortality, particularly after a period of stable anticyclonic conditions (Hisdale, 1953; Kuhnke, 1958). For the north-central and northeast United States mortality rates increase with pre-frontal weather, according to Driscoll (1971), in spite of previously reported negative results (Driscoll and Landsberg, 1967). One of the most important synoptic situations affecting mortality rate is the winter anticyclonic type accompanied by a persistent low level inversion which allows urban concentrations of atmospheric pollution to increase sharply. The infamous London smog of December 1952 is attributed with an 'excess mortality' of more than 4000 people in one week alone through its effects on the aged and persons suffering from respiratory diseases such as bronchitis-emphysema. During the 1960s the incidence of such smogs has been markedly reduced in Britain as a result of the smokeless zones established under the 1956 Clean Air Act.

Obviously broad empirical studies, while useful in highlighting some important problems, leave many questions unanswered as to what meteorological factors are critical. The following seem to be significantly health-related:

1 Solar radiation especially in the ultraviolet
2 Extremes of temperature
3 Extremes of humidity
4 Wind
5 Air ions and pollutants
6 Large fluctuations of air pressure.

The reported effects in connection with specific diseases and ailments have been summarized by Sargent and Tromp (1964). An abridged list is given in table 5.25. It must, of course, be emphasized that weather effects are one small element in a complex of factors involved in man's internal physiological processes. Nevertheless, the development of synoptic climatological analyses would seem to be worthwhile. The majority of medical biometeorological studies do not go beyond correlation between selected weather parameters and particular ailments or

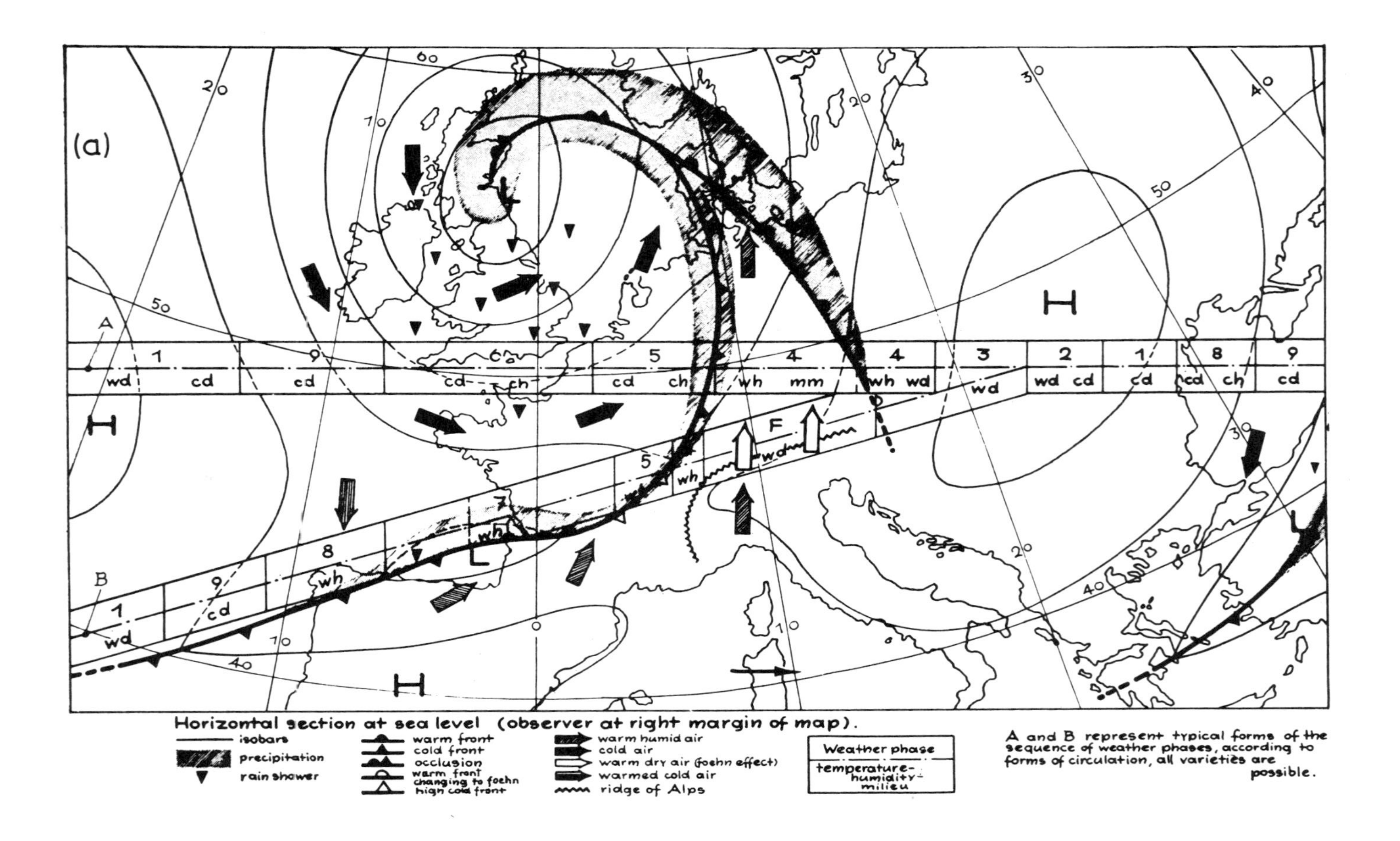
(a)
A
B
H
H
H
L
F
1
9
6
5
4
4
3
2
1
8
9
wd
cd
cd
cd
ch
cd
ch
wh
mm
wh
wd
wd
wd
cd
cd
cd
ch
cd
1
9
8
7
5
wd
cd
wh
wh
wh
wd
Horizontal section at sea level (observer at right margin of map).
isobars
precipitation
rain shower
warm front
cold front
occlusion
warm front changing to foehn
high cold front
warm humid air
cold air
warm dry air (foehn effect)
warmed cold air
ridge of Alps
Weather phase
temperature-humidity-milieu
A and B represent typical forms of the sequence of weather phases, according to forms of circulation, all varieties are possible.

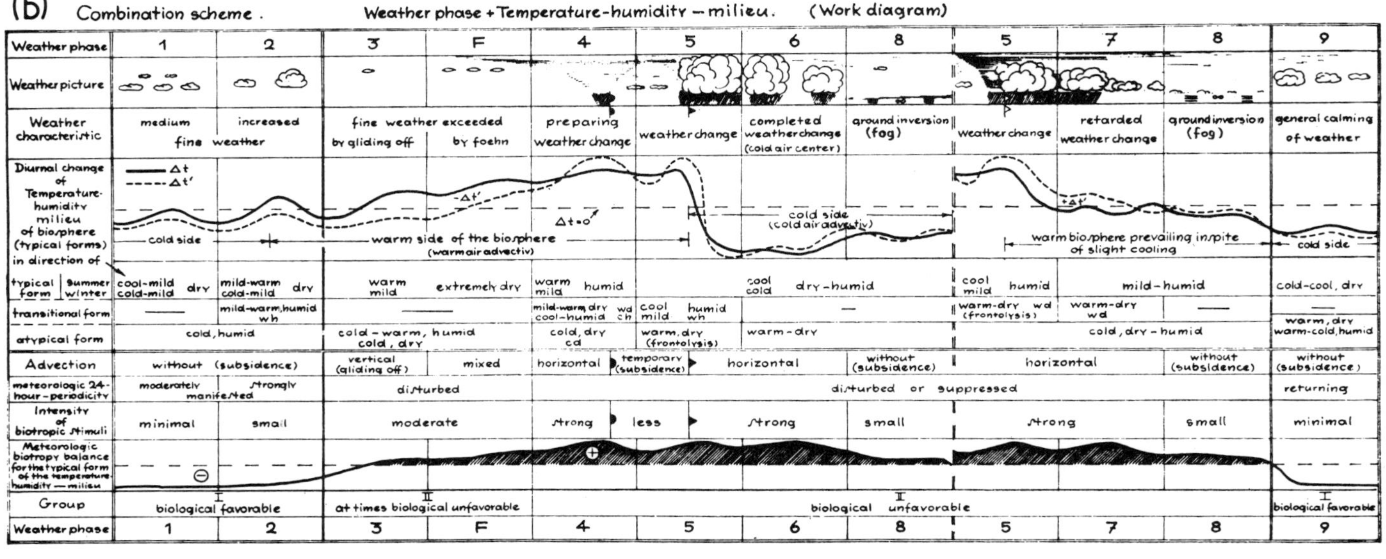

Fig. 5.65 Weather phases and their postulated biological effects (after Brezowsky, 1964). (*a*) Synoptic occurrence of typical phases. (*b*) Weather phases and biological response.

diseases (Tromp and Weihe, 1957, for example). Nevertheless, if weather effects can be delimited by these empirical means, or by experiments using controlled chambers (Hollander and Yeostros, 1963), then for prediction purposes a synoptic approach is essential.

A substantial contribution to a synoptic climatological view of biometeorology has been provided by Brezowsky (1964). From a synoptic identification of 'weather phases' and the associated temperatures and humidity conditions, especially their diurnal change (fig. 5.65(*a*)), he

Table 5.25 Weather effects on diseases and ailments (after Sargent and Tromp, 1964, and others)

Disease	*Weather factors*
Cancer	
Skin cancer	Increased u-v radiation exposure.
Headaches	Frontal and cyclonic cold air situations.
Heart diseases	
Coronary thrombosis, angina pectoris	More frequent soon after a period of strong cooling.
Infectious diseases	
Common cold	Sharp weather changes, such as a warming after a very cold period, which affect thermoregulation mechanism, membrane permeability, and virus transmission.
Influenza – development and spread of virus	Relative humidity below 50% and light winds.
Poliomyelitis – development and spread of virus	Warm, humid air.
Lung diseases	
Bronchial asthma	Increases with rapid cooling.
Bronchitis	Frequent with fog/smog episodes especially if cold.
Tuberculosis	Haemoptysis increase following oppressive warm weather, föhn, humid cold foggy weather, or sudden heat waves.
Rheumatic diseases	
Arthritis (most forms), sciatic pains	Strong cooling (with wind). Humidity seems to an indirect effect only.
Ulcers	
Perforations of peptic ulcers Haemorrhaging of duodenal ulcers	Marked temperature variations (e.g. cold waves).

assigns the categories biologically favourable or unfavourable (fig. 5.65(*b*)) on the basis of the response of the body system to meteorological stimuli ('biotropic stimuli').

In central Europe the proportion of unfavourable/favourable phases is estimated to be 2:1. Brezowsky also illustrates the use of frequency distributions of the reaction to warm and cold 'biospheres'. Extension of this approach to other regions and other climatic zones is overdue.

6 Synoptic climatology: status and prospects

Only when the merits of long standing methods are seen in their original setting can their present utility be adequately judged. The most complete prisoners of the past are those who are unconscious of it. (*E. A. Wrigley, 1965*)

The field which we have identified as synoptic climatology is manifestly varied in its objectives. Many topics, familiar in themselves to meteorologists and climatologists, have been synthesized in what is perhaps an unfamiliar setting, but it is hoped that this very process will have served to highlight their coherence and their utility. Synoptic climatology studies possess at least one common denominator; their aim is to interpret spatial and temporal climatic patterns in terms of large-scale weather processes and events, and clearly this is now leading to analysis of the patterns of horizontal and vertical transfer of energy and moisture. This broader purpose, which complements that of research in dynamic climatology, has replaced the original primary concern of synoptic climatology with long-range forecasting, although it still encompasses direct or indirect contributions to such forecasting.

This shift of emphasis reflects greater appreciation of the forecasting problem and the limitation of analogues. Nevertheless, optimism over the future role of numerical prediction must at present be tempered with caution for the longer time scales. It seems likely that, for some time to come, a judicious combination of physical and statistical procedures will be required for long-range forecasts and that synoptic climatological information on a regional to hemispheric scale will constitute a valuable input to and check upon numerical prediction methods. Pokrovskaia and Esakova (1967) note that attention is being given to these problems in the Soviet Union and Kudashkin and Iudin (1968), for example, discuss the use of analogues, determined from eigenvectors, in making statistical adjustments to two to four day hydrodynamic forecasts.

The topic of synoptic classification has been discussed extensively because this constitutes a central problem to which much research effort has been devoted. It is our primary means of stratifying highly variable synoptic data with respect to potentially significant combinations of parameters. Depending on the precise objective of a particular study, it may or may not be appropriate to work out a refined classification. In some cases a rough and ready scheme is quite adequate, but if the intention is to formulate a long-term catalogue of circulation patterns then it is essential to specify the categories as carefully as possible and to ensure consistency in their identification from the weather map record. This may or may not necessitate objective specification techniques. Where more than one classification system has been developed it is useful to determine which is the most generally effective in characterizing regional weather conditions, as Fliri has done for the Alps (see p. 156). It would be interesting to see such a comparison made for the Vangengeim and Dzerdzeevski hemispheric schemes – something which, to the knowledge of the authors, has not been performed.

The primary object of meteorological analysis up to now has been the synoptic weather map. Analysis of the pressure field in conjunction with the geostrophic wind relationship has generally served as a surrogate for direct analysis of the wind field, except for tropical areas, but with modern computers this situation is changing. It is now feasible, for example, to prepare streamline maps and determine kinematic properties of the motion by objective automatic procedures. In addition, new types of data are becoming available through satellite imagery and other forms of remote sensing. Actual and potential developments in these areas were discussed in Chapter 2, but it largely remains for climatologists to employ these new parameters extensively.

A key question we have not so far specifically considered pertains to the global generality of synoptic classification methods. The regions for which such classifications have been developed are mainly, though not exclusively, in middle latitudes. In part, at least, this reflects the greater concentration of research in these parts of the world, although certain of the approaches that have been discussed are, for one reason or another, less suitable in low latitudes. The traditional air mass concept, for example, was found to be inapplicable in the tropics owing to the characteristically small temperature gradients, and even a modified approach based on airstreams proved inadequate as a basis for understanding tropical weather regimes because boundaries between air streams lack continuity. Earlier, Depperman (1937) related weather types[1] and cloud conditions for the Philippines and thereby, in a sense, anticipated current interest in tropical cloud patterns as recorded by satellite imagery. Some studies of circulation and weather types in the tropics have, nevertheless, been pursued along similar lines to those in middle latitudes. Vowinckel (1956), for example, analysed conditions in

[1] The seven basic categories were Pure Trade, Trade, Convective Trade, Northern, Mild Southwest Monsoon, Frontal Southwest Monsoon and Typhoon.

southern Mozambique in terms of a combination of weather and circulation types. The majority refer to weather situations and are non-directional, but at least the major seasonal differences are clearly indicated (fig. 6.1). However, this approach gives little basis for dynamic explanation; the situations do not demonstrate any tendency for recurrent sequences, for example. In contrast Casanova's (1968) work for west

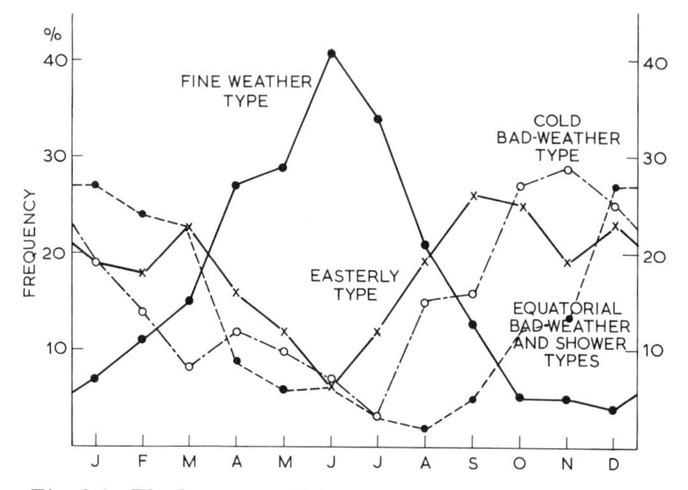

Fig. 6.1 The frequency of the major weather situations over southern Mozambique (after Vowinckel, 1956).

Africa used broad surfaces and 500 mb circulation patterns and it is evident that more detail could be obtained if the synoptic maps for this area were classified on the basis of the schematic equatorial flow patterns of Johnson and Mörth (1960) illustrated in fig. 6.2. At the same time we must be aware of the problems created in low latitudes by random errors in pressure measurements. As Ramage (1971, p. 89) remarks, 'day-to-day weather variations in transequatorial flow can seldom be related to variations in the synoptic fields of surface pressure and surface wind'. Averaging pressure over several days of similar conditions (i.e. short-term synoptic climatological averages) is essential if weather conditions are to be related to the airflow.

Following from this, the possibility of formulating a classificatory approach suitable for application in any climatic region also needs to be considered. To answer this question fully we should know, among other things, the temporal and spatial scale characteristics of weather systems in different climatic regions. At the present time this information is still incomplete. For example, previously undetected synoptic-scale *cloud clusters* over the tropical Pacific Ocean are now being studied from satellite photographs. In studies of monsoon weather systems in south-

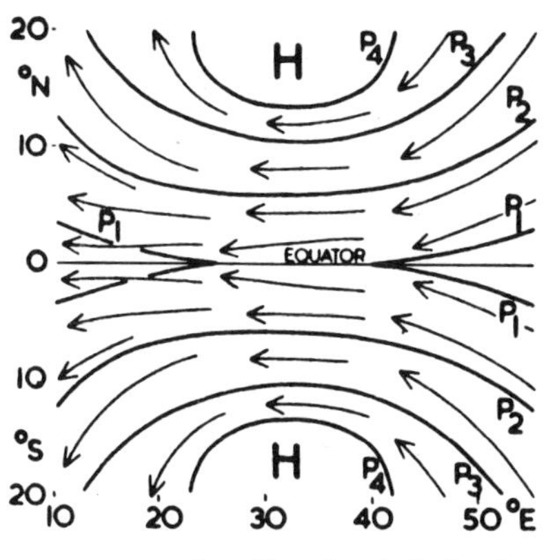

A - Equatorial duct

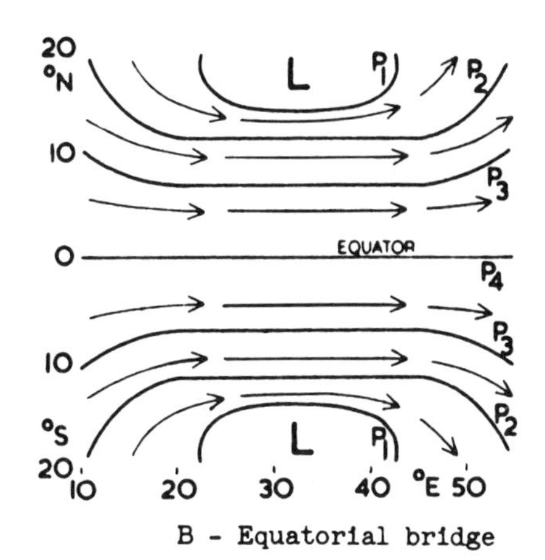

B - Equatorial bridge

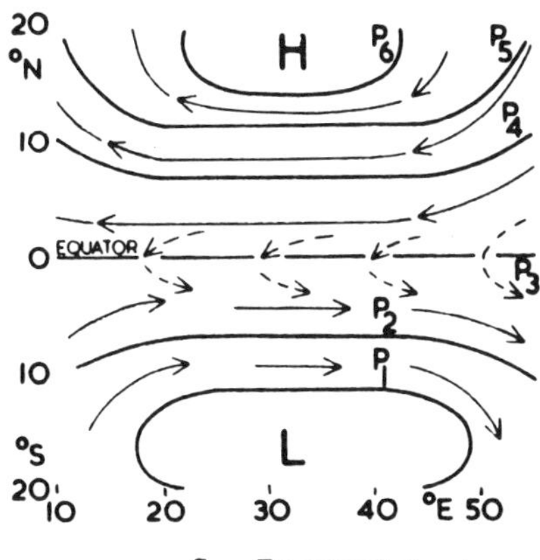

C - Equatorial step

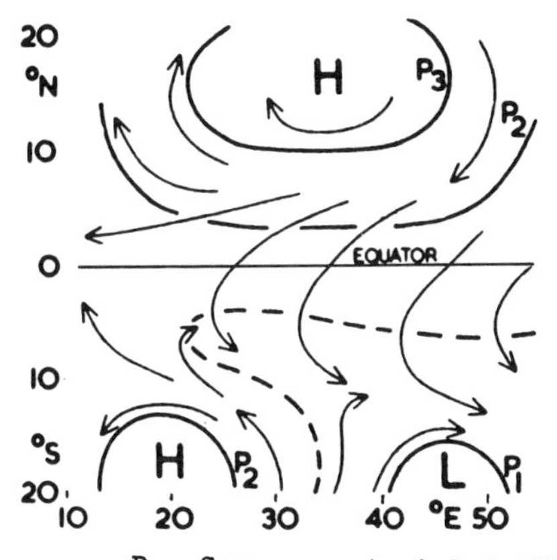

D - Cross-equatorial drift

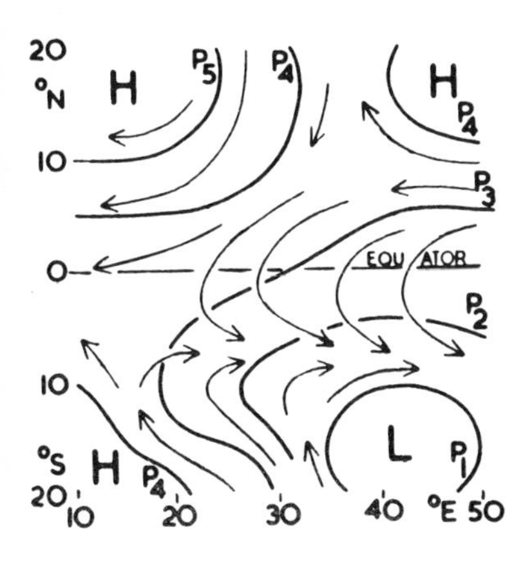

E - Cross-equatorial drift

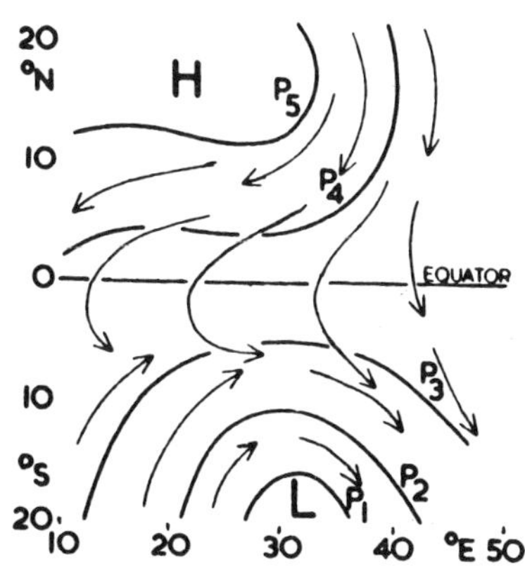

F - Cross-equatorial drift

Fig. 6.2 Equatorial flow patterns (from Johnson, 1965).

east Asia (Riehl, 1967) it was initially assumed that the 700 (or 850) mb level could be taken as representative of disturbances in the lower trades and monsoon flow, and the 200 mb level as representative of upper tropospheric systems, but it was found that the maximum intensity of disturbances was commonly at about 500 mb. Over India, Keshavamurty (1971) has shown that during the summer monsoon disturbances at 850 mb having a period of five to six days interact with others at 500 mb having a period of seven to eight days, thereby giving rise to monsoon depressions with a period of fifteen to twenty days. In the Canadian Arctic Archipelago it seems that local (subsynoptic) scale systems, related perhaps to open water areas, are of great significance in local weather conditions. These examples suggest that there is probably no single time and space scale of analysis suitable for all regions.

Apart from the scale question, there is the further consideration of the role of advective and non-advective elements in the regional or local weather. While travelling systems of several types occur in the monsoon season over southeast Asia, many other systems develop *in situ* and their structure is difficult to characterize adequately in terms of map composites (Riehl, 1967). The diverse vertical structure of wind profiles in the tropics compared with middle and high latitudes should certainly caution against expecting to find simple circulation types appropriate in different seasons and different locations. Mosiño's (1964) analysis of upper flow patterns over Mexico emphasizes the marked nature of the seasonal change-over in circulation in the tropics (fig. 6.3).

For essentially similar climates, in terms of wind regime, it is quite feasible, at least hypothetically, to adopt an objective classification approach such as the correlation method or eigenvector extraction. Direct synthesis of circulation and weather types along the lines of Christensen and Bryson (1966) could also be employed, and further exploration of this possibility is desirable. The main obstacle to an extensive development of objective classifications is the inadequacy or inaccessibility of available data. The increased storage of grid point pressure and height data on magnetic tape will greatly accelerate the objective classification of pressure fields, although the expense of obtaining limited quantities of data from large tape files is often a serious problem as a result of the way in which the records are stored. This is also true of weather map series on microfilm and satellite imagery. Regrettably, synoptic records are still primarily viewed with respect to their forecasting and research applications and not as data sources for climatology. Nevertheless, Vowinckel (1964) considers the task of regional climatology to be that of assessing how far physical meteorology corresponds to reality!

The benefits that could derive from improved climatological knowledge are insufficiently recognized, although awareness is growing

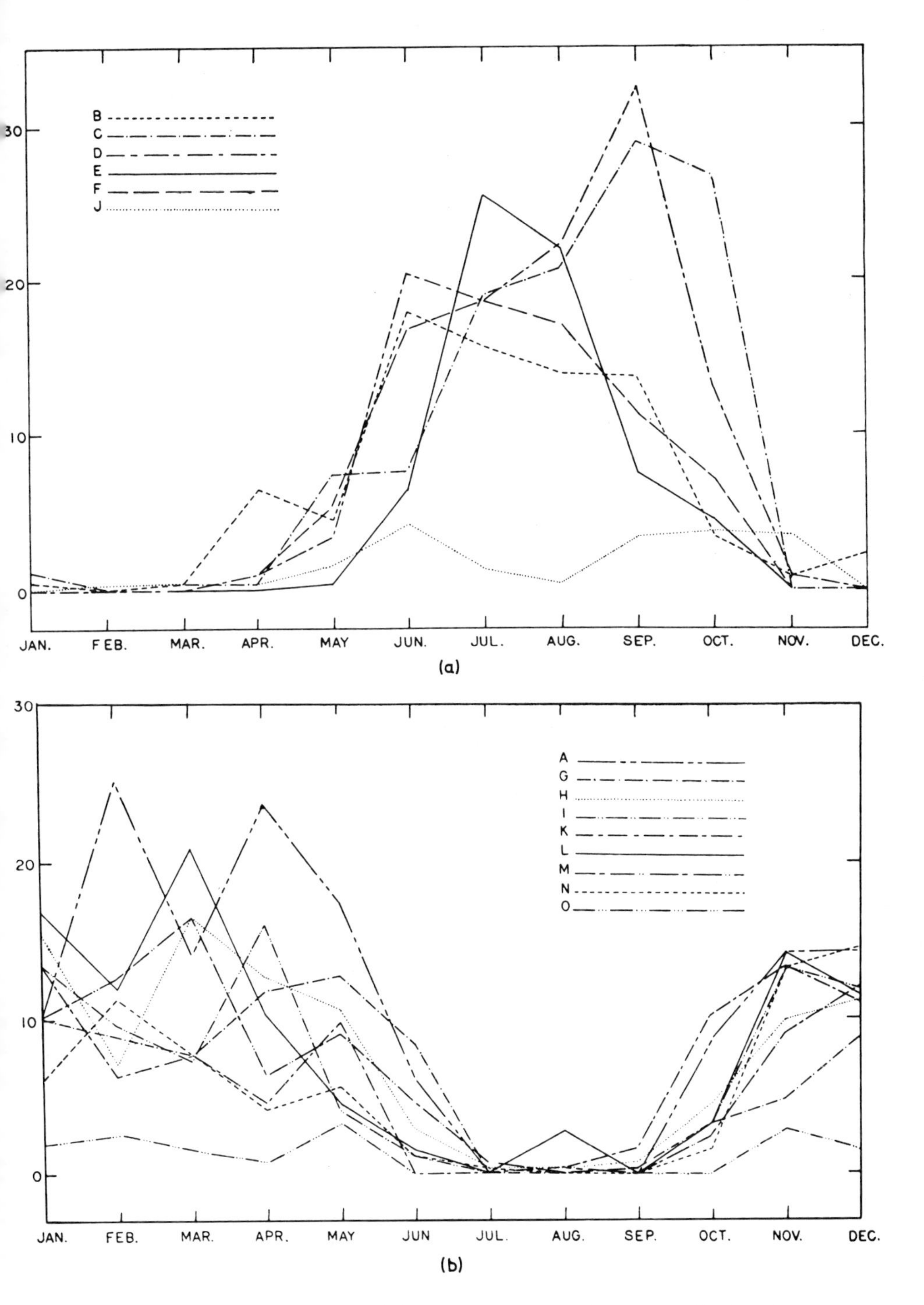

Fig. 6.3 The seasonal variations of airflow patterns at 700 mb over Mexico (from Mosino, 1964). (*a*) Easterly. (*b*) Westerly.

(Maunder, 1970). There is, for example, a general tendency for increased use of cost–benefit analysis in meteorological investigations. Also, there is recognition of the possible crucial significance of even minor climatic fluctuations and the need to monitor climatic parameters and related environmental factors. This concern for long-term planning has stimulated a demand for super long-range forecasts and it is in this respect that greater awareness of the potential value of synoptic climatology is likely to develop. In addition to some of the general climatic forecasts already referred to (p. 378), Weiss and Lamb (1970) suggest that the following trends are likely in the British Isles:

1 Continuation of the present reduced frequency of W and SW winds for 80–100 years.
2 Continuation of the present enhanced frequency of N wind components especially in the 1970s with a possibility of some reduction for ten to twenty years about the 1990s.

These predictions are based primarily on extrapolation of analysed trends in airflow over the country. In the Soviet Union similar forecasts have been made of circulation characteristics taking into account the supposed dependence of the Vangengeim types on sunspot activity (Bolotinskaia, 1965; Bolotinskaia and Beliazo, 1969). Fig. 6.4 shows the calculated changes up to 1996. In view of the importance attaching to the accuracy of such attempted forecasts it is vital that every effort be made to improve our understanding of synoptic and dynamic climatology. Forecasts of the development of circulation tendencies made in the late 1950s and early 1960s based on the expected solar activity have not in general been particularly successful. It is not yet clear whether this reflects impersistent heliogeophysical relationships or the dominance of other factors. An intensification of solar activity began in the 1930s and reached a peak in 1957. A general lowering of the level of activity is now anticipated, although this may be delayed until after A.D. 2000. Super long-range forecasts must take account of terrestrial variables such as possible volcanic activity, levels of man-made dust and carbon dioxide content in the atmosphere, since there is reason to suppose that these variables are more critical than solar activity. However, the nature of their interactions and effect on the circulation have yet to be determined with precision (Lamb, 1970; Mitchell, 1971).

Finally, it is worth reiterating the potential value of a synoptic climatological approach in applied climatology. The examples given in Chapter 5.E represent areas where such usage is fairly well developed. Other illustrations exist in a variety of research areas. The work of Schroeder *et al.* (1964) on critical fire-weather conditions is but one specific example of a directly weather-related problem of great practical concern. The analysis of the occurrence of stagnating anticyclones over the eastern United States, which are critical for pollution episodes (Korshover, 1967), is another. It is essential for prediction purposes, however, to determine that the synoptic events are both *necessary and sufficient conditions* for the problem being investigated. In the two examples cited this may well be the case but in biometeorological research it is unlikely that the sufficient conditions will be determined

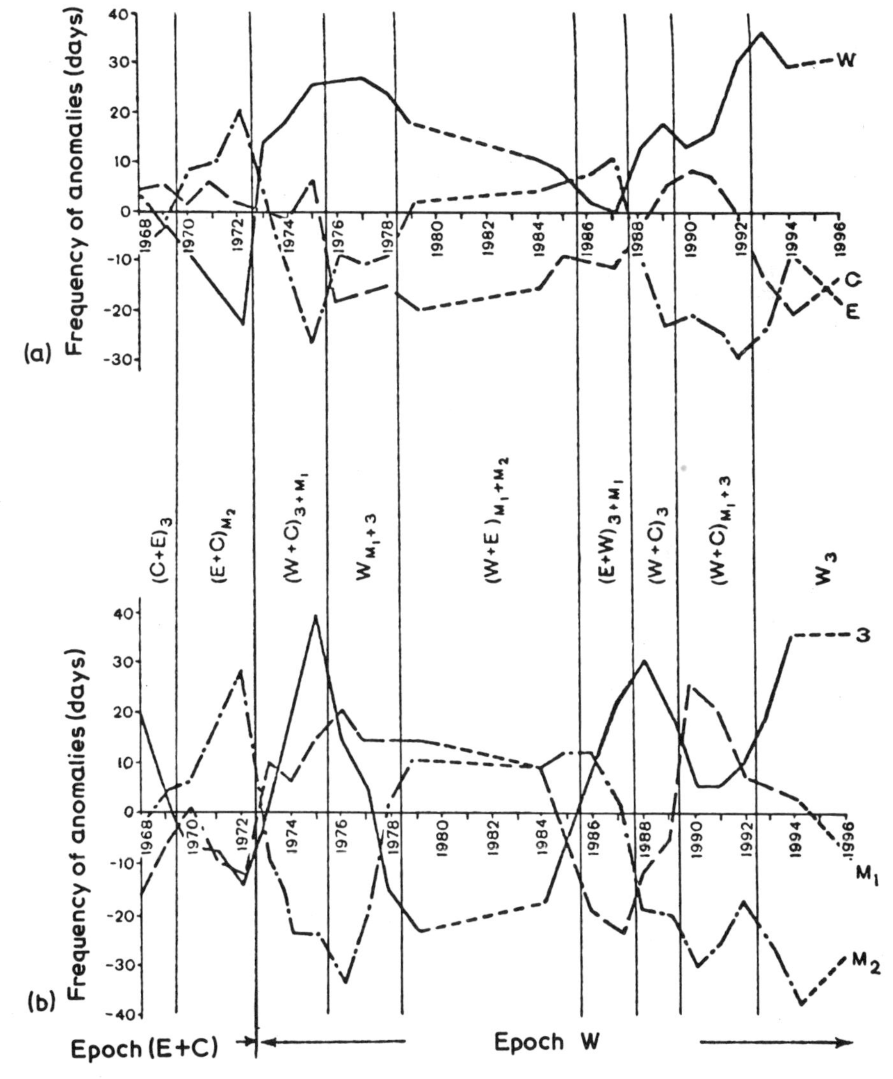

Fig. 6.4 Predicted circulation changes (frequency of departures from normal in days) up to 1996, adjusted for changes in solar activity (from Bolotinskaia, 1965). (*a*) W, E and C types. (*b*) Z, M_1 and M_2 types.

solely from meteorological events. In this respect synoptic climatological studies concerned with forecasting, climatic change, and so on, have a much simpler task than those involving biometeorological problems.

In this book the vast literature of synoptic climatology has been presented in a systematic manner for the first time. In showing what has been accomplished so far in synoptic climatology, it is the author's hope that others may be encouraged to develop the field further and improve the techniques in present use. Certainly greater attention needs to be given to synoptic climatology in the teaching context since considerable benefit will derive from the cross-fertilization of ideas and methodology when the types of theme presented in this book are considered under one broad heading. Equally, numerous research challenges remain. The opportunities presented by modern computer facilities and data banks afford a basis for thoroughgoing analyses not available to many of the researchers whose results provided the material presented here.

References

Chapter 1

BACON, F. 1662. *Historia naturalis et experimentalis de ventis*. Amsterdam: Lugd. Bat. (1638 and 1648). 58–60.

BERGERON, T. 1930. Richtlinien einer dynamischen Klimatologie. *Met. Zeit.* 47, 246–62.

CHU, C.-C. 1962. The pulsation of world climate during historical times. *Acta Met. Sinica* 31, 275–88.

CLARIDGE, J. 1748. *The Shepherd of Banbury's rules to judge the changes of the weather, grounded on forty years experience*. 2nd ed. London. 54 pp.

COURT, A. A. 1957. Climatology: complex, dynamic and synoptic. *Ann. Assoc. Amer. Geogr.* 47, 125–36.

D'ARCY THOMPSON, W. 1918. The Greek winds. *Classical Rev.* 32, 49–56.

DURST, C. S. 1951. Climate: the synthesis of weather. In Malone, T. F. (ed.) *Compendium of meteorology*. Boston, Mass.: Amer. Met. Soc. 967–75.

GODSKE, C. L. 1966. A statistical approach to climatology. *Arch. Met. Geophys. Biokl.* B, 14, 269–79.

HARE, F. K. 1955. Dynamic and synoptic climatology. *Ann. Assoc. Amer. Geogr.* 45, 152–62.

HARE, F. K. 1957. The dynamic aspects of climatology. *Geogr. Annal.* 39, 87–104.

HELLMANN, G. 1908. Die Anfänge der Meteorologie. *Met. Zeit.* 25, 481–91.

HENINGER, S. K. 1960. *A handbook of Renaissance meteorology*. Durham, N.C.: Duke Univ. Press. 269 pp.

HESSELBERG, T. 1932. Arbeitsmethoden einer dynamischen Klimatologie. *Beitr. Phys. f. Atmos.* 19, 291–305.

HUSCHKE, R. E. (ed.) 1959. *Glossary of meteorology*. Boston, Mass.: Amer. Met. Soc. 638 pp.

INWARDS, R. L. 1950. *Weather lore*. 4th ed. London: Rider and Co. 251 pp.

JACOBS, W. J. 1946. Synoptic climatology. *Bull. Amer. Met. Soc.* 27, 306–11.

JACOBS, W. J. 1947. *Wartime developments in applied climatology*. Met. Monogr. (Amer. Met. Soc.) 1 (1). 52 pp.

KARAPIPERIS, P. P. 1951. The Tower of the Winds. *Weatherwise* 4, 112–13.

LANDSBERG, H. E. 1957. *Review of climatology, 1951–55.* Met. Monogr. (Amer. Met. Soc.) 3 (12). 1–43.

LYDOLPH, P. E. 1957. How many climatologies are there? *Prof. Geogr.* 9 (6), 5–7.

POINTER, J. 1723. *A rational account of the weather showing the signs of its several changes and alterations together with the philosophical reasons of them.* Oxford. 88 pp.

SCHNEIDER-CARIUS, K. 1955. *Wetterkunde, Wetterforschung.* Freiburg/Munich: K. Alber. 423 pp.

SHAW, W. N. 1926. *Manual of meteorology,* vol. 1: *Meteorology in history.* Cambridge: Cambridge University Press. 343 pp.

SUTCLIFFE, R. C. 1964. Expansion of Meteorological Office research in dynamic climatology. *Met. Mag.* 93, 3–4.

Chapter 2

2.A SYNOPTIC DATA

2.B CLIMATOLOGICAL MAPS

ANDERSON, R., BOVILLE, B. W. and MCCLELLAN, D. E. 1955. An operational frontal contour-analysis model. *Quart. J. R. Met. Soc.* 81, 588–99.

ANGELL, J. K. 1961. Use of constant level balloons in meteorology. *Adv. Geophys.* 8, 137–219.

ASSELIN, R. 1966. *A technique for designing smoothing operators.* Met. Branch, CIR-4417, TEC-615. Toronto: Dept of Transport. 21 pp.

ATKINSON, B. W. 1968. *The weather business.* London: Aldus. 192 pp.

BARRETT, E. C. 1970a. A contribution to the dynamic climatology of the equatorial eastern Pacific and central America, based on meteorological satellite data. *Trans. Inst. Brit. Geogr.* No. 50, 25–53.

BARRETT, E. C. 1970b. Rethinking climatology, an introduction to the uses of weather satellite photographic data in climatological studies. *Progress in Geogr.* 2, 153–206.

BARRY, R. G. 1967. Seasonal location of the arctic front over North America. *Geogr. Bull.* 9, 79–95.

BARRY, R. G. and CHORLEY, R. J. 1971. *Atmosphere, weather and climate.* 2nd ed. London: Methuen. 379 pp.

BELLAMY, J. C. 1949. Objective calculations of divergence, vertical velocity and vorticity. *Bull. Amer. Met. Soc.* 30, 45–9.

BERGERON, T. 1928. Über die dreidimensional verknupfende Wetteranalyse, Part 1: Prinzipelle Einführung in das Problem der Luftmassen- und Frontenbildung. *Geofys. Publik.* (Oslo), 5 (6), 1–111.

BERGERON, T. 1930. Richtlinien einer dynamischen Klimatologie. *Met. Zeit.* 47, 246–62.

BERGERON, T. 1951. A general survey in the field of cloud physics. *Int. Assoc. Met. I.U.G.G., 9th Gen. Assembly* (Brussels), 120–34.

BERRY, F. A. JR, BOLLAY, E. and BEERS, N. R. 1945. *Handbook of meteorology.* New York: McGraw-Hill. 1068 pp.

BERRY, F. A., HAGGARD, W. H. and WOLFF, P. M. 1954. *Description of contour patterns at 500 mb.* Norfolk, Va: Bureau of Aeronautics Project AROWA, U.S. Naval Air Station.

BJERKNES, J. and PALMÉN, E. 1937. Investigation of selected European cyclones by means of serial ascents, Case 4: February 15–17, 1935; *Geofys. Publik.* (Oslo) 12 (2). 62 pp.

BJERKNES, J. and SOLBERG, H. 1922. Life cycle of cyclones and the polar front theory of atmospheric circulation; *Geofys. Publik.* (Oslo) 3 (1). 18 pp.

BLEEKER, W. 1960. Some climatological remarks about the field of flow and temperature over the Mediterranean Sea. *Met. Abhand.* 9 (1), 51–76.

BOUCHER, R. J. and NEWCOMB, R. J. 1962. Synoptic interpretation of some TIROS vortex patterns: a preliminary cyclone model. *J. Appl. Met.* 1, 127–36.

BROWNING, K. A. and HARROLD, T. W. 1969. Air motion and precipitation growth in a wave depression. *Quart. J. R. Met. Soc.* 95, 288–309.

BROWNING, K. A. and WEXLER, R. 1968. A determination of kinematic properties of a wind field using Doppler radar. *J. Appl. Met.* 7, 105–13.

BRYSON, R. A. 1966. Air masses, streamlines and the boreal forest. *Geogr. Bull.* 8, 228–69.

BRYSON, R. A. and KUHN, P. M. 1961. Stress-differential induced divergence with application to littoral precipitation. *Erdkunde* 15, 287–94.

BULINSKAIA, N. A. 1963. *Atlas of the pressure characteristics of depressions and anticyclones.* Moscow: Inst. Prikladnoi Geofiz., Akad. Nauk. 194 pp. (in Russian).

BURBIDGE, F. E. 1951. The modification of continental polar air over Hudson Bay. *Quart. J. R. Met. Soc.* 77, 365–74.

BUSHBY, F. H. 1952. The evaluation of vertical velocity and thickness tendency from Sutcliffe's theory. *Quart. J. R. Met. Soc.* 78, 354–61.

CHRISTIE, A. D. and RITCHIE, J. C. 1969. On the use of isentropic trajectories in the study of pollen transports. *Naturaliste canadien* 96, 531–49.

CLARKE, L. C. and RENARD, R. J. 1966. The U.S. Navy numerical frontal analysis scheme: further developments and a limited evaluation. *J. Appl. Met.* 5, 764–77.

CLODMAN, S. and JARVIS, E. C. 1966. *Some possible applications of climatological distributions to local forecasting.* Met. Branch, CIR-4412, TEC-610. Toronto: Dept of Transport. 10 pp.

COLLINS, G. O. and KUHN, P. M. 1954. Computation of precipitation resulting from vertical velocities deduced from vorticity changes. *Mon. Wea. Rev.* 82, 173–82.

CONOVER, J. H. 1962. *Cloud interpretation from satellite altitudes.* Air Force Cambridge Res. Lab., Research Note No. 81. 77 pp.

CRESWICK, W. S. 1967. Experiments in objective frontal analysis. *J. Appl. Met.* 6, 774–81.

CROCKER, A. M., GODSON, W. L. and PENNER, C. M. 1947. Frontal contour charts. *J. Met.* 4, 95–9.

CROWE, P. R. 1965. The geographer and the atmosphere. *Trans. Inst. Brit. Geogr.* No. 36, 1–19.

DAMMAN, W. 1960. Klimatologie der atmosphärischen Störungen über Europa. *Erdkunde* 14, 204–21.

DANARD, M. B. 1964. On the influence of released latent heat on cyclone development. *J. Appl. Met.* 3, 27–37.

DANIELSEN, E. F. 1959. The laminar structure of the atmosphere and its relation to the concept of a tropopause. *Arch. Met. Geophys. Biokl.* A, 11, 293–32.

DANIELSEN, E. F. 1961. Trajectories: isobaric, isentropic and actual. *J. Met.* 18, 479–86.

DANIELSEN, E. F. and BLECK, R. 1966. *Research in four-dimensional diagnosis in cyclonic storm cloud systems.* Final Rep., Contract AF 19(628)-4762, Pennsylvania State University. 96 pp.

DODDS, I. 1971. A comparison of depressions over Europe and the northeast Atlantic. *Weather* 26, 210–16.

DOVE, H. W. 1852. *Das Gesetz der Stürme.* Berlin. 117 pp. (trans. 1862 *The law of storms considered in connexion with the ordinary movements of the atmosphere*; 2nd ed., Longmans, London).

DUNN, G. E. 1940. Cyclogenesis in the tropical Atlantic. *Bull. Amer. Met. Soc.* 21, 215–29.

DUQUET, R. T. and SPAR, J. 1957. Some statistical characteristics of cyclones at 500 mb. *J. Met.* 3, 27–37.

EDDY, A. 1964. The objective analysis of horizontal wind divergence fields. *Quart. J. R. Met. Soc.* 90, 424–40.

EKHOLM, N. 1904. Wetterkarten der Luftdruckschwankungen. *Met. Zeit.* 21, 345–57.

ENDLICH, R. M. 1967. An iterative method for altering the kinematic properties of wind fields. *J. Appl. Met.* 6, 837–44.

EVJEN, S. 1954. Number of cyclones and anticyclones in northwest and middle Europe. *Met. Annal.* 3, 459–85.

FERGUSON, H. L. 1961. *A geostrophic advection scale for constant pressure surfaces.* Met. Branch, CIR-3516, TEC-363. Toronto; Dept of Transport. 9 pp. (and Amendment, CIR-3668, TEC-411).

FERGUSON, H. L. 1963. *A geostrophic advection scale for polar stereographic charts.* Met. Branch, CIR-3957, TEC-473, Toronto: Dept of Transport. 5 pp.

FETT, R. W. 1964. Aspects of hurricane structure: new model considerations suggested by TIROS and Project Mercury observations. *Mon. Wea. Rev.* 92, 43–59.

FINDLATER, J. 1967. A note on analysis of anticyclones. *Met. Mag.* 96, 69–73.

FJØRTOFT, R. 1952. On a numerical method of integrating the barotropic vorticity equation. *Tellus* 4, 179–94.

FLEAGLE, R. G. 1958. On the mechanism of large-scale vertial motion. *J. Met.* 15, 249–58.

FLOHN, H. and HUTTARY, J. 1950. Über die Bedeutung der Vb-Lagen für das Niederschlagsregime Mitteleuropas. *Met. Rund.* 3, 167–70.

FROST, R. and STEPHENSON, P. M. 1965. Mean streamlines and isotachs at standard pressure levels over the Indian and west Pacific oceans and adjacent land areas. *Geophys. Mem.* (London) 14 (109). 24 pp.

FULKS, J. R. 1945. Constant pressure maps – methods of preparation and advantages in their use. *Bull. Amer. Met. Soc.* 26, 133–46.

GALLOWAY, J. L. 1958. The three-front model: its philosophy, nature, construction and use. *Weather* 13, 395–403.

GODSKE, C. L., BERGERON, T., BJERKNES, J. and BUNDGAARD, R. C. 1957. *Dynamic meteorology and weather forecasting*. Boston, Mass.: Amer. Met. Soc. 800 pp.

GODSON, W. L. 1950. The structure of North American weather systems. In *Centen. Proc. Roy. Met. Soc.* (London). 89–106.

GODSON, W. L. 1951. Synoptic properties of frontal surfaces. *Quart. J. R. Met. Soc.* 77, 633–53.

GRAHAM, R. D. 1953. A new method of computing vorticity and divergence. *Bull. Amer. Met. Soc.* 34, 68–74.

GRAYSTONE, P. 1962. The introduction of topographic and frictional effects in a baroclimic model. *Quart. J. R. Met. Soc.* 88, 256–70.

GREEN, J. S. A., LUDLAM, F. H. and MCILVEEN, J. F. R. 1966. Isentropic relative-flow analysis and the parcel theory. *Quart. J. R. Met. Soc.* 92, 210–19.

HALTINER, G. J., CLARKE, L. C. and LAWNICZAK, G. E. JR. 1963. Computation of the large scale vertical velocity. *J. Appl. Met.* 2, 242–59.

HALTINER, G. J. and MARTIN, F. L. 1957. *Dynamical and physical meteorology*. New York: McGraw-Hill. 470 pp.

HARE, F. K. 1955. Dynamic and synoptic climatology. *Ann. Assoc. Amer. Geogr.* 45, 152–62.

HARE, F. K. 1957. The dynamic aspects of climatology. *Geogr. Annal.* 39, 87–104.

HARLEY, W. S. 1962. *Distributions of wet-bulb potential temperature in North American air masses (1957) and a statistical analysis of results*. Met. Branch, CIR-3622, TEC-400. Toronto: Dept of Transport. 46 pp.

HARLEY, W. S. 1964. *Use of the advection scale in the determination of vertical velocities*. Met. Branch, CIR-4052, TEC-529. Toronto: Dept of Transport. 5 pp.

HARLEY, W. S. 1965a. An operational method of quantitative precipitation forecasting. *J. Appl. Met.* 4, 305–19.

HARLEY, W. S. 1965b. *Determination of spot values of vertical velocity and precipitation rates from values of absolute vorticity advection, 1000–500 mb thickness advection and precipitable water*. Met. Branch, CIR-4303,TEC-580. Toronto: Dept of Transport. 4 pp.

HARLEY, W. S., DRAGERT, H. and RUTHERFORD, I. D. 1964. *The determination of spot values of vertical velocity and precipitation*. Met. Branch, CIR-4139, TEC-544. Toronto: Dept of Transport. 17 pp.

HARROLD, T. W. and BROWNING, K. A. 1969. The polar low as a baroclinic disturbance. *Quart. J. R. Met. Soc.* 95, 710–23.

HEASTIE, H. and STEPHENSON, P. M. 1960. Upper winds over the world, Parts I and II. *Geophys. Mem.* (London) 13 (3). 217 pp.

HESSELBERG, T. 1962. Climatological deviation maps. *Geofys. Publik.* (Oslo) 24, 39–66.

HEWSON, E. W. 1936, 1937, 1938. The application of wet-bulb potential temperature to air mass analysis. *Quart. J. R. Met. Soc.* 62, 387–420; 63, 323–37; 64, 407–18.

HOGBEN, G. L. 1946. A theoretical note on some errors in estimating the curvature of air trajectories and streamlines. *Quart. J. R. Met. Soc.* 72, 318–22.

HOLLOWAY, J. L. JR. 1958. Smoothing and filtering of time series and space fields. *Adv. Geophys.* 4, 351–89.

HOPKINS, M. M. JR. 1967. An approach to the classification of meteorological satellite data. *J. Appl. Met.* 6, 164–78.

HUBERT, L. F., KRUEGER, A. F. and WINSTON, J. S. 1969. The double intertropical convergence zone – fact or fiction? *J. Atm. Sci.* 27, 771–3.

JACOBS, I. 1958. 5-bzw 40 jährige Monatsmittel der absoluten Topographien der 1000 mb, 850 mb, 500 mb und 300 mb Flachen sowie der relativen Topographie 500/100 mb und 300/500 mb über der Nordhemisphäre und ihre monatlichen Änderungen. *Met. Abhand.* (Berlin) 4 (2), Part 1, 47 pp.; Part 2, 121 pp.

JARVIS, E. C. 1966. *The analysis and interpretation of precipitation patterns.* Met. Branch, CIR-4510, TEC-632. Toronto: Dept of Transport. 11 pp.

JARVIS, E. C. 1967. *The use of streamline analysis in short-range forecasting.* Met. Branch, CIR-4534, TEC-641. Toronto: Dept of Transport. 12 pp.

JARVIS, E. C. and AGNEW, T. 1970. A note on the computation of terrain and frictionally induced vertical velocities. *J. Appl. Met.* 9, 942–6.

JARVIS, E. C. and LEONARD, R. 1969. *Vertical velocities induced by smoothed topography and their use in areal forecasting.* Met. Branch Tech. Mem. 728. Toronto: Dept of Transport. 17 pp.

JOHNSON, D. H. 1962. Rain in east Africa. *Quart. J. R. Met. Soc.* 88, 1–19.

JOHNSON, D. H. 1965. The meterological implications of downwind movement. In *Meteorology and the desert locust.* Tech. Note No. 69 (W.M.O. No. 171, TP 85). Geneva: World Meteorological Organization. 6–19.

JOHNSON, D. H. and MÖRTH, H. T. 1960. Forecasting research in east Africa. In Bargman, D. J. (ed.) *Tropical meteorology in Africa.* Nairobi: Munitalp Foundation. 56–132.

KAGAWA, H. and LEE, R. 1967. *Kinematic interpretation of pressure and height change fields.* Met. Branch Tech. Mem. 656. Toronto: Dept of Transport. 21 pp.

KASAHARA, A. and WASHINGTON, W. M. 1967. NCAR global general circulation model of the atmosphere. *Mon. Wea. Rev.* 95, 389–402.

KEEGAN, T. J. 1958a. Arctic synoptic activity in winter. *J. Met.* 15, 513–21.

KEEGAN, T. J. 1958b. The wintertime circulation in the arctic troposphere. In Wilson, C. V. (ed.) *Contributions to the study of the arctic circulation.* Montreal: Arctic Met. Res. Group, McGill Univ. 22–47.

KIRK, T. H. 1966a. Some aspects of the theory of fronts and frontal analysis. *Quart. J. R. Met. Soc.* 92, 374–81.

KIRK, T. H. 1966b. A parameter for the objective location of frontal zones. *Met. Mag.* 94, 351–3.

KIRK, T. H. 1970. The Laplacian and its relevance for analysis. *Met. Mag.* 99, 151–2.

KLEIN, W. H. 1957. *Principal tracks and mean frequencies of cyclones and anticyclones in the northern hemisphere.* Research Paper No. 40. Washington, D.C.: U.S. Weather Bureau. 22 pp.

KLEIN, W. H. 1958. The frequency of cyclones and anticyclones in relation to the mean circulation. *J. Met.* 15, 98–102.

KLEIN, W. H. and WINSTON, J. S. 1958. Geographical frequency of troughs and ridges on mean 700 mb charts. *Mon. Wea. Rev.* 86, 344–58.

KNIGHTING, E. 1960. Some computations of the variation of vertical velocity with pressure on a synoptic scale. *Quart. J. R. Met. Soc.* 86, 318–25.

KÖPPEN, W. 1880. Die Zugstrassen der barometrischen Minima in Europa und auf dem nordatlantischen Ozean und ihr Einfluss auf Wind und Wetter bie uns. *Mitt. Geogr. Ges. Hamburg* 1880–1, 76–97.

KREBS, S. and BARRY, R. G. 1970. The arctic front and the tundra-taiga boundary in Eurasia. *Geog. Rev.* 60, 548–54.

KREITZBERG, C. W. and BROWN, H. A. 1970. Mesoscale weather systems within an occlusion. *J. Appl. Met.* 9, 417–32.

KRISHNAMURTI, T. N. 1969. An experiment in numerical prediction in equatorial latitudes. *Quart. J. R. Met. Soc.* 95, 594–620.

KRISHNAMURTI, T. N. 1971. Observational study of the upper tropospheric motion field during the northern hemisphere summer. *J. Appl. Met.* 10, 1066–1096.

KRYZHANOVSKAIA, A. P. 1968. Cyclone and anticyclone trajectories at sea level in the northern hemisphere. *Trudy Gidromet. Nauch. Issled. Tsentr. SSSR* (Leningrad) 26, 79–91 (in Russian).

KRYZHANOVSKAIA, A. P. 1969. Seasonal features of motion of depressions and anticyclones over the northern hemisphere. *Trudy Gidromet. Nauch. Issled. Tsentr. SSSR* (Leningrad) 41, 71–87 (in Russian).

KURASHIMA, A. 1968. Studies on the winter and summer monsoons in east Asia based on dynamic concept. *Geophys. Mag.* (Tokyo) 34, 165–236.

LALLY, V. E., LITCHFIELD, E. W. and SOLOT, S. B. 1966. The southern hemisphere GHOST experiment (the first seventy days). *W.M.O. Bull.* 15, 124–8.

LAMB, H. H. 1959. The southern westerlies: a preliminary survey; main characteristics and apparent associations. *Quart. J. R. Met. Soc.* 85, 1–23.

LANDERS, H. 1955. A three-dimensional study of the horizontal velocity divergence. *J. Met.* 12, 415–27.

LENOVA, G. V. 1970. Description of the synoptic situation by means of centres of gravity of sets of pressure systems. *Trudy Gidromet. Nauch. Issled. Tsentr. SSSR* (Leningrad) 162, 3–17 (in Russian).

LOCKWOOD, J. G. 1968. Some uses of profiles and charts of total energy content. *Met. Mag.* 97, 145–55.

LORENZ, E. N. 1967. *The nature and theory of the general circulation of the atmosphere* (W.M.O. No. 218, TP 115). Geneva: World Meteorological Organization. 161 pp.

LUDLAM, F. H. 1966. *The cyclone problem: a history of models of the cyclonic storm.* Inaugural Lecture, Imperial Coll. Sci. Technol., London. 19–49.

LUDLAM, F. H. and MILLER, L. I. 1959. *Research on the properties of cloud systems.* Final Report, Contract AF-61(514)-1292. London: Met. Dept, Imperial College.

MCINTYRE, D. P. 1958. The Canadian three-front, three-jet stream model. *Geophysica* (Helsinki) 6, 309–24.

MCKAY, G. A., FINDLAY, B. F. and THOMPSON, H. A. 1970. A climatic perspective of tundra areas. In Fuller, W. A. and Kevan, P. G. (eds.) *Productivity and conservation in northern circumpolar lands.* Pub. No. 16. Morges, Switzerland: Int. Union Conserv. Nature, Natural Resources. 10–33.

MCPHERSON, G. A., THOMPSON, F. D., TIBBLES, L. G. and TREIDL, R. A. 1969. *The meaning and application of advection fields in analysis and forecasting.* Met. Branch. Tech. Mem. 715. Toronto: Dept of Transport. 20 pp.

MASON, B. J. 1969. Some outstanding problems in cloud physics – the interaction of micro-physical and dynamical processes. *Quart. J. R. Met. Soc.* 95, 449–85.

MERRITT, E. S. 1964. Easterly waves and perturbations, a reappraisal. *J. Appl. Met.* 3, 367–82.

MILLER, J. E. 1948. Studies of large-scale vertical motions of the atmosphere. *Met. Papers* No. 1, New York Univ. 1–48.

MINTZ, T. and DEAN, G. 1952. *The observed mean field of motion in the atmosphere.* Geophys. Res. Pap. No. 17. Cambridge, Mass,: U.S. Air Force, Cambridge Res. Cen. 65 pp.

MONTGOMERY, R. B. 1937. A suggested method for representing gradient flow in isentropic surfaces. *Bull. Amer. Met. Soc.* 18, 210–12.

NAMIAS, J. 1940. *An introduction to the study of air mass and isentropic analysis.* 5th ed. Boston, Mass.: Amer. Met. Soc. 232 pp.

NAMIAS, J. and CLAPP, P. F. 1946. Normal fields of convergence and divergence at the 10,000 foot level. *J. Met.* 3, 14–22.

NEWTON, C. W. 1966. Severe convective storms. *Adv. Geophys.* 12, 257–308.

O'CONNOR, J. F. 1964. Hemispheric distribution of five-day mean 700 mb circulation centers. *Mon. Wea. Rev.* 92, 303–15.

OLIGER, J. E., WELLCK, R. E., KASAHARA, A. and WASHINGTON, W. M. 1970. *Description of NCAR global circulation model.* Boulder, Colo.: Nat. Cen. Atm. Res. 94 pp.

PALMÉN, E. and HOLOPAINEN, E. O. 1962. Divergence, vertical velocity and conversion between potential and kinetic energy in an extratropical disturbance. *Geophysica* 8, 89–113.

PALMÉN, E. and NEWTON, C. W. 1969. *Atmospheric circulation systems: their structure and physical interpretation.* New York: Academic Press. 603 pp.

PALMER, C. E. 1952. Tropical meteorology. *Quart. J. R. Met. Soc.* 78, 126–64.

PANOFSKY, H. A. 1951. Large-scale vertical velocity and divergence. In Malone, T. F. (ed.) *Compendium of meteorology.* Boston, Mass.: Amer. Met. Soc. 639–46.

PENNER, C. M. 1955. A three-front model for synoptic analysis. *Quart. J. R. Met. Soc.* 81, 89–91.

PENNER, C. M. 1963. An operational method for determination of vertical velocities. *J. Appl. Met.* 2, 235–41.

PETTERSSEN, S. 1950. Some aspects of the general circulation of the atmosphere. *Centen. Proc. R. Met. Soc.* (London). 120–55.

PETTERSSEN, S. 1956. *Weather analysis and forecasting*, vol. 1. New York: McGraw-Hill. 428 pp.

PETTERSSEN, S., BRADBURY, D. L. and PEDERSEN, K. 1962. The Norwegian cyclone models in relation to heat and cold sources. *Geophys. Publik.* (Oslo) 24, 243–80.

PETTERSSEN, S. and CALABRESE, P. A. 1959. On some weather influences due to warming of the air by the Great Lakes in winter. *J. Met.* 16, 646–52.

REED, R. J. 1960. Principal frontal zones of the northern hemisphere in winter and summer. *Bull. Amer. Met. Soc.* 41, 591–8.

REED, R. J. and KUNKEL, B. A. 1960. The arctic circulation in summer. *J. Met.* 17, 489–506.

REINEL, H. 1960. Die Zugbahnen der Hochdruckgebiete über Europa als klimatologisches Problem. *Mitt. Fränk. Geogr. Ges.* 6, 1–73.

RENARD, R. J. and CLARKE, L. C. 1965. Experiments in numerical objective frontal analysis. *Mon. Wea. Rev.* 93, 547–56.

REX, D. F. 1958. Vertical atmospheric motion in the equatorial Pacific. *Geophysica* 6, 479–500.

RICHARDSON, L. F. 1922. *Weather prediction by numerical process.* Cambridge: Cambridge Univ. Press. 236 pp.

ROMANOV, Y. A. 1965. Mean resultant wind divergence and vorticity charts for the Indian Ocean. *Trudy Inst. Okeanol., Akad. Nauk SSSR* 78, 119–27 (in Russian).

ROSSBY, C. G. and collaborators. 1937. Isentropic analysis; *Bull. Amer. Met. Soc.* 18, 201–9.

SADLER, J. C. 1965. The feasibility of global tropical analysis. *Bull. Amer. Met. Soc.* 46, 118–30.

SALTZMANN, B. and PEIXOTO, J. 1957. Harmonic analysis of the mean northern hemisphere wind field for the year 1950. *Quart. J. R. Met. Soc.* 83, 360–4.

SAUCIER, W. M. 1955. *Principles of meteorological analysis.* Chicago: Univ. of Chicago Press. 438 pp.

SAWYER, J. S. 1949. Large-scale vertical motion in the atmosphere: a discussion. *Quart. J. R. Met. Soc.* 75, 185–8.

SAWYER, J. S. 1964. Meteorological analysis: a challenge for the future. *Quart. J. R. Met. Soc.* 90, 227–47.

SCHÜEPP, M. 1963. Die Häufigkeit der starken Höhenwinde bei verschiedenen Wetter- und Witterungslagen. *Geofis. e Met.* 11, 94–6.

SCHWARZL, S. 1965. Die Häufigkeit von Vb-Lagen. *Carinthia II* (Vienna) Sonderheft 24, 101–6.

SCORER, R. S. 1957. Vorticity. *Weather* 12, 72–83.

SCORER, R. S. 1958. *Natural aerodynamics.* Oxford: Pergamon. 312 pp.

SELA, J. and CLAPP, P. F. 1971. Computation of climatological vertical velocities by the thermodynamic method. *Mon Wea. Rev.* 99, 524–31.

SHAW, W. N. 1930. *Manual of meteorology*, vol. 3: *The physical processes of weather.* Cambridge: Cambridge Univ. Press, 259–66.

SHAW, W. N. and LEMPFERT, R. K. G. 1906. *The life history of surface air currents and a study of surface traejctories of moving air.* Met. Office 174. London. 107 pp.

SHEPPARD, P. A. 1949. Large-scale vertical motion in the atmosphere: a discussion. *Quart. J. R. Met. Soc.* 75, 188–92.

SMAGORINSKY, J. 1953. The dynamic influence of large scale heat sources and sinks on the quasi-stationary mean motions of the atmosphere. *Quart. J. R. Met. Soc.* 79, 342–366.

SMAGORINSKY, J. 1970. Numerical simulation of the global atmosphere. In CORBY, G. A. (ed.) *The global circulation of the atmosphere.* London: Roy. Met. Soc. 24–41.

SPAR, J. 1950. On the theory of annual pressure variations. *J. Met.* 7, 167–80.

STALEY, D. O. 1966. The lapse rate of air temperature following an air parcel. *Quart. J. R. Met. Soc.* 92, 147–50.

STARK, L. P. 1965. Positions of monthly mean troughs and ridges in the northern hemisphere, 1949–1963. *Mon. Wea. Rev.* 93, 705–20.

SUTCLIFFE, R. C. 1948. The use of upper air thickness patterns in forecasting *Met. Mag.* 77, 145–52.

SUTCLIFFE, R. C. 1956. Water balance and the general circulation of the atmosphere. *Quart. J. R. Met. Soc.* 82, 385–95.

SUTCLIFFE, R. C. and FORSDYKE, A. G. 1950. The theory and use of upper air thickness patterns in forecasting. *Quart. J. R. Met. Soc.* 76, 189–217.

TALJAARD, J. J. 1967. Development, distribution and movement of cyclones and anticyclones in the southern hemisphere during the IGY. *J. Appl. Met.* 6, 973–87.

TALJAARD, J. J., SCHMITT, W. and VAN LOON, H. 1961. Frontal analysis with application to the southern hemisphere. *Notos* 10, 25–57.

TAYLOR, J. A. and YATES, R. A. 1967. *British weather in maps.* 2nd ed. London: Macmillan. 315 pp.

TAYLOR, V. R. and WINSTON, J. S. 1969. *Monthly and seasonal mean global charts of brightness from ESSA 3 and 5 digitized pictures, Feb. 1967–Feb. 1968.* ESSA Tech. Rep. No. 46. Washington, D.C.; Nat. Env. Satellite Center.

THOMPSON, B. W. 1951. An essay on the general circulation over southeast Asia and the west Pacific. *Quart. J. R. Met. Soc.* 77, 569–97.

THOMPSON, F. D. 1969. *Lake-effect snowstorms in southern Ontario during November and December, 1968.* Met. Branch Tech. Mem. 726. Toronto: Dept of Transport. 20 pp.

TREWARTHA, G. T. 1968. *An introduction to climate.* 4th ed. New York: McGraw-Hill. 334 pp.

TUCKER, G. B. 1960. Upper winds over the world, Part 3. *Geophys. Mem.* (London) 13 (5). 101 pp.

VAISANEN, A. 1961. *Investigation of the vertical air movement and related phenomena in selected synoptic situations.* Inst. of Met. (Helsinki) Paper No. 93. 72 pp.

VAN BEBBER, J. 1891. Die Zugstrassen der barometrischen Minima nach den Bahnenkarten der Deutschen Seewarte für den Zeitraum 1875–1890. *Met. Zeit.* 8, 361–6.

VAN LOON, H. 1965. A climatological study of the atmospheric circulation in the southern hemisphere during the IGY, Part 1: 1 July 1957–31 March 1958. *J. Appl. Met.* 4, 479–91.

VAN LOON, H. and TALJAARD, J. J. 1958. A study of the 1000–500 mb thickness distribution in the southern hemisphere. *Notos* 7, 123–58.

VIAUT, M. 1961. *Les nuages: système nuageux et types de ciel.* Paris: Meteorol. Nat. 158 pp.

VITELS, L. A. 1968. *The monthly, seasonal and yearly characteristics of the pressure circulation regime of the European synoptic region, 1900–1964.* Leningrad: Gidromet. Izdat. 128 pp. (in Russian).

WAHL, E. W. and LAHEY, J. F. 1969. *A 700 mb atlas for the northern hemisphere: five-day mean heights, standard deviations, and changes for the 700 mb pressure surface.* Madison: Univ. of Wisconsin Press. 151 pp.

WALKER, J. M. 1967. Subterranean isobars. *Weather* 22, 296–7.

WATTS, I. E. M. 1955. *Equatorial meteorology, with particular reference to southeast Asia.* London: Univ. of London Press. 223 pp.

WEICKMANN, L. 1960. Haufigkeitsverteilung und Zugbahnen von Depressionen im mittleren Osten. *Met. Rund.* 13, 33–8.

WEXLER, H. and NAMIAS, J. 1938. Mean monthly isentropic charts and their relation to departures of summer rainfall. *Trans. Amer. Geophys. Union* 19, 164–70.

WILSON, H. P., BEAUDOIN, B. and DONAIS, R. 1968. *Slopes of significant surfaces in the atmosphere.* Met. Branch Tech. Mem. 701. Toronto: Dept of Transport. 8 pp.

WORLD METEOROLOGICAL ORGANIZATION. 1961. *Techniques for high-level wind analysis and forecasting of wind and temperature fields.* Tech. Note No. 35. Geneva: W.M.O. 187 pp.

WORLD METEOROLOGICAL ORGANIZATION. 1965. *Catalogue of meteorological data for research* (W.M.O. No. 174, TP 86). Geneva: W.M.O.

YOSHIMURA, M. 1969. Annual change in frontal zones in the northern hemisphere. *Japanese Progr. in Climatol.* 69–71.

YOSHINO, M. M. 1969. *Climatological studies on the polar frontal zones and the intertropical convergence zones over south, southeast and east Asia.* Climatol. Notes No. 1. Tokyo: Hosei Univ. 71 pp.

ZIPSER, E. J. and COLON, J. A. 1962. Mean layer wind charts in tropical analysis. *Mon. Wea. Rev.* 90, 465–70.

2.C DATA SYNTHESIS – THE GENERAL CIRCULATION

BARRY, R. G. 1970. A framework for climatological research with particular reference to scale concepts. *Trans. Inst. Brit. Geogr.* No. 49, 61–70.

BARRY, R. G. and CHORLEY, R. J. 1971. *Atmosphere, weather and climate.* 2nd ed. London: Methuen. 379 pp.

BAUR, F. 1951. Extended-range weather forecasting. In Malone, T. F. (ed.) *Compendium of meteorology.* Boston, Mass.: Amer. Met. Soc. 814–33.

BLÜTHGEN, J. 1966. *Allgemeine Klimageographie.* 2nd ed. Berlin: de Gruyter. 720 pp.

CORBY, G. A. 1970. Editor's foreword in *The global circulation of the atmosphere.* London: Roy. Met. Soc. 1–2.

DEFANT, F. and TABA, H. 1957. The threefold structure of the atmosphere and the characteristics of the tropopause. *Tellus* 9, 259–74.

FIEDLER, F. and PANOFSKY, H. A. 1970. Atmospheric scale and spectral gaps. *Bull. Amer. Met. Soc.* 51, 1114–19.

FLOHN, H. 1965. *Research aspects of long range forecasting*. Tech. Note No. 66 (W.M.O. 162, TP 79). Geneva: World Meteorological Organization. 1–10.

GOMMEL, W. R. 1963. Mean distribution of 500 mb topography and sea-level pressure in middle and high latitudes of the northern hemisphere during the 1950–9 decade, January and July. *J. Appl. Met.* 2, 105–13.

KLEIN, W. H. and WINSTON, J. S. 1958. Geographical frequency of troughs and ridges on mean 700 mb charts. *Mon. Wea. Rev.* 86, 344–58.

LORENZ, E. N. 1967. *The nature and theory of the general circulation of the atmosphere* (W.M.O. No. 218, TP 115). Geneva: World Meteorological Organization. 161 pp.

MASON, B. J. 1970. Future developments in meteorology: an outlook to the year 2000. *Quart. J. R. Met. Soc.* 96, 349–68.

NAMIAS, J. 1953. *Thirty-day forecasting: a review of a ten-year experiment.* Met. Monogr. (Amer. Met. Soc.) 2 (6). 83 pp.

O'CONNOR, J. F. 1964. Hemispheric distribution of five-day mean 700 mb circulation centers. *Mon. Wea. Rev.* 92, 303–15.

PALMÉN, E. and NEWTON, C. W. 1969. *Atmospheric circulation systems: their structure and physical interpretation.* New York: Academic Press. 603 pp.

PETTERSSEN, S. 1950. Some aspects of the general circulation of the atmosphere. *Centen. Proc. R. Met. Soc.* (London). 120–55.

PRIESTLEY, C. H. B. 1949. Heat transport and zonal stress between latitudes. *Quart. J. R. Met. Soc.* 75, 28–40.

PUTNINS, P. 1962. Correlation between pressure changes aloft and at the surface in the Greenland area and some aspects of the 'steering problem'. *Arch. Met. Geophys. Biokl.* A, 13, 218–40.

REITER, E. R. 1969. *Atmospheric transport processes*, Part 1: *Energy transfers and transformations*. U.S. Atomic Energy Commission, Div. Tech. Information. 253 pp.

REX, D. F. (ed.) 1969. *Climate of the free atmosphere. World survey of climatology* (ed. H. E. Landsberg), vol. 4. Amsterdam: Elsevier. 450 pp.

RIEHL, H. 1950. On the role of the tropics in the general circulation of the atmosphere. *Tellus* 2, 1–17.

RIEHL, H. 1970. Mechanisms of the general circulation of the troposphere. In Flohn, H. (ed.) *General climatology. World survey of climatology* (ed. H. E. Landsberg), vol. 2. Amsterdam: Elsevier. 1–36.

SALTZMANN, B. and PEIXOTO, J. 1957. Harmonic analysis of the mean northern hemisphere wind field for the year 1950. *Quart. J. R. Met. Soc.* 83, 360–4.

STARK, L. P. 1965. Positions of monthly mean troughs and ridges in the northern hemisphere, 1949–1963. *Mon. Wea. Rev.* 93, 705–20.

STARR, V. P. 1968. *Physics of negative viscosity phenomena.* New York: McGraw-Hill. 256 pp.

STARR, V. P. and WHITE, R. M. 1951. A hemispherical study of the atmospheric angular-momentum balance. *Quart. J. R. Met. Soc.* 77, 215–25.

STARR, V. P. and WHITE, R. M. 1952. Schemes for the study of hemispheric exchange processes. *Quart. J. R. Met. Soc.* 78, 407–10.

STARR, V. P. and WHITE, R. M. 1954. *Balance requirements of the general circulation.* Geophys. Res. Paper No. 35. Cambridge, Mass. 57 pp.

SUTTON, G. 1965. The resurgence of interest in the observational sciences. *Weather* 20, 173–82.

TALJAARD, J. J., VAN LOON, J., CRUTCHER, H. L. and JENNE, R. L. 1969. *Climate of the upper air*, Part 1: *Southern hemisphere*, vol. 1: *Temperatures, dew points and heights at selected levels.* Washington, D.C.: Government Printing Office.

TEISSERENC DE BORT, L. 1883. Étude sur l'hiver de 1879–80 et recherches sur la position des centres d'action de l'atmosphère dans les hivers anormaux. *Ann. Bureau Central Met. de France*, 1881, pt 4, 17–62.

TYSON, P. D. 1968. Velocity fluctuations in the mountain wind. *J. Atm. Sci.* 25, 381–4.

VEIGAS, K. W. and OSTBY, F. P. JR. 1963. Application of a moving coordinate prediction model to east coast cyclones. *J. Appl. Met.* 2, 24–38.

VUORELA, L. A. and TUOMINEN, I. 1964. On the mean zonal and meridional circulations and the flux of moisture in the northern hemisphere during the summer seasons. *Pure Appl. Geophys.* 57, 167–80.

WALKER, J. M. 1967. Subterranean isobars. *Weather* 22, 296–7.

ZIPSER, E. J. 1970. The Line Islands experiment, its place in tropical meteorology and the rise of the fourth school of thought. *Bull. Amer. Met. Soc.* 51, 1136–46.

Chapter 3

3.A INTRODUCTION

ABERCROMBY, R. 1883. On certain types of British weather. *Quart. J. R. Met. Soc.* 9, 1–25.

ABERCROMBY, R. 1887. *Weather: a popular exposition of the nature of weather changes from day to day.* London: Kegan Paul. 472 pp.

ANDERSON, D. V. 1967. Review of basic statistical concepts in hydrology. In *Statistical methods in hydrology. Proc. Hydrol. Sympos. No. 5.* Ottawa: Inland Waters Branch, Dept of Energy, Mines & Resources. 3–27.

ARAKAWA, H. 1937. Die Luftmassen in den japanischen Gebieten. *Met. Zeit.* 54, 169–74 (also in *Bull. Amer. Met. Soc.* 18, 407–10).

BAUR, F. 1931. Die Formen der atmosphärischen Zirkulation in der gemässigten Zone. *Gerlands Beitr. Geophys.* 34, 264–309.

BAUR, F. 1936a. Die Bedeutung der Stratosphäre für die Grosswetterlage. *Met. Zeit.*, 53, 237–47.

BAUR, F. 1936b. Wetter, Witterung, Grosswetter und Weltwetter. *Z. angew. Met.* 53, 377–81.

BAUR, F. 1944. Über die grundsätzliche Möglichkeit langfristiger Witterungsvorhersagen. *Ann. Hydrogr. marit. Met.* (Hamburg) 72, 15–25.

BAUR, F. 1947. *Musterbeispiele europäischer Grosswetterlagen.* Wiesbaden: Dieterich. 35 pp.

BAUR, F. 1948. *Einführung in die Grosswetterkunde.* Wiesbaden: Dieterich. 165 pp.

BAUR, F. 1963. *Grosswetterkunde und langfristige Witterungsvorhersage.* Frankfurt am-Main: Akad. Verlagsgesellschaft. 91 pp.

BERGERON, T. 1928. Über die dreidimensional verknüpfende Wetteranalyse, Part 1: Prinzipelle Einführung in das Problem der Luftmassen- und Frontenbildung. *Geofys. Publik.* (Oslo), 5 (6), 1–111.

BERGERON, T. 1930. Richtlinien einer dynamischen Klimatologie. *Met. Zeit.* 47, 246–62.

DINIES, E. 1932. Luftkörper Klimatologie. *Arch. dtsch. Seewarte* 50 (6), 21 pp.

ELLIOTT, R. D. 1949. The weather types of North America. *Weatherwise* 2, 15–18, 40–3, 64–7, 86–8, 110–13, 136–8.

ELLIOTT, R. D. 1951. Extended-range forecasting by weather types. In Malone, T. F. (ed.) *Compendium of meteorology.* Boston, Mass.: Amer. Met Soc. 834–40.

GOLD, E. 1920. Aids to forecasting: types of pressure distribution, with notes and tables for the fourteen years 1905–1918. *Geophys. Mem.* (London) 2 (16), 149–74.

JACOBS, W. C. 1947. *Wartime developments in applied climatology.* Met. Monogr. (Amer. Met. Soc.) 1 (1). 52 pp.

KÖPPEN, W. 1874. Über die Abhängigkeit des klimatischen Charakters der Winde von ihrem Ursprunge. *Rep. Met.* (St Petersburg) 4 (4). 15 pp.

LAMB, H. H. 1950. Types and spells of weather around the year in the British Isles: annual trends, seasonal structure of the year, singularities. *Quart. J. R. Met. Soc.* 76, 393–429.

LEIGHLY, J. B. 1949. Climatology since 1800. *Trans. Amer. Geophys. Union* 30, 658–72.

MÜLLER-ANNEN, H. 1955. Untersuchungen zur Frage der Realitat der Singularitäten. *Zeit. Met.* 9, 97–110.

RIETZ, H. L. (ed.) 1924. *Handbook of mathematical statistics.* Boston, Mass.: Houghton. 82–8.

SHOWALTER, A. K. 1939. Further studies of American air mass properties. *Mon. Wea. Rev.* 67, 204–18.

THOM, H. C. S. 1970. The analytical foundations of climatology. *Arch. Met. Geophys. Biokl.* B, 18, 205–20.

TU, C.-W. 1939. Chinese air mass properties. *Quart. J. R. Met. Soc.* 65, 33–51.

WILLETT, H. C. 1931. Ground plan of a dynamic climatology. *Mon. Wea. Rev.* 59, 219–23.

WILLETT, H. C. 1933. *American air mass properties.* Pap. Phys. Oceanog. Met. 2 (2). Cambridge, Mass.: Mass. Inst. Technol., Woods Hole Oceanog. Inst. 116 pp.

3.B STATIC VIEW OF THE WEATHER MAP

ANDERSEN, P. 1967. An investigation of European statistical weather types. In Godske, C. L. (ed.) *Further studies of statistical meteorology.* Contract AF 61(052)-760 (ed. C. L. Godske), University of Bergen. 58–80.

BARRY, R. G. 1960. A note on the synoptic climatology of Labrador–Ungava. *Quart. J. R. Met. Soc.* 86, 557–65.

BARRY, R. G. 1972. *Further climatological studies of Baffin Island, Northwest Territories.* Tech. Rep. 65. Ottawa: Inland Waters Directorate, in press.

BILHAM, E. G. 1938. *The climate of the British Isles.* London: Macmillan. 347 pp.

BOVILLE, B. W. and KWIZAK, M. 1959. *Fourier analysis applied to hemispheric waves of the atmosphere.* Met. Branch, CIR-3155, TEC-292. Toronto: Dept of Transport. 21 pp.

GOLD, E. 1920. Aids to forecasting: types of pressure distribution, with notes and tables for the fourteen years 1905–1918. *Geophys. Mem.* (London) 2 (16), 149–74.

GRAHAM, R. D. 1955. An empirical study of planetary waves by means of harmonic analysis. *J. Met.* 12, 298–307.

HARE, F. K. 1958. The quantitative representation of the north polar pressure field. In Sutcliffe, R. C. (ed.) *The polar atmosphere symposium*, part I. Oxford: Pergamon Press. 137–50.

HARE, F. K., GODSON, W. L., MacFARLANE, M. A. and WILSON, C. V. 1957. *Specification of pressure fields and flow patterns in polar regions: theories and techniques.* Pub. in Met. No. 5. Montreal: Arctic Met. Res. Group, McGill University. 72 pp.

HARTRANFT, F. R., RESTIVO, J. S. and SABIN, R. C. 1970. Computerized map typing procedures and their applications in the development of forecasting aids; *4th Weather Wing, Aerospace Sciences Division, Tech. Paper* 70–2, Ent Air Force Base, Colo, 57 pp.

HORN, L. H., ESSENWANGER, O and BRYSON, R. A. 1958. *Half-hemispheric 500 mb topography description by means of orthogonal polynomials*, Part II: *Specification.* Sci. Rep. No. 11, Contract AF 19(604)-992. Madison, Wis.: Met. Dept, University of Wisconsin.

HOUGHTON, J. G. 1969. *Characteristics of rainfall in the Great Basin.* Reno: Desert Res. Inst., Univ. of Nevada. 205 pp.

KUIPERS, W. J. A. 1970. An experiment on numerical classification of scalar fields. *Idojaras* 74, 296–306.

LAMB, H. H. 1950. Types and spells of weather around the year in the British Isles: annual trends, seasonal structure of the year, singularities. *Quart. J. R. Met. Soc.* 76, 393–429.

LONG, C. S. 1967. *A classification of synoptic weather map pressure patterns over the British Isles by statistical methods and some related weather characteristics.* Unpub. B.Sc. dissertation, Univ. of Southampton. 38 pp.

LUND, I. A. 1963. Map-pattern classification by statistical methods. *J. Appl. Met.* 2, 56–65.

NEWNHAM, E. V. 1925. Classification of synoptic charts for the North Atlantic for 1896–1910. *Geophys. Mem.* (London) 4 (26), 181–200.

PETTERSSEN, S. 1958. *Weather analysis and forecasting*, vol. 2. New York: McGraw-Hill. 266 pp.

PICK, W. H. 1929. The persistence of types of pressure distribution over the British Isles in winter. *Quart. J. R. Met. Soc.*, 55, 403–4.

PICK, W. H. 1930. The persistence of types of pressure distribution over the British Isles in summer. *Quart. J. R. Met. Soc.* 56, 82–4.

PUTNINS, P. 1966. *The sequence of baric pressure patterns over Alaska.* Studies on the meteorology of Alaska, 1st Interim Rep. Washington, D.C.: Environmental Data Service, ESSA. 81 pp.

SALTZMANN, B. 1958. Some hemispheric spectral statistics. *J. Met.* 15, 259–63.

SALTZMANN, B. and FLEISHER, A. 1962. Spectral statistics of the wind at 500 mb. *J. Atm. Sci.* 19, 195–204.

SALTZMANN, B. and PEIXOTO, J. 1957. Harmonic analysis of the mean northern hemisphere wind field for the year 1950. *Quart. J. R. Met. Soc.* 83, 360–4.

SANDS, R. D. 1966. *A feature-of-circulation approach to synoptic climatology applied to western United States.* Tech. Paper 66-2. Geography Dept, Univ. of Denver. 332 pp.

VAN BEBBER, W. J. and KÖPPEN, W. 1895. Die Isobarentypen des Nordatlantischen Ozeans und Westeuropas, ihre Beziehung zur Lage und Bewegung der Barometrischen Maxima und Minima. *Arch. dtsch. Seewarte* (Hamburg), 18 (4), 27 pp.

WADSWORTH, G. P., GORDON, C. H. and BRYAN, J. G. 1948. *Short-range and extended forecasting by statistical methods.* Air Weather Service Tech. Rep. 105-38. Washington, D.C. 207 pp.

WILSON, C. V. 1958. *Synoptic regimes in the lower arctic troposphere during 1955.* Pub. in Met. No. 8. Montreal: Arctic Met. Res. Group, McGill Univ. 58 pp.

3.C KINEMATIC VIEW OF THE WEATHER MAP

ABERCROMBY, R. 1883. On certain types of British weather. *Quart. J. R. Met. Soc.* 9, 1–25.

AIR WEATHER SERVICE. 1944. Preparation of a classification graph east Asia–west Pacific synoptic region. January 1899 through June 1939 (Part I). *Air Weather Service Tech. Rep. 105–12A*, Washington, D.C.

AIR WEATHER SERVICE. 1955. Atlantic–European weather types, 1899–1945: Classification, calendar, uses and climatology. *Air Weather Service Tech. Rep. 105-137.* Washington, D.C. 40 pp.

ALLEN, R. A., FLETCHER, R., HOLMBOE, J., NAMIAS, J. and WILLETT, H. C. 1940. *Report on an experiment in five-day weather forecasting.* Pap. Phys. Oceanog. Met. 8 (3). Cambridge, Mass.: Mass. Inst. Tech., Woods Hole Oceanog. Inst. 94 pp.

ALTYKIS, E. V. 1966. The quantitative estimation of atmospheric circulation forms. In *Contributions to long-range weather forecasting in the Arctic* (*Trudy Arkt. Antarkt. Nauch. Issled. Inst.* 225, 1963). Jerusalem: Israel Prog. Sci. Trans. 118–28.

ARAI, Y. 1964. Classification of 500 mb patterns related to the abnormal January 1963. *Pap. Met. Geophys.* (Tokyo) 15, 93–118.

ARISTOV, N. A. and BLUMINA, L. I. 1969. Fifty years development of the B. P. Multanovski school's synoptic method of long-range weather forecasting. *Trudy Gidromet. Nauch. Issled. Tsentr. SSSR* (Leningrad) 43, 3–22 (in Russian).

BARBER, A. 1970. An international weather corporation: a projected world utility. In Burnell, E. H. and von Simson, P. (eds.) *Pacem in Maribus, Ocean Enterprises.* Occas. Pap. 2 (4). Santa Barbara, Calif.: Center for the Study of Democratic Institutions. 75–83.

BARRY, R. G. 1959. *A synoptic climatology for Labrador–Ungava.* Pub. in Met. No. 17. Montreal: Arctic Met. Res. Group, McGill Univ. 168 pp.

BARRY, R. G. 1960. A note on the synoptic climatology of Labrador–Ungava. *Quart. J. R. Met. Soc.* 86, 557–65.

BARRY, R. G. 1963. Aspects of the synoptic climatology of central south England. *Met. Mag.* 92, 300–8.

BAUR, F. 1936a. Wetter, Witterung, Grosswetter und Weltwetter. *Z. angew. Met.* 53, 377–81.

BAUR, F. 1936b. Die Bedeutung der Stratosphäre für die Grosswetterlage. *Met. Zeit.* 53, 237–47.

BAUR, F. 1947. *Musterbeispiele europäischer Grosswetterlagen.* Wiesbaden: Dieterich. 35 pp.

BAUR, F. 1948. *Einführung in die Grosswetterkunde.* Wiesbaden: Dieterich. 165 pp.

BAUR, F. 1951. Extended-range weather forecasting. In Malone, T. F. (ed.) *Compendium of meteorology.* Boston, Mass.: Amer. Met. Soc. 814–33.

BAUR, F. 1963. *Grosswetterkunde und langfristige Witterungsvorhersage.* Frankfurt-am-Main: Akad. Verlagsgesellschaft. 91 pp.

BELASCO, J. E. 1952. Characteristics of air masses over the British Isles. *Geophys. Mem.* 11 (87). 34 pp.

BLEWITT, S. E. and PAULHUS, J. L. 1942. *Six-day weather types of North America.* Pasedena, Calif.: Met. Dept. California Inst. Tech. 297 pp.

BLÜTHGEN, J. 1966. *Allgemeine Klimageographie.* 2nd ed. Berlin: de Gruyter. 720 pp.

BORSOS, J. 1952. Types of cyclone and anticyclone paths and their frequencies. *Idojaras* 56, 279–84 (in Hungarian).

BRADBURY, D. L. 1958. On the behaviour patterns of cyclones and anticyclones as related to zonal index. *Bull. Amer. Met. Soc.* 39, 149–51.

BUTZER, K. W. 1960. Dynamic climatology of large-scale circulation patterns in the Mediterranean area. *Met. Rund.* 13, 97–105.

BUTZER, K. W. 1961. Climatic change in arid regions since the Pliocene. In Stamp, L. D. (ed.) *A history of land use in arid regions.* Arid Zone Research. Paris: UNESCO. 31–56.

CADEZ, M. 1957. Sur une classification des types de temps. *Météorologie* ser. 4, 45–56, 317–23.

CALIFORNIA INSTITUTE OF TECHNOLOGY. 1943. *Synoptic weather types of North America.* Pasadena, Calif.: Met. Dept, Calif. Inst. Tech. 243 pp.

CALIFORNIA INSTITUTE OF TECHNOLOGY. 1945. *Preparation of a classification graph of synoptic weather sequences for North America, January 1899 through June 1939.* Rep. No. 943, Weather Division, Army Air Force HQ.

CHARNEY, J. 1949. On a physical basis for numerical prediction of large-scale motions in the atmosphere. *J. Met.* 6, 371–85.

CLAYTON, H. H. 1923. *World weather: including a discussion of the influence of variations of solar radiation on the weather and the meteorology of the sun.* New York: MacMillan. 393 pp.

COURT, A. A. 1957. Climatology: complex, dynamic and synoptic. *Ann. Assoc. Amer. Geogr.* 47, 125–36.

DAVIDOVA, N. G. 1967. Types of synoptic process and associated wind fields in oceanic regions of the southern hemisphere. In *Polar Meteorology*. Tech. Note No. 87 (W.M.O. No. 211, TP 111), Geneva: World Meteorological Organization. 263–91.

DINIES, E. 1968. Monthly and annual means of atmospheric pressure for parallels of latitude in the northern hemisphere, 1899–1967. *Ber. dtsch Wetterd.* (Offenbach) No. 109. 18 pp.

DOVE, H. W. 1827. Einige meteorologische Untersuchungen über den Wind. *Ann. Phys. u. Chem.* 11, 545–90.

DZERDZEEVSKI, B. L. 1945. Typification of atmospheric processes over the northern hemisphere as a method of characterizing seasons. *Dokl. Gos. Okean. Inst.* (Moscow) No. 42. 24 pp. (in Russian).

DZERDZEEVSKI, B. L. 1962. Fluctuations of climate and of general circulation of the atmosphere in extratropical latitudes of the northern hemisphere and some problems of dynamic climatology. *Tellus* 14, 328–36.

DZERDZEEVSKI, B. L. 1963. Fluctuations of general circulation of the atmosphere and climate in the twentieth century. In *Changes of climate*. Arid Zone Res. 20. Paris: UNESCO. 285–95.

DZERDZEEVSKI, B. L. 1966. Some aspects of dynamic climatology. *Tellus* 18, 751–60.

DZERDZEEVSKI, B. L. 1968. Circulation of the atmosphere – circulation mechanisms of the atmosphere in the northern hemisphere in the twentieth century. *Results of Meteorological Investigations, IGY Committee.* Moscow: Inst. of Geography, Akad. Nauk SSSR. 240 pp. (in Russian).

DZERDZEEVSKI, B. L. 1970. Calendar of changes of ECM for 1967, 1968, 1969. In Caplygina, A. S. (ed.) The comparison of the characteristics of the atmospheric circulation over the northern hemisphere with the analogous characteristics for its sectors: circulation of the atmosphere. *Results of meteorological investigations, IGY Committee.* Moscow: Inst. of Geography, Akad. Nauk SSSR. 19–22 (in Russian).

DZERDZEEVSKI, B. L., KURGANSAKAIA, V. M. and VITVITSKAIA, Z. M. 1946. Classification of circulation mechanisms over the northern hemisphere and characteristics of synoptic seasons. *Trudy Nauk. Issled. Uchrezden* (Glav. Uprav. Gidromet. Sluzhby, Moscow) ser. 2, 21. 80 pp. (in Russian).

DZERDZEEVSKI, B. L. and MONIN, A. S. 1954. Typical schemes of the general circulation of the atmosphere in the northern hemisphere and circulation indices. *Izv. Akad. Nauk. SSSR, Geophys. Ser.* No. 6, 562–74 (in Russian).

ELLIOTT, R. D. 1949. The weather types of North America. *Weatherwise* 2, 15–18, 40–3, 64–7, 86–8, 110–13, 136–9.

ELLIOTT, R. D. 1951. Extended range forecasting by weather types. In Malone, T. F. (ed.) *Compendium of meteorology*. Boston, Mass.: Amer. Met. Soc. 834–40.

FICKER, H. V. 1938. Zur Frage der 'Steuerung' in der Atmosphäre. *Met. Zeit* 55, 8–12.

FLIRI, F. 1962. *Wetterlagenkunde von Tirol*. Tiroler Wirtschaftstudien No. 13 (Innsbruck). 436 pp.

FLIRI, F. 1965. Über Signifikanzen synoptisch-klimatologischer Mittelwerte in verschiedenen Alpinen Wetterlagensystemen. *Carinthia II* (Vienna) Sonderheft 24, 36–48.

FLOHN, H. 1954. Discussion on climatic change. *Proc. Toronto Met. Conf. 1953*. London: Roy. Met. Soc. 223.

FORSDYKE, A. G. 1955. *Depressions crossing Labrador and the St Lawrence basin*. Met. Office Prof. Notes No. 113. 44 pp.

GAZZOLA, A. 1969. First results of an investigation of precipitation in Italy in relation to the meteorological situation. *Riv. Met. Aeronaut.* (Rome) 29, suppl. to No. 4, 84–114 (in Italian).

GAZZOLA, A. and MONTALTO, M. 1960. Synoptic patterns, upper winds and precipitation over Italy in the months of November and December. *Met. Abhand.* 9 (1), 105–17.

GIRS, A. A. 1948. On the problem of investigation of the main types of atmospheric circulation. *Met. i Gidrol.* (Moscow) No. 3 (in Russian).

GIRS, A. A. 1960. *Principles of long-range weather forecasting*. Leningrad: Gidrometeoizdat. 560 pp.

GREGORY, S. 1964. Climate. In Watson, J. W. and Sissons, J. B. (eds.) *The British Isles: a systematic geography*. London: Nelson. 53–73.

GRESSEL, W. 1954. Zur Aufstellung eines synoptischen Kalendariums für den Alpenraum. *Met. Rund.* 7, 170–4.

GRESSEL, W. 1959. Zur Klassifikation des alpinen Wettergeschehens. *Met. Rund.* 12, 150–2.

GRUNOW, J. 1965. Über die Eignung von Klassifikationssystemen über alpinen Wetterlagen. *Carinthia II* (Vienna) Sonderheft 24, 7–25.

HAY, R. F. M. 1949. *The rainfall in eastern Scotland in relation to the synoptic situation*. Met. Office Prof. Notes No. 98.

HESS, P. and BREZOWSKY, H. 1952. Katalog der Grosswetterlagen Europas. *Ber. dtsch Wetterd., U.S. Zone* (Bad Kissingen) No. 33. 39 pp.

HESS, P. and BREZOWSKY, H. 1969. Katalog der Grosswetterlagen Europas. *Ber. dtsch. Wetterd.* (Offenbach) 15, (113). 56 pp.

HOUGHTON, J. G. 1969. *Characteristics of rainfall in the Great Basin*. Reno: Desert Res. Inst., Univ. of Nevada. 205 pp.

IROSTNIKOV, M. B. 1957. Problem of the rhythm of ultrapolar synoptic processes over Siberia and the Far East. *Trudy Tsen. Inst. Prog.* (Moscow) 51, 124–49 (in Russian).

JACKSON, I. J. 1969. *Pressure types and precipitation over north-east England*. Res. Ser. No. 5. Dept of Geography, Univ. of Newcastle upon Tyne. 81 pp.

JACOBS, W. C. 1947. *Wartime developments in applied climatology*. Met. Monogr. (Amer. Met. Soc.) 1 (1). 52 pp.

JENKINSON, A. F. 1957. Discussion. The relation between standard deviation of contour heights and standard vector deviation of wind. *Quart. J. R. Met. Soc.* 83, 554.

JULIAN, P. R. 1966. The index cycle: a cross-spectral analysis of zonal index data. *Mon. Wea. Rev.* 94, 283–93.

KINGTON, J. A. 1970. Preparation of weather charts from 1780. *Swansea Geogr.* 8, 11–14.

KLEIN, W. H. 1958. The frequency of cyclones and anticyclones in relation to the mean circulation. *J. Met.* 15, 98–102.

KLETTER, L. 1959. Charakteristische Zirkulationstypen in mittleren Breiten der nordlichen Hemisphare. *Arch. Met. Geophys. Biokl.* A, 11. 191–6.

KLETTER, L. 1962. Die Aufeinanderfolge charakteristische Zirkulationstypen in mittleren Breiten der nordlichen Hemisphare. *Arch. Met. Geophys. Biokl.* A, 13, 1–33.

KURASHIMA, A. 1957. Investigation on broad-scale weather types, Part 2. *J. Met. Res.* (Tokyo) 9, 322–34 (in Japanese).

LAMB, H. H. 1950. Types and spells of weather around the year in the British Isles: annual trends, seasonal structure of the year, singularities. *Quart. J. R. Met. Soc.* 76, 393–429.

LAMB, H. H. 1951. *Hemisphere-wide circulation patterns and types classification.* Met. Res. Pap. No. 632 (London). 7 pp.

LAMB, H. H. 1964. *The English Climate.* 2nd ed. London: English Universities Press. 212 pp.

LAMB, H. H. 1972. British Isles weather types and a register of the daily sequence of circulation patterns, 1861–1971. *Geophys. Mem.* (London). 16 (116), 85 pp.

LA SEUR, N. E. 1954. On the asymmetry of the middle latitude circumpolar current. *J. Met.* 11, 43–57.

LAUSCHER, F. 1954. Dynamische Klimaskizze von Österreich. In Flohn, H. (ed.) *Witterung und Klima in Mitteleuropa, Forsch. dt. Landeskunde* (Stuttgart) 88, 145–58.

LAUSCHER, F. 1958. Studien zur Wetterlagenklimatologie der Ostalpenländer. *Wetter u. Leben* 10, 79–83.

LEVICK, R. B. M. 1949. Fifty years of English weather. *Weather* 4, 206–11.

LEVICK, R. B. M. 1950. Fifty years of British weather. *Weather* 5, 245–7.

LEVICK, R. B. M. 1955. Mapping the weather. *Weather* 10, plate 47, facing 412.

LITYNSKI, J. K. 1970. Classification numerique des types de circulation et des types de temps en Pologne. *Cah. Géogr. Quebec* 33, 329–38.

MAEDE, H. 1965. Der jahreszeitliche Gang der Höhenwetterlagenhaufigkeit in den Gebieten Ostatlantik und Mitteleuropa. *Zeit. Met.* suppl. to 17, 127–33.

MERTZ, J. 1957. Essai de classification des types de temps sur les Alpes d'après la disposition des isohypses à 500 mb. *Météorologie* ser. 4, 45–6, 305–15.

MOSIÑO, P. A. 1964. Surface weather and upper-air flow patterns in Mexico. *Geofis. Internac.* 4, 117–68.

MÜGGE, R. 1938. Über das Wesen der Steuerung. *Zeit Met.* 55, 197–205.

MURRAY, R. and LEWIS, R. P. W. 1966. Some aspects of the synoptic climatology of the British Isles as measured by simple indices. *Met. Mag.* 95, 193–203.

NAMIAS, J. 1947. Physical nature of some fluctuations in the speed of the zonal circulation. *J. Met.* 4, 125–33.

NAMIAS, J. 1950. The index cycle and its role in the general circulation. *J. Met.* 7, 130–9.

NAMIAS, J. 1951. Extended-range weather forecasting. In Malone T. F. (ed.), *Compendium of meteorology*. Boston, Mass.: Amer. Met. Soc. 802–13.

NAMIAS, J. and CLAPP, P. F. 1944. Studies of the motion and development of long waves in the westerlies. *J. Met.* 1, 57–77.

NEIS, B. 1949. Die synoptische Methode der Langfristprognose der Schule B. P. Multanowsky. *Zeit. Met.* 3, 304–9, 329–35.

NEMOTO, J. and KUBOKI, K. 1968. Studies on the seasonal forecasting of summer season in Japan, Part 2. *J. Met. Res.* (Tokyo) 20, 248–92 (in Japanese).

PALMÉN, E. and NEWTON, C. W. 1969. *Atmospheric circulation systems: their structure and physical interpretation.* New York: Academic Press. 603 pp.

PERRY, A. H. 1970. Changes in duration and frequency of synoptic types over the British Isles. *Weather* 25, 123–6.

PUTNINS, P. 1962. Correlation between pressure changes aloft and at the surface in the Greenland area and some aspects of the 'steering' problem. *Arch. Met. Geophys. Biokl.* A, 13, 218–40.

PUTNINS, P. 1966. *The sequences of baric pressure patterns over Alaska.* Studies on the meteorology of Alaska, 1st Interim Rep. Washington D.C.: Environmental Data Service, ESSA. 81 pp.

PUTNINS, P. 1968. *Some aspects of the atmospheric circulation over the Alaska area.* Studies on the meteorology of Alaska, 3rd InterimRep.Washington, D.C.: Environmental Data Service, ESSA. 57 pp.

RAGOZIN, A. I. and CHUKANIN, K. K. 1966. Prevailing trajectories of cyclones and anticyclones during standard synoptic processes in the Arctic. In *Contributions to long-range weather forecasting in the Arctic* (*Trudy Arkt. Antarkt. Nauch. Issled. Inst.* 225, 1963). Jerusalem: Israel Prog. Sci. Trans. 143–57.

REED, T. R. 1932. Weather types of the northeast Pacific Ocean as related to the weather on the North Pacific coast. *Mon. Wea. Rev.* 60, 246–52.

REINEL, H. 1960. Die Zugbahnen der Hochdruckgebiete über Europa als klimatologisches Problem. *Mitt. Fränkischen Geogr. Ges.* (Erlangen) 6, 1–73.

REITER, E. R. 1963. *Jet-stream meteorology.* Univ. of Chicago Press, 515 pp.

REMPEL, W. H. and STONE, N. C. 1945. *Preparation of a classification graph east Pacific synoptic region January 1899 through June 1939.* Report No. 943, Met. Dept., Calif. Inst. Tech., Pasadena, Calif. (revised Weather Division, HQ Army Air Forces, 93 pp.).

RIEHL, H., ALAKA, M. A., JORDAN, C. L. and RENARD, R. J. 1954. *The jet stream.* Met. Monogr. 2 (7). 100 pp.

RIEHL, H., LA SEUR, N. E. *et al.* 1952. *Forecasting in middle latitudes.* Met. Monogr. 1 (5). 80 pp.

ROSSBY, C. G. 1940. Planetary flow patterns in the atmosphere. *Quart. J. R. Met. Soc.*, suppl. to No. 66, 68–87.

ROSSBY, C. G. *et al.* 1939. Relations between variations in the intensity of the zonal circulation of the atmosphere and displacements of the semi-permanent centres of action. *J. Mar. Res.* 2, 38–55.

ROSSBY, C. G. and WILLETT, H. C. 1948. The cirulation of the upper troposphere and the lower stratosphere. *Science* 108, 643–52.

RUDLOFF, H. 1967. *Die Schwankungen und Pendelungen des Klimas in Europa seit dem Beginn der regelmassigen instrumenten Beobachtungen 1670.* Braunschweig: F. Vieveg. 370 pp.

SANDS, R. D. 1966. *A feature of circulation approach to synoptic climatology applied to western United States.* Tech Paper 66-2, Dept of Geography, Univ. of Denver. 332 pp.

SAWYER, J. S. 1956. *Rainfall of depressions which pass eastward over or near the British Isles.* Met Off. Prof. Notes No. 118.

SCHELL, I. I. 1940. A preliminary study of the Multanovski school of long-range weather forecasting. *Mon. Wea. Rev.*, suppl. No. 39, 92–6.

SCHÜEPP, M. 1957. Klassifikationschema, Beispiele und Probleme der Alpenwetterstatistik. *Météorologie.* ser. No. 4, 45–6, 291–9.

SCHÜEPP, M. 1959. Die Klassifikation der Witterungslagen. *Geof. pura e appl.* 44, 242–8.

SCHÜEPP, M. 1968. Kalendarien der Wetter- und Witterungslagen von 1955 bis 1967. *Veröff. Schweiz. Met. Zentr. Anst.* (Zurich) 11. 43 pp.

SUSSEBACH, J. 1968. Synoptische-statistische Untersuchungen zur Sturmwetterlagen im Mittel und Osteuropa. *Met. Abhand.* 86 (2). 105 pp.

THOMAS, T. M. 1960a. Some observations on the tracks of depressions over the eastern half of the North Atlantic. *Weather* 15, 325–36.

THOMAS, T. M. 1960b. Precipitation within the British Isles in relation to depression tracks. *Weather* 15, 361–73.

VAN BEBBER, W. J. 1882. Typische Witterungs-Erscheinungen. *Arch. dtsch. Seewarte* (Hamburg) 5 (3). 45 pp.

VANGENGEIM, G. I. 1935. *Application of synoptic methods to the study and characterization of climate.* Moscow: Gidromet. Sluzhby. 109 pp. (in Russian).

VANGENGEIM, G. I. 1946. On variations in the atmospheric circulation over the northern hemisphere. *Izv. Akad. Nauk SSSR, Ser. Geogr. i Geofiz.* 5 (in Russian).

VAN LOON, H. 1958. On the synoptic climatology of the Tristan da Cunha region. *Arch. Met. Geophys. Biokl.* B, 9, 313–22.

VERNON, E. M. 1947. An objective method of forecasting precipitation 24–48 hours in advance at San Francisco, California. *Mon. Wea. Rev.* 75, 211–20.

VON BUCH, L. 1820. Über barometrische Wind-Rosen. *Abh. Akad. Wiss.* (Berlin), 1818–1819, 104–10.

WALDEN, H. 1959. Statistisch-synoptische Untersuchung über das Verhalten von Tiefdruckgebieten im Bereich von Grönland. *Dt. Wetterd. Seewetteramt* (Hamburg) No. 20. 69 pp.

WALKER, G. T. 1924. Correlation in seasonal variations of weather, IX: A further study of world weather. *Mem. Ind. Met. Dept.* (Poona) 24, 275–332.

WALLÉN, C. C. 1953. The variability of summer temperature in Sweden and connection with changes in the general circulation. *Tellus* 5, 157–78.

WILLFARTH, J. 1959. Einfahrungsbericht zur Schüepp'schen Wetterstatistik im Ostalpenraum. *Ber. dtsch Wetterd.* 8 (54), 174–6.

YOSHINO, M. M. 1968. Pressure pattern calendar of east Asia. *Met. Rund.* 21, 162–9.

3.D WEATHER ELEMENTS

3.D.1 *Complex climatology*

CALEF, W. *et al.* 1957. *Winter weather type frequencies of the northern Great Plains.* Tech. Rep. E-64. Natick, Mass.: Environment Protection Research Division, HQ Quartermaster Research and Engineering Command, U.S. Army. 260 pp.

CHUBUKOV, L. A. 1949. *Kompleksnaia Klimatologiia.* Moscow: Akad. Nauk. 94 pp. (in Russian).

CHUBUKOV, L. A. and SHVAREVA, YU, N. 1964. Structure of climate by weather types. In Gerasimov, I. P. *et al.* (eds.) *Physico-geographic atlas of the world.* Moscow. 204–5 (in Russian).

FEDEROV, E. E. 1927. Climate as totality of weather (trans. E. S. Nichols). *Mon. Wea. Rev.* 55, 401–3.

LYDOLPH, P. E. 1957. How many climatologies are there? *Prof. Geogr.* 9 (6), 5–7.

LYDOLPH, P. E. 1959. Federov's complex method in climatology. *Ann. Assoc. Amer. Geogr.* 49, 120–44.

NICHOLS, E. S. 1925. A classification of weather types *Mon. Wea. Rev.* 53, 431–3.

PELZL, E. 1955. Komplex-Klimatologie als witterungsklimatologische Untersuchungsmethode. *Ann. Met.* 7, 35–8.

PETROVIC, S. 1967. Complex climatology of Strba Lake. *Met. Zpravy* 20, 67–70 (in Czech).

PETROVIC, S. 1969. Comparison of weather types at Strba Lake as a function of dynamic and complex climatic evaluation. *Met. Zpravy* 21, 12–15 (in Czech).

POTAPOVA, L. A. 1968. Relationship between elementary circulation mechanisms and the local weather conditions during January in the region of Moscow. In Lyakhov, M. E. (ed.) *Climatologic and circulation epochs in the northern hemisphere in the first half of the twentieth century.* Met. Researches No. 13. Moscow (trans. 1970, Israel Prog. Sci. Trans., Jerusalem, 139–49).

QUITT, E. 1968. Contribution to the method of climatic regionalizing of Czechoslovakia. *Met. Zpravy*, 21, 65–9 (in Czech).

SWITZER, J. E. 1925. Weather types in the climates of Mexico, the Canal Zone and Cuba. *Mon. Wea. Rev.* 53, 434–7.

WOS, A. 1970. The climate of selected places in Yugoslavia in a complex aspect. *Przeglad Geogr.* (Warsaw) 42, 69–92 (in Polish).

3.D.2 *Air mass climatology*

(*Some additional references on air mass not specifically cited in the text are included here.*)

ARAKAWA, H. 1937. Die Luftmassen in den japanischen Gebieten. *Met. Zeit.* 54, 169–74 (also in *Bull. Amer. Met. Soc.* 18, 407–10).

BAKALOW, D. 1939. Über die Transformation der Luftmassen. *Beitr. Phys. Atmos.* 26, 1–22.

BARRY, R. G. 1967. Seasonal location of the arctic front over North America. *Geogr. Bull.* 9, 79–95.

BELASCO, J. E. 1952. Characteristics of air masses over the British Isles. *Geophys. Mem.* 11 (87). 34 pp.

BERGERON, T. 1928. Über die dreidimensional verknüpfende Wetteranalyse, Part 1: Prinzipelle Einführung in das Problem der Luftmassen- und Frontenbildung. *Geofys. Publik.* (Oslo) 5 (6), 1–111.

BERGERON, T. 1930. Richtlinien einer dynamischen Klimatologie. *Met. Zeit.* 47, 246–62.

BORISOV, A. A. 1965. *Climates of the USSR.* 2nd ed. (trans. R. A. Ledwood, ed. C. A. Halstead). Edinburgh: Oliver and Boyd. 255 pp.

BRADBURY, D. L. and PALMÉN, E. 1953. On the existence of a polar-front zone at the 500 mb level. *Bull. Amer. Met. Soc.* 34, 56–62.

BRUNNSCHWEILER, D. H. 1957. Die Luftmassen der Nordhemisphäre. Versuch einer genetischen Klimaklassifikation auf aerosomatischer Grundlage. *Geogr. Helv.* 12, 164–95.

BRYSON, R. A. 1966. Air masses, streamlines and the boreal forest. *Geogr. Bull.* 8, 228–69.

BUNKER, A. F. 1960. Heat and water vapour fluxes in air flowing south over the western North Atlantic. *J. Met.* 17, 52–63.

BURBIDGE, F. E. 1951. The modification of continental polar air over Hudson Bay. *Quart. J. R. Met. Soc.* 77, 365–74.

BURKE, C. J. 1945. Modification of continental polar air over a water surface. *J. Met.* 2, 94–112.

CHARNOCK, H. 1951. Energy transfer between the atmosphere and the ocean. *Sci. Progr.* 39, 80–95.

CHRISTIE, A. D. 1965. *Atmospheric tracers: a review.* Met. Branch CIR-4214, TEC-564. Toronto: Dept of Transport. 29 pp.

CHROMOV, S. P. and KONCEK, N. 1940. *Einführung in die synoptische Wetteranalyse* (trans. G. Swoboda). Vienna: Springer. 532 pp.

COLLMANN, T. 1951. Beispiele zur Frage der Luftmassentransformation. *Ann. Met.* 4, 444–64.

CRADDOCK, J. M. 1951. The warming of arctic air masses over the eastern North Atlantic. *Quart. J. R. Met. Soc.* 77, 351–64.

CRESWICK, W. S. 1967. Experiments in objective frontal analysis. *J. Appl. Met.* 6, 774–81.

CROWE, P. R. 1965. The geographer and the atmosphere. *Trans. Inst. Brit. Geogr.* No. 36, 1–19.

DEFANT, F. and TABA, H. 1957. The threefold structure of the atmosphere and the characteristics of the tropopause. *Tellus* 9, 259–74.

DINIES, E. 1932. Luftkörper Klimatologie. *Arch. dtsch. Seewarte* 50 (6), 21 pp.

FLOHN, H. 1952. Zur Aerologie der Polargebiete. *Met. Rund.* 5, 121–8.

FLOHN, H. 1958. Luftmassen, Fronten und Strahlstrome. *Met. Rund.* 11, 7–13.

FRISBY, E. M. and GREEN, F. H. 1949. Comparison of the regional and seasonal frequency of air masses. *Comptes Rendus, Congrés Internat. de Geogr.* (Lisbon) vol. 2, 307–14.

FRITZ, S. 1958. Seasonal heat storage in the ocean and heating of the atmosphere. *Arch. Met. Geophys. Biokl.* 10, 291–300.

GENTILLI, J. 1949. Air masses of the southern hemisphere. *Weather* 4, 258–61.

GODSON, W. L. 1950. The structure of North American weather systems. *Centen. Proc. Roy. Met. Soc.* (London). 89–106.

GOLDIE, A. H. R. 1923. Circumstances determining the distribution of temperature in the upper air under conditions of high and low barometric pressure. *Quart. J. R. Met. Soc.* 49, 6–16.

GREEN, J. A. S., LUDLAM, F. H. and MCILVEEN, J. F. R. 1966. Isentropic relative-flow analysis and the parcel theory. *Quart. J. R. Met. Soc.* 92, 210–19.

HARE, F. K. and MONTGOMERY, M. R. 1949. Ice, open water and winter climate in the eastern Arctic of North America. *Arctic* 2, 78–89, 140–64.

HARLEY, W. S. 1962. *Distributions of wet-bulb potential temperature in North American air masses (1957) and a statistical analysis of results.* Met. Branch CIR-3622, TEC-400. Toronto: Dept of Transport. 46 pp.

JAMES, R. W. 1969. Elementary air mass analysis. *Met. Rund.* 22, 75–9.

JAMES, R. W. 1970. Air mass climatology. *Met. Rund.* 23, 65–70.

JOHN, I. G. 1949. The properties of the upper air over Singapore. *Mem. Malayan Met. Services* No. 4. 35 pp.

KLEIN, W. H. 1946. Modifications of the continental polar air over a water surface. *J. Met.* 3, 100–1.

KUHLBROOKA, H. 1957. Modification of Siberian air mass caused by flowing out over the open sea surface of north Japan. *J. Met. Soc. Japan* ser. 2, 35, 52–9.

LINKE, F. and DINIES, E. 1930. Uber Luftkörperbestimmungen. *Zeit. angew. Met.* 47, 1–5.

LONGLEY, R. W. 1959. *The three-front model: a critical analysis.* Met. Branch CIR-3245, TEC-309. Toronto: Dept of Transport. 14 pp.

LU, A. 1945. Winter frontology of China. *Bull. Amer. Met. Soc.* 8, 310–14.

MCINTYRE, D. P. 1950. On the air mass temperature distribution in the middle and high troposphere in winter. *J. Met.* 7, 101–7.

MANABE, S. 1957. On the modification of air mass over the Japan Sea when the outburst of cold air predominates. *J. Met. Soc. Japan* ser. 2, 35, 311–25.

MILLER, A. A. 1953. Air mass climatology. *Geography* 38, 55–67.

MOESE, O. and SCHINZE, G. 1929. Zur Analyse von Neubildungen. *Ann. Hydrogr. marit. Met.* 57, 76–81.

MÖLLER, F. 1929. Statische Untersuchungen über die Konstanz der Luftkörper. *Gerlands Beitr. Geophys.* 21, 387–435.

OZORAI, Z. 1963. An assessment of ideas relation to air masses. *Idojaras* 67, 193–203.

PALMÉN, E. and NEWTON, C. W. 1969. *Atmospheric circulation systems: their structure and physical interpretation.* New York: Academic Press. 603 pp.

PETTERSSEN, S. 1940. *Weather analysis and forecasting.* New York: McGraw-Hill. Chapter 3.

PETTERSSEN, S. 1958. *Weather analysis and forecasting.* 2nd ed. New York: McGraw-Hill. vol. 2, 1–34.

PETTERSSEN, S. and CALABRESE, P. A. 1959. On some weather influences due to the warming of the air by the Great Lakes in winter. *J. Met.* 16, 646–52.

PÔNE, R. 1947. L'analyse des masses d'air à l'aide des températures caractèristiques. *Météorologie* ser. 4, No. 7, 204–15.

POTHECARY, T. J. 1960. Modifications of warm maritime air streams over Europe in winter. *Met. Mag.* 89, 1–7.

PRIESTLEY, C. H. B. 1965. *Turbulent transfer in the lower atmosphere.* Chicago: Univ. of Chicago Press. 130 pp.

ROY, A. K. 1949. Study of air mass characteristics over India and the use of upper air charts. *Proc. Nat. Inst. Sci. India* 15, 301–5.

SCHAMP, H. 1939. Luftkörperklimatologie des grieschischen-Mittelmeergebietes. *Frankfurt Geogr.* 13, 79 pp.

SCHERHAG, R. 1948. *Neue Methoden der Wetteranalyse und Wetterprognose.* Berlin: Springer. 424 pp.

SCHINZE, G. 1932a. Troposphärische Luftmassen und Vertikaler Temperaturgradient. *Beitr. Physik. f. Atmos.* 19, 79–90.

SCHINZE, G. 1932b. Die Erkennung der troposphärische Luftmassen aus ihren Einzelfeldern. *Met. Zeit.* 49, 169–79.

SCHINZE, G. and SIEGEL, R. 1943. Die luftenmässige Arbeitsweise. *Reichsamt Wetterdienst, Sonderband, Wiss. Abhand.* 99 pp.

SERRA, A. 1949. Sulle caratteristiche fisiche delle principali masse d'ara nel Mediterraneo Occidentale. *Rivista Met. Aeron.* 9, 3–18.

SHOWALTER, A. K. 1939. Further studies of American air mass properties. *Mon. Wea. Rev.* 67, 204–18.

STONE, R. G. (ed.) *et al.* 1940. *Air mass and isentropic analysis.* Boston. Mass.: Amer. Met. Soc. 232 pp.

TALJAARD, J. J. 1969. Air masses of the southern hemisphere. *Notos* 18, 79–104.

TU, C.-W. 1939. Chinese air mass properties. *Quart. J. R. Met. Soc.* 65, 33–51.

WATSON, A. G. D. 1968. Air-sea interaction. *Sci. Progr.* 56, 303–23.

WATTS, I. E. M. 1955. *Equatorial meteorology, with particular reference to south-east Asia.* London: Univ. of London Press. 223 pp.

WHITING, G. 1959. Frontal passages over the Atlantic Ocean. *Mon. Wea. Rev.* 87, 409–16.

WILLETT, H. C. 1931. Ground plan for a dynamic climatology. *Mon. Wea. Rev.* 59, 219–23.

WILLETT, H. C. 1933. *American air mass properties*. Pap. Phys. Oceanog. Met. 2 (2). Cambridge, Mass.: Mass. Inst. Tech., Woods Hole Oceanog. Inst. 116 pp.

3.D.3 *Relationships between weather conditions and synoptic features*

BARRY, R. G. and FOGARASI, S. 1968. *Climatology studies of Baffin Island, Northwest Territories*. Tech. Bull. No. 13. Ottawa: Dept Energy, Mines & Resources, Inland Waters Branch. 106 pp.

BEEBE, R. G. 1956. Tornado composite charts. *Mon. Wea. Rev.* 84, 127–42.

BOUCHER, R. J. and NEWCOMB, R. J. 1962. Synoptic investigation of some TIROS vortex patterns: a preliminary cyclone model. *J. Appl. Met.* 1, 127–36.

BROWNING, K. A. and HARROLD, T. W. 1969. Air motion and precipitation growth in a wave depression. *Quart. J. R. Met. Soc.* 95, 288–309.

CHRISTENSEN, W. I. JR and BRYSON, R. A. 1966. An investigation of the potential of component analysis for weather classification. *Mon Wea. Rev.* 94, 697–709.

FETT, R. W. 1964. Aspects of hurricane structure: new model considerations suggested by TIROS and Project Mercury observations. *Mon. Wea. Rev.* 92, 43–59.

FRISBY, E. M. and GREEN, F. H. W. 1949. Further notes on comparative regional climatology. *Trans. Inst. Brit. Geogr.* No. 15, 143–51.

GOREE, P. A. and YOUNKIN, R. J. 1966. Synoptic climatology of heavy snowfall over the central and eastern United States. *Mon. Wea. Rev.* 94, 663–8.

HISER, H. W. 1956. Type distributions of precipitation at selected stations in Illinois. *Trans. Amer. Geophys. Union* 37, 421–4.

JACOBS, W. C. 1947. *Wartime developments in applied climatology*. Met. Monogr. (Amer. Met. Soc.) 1 (1). 52 pp.

JORGENSEN, D. L. 1963. A computer derived synoptic climatology of precipitation from winter storms. *J. Appl. Met.* 2, 226–34.

JORGENSEN, D. L., KLEIN, W. H. and KORTE, A. F. 1967. A synoptic climatology of winter precipitation from 700 mb lows for intermontane areas of the West. *J. Appl. Met.* 6, 782–90.

KLEIN, W. H., JORGENSEN, D. L. and KORTE, A. F. 1968. Relation between upper airflows and winter precipitation in the Western Plateau States. *Mon. Wea. Rev.* 96, 162–8.

LOWE, A. B. and MCKAY, G. A. 1962. Tornado composite charts for the Canadian Prairies. *J. Appl. Met.* 1, 157–62.

LUDLAM, F. H. 1966. *The cyclone problem: a history of models of the cyclonic storm*. Inaugural lecture, Imperial Coll. Sci. Tech., London. 19–49.

MATTHEWS, R. P. 1972. Variation of precipitation intensity with synoptic type over the Midlands. *Weather* 27, 63–72.

MOKOSCH, E. 1962. *Mean pressure maps for summer rain in southern Alberta*. Met. Branch CIR-3594, TEC-393. Toronto: Dept of Transport. 13 pp.

MURRAY, R. and DANIELS, S. M. 1953. Transverse flow at entrance and exit to jet streams. *Quart. J. R. Met. Soc.* 79, 236–41.

NAMIAS, J. 1960. Snowfall over eastern United States: factors leading to its monthly and seasonal variations. *Weatherwise* 13, 238–47.

ORGILL, M. M. 1967. *Rainfall patterns associated with monsoon disturbances over southeastern Asia.* Southeast Asia Monsoon Study Rep. No. 3. Fort Collins: Atmos. Sci. Dept, Colo. State Univ. 51 pp.

RICHTER, D. A. and DAHL, R. A. 1958. Relationship of heavy precipitation to the jet maximum in the eastern United States. *Mon. Wea. Rev.* 86, 368–76.

ROSSI, V. 1948. On the effect of the Scandinavian mountains on the precipitation fronts approaching from the sea. *Fennia* (Helsinki) 70 (4), 1–23.

SHAW, E. M. 1962. An analysis of the origins of precipitation in northern England, 1956–1960. *Quart. J. R. Met. Soc.* 88, 539–47.

SMITHSON, P. A. 1969. Regional variations in the synoptic origin of rainfall across Scotland. *Scot. Geogr. Mag.* 85, 182–95.

SOLOT, S. B. 1948. *Possibility of long-range precipitation forecasting for the Hawaiian Islands.* Res. Pap. No. 28. Washington, D.C.: U.S. Weather Bureau. 52 pp.

SPINNANGR, E. 1942. Synoptic studies on precipitation in southern Norway, II: Frontal precipitation. *Met. Ann.* (Oslo) No. 17, 323–56.

STIDD, C. K. 1954. The use of correlation fields in relating precipitation to circulation. *J. Met.* 11, 202–13.

WALLINGTON, C. E. 1963. Mesoscale patterns of frontal rain and cloud. *Weather* 18, 171–81.

3.D.4 *Anomaly patterns*

BAUMAN, I. A. and KANAEVA, A. P. 1967. Distribution of the anomalies of mean monthly precipitation totals in connection with the properties of the basic forms of the atmospheric circulation and with their long-term transformation. *Trudy. Glav. Geofiz. Obs.* (Leningrad) 211, 94–103 (in Russian).

BKERKNES, J. 1962. Synoptic survey of the interaction of sea and atmosphere in the north Atlantic. *Geophys. Norvegica* (V. Bjerknes Centen. vol.) 24, 115–57.

BLAIR, T. A. 1933. Weather types and pressure anomalies. *Mon. Wea. Rev.* 61, 196–7.

BROOKS, C. E. P. and QUENNELL, W. A. 1926. Classification of monthly charts of pressure anomalies. *Geophys. Mem.* (London) 4 (31). 12 pp.

CAPLYGINA, A. S. 1968. Fluctuations in circulation and temperature conditions in the atmosphere over the northern hemisphere in the first half of the 20th century. *Izv. Akad. Nauk. Ser. Geog.* (Moscow) 4, 5–14 (in Russian).

CRADDOCK, J. M. 1957. An analysis of the slower temperature variations at Kew Observatory by means of mutually exclusive band pass filters. *J. R. Stat. Soc.* A, 120, 387–97.

CRADDOCK, J. M. and LOWNDES, C. A. S. 1958. *A synoptic study of anomalies of surface air temperatures over the Atlantic half of the northern hemisphere.* Met. Office Prof. Notes 8, No. 126. 6 pp.

CRADDOCK, J. M. and WARD, R. 1962. *Some statistical relationships between the temperature anomalies in neighbouring months in Europe and western Siberia.* Sci. Pap. No. 12. Met. Office, Bracknell Berks. 31 pp.

DEFANT, A. 1924. Die Schwankungen der atmosphärischen Zirkulation über dem nordatlantischen Ozean in 25 jahrigen Zeitraum 1881–1905. *Geogr. Annal.* 6, 13–41.

DROGAITSEV, D. A. 1966. Investigation of historical fluctuations of air temperature anomaly with the purpose of forecasting it. *Trudy Tsentr. Inst. Progr.* (Moscow) 147, 10–26. (in Russian).

GEDEONOV, A. D. 1967. Areas of large anomalies of mean monthly air temperature from the long period averages in the northern hemisphere. *Trudy Glav. Geofiz. Obs.* (Leningrad) 211, 68–74 (in Russian).

GIRSKAIA, E. I. 1968a. Methods of finding conjugate areas of temperature anomalies in the northern hemisphere. *Trudy Glav. Geofiz. Obs.* (Leningrad) 201, 106–10 (in Russian).

GIRSKAIA, E. I. 1968b. Correlation of temperature anomalies in the northern hemisphere during winter. *Trudy Glav. Geofiz. Obs.* (Leningrad) 247, 97–102 (in Russian).

GIRSKAIA, E. I. 1969. Correlation of temperature anomalies in the northern hemisphere during summer. *Trudy Glav. Geofiz. Obs.* (Leningrad) 245, 72–6 (in Russian).

GIRSKAIA, E. I. and KLEBANER, L. B. 1969. Anomalies and variations of mean monthly air temperatures. *Trudy Glav. Geofiz. Obs.* (Leningrad) 247, 97–102 (in Russian).

GRIMMER, M. 1963. The space filtering of monthly surface anomaly data in terms of pattern, using empirical orthogonal functions. *Quart. J. R. Met. Soc.* 89, 395–408.

HAWKINS, H. F. JR. 1956. The weather and circulation of October 1956 including a discussion of the relationship of mean 700 mb height anomalies to sea level flow. *Mon. Wea. Rev.* 84, 363–70.

HAY, R. F. M. 1960. *Winter temperatures in Britain related to monthly mean depression centres in the Atlantic–European sectors.* Syn. Clim. Branch Mem. No. 14. Bracknell Berks: Met. Office. 7 pp.

HESSELBERG, T. 1962. Climatological deviation maps. *Geofys. Publik.* (Oslo) 24, 39–66.

HOVMÖLLER, E. 1949. The trough-and-ridge diagram. *Tellus* 62–6.

IAKOVLEVA, N. I. and TURIKOV, V. G. 1968. Synchronous relationship between pressure anomalies (geopotential) over various sections of the northern hemisphere. *Trudy Glav. Geofiz. Obs.* (Leningrad) 201, 79–89 (in Russian).

KATS, A. L., MORSKOI, G. I. and SEMENOV, V. G. 1957. The formation of large air temperature anomalies over the territory of the U.S.S.R. during the winter months. *Trudy Tsentr. Inst. Prog.* (Moscow) 49, 3–180 (in Russian).

KLEIN, W. H. 1956. The central role of the height anomaly in the outlook for long-range forecasting. *Trans. N.Y. Acad. Sci.* ser. 2, 18, 375–87.

KLIMENKO, L. V. 1962. Estimation of the values of temperature anomalies. *Met. i. Gidrol.* (Moscow) No. 9, 33–6 (in Russian).

KURSHINOVA, K. V. 1970. Relationship between circulation types and anomalies of monthly average temperatures and precipitation over the European U.S.S.R. In Lyakhov, M. E. (ed.) *Climatologic and circulation epochs in the northern hemisphere in the first half of the 20th century.* Met. Researches No. 13 (Moscow, 1968). Jerusalem: Israel Prog. Sci. Trans. 124–38.

KUTZBACH, J. E. 1967. Empirical eigenvectors of sea level pressure, surface temperature and precipitation complexes over North America. *J. Appl. Met.* 6, 791–802.

KUTZBACH, J. E. 1970. Large-scale features of monthly mean northern sphere anomaly maps of sea level pressure. *Mon. Wea. Rev.* 98, 708–16.

LOEWE, F. 1966. Temperature see-saw between western Greenland and Europe. *Weather* 21, 241–6.

MARTIN, D. E. and HAWKINS, H. F. JR. 1950. Forecasting the weather: the relationship of temperature and precipitation over the United States to the circulation aloft. *Weatherwise* 3, 16–19, 40–3, 65–7, 89–92, 113–16, 138–41.

MESHCHERSKAIA, A. V. and KLIUKVIN, L. N. 1968. Variability of mean monthly temperatures. *Trudy Glav. Geofiz. Obs.* (Leningrad) 201, 97–105 (in Russian).

MILES, M. K. 1963. Lecture discussion. In Baur, F. *Grosswetterkunde und langfristige Witterungsvorhersage.* Frankfurt-am-Main: Akad. Verlags-gessellschaft. 85.

MITCHELL, M. J. JR. 1968. Some remarks on the predictability of drought and related climatic anomalies. *Proc. conf. on drought in the northeastern United States.* Rep. TR-68-3. Geophys. Sci. Lab., New York Univ. 129–59.

MURRAY, R. 1966a. Some features of the large-scale circulation anomalies and the weather over the British Isles in autumn 1965. *Met. Mag.* 95, 225–36.

MURRAY, R. 1966b. A note on the large-scale features of the 1962–63 winter. *Met. Mag.* 95, 339–48.

NAMIAS, J. 1947. Characteristics of the general circulation over the northern hemisphere during the abnormal winter 1946–47. *Mon. Wea. Rev.* 75, 145–52.

NAMIAS, J. 1951. The great Pacific anticyclone of the winter of 1949–50: a case study in the evolution of climatic anomalies. *J. Met.* 8, 251–61.

NAMIAS, J. 1952. The annual course of month-to-month persistence in climatic anomalies. *Bull. Amer. Met. Soc.* 32, 279–85.

NAMIAS, J. 1953. *Thirty-day forecasting: a review of a ten-year experiment.* Met. Monogr. (Amer. Met. Soc.) 2 (6). 83 pp.

NAMIAS, J. 1962. Influences of abnormal surface heat sources and sinks on atmospheric behaviour In *Proc. internat. sympos. on numerical weather prediction.* Tokyo: Met. Soc. Japan. 615–27.

NAMIAS, J. 1966. Nature and possible causes of the northeastern United States drought during 1962–65. *Mon. Wea. Rev.* 94, 543–54.

OSUCHOWSKA, B. 1970. On the persistence of meteorological patterns. *Geophys. Polon.* (Warsaw) 18, 23–41 (in Polish).

PERRY, A. H. 1969. Sensible and latent heat transfer over the North Atlantic during some recent winter months. *Ann. Met.* new ser., 4 (4), 40–6.

PERRY, A. H. 1970. Filtering climatic anomaly fields using principal component analysis. *Trans. Inst. Brit. Geogr.* No. 50, 55–72.

RAFAILOVA, K. K. 1968. Continuity and conjugation of fields of isonomals over individual sectors of the northern hemisphere and forecasting mean monthly atmospheric temperature over the U.S.S.R. *Trudy Gidromet. Nauch. Issled. Tsentr. SSSR* (Moscow) 21, 3–19 (in Russian).

RAGOZINA, V. S. 1967. Air temperature regime in the northern hemisphere in autumn in relation to circulation conditions. *Probl. Arkt. i Antarkt.* (Leningrad) 26, 12–20 (in Russian).

RAHAU, L. *et al.* 1969. Some features of the monthly mean temperature departures for the Socialist Republic of Rumania in January. *Inst. Met. Culegere de Lucrari* (Bucharest), 33–44 (in Rumanian).

RAKIPOVA, L. R. 1959. The formation of mean monthly anomaly in atmospheric temperature. *Trudy Glav. Geofiz. Obs.* (Leningrad) 87 (in Russian).

SAWYER, J. S. 1970. Observational characteristics of atmospheric fluctuations with a time scale of a month. *Quart. J. R. Met. Soc.* 96, 610–25.

SAZONOV, B. J. and GIRSKAIA, E. I. 1969. Temperature anomalies and the persistence of the atmospheric circulation. *Trudy Glav. Geofiz. Obs.* (Leningrad) 245, 77–87 (in Russian).

SISKOV, J. 1962. The meridional heat transport in the lower troposphere and anomalies of the temperature regime in the north part of the North Atlantic Ocean. *Trudy Akad. Nauk Inst. Okeanol.* (Moscow), 156–99 (in Russian).

SMITH, K. 1947. *Monthly temperature anomaly types.* Unpub. research report. Washington, D.C.: Extended Forecast Section, U.S. Weather Bureau. 3 pp.

ZAVIALOVA, I. N. 1965. On the frequency of deviations of the mean monthly values of temperature and pressure from the normal in the Arctic. *Trudy Arkt. Antarkt. Nauch. Issled. Inst.* (Leningrad) 273, 46–63 (in Russian).

ZORINA, V. P. 1967. Spatial distribution of large air temperature anomalies in the northern hemisphere. *Probl. Arkt. Antarkt.* (Leningrad) No. 24, 46–52 (in Russian).

Chapter 4

GENERAL REFERENCES

BAILEY, N. T. 1964. *The elements of stochastic processes with applications to the natural sciences.* New York: Wiley. 249 pp.

BROOKS, C. E. P. and CARRUTHERS, N. 1953. *Handbook of statistical methods in meteorology.* London: H.M.S.O. 412 pp.

FELLER, W. 1957. *An introduction to probability theory and its applications.* vol. 1. 2nd ed. New York: Wiley. 461 pp.

FOSTER, H. A. 1924. Theoretical frequency curves. *Trans. Amer. Soc. Civ. Engrs.* 89, 142–203.

FRASER, D. A. S. 1957. *Non-parametric methods in statistics.* New York: Wiley. 299 pp.

HODGES, J. L. and LEHMAN, E. L. 1965. *Elements of finite probability.* San Francisco: Holden-Day. 230 pp.

KENDALL, M. G. 1957. *A course in multivariate analysis*. London: Griffin. 185 pp.

PANOFSKY, H. A. and BRIER, G. W. 1958. *Some applications of statistics to meteorology*. Pennsylvania State Univ. College. 224 pp.

PARZEN, E. 1960. *Modern probability theory and its applications*. New York: Wiley. 464 pp.

THOM, H. C. S. 1966. *Some methods of climatological analysis*. Tech. Note No. 81. Geneva: World Meteorological Organization. 53 pp.

WEATHERBURN, C. E. 1949. *A first course in mathematical statistics*. 2nd ed. Cambridge: Cambridge Univ. Press. 277 pp.

YULE, G. U. and KENDALL, M. G. 1950. *An introduction to the theory of statistics*. 14th ed. London: Griffin. 701 pp.

4.A FREQUENCY AND PROBABILITY ANALYSIS

BENNETTS, P. 1967. *The distribution of departures of Manley's central England seasonal and monthly mean temperatures from 24-year running means.* Mem. No. 16, Synoptic Climatology Branch. Bracknell, Berks.: Met. Office. 3 pp.

BRYSON, R. A. 1966. Air masses, streamlines and the boreal forest. *Geogr. Bull.* 8, 228–69.

CHOW, V. T. 1955. The log-probability law and its engineering applications. *Proc. Amer. Soc. Civ. Engrs*, 80, 1–25.

CONRAD, V. and POLLAK, L. W. 1950. *Methods in climatology*. Cambridge, Mass.: Harvard Univ. Press. 459 pp.

COX, D. R. 1962. *Renewal Theory*. London: Methuen. 142 pp.

CRADDOCK, J. M. 1963. Persistent temperature regimes at Kew. *Quart. J. R. Met. Soc.* 89, 461–8.

DECKER, W. L. 1952. Hail damage frequencies for Iowa and a method of evaluating the probability of a specified amount of hail damage. *Trans. Amer. Geophys. Union* 33, 204–10.

ESSENWANGER, O. 1954. Neue Methoden der Zerlegung von Häufigkeitsverteilungen. *Ber. dtsch. Wetterdienst* U.S. Zone (Bad Kissingen) No. 10. 11 pp.

ESSENWANGER, O. 1955. Zur Realitat der Zerlegung von Häufigkeitsverteilungen in Normalkurven. *Arch. Met. Geophys. Biokl.* B, 7, 49–59.

ESSENWANGER, O. 1960a. Frequency distributions of precipitation. In Weickmann, H. (ed.) *Physics of Precipitation*. Geophys. Monogr. No. 5. Washington, D.C.: Amer. Geophys. Union. 271–8.

ESSENWANGER, O. 1960b. Linear and logarithmic scale for frequency distribution of precipitation. *Geof. pura e appl.* 45, 199–214.

FIEDLER, V. F. 1965. Woraus bestehen die Haufigkeitsverteilungen der Tagestemperaturen von Frankfurt. *Zeit. Met.* 17, 305–10.

GODSKE, C. L. 1966. Methods of statistics and some applications to climatology. In *Statistical analysis and prognosis in meteorology*. Tech. Note No. 71 (W.M.O. No. 178, TP 88). Geneva: World Meteorological Organization. 9–86.

GRINGORTEN, I. I. 1966. A stochastic model of the duration and frequency of weather events. *J. Appl. Met.* 5, 606–24.

GRUNOW, J. 1956. Zur Erfassung und Statistik der kleinsten Niederschlage. *Geof. pura e appl.* 33, 251–61.

HAY, R. F. M. 1968. Relationships between autumn rainfall and winter temperatures. *Met. Mag.* 97, 278–82.

HILTIER, F. S. and LIEBERMAN, G. J. 1967. *Introduction to operations research.* San Fransisco: Holden-Day. 452–71.

HUFF, F. A. and NEILL, J. C. 1959. Comparison of several methods for rainfall frequency analysis. *J. Geophys. Res.* 64, 541–7.

JORDAN, C. L. and HO, T.-C. 1962. Variations in the annual frequency of tropical cyclones. *Mon. Wea. Rev.* 90, 157–64.

KAPLANSKY, I. 1945. A common error concerning kurtosis. *J. Amer. Stat. Ass.* 40, 259.

KENDALL, G. R. 1960. The cube-root normal distribution applied to Canadian monthly rainfall totals. *Int. Ass. Sci. Hydrol.* (I.U.G.G., Helsinki) Pub. 53. 250–60.

LUND, I. A. 1970. A Monte Carlo method for testing the statistical significance of a regression equation. *J. Appl. Met.* 9, 330–2.

MURRAY, R. 1968. Some predictive relationships concerning seasonal rainfall over England and Wales and seasonal temperature in central England. *Met. Mag.* 97, 303–10.

MURRAY, R. and LEWIS, R. P. W. 1966. Some aspects of the synoptic climatology of the British Isles as measured by simple indices. *Met. Mag.* 95, 193–203.

PALMIERI, G., WANKE, E. and PALMIERI, S. 1969. Meteorological application of a probability classification machine. *Rivista Met. Aeronaut.* (Rome) 29, suppl. to No. 6, 52–62 (in Italian).

PEARSON, K. *et al.* 1951. *Tables of the incomplete gamma function.* Cambridge: Cambridge Univ. Press.

STIDD, C. K. 1953. The cube-root normal precipitation distribution. *Trans. Amer. Geophys. Union* 34, 31–4.

SUZUKI, E. 1964. Hyper gamma distribution and its fitting to rainfall data. *Pap. Met. Geophys.* (Tokyo) 15, 31–51.

SUZUKI, E. 1967. A statistical and climatological study on the rainfall in Japan. *Pap. Met. Geophys.* (Tokyo) 18, 103–82.

THOM, H. C. S. 1958. A note on the gamma distribution. *Mon. Wea. Rev.* 86, 117–22.

THOM, H. C. S. 1966a. *Some methods of climatological analysis.* Tech. Note No. 81. Geneva: World Meteorological Organization. 53 pp.

THOM, H. C. S. 1966b. The distribution of annual tropical cyclone frequency. *J. Geophys. Res.* 65, 213–22.

WACHTER, H. 1968. Häufigkeitsverteilung klimatologischer Grossen. *Ber. dtsch. Wetterdienst* (Offenbach) 15 (107). 35 pp.

4.B TIME SERIES

ALTER, D. 1933. Correlation periodogram investigation of English rainfall. *Mon. Wea. Rev.* 61, 345–52.

BELASCO, J. E. 1948. The incidence of anticyclonic days and spells over the British Isles. *Weather* 3, 233–42.
BLACKMAN, R. B. and TUKEY, J. W. 1958. *The measurement of power spectra from the point of view of communications engineering*. New York: Dover. 190 pp.
BROOKS, C. E. P. and MIRRLEES, S. T. A. 1930. Irregularities in the annual variation of the temperature of London. *Quart. J. R. Met. Soc.* 56, 375–88.
BRUCE, J. P. and CLARK, R. H. 1966. *Introduction to hydrometeorology*. Oxford: Pergamon. 319 pp.
BRYON, R. A. and DUTTON, J. A. 1961. Some aspects of the variance spectra in tree rings and varves. *Ann. N.Y. Acad. Sci.* 95, 580–604.
CASKEY, J. E. JR. 1963. A Markov chain model for the probability of precipitation occurrence in intervals of various lengths. *Mon. Wea. Rev.* 91, 298–301.
CASKEY, J. E. JR. 1964. Markov chain model of cold spells at London. *Met. Mag.* 93, 136–8.
CAVADIAS, G. S. 1967. River flow as a stochastic process. In *Proc. hydrol. sympos. No. 5: Statistical methods in hydrology*. Ottawa: Inland Waters Branch. 315–51.
COCHRAN, W. G. 1938. An extension of Gold's method of examining the apparent persistence of one type of weather. *Quart. J. R. Met. Soc.* 64, 631–4.
COX, D. R. and LEWIS, P. A. W. 1966. *The statistical analysis of series of events*. London: Methuen. 285 pp.
CRADDOCK, J. M. 1957. An analysis of the slower temperature variations at Kew Observatory by means of mutually exclusive band pass filters. *J. R. Stat. Soc.* A, 120, 387–97.
CRADDOCK, J. M. 1965. The analysis of meteorological time series for use in forecasting. *The Statistician* 15, 167–90.
CRADDOCK, J. M. 1968. *Statistics in the computer age*. London: English Universities Press. 214 pp.
DOBERITZ, R. 1967. Spectrum and filter analysis of rainfall for equatorial Pacific Islands. *Bonner. Met. Abh.* 7, 9–51.
DOBERITZ, R. 1968. Kohärenzanalyse von Niederschlag und Wassertemperatur im tropischen Pazifischen Ozean. *Ber. dtsch. Wetterd.* (Offenbach) 15 (112). 22 pp.
FEYERHERM, A. M. and BARK, L. D. 1965. Statistical methods for persistent precipitation patterns. *J. Appl. Met.* 4, 320–8.
FEYERHERM, A. M. and BARK, L. D. 1967. Goodness of fit of a Markov chain model for sequences of wet and dry days. *J. Appl. Met.* 6, 770–3.
FITZPATRICK, E. A. and KRISHNAN, A. 1967. A first-order Markov model for assessing rainfall discontinuity in central Australia. *Arch. Met. Geophys. Biokl.* B, 15, 242–59.
GABRIEL, K. R. and NEWMANN, J. 1957. On a distribution of weather cycles by length. *Quart. J. R. Met. Soc.* 88, 90–5.
GABRIEL, K. R. and NEWMANN, J. 1962. A Markov chain model for daily rainfall occurrence at Tel Aviv. *Quart. J. R. Met. Soc.* 88, 90–5.

GILMAN, D. I., FUGLISTER, F. J. and MITCHELL, J. M. JR. 1963. On the power spectrum of 'red noise'. *J. Atm. Sci.* 20, 182–4.

GODSKE, C. L. 1966. Methods of statistics and some applications to climatology. In *Statistical analysis and prognosis in meteorology.* Tech. Note No. 71 (W.M.O. No. 178, TP 88). Geneva: World Meteorological Organization. 9–86.

GOLD, E. 1929. Note on the frequency of occurrence of sequences in a series of events of two types. *Quart. J. R. Met. Soc.* 55, 307–9.

GREEN, J. R. 1965. Two probability models for the sequences of wet or dry days. *Mon. Wea. Rev.* 93, 155–6.

GREEN, J. R. 1970. A generalized probability model for sequences of wet and dry days. *Mon. Wea. Rev.* 98, 238–41.

GRINGORTEN, I. I. 1966. A stochastic model of the frequency and duration of weather events. *J. Appl. Met.* 5, 606–24.

HAMON, B. V. and HANNAN, E. J. 1963. Estimating relationships between time series. *J. Geophys. Res.* 68, 6033–42.

HERSHFIELD, D. M. 1970. A comparison of conditional and unconditional probabilities for wet and dry day sequences. *J. Appl. Met.* 9, 825–7.

HINICH, M. J. and CLAY, C. S. 1968. The application of the discrete Fourier transform in the estimation of power spectra, coherence, and bispectra of geophysical data. *Rev. Geophys.* 6, 347–63.

HOLLOWAY, J. L. JR. 1958. Smoothing and filtering of time series and space fields. *Adv. Geophys.* 4, 351–89.

HOPKINS, J. W. and ROBILLARD, P. 1964. Some statistics of daily rainfall occurrence for the Canadian prairie provinces. *J. Appl. Met.* 3, 600–2.

HORN, L. H. and BRYSON, R. A. 1960. Harmonic analysis of the annual march of precipitation over the United States. *Ann. Assoc. Amer. Geogr.* 50, 157–71.

JENKINS, G. H. 1961. General considerations in the analysis of spectra. *Technometrics* 3, 133–66.

JENKINS, G. H. and WATTS, D. G. 1968. *Spectral analysis and its applications.* San Francisco: Holden Day. 525 pp.

JULIAN, P. R. 1966. The index cycle: a cross-spectral analysis of zonal index data. *Mon. Wea. Rev.* 94, 283–93.

JULIAN, P. R. 1967. Variance spectrum analysis. *Water Resources Res.* 3, 831–45.

LANDSBERG, H. E., MITCHELL, J. M. JR. and CRUTCHER, H. L. 1959. Power spectrum analysis of climatological data for Woodstock College, Maryland. *Mon Wea. Rev.* 87, 283–98.

LANDSBERG, H. E., MITCHELL, J. M. JR, CRUTCHER, H. L. and QUINLAN, F. T. 1963. Surface signs of the biennial atmospheric pulse. *Mon. Wea. Rev.* 91, 549–56.

LAWRENCE, E. N. 1957. Estimation of the frequency of 'runs of dry days'. *Met. Mag.* 86, 257–69, 301–4.

LETTAU, K. and WHITE, F. 1964. Fourier analysis of Indian rainfall. *Indian J. Met. Geophys.* 15, 27–8.

LONGLEY, R. W. 1953. The length of wet and dry periods. *Quart. J. R. Met. Soc.* 79, 520–7.

LOWRY, W. P. and GUTHRIE, D. 1968. Markov chains of order greater than one. *Mon. Wea. Rev.* 96, 798–801.

MCGEE, O. S. and HASTENRATH, S. L. 1966. Harmonic analysis of the rainfall over South Africa. *Notos* 15, 79–90.

MANDELBROT, B. B. and WALLIS, J. R. 1968. Noah, Joseph and operational hydrology. *Water Resources Res.* 4, 909–18.

MANDELBROT, B. B. and WALLIS, J. R. 1969a. Computer experiments with fractional Gaussian noises, Part I: Averages and variances. *Water Resources Res.* 5, 228–41.

MANDELBROT, B. B. and WALLIS, J. R. 1969b. Computer experiments with fractional Gaussian noises, Part 2: Rescaled ranges and spectra. *Water Resources Res.* 5, 242–59.

MANDELBROT, B. B. and WALLIS, J. R. 1969c. Some long-run properties of geophysical records. *Water Resources Res.* 5, 321–40.

MANDELBROT, B. B. and WALLIS, J. R. 1969d. Robustness of the rescaled range R/S in the measurement of non-cyclic long-run statistical dependence. *Water Resources Res.* 5, 967–88.

MATALAS, N. C. 1967. Some aspects of time series analysis in hydrologic studies. In *Proc. hydrol. sympos. No. 5: Statistical methods in hydrology.* Ottawa: Dept of Energy, Mines & Resources, Inland Waters Branch. 271–309.

MATALAS, N. C. and BENSON, M. A. 1961. Effect of interstation correlation on regression analysis. *J. Geophys. Res.* 66, 3285–93.

MITCHELL, J. M. JR. 1966. Stochastic models of air–sea interaction and climatic fluctuation. In Fletcher, J. O. (ed.) *Proc. sympos. on the arctic heat budget and atmospheric circulation.* Mem. RM-5233-NSF. Santa Monica, Calif.: Rand Corp. 46–74.

NEWNHAM, E. V. 1916. The persistence of wet and dry weather. *Quart. J. R. Met. Soc.* 42, 153–62.

PANOFSKY, H. A. 1949. Significance of meteorological correlation coefficients. *Bull. Amer. Met. Soc.* 30, 326–7.

PANOFSKY, H. A. 1955. Meteorological applications of power spectrum analysis. *Bull. Amer. Met. Soc.* 36, 163–6.

PANOFSKY, H. A. and WOLFF, P. 1957. Spectrum and cross-spectrum analysis of hemispheric westerly index. *Tellus* 9, 195–200.

POLOWCHAK, V. M. and PANOFSKY, H. A. 1968. The spectrum of daily temperatures as a climatic indicator. *Mon. Wea. Rev.* 96, 596–600.

QUENOUILLE, M. H. 1952. *Associated measurements.* London: Butterworths.

QUENOUILLE, M. H. 1957. *The analysis of multiple time series.* London: Griffin. 105 pp.

RAYMENT, R. 1970. Introduction to the fast Fourier transform (FFT) in the production of spectra. *Met. Mag.* 99, 261–70.

RAYNER, J. N. 1965. *An approach to the dynamic climatology of New Zealand.* Unpub. Ph.D. thesis, Univ. of Canterbury, Christchurch. 173 pp.

RAYNER, J. N. 1967. A statistical model for the explanatory description of large-scale time and spatial climate. *Canad. Geogr.* 11, 67–86.

RAYNER, J. N. 1971. *An introduction to spectral analysis.* London: Pion Press. 174 pp.

RODRIGUEZ-ITURBÉ, I. and NORDIN, C. F. 1969. Some applications of cross-spectral analysis in hydrology: rainfall and runoff. *Water Resources Res.* 5, 608–21.

ROSENBLATT, M. (ed.) 1963. *Proceedings of the symposium on time series analysis.* New York: Wiley. 497 pp.

SABBAGH, M. E. and BRYSON, R. A. 1962. Aspects of the precipitation climatology of Canada investigated by the method of harmonic analysis. *Ann. Assoc. Amer. Geogr.* 52, 426–40.

SCHUMANN, T. E. W. and HOFMEYR, W. L. 1942. The problem of autocorrelation of meteorological time series. *Quart. J. R. Met. Soc.* 68, 177–88.

SRINIVASAN, T. R. 1964. Rainfall persistence in India during May–October. *Indian J. Met. Geophys.* 15, 163–74.

TAKAHASHI, U. 1965. *A review of the method of long-range weather forecasting especially used in Japan.* Tech. Note No. 66. Geneva: World Meteorological Organization. 31–45.

THOM, H. C. S. 1966. *Some methods of climatological analysis.* Tech. Note No. 81. Geneva: World Meteorological Organization. 53 pp.

TUKEY, J. W. 1967. Spectrum calculations in the new world of the fast Fourier transform. In Harris, B. (ed.) *Spectral analysis of time series.* New York: Wiley. 25–46.

TYSON, P. D. 1969. *Time series: a problem of numerical analysis in geography.* Occasional paper No. 1. Johannesburg: Dept of Geography & Environmental Studies, Univ. of Witwatersrand. 13 pp.

WALKER, E. R. 1964. *Analysis of normal monthly precipitation over Alaska and western Canada.* Met. Branch CIR-4043, TEC-522. Toronto: Dept of Transport. 8 pp.

WALLACE, J. M. and CHANG, C.-P. 1969. Spectral analysis of large-scale wave disturbances in the tropical lower troposphere. *J. Atm. Sci.* 25, 1010–25.

WARD, F. and SHAPIRO, R. 1961. Meteorological periodicities. *J. Met.* 8, 635–56.

WATTERSON, G. A. and LEGG, M. P. C. 1967. Daily rainfall patterns at Melbourne. *Austral. Met. Mag.* 15, 1–12.

WEISS, L. L. 1964. Sequences of wet or dry days described by a Markov chain probability model. *Mon. Wea. Rev.* 92, 169–76.

WEISS, L. L. and WILSON, W. T. 1953. Evaluation of significance of slope changes in double mass curves. *Trans. Amer. Geogr. Union* 34.

WILLIAMS, C. B. 1952. Sequences of wet and dry days considered in relation to the logarithmic series. *Quart. J. R. Met. Soc.* 78, 91–6.

WISER, E. H. 1965. Modified Markov probability models of sequences of precipitation events. *Mon. Wea. Rev.* 93, 511–16.

4.C SPATIAL SERIES

ANDERSON, P. M. 1970. The uses and limitations of trend surface analysis in studies of urban air pollution. *Atm. Environment* 4, 129–47.

BAGROV, N. A. 1959. The analytical representation of the sequence of meteorological fields by means of natural orthogonal polynomials. *Trudy Tsentr. Inst. Prog.* (Moscow) 7, 3–24 (in Russian).

BERTONI, E. A. and LUND, I. A. 1963. Space correlations of the height of constant pressure surfaces. *J. Appl. Met.* 2, 539–45.

CAFFEY, J. E. 1965. *Interstation correlations in annual precipitation and in annual effective precipitation.* Hydrol. Pap. No. 6. Fort Collins: Colo. State Univ. 47 pp.

CATTELL, R. B. 1952. *Factor analysis: an introduction and manual for the psychologist and social scientist.* New York: Harper. 462 pp.

CATTELL, R. B. 1965. Factor analysis: an introduction to essentials I and II. *Biometrics* 21, 190–215, 405–35.

CEHAK, K. 1962. Die Verwendungen von orthogonalen Polynomen in der Meteorologie, I: Mitteilung. *Arch. Met. Geophys. Biokl.* A, 12, 40–61.

CHORLEY, R. J. and HAGGETT, P. 1965. Trend surface mapping in geographical research. *Trans. Inst. Brit. Geogr.* No. 37, 47–67.

CORNISH, E. A., HILL, G. W. and EVANS, M. J. 1961. *Inter-station correlations of rainfall in southern Australia.* Tech. Pap. No. 10. Melbourne: Division of Math. Statistics, C.S.I.R.O. 16 pp.

CRADDOCK, J. M. 1968. A meteorological application of principal components analysis. *The Statistician* 15, 143–56.

CRADDOCK, J. M. 1969. *The use of eigenvector analysis in statistical meteorology.* Met. O.13, Branch Mem. No. 27. Bracknell, Berks: Met. Office. 20 pp.

CRADDOCK, J. M. and FLINTOFF, S. 1970. Eigenvector representations of northern hemispheric fields. *Quart. J. R. Met. Soc.* 96, 124–9.

CRADDOCK, J. M. and FLOOD, C. R. 1969. Eigenvectors for representing the 500 mb geopotential surface over the northern hemisphere. *Quart. J. R. Met. Soc.* 95, 576–93.

CURRY, L. 1966. A note on spatial association. *Prof. Geogr.* 18, 97–9.

DELAND, R. J. 1965. Some observations on the behavior of spherical harmonic waves. *Mon. Wea. Rev.* 93, 307–12, 494.

DICKSON, R. R. 1971. On the relationship of variance spectra of temperature to the large-scale atmospheric circulation. *J. Appl. Met.* 10, 186–93.

DIXON, R. 1969. *Orthogonal polynomials as a basis for objective analysis.* Sci. Pap. No. 30. Bracknell, Berks: Met. Office. 20 pp.

DRAPER, N. R. and SMITH, H. 1966. *Applied regresion analysis.* New York: Wiley. 407 pp.

ELIASEN, E. and MACHENHAUER, B. 1965. A study of the fluctuations of the atmospheric flow patterns represented by spherical harmonics. *Tellus* 17, 220–38.

ELSAESSER, H. W. 1966. Expansion of hemispheric meteorological data in antisymmetric surface spherical harmonic (Laplace) series. *J. Appl. Met.* 5, 263–76.

EL THOM, M. A. 1969. A statistical analysis of the rainfall over the Sudan *Geogr. J.* 135, 378–87.

FLIRI, F. 1967. Beiträge zur Kenntnis der Zeit-Raum-Struktur des Niederschlags in den Alpen. *Wetter u. Leben* 19, 241–68.

FRIEDMAN, D. G. 1955. Specification of temperature and precipitation in terms of circulation patterns. *J. Met.* 12, 428–35.

GANDIN, L. S. 1965. *Objective analysis of meteorological fields* (Leningrad, Gidromet. Izdat. 1963). Jerusalem: Israel Prog. Sci. Trans. 242 pp.

GILMAN, D. L. 1957. *Empirical orthogonal functions applied to thirty-day forecasting.* Sci. Rep. No. 1, Contract No. AF 19(604)-1283. Cambridge, Mass.: Met. Dept. Mass. Inst. Tech. 129 pp.

GODSKE, C. L. 1952. *Studies in local meteorology and representativeness, 1: The distribution of precipitation in Hardanger.* Univ. Bergen Arbok, Naturvitenskapelig Rekke No. 10. 99 pp.

GODSKE, C. L. 1965. *Statistics of meteorological variables.* Report, Contract AF 61(052)-416. Geophys. Inst., Univ. of Bergen.

GOULD, P. R. 1967. On the geographical interpretation of eigenvalues. *Trans. Inst. Brit. Geogr.* No. 42, 53–86.

GREGORY, S. 1968. The orographic component in rainfall distribution patterns. In Sporck, J. A. (ed.) *Mélanges de Géographie offerts à M. Omer Tulippe, 1: Géographie physique et géographie humaine.* Belgium: Gembloux. 234–52.

GRIMMER, M. 1963. The space filtering of monthly surface temperature anomaly data in terms of pattern, using empirical orthogonal functions. *Quart. J. R. Met. Soc.* 89, 395–408.

HARE, F. K. 1958. The quantitative representation of the north polar pressure field. In Sutcliffe, R. C. (ed.) *The polar atmosphere symposium,* part 1. Oxford: Pergamon. 137–50.

HARE, F. K., GODSON, W. L., MACFARLANE, M. A. and WILSON, C. V. 1957. *Specification of pressure fields and flow patterns in polar regions: theories and techniques.* Pub. in Met. No. 5. Montreal: Arctic Met. Res. Group McGill Univ. 72 pp.

HAURWITZ, B. and CRAIG, R. A. 1952. *Atmospheric flow patterns and their representation by spherical surface harmonics.* Geophys. Res. Paper No. 14. Cambridge, Mass.: Air Force Res. Center. 78 pp.

HUFF, F. A. and SHIPP, W. L. 1968. Mesoscale spatial variability in midwestern precipitation. *J. Appl. Met.* 7, 886–91.

HUFF, F. A. and SHIPP, W. L. 1969. Spatial correlations of storm, monthly and seasonal precipitation. *J. Appl. Met.* 8, 562–50.

IAKOVELVA, N. I., CHUVASHINA, I. E. and KUDASHIN, G. D. 1968. Improved natural orthogonal functions of the pressure field (geopotential) over the northern hemisphere. *Trudy Glav. Geofiz. Obs.* (Leningrad) 201, 60–71 (in Russian).

IUDIN, M. I. (ed.) 1969. Physico-statistical investigations of the atmospheric circulation. *Trudy Glav. Geofiz. Obs.* (Leningrad) 236, 83–164 (in Russian).

IZAWA, T. 1965. Two- or multi-dimensional gamma-type distribution and its application. *Pap. Met. Geophys.* (Tokyo) 15, 167–200.

JACKSON, I. J. 1969. *Pressure types and precipitation over north-east England.* Res. Ser. No. 5. Newcastle-upon-Tyne: Dept. of Geography, Univ. of Newcastle-upon-Tyne. 81 pp.

JEFFERS, J. M. R. 1967. Two case studies in the application of principal component analysis. *Appl. Stat.* 16, 225–36.

JULIAN, P. R. 1970. An application of rank-order statistics to the joint spatial and temporal variations of meteorological elements. *Mon. Wea. Rev.* 98, 142–53.

KING, L. J. 1969. *Statistical analysis in geography*. Englewood Cliffs: Prentice-Hall. 288 pp.

KLEIN, W. H. 1963. Specification of precipitation from the 700 mb circulation. *Mon. Wea. Rev.* 91, 527–36.

KLEIN, W. H. 1965. Five-day precipitation patterns derived from circulation and moisture. In Amdur, E. J. (ed.) *Humidity and moisture*, vol. 2: *Applications*. New York: Reinhold. 532–49.

KLEIN, W. H., LEWIS, B. M. and ENGER, I. 1959. Objective prediction of five-day mean temperatures during winter. *J. Met.* 16, 672–82.

KUBOTE, S., HIROSE, M., KIBUCHI, Y. and KURIHARA, Y. 1960. Barotropic forecasting with the use of surface spherical harmonic representations. *Pap. Met. Geophys.* (Tokyo) 12, 199–215.

KUTZBACH, J. E. 1967. Empirical eigenvectors of sea-level pressure, surface temperature and precipitation complexes over North America. *J. Appl. Met.* 6, 791–802.

LEE, P. J. 1969. The theory and application of canonical trend surfaces. *J. Geol.* 77, 303–18.

LEESE, J. A. and EPSTEIN, E. S. 1963. Application of two-dimensional spectral analysis to the quantification of satellite cloud photographs. *J. Appl. Met.* 2, 629–44.

LORENZ, E. N. 1956. *Empirical orthogonal functions and statistical weather prediction*. Sci. Rep. No. 1. Contract AF 19 (604)-1566. Cambridge, Mass.: Met. Dept, Mass. Inst. Tech. 49 pp.

MCDONALD, J. E. 1960. Remarks on correlation methods in geophysics. *Tellus* 12, 176–83.

MARTIN, D. E. and LEIGHT, W. G. 1949. Objective temperature estimates from mean circulation patterns. *Mon. Wea. Rev.* 77, 275–83.

MATALAS, N. C. and BENSON, M. A. 1961. Effect of interstation correlation on regression analysis. *J. Geophys. Res.* 66, 3285–93.

MDINARADZE, D. 1968. Pecularities of the main terms of analysis of the 500 mb geopotential field according to natural functions. *Trudy Gidromet. Nauch. Issled. Tsentr SSSR* (Moscow) No. 28, 51–63 (in Russian).

MILLER, R. G. 1962. *Statistical prediction by discriminant analysis*. Met. Monogr. (Amer. Met. Soc.) 4 (25), 54 pp.

NORCLIFFE, G. B. 1969. On the uses and limitations of trend surface models. *Canad. Geogr.* 13, 338–48.

NORDØ, J. and HJORTNAES, K. 1967. Statistical studies of precipitation on local, national and continental scales. *Geofys. Publik.* (Oslo) 26. 46 pp.

PAGE, P. 1968. *Climatological aspects of air pollution in the West Midlands conurbation*. Unpub. Ph.D. thesis, Univ. of London.

PALMIERI, S. 1968. Study of the space relationships of meteorological fields over northern Italy by means of principal component analysis. *Rivista Met. Aeronaut.* 28, 5–19 (in Italian).

PERRY, A. H. 1970. Filtering climatic anomaly fields using principal component analysis. *Trans. Inst. Brit. Geogr.* No. 50, 55–72.

PETERSON, J. T. 1970. Description of sulphur dioxide over metropolitan St Louis as described by empirical eigenvectors and its relation to meteorological parameters. *Atm. Environment* 4, 501–18.

POOLE, M. A. and O'FARRELL, P. N. 1971. The assumptions of the linear regression model. *Trans. Inst. Brit. Geogr.* No. 52, 145–58.

RAYNER, J. N. 1967a. A statistical model for the explanatory description of large-scale time and spatial climate. *Canad. Geogr.* 11, 67–86.

RAYNER, J. N. 1967b. Correlation of surfaces by spectral methods. *Computer Contrib.* 12 (Kansas Geol. Survey, Lawrence, Kan.) 31–7.

RAYNER, J. N. 1971. *An introduction to spectral analysis.* London: Pion Press. 174 pp.

ROBINSON, G. 1970. Some comments on trend-surface analysis. *Area* No. 13, 31–6.

SELLERS, W. D. 1968. Climatology of monthly precipitation patterns in the western United States, 1931–66. *Mon. Wea. Rev.* 96, 585–95.

SMITHSON, P. A. 1969. *Rainfall in Scotland in relation to the synoptic situation.* Unpub. Ph.D. thesis, Univ. of Liverpool.

STIDD, C. K. 1954. The use of correlation fields in relating precipitation to circulation. *J. Met.* 11, 202–13.

STIDD, C. K. 1967. The use of eigenvectors for climatic estimates. *J. Appl. Met.* 6, 255.

TRISCHLER, F. 1956. Untersuchung über räumlich kohärente und singulare Tagesniederschläge. *Arch. Met. Geophys. Biokl.* B, 7, 370–405.

UNWIN, D. J. 1969. The areal extension of rainfall records: alternative model. *J. Hydrol.* 7, 404–14.

WACHTER, H. 1963. On the frequency distribution of concurrent rainfall in a large area. *Geof. pura e appl.* 55, 126–30.

WACHTER, H. 1965. Gesetzmässigkeiten gleichzeitigen Niederschlags an vielen synoptischen Stationen. *Zeit. Met.* 17, 297–304.

WACHTER, H. 1968. Häufigkeitsverteilung klimatologischer Grossen. *Ber. dtsch. Wetterdienst.* (Offenbach) 15 (107). 35 pp.

WALLACE, J. M. 1971. Spectral studies of tropospheric wave disturbances in the tropical western Pacific. *Rev. Geophys. and Space Phys.* 9, 557–612.

WALLIS, J. R. 1965. Multivariate statistical methods in hydrology: comparison using data of known functional relationship. *Water Resources Res.* 1, 447–61.

WHITE, R. M., DERBY, R. C., COOLEY, D. S. and SEAVER, F. A. 1957. Hemispherical prediction by statistical techniques. *J. Met.* 14, 448–57.

WHITE, R. M., DERBY, R. C., COOLEY, D. S. and SEAVER, F. A. 1958. The development of efficient linear statistical operators for the prediction of sea-level pressure. *J. Met.* 15, 426–34.

4.D CLASSIFICATORY METHODS

ANDREWS, J. T. and ESTABROOK, G. F. 1971. Applications of information and graph theory to multivariate geomorphological analyses. *J. Geol.* 79, 207–21.

ARMSTRONG, J. S. 1967. Derivation of theory by means of factor analysis, or Tom Swift and his electric factor analysis machine. *Amer. Stat.* 21, 17–21.

BOYCE, A. J. 1969. Mapping diversity: a comparative study of some numerical methods. In Cole, A. J. (ed.) *Numerical taxonomy*. London: Academic Press. 1–31.

BRINKMANN, W. A. R. 1970. The chinook at Calgary (Canada). *Arch. Met. Geophys. Biokl.* B, 18, 269–98.

CASETTI, E. 1964. *Classificatory and regional analysis by discriminant iteration*. Tech. Rep. No. 12. Geog. Branch, Office of Naval Res. Contract 1228(26), Task 389–135. Evanston, Ill.: Northwestern Univ. 95 pp.

CATTELL, R. B. 1965. Factor analysis: an introduction to essentials I and II. *Biometrics* 21, 190–215, 405–35.

CHORLEY, R. J. and HAGGETT, P. 1970. *Network analysis in geography*. London: Arnold. 358 pp.

CHRISTENSEN, W. I. JR and BRYSON, R. A. 1966. An investigation of the potential of component analysis for weather classification. *Mon. Wea. Rev.* 94, 697–709.

COOLEY, W. W. and LOHNES, P. R. 1962. *Multivariate procedures for the behavioural sciences*. New York: Wiley. 211 pp.

ESTABROOK, G. F. 1966. A mathematical model in graph theory for biological classification. *J. Theoret. Biol.* 12, 297–310.

ESTABROOK, G. F. 1967. An information theory model for character analysis. *Taxon*. 16, 86–97.

FLIRI, F. 1967. Beiträge zur Kenntnis der Zeit-Raum-Struktur des Niederschlags in den Alpen. *Wetter u. Leben* 19, 241–68.

FRITTS, H. C., BLASING, T. J., HAYDEN, B. P. and KUTZBACH, J. E. 1971. Multivariate techniques for specifying tree-growth and climate relationships and for reconstructing anomalies in palaeoclimate. *J. Appl. Met.* 10, 845–64.

GLAHN, H. R. 1968. Canonical correlation and its relationship to discriminant analysis and multiple regression. *J. Atm. Sci.* 25, 23–31.

GLASSPOOLE, J. 1925. Relation between annual rainfall over Europe, Oxford and Glenquoich. *Brit. Rainfall* 65, 254–69.

GODSKE, C. L. 1959. Information, climatology and statistics. *Geogr. Annal.* 61, 85–93.

GOWER, J. C. 1966. Some distance properties of latent root and vector methods used in multivariate analysis. *Biometrika* 53, 325–38.

GOWER, J. C. 1967. Multivariate analysis and multidimensional geometry. *The Statistician* 17, 13–28.

GREGORY, S. 1964. Climate. In Watson, J. W. and Sissons, J. B. (eds.) *The British Isles: a systematic geography*. London: Nelson. 53–73.

HAGGETT, P. and CHORLEY, R. J. 1969. *Network analysis in geography*. London: Arnold. 348 pp.

HARARAY, F., NORMAN, R. Z. and CARTWRIGHT, D. 1965. *Structural models: an introduction to the theory of directed graphs*. New York: Wiley.

HARBAUGH, J. W. and MERRIAM, D. F. 1968. *Computer applications in stratigraphic analysis*. New York: Wiley. 282 pp.

IAKOVLEVA, N. I. and GURLEVA, K. A. 1969. Objective division into regions by expansion in terms of empirical function. *Trudy Glav. Geofiz. Obs.* (Leningrad) 236, 155–64 (in Russian).

JOHNSON, R. J. 1968. Choice in classification: the subjectivity of objective methods. *Ann. Assoc. Amer. Geogr.* 58, 575–89.

KING, L. J. 1969. *Statistical analysis in geography.* London: Prentice-Hall. 288 pp.

KUIPERS, W. J. A. 1970. An experiment on numerical classification of scalar fields. *Idojaras* 74, 296–306.

KUTZBACH, J. E. 1967. Empirical eigenvectors of sea-level pressure, surface temperature and precipitation complexes over North America. *J. Appl. Met.* 6, 791–802.

LAMBERT, J. M. and WILLIAMS, W. T. 1962. Multivariate methods in plant ecology, IV: Nodal analysis. *J. Ecol.* 50, 775–802.

LANCE, G. N. and WILLIAMS, W. T. 1967a. A general theory of classificatory sorting strategies, I: Hierarchical systems. *Computer J.* 9, 373–80.

LANCE, G. N. and WILLIAMS, W. T. 1967b. A general theory of classificatory sorting strategies, II: Clustering systems. *Computer J.* 10, 271–7.

LUND, I. A. 1963. Map pattern classification by statistical methods. *J. Appl. Met.* 2, 56–65.

MACNAUGHTON-SMITH, P., WILLIAMS, W. T., DALE, M. B. and MOCKETT, L. G. 1964. Dissimilarity analysis: a new technique of hierarchical subdivision. *Nature* 202, 1034–5.

MCQUITTY, L. L. 1957. Elementary linkage analysis for orthogonal and oblique types and typal relevancies. *Educ. Psychol. Measurement* 17, 207–29.

MCQUITTY, L. L. 1960. Hierarchical syndrome analysis. *Educ. Psychol. Measurement* 20, 293–304.

MCQUITTY, L. L. 1966. Single and multiple classification by reciprocal pairs and rank order types. *Educ. Psychol. Measurement* 26, 253–65.

MATHER, P. M. 1969a. Cluster analysis. *Computer Applications in the Natural and Social Sciences* No. 1, Geog. Dept, Univ. of Nottingham. 19 pp.

MATHER, P. M. 1969b. Multiple discriminant analysis. *Computer Applications in the Natural and Social Sciences* No. 6. Geog. Dept, Univ. of Nottingham. 17 pp.

MATHER, P. M. and DOORNKAMP, J. C. 1970. Multivariate analysis in geography. *Trans. Inst. Brit. Geogr.* No. 51, 163–87.

MILLER, R. G. 1962. *Statistical prediction by discriminant analysis.* Met. Monogr. (Amer. Met. Soc.) 4 (25). 54 pp.

NOSEK, M. 1967. Varianzanalyse und Signifikanzteste in der dynamischen Klimatologie. *Ann. Met.* 20, 211–16.

PANOFSKY, H. A. 1949. Significance of meteorological correlation coefficients. *Bull. Amer. Met. Soc.* 30, 327–7.

PANOFSKY, H. A. and BRIER, G. W. 1958. *Some applications of statistics to meteorology.* Pennsylvania State University. 224 pp.

PERRY, A. H. 1968. The regional variation of climatological characteristics with synoptic indices. *Weather* 23, 325–30.

PERRY, A. H. 1969. Sensible and latent heat transfer over the North Atlantic during some recent winter months; *Ann. Met.* new ser. 4 (4), 40–6.

RAJSKI, C. 1961. Entropy and metric spaces. In Cherry, C. (ed.) *Information theory.* London: Butterworths. 41–5.

RAO, C. R. 1952. *Advanced statistical methods in biometric research.* London: Chapman & Hall. 390 pp.

RUSSELL, J. S. and MOORE, A. W. 1970. Detection of homoclimate by numerical analysis with reference to the Brigalow region (eastern Australia). *Agric. Met.* 7, 455–79.

SEKIGUTI, T. 1952. Some problems of climatic classification: a new classification of climates of Japan. *Proc. 8th gen. assembly and 17th internat. congress, Internat. Geogr. Union* (Washington). 285–90.

SNEATH, P. H. A. 1969. Evaluation of clustering methods. In Cole, A. J. (ed.) *Numerical taxonomy.* London: Academic Press. 257–71.

SOKAL, R. R. and MICHENER, C. D. 1958. A statistical method of evaluating systemic relationships. *Univ. Kansas Sci. Bull.* 38, 1409–38.

SOKAL, R. R. and SNEATH, P. H. A. 1963. *Principles of numerical taxonomy.* San Francisco: Freeman. 359 pp.

SØRENSEN, T. 1948. A method of establishing groups of equal amplitude in plant sociology based on similarity of species content. *K. danske vidensk. Selsk.* 5, 1–34.

SPENCE, N. A. and TAYLOR, P. J. 1970. Quantitative methods in regional taxonomy. *Prog. in Geogr.* 2, 1–64.

STEINER, D. 1965. A multivariate statistical approach to climatic regionalization and classification. *Tijd. Kon. Ned. Aardrijksk. Genoot.* 82, 329–47.

SUZUKI, E. 1964. Categorical prediction schemes of rainfall types by discriminant analysis. *Papers Met. Geophys.* (Tokyo) 15, 119–30.

SUZUKI, E. 1969. A discrimination theory based on categorical variables and its application to meteorological variables. *J. Met. Soc. Japan* ser. 2, 47, 145–58.

WILLIAMS, W. T. and LAMBERT, J. M. 1959. Multivariate methods in plant ecology, II: The use of an electronic digital computer for association-analysis. *J. E. Ecol.* 48, 689–710.

WILLIAMS, W. T. and LAMBERT, J. M. 1966. A comparison of 'information analysis' and similarity analyses. *J. Ecol.* 54, 635–64.

WIRTH, M., ESTABROOK, G. F. and ROGERS, D. J. 1966. A graph theory model for systematic biology, with an example for the Oncidiinae (Orthidaceae). *Syst. Zool.* 15, 59–69.

WISHART, D. 1969. Mode analysis: a reduction of nearest neighbour which reduces chaining effects. In Cole, A. J. (ed.) *Numerical analysis.* London: Academic Press. 282–311.

Chapter 5

5.A DESCRIPTIONS AND ANALYSIS OF CLIMATIC REGIONS

ALLEN, R. A., FLETCHER, R., HOLMBOE, J., NAMIAS, J. and WILLETT, H. C. 1940. *Report on an experiment in five-day weather forecasting.* Pap. Phys. Oceanog. Met. (Mass. Inst. Tech., Woods Hole Oceanog. Inst.) 8 (3). 94 pp.

ASAKURA, T. 1968. Dynamic climatology of atmospheric circulation over East Asia centered in Japan. *Pap. Met. Geophys.* (Tokyo) 19, 1–69.

BARRY, R. G. and CHORLEY, R. J. 1971. *Atmosphere, weather and climate.* 2nd ed. London: Methuen. 379 pp.

BARTELS, J. 1948. Anschauliches über den statistischen Hintergrund der sogenannten Singularität im Jahresgang der Witterung. *Ann. Met.* 1, 106–27.

BAUR, F. 1947. *Musterbeispiele europäischer Grosswetterlagen.* Wiesbaden: Dieterich. 35 pp.

BAUR, F. 1948. Zur Frage der Echtheit der sogenannten Singularitäten im Jahresgang der Witterung. *Ann. Met.* 1, 372–78.

BAUR, F. 1958. *Physikalische-statistische Regeln als Grundlagen für Wetter und Witterungsvorhersagen.* vol. 2. Frankfurt-am-Main: Akad. Verlagsgesellschaft. 152 pp.

BAYER, K. 1959. Witterungssingularitäten und allgemeine Zirkulation der Erdatmosphere. *Geofys. Sbornik* (Prague) No. 125, 521–634.

BIGG, E. K. 1957. January anomalies in cirriform cloud coverage over Australia. *J. Met.* 14, 525–6.

BILWILLER, R. 1884. Die Kälterückfälle im Mai. *Zeit. österreich Gessell. f. Met.* (Vienna) 19, 245–6.

BOWEN, E. G. 1956. A relation between meteor showers and the rainfall of November and December. *Tellus* 8, 394–402.

BOWEN, E. G. 1957. Relation between meteor showers and the rainfall. *Aust. J. Phys.* 10, 402–17.

BRADKA, J. 1966. Natural seasons in the northern hemisphere. *Geofys. Sbornik* (Prague) 14, 597–648.

BRANDES, H. W. 1820. Untersuchungen über den mittleren Gange der warme Änderungen durchs ganze Jahre. *Beiträge zur Witterungkunde.* Leipzig.

BRANDES, H. W. 1826. Bemerkungen über die Zeitpunkte grösser Kälte nach der Mitte des Winters. *Unterhaltungen f. Freunde d. Physik u. Astron.* 2, 148–59.

BRIER, G. W., SHAPIRO, R. and MACDONALD, N. J. 1963. A search for rainfall calendaricities. *J. Atm. Sci.* 20, 529–32.

BROOKS, C. E. P. 1946. Annual recurrences of weather: singularities. *Weather* 1, 107–13, 130–4.

BROOKS, C. E. P. 1954. *The English climate.* London: English Universities Press. ch. 9.

BRYSON, R. A. and LAHEY, J. F. 1958. *The march of the seasons.* Final Rep., Contract AF 19-(604)-992. Madison: Met. Dept, Univ. of Wisconsin. 41 pp.

BRYSON, R. A. and LOWRY, W. P. 1955. Synoptic climatology of the Arizona summer precipitation singularity. *Bull. Amer. Met. Soc.* 36, 329–39.

BUCHAN, A. 1867–9. Interruptions in the regular rise and fall of temperature in the course of the year. *J. Scot. Met. Soc.* new ser. 2, 3–15, 41–51, 107–12.

CHU, C.-C. 1962. The pulsation of world climate during historical times. *Acta Met. Sinica* 31, 275–88.

CRADDOCK, J. M. 1956. The representation of the annual temperature variation over central and northern Europe by a two-term harmonic form. *Quart. J. R. Met. Soc.* 82, 275–88.

CRADDOCK, J. M. 1957. The serial correlations of daily mean temperatures at Kew Observatory. *J. Met. Soc. Japan* 75, 350–64.

CRADDOCK, J. M. 1963. Persistent temperature regimes at Kew. *Quart. J. R. Met. Soc.* 89, 461–8.

DE LA MOTHE, P. D. 1968. Middle latitude wavelength variation at 500 mb. *Met. Mag.* 97, 333–9.

DOVE, H. W. 1857. Über die Rückfälle der Kälte in Mai. *Abh. Königl. Preuss. Akad. Wissenschaften* (Berlin) 121–92.

DUQUET, R. T. 1963. The January warm spell and associated large-scale circulation changes. *Mon. Wea. Rev.* 91, 47–60.

DZERDZEEVSKI, B. L. 1957. Circulation schemes for the seasons in the northern hemisphere. *Izv. Akad. Nauk. SSSR Ser. Geogr.* No. 1, 36 (in Russian).

EHRLICH, A. 1954. Synoptic patterns associated with singularities in the northeastern Atlantic. *Bull. Amer. Met. Soc.* 35, 215–19.

FLOHN, H. 1941. Über Begriff und Wesen der Singularitäten der Witterung. *Met. Zeit.* 58, 229–33.

FLOHN, H. 1947. Indiansommer-Altweibersommer. *Met. Rund.* 1, 282–7.

FLOHN, H. and HESS, P. 1949. Grosswettersingularitäten in jährlichen Witterungsverlauf Mitteleuropas. *Met. Rund.* 2, 258–63.

FREDERICK, R. H. 1966. Geographical distribution of a 30-year mean January warm spell. *Weather* 21, 9–15.

FULTZ, D. *et al.* 1959. *Studies of thermal convection in a rotating cylinder with some implications for large-scale atmospheric motions.* Met. Monogr. (Amer. Met. Soc.) 4 (21). 104 pp.

FULTZ, D. *et al.* 1964. *Experimental investigations of the spectrum of thermal convective motions in a rotating annulus.* Final Report, Contract AF 19(604)-8361, Dept Geophys. Sciences, Univ. of Chicago, Article 2B.

HOUGHTON, D. H. 1958. Heat sources and sinks at the earth's surface. *Met. Mag.* 87, 132–43.

INWARDS, R. L. 1950. *Weather Lore.* 4th ed. London: Rider and Co. 251 pp.

JAMES, R. W. 1970. Air mass climatology. *Met. Rund.* 23, 65–70.

KARAPIPERIS, L. N. 1953. On the spells of cold weather over the eastern Mediterranean during the autumn. *Arch. Met. Geophys. Biokl.* B, 41, 420–30.

KLINE, D. B. and BRIER, G. W. 1958. A note on freezing nuclei anomalies. *Mon. Wea. Rev.* 86, 329–33.

KÖPPEN, W. 1884. Zur Frage der 'gestrengen Herren'. *Zeit. österreich. Gesell. f. Met.* (Vienna) 18, 447–58.

KUPFER, E. 1948. Mittlere Hohenwetterkarten von Europa an Singularen Daten. *Zeit. Met.* 269–75.

LAMB, H. H. 1950. Types and spells of weather around the year in the British Isles: annual trends, seasonal structure of the year, singularities. *Quart. J. R. Met. Soc.* 76, 393–429.

LAMB, H. H. 1955. Two-way relationships between the snow or ice-limit and 1000–500 mb thickness in the overlying atmosphere. *Quart. J. R. Met. Soc.* 81, 172–89.

LAMB, H. H. 1964. *The English climate.* 2nd ed. London: English Universities Press. 212 pp.

LAMB, H. H. and JOHNSON, A. I. 1959. Climatic variation and observed changes in the general circulation, I and II. *Geogr. Annal.* 41, 94–134.

LAMB, H. H. and JOHNSON, A. I. 1961. Climatic variation and observed changes in the general circulation, III. *Geogr. Annal.* 43, 363–400.

LANCASTER, A. 1886–7. L'été de la Saint Martin. *Ciel et Terre* (Brussels) 2, 447–54.

LAWRENCE, E. N. 1954. Application of mathematical series to the frequency of weather spells. *Met. Mag.* 83, 195–200.

LAWRENCE, E. N. 1957. Estimation of the frequency of 'runs of dry days'. *Met. Mag.* 86, 257–69, 301–4.

LEHMANN, A. 1911. Altweibersommer: die Wärmerückfälle des Herbstes in Mitteleuropa. *Landwirtschaftliche Jährbücher, Zeit. f. wiss. Landwirtschaft* (Berlin) 41, 57–129.

LEHMANN, M. 1950. Die Hochdruckslagen als Singularitäten im Witterungsablauf von Mitteldeutschland. *Zeit. Met.* 4, 21–31.

LETTAU, H. 1947. Spezifische Singularitäten. *Met. Rund.* 1, 152–5.

LIU, K. and WU, H. 1956. A preliminary study of the determination of the natural synoptic summer season over the Asiatic natural synoptic region and the prevailing weather processes in this season. *Acta Met. Sinica* 27, 219–42.

LONGLEY, R. W. 1953. The length of wet and dry spells. *Quart. J. R. Met. Soc.* 79, 520–7.

LORENZ, E. N. 1967. *The nature and theory of the general circulation of the atmosphere* (W.M.O. No. 218, TP 115). Geneva: World Meteorological Organization. 161 pp.

MCINTOSH, D. H. 1953. Annual recurrences in Edinburgh temperature. *Quart. J. R. Met. Soc.* 79, 262–71.

MADDEN, R. A. and JULIAN, P. R. 1971. Detection of a 40–50 day oscillation in the zonal wind in the tropical Pacific. *J. Atm. Sci.* 28, 702–8.

MÄDLER, J. H. 1843. Über den Gange der Temperature im Laufe des Jahres. *Jahrbuch für 1843.* Stuttgart: Schumacher. 70–122.

MAEJIMA, I. 1967. Natural seasons and weather singularities in Japan. *Geogr. Rep.* No. 2. Tokyo: Metropolitan Univ. 77–103.

MARVIN, C. F. 1919. Normal temperatures (daily): are irregularities in the annual march of temperature persistent? *Mon. Wea. Rev.* 47, 544–55.

NAGAO, T. 1957, 1960. Turbulence theory of singularities. *Geophys. Mag.* (Tokyo) 28, 45–88; 29, 583–632; 30, 287–309, 311–54.

NAMIAS, J. 1952. The annual course of month-to-month persistence in climatic anomalies. *Bull. Amer. Met. Soc.* 32, 279–85.

NAMIAS, J. 1968. A late November singularity. In Court, A. (ed.) *Eclectic Climatology.* Corvallis: Univ. of Oregon. 55–62.

NAMIAS, J. and MORDY, W. A. 1952. The February minimum in Hawaiian rainfall as a manifestation of the primary index-cycle of the general circulation. *J. Met.* 9, 180–186.

PILGRAM, A. 1788. *Untersuchugen uber das Wahrscheinliche in der Witterungskunde durch vieljährige Beobachtungen.* Vienna.

POGADE, G. 1948. Singularitäten im Spiegel der Grosswetterlagen zur Zeit der Sommermonsunbeginns in Europa. *Ann. Met.* 1, 182–90.

REX, D. F. 1950. The effect of Atlantic blocking upon European climate. *Tellus* 2, 196–211, 275–301.

SAKATA, K. 1950, 1953. A new classification of seasons. *J. Met. Res.* (Tokyo) 2, 182–90, 903–12 (in Japanese).

SCHMAUSS, A. 1928. Singularitäten in jährlichen Witterungsverlauf von München. *Dtsch. Met. Jahrbuch* (Bayern) 50. 22 pp.

SCHMAUSS, A. 1932. Der Sinn der Singularitätenforschung. *Z. angew. Met.* 49, 97–107.

SCHMAUSS, A. 1938. Synoptische Singularitäten. *Met. Zeit.* 55, 384–403.

SCHMAUSS, A. 1941. Kalendermässige Bindungen des Wetters (Singularitäten). *Z. angew. Met.* 58, 237–44, 373–6.

TAKAHASHI, K. 1955. *Dynamic climatology.* Tokyo: Iwanami Book Co. 316 pp. (in Japanese).

TALMAN, C. F. 1919. Literature concerning supposed recurrent irregularities in the annual march of temperature. *Mon. Wea. Rev.* 47, 555–65.

VEITCH, L. G. 1965. The description of Australian pressure fields by principal components. *Quart. J. R. Met. Soc.* 91, 184–95.

WAHL, E. W. 1952. The January thaw in New England: an example of a weather singularity. *Bull. Amer. Met. Soc.* 33, 380–6.

WAHL, E. W. 1953. Singularities and the general circulation. *J. Met.* 10, 42–5.

WAHL, E. W. 1954. A weather singularity over the United States in October. *Bull. Amer. Met. Soc.* 35, 351–6.

WINSTON, J. S. 1954. The annual course of zonal wind speed at 700 mb. *Bull. Amer. Met. Soc.* 35, 468–71.

YARHAM, E. R. 1966. Weather lore of the twelve months. *Weather* 21, 433–9.

YEH, T. C., DAO, S.-Y. and LI, M.-T. 1959. The abrupt change of circulation over the northern hemisphere during June and October. In Bolin, B. (ed.) *The atmosphere and the sea in motion.* New York: Rockfeller Inst. Press. 249–62.

YOSHINO, M. M. 1968. Pressure pattern calendar of east Asia. *Met. Rund.* 21, 162–9.

ZAHAROVA, N. M. 1969. The 500 mb height field in the northern hemisphere in natural synoptic seasons. *Trudy Gidromet. Nauk. Issled. Tsentr. SSSR* (Leningrad) 43, 23–56 (in Russian).

ZIMMER, F. 1941. Der Wert der Bauernregel über den jährlichen Temperaturgang und die Witterungsabschnitte des Jahres. *Met. Zeit.* 58, 210–19, 464.

5.B DESCRIPTION AND ANALYSIS OF SPATIAL CHARACTERISTICS OF CLIMATE

5.B.1 *Regional climatic characteristics*

AELION, E. 1958. *A report on weather types causing marked storms in Israel during the cold season.* Misc. Rep. No. 10, ser. C. Hakirya: Israel Met. Service.

BARRY, R. G. 1963. Aspects of the synoptic climatology of central south England. *Met. Mag.* 92, 300–8.

BARRY, R. G. 1967. Variations in the content and flux of water vapour over eastern North America during two winter seasons. *Quart. J. R. Met. Soc.* 93, 535–43.

BARRY, R. G. 1970. A synoptic climatological diagram. *Rev. Géogr. Montréal* 24, 340–1.

BELASCO, J. E. 1952. *Characteristics of air masses over the British Isles.* Geophys. Mem. 11 (87). 34 pp.

BLANCHET, G. 1959. Masses d'air et types de temps dans le Couloir Rhodanien. *Ber. dtsch. Wetterd.* 8 (54), 197–202.

BLÜTHGEN, J. 1942. Kaltlufteinbrüche im Winter des atlantischen Europas. *Geogr. Zeit.* 48, 21–46.

BLÜTHGEN, J. 1966. *Allgemeine Klimageographie.* 2nd ed. Berlin: W. de Gruyter. 720 pp.

BÖER, W. 1954. Über den Zusammenhang zwischen Grosswetterlagen und extremen Abweichung der Monats mitteltemperaturen. *Zeit. Met.* 8, 11–16.

BOLOTINSKAIA, M. and RYZAKOV, L. 1964. *Catalogue of macrosynoptic processes according to the classification of Vangengeim, 1881–1962.* Leningrad: Arkt. i. Antarkt. Nauch.-Issled Inst. 158 pp. (in Russian).

BÜRGER, H. 1952. Über die Temperaturen der Grosswetterlagen. *Ber. dtsch. Wetterd. U.S. Zone* No. 42, 146 pp.

BÜRGER, H. 1958. Zur Klimatologie der Grosswetterlagen: ein Witterungsklimatologischer Beitrag. *Ber. dtsch Wetterd.* 6 (45), 79 pp.

CALIFORNIA INSTITUTE OF TECHNOLOGY. 1943. *Synoptic weather types of North America.* Pasadena, Calif.: Met. Dept, Calif. Inst. Tech. 243 pp.

CLERGET, M. 1937. Les types de temps en Mediterranée. *Ann. de Géogr.* 46, 225–46.

DZERDZEEVSKI, B. L. 1968. Circulation of the atmosphere: circulation mechanisms of the atmosphere in the northern hemisphere in the twentieth century. *Results of Meteorological Investigations, IGY Committee.* Moscow: Inst. of Geography, Akad. Nauk SSSR (in Russian).

FÉNELON, P. 1951. Types de temps australiens. *Ann. de Géogr.* 60, 288–94.

FLIRI, F. 1962. *Wetterlagenkunde von Tirol.* Tiroler Wirtschaftstudien No. 13. Innsbruck. 436 pp.

FLIRI, F. 1965a. Synoptische Klimadiagramme. *Erde* 96, 122–35.

FLIRI, F. 1965b. Über Signifikanzen synoptisch-klimatologischer Mittelwerte in verschiedenen Alpinen Wetterlagensystemen. *Carinthia II* (Vienna) 24, 36–48.

FLOHN, H. 1954. *Witterung und Klima in Mitteleuropa.* Forsch. dt. Landeskunde 78. Stuttgart. 214 pp.

FLOHN, H. and HUTTARY, J. 1950. Über die Bedeutung der Vb-Lagen für das Niederschlagsregime Mitteleuropas. *Met. Rund.* 3, 167–70.

FOJT, W. 1967. Zur verteilung hoher Tagessummen des Niederschlages im Erzgebirgsraum. *Zeit. Met.* 19, 290–300.

GARNIER, B. J. 1958. *The climate of New Zealand.* London: Arnold. 191 pp.

GAZZOLA, A. 1969. First results of an investigation of precipitation in Italy in relation to the meteorological situation. *Rivista Met. Aeronaut.* (Rome) 29, suppl. to No. 4, 84–114 (in Italian).

GRUTTER, M. 1966. Die bemerkenswerten Niederschlage der Jahre 1948–64 in der Schweiz. *Veröff. Schweiz. Met. Zentr. Aust.* (Zurich) 3. 20 pp.

HESSE, W. 1953. Wintertypen: eine synoptisch-klimatologische Studie über die letzten 101 Winter in Nordwestsachsen. *Zeit. Met.* 7, 362–73.

HEYER, E. 1955. Die wetterlagenmässige Betrachtung strenger Winter. *Zeit. Met.* 9, 85–8.

HOWELL, W. E. 1953. A study of the rainfall of central Cuba. *J. Met.* 10, 270–8.

JOHNSON, D. H. and MÖRTH, H. T. 1960. Forecasting research in east Africa. In Bargman, D. J. (ed.) *Tropical meteorology in Africa.* Nairobi: Munitalp Foundation. 56–132.

KERN, H. 1963. Grosse Tagessummen des Niederschlage am bayerischen Alpennordrand in Beziehung zur Grosswetterlage. *Geof. e. Met.* (Genoa) 11, 191–4.

KIRSTEN, H. 1960. Ein Beitrag zur synoptischen Klimatologie von Starkniederschlägen in Thüringen. *Zeit. Met.* 14, 286–96.

KOSSOWSKI, J. 1968. Frequency of occurrence of the principal weather types over Poland. *Przeglad Geofiz.* (Warsaw) 14, 286–96 (in Polish).

LAMB, H. H. 1950. Types and spells of weather around the year in the British Isles: annual trends, seasonal structure of the year, singularities. *Quart. J. R. Met. Soc.* 76, 393–429

LAMB, H. H. 1972. British Isles weather types and a register of the daily sequence of circulation patterns, 1861–1971. *Geophys. Mem.* (London), 16 (116), 85 pp.

LAWRENCE, E. N. 1971. Synoptic type averages over England and Wales. *Met. Mag.* 100, 333–9.

MAEDE, H. 1951. Zur Frage der Abgrenzung des Wirkungsbereiches einer Wetterlage. *Zeit. Met.* 5, 268–73.

METEOROLOGICAL OFFICE. 1949. Aviation meteorology of the Azores. *Met. Rep.* 1 (2), 10.

METEOROLOGICAL OFFICE. 1962. *Weather in the Mediterranean,* vol. 1: *General meteorology.* 2nd ed. M.O. 391. London: H.M.S.O. 352 pp.

METEOROLOGICAL OFFICE. 1963. *Weather in the Black Sea.* M.O. 706. London: H.M.S.O. 294 pp.

MÉZIN, M. 1945. Évolution des types de temps en Europe occidentale. *Congrès de la victoire.* vol. 2. Paris: A.F.A.S. 171–83.

MOSIÑO, P. A. 1964. Surface weather and upper air flow patterns in Mexico. *Geofis. Internac.* 4, 117–68.

MÜLLER, W. 1969. Eine agrarmeteorologische Pentadensprognose. *Wetter u. Leben* 21, 51–67.

MURRAY, R. and LEWIS, R. P. W. 1966. Some aspects of the synoptic climatology of the British Isles as measured by simple indices. *Met. Mag.* 95, 193–203.

NOIN, D. 1963. Types de temps d'été au Maroc. *Ann. de Géogr.* 72, 1–12.

OSMAN, O. E. and HASTENRATH, S. L. 1969. On the synoptic climatology of summer rainfall over central Sudan. *Arch. Met. Geophys. Biokl.* B, 17, 297–324.

PÉDÉLABORDE, P. 1957, 1958. *Le climat du bassin parisien: Essai d'une méthode rationelle de climatologie physique.* 2 vols. Paris: Genin. 534 and 116 pp.

PÉDÉLABORDE, P. and DELANNOY, H. 1958. Recherches sur les types de temps et le mécanisme des pluies in Algerie. *Ann. de Géogr.* 67, 216–44.

PENZAR, B. 1967. Some characteristics of weather types over the Adriatic Sea. *Hydrofski Inst. Godisnajk* (Split), 99–124 (in Serbo-Croat).

PERRY, A. H. 1968. The regional variation of climatological characteristics with synoptic indices. *Weather* 23, 325–30.

PERRY, A. H. 1969. The P.S.C.M. index and regional anomalies of temperature rainfall and sunshine. *Weather* 24, 225–8.

PUTNINS, P. 1966. *The sequences of baric weather patterns over Alaska.* Studies on the meteorology of Alaska, 1st interim rep. Washington, D.C.: Environmental Data Service, ESSA. 81 pp.

PUTNINS, P. and LANGDON, L. C. 1969. *Weather situations in Alaska during the occurrence of specific baric pressure patterns.* Studies on the meteorology of Alaska. Final Rep. Washington, D.C.: Environmental Data Service, ESSA. 267 pp.

RAGHAVAN, J. 1967. A climatological study of severe cold waves in India. *Ind. J. Met. Geophys.* 18, 91–6.

SCHÜEPP, M. 1968. Kalendarien der Wetter- und Witterungslagen von 1955 bis 1967. *Veröff. Schweiz. Met. Zentr. Anst.* (Zurich) 11. 43 pp.

SCHÜEPP, M. and FLIRI, F. 1967. Witterungsklimatologie. *Veröff. Schweiz. Met. Zentr. Anst.* (Zurich) 4, 215–29.

SHKLYAEV, A. S. and ALIKHINA, I. Y. 1968. Characteristics of temperatures in the central and south Urals with the G.Ia. Vangengeim basic forms of circulation. *Collection of Papers, Sverdlovsk Hydromet. Obs.* 7, 63–75.

SNEAD, R. E. 1968. Weather patterns in southern West Pakistan. *Arch. Met. Geophys. Biokl.* B, 16, 316–46.

VOROBIEVA, E. V. 1967. Characteristics of precipitation for the basic forms of atmospheric circulation. *Trudy Glav. Geofiz. Obs.* (Leningrad) 211, 81–93 (in Russian).

VOWINCKEL, E. 1955. Beitrag zur Witterungs Klimatologie Sudafrikas. *Arch. Met. Geophys. Biokl.* B, 7, 11–31.

VULQUIN, A. 1971. Arguments en faveur d'une mousson en Amazonie. *Tellus* 23, 74–81.

WALKER, E. R. 1961. *A synoptic climatology for parts of the Western Cordillera.* Pub. in Met. No. 35. Montreal: McGill Univ. 218 pp.

WORTHLEY, L. S. 1967. *Synoptic climatology of Hawaii, Part 1: Weather phenomena in Hawaii.* Hawaii Inst. Geophysics Pub. No. 67–9. 40 pp.

YEH, T. C., WALLÉN, C. C. and CARSON, J. E. 1951. *A study of rainfall over Oahu.* Met. Monogr. (Amer. Met. Soc.) 1 (3). 34–55.

YOSHINO, M. 1968. Pressure pattern calendar of east Asia. *Met. Rund.* 21, 162–9.

5.B.2 *Local climatic characteristics*

ATKINSON, B. W. 1966a. *Thunder outbreaks over southeast England 1951–60.* Unpub. Ph.D. thesis, Univ. of London.

ATKINSON, B. W. 1966b. Some synoptic aspects of thunder outbreaks over southeast England during 1951–60. *Weather* 21, 203–9.

ATKINSON, B. W. 1968. A preliminary examination of the possible effects of London's urban area on the distribution of thunder rainfall 1951–60. *Trans. Inst. Brit. Geogr.* No. 44, 97–118.

ATKINSON, B. W. 1969. A further examination of the urban maximum of thunder rainfall in London, 1951–60. *Trans. Inst. Brit. Geogr.* No. 48, 97–119.

BARRY, R. G. 1964. Weather and climate. In Monkhouse, F. J. (ed.) *A survey of Southampton and its region.* Univ. of Southampton. 73–92.

BARRY, R. G. 1967. The prospect for synoptic climatology: a case study. In Steel, R. W. and Lawton, R. (eds.) *Liverpool essays in geography.* London: Longmans. 85–106.

ERIKSEN, W. 1964. Das Stadtklima. *Erdkunde* 18, 257.

FLIRI, F. and DIMAI-FEUCHT, B. 1970. Die Witterungslagen in Innsbruck. *Wetter u. Leben* 22, 133–50.

FONTAINE, P. 1959. Nouvelles données sur l'enneigement moyen d'hiver et de printemps dans les Alpes françaises: essai de schématisation des types de temps correspondants. *Ber. dtsch. Wetterd.* 8 (54), 203–8.

FONTAINE, P. 1963. Contribution a l'étude hydrométéorologique des creus de l'haute Durance. *Geofis. e Met.* (Genoa) 11, 163–74.

GOEDECKE, E. 1955. Über die Bedeutung der Ost-Wetterlagen Mitteleuropas für die Vereisung der Deutschen Nord- und Ostseekusten. *Ann. Met.* new ser., 7, 386–403.

HOWE, G. M. 1953. Local climatic conditions in the Aberystwyth area. *Met. Mag.* 82, 270–4.

HUFTY, A. 1962. Influence des conditions météorologiques sur le mésoclimat à Florennes. *Bull. Soc. Belge Étud. Géogr.* 31, 309–20.

JACOBS, W. C. 1947. *Wartime developments in applied climatology.* Met. Monogr. (Amer. Met. Soc.) 1 (1). 52 pp.

LANDSBERG, H. E. 1937. Air mass climate for central Pennsylvania. *Gerlands Beitr. Geophys.* 51, 278–85.

LAWRENCE, E. N. 1966. Sunspots: a clue to bad smog? *Weather* 21, 367–70.

LAWRENCE, E. N. 1971. Synoptic-type rainfall averages over England and Wales. *Met. Mag.* 100, 333–9.

LAWRENCE, E. N. 1972. Westerly-type rainfall and atmospheric mean-sea-level pressure over England and Wales. *Met. Mag.* 101, 129–37.

LOWNDES, C. A. S. 1968. Forecasting large 24-hour rainfall totals in the Dee and Clwyd river authority area from September to February. *Met. Mag.* 97, 226–35.

LOWNDES, C. A. S. 1969. Forecasting large 24-hour rainfall totals in the Dee and Clwyd River Authority area from March to August. *Met. Mag.* 98, 325–40.

MAEDE, H. 1952. Klimatologische Untersuchungen über das Verhalten der Westwetterlagen in Raum der südlichen Ostsee und in norddeutschen Flachland. *Zeit. Met.* 6, 291–303.

MAEDE, H. 1953a. Das Verhalten der Nordwest- und Nordlagen in Raum der südlichen Ostsee und des agrenzenden Flachlandes klimatologisch betrachtet. *Zeit. Met.* 7, 48–57.

MAEDE, H. 1953b. Tiefkern-Vb-, zyklonale Ost- und Südlagen in Raum der südlichen Ostsee und des agrenzenden Flachlandes in klimatologischer Betrachtung. *Zeit. Met.* 7, 117–23.

MAEDE, H. 1953c. Die Hochdruckwetterlagen in Raum der südlichen Ostsee und des angrenzenden Flachlandes in klimatologischer Betrachtung. *Zeit. Met.* 7, 129–46.

MAEDE, H. and MATZKE, H. 1952. Die Globalstrahlung in Zusamnenghang mit typischen Wetterlagen, dargestellt am Beispeil von Greifswald. *Zeit Met.* 6, 15–22.

MUNN, R. E., HIRT, M. S. and FINDLAY, B. F. 1969. A climatological study of the urban temperature anomaly in the lakeshore environment at Toronto. *J. Appl. Met.* 8, 411–22.

PECZELY, G. 1961. *Die klimatologische Charakterisierung der makrosynoptischen Lagen Ungarns.* Budapest: As Orszagos Meteorol. Intezet. 128 pp. (in Hungarian: German abstract 36–49).

PERRY, V. 1969. *Some characteristics of variations in the microclimate of the British coastlands.* Unpub. B.A. dissertation, Univ. of Southampton. 43 pp.

PICK, W. H. and WRIGHT, G. A. 1925. Meteorological characteristics associated with the northeasterly type of weather at Cranwell, Lincs. *Quart. J. R. Met. Soc.* 51, 260.

PICK, W. H. and WRIGHT, G. A. 1926a. Meteorological characteristics associated with the southwesterly type of weather at Cranwell, Lincs. *Quart J. R. Met. Soc.*, 52, 196–7.

PICK, W. H. and WRIGHT, G. A. 1926b. Meteorological characteristics associated with the northerly type of weather at Cranwell, Lincs. *Quart. J. R. Met. Soc.* 52, 197–8.

PICK, W. H. and WRIGHT, G. A. 1927. Meteorological characteristics associated with the northwesterly type of weather at Cranwell, Lincs. *Quart. J. R. Met. Soc.* 53, 39–40.

RICHARDSON, W. E. 1956. Temperature differences in the South Tyne Valley, with special reference to the effects of air mass. *Quart. J. R. Met. Soc.* 82, 342–8.

SCHMIDT, F. H. and VELDS, C. A. 1969. On the relation between changing meteorological circumstances and the decrease of the SO_2 concentration around Rotterdam. *Atmos. Env.* 1, 455–60.

THOMPSON, R. D. 1970. *The dynamic climatology of north-east New South Wales.* Unpub. Ph.D. thesis, Univ. of New England, N.S.W.

VELDS, C. A. 1970. Relation between SO_2 concentration and circulation type in Rotterdam and surroundings. In *Urban climates*. Tech. Note No. 108

(W.M.O. No. 254, TP 141). Geneva: World Meteorological Organization. 280–5.

WILMERS, F. 1968. Wettertypen für mikroklimatische Untersuchungen. *Arch. Met. Geophys. Biokl.* B, 6, 144–50.

5.C SYNOPTIC CLIMATOLOGY AND CLIMATIC CHANGE

ABRAMOV, R. V. 1966. Southern displacement of the North Atlantic low. *Dokl. Akad. Nauk SSSR* (Moscow) 166, 165–6 (in Russian).

ANGELL, J. K., KORSHOVER, J. and COTTEN, G. F. 1969. Quasi-biennial variations in the centres of action. *Mon. Wea. Rev.* 97, 867–72.

BARRY, R. G. 1960. The application of synoptic studies in palaeoclimatology: a case study for Labrador-Ungava. *Geogr. Annal.* A, 42, 36–44.

BARRY, R. G. 1966. Meteorological aspects of the glacial history of Labrador-Ungava with special reference to atmospheric vapour transport. *Geogr. Bull.* 8, 319–40.

BARRY, R. G. and CHORLEY, R. J. 1971. *Atmosphere, weather and climate.* 2nd ed. London: Methuen. 379 pp.

BARRY, R. G. and PERRY, A. H. 1969. 'Weather type' frequencies and the recent temperature fluctuation. *Nature* 222, 463–4.

BARRY, R. G. and PERRY, A. H. 1970. 'Weather type' frequencies and temperature fluctuations: a reply. *Nature* 226, 634.

BERGGREN, R., BOLIN, B. and ROSSBY, C. G. 1949. An aerological study of zonal motion, its perturbations and breakdown. *Tellus* 1, 14–37.

BJERKNES, J. 1966. A possible response of the atmospheric Hadley circulation to equatorial anomalies of ocean temperature. *Tellus* 18, 820–8.

BJERKNES, J. 1969. Atmospheric teleconnections from the equatorial Pacific. *Mon. Wea. Rev.* 97, 163–72.

BOLOTINSKAIA, M. and RYZAKOV, L. 1964. *Catalogue of macrosynoptic processes according to the classification of Vangengeim, 1891–1962.* Leningrad: Arkt. i. Antarkt. Nauch. Issled. Inst. 158 pp. (in Russian).

BREZOWSKY, H., FLOHN, H. and HESS, P. 1951. Some remarks on the climatology of blocking action. *Tellus* 3, 191–4.

BRINKMANN, W. A. R. and BARRY, R. G. 1972. Palaeoclimatological aspects of the synoptic climatology of Keewatin, Northwest Territories. *Palaeogeogr., Palaeoclim., Palaeoecol.* 11, 77–91.

BROWN, P. R. 1963. Climatic fluctuations over the oceans and in the tropical Atlantic. In *Changes of climate.* Arid Zone Res. No. 20. Paris: UNESCO. 109–24.

BRYSON, R. A. 1966. Air masses, streamlines and the boreal forest. *Geogr. Bull.* 8, 228–69.

BRYSON, R. A. and JULIAN, P. R. 1962. *Proceedings of the conference on the climate of the eleventh and sixteenth centuries.* NCAR Tech. Notes 63-1. Boulder, Colo.: Nat. Cen. for Atm. Res. 102 pp.

BRYSON, R. A. and WENDLAND, W. M. 1967. Tentative climatic patterns for some late glacial and postglacial episodes in central North America. In MAYER-OAKES, W. J. (ed.) *Life, land and water.* Occas. Pap. No. 1. Winnipeg: Anthropology Dept., Univ. of Manitoba. 271–98.

BUTZER, K. W. 1957a. Mediterranean pluvials and general circulation of the Pleistocene. *Geogr. Annal.* A, 39, 48–53.

BUTZER, K. W. 1957b. The recent climatic fluctuation in lower latitudes and the general circulation of the Pleistocene. *Geogr. Annal.* A, 39, 105–13.

BUTZER, K. W. 1958. Quaternary stratigraphy and climate in the Near East. *Bonn. Geogr. Abhand.* 24, 157 pp.

BUTZER, K. W. 1961. Climatic change in arid regions since the Pliocene. In Stamp, L. D. (ed.) *A history of land use in arid regions.* Arid Zone Research No. 17. Paris: UNESCO. 31–56.

BUTZER, K. W. 1971. *Environment and archaeology: an ecological approach to prehistory.* 2nd ed. Chicago: Aldine. 703 pp.

CHARNEY, J. G. 1969. On the intertropical convergence zone and the Hadley circulation of the atmosphere. *Proc. WMO/IUGG sympos. on numerical weather prediction.* Tech. Rep. No. 67. Tokyo: Japan Met. Agency. 73–9.

CURRY, L. 1962. Climatic change as a random series. *Ann. Assoc. Amer. Geogr.* 52, 21–31.

DERBYSHIRE, E. 1969. Approche synoptique de la circulation du dernier maximum glaciaire dans le sud-est de l'Australie; *Rev. Géogr. Phys. Geol. Dynam.* 11 (3), 341–62.

DERBYSHIRE, E. 1971. A synoptic approach to the atmospheric circulation of the Last-Glacial maximum in southeastern Australia. *Palaeogeogr., Palaeoclim., Palaeoecol.* 10, 103–24.

DOBERITZ, R. 1969. Kohärenzanalyse von Niederschlag und Wassertemperatur im tropischen Pazifischen Ozean. *Ber. dtsch. Wetterd.* (Offenbach) 15 (112). 22 pp.

DZERDZEEVSKI, B. L. 1963. Fluctuations of general circulation of the atmosphere and climate in the twentieth century. In *Changes of Climate.* Arid Zone Res. No. 20. Paris: UNESCO. 285–95.

DZERDZEEVSKI, B. L. 1968a. Climatic fluctuations and the problem of super long range forecasting. *Izv. Akad. Nauk., Ser. Geog.* (Moscow), 43–55 (in Russian).

DZERDZEEVSKI, B. L. 1968b. Circulation of the atmosphere: circulation mechanisms of the atmosphere of the northern hemisphere in the twentieth century. *Results of Meteorological Investigations, IGY Committee.* Moscow: Inst. of Geog. Akad. Nauk SSSR. 240 pp. (in Russian).

DZERDZEEVSKI, B. L. 1969. Climatic epochs in the 20th century and some comments on the analysis of past climates. In Wright, H. E. Jr. (ed.) *Quaternary Geology and Climate* (*Proc. 7th Cong*) *INQUA* 16. Nat. Acad. Sci. Publ. 1701. Washington 49–60.

DZERDZEEVSKI, B. L. 1970. Calendar of changes of ECM for 1967, 1968, 1969. In Caplygina, A. S. (ed.) The comparison of the characteristics of the atmospheric circulation over the northern hemisphere with the analogous characteristics for its sectors: Circulation of the atmosphere. *Results of Meteorological Investigations, IGY Committee.* Moscow: Inst. of Geog., Akad. Nauk. SSSR. 19–22 (in Russian).

FAEGRI, K. 1950. On the value of palaeoclimatological evidence. *Centen. Proc. R. Met. Soc.* (London), 188–95.

FLOHN, H. 1952. Allgemeime atmospharische Zirkulation und Paläoklimatologie. *Geol. Rund.* 40, 153–78.
FLOHN, H. 1953. Studies über die atmosphärische Zirkulation in der letzten Eiszeit. *Erdkunde* 7, 266–75.
FLOHN, H. 1969. Ein geophysikalisches Eiszeit-Modell. *Eiszeit. u. Gegenwart* 20, 204–31.
FRITTS, H. C., BLASING, T. J., HAYDEN, B. P. and KUTZBACH, J. E. 1971. Multivariate techniques for specifying tree-growth and climate relationships and for reconstructing anomalies in palaeoclimate. *J. Appl. Met.* 10, 845–64.
GIRS, A. A. 1966a. Intra-periodical transformations of the atmospheric circulation and their causes. In Girs, A. A. and Dydina, L. A. (eds.) *Contributions to long-range weather forecasting in the Arctic* (Leningrad, 1963). Jerusalem: Israel Prog. Sci. Trans. 13–45.
GIRS, A. A. 1966b. Heat regime of the Soviet Arctic related to the main atmospheric circulation patterns and their secular variations. In Fletcher, J. O. (ed.) *Proc. sympos. on arctic heat budget and atmospheric circulation.* Mem. No. RM-5233-NSF. Santa Monica, Calif.: Rand Corp. 75–110.
HESS, P. and BREZOWSKY, H. 1969. Katalog der Grosswetterlagen Europas. *Ber. dtsch Wetterd.* (Offenbach) 15 (113) 56 pp.
HOINKES, H. C. 1968. Glacier variations and weather. *J. Glaciol.* 7, 3–19.
IMBRIE, J. and KIPP, N. G. 1971. A new micropalaeontological method for quantitative palaeoclimatology: application to a late Pleistocene Caribbean core. In Turekian, K. K. (ed.) *The late Cenozoic glacial ages.* New Haven, Conn.: Yale Univ. Press 71–181.
JOHNSEN, S. J., DANSGAARD, W., CLAUSEN, H. B. and LANGWAY, C. C. 1970. Climatic oscillations, 1200–200 A.D. *Nature* 227, 482–83.
KINGTON, J. A. 1970. Preparation of weather charts from 1780. *Swansea Geogr.* 8, 11–14.
KRAUS, E. B. 1956. Secular changes of the standing circulation. *Quart. J. R. Met. Soc.* 82, 289–300.
KRAUS, E. B. 1958. Recent climatic changes. *Nature* 181, 666–8.
KUTZBACH, J. E. 1970. Large-scale features of monthly mean northern hemisphere anomaly maps of sea level pressure. *Mon. Wea. Rev.* 98, 708–16.
KUTZBACH, J. E., BRYSON, R. A. and SHEN, W. C. 1966. An evaluation of the thermal Rossby number in the Pleistocene. In Mitchell, J. M. Jr. (ed.) *Causes of climatic change.* Met. Monogr. (Amer. Met. Soc.), 8 (30) 134–8.
LA MARCHE, V. C. Jr. and FRITTS, H. C. 1971. Anomaly patterns of climate over the western United States, 1700–1930, derived from principal components analysis of tree-ring data. *Mon. Wea. Rev.* 99, 138–42.
LAMB, H. H. 1962. The climates of the eleventh and sixteenth centuries A.D. *Weather* 17, 381–9.
LAMB, H. H. 1963. What can we find out about the trend of our climate? *Weather* 18, 194–216.
LAMB, H. H. 1964. Fundamentals of climate. In Nairn, A. E. M. (ed.) *Problems in palaeoclimatology.* New York: Interscience, 8–44.

LAMB, H. H. 1965a. Frequency of weather types. *Weather* 20, 9–12.

LAMB, H. H. 1965b. The history of our climate: Wales. In *Climatic change with special reference to Wales and its agriculture.* Mem. No. 8, Symposia in Agric. Met. Aberystwyth. 1–18.

LAMB, H. H. 1966a. Climate in the 1960s: world's wind circulation reflected in prevailing temperatures, rainfall patterns and the levels of African lakes. *Geogr. J.* 132, 183–212.

LAMB, H. H. 1966b. *The changing climate: selected papers.* London: Methuen. 236 pp.

LAMB, H. H. 1969. The new look of climatology. *Nature* 223, 1209–15.

LAMB, H. H. 1972. British Isles weather types and a register of the daily sequence of circulation patterns, 1861–1971. *Geophys. Mem.* (London) 16 (116), 85 pp.

LAMB, H. H. and JOHNSON, A. I. 1959. Climatic variation and observed changes in the general circulation, I and II. *Geogr. Annal.* 41, 94–134.

LAMB, H. H. and JOHNSON, A. I. 1961. Climatic variation and observed changes in the general circulation, III. *Geogr. Annal.* 43, 363–400.

LAMB, H. H. and JOHNSON, A. I. 1966. *Secular variations of the atmospheric circulation since 1750.* Geophys. Mem. 14 (5), No. 110. 125 pp.

LAMB, H. H., LEWIS, R. P. W. and WOODROFFE, A. 1966. Atmospheric circulation and the mean climatic variables between 8000 and 0 B.C.: meteorological evidence. In Sawyer, J. S. (ed.) *World climate from 8000 to 0 B.C.* London: Roy. Met. Soc. 174–217.

LAMB, H. H. and WOODROFFE, A. 1970. Atmospheric circulation during the last Ice Age. *Quat. Res.* 1, 29–58.

LAWRENCE, E. N. 1970. Variation in weather-type temperature averages. *Nature* 226, 633–4.

LE ROY LADURIE, E. 1967. *Histoire du climat depuis l'an mil.* Paris: Flammarion. 378 pp. (Trans. *Times of feast, times of famine.* New York: Doubleday, 1971. 426 pp.)

LORENZ, E. N. 1965. On the possible reasons for long-period fluctuations of the general circulation. In Tech. Note No. 66 (W.M.O. No. 162, TP 79). Geneva: World Meteorological Organization. 203–11.

LORENZ, E. N. 1970. Climatic change as a mathematical problem. *J. Appl. Met.* 9, 325–9.

MAKSIMOV, I. V. and KARKLIN, V. P. 1969. Seasonal and perennial changes of the geographical position of the Siberian high. *Izv. Vses. Geog. Obsch.* (Moscow) 101, 320–30 (in Russian).

MAKSIMOV, I. V. and KARKLIN, V. P. 1970. Seasonal and long-term changes in the geographical location and the intensity of the Azores atmospheric pressure maximum. *Izv. Akad. Nauk. SSSR, Ser. Geogr.* No. 1, 17–23. (in Russian).

MAKSIMOV, I. V., SARUKHANIAN, E. I. and SMIRNOV, N. P. 1970. On the relation between the force of deformation and the movements of atmospheric centres of action. *Dokl. Akad. Nauk SSSR,* Ser. Matem. Fiz. (Moscow) 190, 1095–7 (in Russian).

MANABE, S. and BRYAN, K. 1969. Climate and the ocean circulation. *Mon. Wea. Rev.* 97, 739–827.

MANDELBROT, B. B. and WALLIS, J. R. 1969. Some long-run properties of geophysical records. *Water Resources Res.*, 5, 321–40.

MASON, B. J. 1970. Future developments in meteorology: an outlook to the year 2000. *Quart. J. R. Met. Soc.* 96, 349–68.

MATHER, J. R. 1954. The present climatic fluctuation and its bearing on a reconstruction of the Pleistocene. *Tellus* 6, 287–301.

MAUNDER, W. J. 1970. *The value of the weather*. London: Methuen. 388 pp.

MILANKOVITCH, M. 1969. *Canon of insolation and the ice-age problem*. Jerusalem: Israel Prog. Sci. Trans. 484 pp.

MURRAY, R. and BENWELL, P. R. 1970. PSCM indices in synoptic climatology and long-range forecasting. *Met. Mag.* 99, 232–44.

NAMIAS, J. 1957. Characteristics of cold winters and warm summers over Scandinavia in relation to the general circulation. *J. Met.* 14, 235–50.

NAMIAS, J. 1970. Climatic anomaly over the United States during the 1960s. *Science* 170, 741–3.

OLIVER, J. and KINGTON, J. A. 1970. The usefulness of ship's log books in the synoptic analysis of past climates. *Weather* 25, 520–7.

PECZELY, C. 1957. *Grosswetterlagen in Ungarn*. Orszagos Met. Intezet. (Budapest) No. 30. 86 pp.

PERRY, A. H. 1969. Circulation anomalies during July 1965, in relation to pressure patterns over the Atlantic-European sector. *Weather* 24, 19–22.

PERRY, A. H. 1970. Changes in duration and frequency of synoptic types over the British Isles. *Weather* 25, 123–6.

PERRY, A. H. and BARRY, R. G. 1973. Recent temperature changes due to changes in the frequency and average temperature of weather types over the British Isles. *Met. Mag.* 102 (in press).

QUINN, W. H. 1971. Late Quaternary meteorological and oceanographical developments in the Equatorial Pacific. *Nature* 229, 330–1.

RASOOL, S. I. and HOGAN, J. S. 1969. Ocean circulation and climatic changes. *Bull. Amer. Met. Soc.* 50, 130–4.

REX, D. F. 1950. Blocking action in the middle troposphere and its effects upon regional climate, II. *Tellus* 275–301.

ROSSBY, C.-G. and WILLETT, H. C. 1948. The circulation of the upper troposphere and the lower stratosphere. *Science* 108, 643, 652.

RUDLOFF, H. 1969. Die Abkühlung im nordpolaren auf die Grosswetterung in Zentraleuropa. *Met. Rund.* 22, 125–6.

SAVINA, S. S. 1970. Comparative description of extremal periods of circulation epoch over the European U.S.S.R. and west Siberia. In Lyakhov, M. E. (ed.) *Climatologic and circulation epochs in the northern hemisphere in the first half of the twentieth century*. Met. Res. No. 13 (Moscow, 1968). Jerusalem: Israel Prog. Sci. Trans. 116–23.

SAWYER, J. S. 1966. Possible variations in the general circulation of the atmosphere. In Sawyer, J. S. (ed.) *World climate from 8000 to 0 B.C.* London: Roy. Met. Soc. 218–29.

SCHERHAG, R. 1939. Die Erwarmung des Polargebiets. *Ann. Hydrogr. marit. Met.* 67, 57–67.

SELLERS, W. D. 1968. Climatology of monthly precipitation patterns in the western United States, 1931–66. *Mon. Wea. Rev.* 96, 585–95.

SIMPSON, G. C. 1957. Further studies in world climate. *Quart. J. R. Met. Soc.* 83, 459–85.

SORKINA, A. I. 1965. Types of atmospheric circulation and wind field over the North Atlantic. *Trudy Gos. Okean. Inst.* (Moscow) 84, 1–133 (in Russian).

TSUCHIYA, I. 1963. An analysis on the relationship between general circulation and climatic fluctuation, Part I. *J. Met. Soc. Japan* 41, 288–98.

TSUCHIYA, I. 1964. An analysis on the relationship between general circulation and climatic fluctuation, Part II. *J. Met. Soc. Japan* 42, 299–308.

WAGNER, A. J. 1971. Long-period variations in seasonal sea-level pressure over the northern hemisphere. *Mon. Wea. Rev.* 99, 49–66.

WALLÉN, C. C. 1953. The variability of summer temperature in Sweden and its connection with changes in the general circulation. *Tellus* 5, 157–78.

WEBB, T. III and BRYSON, R. A. 1972. Late glacial and postglacial climatic change in the northern midwest, U.S.A.: quantitative estimates derived from fossil pollen spectra by multivariate statistical analysis. *Quat. Res.* 2, 70–115.

WEGE, K. 1961. Die Änderung der Luftdruckverhaltnisse zwischen den Zeitraumen 1901–30 und 1931–60 am Beispel der deutschen Climatstationen. *Met. Rund.* 14, 138–42.

WEISS, I. and LAMB, H. H. 1970. *Die Zunahme der Wellenhöhen in jüngster Zeit in der Operationsgebieten der Bundesmarine, ihre vermütlichen Ursachen und ihre voraussichtliche weitere Entwicklung.* Fachliche Mitteilungen No. 160 (Porz-Wahn). 18 pp.

WILLETT, H. C. 1949. Long-period fluctuations of the general circulation of the atmosphere. *J. Met.* 6, 34–50.

WILLETT, H. C. 1950. The general circulation at the last (Würm) glacial maximum. *Geogr. Annal.* 32, 179–87.

WILLETT, H. C. and SANDERS, F. 1959. *Descriptive meteorology.* 2nd ed. New York: Academic Press. 355 pp.

YEVJEVICH, V. 1968. Reply to comments on 'Misconceptions in hydrology and their consequences'. *Water Resources Res.* 4, 1147.

5.D LONG-RANGE FORECASTING

ADEM, J. 1965. Experiments aimed at monthly and seasonal numerical weather prediction. *Mon. Wea. Rev.* 93, 495–503.

ANGELL, J. K. and KORSHOVER, J. 1962. The biennial wind and temperature oscillations of the stratosphere and their possible extension to higher latitudes. *Mon. Wea Rev.* 90, 127–32.

ANGELL, J. K. and KORSHOVER, J. 1968. Additional evidence for quasi-biennial variations in tropospheric parameters. *Mon. Wea. Rev.* 96, 778–84.

ARCHIBALD, D. 1898. *Weather types in relation to long-period forecasting.* London: Eyre & Spottiswoode. 19 pp.

ARCTOWSKI, H. B. 1910. La dynamique des anomalies climatiques: contribution à l'étude de la répartition de la pression atmosphérique aux États – Unis. *Odlitka z. Prac. Mat.-Frzyc.* (Warsaw) 21, 179–96.

BAUR, F. 1936. Die Bedeutung der Stratosphäre für die Grosswetterlage. *Met. Zeit.* 53, 237–47.

BAUR, F. 1951. Extended range weather forecasting. In Malone, T. F. (ed.) *Compendium of meteorology.* Boston, Mass.: Amer. Met. Soc. 814–33.

BAUR, F. 1956, 1958. *Physikalisch-statische Regeln als Grundlagen für Wetter – und Witterungsvorhersagen.* Frankfurt-am-Main: Akad. Verlagagesellschaft. Vol. 1, 138 pp.; Vol. 2, 152 pp.

BAUR, F. 1963. *Grosswetterkunde und langfristige Witterungsvorhersage.* Frankfurt-am-Main: Akad. Verlagsgesellschaft. 91 pp.

BELMONT, A. D. and DART, D. G. 1968. Variation with longitude of the quasi-biennial oscillation. *Mon. Wea. Rev.* 96, 767–77.

BERLAGE, H. P. 1957. Fluctuations of the general atmospheric circulation of more than one year, their nature and prognostic value. *Kon. Ned. Met. Inst., Meded. en Verhandel* (de Bilt) No. 69. 152 pp.

BERLAGE, H. P. 1966. The Southern Oscillation and world weather. *Kon. Ned. Met. Inst., Meded. en Verhandel* (de Bilt) No. 88. 152 pp.

BERLAGE, H. P. and DE BOER, H. J. 1959. On the extension of the Southern Oscillation throughout the world during the period 1 July 1949 up to 1 July 1957. *Geof. pura e appl.* 44, 287–95.

BERLAGE, H. P. and DE BOER, H. J. 1960. On the Southern Oscillation, its way of operation and how it affects pressure patterns in the higher latitudes. *Geof. pura e appl.* 46, 329–51.

BERSON, F. A. and KULKARNI, R. N. 1968. Sunspot cycle and the quasi-biennial oscillation. *Nature* 217, 1133–4.

BILHAM, E. G. 1934. Note on sequences of wet and dry months in England and Wales. *Quart. J. R. Met. Soc.* 60, 514–16.

BJERKNES, J. 1966. A possible response of the atmospheric Hadley circulation to equatorial anomalies of ocian temperature. *Tellus* 18, 820–8.

BJERKNES, J. 1969. Atmospheric teleconnections from the equatorial Pacific. *Mon. Wea. Rev.* 97, 163–72.

BLISS, E. W. and WALKER, G. T. 1932. World weather, V: some applications to seasonal foreshadowing. *Mem. R. Met. Soc.* (London) 4, 53–84.

BÖHME, W. 1967. Eine 26-monatige Schwankung der Haüfigkeit meridionaler Zirkulationsformen über Europe. *Zeit. Met.* 19, 113–15.

BOWEN, E. G. and ADDERLEY, E. E. 1962. Lunar component in precipitation data. *Science* 137, 749–50.

BREZOWSKY, H., FLOHN, H. and HESS, P. 1951. Some remarks on the climatology of blocking action. *Tellus* 3, 191–4.

BRIER, G. W. 1966. Evidence of a longer period tidal oscillation in the tropical atmosphere. *Quart. J. R. Met. Soc.* 92, 284–9.

BRIER, G. W. 1968. Long-range prediction of the zonal westerlies and some problems in data analysis. *Rev. Geophys.* 6, 525–51.

BRIER, G. W. and BRADLEY, D. A. 1964. The lunar synodical period and precipitation in the United States. *J. Atm. Sci.* 21, 386–95.

BRIER, G. W. and CARPENTER, T. 1967. A study of a non-deepening tropical disturbance. *J. Appl. Met.* 6, 425–6.

BRIER, G. W. and SIMPSON, J. 1969. Tropical cloudiness and rainfall related to pressure and tidal variations. *Quart. J. R. Met. Soc.* 95, 120–47.

BROOKS, C. E. P. 1928a. *Sunspots and the distribution of pressure over western Europe.* London: Met. Office Prof. Notes No. 49. 6 pp.

BROOKS, C. E. P. 1928b. *Periodicities in the Nile floods.* Mem. R. Met. Soc. 2 (12). 26 pp.

BROOKS, C. E. P. 1951. The relations of solar and meteorological phenomena. *Commission pour l'étude des relations entre les phénomenès solaires et terrestres.* Rep. 7, Paris: Int. Comm. Sci. Un. 183–98.

BROOKS, C. E. P. and GLASSPOOLE, J. 1922. The drought of 1921. *Quart. J. R. Met. Soc.* 48, 139–66.

BROOKS, C. E. P. and QUENNELL, W. A. 1926. *Classification of monthly charts of pressure anomaly.* Geophys. Mem. (London) 4 (31). 12 pp.

BROOKS, C. E. P. and QUENNELL, W. A. 1928. *The influence of arctic ice on the subsequent distribution of pressure over the eastern north Atlantic and western Europe.* Geophys. Mem. (London) 5 (4). 36 pp.

BRÜCKNER, E. 1890. Klimaschwankungen seit 1700 nebst Bemerkungen über die Klimaschwankungen der Diluvialzeit. *Geogr. Abh.* 4. 324 pp.

BRYSON, R. A. 1948. On a lunar bi-fortnightly tide in the atmosphere. *Trans. Amer. Geophys. Union* 29, 473–5.

BRYSON, R. A., LAHEY, J. F., KUHN, P. M. and HORN, L. H. 1955. *Singularities in the sequence of European 'Grosswetter' types.* Sci. Rep. No. 2, Contract AF 19 (604)-992. Madison: Met. Dept. Univ. of Wisconsin.

CHAPMAN, S. 1951. Atmospheric tides and oscillations. In Malone, T. F. (ed.) *Compendium of meteorology.* Boston, Mass.: Amer. Met. Soc. 510–30.

CLARKE, P. C. 1970. Arctic sea ice in summer. *Weather* 25, 215–18.

CLAYTON, H. H. 1885. A lately discovered meteorological cycle. *Amer. Met. J.* 1, 130, 528.

CLAYTON, H. H. 1936. Long-range weather changes and methods of forecasting. *Mon. Wea. Rev.* 64, 359–76.

CRADDOCK, J. M. 1958. Research in the Meteorological Office concerned with long-range forecasting. *Met. Mag.* 87, 239–49.

CRADDOCK, J. M. 1964. Long-range weather forecasting in Great Britain. *Met. Mag.* 93, 98–105.

CRADDOCK, J. M. 1968. *Atmospheric oscillations with periods of about two years.* Met. Office Syn. Clim. Branch Mem. No. 22. 19 pp.

CRADDOCK, J. M. 1970. Work in synoptic climatology with a digitized data bank. *Met. Mag.* 99, 221–31.

CRADDOCK, J. M., FLOHN, H. and NAMIAS, J. 1962. *The present status of long-range forecasting in the world.* Tech. Note No. 48 (W.M.O. No. 126, TP 56). Geneva: World Meteorological Organization. 23 pp.

CRADDOCK, J. M. and WARD, R. 1962. *Some statistical relationships between the temperature anomalies in neighbouring months in Europe and western Siberia.* London: Met. Office Sci. Pap. No. 12. 31 pp.

DARTT, D. G. and BELMONT, A. D. 1970. A global analysis of the variability of the quasi-biennial oscillation. *Quart. J. R. Met. Soc.* 96, 186–94.

DAVIS, N. E. 1967. The summers of northwest Europe. *Met. Mag.* 96, 178–87.

DE FELICE, P. 1970. Étude de la persistence des situations bariques à l'aide d'un catalogue de la carte quotidienne 500 mb. *Arch. Met. Geoph. Biokl.* A, 19, 227–34.

DEHSARA, M. and CEHAK, K. 1970. A global survey on periodicities in annual mean temperatures and precipitation totals. *Arch. Met. Geophys. Biokl.* B, 18, 253–68.

DICKSON, R. R. 1967. The climatological relationship between temperatures of successive months in the United States. *J. Appl. Met.* 6, 31–8.

DUVANIN, A. I. 1968. On a model of the interaction between large scale processes in the ocean and atmosphere. *Akad. Nauk Okeanograf. Kom. Okeanol.* (Moscow) 8, 571–80 (in Russian).

DYDINA, L. A. 1967. Development of a method of meteorological forecasting for 3–10 days for the Arctic. *Problemy Arkt. i. Antarkt.* No. 27, 114–22 (in Russian).

EASTON, C. 1918. Periodicity of winter temperatures in western Europe since A.D. 760. *Proc. Kon. Akad. v. Wet.* (Amsterdam) 20, 1092–107.

EASTWOOD, J. R. 1965. *An investigation of the latitude positions of the main zones of the circulation in relation to the 80–90 year and 170 year sunspot cycles.* Met. 0.13. Tech. Note No. 10. Bracknell: Met. Office.

ELLIOTT, R. D. 1942. *Studies of persistent regularities in weather phenomena.* Pasadena: Met. Dept, California Inst. Tech.

FLOHN, H. 1942. Kalendermassige Bindungen im Wettergeschehen. *Naturwissenschaften* 30, 718–28.

GEB, M. 1966. Synoptische-statistische Untersuchungen zur Einleitung blockierender Hochdrucklagen über dem Nordostatlantik und Europa *Met. Abhand.* 69 (1). 94 pp.

GIRS, A. A. 1956. Secular development of the form of the atmospheric circulation and fluctuations in solar activity. *Met. i. Gidrol.* (Leningrad) No. 10, 3–13 (in Russian).

GIRS, A. A. 1960a. *Principles of long-range weather forecasting.* Leningrad: Gidrometeoizdat. 560 pp. (in Russian).

GIRS, A. A. 1960b. The main results of the investigation of long-term variations of the general atmospheric circulation applicable to the problem of super long-range hydrometerological forecasts. *Trans. First Sci. Congress on General Circulation of the Atmosphere.* Moscow: Central Inst. Prognosis. 13–22 (in Russian).

GIRS, A. A. and DYDINA, L. A. (eds.). 1966. Contributions to long-range weather forecasting in the Arctic (*Trudy Arkt. i Antarkt. Nauch. Issled. Inst.* 255). Jerusalem: Israel Prog. Sci. Trans. 240 pp.

GROISSMAYR, F. B. 1930. Relations between winters in Manitoba and the following spring in eastern United States. *Mon. Wea. Rev.* 58, 246–7.

GRYTOYR, E. 1950. A method of selecting analogous synoptic situations, and the use of past synoptic situations in forecasting for east Norway. *Geofys. Publik.* (Oslo) 17. 28 pp.

HAURWITZ, B. and COWLEY, A. D. 1965. The lunar and solar tides at six stations in North and Central America. *Mon. Wea. Rev.* 93, 505–9.

HAWKE, E. 1934. Frequency distribution of wet and dry months from 1815 to 1914, and of warm and cold months from 1841 to 1930 at Greenwich. *Quart. J. R. Met. Soc.* 60, 71–3.

HAY, R. F. M. 1956. *Some aspects of air and sea surface temperature in short periods at O.W.S. 'I' and 'J' of significance to the synoptic climatologist.* Met. Res. Pap. (London) No. 949.

HAY, R. F. M. 1966. *Winter temperatures in Britain related to monthly mean depression centres in the Atlantic-European sector.* Mem. No. 14. Bracknell, Berks: Synoptic Climatology Branch, Met. Office. 7 pp.

HAY, R. F. M. 1967. The association between autumn and winter circulations near Britain. *Met. Mag.* 96, 167–78.

HAY, R. F. M. 1968. *Association between northern hemisphere pressure patterns in March and April pressure in Britain.* Mem. No. 18. Bracknell, Berks: Syn. Clim. Branch, Met. Office.

HAY R. F. M. 1969. An analysis of monthly mean pressure patterns near the British Isles with possible applications to seasonal forecasting. *Met. Mag.* 98, 357–64.

HAY, R. F. M. 1970. Further analysis of monthly mean pressure patterns near the British Isles (1874–1968). *Met. Mag.* 99, 189–98.

HELLAND-HANSEN, B. and NANSEN, F. 1920. *Temperature variations in the north Atlantic Ocean and in the atmosphere.* Smithsonian Misc. Coll. 70 (4). 408 pp.

HILDEBRANSSON, H. H. 1897. *Quelques recherches sur les centres d'action de l'atmosphère: écarts des moyennes barométriques mensuelles.* Kongl. Svenska Vetenscaps-akad. Handlinger 29. 36 pp.

HOWARD, L. 1842. *A cycle of 18 years in the seasons of Britain deduced from meteorological observations made at Ackworth in the West Riding of Yorkshire from 1824 to 1841.* London. 22 pp.

HURST, G. W. 1967. Honey production and summer temperature. *Met. Mag.* 96, 116–20.

JOHANSEN, H. 1958. On continental and oceanic influences on the atmosphere. *Met. Ann.* (Oslo) 4, 143–58.

JULIAN, P. R. 1966. The index cycle: a cross-spectral analysis of zonal index data. *Mon. Wea. Rev.* 94, 283–93.

KEELEY, D. A. and OLIEN, T. C. 1961. *Persistence in sequences of wet and dry months at Regina.* Met. Branch CIR-3573, TEC-384. Toronto: Dept of Transport. 12 pp.

KLEIN, W. H. 1948. Winter precipitation as related to the 700 mb circulation. *Bull. Amer. Met. Soc.* 29, 439–53.

KLEIN, W. H. 1963. Specification of precipitation from the 700 mb circulation. *Mon. Wea. Rev.* 91, 527–36.

KLEIN, W. H. 1965a. Synoptic climatological models for the United States. *Weatherwise* 18, 252–9.

KLEIN, W. H. 1965b. Five-day precipitation patterns derived from circulation and moisture. In Wexler, A. (ed.) *Humidity and moisture*, vol. 2: *Applications*. New York: Reinhold. 532–49.

KLEIN, W. H. 1965c. Application of synoptic climatology and existing numerical prediction to medium-range forecasting. In Tech. Note No.

66 (W.M.O. 162, TP 79). Geneva: World Meteorological Organization. 103–25.

KLEIN, W. H. 1967. Specification of sea-level pressure from 700 mb height. *Quart. J. R. Met. Soc.* 93, 214–26.

KLEIN, W. H. 1968. Precipitation and circulation patterns. In *Proceedings of the seventh Stanstead Seminar on the Middle Atmosphere.* Pub. in Met. 90, Montreal: Arctic Met. Res. Group, McGill Univ. 127–39.

KLEIN, W. H., CROCKETT, C. W., and ANDREWS, J. F. 1965. Objective prediction of daily precipitation and cloudiness. *J. Geophys. Res.* 70, 801–13.

KLEIN, W. H., LEWIS, B. M., and ENGER, I. 1959. Objective prediction of five-day mean temperatures during winter. *J. Met.* 16, 672–82.

KLEIN, W. H., LEWIS, F. and CASELY, G. P. 1967. Automated nationwide forecasts of maximum and minimum temperature. *J. Appl. Met.* 6, 216–28.

KONCEK, N. and CEHAK, K. 1969. Untersuchungen über Folgen von Zeiträumen mit unter-oder übernormaler temperature in Mitteleuropa. *Arch. Met. Geoph. Biokl.* B, 17, 429–38.

KÖPPEN, W. 1881. Über mehrjährige Perioden der Witterung. *Zeit Öster. Ges. Met.* 16, 140–83.

KRICK, P. 1942. *A dynamical theory of the atmospheric circulation and its uses in weather forecasting.* Pasadena: Met. Dept, California Inst. Tech. 43 pp.

KROWN, L. 1966. An approach to forecasting seasonal rainfall in Israel. *J. Appl. Met.* 5, 590–4.

KRUEGER, A. F. and GRAY, T. I. JR. 1969. Long-term variations in equatorial circulations. *Mon. Wea. Rev.* 97, 700–11.

KUDASHKIN, G. D. and IUDIN, M. I. 1968. Selection and use of analogues for more accurate numerical weather forecasts. *Trudy Glav. Geofiz. Obs.* (Leningrad) 197, 3–18 (in Russian).

LABITZKE, K. 1965. On the mutual relation between stratosphere and troposphere during periods of stratospheric warming in winter. *J. Appl. Met.* 4, 91–9.

LAMB, H. H. 1972. British Isles weather types and a register of the daily sequence of circulation patterns, 1861–1971. *Geophys. Mem.* (London), 16 (116), 85 pp.

LAMB, H. H. 1972. *Climate: Present, Past and Future,* vol. 1. London: Methuen. 613 pp.

LANDSBERG, H. E. 1962. Biennial pulses in the atmosphere. *Beitr. Phys. Atmos.* 35, 184–94.

LANDSBERG, H. E., MITCHELL, J. M. JR., CRUTCHER, H. L. and QUINLAN, F. T. 1963. Surface signs of the biennial atmospheric pulse. *Mon. Wea. Rev.* 91, 549–56.

LAWRENCE, E. N. 1965. Terrestrial climate and the solar cycle. *Weather* 334–43.

LINDZEN, R. S. and HOLTON, J. R. 1968. A theory of the quasi-biennial oscillation. *J. Atm. Sci.* 25, 1095–107.

LONDON, J. and HAURWITZ, M. W. 1963. Ozone and sunspots. *J. Geophys. Res.* 68, 795–801.

LORENZ, E. N. 1969. Atmospheric predictability as revealed by naturally occurring analogues. *J. Atm. Sci.* 26, 636–46.

LUND, I. A. 1965. Indication of a lunar synodic period in United States observations of sunshine. *J. Atm. Sci.* 22, 24–39.

MACDONALD, N. J. and ROBERTS, W. O. 1960. Further evidence of a solar corpuscular influence on large-scale circulation at 300 mb. *J. Geophys. Res.* 65, 529–34.

MANABE, S. and BRYAN, K. 1969. Climate and the ocean circulation. *Mon. Wea. Rev.* 97, 739–827.

MARTIN, D. E. and HAWKINS, H. F. JR. 1950. Forecasting the weather: the relationship of temperature and precipitation over the United States to the circulation aloft. *Weatherwise* 3, 16–19, 40–3, 65–7, 89–92, 113–16, 138–41.

MARSHALL, N. 1968. Icefields around Iceland. *Weather* 23, 368–76.

MERTZ, J. 1959. Succession et durée des regimes météorologiques sur les Alpes. *Ber. dtsch. Wetterd.* 8 (54), 231–3.

MILLER, R. G. 1958. *The screening procedure: a statistical procedure for screening predictors in multiple regression, Part II.* Contract AF19(604)-1590 (ed. Shorr, B.). Hartford, Conn.: Travelers Weather Res. Cen., 86–136.

MILLS, C. A. 1966. Extraterrestrial factors in the initiation of North American polar fronts. *Mon. Wea Rev.* 94, 313–18.

MITCHELL, J. M. JR. 1966. Stochastic models of air-sea interaction and climatic fluctuation. In Fletcher J. O. (ed.) *Proc. sympos. on the arctic heat budget and atmospheric circulation.* Mem. RM-5233-NSF. Santa Monica, Calif.: Rand Corp. 46–74.

MITCHELL, J. M. JR. 1968. Some remarks on the predictability of drought and related climatic anomalies. In *Proc. conf. on drought in the northeastern United States.* Rep. TR-68-3. Geophys. Sci. Lab., New York Univ. 129–59.

MONTALTO, M., CONTE, M. and URBANI, M. 1971. Climatologia sinnottica delle situazioni di blocco sulla regione Euro-atlantica. *Rivista Met. Aeronaut.* (Rome) 31, 157–67, 275–87.

MONTGOMERY, R. B. 1940. Report on the work of G. T. Walker. *Mon. Wea. Rev.* suppl., 39, 1–22.

MOOK, C. P. 1954. On the normal monthly variation of blocking patterns over the North Atlantic. *Bull. Amer. Met. Soc.* 35, 379–80.

MÜLLER-ANNEN, H 1961 Über die Schwankungen der Zonal-Zirkulation, 4: Die Häufigkeit der West-Grosswetterlagen in den Sonnenflecken-Zyklen. *Met. Rund.* 14, 38–47.

MULTANOVSKI, B. P. 1933. *Basic principles of the synoptic method of long-range weather forecasting.* Glav. Upravlen; Gidromet. Sluzhby SSSR, 129 pp (in Russian).

MURGATROYD, R. J. 1970. Structure and dynamics of the stratosphere. In Corby, G. A. (ed.) *The global circulation of the atmosphere.* London: Roy. Met. Soc. 159–95.

MURRAY, R. 1966. The blocking patterns of February 1965 and a good pressure analogue. *Weather* 21, 66–9.

MURRAY, R. 1967a. Cyclonic Junes over the British Isles and the synoptic character of the following Septembers. *Met. Mag.* 96, 65–9.

MURRAY, R. 1967b. Sequences in monthly rainfall over England and Wales. *Met. Mag.* 96, 129–35.

MURRAY, R. 1967c. Persistence in monthly mean temperature in central England. *Met. Mag.* 96, 356–63.

MURRAY, R. 1968a. Sequences in monthly rainfall over Scotland. *Met. Mag.* 97, 181–3.

MURRAY, R. 1968b. Some predictive relationships concerning seasonal rainfall over England and Wales and seasonal temperature in central England. *Met. Mag.* 97, 303–10.

MURRAY, R. 1968c. On certain large-scale synoptic features of the summer over the British Isles and autumn mean temperatures in central England and autumn rainfall over England and Wales. *Met. Mag.* 97, 321–7.

MURRAY, R. 1969. Prediction of monthly rainfall over England and Wales from 15-day and monthly mean troughs at 500 mb. *Met. Mag.* 98, 141–4.

MURRAY, R. 1970. Recent developments in long-range forecasting in the Meteorological Office. *Quart. J. R. Met. Soc.* 96, 329–36.

MURRAY, R. and BENWELL, P. R. 1970. PSCM indices in synoptic climatology and long-range forecasting. *Met. Mag.* 99, 232–45.

MURRAY, R. and LEWIS, R. P. W. 1966a. The limitations of analogue selected from sequences of daily weather types over the British Isles. *Quart. J. R. Met. Soc.* 92, 402–6.

MURRAY, R. and LEWIS, R. P. W. 1966b. Some aspects of the synoptic climatology of the British Isles as measured by simple indices. *Met. Mag.* 98, 201–19.

MURRAY, R. and MOFFITT, B. J. 1969. Monthly patterns of the quasi-biennial pressure oscillation. *Weather* 24, 382–9.

MURRAY, R. and RATCLIFFE, R. A. S. 1969. The summer of 1968: related atmospheric circulation and sea temperature patterns. *Met. Mag.* 98, 201–19.

NAMIAS, J. 1951. General aspects of extended-range forecasting. In Malone, T. F. (ed.) *Compendium of meteorology*. Boston, Mass.: Amer. Met. Soc. 802–13.

NAMIAS, J. 1953. *Thirty-day forecasting: a review of a ten-year experiment.* Met. Monogr. (Amer. Met. Soc.) 2 (6). 83 pp.

NAMIAS, J. 1964. Seasonal persistence and recurrence of European blocking during 1958–60. *Tellus* 16, 394–407.

NAMIAS, J. 1968. Long-range weather forecasting: history, current status and outlook. *Bull. Amer. Met. Soc.* 49, 438–70.

NAMIAS, J. 1969. Autumnal variations in the North Pacific and North Atlantic anticyclones as manifestations of air–sea interaction. *Deep Sea Res.* 16, 153–64.

NAZAROV, V. S. 1968. The duration and distribution of anomalies in temperature of the ocean surface. *Akad. Nauk Okeanograf.* (Moscow) 8, 23–5 (in Russian).

O'CONNOR, J. F. 1969. Hemispheric teleconnections of mean circulation anomalies at 700 mb. Weather Bureau Tech. Rep. No. 10. Washington, D.C.: ESSA. 103 pp.

O'MAHONY, G. 1965. Rainfall and moon phase. *Quart. J. R. Met. Soc.* 91, 196–208.

PAGE, L. F. 1940. Some statistical tests of solar constant-weather relationships. *Mon. Wea. Rev.* suppl. 39, 121–5.

PERRY, A. H. 1969. *A synoptic climatology of the thermal surface and heat balance of the North Atlantic.* Unpub. Ph.D. thesis, Univ. of Southampton. 388 pp.

PERRY, A. H. 1970. Spatial variations in Irish summer weather during 1968 and 1969. *Irish Nat. J.* 16, 301–5.

PETTERSSEN, S. 1950. Some aspects of the general circulation of the atmosphere. *Centen. Proc. R. Met. Soc.* (London). 120–55.

POKROVSKAIA, T. V. 1969. Solar activity and long-range weather forecasting. In *Synoptic-climatological and solar-geophysical long-range weather forecasting,* Leningrad, Gidrometeoizdat, Part 2, 152–225 (in Russian).

POULTER, R. M. 1962. The next few summers in London. *Weather* 17, 253–5.

QUINN, W. H. and BURT, W. V. 1970. Prediction of abnormally heavy precipitation over the equatorial Pacific dry zone. *J. Appl. Met.* 9, 20–8.

RAFAILOVA, K. K. 1968. Objective method of analogue selection from composite and kinematic charts of natural synoptic periods by means of a computer. *Met. i. Gidrol.* No. 8, 20–8 (in Russian).

RATCLIFFE, R. A. S. 1970. Meteorological Office long-range forecasts: six years of progress. *Met. Mag.* 99, 125–30.

RATCLIFFE, R. A. S. and MURRAY, R. 1970. New lag-associations between North Atlantic sea temperatures and European pressure applied to long-range weather forecasting. *Quart. J. R. Met. Soc.* 96, 226–46.

REED, R. J. 1965. The present status of the 26-month oscillation. *Bull. Amer. Met. Soc.* 46, 374–87.

REUTER, F. 1936. Die synoptische Darstellung der halbjäriegen Druckwelle. *Veröff. Geophys. Inst. Univ. Leipzig* 7, 257–95.

RIEHL, H. 1956. *Sea surface temperatures of the North Atlantic, 1887–1936.* Research Rep., Contract N60ri-02036. Met. Dept, Univ. of Chicago.

ROSSBY, C.-G. 1939. Relation between variations of the zonal circulation of the atmosphere and displacement of the semi-permanent centres of action. *J. Mar. Res.* 2, 38–55.

SAWYER, J. S. 1965. *Notes on possible physical causes of long-term weather anomalies.* Tech. Note No. 66 (W.M.O. No. 162, TP 79). Geneva: World Meteorological Organization. 227–48.

SAWYER, J. S. 1970. Observational characteristics of atmospheric fluctuations with a time scale of a month. *Quart. J. R. Met. Soc.* 96, 610–25.

SAZONOV, B. J. and GIRSKAIA, E. I. 1969. Temperature anomalies and the persistence of the atmospheric circulation. *Trudy Glav. Geofiz. Obs.* (Leningrad) 245, 77–87 (in Russian).

SCHELL, I. I. 1940. Baur's contribution to long-range weather forecasting. *Mon. Wea. Rev.* suppl. 39, 63–91.

SCHELL, I. I. 1947. *Dynamic persistence and its applications to long-range foreshadowing.* Harvard Met. Studies 8. Cambridge, Mass. 80 pp.

SCHELL, I. I. 1956a. Interrelations of arctic ice with the atmosphere and the ocean in the north Atlantic, Arctic and adjacent areas. *J. Met.* 13, 46–58.

SCHELL, I. I. 1956b. Further evidence of dynamic persistence and of its application to foreshadowing. *J. Met.* 13, 471–81.

SCHELL, I. I. 1956c. On the nature and origin of the Southern Oscillation. *J. Met.* 13, 592–8.

SCHELL, I. I. 1970. Arctic ice and sea temperature anomalies in the north-eastern north Atlantic and their significance for seasonal foreshadowing locally and to the eastward. *Mon. Wea. Rev.* 96, 833–50.

SCHMAUSS, A. 1937. *Das Problem der Wettervorhersage.* 2nd ed. Leipzig: Akad. Verlag. 102 pp.

SCHUURMANS, C. J. E. 1969. The influence of solar flares on the tropospheric circulation. *Med. Verhand. Nederlands Met. Inst.* (de Bilt) 92. 122 pp.

SCHWERDTFEGER, W. and PROHASKA, F. 1956. The semi-annual pressure oscillation: its cause and effects. *J. Met.* 13, 217–18.

SEMENOV, V. G. 1960a. *The influence of the Atlantic Ocean on the temperature and precipitation regimes over European Russia.* Moscow: Gidromet. 148 pp (in Russian).

SEMENOV, V. G. 1960b. The role of the ocean surface in the formation of blocking anticyclones. *Met. i. Gidrol.* No. 6, 17–20 (in Russian).

SHAPIRO, R. 1964. A mid-latitude biennial oscillation in the variation of the surface pressure distribution. *Quart. J. R. Met. Soc.* 90, 328–31.

SHAPIRO, R. and STOLOV, H. L. 1970. Surface pressure variations in polar regions. *J. Atm. Sci.* 27, 1021–6.

SHAPIRO, R. and WARD, F. 1962. A neglected cycle in sunspot numbers? *J. Atm. Sci.* 19, 506–8.

SIMPSON, J., GARSTANG, M., ZIPSER, E. J. and DEAN, G. A. 1967. A study of a non-deepening tropical disturbance. *J. Appl. Met.* 6, 237–54.

SMED, J. 1948. Monthly anomalies of the surface temperature in areas of the northeastern North Atlantic during the years 1887–1939 and 1945–1948. *Annal. Biol., Cons. Perm. Internat. Explor. Mer.* (Copenhagen) 5, 10–15.

STOLOV, H. L. 1955. Tidal wind fields in the atmosphere. *J. Met.* 12, 117–40.

SUMNER, E. J. 1954. A study of blocking in the Atlantic-European sector of the northern hemisphere. *Quart. J. R. Met. Soc.* 80, 402–16.

SUMNER, E. J. 1959. Blocking anticyclones in the Atlantic-European sector of the northern hemisphere. *Met. Mag.* 88, 300–11.

TEISSERENC DE BORT, L. 1883. Étude sur l'hiver de 1879–80 et recherches sur la position des centres d'action de l'atmosphère dans les hivers anormaux. *Ann. Bureau Central Met. de France,* 1881, Pt. 4. 17–62.

TROUP, A. J. 1962. A secular change in the relation between the sunspot cycle and temperature in the tropics. *Geof. pura e appl.* 51, 184–98.

TROUP, A. J. 1965. The southern oscillation. *Quart. J. R. Met. Soc.* 91, 490–506.

TUCKER, G. B. 1964. Solar influences on the weather. *Weather* 19, 302–11.

UNITED STATES AIR WEATHER SERVICE. 1954. *A description of some methods of extended period forecasting.* Tech. Rep. 105–93. Washington, D.C.: Air Weather Service. 90 pp.

VALERIANOVA, M. A. 1965. Variability of water temperature in the North Atlantic with different types of circulation. *Trudy Gidromet. Inst.* (Leningrad) 20, 37–42 (in Russian).

VAN LOON, H. 1967. The half-yearly oscillations in middle and high southern latitudes and the coreless winter. *J. Atm. Sci.* 24, 472–86.

VAN LOON, H. 1971. On the interaction between Antarctica and middle latitudes. In Quam, L. A. (ed.) *Research in the Antarctic.* Amer. Assoc. Adv. Sci. 477–87.

VAN LOON, H. and JENNE, R. L. 1969. The half-yearly oscillations in the tropics of the southern hemisphere. *J. Atm. Sci.* 26, 218–32.

VAN LOON, H. and JENNE, R. L. 1970. On the half-yearly oscillations in the tropics. *Tellus* 22, 391–8.

VERYARD, R. G. and EBDON, R. A. 1961. Fluctuations in tropical stratospheric winds. *Met. Mag.* 90, 125–43.

VISSER, S. W. 1959. On the connections between the 11-year sunspot period and periods of about 3 and 7 years in world weather. *Geof. pura e Appl.* 43, 302–18.

VISVANTHAN, T. R. 1966. Heavy rainfall distribution in relation to the phase of the moon. *Nature* 210, 406.

VITELS, L. A. 1948. Method of use of analogues in long-period weather forecasting. *Met. i. Gidrol.* No. 3, 22–32 (in Russian).

VOEIKOV, A. 1895. Die Schneedecke in 'paaren' and 'unpaaren' Wintern. *Met. Zeit.* 12, 77.

WADA, H. 1962. A study of the behaviour of the polar vortex and its application to long-range weather forecasting. *Geophys. Mag.* 31, 411–55.

WALKER, G. T. 1923. Correlation in seasonal variations of weather, VIII: A preliminary study of world weather. *Mem. Ind. Met. Dept.* (Poona) 24, 75–132.

WALKER, G. T. 1924. Correlation in seasonal variations of weather, IX: A further study of world weather. *Mem. Ind. Met. Dept.* (Poona) 24, 275–332.

WALKER, G. T. 1938. On monsoon forecasting in India. *Bull. Amer. Met. Soc.* 19, 297–9.

WALKER, G. T. and BLISS, E. W. 1930. World weather, IV: Some applications to seasonal foreshadowing. *Mem. R. Met. Soc.* (London) 3 (24), 81–95.

WALLACE, J. M. and NEWELL, R. E. 1966. Eddy fluxes and the biennial stratospheric oscillation. *Quart. J. R. Met. Soc.* 92, 481–9.

WEIGHTMAN, R. H. 1941. Preliminary studies in seasonal weather forecasting. *Mon. Wea. Rev.* suppl. 45. 99 pp.

WEXLER, H. 1956. Variations in insolation, atmospheric circulation and climate. *Tellus* 8, 480–94.

WILLETT, H. C. 1940. Review of H. H. Clayton on long-range weather changes and methods of forecasting. *Mon. Wea. Rev.* suppl. 39, 126–30.

WILLETT, H. C. 1962. The relationship of total atmospheric ozone to the sunspot cycle. *J. Geophys. Res.* 67, 661–70.

WILLETT, H. C. 1964. Evidence of solar-climatic relationships. In *Weather and our food supply.* Rep. 20. Ames, Iowa: Center for Agric. and Econ. Developments. 123–51.

WOLFF, P. M. 1966. Numerical synoptic analysis of sea surface temperature. *Int. J. Oceanol. Limnol.* 277–90.

WOLFF, P. M. and LAEVASTU, T. 1968. *The effect of oceanic heat exchange on 500 mb large-scale pattern change.* Tech. Note No. 35. Monterey, Calif.: Fleet Numerical Weather Facility. 15 pp.

WORLD METEOROLOGICAL ORGANIZATION. 1965. *Research and development aspects of long-range forecasting.* Tech. Note No. 66 (W.M.O. No. 162, TP 79). Geneva: World Meteorological Organization. 339 pp.

WRIGHT, P. B. 1968a. A widespread biennial oscillation in the troposphere. *Weather* 23, 50–4.

WRIGHT, P. B. 1968b. Wine harvests in Luxemburg and biennial oscillations in European summers. *Weather* 23, 300–24.

WRIGHT, P. B. 1971. Quasi-biennial oscillations in the atmosphere. *Weather* 26, 69–76.

5.E OTHER APPLICATIONS

5.E.1 *Air–sea interaction*

BJERKNES, J. 1963. Climatic change as an ocean-atmosphere problem. In *Changes of climate.* Arid Zone Res. No. 20. Paris: UNESCO. 297–321.

BUDYKO, M. I. (ed). 1963. *Atlas teplovogo balansa zemnogo shara* (*Atlas of the heat balance of the earth*). 2nd ed. Leningrad: Gidromet. Izdat. 69 pp.

DEACON, E. L. and WEBB, E. K. 1962. Inter-change of properties between sea and air: small-scale interactions. In Hill, M. N. (ed.) *The sea.* vol. 1. New York: Interscience. 43–87.

EDWARDS, P. 1970. Temperature measurements in ships. *Quart. J. R. Met. Soc.* 96, 130–1, 543.

GAGNON, R. M. 1964. *Types of winter energy budgets over the Norwegian Sea.* Pub. in Met. No. 64. Montreal: Arctic Met. Res. Group, McGill Univ. 56 pp.

GAMBO, K. 1963. The role of sensible and latent heat in the baroclinic atmosphere. *J. Met. Soc. Japan* 11 (41), 233–46.

GARSTANG, M. 1967. Sensible and latent heat exchange in low latitude synoptic scale systems. *Tellus* 19, 492–509.

HARE, F. K. 1966. The concept of climate. *Geography* 51, 99–110.

HAY, R. F. M. 1956. *Some aspects of variations of air and sea surface temperature in short periods at O.W.S. 'I' and 'J' of significance to the synoptic climatologist.* Met. Res. Paper (London) No. 949.

KLIMENKO, L. V. and STROKINA, L. A. 1969. Relation between the variation of the heat content of the waters of the North Atlantic and the air temperature in the European U.S.S.R. in winter. *Trudy Glav. Geofiz. Obs.* (Leningrad) 249, 52–7 (in Russian).

LAEVASTU, T. 1965. Daily heat exchange in the North Pacific: its relations to weather and its oceanographic consequences. *Soc. Scient. Fennica, Comm. Phys.-Math.* (Helsinki) 31, 3–53.

MALKUS, J. S. 1962. Large-scale interactions. In Hill, M. N. (ed.) *The sea.* Vol. 1. New York: Interscience. 88–294.

MANABE, S. 1957. On the modification of air mass over the Japan Sea when the outburst of cold air predominates. *J. Met. Soc. Japan* ser. 2, 35, 311–26.

MANABE, S. 1958. On the estimation of energy exchange between the Japan Sea and the atmosphere during winter based upon the energy budget of both the atmosphere and the sea. *J. Met. Soc. Japan* 36, 123–33.

MANABE, S. 1969. Climate and the ocean circulation, I: The atmospheric circulation and the hydrology of the earth's surface. *Mon. Wea. Rev.* 97, 739–74.

MANABE, S. and BRYAN, K. 1969. Climate and the ocean circulation. *Mon. Wea. Rev.* 97, 739–827.

MANIER, G. and MÖLLER, F. 1961. *Determination of the heat balance of the boundary layer over the sea.* Final Rep. Contract AF-61(052)-315. Mainz: Johannes Guttenberg Univ. 87 pp.

MATHESON, K. M. 1967. *The meteorological effect on ice in the Gulf of St. Lawrence.* Pub. in Met. No. 89. Montreal: Arctic Met. Res. Group, McGill Univ. 110 pp.

MITCHELL, J. M. 1966. Stochastic models of air–sea interaction and climatic fluctuation. In Fletcher, J. O. (ed.) *Proc. sympos. on the arctic heat budget and atmospheric circulation.* Mem. RM-5233-NSF. Santa Monica, Calif.: Rand Corp. 46–74.

MURRAY, R. 1966. The blocking pattern of February 1965 and a good pressure analogue. *Weather* 21, 66–9.

NAMIAS, J. 1969. Seasonal interactions between the north Pacific Ocean and the atmosphere during the 1960s. *Mon. Wea. Rev.* 97, 173–92.

PAGAVA, S. T. 1962. On the type of relationship between thermal conditions of the north Atlantic and the air temperature in Europe. *Met. i. Gidrol.* No. 1, 10–18 (in Russian).

PERRY, A. H. 1968. Turbulent heat flux patterns over the north Atlantic during recent winter months. *Met. Mag.* 97, 246–54.

PERRY, A. H. 1969. *A synoptic climatology of the thermal surface and heat balance of the North Atlantic.* Unpub. Ph.D. thesis, Univ. of Southampton. 388 pp.

PETTERSSEN, S., BRADBURY, D. L. and PEDERSEN, K. 1962. The Norwegian cyclone models in relation to heat and cold sources. *Geofys. Publik.* (Oslo) 24, 243–80.

RATCLIFFE, R. A. S. 1971. North Atlantic sea temperature classification, 1877–1970. *Met. Mag.* 100, 225–31.

RATCLIFFE, R. A. S. and MURRAY, R. 1970. New lag-associations between north Atlantic sea temperatures and European pressure applied to long-range weather forecasting. *Quart. J. R. Met. Soc.* 96, 226–46.

RIEHL, H. 1969. Some aspects of cumulonimbus convection in relation to tropical weather disturbances. *Bull. Amer. Met. Soc.* 50, 587–95.

SAUR, J. F. T. 1963. A study of the quality of seawater temperatures reported in logs of ships' weather observations. *J. Appl. Met.* 2, 417–25.

SAWYER, J. S. 1965. *Notes on possible physical causes of long-term weather anomalies.* Tech. Note No. 66 (W.M.O. No. 162, TP 79). Geneva: World Meteorological Organization. 227–48.

SEMENOV, V. G. 1963. The role of the North Atlantic in forming the temperature field over the European U.S.S.R. *Trudy Vsesoyuz. Nauch. Met. Soveshch.* (Leningrad) 3, 105–12 (in Russian).

SIMPSON, J. 1969. On some aspects of sea–air interaction in middle latitudes. *Deep Sea Res.*, suppl. 16, 233–61.

TAUBER, G. M. 1969. The comparative measurements of sea-surface temperature in the U.S.S.R. In *Sea surface temperature.* Tech. Note No. 103 (W.M.O. No. 247, TP 135). Geneva: World Meteorological Organization. 141–51.

VALERIANOVA, M. A. 1965. Variability of water temperature in the North Atlantic with different types of circulation. *Trudy Gidromet. Inst.* (Leningrad) 20, 37–42 (in Russian).

VERPLOEGH, G. 1967. Observation and analysis of the surface wind over the ocean. *Kon. Ned. Met. Inst.* No. 89. 67 pp.

VINOGRADOV, N. D. 1967. The relation between types of atmospheric circulation and the temperature of the surface water in the northern part of the North Atlantic. *Probl. Arkt. i. Antarkt.* (Leningrad) 25, 44–53 (in Russian).

VOWINCKEL, E. 1965. *The energy budget over the North Atlantic – January 1963, Part 1: energy loss at the surface.* Pub. in Met. No. 78. Montreal: Arctic Met. Res. Group, McGill Univ. 38 pp.

WINSTON, J. S. 1955. Physical aspects of rapid cyclogenesis in the Gulf of Alaska. *Tellus* 7, 481–500.

5.E.2 *Bioclimatology and biometeorology*

ASPLIDEN, C. I. H. and RAINEY, R. C. 1961. Desert locust control. *W.M.O. Bull.* 10 (3), 155–61.

BEAUMONT, A. 1947. The dependence on the weather of the dates of outbreak of potato blight epidemics. *Trans. Brit. Mycol. Soc.* 31, 45–53.

BOURKE, P. M. A. 1957. *The use of synoptic weather maps in potato blight epidemiology.* Tech. Notes No. 23. Dublin: Met. Service.

BREZOWSKY, H. 1964. Morbidity and weather. In Licht, S. (ed.) *Medical climatology.* Baltimore, Md: Waverly Press. 358–99.

BREZOWSKY, H. and WEISSER, P. 1961. Der Einfluss der Wettervorgänge auf Beitriebsunfälle und Befindensstörungen in einem Industriewerk. *Zentralbl. f. Arbeitsmed. u. Arbeitsschutz* 11, 81–4.

BROOKS, F. A. and KELLY, C. F. 1951. Instrumentation for recording microclimatological factors. *Trans. Amer. Geophys. Union* 32, 833–46.

BUCHAN, A. and MITCHELL, A. 1878. The influence of weather on mortality from different diseases and at different ages. *J. Scot. Met. Soc.* 5, 171–88.

CHRISTIE, A. D. and RITCHIE, J. C. 1969. On the use of isentropic trajectories in the study of pollen transports. *Naturaliste canadien* 96, 531–49.

DAUBERT, K. 1956. Betriebsunfälle der Bundesbahn in Württemberg und Wettergeschehen. *Ann. Met.* (*Medizin-Met. Heft*) 11, 149–51.

DAVIS, F. K. JR. 1958. Ulcers and temperature changes. *Bull. Amer. Met. Soc.* 39, 652–4.

DRISCOLL, D. M. 1971. The relationship between weather and mortality in ten major metropolitan areas in the United States, 1962–1965. *Int. J. Biomet.* 15, 23–39.

DRISCOLL, D. M. and LANDSBERG, H. E. 1967. Synoptic aspects of mortality: a cast study. *Int. J. Biomet.* 11, 323–8.

FRENCH, R. A. and WHITE, J. H. 1960. The diamond-back moth outbreak of 1958. *Plant Path.* 9, 77–84.

GREEN, G. W. 1955. Temperature relations of ant-lion larvae. *Canad. Entomol.* 87, 441–59.

HARCOURT, D. G. and LE ROUX, E. J. 1967. Population regulation in insects and man. *Amer. Sci.* 55, 400–15.

HAUFE, W. O. 1963. Entomological biometeorology. *Int. J. Biomet.* 7, 129–36.

HENSEN, W. R. 1951. Mass flights of the spruce budworm. *Canad. Entomol.* 83, 240.

HENSEN, W. R., STARK, R. W. and WELLINGTON, W. G. 1954. Effects of the weather of the coldest month on winter mortality of the lodge pole needle miner in Banff National Park. *Canad. Entomol.* 86, 13–19.

HIRST, J. M. and HURST, G. W. 1967. Long-distance spore transport. In *Airborne microbes* (*17th Sympos. Soc. Gen. Microbiol.*). Cambridge: Cambridge Univ. Press. 307–44.

HISDALE, V. 1953. The influence of weather on man. *Met. Annal.* 3, 447–8.

HOGG, W. H. 1967. The use of upper air data in relation to plant disease. In Taylor, J. A. (ed.) *Weather and agriculture.* Oxford: Pergamon. 115–27.

HOGG, W. H. 1970. Weather, climate and plant disease. *Met. Mag.* 99, 317–26.

HOGG, W. H., HOUNAM, C. E., MALLIK, A. K. and ZADOKS, J. C. 1969. *Meteorological factors affecting the epidemiology of wheat rusts.* Tech. Note No. 99 (W.M.O. No. 238, TP 130). Geneva: World Meteorological Organization. 143 pp.

HOLLANDER, J. L. and YEOSTROS, S. J. 1963. The effect of simultaneous variations of humidity and atmospheric pressure on arthritis. *Bull. Amer. Met. Soc.* 44, 489–94.

HURST, G. W. 1963. Small mottled willow moth in southern England. *Met. Mag.* 92, 308–12.

HURST, G. W. 1968. Areal infiltration by windborne insects and spores. In *Agroclimatological methods.* Natural Resources Res. No. 7. Paris: UNESCO. 153–6.

HURST, G. W., COCHRANE, J. and RUMNEY, R. P. 1969. *Trajectories from continental sources to the United Kingdom, 1960.* Agric. Mem. No. 285. Bracknell, Berks: Met. Office. 9 pp.

HURST, G. W. and RUMNEY, R. P. 1969. *Trajectories from continental sources to Great Britain, 1946–68.* Agric. Mem. No. 291. Bracknell, Berks: Met. Office.

KOHN, W. 1956. Hamburger Verkehrsunfälle und Wetter. *Ann. Met.* (*Medizin Met. Heft*) 11, 137–47.

KUHNKE, W. 1958. Kann man von einer Föhnwirkung in Norddeutschland sprechen? *Ann. Met.* (*Medizin-Met. Heft*) 13, 117–26.

MOMIYAMA, M. 1968. Biometeorological study of the seasonal variation of mortality in Japan and other countries on the seasonal disease calendar. *Int. J. Biomet.* 12, 377–93.

MOMIYAMA, M. and KATAYAMA, K. 1967. Seasonal fluctuations of mortality in the U.S.A. (abstract). *Int. J. Biomet.* 11, 223.

NAYA, A. 1967. Insects, insecticides and the weather. *Weather* 22, 211–14.

OBENLAND, E. 1965. Untersuchung über die Wetterabhängigkeit des Kopfschmerzes. *Arch. Met. Geophys. Biokl.* B, 13, 414–37.

PEDGLEY, D. E. and SYMMONS, P. M. 1968. Weather and the locust upsurge. *Weather* 23, 484–92.

PSCHORN-WALCHER, H. 1954. Die Zunahme der Schadlingsauftreten im Lichte der rezenten Klimagestaltung. *Anz. Schädlingskunde* (Berlin) 27, 89–91.

RAINEY, R. C. 1963. *Meteorology and the migration of the desert locust.* Tech. Note No. 54 (W.M.O. No. 138, TP 64). Geneva: World Meteorological Organization. 115 pp.

REIFFERSCHEID, H. 1954. Über den Einfluss des Wetters auf Leistung und Befinden bei Akkordarbeiten. *Zentralbl. f. Arbeitsund. u. Arbeitsschutz* 4, 112–16.

REITER, R. 1960. *Meteorobiologie und Elektrizität der Atmosphäre.* Geest and Portig, Leipzig. 424 pp.

RICHARDS, O. W. 1961. The theoretical and practical study of natural insect population. *Ann. Rev. Entomol.* 6, 147–62.

SARGENT, F., II. and TROMP. S. W. 1964. *A survey of human biometeorology.* Tech. Note No. 65 (W.M.O. No. 160, TP 78). Geneva: World Meteorological Organization. 113 pp.

SAYER, H. J. 1962. The desert locust and tropical convergence. *Nature* 194, 330–6.

SCHIRMER, H. 1952. Über die räumliche Struktur der Niederschlagsverteilung. *Ann. Met.* 5, 248–53.

SHAW, M. W. 1962. The diamond-back moth migration of 1958. *Weather* 17, 221–34.

SHAW, M. W. and HURST, G. W. 1969. A minor immigration of the diamond-back moth. *Plutella Zylostella* (L.) (*Maculipennis Curtis*). *Agric. Met.* 6, 125–32.

SMITH, L. P. 1970. *Weather and animal diseases.* Tech. Note No. 113 (W.M.O. No. 268, TP 152). Geneva: World Meteorological Organization. 49 pp.

SPANN, W. 1956. Verkehrsunfälle und Wetter. *Ann. Met.* (*Medizin-Met. Heft*) 11, 152.

STEVENSON, C. M. 1969. The dust fall and severe storms of 1 July 1968. *Weather* 24, 126–32.

TINLINE, R. 1970. Lee wave hypothesis for the initial pattern of spread during the 1967–68 foot and mouth epidemic. *Nature* 227, 860–2.

TROMP, S. W. 1963. *Medical biometeorology.* Amsterdam: Elsevier. 991 pp.

TROMP, S. W. and WEIHE, W. H. *Biometeorology.* Vol. 2, pt. 1. Oxford: Pergamon. 520 pp.

UVAROV, B. P. 1931. Insects and climate. *Trans. Entomol. Soc.* (London) 79, 1–247.

WELLINGTON, W. G. 1954a. Weather and climate in forest entomology. *Met. Monogr.* 2 (8), 11–18.

WELLINGTON, W. G. 1954b. Air mass climatology of Ontario north of Lake Huron and Lake Superior before outbreaks of the spruce budworm and the forest tent caterpillar. *Canad. J. Zool.* 30, 114–17.

WELLINGTON, W. G. 1954c. Atmospheric circulation processes and insect ecology. *Canad. Entomol.* 86, 312–33.

WELLINGTON, W. G. 1957. The synoptic approach to studies of insects and climate. *Ann. Rev. Entomol.* 2, 143.

WELLINGTON, W. G. 1964. Quantitative changes in populations in unstable environments *Canad. Entomol.* 96, 436–41.

WELLINGTON, W. G. 1965. The use of cloud patterns to outline areas with different climates during population. *Canad. Entomol.* 97, 617–31.

WORLD METEOROLOGICAL ORGANIZATION. 1965. *Meteorology and the desert locust.* Tech. Note No. 69 (W.M.O. No. 171, TP 85). Geneva. 310 pp.

Chapter 6

BOLOTINSKAIA, M. S. 1965. Variations in the frequency of atmospheric circulation types as related to solar activity. *Probl. Arkt. i Antarkt.* (Leningrad) 20, 40–8 (in Russian).

BOLOTINSKAIA, M. S. and BELIAZO, V. A. 1969. Effect of solar activity on the formation of circulation epochs and their stages. *Trudy Arkt. i Antarkt. Nauch. Issled. Inst.* (Leningrad) 289, 132–51 (in Russian).

CASANOVA, H. 1968. *Principaux types de temps en Afrique occidentale.* Met. Publ. No. 6. Dakar: Agence Securité Nav. Aer. en Afrique et Madagascar, Dir. Exploit. 43 pp.

CHRISTENSEN, W. I. JR. and BRYSON, R. A. 1966. An investigation of the potential of component analysis for weather classification. *Mon. Wea. Rev.* 96, 697–709.

DEPPERMAN, C. E. 1937. *The weather and clouds of Manila.* Manila: Bureau of Printing.

JOHNSON, D. H. 1965. African synoptic meteorology. In *Meterology and the desert locust,* Tech. Note No. 69. Geneva: World Meteorological Organization. T.P. 85, 48–90.

JOHNSON, D. H. and MÖRTH, H. T. 1960. Forecasting research in east Africa. In Bargman, D. J. (ed.) *Tropical meteorology in Africa.* Nairobi: Munitalp Foundation. 56–132.

KESHAVAMURTY, R. N. 1971. Vertical coupling in the Indian summer monsoon. *Nature* 232, 169–70.

KORSHOVER, J. 1967. *Climatology of stagnating anticyclones east of the Rocky Mountains,* 1936–1965. Cincinatti, Ohio: Public Health Service Publ. 999-AP-34.

KUDASHKIN, G. D. and IUDIN, M. I. 1968. Selection and use of analogues for more accurate numerical weather forecasts. *Trudy Glav. Geofiz. Obs.* (Leningrad) 197, 3–18 (in Russian).

LAMB, H. H. 1970. Volcanic dust in the atmosphere: with a chronology and an assessment of its meteorological significance. *Phil. Trans. Roy. Soc.* (London) A, 266 (1178), 425–33.

MAUNDER, W. J. 1970. *The value of the weather.* London: Methuen. 388 pp.

MITCHELL, J. M., JR. 1971. The effect of atmospheric aerosols on climate with special reference to temperature near the earth's surface. *J. Appl. Met.* 10, 703–14.

MOSIÑO, P. A. 1964. Surface weather and upper air flow patterns in Mexico. *Geofis. Internac.* 4, 117–68.

POKROVSKAIA, T. V. and ESAKOVA, N. P. 1967. Weather forecasts and synoptic climatological investigations. *Trudy Glav. Geofiz. Obs.* (Leningrad) 218, 29–44 (in Russian).

RAMAGE, C. C. 1971. *Monsoon meteorology.* New York: Academic Press. 296 pp.

RIEHL, H. 1967. *Southeast Asia monsoon study.* Final Rep. DA 28-043-AMC-01202 (E). Fort Collins, Colo.: Dept. of Atm. Sci., Colorado State Univ. 33 pp.

SCHROEDER, M. J. *et al.* 1964. *Synoptic weather types associated with critical fire weather.* Berkeley, Calif.: Pacific Southwest Forest & Range Experiment Station. 492 pp.

VOWINCKEL, E. 1956. Weather type climatology of southern Mozambique and Lourenco Marques. *Misc. Geofis. 10th Anniv. Met. Service Angola* (Luanda). 63–86.

VOWINCKEL, E. 1964. Die Bedeutung regionaler Klimatologie in der Meteorologie. *Met. Rund.* 17, 104–5.

WEISS, I. and LAMB, H. H. 1970. *Die Zunahme der Wellenhöhen in jüngster Zeit in der Operationsgebieten der Bundesmarine, ihre vermütlichen Ursachen und ihre voraussichtliche weitere Entwicklung*, Fachliche Mitteilungen No. 160 (Porz-Wahn). 18 pp.

WRIGLEY, E. A. 1965. Changes in the philosophy of geography. In Chorley, R. J. and Haggett, P. (eds.) *Frontiers in geographical teaching.* London: Methuen.

Addenda

CHAPTER 2.A, B

COCHEMÉ, J. 1965. Assessments of divergence in relation to the desert locust. In *Meteorology and the desert locust.* Tech. Note No. 69 (W.M.O. 171, T.P. 85). Geneva: World Meteorological Organization. 23–41.

REITER, E. R. 1972. *Atmospheric transport processes.* Part 3. *Hydrodynamic tracers.* U.S. Atomic Energy Commission, Div. Tech. Information. 212 pp.

SMITH, P. J. 1971. An analysis of kinematic vertical motions. *Mon. Wea. Rev.* 99, 715–24.

STRETEN, N. A. and TROUP, A. J. 1972. A synoptic climatology of satellite observed cloud vortices over the southern hemisphere. *Quart. J.R. Met. Soc.* 99, 56–72.

WORLD METEOROLOGICAL ORGANIZATION. 1972. Basic synoptic networks of observing stations. W.M.O. 217, T.P. 113, Geneva.

CHAPTER 2.C

ERICKSON, C. O. and WINSTON, J. S. 1972. Tropical storm, mid-latitude cloud band connections and the autumnal buildup of the planetary circulation. *J. Appl. Met.* 11, 23–36.

GODEV, N. 1971. Anticyclonic activity over south Europe and its relationship to orography. *J. Appl. Met.* 10, 1097–1102.

MILLER, D. B. and FEDDES, R. G. 1971. *Global atlas of relative cloud cover 1967–70 based on data from meteorological satellites.* Washington, D.C.; Air Force Environ. Tech. Appl. Cen. 240 pp.

NEWTON, C. W. (ed.) 1972. *Meteorology of the southern hemisphere.* Met. Monogr. (Amer. Met. Soc.) 13 (35). 263 pp.

OORT, A. H. and RAMUSSON, E. M. 1971. *Atmospheric circulation statistics.* Princeton, N.J.: Environ. Res. Labs., NOAA. 335 pp.

CHAPTER 3.D

HUFTY, A. 1972. Méthode descriptive des types de temps. In Adams, W. P. and Helleiner, F. M. (eds.) *International Geography 1972*, Vol. 2. Univ. of Toronto. 236–8.

LITYNSKI, J. 1972. Classification des types de circulation et des types de temps pour le Québec (periode 1963–9). In Adams, W. P. and Helleiner, F. M. (eds.) *International Geography 1972*, Vol. 2. Univ. of Toronto. 1352–4.

NIEUWOLT, S. 1971. Climatic variability and weather types in Lusaka, Zambia. *Arch. Met. Geophys. Biokl.* B 19, 345–66.

CHAPTER 4.A

HAGGARD, W. H., BILTON, T. H. and CRUTCHER, H. L. 1973. Maximum rainfall from tropical cyclone systems which cross the Appalachians. *J. Appl. Met.* 12, 50–61.

MIELKE, P. W., JR. 1973. Another family of distributions for describing and analyzing precipitation data. *J. Appl. Met.* 12, 275–280.

CHAPTER 4.B

COLE, J. A. and SHERRIFF, J. D. F. 1972. Some single- and multi-site models of rainfall with discrete time increments. *J. Hydrol.* 17, 97–114.

CURRY, L. 1967. A wave statistics model for climatic time series. In *Colloquium on time series analysis.* Univ. of Kansas: Computer Contrib. No. 18, 46–50.

CHAPTER 4.C

TYSON, P. D. 1971. Spatial variation of rainfall spectra in South Africa. *Ann. Assoc. Amer. Geogr.* 61, 711–20.

CHAPTER 5.A

WAHL, E. H. 1972. Climatological studies of the large-scale circulation in the northern hemisphere: zonal and meridional indices at the 700-millibar level. *Mon. Wea. Rev.* 100, 553–64.

CHAPTER 5.B

DAVIS, N. E. 1972. Variability of the onset of spring in Britain. *Quart. J. R. Met. Soc.* 98, 763–77.

DESAI, B. N. 1970. Synoptic climatology of the Indian subcontinent. *Met. and Geophys. Rev.* No. 2 (Poona).

MURRAY, R. 1972. Monthly mean temperature related to synoptic types over Britain specified by PSCM indices. *Met. Mag.* 101, 305–11.

CHAPTER 5.C

BARRY, R. G. 1973. The conditions favouring glacierization and deglacierization in North America from a climatological viewpoint. *Arct. Alp. Res.* 5 (3), in press.

DAVIS, N. E. 1972. Some aspects of circulation changes in the northern hemisphere in January in the 20th century. *Met. Mag.* 101, 317–29.

WILLIAMS, JILL, BARRY, R. G. and WASHINGTON, W. M. 1973. Simulation of climate at the last glacial maximum using the NCAR global circulation model. *Occas. Pap.* No. 5. Boulder, Colo: Inst. Arct. Alp. Res., Univ. of Colorado. 39 pp.

CHAPTER 5.D

BORISOVA, L. G. and RUDICHEVA, L. M. 1968. The use of special features of natural synoptic seasons in making monthly weather forecasts. *Trudy Gidromet. Nauch. Issled. Tsentr SSSR* (Leningrad) 12, 12–18 (in Russian).

LABITZKE, K. and COLLABORATORS. 1972. Climatology of the stratosphere in the northern hemisphere. *Met. Abh.* 100 (4). 290 pp.

QUINN, W. H. and BURTS, W. V. 1972. Use of the Southern Oscillation in weather prediction. *J. Appl. Met.* 11, 616–28.

ROBERTS, W. O. and OLSON, R. H. 1973. Geomagnetic storms and wintertime 300 mb trough development in the North Pacific–North America area. *J. Atmos. Sci.* 30, 135–40.

CHAPTER 5.E

KRAUS, E. B. 1972. *Atmosphere–ocean interaction.* Oxford Univ. Press. 284 pp.

Appendix

For the benefit of the reader with little background in meteorology and mathematical analysis, a few concepts referred to in the text are summarized here. For the most part these refer to material discussed in Chapter 2.A. Further details may be found in the reference list.

Geopotential

Geopotential (Φ) is the work done against the local acceleration of gravity (g) in raising a unit mass from mean sea level to a height Z:

$$\Phi = \int_0^Z g\, dz$$

where Φ and g are cm s^{-2} and Z is in cm above MSL. The difference between geopotential metres and geometric metres is for many purposes negligible, viz:

g.p.m.	*m at Equator*	*m at Pole*
1,000	1,002	997
10,000	10,036	9,983

Height and geopotential have the same numerical value at 45° latitude.

Thickness

From the equation of hydrostatic equilibrium:

$$\frac{dp}{dz} = -g\rho$$

where p = pressure
z = height
g = gravity
ρ = density
and from the gas equation

$$p = R\rho T$$

where R = the specific gas constant for dry air we have:

$$dz = -\frac{1}{g\rho}\,dp$$
$$= -\frac{RT}{g}\,d\ln p.$$

On integration,

$$Z_2 - Z_1 = -\frac{R}{g}\int_{P_1}^{P_2} T\,d\ln p$$
$$= -\frac{R\bar{T}_v}{g}\ln(P_2 - P_1)$$
$$= \frac{R\bar{T}_v}{g}\ln\left(\frac{P_1}{P_2}\right)$$

where ln denotes natural logarithms; $\bar{T}_v$ = mean virtual temperature of the layer $Z_2 - Z_1$

$$T_v = T\,\frac{[1 + r/\varepsilon]}{1 + r}$$

where ε = 0·62197
r = humidity mixing ratio

$$T_v \approx T + 0{\cdot}61rT$$

Eulerian and Lagrangian methods

There are two basic analytical approaches to studies of fluid motion. One refers to the fluid motion past fixed points (Eulerian methods) and the other to the movement of individual fluid elements (Lagrangian methods). In the latter system total differentials are involved since our reference frame is the trajectory of an individual particle. For example, the speed components are:

$$u = \frac{dx}{dt}, \qquad v = \frac{dy}{dt}, \qquad w = \frac{dz}{dt}.$$

In the Eulerian system we consider the instantaneous change at individual points as the fluid passes. The acceleration components are now partial derivatives (i.e. functions of more than one variable):

$$a_x = \frac{\partial u}{\partial x}u + \frac{\partial u}{\partial y}v + \frac{\partial u}{\partial z}w + \frac{\partial u}{\partial t} \equiv \frac{Du}{Dt}$$
$$a_y = \frac{\partial v}{\partial x}u + \frac{\partial v}{\partial y}v + \frac{\partial v}{\partial z}w + \frac{\partial v}{\partial t} \equiv \frac{Dv}{Dt}$$
$$a_z = \frac{\partial w}{\partial x}u + \frac{\partial w}{\partial y}v + \frac{\partial w}{\partial z}w + \frac{\partial w}{\partial t} \equiv \frac{Dw}{Dt}$$

where $$\frac{D}{Dt} = \frac{\partial}{\partial y}\, u + \frac{\partial}{\partial y}\, v + \frac{\partial}{\partial z}\, w + \frac{\partial}{\partial t}$$ (Stoke's operator).

Finite differences of partial derivatives

$$\frac{\partial S}{\partial x} \approx \frac{(S_x - S_{-x})}{L}$$

where S is some field variable.

$$\frac{\partial^2 S}{\partial x^2} = \frac{\partial}{\partial x}\left(\frac{\partial S}{\partial x}\right)$$

$$= (S_{2x} - 2S_x + S_{-x})/L^2$$

$$\nabla^2 S = \frac{\partial^2 S}{\partial x^2} + \frac{\partial^2 S}{\partial y^2}.$$

$$\approx (S_x + S_{-x} + S_y + S_{-y} - 4S_0)/L^2.$$

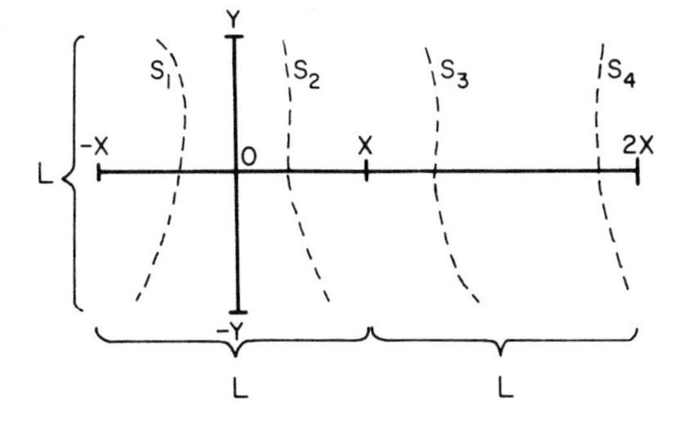

Fig. A.1
Grid framework for finite difference evaluation of partial derivatives.

Scalars and vectors

A *scalar* quantity has only magnitude; for example, temperature or pressure.

A *vector* quantity has magnitude and direction; for example, wind velocity, the gradient of a temperature field.

A wind vector (**V**) can be resolved into components in a cartesian co-ordinate system (x, y). The scalars u and v are the lengths of the projections of **V** onto the x and y axes, respectively. In three dimensions a similar projection for the w component may be made on the vertical z axis.

$$\mathbf{V} = \mathbf{i}u + \mathbf{j}v + \mathbf{k}w$$

where u and v = horizontal wind components (in x and y directions, respectively)

w = vertical wind component (in z direction)

i, **j** and **k** = unit vectors denoting the directions of u, v, w respectively.

The magnitude of $\mathbf{V}=(u^2+v^2+w^2)^{1/2}$. The direction of the vector is measured by cosines of the angles between the vector $\mathbf{V}$ and the three axes x, y, z.

The sum of two vectors is illustrated in fig. A.2 for positive and negative cases. From the definition of $\mathbf{V}$ above it follows that

$$\mathbf{V}_1 \pm \mathbf{V}_2 = \mathbf{i}(u_1 \pm u_2)+\mathbf{j}(v_1 \pm v_2)+\mathbf{k}(w_1 \pm w_2).$$

The two principal multiplication operations are the dot product and the cross-product of two vectors. The definitions are:

$$\mathbf{V}_1 . \mathbf{V}_2 = V_1 V_2 \cos \theta$$

where θ is the angle $\leqslant 180°$ between the vectors $\mathbf{V}_1$ and $\mathbf{V}_2$. Also $\mathbf{V}_1 . \mathbf{V}_2 = \mathbf{V}_2 . \mathbf{V}_1$. The dot product is a scalar.

For unit vectors:

$$\mathbf{i}.\mathbf{i} = \mathbf{j}.\mathbf{j} = \mathbf{k}.\mathbf{k} = 1$$

and

$$\mathbf{i}.\mathbf{j} = \mathbf{j}.\mathbf{k} = \mathbf{k}.\mathbf{i} = 0.$$

The cross product gives a vector of magnitude:

$$\mathbf{V}_1 \times \mathbf{V}_2 = V_1 V_2 \sin \theta$$

and direction normal to both vectors, positive in the sense of a right-hand screw when turned from direction $\mathbf{V}_1$ to $\mathbf{V}_2$ through an angle $\theta \leqslant 180°$ (fig. A.2(*d*)).

For unit vectors:

$$\mathbf{i} \times \mathbf{j} = \mathbf{k}$$
$$\mathbf{j} \times \mathbf{k} = \mathbf{i}$$
$$\mathbf{k} \times \mathbf{i} = -\mathbf{j}$$
$$\mathbf{j} \times \mathbf{i} = -\mathbf{k}$$
$$\mathbf{i} \times \mathbf{i} = \mathbf{j} \times \mathbf{j} = \mathbf{k} \times \mathbf{k} = 0.$$

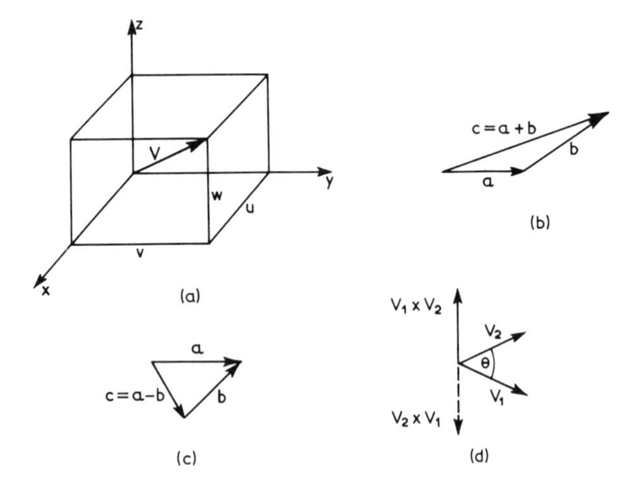

Fig. A.2 (*a*) Vector and components in cartesian coordinate. (*b*) Vector addition. (*c*) Vector subtraction. (*d*) Vector cross-product.

Vector operators

1 Gradient:[1] $$\text{grad}\, S = \nabla S = \mathbf{i}\,\frac{\partial S}{\partial x} + \mathbf{j}\,\frac{\partial S}{\partial y} + \mathbf{k}\,\frac{\partial S}{\partial z}$$

where **i**, **j**, **k** are unit vectors in the directions of x, y, z respectively. $S =$ any scalar.

The gradient is defined as positive towards lower values of S.

2 Divergence: $$\text{div}\, S = \nabla . S = \frac{\partial S_x}{\partial x} + \frac{\partial S_y}{\partial y} + \frac{\partial S_z}{\partial z}.$$

Divergence is the net outflow of fluid from unit volume in unit time.

3 Laplacian operator: $$\nabla^2 = \nabla . \nabla = \frac{\partial^2}{\partial x^2} + \frac{\partial^2}{\partial y^2} + \frac{\partial^2}{\partial z^2}.$$

4 Curl: $$\text{curl}\, S = \nabla \times S = \mathbf{i}\left(\frac{\partial S_z}{\partial y} - \frac{\partial S_y}{\partial z}\right) + \mathbf{j}\left(\frac{\partial S_x}{\partial z} - \frac{\partial S_z}{\partial x}\right) + \mathbf{k}\left(\frac{\partial S_y}{\partial x} - \frac{\partial S_x}{\partial y}\right).$$

Note that the vertical relative vorticity, ζ, is:

$$\zeta = \left(\frac{\partial v}{\partial x} - \frac{\partial u}{\partial y}\right)$$

corresponding to the last term of the complete three-dimensional curl.

References: Appendix

BERRY, F. A. JR., BOLLAY, E. and BEERS, N. R. 1945. *Handbook of meteorology*. New York: McGraw-Hill. 1068 pp.

†GODSKE, C. L., BERGERON, T., BJERKNES, J. and BUNDGAARD, R. C. 1957. *Dynamic meteorology and weather forecasting*. Boston, Mass.: Amer. Met. Soc. 800 pp.

GORDON, A. H. 1962. *Elements of dynamic meteorology*. Princeton: Van Nostrand. 217 pp.

*HESS, S. L. 1959. *Introduction to theoretical meteorology*. New York: H. Holt. 362 pp.

LIST, R. J. 1951. *Smithsonian meteorological tables*. 6th ed. Washington, D.C.: Smithsonian Institution. 527 pp.

[1] Strictly ∇S is the ascendant. In meteorological usage, however, it is referred to as the gradient.

* Most useful for the beginner.

† Advanced.

Author Index

Subject Index